IGMacdonald

1972

Die Grundlehren der mathematischen Wissenschaften

in Einzeldarstellungen
mit besonderer Berücksichtigung
der Anwendungsgebiete

Band 189

Herausgegeben von

J. L. Doob · A. Grothendieck · E. Heinz · F. Hirzebruch
E. Hopf · W. Maak · S. MacLane · W. Magnus · J. K. Moser
M. M. Postnikov · F. K. Schmidt · D. S. Scott · K. Stein

Geschäftsführende Herausgeber

B. Eckmann und B. L. van der Waerden

Garth Warner

Harmonic Analysis on Semi-Simple Lie Groups II

Springer-Verlag Berlin Heidelberg New York 1972

Garth Warner

Department of Mathematics, University of Washington
Seattle, Washington

AMS Subject Classifications (1970)
22 E 45, 22 D 10, 43 A 90, 43 A 85, 22 D 30

ISBN 3-540-05469-3 Springer-Verlag Berlin Heidelberg New York
ISBN 0-387-05469-3 Springer-Verlag New York Heidelberg Berlin

 Library of Congress Catalog Card Number 70-160590. Printed in Germany. Typesetting: Dai Nippon Printing Co., Ltd., Tokyo. Printing and binding: Brühlsche Universitätsdruckerei, Gießen

Contents

Chapter 6

Spherical Functions — The General Theory

Chapter 7

Topology on the Dual   Plancherel Measure

Chapter 8

Analysis on a Semi-Simple Lie Group

Chapter 9

Spherical Functions on a Semi-Simple Lie Group

Chapter 10

The Discrete Series for a Semi-Simple Lie Group — Existence and Exhaustion

Appendix

Contents Volume I

Chapter 6

Spherical Functions—The General Theory

Introduction

Historically the theory of spherical functions dates to the classical papers of É. Cartan and H. Weyl; they showed that spherical harmonics arise in a natural way from a study of functions on G/K where G is the orthogonal group in n-space and where K consists of those transformations in G which leave a given vector invariant – this study is carried out by the methods of group representations. However in order to get a theory applying to larger classes of 'special functions' it is necessary to drop the assumption that G is compact and also to consider functions not just on G/K but also on G. In 1947 Bargmann studied the pair (G, K) where $G = \mathbf{SL}(2, \mathbf{R})$ and $K = \mathbf{SO}(2)$; there functions Φ on G were considered which, for a given character χ of K, verify the relation

$$\Phi(k_1 x k_2) = \chi(k_1)\Phi(x)\chi(k_2)\,(k_1, k_2 \in K; x \in G);$$

such functions arise upon considering finite or infinite dimensional irreducible representations of G; these representations have coefficients satisfying the above relation and, when G is suitably 'parameterized', it turns out that these functions can be identified with hypergeometric functions; in particular when χ is the trivial character of K one is led to Legendre functions of arbitrary index and to group theoretical explanations of three important properties of these functions, namely their differential equation, their representation by integral formulas and their functional equation. [There is a certain procedure, due to Langlands, by means of which one can obtain all hypergeometric functions from the group theoretic point of view; it is necessary, however, to also work with the universal covering group of $\mathbf{SL}(2, \mathbf{R})$.] At the same time Gelfand and Naimark where led to similar questions in their study of the pair (G, K) where $G = \mathbf{SL}(n, \mathbf{C})$ and $K = \mathbf{SU}(n)$; among other things they were able to obtain an explicit formula for the zonal spherical functions on $\mathbf{SL}(n, \mathbf{C})$, a formula which is very similar to that of Weyl's for the characters of compact (connected) semi-simple groups (Harish-Chandra's generalization of this to arbitrary complex (connected) semi-simple Lie groups will be presented in Chapter 9). The first general results were obtained in 1950 by Gelfand who considered zonal spherical functions on a Riemannian symmetric pair (G, K); a short time thereafter the fundamental papers of Godement and Harish-Chandra appeared.

Let G be a locally compact unimodular group, K a large compact subgroup of G; let U be a TCI Banach representation of G on E; fix a class $\delta \in \hat{K}$ which occurs $m(\geqslant 1)$ times in $U|K$ – then the function ψ_δ^U defined by the prescription

$$x \mapsto \operatorname{tr}(\mathsf{P}(\delta)U(x)\mathsf{P}(\delta)) \qquad (x \in G)$$

is said to be a *spherical trace function* on G of type δ and height m. These are the functions which have been studied traditionally in the 'theory of spherical functions'. An important theorem due to Godement (cf. Theorem 6.1.1.2) establishes a close connection between the irreducible representations of a given algebra $I_{c,\delta}(G)$ $(\delta \in \hat{K})$ and the spherical trace functions on G. Thus the following characterization of spherical trace functions of type δ obtains $(\delta \in \hat{K})$: Let ψ be a K-central continuous function on G such that $\chi_\delta * \psi = \psi$ and which does not 'grow too fast at infinity' – then ψ is a spherical trace function on G of type δ and height m iff there exists an m-dimensional irreducible representation μ_δ of $I_{c,\delta}(G)$ such that

$$\int_G f(x)\psi(x)d_G(x) = d(\delta)\mathrm{tr}(\mu_\delta(f))$$

for every $f \in I_{c,\delta}(G)$. [For certain purposes it is best not to work with the scalar valued spherical trace functions but rather directly with the irreducible representations of the algebra $I_{c,\delta}(G)$ $(\delta \in \hat{K})$; this leads to the important notion of 'spherical function of type δ' which is explored systematically in 6. 1.] From this result it follows that spherical trace functions are similar to the characters of abelian and compact groups; in particular it will be shown that a spherical trace function of type δ (say) satisfies the functional equation

$$\psi(1)\int_K \psi(kxk^{-1}y)dk = \psi(x)\psi(y) \qquad (x, y \in G)$$

iff δ occurs exactly once in the given representation of G. In the event that G is a Lie group, the spherical trace functions are actually analytic on G; moreover they admit an 'infinitesimal' characterization as, roughly speaking, eigenfunctions of $\mathfrak{X}$ (= the centralizer of $\mathfrak{k}_c$ in $\mathfrak{G}$ . . .).

Besides being of interest in its own right, the theory of spherical functions will provide us with useful tools in the representation theory of G; for instance we shall prove that two irreducible unitary representations of G are unitarily equivalent iff they have a common spherical trace function (cf. Corollary 6.1.1.9). On the other hand the theory of spherical functions also offers new interpretations of known results in classical analysis, in particular in the theory of 'special functions'. [In 6.2 we compute spherical functions on motion groups and semi-simple groups; roughly speaking the spherical functions which arise from the study of motion groups are 'generalized Bessel functions' while those which arise in the theory of semi-simple groups are 'hypergeometric' in character (cf. Chapter 9).]

The purpose of this chapter, then, is to present in a systematic fashion the generalities which lie at the basis of the theory of spherical functions on those pairs (G, K) where G is a locally compact unimodular group, K a large compact subgroup of G.

6.1 Fundamentals

6.1.1 Spherical Functions – Functional Properties

Let G be a locally compact unimodular group countable at infinity, K a compact subgroup of G – we shall assume in what follows that K is *large* in G. [Accordingly the theory developed in Vol. I, section 4.5 is then available; all unexplained notations and conventions are as there.]

Definition To say that a pair $\mu = (\mu_1, \mu_2)$ is a *double* (*Banach*) *representation* of K on a Banach space E means that E is a left Banach K-module under μ_1 and a right Banach K-module under μ_2, the operations of K on the left commuting with those of K on the right.

Example Consider $C(K \times K)$ as a Banach space under the norm:

$$\|f\| = \sup_{k_1, k_2 \in K} |f(k_1, k_2)| \, (f \in C(K \times K)).$$

Let $\mu_1(k)f, f\mu_2(k)$ $(k \in K)$ denote, respectively, the functions

$$(k_1, k_2) \mapsto f(k^{-1}k_1, k_2), \, (k_1, k_2) \mapsto f(k_1, k_2 k^{-1}) \qquad (k_1, k_2 \in K).$$

Plainly $\mu = (\mu_1, \mu_2)$ is a double Banach representation of K on $C(K \times K)$.

We shall now give a rather general definition of the term 'spherical function' and then show how such functions occur in practice.

Definition Let $\mu = (\mu_1, \mu_2)$ be a double Banach representation of K on a finite dimensional Banach space E. By a *μ-spherical function* Ψ we shall understand a continuous function Ψ from G to E such that

$$\Psi(k_1 x k_2) = \mu_1(k_1)\Psi(x)\mu_2(k_2) \qquad (k_1, k_2 \in K; x \in G).$$

Let $\hat{K}$ denote the set of all equivalence classes of finite dimensional irreducible representations of K; for each $\delta \in \hat{K}$, let ξ_δ denote the character of δ, $d(\delta)$ the degree of δ and $\chi_\delta = d(\delta)\xi_\delta$.

Examples (1) Consider the algebra $I_{c,\delta}(G)$ (cf. Vol. I, number 4.5.1); then we claim that $I_{c,\delta}(G)$ is isomorphic (as an algebra) to a certain algebra $\mathfrak{A}_{c,\delta}(G)$ of μ-spherical functions on G. Denote by $\mu_{\check{\delta}}$ an irreducible unitary representation of K in the dual class $\check{\delta}$ on a space $E_{\check{\delta}}$ (say); for an arbitrary endomorphism T of $E_{\check{\delta}}$ define the number $\sigma(T)$ by $\sigma(T) = d(\check{\delta})\mathrm{tr}(T)$ – then, thanks to the Schur Orthogonality Relations, for any $T \in F_{\check{\delta}} = \mathrm{Hom}_{\mathbf{C}}(E_{\check{\delta}}, E_{\check{\delta}})$ one has

$$(\sigma) \quad T = \int_K \mu_{\check{\delta}}(k^{-1})\sigma(\mu_{\check{\delta}}(k)T)dk.$$

Given $f \in C_c(G)$ define an element $\Psi_f^\delta \in C_c(G; F_{\check{\delta}})$ by

$$\Psi_f^\delta(x) = \int_K \mu_{\check{\delta}}(k^{-1})f(kx)dk \qquad (x \in G).$$

Claim: Restricted to $I_{c,\delta}(G)$, the map $f \mapsto \Psi_f^\delta$ is a linear bijection onto the space $\mathfrak{A}_{c,\delta}(G)$ of all compactly supported continuous functions $\Psi: G \to F_{\check{\delta}}$ which verify the relation

$$(\delta) \quad \Psi(k_1 x k_2) = \mu_{\check{\delta}}(k_1)\Psi(x)\mu_{\check{\delta}}(k_2) \qquad (k_1, k_2 \in K; x \in G).$$

Thus given $f \in I_{c,\delta}(G)$, it is clear that Ψ_f^δ satisfies (δ) above; moreover $\bar{\chi}_\delta * f = f$ – hence

$$\begin{aligned} f(x) &= \int_K \chi_{\check{\delta}}(k)f(k^{-1}x)dk \\ &= \int_K \sigma(\mu_{\check{\delta}}(k^{-1}))f(kx)dk = \sigma(\Psi_f^\delta(x)) \qquad (x \in G) \end{aligned}$$

and so the map $f \mapsto \Psi_f^\delta$ is injective. To show that the map $f \mapsto \Psi_f^\delta$ is surjective $(f \in I_{c,\delta}(G))$, take any Ψ in the space under consideration and set $f(x) =$

$\sigma(\Psi(x))$ $(x \in G)$ – obviously $f \in C_c(G)$. In addition, for $k \in K, x \in G$, $f(kxk^{-1}) = \sigma(\Psi(kxk^{-1})) = \sigma(\Psi(x)) = f(x)$ and

$$\begin{aligned}\bar{\chi}_\delta * f(x) &= d(\check{\delta}) \int_K \chi_{\check{\delta}}(k)\mathrm{tr}(\Psi(k^{-1}x))dk \\ &= d(\check{\delta}) \int_K \mathrm{tr}(\mu_{\check{\delta}}(k))\sigma(\mu_{\check{\delta}}(k^{-1}\Psi(x))dk \\ &= d(\check{\delta})\mathrm{tr}(\Psi(x)) = f(x) \qquad ((\sigma) \text{ above});\end{aligned}$$

therefore $f \in I_{c,\delta}(G)$. Since $\Psi_f^\delta = \Psi$, the claim follows. Now the space $\mathfrak{A}_{c,\delta}(G)$ actually is an algebra under convolution

$$\Psi_1 * \Psi_2(x) = \int_G \Psi_1(xy)\Psi_2(y^{-1})d_G(y),$$

the product $\Psi_1(xy)\Psi_2(y^{-1})$ being, of course, matrix multiplication $(x, y \in G)$. So, to wrap up the present discussion, there remains only to verify that $\Psi_{f*g}^\delta = \Psi_f^\delta * \Psi_g^\delta$ $(f, g \in I_{c,\delta}(G))$; for this purpose, fix an x in G – then

$$\begin{aligned}\Psi_f^\delta * \Psi_g^\delta(x) &= \int_G \Psi_f^\delta(xy^{-1})\Psi_g^\delta(y)d_G(y) \\ &= \int_G \left(\int_K \mu_{\check{\delta}}(k_1^{-1})f(k_1xy^{-1})dk_1\right)\left(\int_K \mu_{\check{\delta}}(k_2^{-1})g(k_2y)dk_2\right) d_G(y) \\ &= \int_G \int_K \int_K \mu_{\check{\delta}}(k_1^{-1}k_2^{-1})f(k_1xy^{-1})g(k_2y)dk_1dk_2d_G(y) \\ &= \int_K \int_G \int_K \mu_{\check{\delta}}(k_1^{-1}k_2^{-1})f(k_1xy^{-1}k_2)g(y)dk_1d_G(y)dk_2 \\ &= \int_K \int_G \int_K \mu_{\check{\delta}}(k^{-1})f(k_2^{-1}kxy^{-1}k_2)g(y)dkd_G(y)dk_2 \\ &= \int_K \int_G \mu_{\check{\delta}}(k^{-1})f(kxy^{-1})g(y)d_G(y)dk \qquad (f = f_K) \\ &= \Psi_{f*g}^\delta(x),\end{aligned}$$

as desired.

(2) Consider a TCI Banach representation U of G on a Banach space E. For each $\delta \in \hat{K}$ we then have, as usual, a continuous projection $\mathsf{P}(\delta): E \to E$ with $\mathsf{P}(\delta)E = E(\delta)$, $E(\delta)$ the closed subspace of E consisting of those vectors in E which transform under K according to δ – of course $\dim(E(\delta)) < \infty$, K being large. Now suppose that $\delta_1, \delta_2 \in \hat{K}$ are such that $E(\delta_i) = \mathsf{P}(\delta_i)E \neq \{0\}$; let $U_{\delta_i}(k)$ denote the restriction of $U(k)$ to $E(\delta_i)$ $(k \in K; i = 1, 2)$ – then we obtain a double representation $\mu_U = (\mu_1, \mu_2)$ of K on $\mathrm{Hom}_{\mathbf{C}}(E(\delta_2), E(\delta_1))$ by writing:

$$\mu_1(k_1)T\mu_2(k_2) = U_{\delta_1}(k_1)TU_{\delta_2}(k_2) \, (T \in \mathrm{Hom}_{\mathbf{C}}(E(\delta_2), E(\delta_1)); \, k_1, k_2 \in K).$$

Put $\Psi_{\delta_1,\delta_2}^U(x) = \mathsf{P}(\delta_1)U(x)\mathsf{P}(\delta_2)$ $(x \in G)$ – then $\Psi_{\delta_1,\delta_2}^U$ can be regarded as a spherical function on G of type μ_U. The case when $\delta_1 = \delta_2 = \delta$ (say) is of particular importance – we shall write Ψ_δ^U for $\Psi_{\delta,\delta}^U$ in this situation. Notice that the μ_U-spherical function Ψ_δ^U satisfies the relation $\chi_\delta * \Psi_\delta^U = \Psi_\delta^U * \chi_\delta = \Psi_\delta^U$ (use the formula (σ) in Example 1 above) and has the important property that the map $f \mapsto \int_G f(x)\Psi_\delta^U(x)d_G(x)$ is an irreducible representation of the algebra $C_{c,\delta}(G)$ on $E(\delta)$ – moreover the growth of Ψ_δ^U at infinity is not arbitrary since there will exist a constant $M > 0$ such that

$$\|\Psi_\delta^U(x)\| \leqslant M\rho(x) \quad (\rho(x) = \|U(x)\|;\ x \in G).$$

Our next task will be to establish the basic properties of spherical trace functions – these are what Godement [7] and Harish-Chandra [7] call spherical functions, but we shall reserve this latter term for a somewhat different concept.

Let U be a TCI Banach representation of G on E.

Definition Let $\delta \in \hat{K}$. The function ψ_δ^U on G given by

$$x \mapsto \operatorname{tr}(\mathsf{P}(\delta)U(x)\mathsf{P}(\delta)) = \operatorname{tr}(\Psi_\delta^U(x))$$

will be called the *spherical trace function of type* δ of the given representation U.

If δ occurs m times in $U|K$, then ψ_δ^U is said to have *height* m.

Proposition 6.1.1.1 Let ψ_δ^U be a spherical trace function on G of type δ – then:

(i) For all $x \in G$, $k \in K$, $\psi_\delta^U(kxk^{-1}) = \psi_\delta^U(x)$;

(ii) For all $x \in G$, $\chi_\delta * \psi_\delta^U(x) = \psi_\delta^U * \chi_\delta(x) = \psi_\delta^U(x)$;

(iii) For all $x, y \in G$, $\int_K \psi_\delta^U(xky)\bar{\chi}_\delta(k)dk = \int_K \psi_\delta^U(ykx)\bar{\chi}_\delta(k)dk$;

(iv) For all $x, y \in G$, $\int_K \psi_\delta^U(kxk^{-1}y)dk = \int_K \psi_\delta^U(kyk^{-1}x)dk$.

Proof To prove (i), first note that $\mathsf{P}(\delta)U(k) = U(k)\mathsf{P}(\delta)$ $(k \in K)$; hence

$$\psi_\delta^U(kxk^{-1}) = \operatorname{tr}(\mathsf{P}(\delta)U(k)U(x)U(k^{-1})\mathsf{P}(\delta)) = \operatorname{tr}(\mathsf{P}(\delta)U(x)\mathsf{P}(\delta)) = \psi_\delta^U(x) \quad (x \in G).$$

Turning to (ii), let $\mu \in M_c(G)$ – then

$$\mu' * \psi_\delta^U(x) = \int_G \psi_\delta^U(yx)d\mu(y) = \operatorname{tr}(\mathsf{P}(\delta)U(\mu)U(x)\mathsf{P}(\delta)),$$

where, as usual (A 2.4), μ' denotes the transform of μ under the map $x \mapsto x^{-1}$ $(x \in G)$. Specialize and take $\mu = \bar{\chi}_\delta$. Since $(\bar{\chi}_\delta)' = \chi_\delta$, we find that $\chi_\delta * \psi_\delta^U = \psi_\delta^U(U(\bar{\chi}_\delta) = \mathsf{P}(\delta))$; in a similar way, one gets the other half of (ii). As for (iii), observe that

$$\begin{aligned}\int_K \psi_\delta^U(xky)\bar{\chi}_\delta(k)dk &= \int_K \operatorname{tr}(\mathsf{P}(\delta)U(x)U(k)U(y)\mathsf{P}(\delta))\bar{\chi}_\delta(k)dk \\ &= \operatorname{tr}(\mathsf{P}(\delta)U(x)\mathsf{P}(\delta)U(y)\mathsf{P}(\delta)).\end{aligned}$$

Because this last expression is symmetric with respect to x and y, (iii) follows. The proof of (iv) (which is equivalent to the relation $f * \psi_\delta^U = \psi_\delta^U * f$, $f \in \mathfrak{J}_c(G)$) may be left to the reader. □

It should be pointed out that spherical trace functions ψ_δ^U may not belong to $\mathfrak{J}_c(G)$ (their support need not be compact) – in fact are not in $L^1(G)$ in general.

A Banach space valued function f on G is said to be *quasi-bounded* if there exists a semi-norm ρ on G such that

$$\sup_{x \in G} \frac{\| f(x) \|}{\rho(x)} < \infty.$$

Plainly every spherical trace function is quasi-bounded.

We shall now take up a very important result, a result which establishes a close connection between spherical trace functions of type δ and 'characters' of the (in general non-commutative) algebra $I_{c,\delta}(G)$.

Theorem 6.1.1.2 (Godement) Let ψ be a quasi-bounded continuous function on G satisfying $\psi(kxk^{-1}) = \psi(x)$ $(k \in K, x \in G)$, $\chi_\delta * \psi = \psi$ – then ψ is a spherical trace function of type δ and height m iff there exists an m-dimensional irreducible representation μ_δ of $I_{c,\delta}(G)$ such that

$$\psi(f) = \int_G f(x)\psi(x)d_G(x) = d(\delta)\mathrm{tr}(\mu_\delta(f))$$

for every $f \in I_{c,\delta}(G)$.

Proof Necessity Let ψ_δ^U be a spherical trace function of type δ and height m. As was observed in the Remark following Proposition 4.5.1.7, there exists an isomorphism $U_\delta(f) \mapsto \mu_\delta(f), f \in I_{c,\delta}(G)$, onto $\mathfrak{gl}(m, \mathbf{C})$ – moreover μ_δ is an m-dimensional irreducible representation of the algebra $I_{c,\delta}(G)$ with

$$\mathrm{tr}(U_\delta(f)) = d(\delta)\mathrm{tr}(\mu_\delta(f)) \qquad (f \in I_{c,\delta}(G)).$$

As

$$\psi_\delta^U(f) = \mathrm{tr}(\mathsf{P}(\delta)U(f)\mathsf{P}(\delta)) = \mathrm{tr}(U_\delta(f)) = d(\delta)\mathrm{tr}(\mu_\delta(f)) \qquad (f \in I_{c,\delta}(G)),$$

the necessity follows. □

To prove the sufficiency, three lemmas will be needed. First let us review some basic terminology. A left ideal J in an associative algebra A is said to be *regular* if A admits a right identity mod J, i.e. if there exists an element $u \in A$ such that $xu \equiv x \bmod J$ for every $x \in A$. Every regular left ideal is contained in at least one regular maximal left ideal; in a complete normed algebra (i.e. a Banach algebra) every regular maximal left ideal is closed.

There is a canonical way to manufacture TCI Banach representations of G. Thus let ρ be a semi-norm on G, J a regular maximal left ideal in the Banach algebra $C_\rho(G)$; since J is closed it is invariant under left translation by elements of G – otherwise said there is induced on the Banach space $C_\rho(G)/J$ a natural representation U of G. The corresponding lift of U to $C_\rho(G)$ associates with each $f \in C_\rho(G)$ the operator which transforms the class of $g \in C_\rho(G)$ (mod J) into the class of $f*g$ (mod J) – thus this representation of $C_\rho(G)$ is algebraically irreducible (J is maximal) and, since it is a question of bounded operators, is in fact TCI (Proposition 4.2.1.4). Hence the representation U of G is TCI. Notice that if u is a right identity mod J and if $\dot{u}$ is the class of u in $C_\rho(G)/J$, then $f \in J$ iff $U(f)\dot{u} = 0$.

[Conversely, let U be a TCI Banach representation of G on E. Taking $\rho(.) = \| U(.)\|$, fix a non-zero vector $a \in E(\delta)$ $(\delta \in \hat{K})$ and consider the set J of all f in $C_\rho(G)$ such that $U(f)a = 0$. It may then be shown, using the following lemma, that J is a regular maximal left ideal in $C_\rho(G)$ (for details see Godement [7, p. 514]).]

Let us turn now to the formulation and proofs of the lemmas referred to above.

Lemma 6.1.1.3 Let ρ be a semi-norm on G; let I be a regular maximal left ideal in $C_{\rho,\delta}(G)$. If J is the set of all $f \in C_\rho(G)$ such that

$$\bar{\chi}_\delta * g * f * \bar{\chi}_\delta \in I \quad \text{for every} \quad g \in C_\rho(G),$$

then J is a regular maximal left ideal in $C_\rho(G)$, $I = J \cap C_{\rho,\delta}(G)$, and we have $f*\bar{\chi}_\delta \equiv f \bmod J$ for all $f \in C_\rho(G)$.

Proof Clearly J is a left ideal; to see that J is regular, let $u \in C_{\rho,\delta}(G)$ be a right identity mod I and observe that for arbitrary $f, g \in C_\rho(G)$ one has $\bar{\chi}_\delta * g * (f*u - f) * \bar{\chi}_\delta = h*u - h$ where $h = \bar{\chi}_\delta * g * f * \bar{\chi}_\delta \in C_{\rho,\delta}(G)$ – of course this implies that u is a right identity mod J. Let us show next that $I = J \cap C_{\rho,\delta}(G)$. Owing to the maximality of I, it is enough to verify that $J \cap C_{\rho,\delta}(G)$ is a non-trivial ideal containing I. Claim: $u \notin J$. For if so, then $\bar{\chi}_\delta * f * u * \bar{\chi}_\delta \in I$ for every $f \in C_\rho(G)$; in turn this would entail that $f \in I$ for every $f \in C_{\rho,\delta}(G)$, a contradiction. Since it is obvious that $I \subset J$, we conclude that $I = J \cap C_{\rho,\delta}(G)$. The fact that $f*\bar{\chi}_\delta - f \in J$ $(f \in C_\rho(G))$ follows from the relation $\bar{\chi}_\delta * g * (f*\bar{\chi}_\delta - f) * \bar{\chi}_\delta = 0$ $(g \in C_\rho(G))$. It remains to show that J is maximal. Let $\tilde{J}$ be a left ideal containing J and assume that $\tilde{J} \neq C_\rho(G)$ (thus $u \notin \tilde{J}$). Reasoning as before, we see that $I = \tilde{J} \cap C_{\rho,\delta}(G)$. Let $f \in \tilde{J}$ – then $f \in J$. For we have:

$$\begin{aligned} f*\bar{\chi}_\delta - f \in J \subset \tilde{J} &\Rightarrow f*\bar{\chi}_\delta \in \tilde{J} \Rightarrow \bar{\chi}_\delta * g * f * \bar{\chi}_\delta \in \tilde{J} \qquad (\text{all } g \in C_\rho(G)) \\ &\Rightarrow \bar{\chi}_\delta * g * f * \bar{\chi}_\delta \in \tilde{J} \cap C_{\rho,\delta}(G) = I \Rightarrow f \in J. \end{aligned}$$

Hence the lemma. □

Remark Lemma 6.1.1.3 is valid without any change in the following more general setting. Call an associative algebra A (over **C**) a *group algebra* of G if A is a two-sided module over $M_c(G)$ – thus, for every $x \in A$, $\mu \in M_c(G)$, it makes sense to consider $x\mu$, μx and the usual rules of computation are satisfied. In such a group algebra A one may then define subalgebras A_δ by the condition: $\bar{\chi}_\delta x = x\bar{\chi}_\delta = x (x \in A, \delta \in \hat{K})$.

As usual, if f is a function on G, then f' will denote the transform of f under the map $x \mapsto x^{-1}$ (hence $f'(x) = f(x^{-1})$, $x \in G$).

Lemma 6.1.1.4 Let ψ be a continuous function on G such that $\psi = \psi_K$, $\chi_\delta * \psi = \psi$. The following conditions are equivalent:
(i) $\psi(f*g) = \psi(g*f)$ for all $f, g \in I_{c,\delta}(G)$;
(ii) $f' * \psi = \psi * f'$ for all $f \in I_{c,\delta}(G)$.

Proof In general we have $\psi(f) = f' * \psi(1) = \psi * f'(1)$ $(f \in C_c(G))$. Now (i) implies that

$$g' * f' * \psi(1) = \psi * f' * g'(1),$$

i.e. (i) implies that $f' * \psi(g) = \psi * f'(g)$ for every $f, g \in I_{c,\delta}(G)$ – thus $f' * \psi(g) = \psi * f'(g)$ for all $g \in C_c(G)$, which gives (ii). Likewise one may reverse these steps and deduce that (ii) $\Rightarrow$ (i). □

Lemma 6.1.1.5 Let ψ be a continuous function on G such that $\psi = \psi_K$, $\chi_\delta * \psi = \psi$; let ρ be a semi-norm on G such that $|\psi(x)| \leqslant M\rho(x)$ for all $x \in G (M > 0)$, $C_\rho(G)$ the Banach algebra corresponding to ρ. If there exists a finite dimensional irreducible representation μ_δ of $I_{c,\delta}(G)$ such that $\psi(f) = d(\delta)\mathrm{tr}(\mu_\delta(f))$ (all $f \in I_{c,\delta}(G)$), then:
(i) $f' * \psi = \psi * f'$ for every $f \in C_{\rho,\delta}(G)$;
(ii) $I_\psi = \{f \in C_{\rho,\delta}(G) : f' * \psi = 0\}$ is a regular two-sided ideal in $C_{\rho,\delta}(G)$;
(iii) $f \in I_\psi \cap I_{c,\delta}(G)$ iff $\mu_\delta(f) = 0$.

Proof Since $\psi(f) = d(\delta)\mathrm{tr}(\mu_\delta(f))$, $\psi(f*g) = \psi(g*f)$ for every $f, g \in I_{c,\delta}(G)$ and so $f' * \psi = \psi * f'$ for all $f \in I_{c,\delta}(G)$ (Lemma 6.1.1.4). Now $C_c(G)$ is dense in $C_\rho(G)$ and the map $f \mapsto \bar{\chi}_\delta * f_K$ is a continuous projection of $C_\rho(G)$ onto $I_{\rho,\delta}(G)$; thus $I_{c,\delta}(G)$ is dense in $I_{\rho,\delta}(G)$ and so $f' * \psi = \psi * f'$ for all $f \in I_{\rho,\delta}(G)$. Plainly this fact persists if we replace f by $\delta_k * f$ $(k \in K)$ – therefore (i) follows from Lemma 4.5.1.11. Clearly (i) in conjunction with the general rule $(f*g)' = g' * f'$ tells us that I_ψ is a two-sided ideal in $C_{\rho,\delta}(G)$. Next observe that $f \in I_\psi \cap I_{c,\delta}(G) \Rightarrow f' * \psi = 0 \Rightarrow \psi(f*g) = 0$ (all $g \in I_{c,\delta}(G)) \Rightarrow \mathrm{tr}(\mu_\delta(f)\mu_\delta(g)) = 0$ (all $g \in I_{c,\delta}(G)) \Rightarrow \mu_\delta(f) = 0$ (Burnside's Theorem); the converse is clear too – hence (iii). Finally, in order to see that I_ψ is regular, take $u \in I_{c,\delta}(G)$ such that $\mu_\delta(u)$ is the identity operator in the representation space of μ_δ – thus $\psi * u' = u' * \psi = \psi$ and so, if $f \in C_{\rho,\delta}(G)$, then

$$(f*u - f)' * \psi = u' * f' * \psi - f' * \psi = \psi * f' - f' * \psi = 0. \quad \square$$

We can now conclude the proof of Theorem 6.1.1.2.

Proof *Sufficiency* (1) Retain the notations of Lemma 6.1.1.5; extend I_ψ to a regular maximal left ideal I in $C_{\rho,\delta}(G)$ ($I_\psi \neq C_{\rho,\delta}(G)$ if $\psi \neq 0$) and then, using the procedure of Lemma 6.1.1.3, construct the regular maximal left ideal J in $C_\rho(G)$ – thus J consists of those $f \in C_\rho(G)$ such that $\bar{\chi}_\delta * g * f * \bar{\chi}_\delta \in I$ for all $g \in C_\rho(G)$. Consider the TCI Banach representation U of G on $E = C_\rho(G)/J$. Claim: The natural representation $f \mapsto U_I(f)$

of $C_{p,\delta}(G)$ on $C_{p,\delta}(G)/I$ is equivalent to the representation $f \mapsto U_\delta(f)$ of $C_{p,\delta}(G)$ on $E(\delta)$ (of course $E(\delta)$ is finite dimensional, K being large). To see this, note that the projection operator $\mathsf{P}(\delta)$ is given by $\mathsf{P}(\delta)(f+J) = \bar{\chi}_\delta * f + J$; on the other hand $f * \bar{\chi}_\delta \equiv f \bmod J$ for all $f \in C_p(G)$ – hence $f \mapsto f + J$ is a mapping of $C_{p,\delta}(G)$ onto $E(\delta)$. Since $I = J \cap C_{p,\delta}(G)$, the claim is clear. Let ψ_δ^U denote the spherical trace function of type δ for the representation U of G on E – then $\psi_\delta^U(f) = \mathrm{tr}(U_I(f))$ for all $f \in C_{p,\delta}(G)$.

(2) Owing to the fact that I_ψ is a two-sided ideal in I, it is plain that $f \in I_\psi \Rightarrow U_I(f) = 0$; in particular, $f \in I_\psi \cap I_{c,\delta}(G) \Rightarrow U_I(f) = 0$ (Lemma 6.1.1.5 (iii)).

(3) Let n be the height of ψ_δ^U – then, as has been seen during the proof of the necessity, there exists an n-dimensional irreducible representation $f \mapsto \nu_\delta(f)$ of $I_{c,\delta}(G)$ such that $\psi_\delta^U(f) = d(\delta)\mathrm{tr}(\nu_\delta(f))$ for every $f \in I_{c,\delta}(G)$. In addition: $\nu_\delta(f) = 0$ iff $U_\delta(f) = 0$ iff $U_I(f) = 0$ $(f \in I_{c,\delta}(G))$; therefore, $f \in I_\psi \cap I_{c,\delta}(G)$ iff $\mu_\delta(f) = 0 \Rightarrow U_I(f) = 0 \Rightarrow \nu_\delta(f) = 0$. Consequently, since it is a question of finite dimensional irreducible representations of an associative algebra, it follows that μ_δ and ν_δ are equivalent. Thus $m = n$ and so

$$\psi(f) = d(\delta)\mathrm{tr}(\mu_\delta(f)) = d(\delta)\mathrm{tr}(\nu_\delta(f)) = \psi_\delta^U(f) \qquad (f \in I_{c,\delta}(G))$$

which implies that $\psi = \psi_\delta^U$.

This completes the proof of Theorem 6.1.1.2. □

Remark The characterization of spherical trace functions as (essentially) the traces of certain 'characters' (i.e. certain finite dimensional irreducible representations) of the various algebras $I_{c,\delta}(G)$ reflects the definition of these functions in terms of the TCI Banach representations of G through the requirement of quasi-boundedness. It has been shown by Dieudonné [3] that it is possible to prove Theorem 6.1.1.2 without assuming any 'condition of increase' on ψ provided representations on a locally convex space are admitted. In fact it is possible (and not difficult) to carry through the whole theory at the level of TCI locally convex representations rather than just TCI Banach representations – however there exists reasonably suggestive evidence to show that such a generalization is not really necessary (cf. Fell [4], Rader [1]). It is for this reason that we have elected to stay within the category of Banach spaces.

Important information concerning the structure of the algebras $I_{c,\delta}(G)$ and $C_{c,\delta}(G)$ is contained in:

Proposition 6.1.1.6 The following properties are equivalent:
(i) $I_{c,\delta}(G)$ is commutative;
(ii) δ occurs at most once in every TCI Banach representation of G;
(iii) $I_{c,\delta}(G)$ is the center of $C_{c,\delta}(G)$.

Proof Assume (ii). Then $I_{c,\delta}(G)$ admits sufficiently many one dimensional representations (cf. Proposition 4.5.1.10) – hence (i). Conversely, if $I_{c,\delta}(G)$ is commutative, then every finite dimensional irreducible representation of $I_{c,\delta}(G)$ is one dimensional so that every spherical trace function of type δ is of height one – thus (i) iff (ii). Plainly (iii) $\Rightarrow$ (i); now assume (i) – then (ii) also is in force. Therefore, for every TCI Banach representation U of G on E, the representation $k \mapsto U_\delta(k)$ of K on $E(\delta)$ is irreducible which implies that $U_\delta(f)$ reduces to a scalar for every $f \in I_{c,\delta}(G)$ – thus $U_\delta(g*f - f*g) = 0$ for all $f \in I_{c,\delta}(G)$, $g \in C_{c,\delta}(G)$ and so $I_{c,\delta}(G)$ is contained in the center of $C_{c,\delta}(G)$, U being arbitrary. Furthermore, if f belongs to the center of $C_{c,\delta}(G)$, then it is evident that $U_\delta(f)$ is a scalar in every TCI Banach representation of G, hence that

$$U_\delta(f) = \int_K U(k)U_\delta(f)U(k^{-1})dk = U_\delta(f_K) \Rightarrow f = f_K \Rightarrow f \in I_{c,\delta}(G). \quad \square$$

Examples (1) Take for δ the class corresponding to the trivial ($k \mapsto 1$) one dimensional representation of K and ask: When is the algebra $I_{c,\delta}(G) = I_c(G)$ commutative? We recall that a sufficient condition for this is that (G, K) be a Riemannian symmetric pair (compare, e.g. Helgason [2, p. 408]). [In fact the involutive automorphism σ corresponding to the pair (G, K) evidently preserves the Haar measure on G; let $f \mapsto f^\sigma$ denote the corresponding automorphism of $I_c(G)$. Since each $x \in G$ can be written $x = kp$ where $\sigma(k) = k$, $\sigma(p) = p^{-1}$, it follows that $f^\sigma = f'$; so, on the one hand, $(f*g)' = g'*f'$, while on the other, $(f*g)^\sigma = f^\sigma * g^\sigma$ – hence $f*g = g*f$ ($f, g \in I_c(G)$). Notice, too, that the same proof applies more generally to the canonical algebras of biinvariant measures and distributions on G.] In particular this is the case when G is a connected semi-simple Lie group with finite center, K a maximal compact subgroup.

(2) Suppose that (G, K) is a pair with the property that δ occurs no more than $d(\delta)$ times in any TCI Banach representation of G – then it is clear, in view of Proposition 6.1.1.6, that the algebra $I_{c,\delta}(G)$ is commutative whenever δ is the class of some one dimensional representation of K. Examples of pairs (G, K) which verify this assumption were given in Vol. I, number 4.5.2.

Spherical trace functions of height one may be characterized by a functional equation.

Theorem 6.1.1.7 (Godement) Let ψ be a quasi-bounded continuous function on G – then ψ is proportional to a spherical trace function of height one iff

$$(*) \quad \psi(1)\int_K \psi(kxk^{-1}y)dk = \psi(x)\psi(y)$$

for arbitrary $x, y \in G$.

Proof (1) Let ψ_δ^U be a spherical trace function of type δ and height one (U a TCI Banach representation of G on E). Put $\psi = \psi_\delta^U/d(\delta)$ – then Theorem 6.1.1.2 implies that the map $f \mapsto \psi(f)$ is a homomorphism of $I_{c,\delta}(G)$ onto $\mathbf{C}$. Since $\psi = \psi_K$, $\chi_\delta * \psi = \psi$, we have, for arbitrary $f, g \in C_c(G)$,

$$\psi(f_K*g) = \psi((f_K*g)_K) = \psi(f_K*g_K) = \psi(\bar{\chi}_\delta*f_K*g_K*\bar{\chi}_\delta)$$
$$= \psi(\bar{\chi}_\delta*f_K)\psi(g_K*\bar{\chi}_\delta) = \psi(f)\psi(g).$$

Now $C_c(G)$ is weakly dense in $M_c(G)$ – therefore, for all $\mu, \nu \in M_c(G)$, we have $\psi(\mu_K * \nu) = \psi(\mu)\psi(\nu)$. In particular, upon taking $\mu = \delta_x, \nu = \delta_y$, the desired functional equation (*) follows at once.

(2) Conversely, if ψ is a quasi-bounded continuous function on G verifying (*), then $\psi((\delta_x)_K * \delta_y) = \psi(\delta_x)\psi(\delta_y)$ for all $x, y \in G$ (we may assume that $\psi(1) = 1$); since every element in $M_c(G)$ is the weak limit of linear combinations of Dirac measures, we see that $\psi(\mu_K * \nu) = \psi(\mu)\psi(\nu)$ for all $\mu, \nu \in M_c(G)$. Consequently, for any $\delta \in \hat{K}$, the rule $f \mapsto U(f)$ given by

$$U(f)(r + \sqrt{-1}s) = \psi(f)(r + \sqrt{-1}s) \qquad (r, s \in \mathbf{R}; f \in I_{c,\delta}(G)),$$

defines a one or zero dimensional representation of $I_{c,\delta}(G)$ such that $\mathrm{tr}(U(f)) = \psi(f)$. So, in order to invoke Theorem 6.1.1.2 to conclude that ψ is proportional to a spherical trace function of type δ, it is enough to exhibit a $\delta \in \hat{K}$ such that $\chi_\delta * \psi = \psi$. Put $\phi = \psi | K$ – then, for all $k_1, k_2 \in K$,

$$\int_K \phi(kk_1k^{-1}k_2)dk = \phi(k_1)\phi(k_2).$$

In other words ϕ satisfies the functional equation for the characters of compact groups. Therefore there exists a $\delta \in \hat{K}$ such that $\phi = \xi_\delta / d(\delta)$, from which it follows that $\psi | K = \chi_\delta / d(\delta)^2$ – but then

$$\begin{aligned}\chi_\delta * \psi(x) &= \int_K \psi(k^{-1}x)\chi_\delta(k)dk = \int_K \int_K \psi(k^{-1}x)\chi_\delta(\tilde{k}^{-1}k\tilde{k})dkd\tilde{k} \\ &= \int_K \int_K \psi(\tilde{k}k^{-1}\tilde{k}^{-1}x)\chi_\delta(k)d\tilde{k}dk = \left(\int_K \psi(k)\overline{\chi_\delta(k)}dk\right)\psi(x) \\ &\qquad (x \in G).\end{aligned}$$

Put $M = \int_K \psi(k)\overline{\chi_\delta(k)}dk$ – then, since $\psi | K = \chi_\delta / d(\delta)^2$, M must actually be equal to one, whence $\chi_\delta * \psi = \psi$, as desired.

The proof of Theorem 6.1.1.7 is now complete. □

Remark Functional equations for arbitrary spherical trace functions do not seem to be known – for some special cases see 6.2.

It is well-known that if two topologically irreducible unitary representations of a group have a common coefficient, then they are unitarily equivalent. This result may be generalized to our setting as follows. Suppose that U is a TCI Banach representation of G on E containing a class $\delta \in \hat{K}$ – then the spherical trace function ψ_δ^U determines U to within Naimark equivalence. To see this, simply observe that for $f \in C_c(G)$,

$$\psi_\delta^U(f) = \mathrm{tr}(\mathsf{P}(\delta)U(f)\mathsf{P}(\delta)) = \int_G f(x)\psi_\delta^U(x)d_G(x)$$

and so the knowledge of ψ_δ^U determines the trace of the finite dimensional algebraically irreducible representation $f \mapsto U_\delta(f)$ of $C_{c,\delta}(G)$ on $E(\delta)$ – in turn the knowledge of the trace of U_δ determines U_δ to within equivalence and thus, as we know (cf. Proposition 4.5.1.6), U to within Naimark equivalence. Hence:

Theorem 6.1.1.8 (Harish-Chandra) Let U and V be TCI Banach representations of G. Suppose that, for some $\delta \in \hat{K}$, the corresponding spherical trace functions of type δ are equal and do not vanish – then U and V are Naimark equivalent. Conversely, if U and V are Naimark equivalent, then $\psi_\delta^U = \psi_\delta^V$ for every $\delta \in \hat{K}$.

Corollary 6.1.1.9 Two irreducible unitary representations of G are unitarily equivalent iff they have a common spherical trace function.

Proof This is an immediate consequence of Theorem 6.1.1.8 and Proposition 4.3.1.4. □

The preceding theorem admits the following important generalization.

Theorem 6.1.1.10 (Harish-Chandra) Let $U^1, \ldots, U^r$ be a finite set of TCI Banach representations of G; suppose that no two of them are Naimark equivalent – then the non-zero functions in the set

$$\psi_{\delta_1}^{U^1}, \ldots, \psi_{\delta_r}^{U^r} \qquad (\delta_i \in \hat{K}, i = 1, \ldots, r)$$

are linearly independent.

Proof (Rader) Without loss of generality, we may assume that all the $\psi_{\delta_i}^{U^i}$ are non-zero; this being so, suppose, if possible, that we have a non-trivial dependence relation of the form

$$c_1\psi_{\delta_1}^{U^1} + \cdots + c_r\psi_{\delta_r}^{U^r} = 0 \qquad (c_i \in \mathbf{C}, i = 1, \ldots, r).$$

By grouping together the terms which have δ_i (say) in common, and then multiplying through by χ_{δ_i}, we see that we might just as well assume that $\delta_1 = \delta_2 = \ldots = \delta_r = \delta$ (say). Our dependence relation then implies that $\sum_i c_i \mathrm{tr}(U_\delta^i(f)) = 0$ for every $f \in C_{c,\delta}(G)$; however the representations $f \mapsto U_\delta^i(f)$ $(i = 1, \ldots, r)$ are finite dimensional algebraically irreducible representations of the algebra $C_{c,\delta}(G)$ which are, moreover, pairwise inequivalent (by hypothesis) – therefore, in view of well-known algebraic facts (cf. the Appendix to Vol. I, number 4.5.8), the functions $f \mapsto \mathrm{tr}(U_\delta^i(f))$ $(i = 1, \ldots, r)$ are linearly independent, a contradiction. □

We shall now take up some questions which were left open in Vol. I, number 4.5.1.

Let us return to the discussion following Proposition 4.5.1.6; there, for each $\delta \in \hat{K}$, we constructed an injection from $\bar{G}(\delta)$ into $\bar{C}_{c,\delta}(G)$ – we wish to consider the range of this injection.

First of all, notice that the elements of $\bar{C}_{c,\delta}(G)$ are finite dimensional; in fact are uniformly bounded in dimension – this is plain in view of our standing suppositions (in particular, the largeness of K in G) and Proposition 4.5.1.5. Suppose now that V_δ is an irreducible finite dimensional representation of $C_{c,\delta}(G)$; let ρ be a semi-norm on G – then the following three conditions are equivalent:

(i) The representation V_δ is continuous on $C_{c,\delta}(G)$ with respect to $\|\cdot\|_\rho$;

(ii) There exists a TCI Banach representation U of G such that $U_\delta \sim V_\delta$ and $\|U(x)\| \leqslant \rho(x)$ for all x in G;

(iii) There exists a TCI Banach representation U of G such that $U_\delta \sim V_\delta$ and $\|U(x)\| \leqslant M\rho(x)$ for all x in G, M some positive constant.

[It is clear that (ii) $\Rightarrow$ (iii) and (iii) $\Rightarrow$ (i); that (i) $\Rightarrow$ (ii) is contained in the proof of Theorem 6.1.1.2 (i.e. in the sufficiency).]

Our main interest is in the finite dimensional irreducible representations of the various algebras $I_{c,\delta}(G)$ ($\delta \in \hat{K}$). As has been observed in Vol. I, number 4.5.1, there exists a natural one-to-one correspondence between the elements of $\bar{C}_{c,\delta}(G)$ and $\bar{I}_{c,\delta}(G)$ and, as a consequence, a natural injection of $\bar{G}(\delta)$ into $\bar{I}_{c,\delta}(G)$. The preceding discussion then tells us that a finite dimensional irreducible representation of $I_{c,\delta}(G)$ is in the range of this injection iff it is continuous with respect to some $\|\cdot\|_\rho$, ρ a semi-norm on G.

Let ρ be a semi-norm on G. Denote by $\bar{G}_\rho$ the subset of $\bar{G}$ consisting of those classes which contain some TCI Banach representation U of G with the property that, for some constant $M > 0$, $\|U(x)\| \leqslant M\rho(x)$ for all $x \in G$ – of course $\bar{G} = \bigcup_\rho \bar{G}_\rho$ (ρ running over all semi-norms on G). [Suppose in particular that $\rho = 1$ – then $\bar{G}_1$ may be viewed as the set of all Naimark equivalence classes of uniformly bounded TCI Banach representations of G. Notice that $\bar{G}_1$ contains $\hat{G}$ as a subset – for a topologically irreducible unitary representation of G is necessarily TCI (Proposition 4.3.1.7) and two such are unitarily equivalent iff they are Naimark equivalent (Proposition 4.3.1.4).]

Let $\delta \in \hat{K}$. If $U \in \bar{U} \in \bar{G}(\delta)$, then the representation $f \mapsto U_\delta(f)$ of $C_{c,\delta}(G)$ is finite dimensional and irreducible and, to within equivalence, depends only on $\bar{U}$ – denote this latter equivalence class by $\bar{U}_\delta$. Similarly the representation $f \mapsto \mu_\delta(f)$ of $I_{c,\delta}(G)$, which corresponds to U_δ in the canonical way, is finite dimensional and irreducible and, to within equivalence, depends only on $\bar{U}$ – denote this latter equivalence class by $\bar{\mu}_\delta$. With these notations, the preceding considerations may be summarized in:

Theorem 6.1.1.11 Fix $\delta \in \hat{K}$ – then:

(1) The map $\bar{U} \mapsto \bar{U}_\delta$ is a bijection of $\bar{G}(\delta) \cap \bar{G}_\rho$ onto the family of all equivalence classes of finite dimensional irreducible representations of $C_{c,\delta}(G)$ which are continuous with respect to $\|\cdot\|_\rho$;

(2) The map $\bar{U} \mapsto \bar{\mu}_\delta$ is a bijection of $\bar{G}(\delta) \cap \bar{G}_\rho$ onto the family of all equivalence classes of finite dimensional irreducible representations of $I_{c,\delta}(G)$ which are continuous with respect to $\|\,.\,\|_\rho$.

To a certain extent, at least, Theorem 6.1.1.11 reduces the problem of classifying the elements of $\bar{G}$ to that of classifying finite dimensional irreducible representations of subalgebras of $C_c(G)$. In turn this latter problem is by no means easy – for one must, evidently, study the finite dimensional irreducible representations of the, in general, *non-commutative* Banach algebras $I_{\rho,\delta}(G)$. [It should be noted that any such representation is automatically continuous. In fact, generalizing the well-known statement concerning the continuity of the homomorphisms of a commutative Banach algebra, the following result holds: Let A be a Banach algebra over **C** – then any finite dimensional irreducible representation of A is continuous (simple examples show that the assumption of irreducibility cannot be dropped). For a proof, see Fell [3, p. 240].]

Remark Retain the preceding notations – then, in general, we have the inclusions $\hat{G} \subset \bar{G}_1 \subset \bar{G}$, and so it is only natural to enquire as to the various combinations of equality and inequality here. Examples of the case in which $\bar{G} = \bar{G}_1 \neq \hat{G}$ do not seem to be known; the non-compact connected semi-simple Lie groups (with finite center) will illustrate the case $\bar{G} \neq \bar{G}_1 \neq \hat{G}$; the additive group of the reals satisfies $\bar{G} \neq \bar{G}_1 = \hat{G}$; finally all compact groups have the property that $\bar{G} = \hat{G}$.

Consider the algebra $I_c(G)$. By definition, a *zonal spherical function* ϕ on G is a continuous, complex valued function which is biinvariant under K and has the property that the map $f \mapsto \int_G f(x)\phi(x)d_G(x)$ is a homomorphism of $I_c(G)$ (cf. Helgason [2, p. 409]). This suggests the following generalization.

Definition Let $\delta \in \hat{K}$. A *spherical function* Φ *(on G) of type* δ is a quasi-bounded continuous function on G with values in $\mathrm{Hom}_{\mathbf{C}}(E, E)$ (E a finite dimensional linear space) such that:

(i) $\Phi(kxk^{-1}) = \Phi(x)$ $(x \in G, k \in K)$;

(ii) $\chi_\delta * \Phi = \Phi (= \Phi * \chi_\delta)$;

(iii) The map $\mu_\Phi, f \mapsto \int_G f(x)\Phi(x)d_G(x)$, is an irreducible representation of the algebra $I_{c,\delta}(G)$.

The dimension of E is called the *height* of Φ.

Notice that the representation μ_Φ of $I_{c,\delta}(G)$ determined by a spherical function Φ of type δ is necessarily continuous with respect to $\|\,.\,\|_\rho$, ρ some semi-norm on G, i.e., is not entirely 'arbitrary'.

Call two spherical functions $\Phi_i (i = 1, 2)$ of type δ with values in $\mathrm{Hom}_{\mathbf{C}}(E_i, E_i)$ $(i = 1, 2)$ *equivalent* if there exists a linear bijection $Q: E_1 \rightarrow E_2$ such that $\Phi_2(x) = Q\Phi_1(x)Q^{-1}$ for all x in G. Claim: The equivalence class of Φ is determined by the equivalence class of μ_Φ.

Thus suppose that $\mu_{\Phi_1} = \mu_{\Phi_2}$ where Φ_1 and Φ_2 are spherical functions on G of type δ – then the claim will follow when it is shown that $\Phi_1 = \Phi_2$. But since

$$\begin{aligned}\int_G f(x)\Phi_1(x)d_G(x) &= \int_G (\bar{\chi}_\delta * f_K * \bar{\chi}_\delta)(x)\Phi_1(x)d_G(x)\\ &= \int_G (\bar{\chi}_\delta * f_K * \bar{\chi}_\delta)(x)\Phi_2(x)d_G(x)\\ &= \int_G f(x)\Phi_2(x)d_G(x)\end{aligned}$$

for all $f \in C_c(G)$, this is clear. It is now also evident that Φ is determined to within equivalence by the scalar function $x \mapsto \operatorname{tr}(\Phi(x))$ $(x \in G)$.

Spherical functions of type δ arise in a natural way upon consideration of TCI Banach representations of G. Thus suppose that U is a TCI Banach representation of G on a space E such that δ occurs in $U|K$. Earlier discussion tells us that we may express $E(\delta)$ as the tensor product of a simple K-module E_δ in the class δ together with the representation space E_δ (say) of the irreducible representation $f \mapsto \mu_\delta(f)$ of $I_{c,\delta}(G)$ – symbolically: $E(\delta) = E_\delta \otimes_{\mathbf{C}} \mathsf{E}_\delta$. Introduce the μ_U-spherical function Ψ_δ^U on G given by $\Psi_\delta^U(x) = \mathsf{P}(\delta)U(x)\mathsf{P}(\delta)$ $(x \in G)$ and set

$$\Psi_{\delta,K}^U(x) = \int_K U_\delta(k)\Psi_\delta^U(x)U_\delta(k^{-1})dk \qquad (x \in G).$$

Since $\Psi_{\delta,K}^U(x)$ commutes with all the $U_\delta(k)$, one has $\Psi_{\delta,K}^U(x) = 1 \otimes \Phi_\delta^U(x)$, Φ_δ^U a quasi-bounded continuous function on G with values in $\operatorname{Hom}_{\mathbf{C}}(\mathsf{E}_\delta, \mathsf{E}_\delta)$ and verifying conditions (i) and (ii) above in the definition of spherical function of type δ. Furthermore, for $f \in I_{c,\delta}(G)$,

$$\begin{aligned}1 \otimes \int_G f(x)\Phi_\delta^U(x)d_G(x) &= \int_G f(x)\Psi_{\delta,K}^U(x)d_G(x)\\ &= \int_G f(x)\Psi_\delta^U(x)d_G(x) = U_\delta(f) = 1 \otimes \mu_\delta(f)\end{aligned}$$

– thus $\mu_\delta(f) = \int_G f(x)\Phi_\delta^U(x)d_G(x)$ and so Φ_δ^U is a spherical function on G of type δ. Plainly $\psi_\delta^U(x) = d(\delta)\operatorname{tr}(\Phi_\delta^U(x))$ $(x \in G)$, where ψ_δ^U denotes the spherical trace function on G of type δ associated with the given representation U. Conversely, if Φ is a spherical function on G of type δ, then there exists a TCI Banach representation U of G such that Φ arises per the above process – this fact is contained in the sufficiency part of the proof of Theorem 6.1.1.2.

Remark Let U be a TCI Banach representation of G on E, δ a class in $\hat{K}$ which occurs in $U|K$; let Φ_δ^U denote the corresponding spherical function on G of type δ – then the following conditions are equivalent: (1) The representation U is Naimark equivalent to an (irreducible) unitary representation of G; (2) The function $x \mapsto \operatorname{tr}(\Phi_\delta^U(x))$ is of positive type on G; (3) The function $x \mapsto \Phi_\delta^U(x)$ is of positive type in the sense that E_δ admits the structure of a Hilbert space such that for all positive integers n, all $x_1, \ldots, x_n$ in G, and all $c_1, \ldots, c_n$ in $\mathbf{C}$, $\sum_{i,j} c_i\bar{c}_j\Phi_\delta^U(x_i^{-1}x_j)$ is a positive operator. [As we shall have no real need of this result, we shall omit its proof (cf. Sakai [1, p. 17]); it should be pointed out that conditions (1), (2), and (3) are *not* equivalent to the condition that $\mu_{\Phi}U_{\delta}$ be a *-representation of $I_{c,\delta}(G)$.]

Let Φ be a spherical function on G of type δ – then Φ verifies the functional equation

$$(\Phi) \quad \int_K \Phi(kxk^{-1}y) = \Phi(x)\Phi(y) \qquad (x, y \in G).$$

The proof of this fact is the same as part (1) of the proof of Theorem 6.1.1.7. Conversely, suppose that we are given a non-zero quasi-bounded continuous function Φ on G assuming values in $\mathrm{Hom}_{\mathbf{C}}(E, E)$ (E a finite dimensional linear space) which satisfies the functional equation (Φ) together with conditions (i) and (ii) in the definition of spherical function of type δ. It is then an easy matter to show that the map

$$f \mapsto \int_G f(x)\Phi(x)d_G(x)$$

is a representation of $I_{c,\delta}(G)$ on E – hence Φ is a spherical function of type δ provided that this representation is irreducible.

Note Let Φ be a non-trivial spherical function on G of type δ – then it follows without difficulty from equation (Φ) and the irreducibility of the representation

$$f \mapsto \int_G f(x)\Phi(x)d_G(x) \qquad (f \in I_{c,\delta}(G))$$

that $\Phi(1) = 1$.

Remark Spherical functions of type δ generalize the characters of compact or abelian groups. For if $K = \{1\}$, then the functional equation (Φ) tells us that a spherical function of type δ and height m (say) is just an m-dimensional irreducible representation of G while on the other hand, if $K = G$ (so that G is compact), then a spherical function Φ of type δ is necessarily of height 1 and is proportional to a character of G.

6.1.2 Spherical Functions – Differential Properties

Throughout the present number, G will denote a *connected* unimodular Lie group countable at infinity, K a uniformly large compact subgroup of G, which, without essential loss of generality, may be taken connected (cf. Corollary 4.5.3.2).

Let $\mathfrak{G}$ be the universal enveloping algebra of $\mathfrak{g}_c$, $\mathfrak{g}_c$ the complexification of the Lie algebra $\mathfrak{g}$ of G; let $\mathfrak{Z}$ be the center of $\mathfrak{G}$; let $\mathfrak{X}$ denote the centralizer of $\mathfrak{k}_c$ in $\mathfrak{G}$, $\mathfrak{k}_c$ the complexification of the Lie algebra $\mathfrak{k}$ of K – the canonical projection of $\mathfrak{G}$ onto $\mathfrak{X}$ will then be denoted by $D \mapsto D_K$ (so that $D_K = \int_K \mathrm{Ad}(k)\cdot D dk$, $D \in \mathfrak{G}$).

Note In what follows the elements of $\mathfrak{G}$ are to be regarded as left invariant differential operators on G.

Let U be a TCI Banach representation of G on E, E_ω the space of analytic vectors in E (for U) – then, as we know (Corollary 4.4.5.17), $E_K = \sum_{\delta \in \hat{K}} E(\delta) \subset E_\omega$. This fact has the following important consequence: Fix a class $\delta \in \hat{K}$ which occurs in $U|K$ – then the functions Ψ_δ^U, ψ_δ^U, Φ_δ^U are *analytic* functions on G.

Proposition 6.1.2.1 Let U be a TCI Banach representation of G on E, κ_U the infinitesimal character of U (cf. Proposition 4.4.1.5); let ψ_δ^U be a spherical trace function on G of type δ for U – then $Z\psi_\delta^U = \kappa_U(Z)\psi_\delta^U$ $(Z \in \mathfrak{Z})$.

Proof Let $a_1, \ldots, a_r$ be a basis for $E(\delta)$; let $\mathbf{z}_i$ be the linear function on $E(\delta)$ which takes the value 1 at a_i and zero at a_j $(i \neq j;\ 1 \leqslant i, j \leqslant r)$; extend $\mathbf{z}_i$ to all of E in the obvious way $(1 \leqslant i \leqslant r)$ – then it is clear that

$$\psi_\delta^U(x) = \sum_i \langle U(x)a_i, \mathbf{z}_i\rangle \quad (x \in G).$$

Since a_i lies in $E(\delta)$, it is analytic, hence differentiable, that is, the function $\tilde{a}_i$, $x \mapsto U(x)a_i$ $(x \in G)$, belongs to the space $C^\infty(G; E)$; since

$$D\tilde{a}_i(x) = U(x)U_\infty(D)a_i \quad (\text{all } D \in \mathfrak{G}),$$

it follows that

$$\begin{aligned} Z\psi_\delta^U &= Z(\sum_i \mathbf{z}_i \circ \tilde{a}_i) \\ &= \sum_i \mathbf{z}_i \circ (Z\tilde{a}_i) = \kappa_U(Z)\psi_\delta^U \end{aligned}$$

for all $Z \in \mathfrak{Z}$, as desired. □

Note Keeping to the notations of the preceding proposition, observe that the argument thereof actually serves to show that $Z\langle U(.)a_i, \mathbf{z}_j\rangle = \kappa_U(Z)\langle U(.)a_i, \mathbf{z}_j\rangle$ (all $Z \in \mathfrak{Z}$); in particular, then, $Z\Psi_\delta^U = \kappa_U(Z)\Psi_\delta^U$, $Z\Phi_\delta^U = \kappa_U(Z)\Phi_\delta^U$ (all $Z \in \mathfrak{Z}$).

Proposition 6.1.2.2 Let U and V be TCI Banach representations of G on spaces E and F, respectively, such that δ $(\delta \in \hat{K})$ occurs in both $U|K$ and $V|K$; let U_δ and V_δ be the representations of $\mathfrak{X}$ on $E(\delta)$ and $F(\delta)$ obtained from U_K and V_K, respectively, by restriction – then, if U_δ and V_δ are algebraically equivalent, U and V are Naimark equivalent.

Proof In view of Theorem 6.1.1.8, the Naimark equivalence of U and V will follow when it is shown that the spherical trace functions ψ_δ^U and ψ_δ^V are equal. Since the functions in question are analytic, it will be enough to show that $D\psi_\delta^U(1) = D\psi_\delta^V(1)$ (all $D \in \mathfrak{G}$) (G being connected); in turn, thanks to the K-centrality of spherical trace functions, we need only check the equality of the derivatives at 1 for $D \in \mathfrak{X}$ – but, for $D \in \mathfrak{X}$, we have $D\psi_\delta^U(1) = \mathrm{tr}(U_\delta(D)) = \mathrm{tr}(V_\delta(D)) = D\psi_\delta^V(1) \ldots.$ □

Remark Let U be a TCI Hilbert representation of G on E, T_U its (distribution) character (Vol. I, number 4.5.8). Consider the Fourier components $T_{U,\delta}$ of $T_U(\delta \in \hat{K})$ – thus $T_{U,\delta}$ is defined by the rule

$$T_{U,\delta}(f) = T_U(f_\delta) = T_U(f*\bar{\chi}_\delta) \qquad (f \in C_c^\infty(G))$$

(cf. Vol. I, number 4.4.3). It is not difficult to relate the $T_{U,\delta}$ and the ψ_δ^U; indeed, assuming, as we may, that the $E(\delta)$ $(\delta \in \hat{K})$ are mutually orthogonal, we find, for a fixed $\delta_0 \in \hat{K}$, that

$$\begin{aligned} T_{U,\delta_0}(f) = T_U(f*\bar{\chi}_{\delta_0}) &= \sum_{\delta \in \hat{K}} \mathrm{tr}(\mathsf{P}(\delta)U(f)\mathsf{P}(\delta_0)\mathsf{P}(\delta)) = \mathrm{tr}(\mathsf{P}(\delta_0)U(f)\mathsf{P}(\delta_0)) \\ &= \int_G f(x)\psi_{\delta_0}^U(x)d_G(x) \qquad (\text{all } f \in C_c^\infty(G)). \end{aligned}$$

Hence, in the sense of distributions, the Fourier components $T_{U,\delta}$ of the character T_U are the (necessarily analytic) functions ψ_δ^U $(\delta \in \hat{K})$.

Consider the class of C^∞ complex valued functions ϕ on G (normalized by the requirement that $\phi(1) = 1$) which are biinvariant under K and are eigenfunctions of $\mathfrak{X}$ (i.e. which have the property that for each $D \in \mathfrak{X}$, there exists a complex number c_D (depending on ϕ, of course) such that $D\phi = c_D\phi$ (indeed, $c_D = D\phi(1)$)); it is a standard, well-known fact that this class of functions coincides with the class of zonal spherical functions on G (compare, e.g. Helgason [2]). Now a priori such functions need not be quasi-bounded (although this will always be the case in the applications); those that are, however, may be regarded as comprising the set of spherical functions on G of type δ, δ the class of the trivial $(k \mapsto 1)$ one dimensional representation of K; this being so, we shall now proceed to extend these considerations to spherical functions on G of type δ, δ an arbitrary class in $\hat{K}$.

Let Φ be a spherical function on G of type δ; let $D \in \mathfrak{X}$ – then $D\Phi = \Phi[D\Phi(1)]$. [To see this, bear in mind that Φ satisfies the functional equation $\int_K \Phi(k.k^{-1}.)dk = \Phi(.)\Phi(.)$; consequently, if T is a distribution on G with compact support which commutes, moreover, with all the δ_k (δ_k the unit mass at the point $k \in K$), then, upon 'multiplying' the functional equation through by T and integrating, we find that

$$\int_G \Phi(xy)dT(y) = \Phi(x)\left(\int_G \Phi(y)dT(y)\right) \qquad (x, y \in G).$$

As in A 2.4, let $D \mapsto T_D$ denote the identification of $\mathfrak{G}$ with the algebra of distributions on G with support $\{1\}$; if $D \in \mathfrak{X}$, then T_D commutes with the δ_k $(k \in K)$ – therefore

$$\begin{aligned} D\Phi(x) &= \Phi * T_D'(x) \\ &= \int_G \Phi(xy^{-1})dT_D'(y) = \Phi(x)[D\Phi(1)] \qquad (x \in G), \end{aligned}$$

which serves to establish our contention.] Furthermore, the mapping $D \mapsto D\Phi(1)$ $(D \in \mathfrak{X})$ is a representation μ_Φ (say) of $\mathfrak{X}$ (direct verification) which is, in fact, irreducible; in particular, therefore, Φ is a $\mathfrak{Z}$-

eigenfunction. [Suppose that $\Phi: G \to \mathrm{Hom}_{\mathbf{C}}(E, E)$ – then the irreducibility of μ_Φ will follow when it is shown that $\mu_\Phi(\mathfrak{X}) = \mathrm{Hom}_{\mathbf{C}}(E, E)$. If the contrary were true, then there would exist a non-trivial linear function $\mathbf{z}$: $\mathrm{Hom}_{\mathbf{C}}(E, E) \to \mathbf{C}$ which vanishes identically on $\mu_\Phi(\mathfrak{X})$. Because

$$\mathrm{Hom}_{\mathbf{C}}(E, E) = \Phi(I_{c,\delta}(G)),$$

there would then be an $f \in I_{c,\delta}(G)$ such that

$$\int_G f(x) < \Phi(x), \mathbf{z} > d_G(x) \neq 0,$$

whence, $\mathbf{z}\circ\Phi$ being a non-zero analytic function, there would be a $D \in \mathfrak{X}$ such that $< D\Phi(1), \mathbf{z} > \neq 0$, a contradiction. . . .]

Here is the to be expected complement to the above remarks.

Theorem 6.1.2.3 (Rader) Let E be a finite dimensional vector space over $\mathbf{C}$, $\Phi : G \to \mathrm{Hom}_{\mathbf{C}}(E, E)$ a quasi-bounded K-central C^∞ function on G; suppose that there exists an irreducible representation μ_Φ of $\mathfrak{X}$ on E such that $D\Phi = \Phi\mu_\Phi(D)$, where $\mu_\Phi(D) = D\Phi(1)$ $(D \in \mathfrak{X})$ – then Φ is a spherical function on G of type δ (for some $\delta \in \hat{K}$).

Proof We shall prove that there exists a $\delta \in \hat{K}$ with the property that $\Phi * \chi_\delta = \Phi$; once this is established, it will be an easy matter to check that the map $f \mapsto \int_G f(x)\Phi(x)d_G(x)$ defines an irreducible representation of $I_{c,\delta}(G)$ on E, thus completing the proof of the present theorem.

(1) The function Φ is analytic. [According to our hypotheses, the space $\mathfrak{X}\Phi$ is finite dimensional; in addition $\mathfrak{X}$ contains an elliptic element Δ (say) (cf. Lemma 4.4.5.20). But then there exist complex numbers c_k $(0 \leqslant k \leqslant n,\ c_n = 1)$ such that $\sum_k c_k \Delta^k \Phi = 0$; as the analytic differential operator $\sum_k c_k \Delta^k$ is elliptic, it follows that Φ is analytic.] This being the case, we shall now show that Φ verifies the functional equation $\int_K \Phi(.k.k^{-1})dk = \Phi(.)\Phi(.)$; thus fix $x \in G$ – then there exists a small neighborhood $\mathcal{O}$ of zero in $\mathfrak{g}$ such that for all $Y \in \mathcal{O}$, we have

$$\begin{aligned}\int_K \Phi(xk \exp Yk^{-1})dk &= \int_K \Phi(x \exp(\mathrm{Ad}(k)Y))dk \\ &= \int_K \sum_{m=0}^{\infty} \frac{1}{m!}(\mathrm{Ad}(k)Y)^m \Phi(x)dk \\ &= \sum_{m=0}^{\infty} \frac{1}{m!}(Y_K)^m \Phi(x) \\ &= \Phi(x)\Big(\sum_{m=0}^{\infty} \frac{1}{m!}(Y_K)^m \Phi(1)\Big) = \Phi(x)\Phi(\exp Y),\end{aligned}$$

where we have tacitly used the fact that $D\Phi(1) = D_K\Phi(1)$ (all $D \in \mathfrak{G}$) (Φ is K-central). The functional equation in question thus results from the analyticity of Φ.

(2) We may assume that the function Φ is not identically zero. Claim: There exists a $\delta \in \hat{K}$ such that $\Phi * \chi_\delta(1) \neq 0$. To see this, note that for any $x \in G$,

$$\begin{aligned}\Phi * \chi_\delta(x) &= \int_K \Phi(xk)\overline{\chi_\delta(k)}dk \\ &= \int_K \left(\int_K \Phi(x\tilde{k}k\tilde{k}^{-1})d\tilde{k}\right)\overline{\chi_\delta(k)}dk \\ &= \Phi(x)(\Phi * \chi_\delta(1)),\end{aligned}$$

which shows that if $\Phi * \chi_\delta(1)$ did vanish for all $\delta \in \hat{K}$, then, of necessity, the same would be true of $\Phi * \chi_\delta$, in contradiction with our supposition on Φ. So fix $\delta \in \hat{K}$ such that $\Phi * \chi_\delta(1) \neq 0$ – then, thanks to the K-centrality of χ_δ, it is clear that $\Phi * \chi_\delta(1)$ commutes with the $D\Phi(1)$ $(D \in \mathfrak{X})$, hence is a non-zero scalar operator M_δ (say) (Schur's Lemma). Consequently,

$$\Phi M_\delta = \Phi * \chi_\delta = (\Phi * \chi_\delta) * \chi_\delta = \Phi M_\delta^2 ,$$

whence $\Phi * \chi_\delta = \Phi$.

(3) Because $\Phi = \Phi_K$ and $\Phi * \chi_\delta = \Phi$, it is clear that the map $f \mapsto \int_G f(x)\Phi(x)d_G(x)$ defines a representation of $I_{c,\delta}(G)$ on E – that this representation is actually irreducible may be seen as follows. The space $C_c^\infty(G)$ is weakly dense in the algebra of distributions on G with compact support; accordingly any operator of the form $D\Phi(1)$ $(D \in \mathfrak{X})$ can be approximated by operators $\Phi(f) = \int_G f(x)\Phi(x)d_G(x)$ where, without loss of generality, it may be supposed that $f \in I_{c,\delta}^\infty(G)$ – otherwise said, $\text{Hom}_{\mathbf{C}}(E, E) = \{\int_G f(x)\Phi(x)d_G(x) : f \in I_{c,\delta}(G)\}$.

This completes the proof of theorem. □

Fix $\delta \in \hat{K}$; consider the class of quasi-bounded K-central analytic functions $\Phi: G \to \text{Hom}_{\mathbf{C}}(E, E)$ (E a finite dimensional linear space depending on Φ) such that $\chi_\delta * \Phi = \Phi$ and with the property that $D\Phi = \Phi\mu_\Phi(D)$, where the map $D \mapsto \mu_\Phi(D)$, $\mu_\Phi(D) = D\Phi(1)$, defines an irreducible representation of $\mathfrak{X}$ on E – then, in view of what has been said above, this class of functions is the set of spherical functions on G of type δ. The possible dimensions of the various E are not entirely arbitrary; thus if $m(\delta)$ is an integer $\geqslant 1$ such that δ occurs no more than $m(\delta)$ times in any TCI Banach representation of G, then $\dim(E) \leqslant m(\delta)$ – of course the cases when $m(\delta) = 1$ are especially interesting (cf. 6.2).

Remark One may also prove an analogue of Theorem 6.1.2.3 for spherical trace functions on G (cf. Godement [7]) – however this result appears to be of limited utility.

6.2 Examples

6.2.1 Spherical Functions on Motion Groups

Let G be a locally compact group which can be written as the semi-direct product of a closed normal abelian subgroup H and a compact subgroup K, i.e. suppose that G is a motion group (see Vol. I, number 4.5.2).

[We recall that the compact subgroup K is then large in G – in fact that a given $\delta \in \hat{K}$ occurs no more than $d(\delta)$ times in any TCI Banach representation U of G (Theorem 4.5.2.1).] Fix a class $\delta \in \hat{K}$ – then the first thing that shall be done in the present number is to obtain integral formulas for spherical trace functions on G of type δ (and even for spherical functions of type δ). The methods which lead to these results yield, more or less directly, a theorem (due to Fell) which essentially serves to classify $\bar{G}$ ($\bar{G}$ denoting, as usual, the set of Naimark equivalence classes of TCI Banach representations of G).

Let dk denote normalized Haar measure on K; normalize the Haar measures on G and H so that $d_G(x) = dk\,d_H(\xi)(= d_H(\xi)dk)$ if $x = k\xi (k \in K, \xi \in H)$.

Fix an element $\delta \in \hat{K}$. According to an Example in 6.1.1, the algebra $I_{c,\delta}(G)$ is isomorphic, via the map

$$f \mapsto \Psi_f^\delta, \ \Psi_f^\delta(x) = \int_K \mu_\delta(k^{-1}) f(kx)dk \qquad (x \in G),$$

to an algebra $\mathfrak{A}_{c,\delta}(G)$ of (μ_δ, μ_δ)-spherical functions on G. Notice, in the case at hand, that an element of $\mathfrak{A}_{c,\delta}(G)$ is completely determined by its restriction to H – in fact that the restriction map $\Psi \mapsto F = \Psi | H$ maps $\mathfrak{A}_{c,\delta}(G)$ bijectively onto the algebra $\mathfrak{A}_{c,\delta}(H)$ of all compactly supported continuous functions $F: H \rightarrow \mathrm{Hom}_{\mathbf{C}}(E_\delta, E_\delta) = F_\delta$ which verify the relation $F(k\xi k^{-1}) = \mu_\delta(k)F(\xi)\mu_\delta(k^{-1})$ $(k \in K, \xi \in H)$. Conclusion: The map $f \mapsto \Psi_f^\delta \mapsto F_f^\delta = \Psi_f^\delta | H$ is a linear bijection of $I_{c,\delta}(G)$ onto $\mathfrak{A}_{c,\delta}(H)$. As we shall now see, this map is very well behaved both multiplicatively and topologically.

Let $f, g \in I_{c,\delta}(G)$ – then, for $\xi \in H$,

$$F_{f*g}^\delta(\xi) = \int_H F_f^\delta(\xi\eta)F_g^\delta(\eta^{-1})d_H(\eta),$$

i.e. the map $f \mapsto F_f^\delta$ carries G-convolution on $I_{c,\delta}(G)$ onto H-convolution on $\mathfrak{A}_{c,\delta}(H)$. In order to prove this it will be necessary to bear in mind the rule (σ) which was employed in the Example referred to above: Let $T \in F_\delta$ – then

$$(\sigma) \quad T = \int_K \mu_\delta(k^{-1})\sigma(\mu_\delta(k)T)dk \qquad (\sigma(T) = d(\check{\delta})\mathrm{tr}(T)).$$

Using the relation $f(x) = \sigma(\Psi_f^\delta(x))$ $(x \in G, f \in I_{c,\delta}(G))$, we find that

$$\begin{aligned} F_{f*g}^\delta(\xi) &= \int_K \mu_\delta(k^{-1})(f*g)(k\xi)dk \\ &= \int_K\int_K\int_H \mu_\delta(k^{-1})f(k\xi\eta\tilde{k})g(\tilde{k}^{-1}\eta^{-1})d_H(\eta)d\tilde{k}dk \\ &= \int_K\int_K\int_H \mu_\delta(k^{-1})\sigma(\mu_\delta(k)\Psi_f^\delta(\xi\eta)\mu_\delta(\tilde{k})) \\ &\quad \times \sigma(\mu_\delta(\tilde{k}^{-1}))\Psi_g^\delta(\eta^{-1}))d_H(\eta)d\tilde{k}dk \end{aligned}$$

$$= \int_H \int_K \Psi_f^\delta(\xi\eta)\mu_\delta(k)\sigma(\mu_\delta(k^{-1})\Psi_g^\delta(\eta^{-1}))dkd_H(\eta)$$

$$= \int_H F_f^\delta(\xi\eta)F_g^\delta(\eta^{-1})d_H(\eta) \Rightarrow F_{f*g}^\delta = F_f^\delta * F_g^\delta, \text{ as desired.}$$

The space $C_c(H; F_\delta)$ is an algebra under convolution. Introduce a norm on the space E_δ relative to which the representation μ_δ is unitary; if $T \in F_\delta$, then $\|T\|$ will denote the operator norm derived from this particular norm on E_δ. This being so, let ρ_H be a semi-norm on H – then $C_c(H; F_\delta)$ acquires the structure of a normed algebra when we write:

$$\|f\|_{\rho_H} = \int_H \|f(\xi)\|\rho_H(\xi)d_H(\xi) \qquad (f \in C_c(H; F_\delta)).$$

Claim: Let ρ be a semi-norm on G – then the isomorphism

$$f \mapsto F_f^\delta \qquad (f \in I_{c,\delta}(G))$$

is a topological isomorphism with respect to the norms $\|\,.\,\|_\rho$ and $\|\,.\,\|_{\rho_H}(\rho_H = \rho\,|\,H)$ on $I_{c,\delta}(G)$ and $\mathfrak{A}_{c,\delta}(H)$, respectively. Thus set $M = \sup_{k\in K}\rho(k)$; using the inequality

$$\rho(\xi) \leqslant \rho(k\xi)\rho(k^{-1}) \leqslant M\rho(k\xi) \qquad (k \in K, \xi \in H),$$

we have that $\|F_f^\delta\|_{\rho_H} \leqslant M\|f\|_\rho$ – on the other hand

$$\|f\|_\rho = \int_H \int_K |f(k\xi)|\,\rho(k\xi)dkd_H(\xi)$$

$$\leqslant M \int_H \int_K |\sigma(\Psi_f^\delta(k\xi))|\rho(\xi)dkd_H(\xi)$$

$$\leqslant Md(\check{\delta})^2 \int_H \int_K \|\Psi_f^\delta(k\xi)\|\,\rho(\xi)dkd_H(\xi)$$

$$= Md(\check{\delta})^2 \int_H \|F_f^\delta(\xi)\|\,\rho_H(\xi)d_H(\xi) = Md(\check{\delta})^2\,\|F_f^\delta\|_{\rho_H}.$$

Hence the claim.

To find all spherical trace functions of type δ, we must construct all finite dimensional irreducible representations of $I_{c,\delta}(G)$ satisfying obvious continuity conditions (Theorem 6.1.1.11). [Actually, in view of Proposition 4.5.1.10, every TCI Banach representation of $I_{c,\delta}(G)$ is finite dimensional – in fact is bounded in dimension by $d(\delta) = d(\check{\delta})$.] The discussion in the preceding paragraphs reduces this problem to a similar question for the algebra $\mathfrak{A}_{c,\delta}(H)$. Now the 'reasonable' (cf. infra) irreducible representations of the algebra $C_c(H)$ are well-known; since $C_c(H; F_\delta) = C_c(H) \otimes_{\mathbf{C}} F_\delta$, the 'reasonable' irreducible representations of $C_c(H; F_\delta)$ are likewise easy to describe – a simple algebraic device will then enable us to relate the irreducible representations of $\mathfrak{A}_{c,\delta}(H)$ to those of $C_c(H; F_\delta)$.

The TCI Banach representations of $C_c(H)$ are all one dimensional ($C_c(H)$ is abelian). If ρ_H is a semi-norm on H and $f \mapsto U(f)$ $(f \in C_c(H))$

is such a representation which is, moreover, continuous with respect to ρ_H, then there exists a (continuous) character $\hat{\xi}$ of H, verifying $|\langle \xi, \hat{\xi} \rangle| \leqslant M\rho_H(\xi)$ $(\xi \in H)$ for some $M > 0$, such that U is given by the rule

$$f \mapsto \int_H \langle \xi, \hat{\xi} \rangle f(\xi) d_H(\xi) \qquad (f \in C_c(H));$$

there is an evident converse statement too. Consider now the TCI Banach representations of the algebra $C_c(H; F_\delta)$ which are continuous with respect to some semi-norm on H. In view of what has just been said and the fact that $C_c(H; F_\delta) = C_c(H) \otimes_{\mathbf{C}} F_\delta$, any such representation is defined by the prescription $F \mapsto \int_H \langle \xi, \hat{\xi} \rangle F(\xi) d_H(\xi)$ $(F \in C_c(H; F_\delta))$, where $\hat{\xi}$ is a quasi-bounded character of H. Having described the irreducible representations of $C_c(H; F_\delta)$ in which we are interested, it is now possible to formulate the following result.

Proposition 6.2.1.1 Let ρ be a semi-norm on G; set $\rho_H = \rho | H$. Let V be an irreducible finite dimensional representation of $\mathfrak{A}_{c,\delta}(H)$ which is ρ_H-continuous – then there exists a character $\hat{\xi}$ of H, verifying $|\langle \xi, \hat{\xi} \rangle| \leqslant M\rho_H(\xi)$ $(\xi \in H)$ for some $M > 0$, such that V is equivalent to a subrepresentation of the representation

$$F \mapsto \int_H \langle \xi, \hat{\xi} \rangle F(\xi) d_H(\xi) \qquad (F \in C_c(H; F_\delta))$$

of $C_c(H; F_\delta)$.

The proof of this result rests on an algebraic fact. Let A be an associative algebra over $\mathbf{C}$; let $x \mapsto x^0$ be a linear idempotent operator on A satisfying $(xy^0)^0 = x^0y^0 = (x^0y)^0$ $(x, y \in A)$ – put

$$A_I = \{x \in A : x^0 = x\}.$$

Of course A_I is a subalgebra of A.

Lemma 6.2.1.2 Let V be a finite dimensional algebraically irreducible representation of A_I – then there exists an algebraically irreducible representation U of A on a space E such that E contains an A_I-stable subspace F with the property that the representation $x \mapsto U(x) | F$ $(x \in A_I)$ of A_I on F is equivalent to V. Furthermore, if A is a Banach algebra and if the map $x \mapsto x^0$ is continuous, then U can be taken to be a norm decreasing algebraically completely irreducible Banach representation of A.

Proof The proof of the present lemma is similar to that of Lemma 6.1.1.3. Choose a non-zero vector in the representation space for V and let I denote its annihilator in A_I; plainly I is a regular maximal left ideal in A_I. Consider the set J of all $x \in A$ such that $(yx)^0 \in I$ for every $y \in A$; it is obvious that J is a left ideal in A and, by using the relations $(xy^0)^0 = x^0y^0 = (x^0y)^0$ $(x, y \in A)$, it is easy to see that every right identity mod I is also a right identity mod J, i.e. J is regular. Since $I \subset J$, $I = J \cap A_I$ by the maximality of I. But, as J is regular, J is contained in some regular maximal left ideal in A. Conclusion: I is the intersection of A_I with some regular maximal left ideal in A–this gives the first contention of the lemma. The second is also easy and may be left to the reader. □

Proof of Proposition 6.2.1.1 (1) If $f \in C_c(H; F_\delta)$, then define $f^0 \in C_c(H; F_\delta)$ by the rule

$$f^0(\xi) = \int_K \mu_\delta(k^{-1}) f(k\xi k^{-1}) \mu_\delta(k) dk \qquad (\xi \in H).$$

Evidently the map $f \mapsto f^0$ is an idempotent linear operator on $C_c(H; F_\delta)$ whose range is $\mathfrak{A}_{c,\delta}(H)$. Let us check the properties:

$$(f * g^0)^0 = f^0 * g^0 = (f^0 * g)^0 \qquad (f, g \in C_c(H; F_\delta)).$$

Thus, for instance,

$$\begin{aligned}(f * g^0)^0(\xi) &= \int_K \mu_\delta(k^{-1}) \left(\int_H f(k\xi k^{-1}\eta) g^0(\eta^{-1}) d_H(\eta) \right) \mu_\delta(k) dk \\ &= \int_H \left(\int_K \mu_\delta(k^{-1}) f(k\xi\eta k^{-1}) \mu_\delta(k) dk \right) g^0(\eta^{-1}) d_H(\eta) \\ &= \int_H f^0(\xi\eta) g^0(\eta^{-1}) d_H(\eta) = (f^0 * g^0)(\xi) \qquad (\xi \in H);\end{aligned}$$

in a similar way one finds that $f^0 * g^0 = (f^0 * g)^0$.

(2) The projection $f \mapsto f^0$ of $C_c(H; F_\delta)$ onto $\mathfrak{A}_{c,\delta}(H)$ is ρ_H-continuous – for $\|f^0\|_{\rho_H} \leqslant M^2 \|f\|_{\rho_H}$, as a simple computation quickly reveals ($M = \sup_{k \in K} \rho(k)$).

(3) Let A be the Banach algebra completion of $C_c(H; F_\delta)$ with respect to the norm $\| \cdot \|_{\rho_H}$. By (2) the map $f \mapsto f^0$ extends to a continuous projection of A onto the ρ_H-completion A_I of $\mathfrak{A}_{c,\delta}(H)$. Since the given representation V is ρ_H-continuous, it extends to an irreducible representation of A_I. Lemma 6.2.1.2 then provides us with a norm decreasing algebraically completely irreducible representation U of A such that V is equivalent to a subrepresentation of $U | A_I$. The restriction of U to $C_c(H; F_\delta)$ is topologically completely irreducible and ρ_H-continuous, whence there exists a character $\hat{\xi}$ of H, verifying $|\langle \xi, \hat{\xi} \rangle| \leqslant M \rho_H(\xi)$ $(\xi \in H)$ for some $M > 0$, such that $U | C_c(H; F_\delta)$ is given by the prescription

$$F \mapsto \int_H \langle \xi, \hat{\xi} \rangle F(\xi) d_H(\xi) \qquad (F \in C_c(H; F_\delta)).$$

The assertion of the proposition is now clear. □

We may paraphrase Proposition 6.2.1.1 in the following way. Fix a class $\delta \in \hat{K}$. Let ρ be a semi-norm on G. If μ_δ is any irreducible finite dimensional ρ-continuous representation of $I_{c,\delta}(G)$, then there exists a character $\hat{\xi}$ of H, verifying $|\langle \xi, \hat{\xi} \rangle| \leqslant M \rho(\xi)$ $(\xi \in H)$ for some $M > 0$, such that μ_δ is equivalent to a subrepresentation of the $d(\check{\delta})$-dimensional representation

$$f \mapsto \int_H \langle \xi, \hat{\xi} \rangle F_f^\delta(\xi) d_H(\xi)$$

of $I_{c,\delta}(G)$ where

$$F_f^\delta(\xi) = \int_K \mu_\delta(k^{-1}) f(k\xi) dk \qquad (f \in I_{c,\delta}(G);\, \xi \in H).$$

Let us turn to the calculation of the spherical trace functions of type δ on our motion group G. Suppose first that the representation $\mu_{\hat{\xi}}$ of $I_{c,\delta}(G)$ defined by the prescription

$$f \mapsto \int_H \langle \xi, \hat{\xi} \rangle F_f^\delta(\xi) d_H(\xi)$$

is irreducible; Theorems 6.1.1.2, 6.1.1.11 guarantee us the existence of a spherical trace function ψ of type δ corresponding to $\mu_{\hat{\xi}}$ – in fact ψ is given by

$$\psi(f) = d(\check{\delta}) \mathrm{tr}\Big(\int_H \langle \xi, \hat{\xi} \rangle F_f^\delta(\xi) d_H(\xi)\Big) = \int_H \langle \xi, \hat{\xi} \rangle f(\xi) d_H(\xi)$$

for all $f \in I_{c,\delta}(G)$. To calculate ψ explicitly, take an $f \in C_c(G)$ and observe that

$$\begin{aligned} \psi(f) &= \psi(\bar{\chi}_\delta * f_K) \\ &= \int_H \int_K \int_K \langle \xi, \hat{\xi} \rangle f(\tilde{k} k \xi \tilde{k}^{-1}) \chi_\delta(k) d\tilde{k}\, dk\, d_H(\xi) \\ &= \int_H \int_K \int_K \langle \tilde{k} \xi \tilde{k}^{-1}, \hat{\xi} \rangle f(k\xi) \chi_\delta(k) d\tilde{k}\, dk\, d_H(\xi); \end{aligned}$$

as $\psi(f) = \int_G f(x)\psi(x) d_G(x)$ and f is arbitrary, we see that

$$(\psi) \quad \psi(k\xi) = \chi_\delta(k) \int_K \langle \tilde{k} \xi \tilde{k}^{-1}, \hat{\xi} \rangle \, d\tilde{k}.$$

Consider the special case when K is abelian – then it is clear, for a suitable choice of $\hat{\xi}$, that *every* spherical trace function of type δ is represented by an integral formula (ψ). Even in this case, our formula leads to non-trivial 'special' functions – e.g. Bessel functions (cf. Helgason [2, p. 402] and infra).

[Retain the immediately preceding assumptions and notations; let us calculate the spherical function Φ of type δ corresponding to $\mu_{\hat{\xi}}$. Thus take an arbitrary $f \in C_c(G)$ – then

$$\begin{aligned} \Phi(f) &= \Phi(\bar{\chi}_\delta * f_K) \\ &= \int_H \int_K \int_K \int_K \langle \xi, \hat{\xi} \rangle \chi_\delta(k) f(k_1 k k_2 \xi k_1^{-1}) \\ &\qquad \mu_\delta(k_2^{-1}) dk_1\, dk_2\, dk\, d_H(\xi) \end{aligned}$$

$$= \int_H \int_K \int_K \int_K \langle k_1^{-1}\xi k_1, \hat{\xi}\rangle \chi_\delta(k) f(k_1 k k_2 k_1^{-1}\xi) \cdot \mu_\delta(k_2^{-1}) dk_1 dk_2 dk d_H(\xi)$$

$$= \int_H \int_K \int_K \int_K \langle k_1^{-1}\xi k_1, \hat{\xi}\rangle \chi_\delta(k k_2^{-1}) f(k_1 k k_1^{-1}\xi) \cdot \mu_\delta(k_2^{-1}) dk_1 dk_2 dk d_H(\xi)$$

$$= \int_H \int_K \int_K \int_K \langle k_1^{-1}\xi k_1, \hat{\xi}\rangle \chi_\delta(k_1^{-1} k k_1 k_2^{-1}) f(k\xi) \cdot \mu_\delta(k_2^{-1}) dk_1 dk_2 dk d_H(\xi)$$

$$= \int_H \int_K \int_K \langle \tilde{k}^{-1}\xi\tilde{k}, \hat{\xi}\rangle f(k\xi)\mu_\delta(\tilde{k}^{-1}k^{-1}\tilde{k}) d\tilde{k} dk d_H(\xi)$$

$$= \int_G f(x)\Phi(x) d_G(x),$$

where

$$(\Phi) \quad \Phi(k\xi) = \int_K \langle \tilde{k}\xi\tilde{k}^{-1}, \hat{\xi}\rangle\, \mu_\delta(\tilde{k}k^{-1}\tilde{k}^{-1}) d\tilde{k}.]$$

We proceed now to the general case in which the representation μ_ξ of $I_{c,\delta}(G)$ may not be irreducible. Let $E \subset E_\delta$ be a minimal subspace invariant under the operators

$$\int_H \langle \xi, \hat{\xi}\rangle\, F_f^\delta(\xi) d_H(\xi) \qquad (f \in I_{c,\delta}(G)).$$

The representation $f \mapsto \mu_\xi(f)\,|\,E$ of $I_{c,\delta}(G)$ on E is irreducible, hence defines a spherical trace function ψ on G of type δ. Choose an orthonormal basis $a_1, \ldots, a_m$ for E – then, for $f \in I_{c,\delta}(G)$, we have that

$$\psi(f) = d(\check{\delta})\Sigma \int_H \langle \xi, \hat{\xi}\rangle (F_f^\delta(\xi)a_i, a_i) d_H(\xi).$$

Fix an arbitrary $f \in C_c(G)$ – then clearly

$$F^\delta_{\bar{\chi}_\delta * f_K}(\xi) = \int_K \int_K \int_K \mu_\delta(k_1^{-1}) f(k k_2 k_1 \xi k^{-1}) \chi_\delta(k_2) dk_1 dk_2 dk$$

and so

$$\psi(f) = \psi(\bar{\chi}_\delta * f_K)$$

$$= d(\check{\delta})\Sigma \int_H \int_K \int_K \int_K \langle \xi, \hat{\xi}\rangle (a_i, \mu_\delta(k_1)a_i) \cdot f(k k_2 k_1 \xi k^{-1}) \chi_\delta(k_2) dk_1 dk_2 dk d_H(\xi)$$

$$= d(\check{\delta})\Sigma \int_H \int_K \int_K \int_K \langle \xi, \hat{\xi}\rangle (\mu_\delta(k k_2)a_i, \mu_\delta(k_1)a_i) \cdot f(k_1 \xi k^{-1}) \chi_\delta(k_2) dk_1 dk_2 dk d_H(\xi)$$

$$= d(\check{\delta})\Sigma \int_H \int_K \int_K \langle \xi, \hat{\xi}\rangle (\mu_\delta(k)a_i, \mu_\delta(\tilde{k})a_i) \cdot f(\tilde{k}\xi k^{-1}) d\tilde{k} dk d_H(\xi) \quad \left(\int_K \mu_\delta(k_2)\chi_\delta(k_2) dk_2 = 1\right)$$

$$= d(\check{\delta})\Sigma \int_H \int_K \int_K \langle \xi, \hat{\xi} \rangle (\mu_\delta(\tilde{k}^{-1}k\tilde{k})a_i, a_i) f(\tilde{k}\xi\tilde{k}^{-1}k^{-1}) d\tilde{k} dk d_H(\xi)$$

$$= d(\check{\delta})\Sigma \int_H \int_K \int_K \langle \tilde{k}^{-1}\xi\tilde{k}, \hat{\xi} \rangle (\mu_\delta(\tilde{k}^{-1}k^{-1}\tilde{k})a_i, a_i)$$
$$\cdot f(k\xi) d\tilde{k} dk d_H(\xi).$$

Therefore we obtain

$$(\psi) \quad \psi(k\xi) = \int_K \theta(\tilde{k}k\tilde{k}^{-1}) \langle \tilde{k}\xi\tilde{k}^{-1}, \hat{\xi} \rangle d\tilde{k}$$

where the function θ is given by $\theta(k) = d(\check{\delta})\Sigma\, (a_i, \mu_\delta(k)a_i)$ $(k \in K)$. Conclusion: Every spherical trace function on G of type δ may be represented by an integral formula (ψ). [Proceeding as above, the reader will have no difficulty in working out a formula for the spherical function Φ on G of type δ corresponding to $\mu_\xi | E$.]

In the case of motion groups every spherical trace function on G satisfies a functional equation, viz.

$$\psi(1) \int_K \psi(xk\xi k^{-1}) dk = \psi(x)\psi(\xi) \qquad (x \in G, \xi \in H).$$

Thus let U be a TCI Banach representation of G; for every $\xi \in H$, the operator $\int_K U(k\xi k^{-1}) dk$ commutes with the $U(k)$ $(k \in K)$ – on the other hand, since $k\xi k^{-1} \in H$ and since H is abelian, this operator commutes with every $U(\xi)$ too, hence with every $U(x)$, hence reduces to a scalar (Proposition 4.2.2.3), say: $\int_K U(k\xi k^{-1}) dk = \lambda(\xi)$. Clearly, then,

$$\int_K \psi_\delta^U(xk\xi k^{-1}) dk = \int_K \operatorname{tr}(\mathsf{P}(\delta)U(x)U(k\xi k^{-1})\mathsf{P}(\delta)) dk = \lambda(\xi)\psi_\delta^U(x);$$

taking $x = 1 \Rightarrow \psi_\delta^U(\xi) = \lambda(\xi)\psi_\delta^U(1)$. The functional equation in question is now obvious.

Example Let J_0 be the classical Bessel function of order zero. In the complex Euclidean 3-space H, consider the compact abelian group of rotations represented by the matrices

$$k = \begin{pmatrix} e^{\sqrt{-1}\alpha} & 0 & 0 \\ 0 & e^{\sqrt{-1}\beta} & 0 \\ 0 & 0 & e^{\sqrt{-1}(-\alpha-\beta)} \end{pmatrix} \qquad (\alpha, \beta \in \mathbf{R}).$$

Letting G denote the motion group generated by these rotations and the translations on H, the above theory shows that the spherical trace functions, restricted to H, are given by

$$(2\pi)^{-2} \int_0^{2\pi} \int_0^{2\pi} \exp\,(c_1 x e^{\sqrt{-1}\alpha} + c_2 \bar{x} e^{-\sqrt{-1}\alpha} + c_3 y e^{\sqrt{-1}\beta} + c_4 \bar{y} e^{-\sqrt{-1}\beta}$$
$$+ c_5 z e^{\sqrt{-1}(-\alpha-\beta)} + c_6 \bar{z} e^{\sqrt{-1}(\alpha+\beta)}) d\alpha d\beta$$

where $c_1, \ldots, c_6$ are arbitrary complex numbers, that is, the spherical trace functions can be expressed in terms of the function

$$(2\pi)^{-2}$$

$$\times \int_0^{2\pi} \int_0^{2\pi} \exp(\sqrt{-1}(x\cos\alpha + y\sin\alpha + z\cos\beta + t\sin\beta + u\cos(\alpha+\beta) + v\sin(\alpha+\beta)))d\alpha d\beta$$

$$= (2\pi)^{-1}$$

$$\times \int_0^{2\pi} J_0([(x + u\cos\beta + v\sin\beta)^2 + (y - u\sin\beta + v\cos\beta)^2]^{1/2}) \times \exp(\sqrt{-1}(z\cos\beta + t\sin\beta))d\beta.$$

As above, let $G = KH$ be a motion group. The reader will recall that Mackey's method gives a complete analysis of the irreducible unitary representations of G in terms of the unitary character group $\hat{H}$ of H and the unitary duals of certain subgroups of K (see Vol. I, number 5.4.1). We shall now see what can be said about $\bar{G}$ (the Naimark equivalence classes of TCI Banach representations of G).

Let $\hat{\xi}$ be a (not necessarily unitary) character of H, $U^{\hat{\xi}}$ the continuous representation of G continuously induced by $\hat{\xi}$; as has been noted earlier $U^{\hat{\xi}}$ is, in the case at hand, a K-finite Banach representation of G, the representation space of which may be taken to be $C(K)$ (in the supremum norm).

Theorem 6.2.1.3 (Fell) Let U be a TCI Banach representation of G on E; put $\rho(x) = \|U(x)\|$ $(x \in G)$ – then there exists a (not necessarily unitary) character $\hat{\xi}$ of H, verifying $|\langle \xi, \hat{\xi}\rangle| \leqslant M\rho(\xi)$ $(\xi \in H)$ for some $M > 0$, such that U occurs in $U^{\hat{\xi}}$.

Proof We recall that, by definition, U is said to occur in $U^{\hat{\xi}}$ provided some irreducible component (i.e. topologically irreducible subquotient) of $U^{\hat{\xi}}$ is Naimark equivalent to U (see Vol. I, number 4.5.6). This being so, fix a class $\delta \in \hat{K}$ which occurs in $U|K$ – then, in the usual notations, the map $f \mapsto \mu_\delta(f)$ defines a ρ-continuous irreducible representation of $I_{c,\delta}(G)$ on E_δ $(E(\delta) = E_\delta \otimes_C \mathsf{E}_\delta)$. Thanks to the discussion above which led to the integral formulas for the spherical trace functions on G, there exists a character $\hat{\xi}$ of H, verifying $|\langle \xi, \hat{\xi}\rangle| \leqslant M\rho(\xi)$ $(\xi \in H)$ for some $M > 0$, such that μ_δ is equivalent to a subrepresentation of $\mu_{\hat{\xi}}$, $\mu_{\hat{\xi}}$ the representation of $I_{c,\delta}(G)$ defined by the prescription

$$f \mapsto \int_H \langle \xi, \hat{\xi}\rangle F_f^\delta(\xi) d_H(\xi) \qquad (f \in I_{c,\delta}(G)).$$

From this it follows that U occurs in $U^{\hat{\xi}}$. [The K-finite Banach representation $U^{\hat{\xi}}$ of G operates on $C(K)$ by the rule

$$U^{\hat{\xi}}(\xi\tilde{k})\phi(k) = \langle k^{-1}\xi k, \hat{\xi}\rangle \phi(\tilde{k}^{-1}k) \qquad (k, \tilde{k} \in K; \xi \in H; \phi \in C(K));$$

there is defined, in a natural way, a $d(\delta)$-dimensional representation

ν_ξ (say) of $I_{c,\delta}(G)$ on a certain subspace of $C(K)(\delta)$ – our contention will follow when it is shown that μ_ξ and ν_ξ are equivalent. Let us begin by computing the action of $U^\xi(f)$ ($f \in I_{c,\delta}(G)$) on functions ϕ in $C(K)$ – thus, bearing in mind that $f(x) = \sigma(\Psi_f^\delta(x))$ ($x \in G$), we have

$$\begin{aligned} U^\xi(f)\phi(k) &= \int_H \int_K f(\xi\tilde{k}) \langle k^{-1}\xi k, \hat{\xi}\rangle \phi(\tilde{k}^{-1}k) d\tilde{k} d_H(\xi) \\ &= \int_K \phi(\tilde{k}) d\tilde{k} \int_H \langle \xi, \hat{\xi}\rangle f(k\xi\tilde{k}^{-1}) d_H(\xi) \\ &= \sigma\left[\int_K \phi(\tilde{k}) d\tilde{k} \int_H \langle \xi, \hat{\xi}\rangle \mu_\delta(k) F_f^\delta(\xi) \mu_\delta(\tilde{k}^{-1}) d_H(\xi)\right] \\ &= \sigma[\mu_\delta(k)\mu_\xi(f)\mu_\delta(\phi')] \qquad (k \in K). \end{aligned}$$

Agreeing to write

$$\nu_\delta(\phi) \text{ for } \mu_\delta(\phi') \ (\phi \in C(K), \phi'(k) = \phi(k^{-1})(k \in K)),$$

we then derive that

$$\nu_\delta(U^\xi(f)\phi) = \mu_\xi(f)\nu_\delta(\phi) \qquad (f \in I_{c,\delta}(G)).$$

Consequently the representations μ_ξ and ν_ξ of $I_{c,\delta}(G)$ are indeed equivalent. . . .] □

The preceding theorem thus serves to locate the elements of $\bar{G}$ in a rather precise fashion: Each one will be found somewhere inside one of the U^ξ. Therefore, in order to obtain a complete classification of $\bar{G}$, it is enough to find a Jordan-Hölder series for each of the U^ξ; this, however, is a rather difficult problem which, at the time of this writing, is not yet completely solved (see Fell [8] for some examples).

Needless to say the U^ξ play the same role for motion groups as do the standard representations in the case of semi-simple Lie groups; accordingly it is not unreasonable to expect an analogue of Theorem 5.5.1.5 to hold in the present situation.

Fix a (continuous) character $\hat{\xi}$ of H and let H_ξ denote its stabilizer in G; put $M_\xi = H_\xi \cap K$. If $\sigma \in \hat{M}_\xi$ and if μ_σ is an irreducible unitary matrix representation in the class σ, then the pair $(\mu_\sigma, \hat{\xi})$ determines, in the obvious way, a representation of the group $M_\xi H$; let $U^{\sigma,\xi}$ denote the corresponding continuously induced representation of G – then:

Theorem 6.2.1.4 (Fell) Let U be a TCI Banach representation of G on E; put $\rho(x) = \|U(x)\|$ ($x \in G$) – then there exists a (not necessarily unitary) character $\hat{\xi}$ of H, verifying $|\langle \xi, \hat{\xi}\rangle| \leqslant M\rho(\xi)$ ($\xi \in H$) for some $M > 0$, and an element $\sigma \in \hat{M}_\xi$ such that U occurs in $U^{\sigma,\xi}$.

[The proof proceeds along the to be expected lines; thus the isotypic components which occur in the restriction to M_ξ of the right regular representation of K on $C(K)$ are U^ξ-stable; each of them breaks up into finitely many U^ξ-stable subspaces . . . etc.]

Unfortunately Theorem 6.2.1.4 still does not serve to classify the elements of $\bar{G}$; for it can be shown by examples that a given $U^{\sigma,\xi}$ may very well be *reducible*. . . .

Remark (Fell) Let us agree to adopt the notations of the Remark following Theorem 6.1.1.11 – then for all motion groups G we have $\bar{G}_1 = \hat{G}$. [To see this note first that a bounded continuous character of H is necessarily a unitary character; consequently, by Theorem 6.2.1.3, an element of $\bar{G}_1$ occurs in the continuous representation of G continuously induced by some unitary character $\hat{\xi}$, hence, as is easily seen, in the unitary representation of G unitarily induced by $\hat{\xi}$ – but elements of $\bar{G}$ occurring in a K-finite unitary representation must themselves be unitary, i.e. must lie in $\hat{G}$.]

6.2.2 Spherical Functions on Semi-Simple Lie Groups

Let G be a connected semi-simple Lie group with finite center, K a maximal compact subgroup. In the present number we shall investigate the representation theory of the algebras $I_{c,\delta}(G)$ $(\delta \in \hat{K})$; as we shall see, it is always possible to construct a certain class of 'elementary' representations of a given $I_{c,\delta}(G)$ such that every 'reasonable' TCI Banach representation of our algebra is equivalent to some irreducible component of one of its 'elementary' representations (for the analogous result on motion groups, see Proposition 6.2.1.1); the key for this investigation is a certain integral transform which will in fact be our main object of study.

Before taking up the theory in its general form, let us make a few simple observations concerning zonal spherical functions.

So let G be a connected semi-simple Lie group with finite center, $G = KA_{\mathfrak{p}}N^+$ an Iwasawa decomposition for G. By definition, a continuous complex valued function ϕ on G, which is biinvariant under K, is said to be a *zonal spherical function* provided the mapping $f \mapsto \int_G f(x)\phi(x)d_G(x)$ is a homomorphism of $I_c(G)$ (no continuity requirement). In the well-known way, the zonal spherical functions on G are parameterized by (possibly complex valued) linear functions ν on $\mathfrak{a}_{\mathfrak{p}}$; thus the function

$$x \mapsto \phi_\nu(x) = \int_K e^{(\sqrt{-1}\nu-\rho)(H(xk))}\,dk \qquad (x \in G)$$

is a zonal spherical function on G and all such arise in this way for a suitable choice of ν – moreover $\phi_{\nu_1} = \phi_{\nu_2}$ iff there exists a $w \in W$ (W the Weyl group of the pair $(\mathfrak{g}, \mathfrak{a}_{\mathfrak{p}})$) such that $\nu_1 = w\nu_2$ (these results will be discussed in greater detail below). We intend to prove that any such function ϕ_ν is automatically quasi-bounded. As a consequence, then, there exists a TCI Banach representation U of G in which the trivial representation of K occurs exactly once and such that ϕ_ν is the corresponding spherical trace function (Theorem 6.1.1.7 – recall that ϕ_ν verifies the functional equation $\int_K \phi_\nu(xky)dk = \phi_\nu(x)\phi_\nu(y)$ $(x, y \in G)$). Furthermore this observation tells us, in the case at hand, that the set of spherical functions of type δ, δ the class of the trivial $(k \mapsto 1)$ one dimensional representation of K, coincides with the set of zonal spherical functions on G.

Let us first recall the methods which lead to the 'formula' for zonal spherical functions on G (cf. Helgason [2, Chapter 10]).

Thus let $\mathfrak{X}$ be the centralizer of $\mathfrak{k}_c$ in $\mathfrak{G}$ ($\mathfrak{G}$ the universal enveloping algebra of $\mathfrak{g}_c$) – then $\mathfrak{X}$ can be identified with the algebra of differential operators on G which are invariant under left translation by elements of G and right translation by elements of K. Let $I(\mathfrak{a}_{\mathfrak{p}_c})$ denote the set of W-invariants in $S(\mathfrak{a}_{\mathfrak{p}_c})$, $\gamma_{A_\mathfrak{p}}$ the canonical homomorphism of $\mathfrak{X}$ onto $I(\mathfrak{a}_{\mathfrak{p}_c})$. [Let $\mathfrak{A}_\mathfrak{p}$ denote the universal enveloping algebra of $\mathfrak{a}_{\mathfrak{p}_c}$ – then, for each $D \in \mathfrak{G}$, there exists a unique element $\gamma^*_{A_\mathfrak{p}}(D) \in \mathfrak{A}_\mathfrak{p}$ such that $D - \gamma^*_{A_\mathfrak{p}}(D) \in \mathfrak{k}_c\mathfrak{G} + \mathfrak{G}\mathfrak{n}_c^+$. This being so, the homomorphism $\gamma_{A_\mathfrak{p}}$ is the mapping (of differential operators) $D \mapsto e^\rho \gamma^*_{A_\mathfrak{p}}(D)\circ e^{-\rho}$ $(D \in \mathfrak{X})$. In passing it should be noted that the use of the term 'canonical' is legitimate here – for it can be shown that the definition of $\gamma_{A_\mathfrak{p}}$ is independent of the choice for the ordering of the roots of the pair $(\mathfrak{g}, \mathfrak{a}_\mathfrak{p})$.] It is not difficult to see that the kernel of $\gamma_{A_\mathfrak{p}}$ is $\mathfrak{X} \cap \mathfrak{G}\mathfrak{k}_c$ $(= \mathfrak{X} \cap \mathfrak{k}_c\mathfrak{G})$; hence $\mathfrak{X}/\mathfrak{X} \cap \mathfrak{G}\mathfrak{k}_c$ is a polynomial ring in l generators ($l = \dim(\mathfrak{a}_{\mathfrak{p}_c})$).

Example For sake of illustration, we shall compute $\gamma_{A_\mathfrak{p}}(\omega)$ (ω the Casimir element in $\mathfrak{G}$). So let Φ denote the set of roots of the pair $(\mathfrak{g}_c, \mathfrak{a}_c)$; given $\alpha \in \Phi$, let X_α and H_α be as usual – thus $B(X_\alpha, X_{-\alpha}) = 1$, whence $[X_\alpha, X_{-\alpha}] = H_\alpha$. Choose a basis $H_1, \ldots, H_l$ for $\mathfrak{a}_\mathfrak{p}$ such that $B(H_i, H_j) = \delta_{ij} (1 \leqslant i, j \leqslant l)$; choose a similar basis for $\mathfrak{a}_\mathfrak{k}$ – then, since X_α, $X_{-\alpha}$ and H_α lie in $\mathfrak{k}_c$ if α vanishes on $\mathfrak{a}_\mathfrak{p}$, the very definition of the Casimir element ω tells us that

$$\omega \equiv H_1^2 + \cdots + H_l^2 + 2Q_\rho + 2 \sum_{\alpha \in P_+} X_{-\alpha}X_\alpha \bmod \mathfrak{k}_c\mathfrak{G} \cap \mathfrak{G}\mathfrak{k}_c.$$

This shows that $\omega \equiv H_1^2 + \cdots + H_l^2 + 2Q_\rho \bmod (\mathfrak{k}_c\mathfrak{G} + \mathfrak{G}\mathfrak{n}_c^+)$, whence $\gamma^*_{A_\mathfrak{p}}(\omega) = \tilde{\omega} + 2Q_\rho$ ($\tilde{\omega}$ the restriction to $\mathfrak{a}_{\mathfrak{p}_c}$ of the Casimir polynomial of $\mathfrak{g}_c$); finally, then, it is clear that $\gamma_{A_\mathfrak{p}}(\omega) = \tilde{\omega} - (\rho, \rho)$.

As we know, the zonal spherical functions in the present case admit an 'infinitesimal' characterization. Thus a C^∞ function ϕ on G which is biinvariant under K is a zonal spherical function on G iff $\phi(1) = 1$ and ϕ is an eigenfunction of every differential operator $D \in \mathfrak{X}$ (i.e., $D\phi = c_D\phi$ $(c_D \in \mathbf{C})$ for all $D \in \mathfrak{X}$). [In passing let us recall that a zonal spherical function ϕ is completely determined by its system of eigenvalues, that is $c_D = 0$ (all $D \in \mathfrak{X}$) $\Rightarrow \phi = 0$ – this follows from the fact that for $D \in \mathfrak{X}$ one has $\phi(x; D) = \phi(1; D)\phi(x)$ $(x \in G)$.] Given a linear function $\nu: \mathfrak{a}_\mathfrak{p} \to \mathbf{C}$, consider the function

$$x \mapsto \phi_\nu(x) = \int_K e^{(\sqrt{-1}\nu - \rho)(H(xk))}\, dk \qquad (x \in G)$$

on G. As is shown in Helgason [2, p. 428], ϕ_ν is a zonal spherical function on G which verifies the relation

$$D\phi_\nu = \gamma_{A_\mathfrak{p}}(D\colon \sqrt{-1}\nu)\phi_\nu \quad (\text{all } D \in \mathfrak{X})$$

($\gamma_{A_\mathfrak{p}}(D\colon \sqrt{-1}\nu)$ the value of the polynomial function $\gamma_{A_\mathfrak{p}}(D)$ at the point $\sqrt{-1}\nu$ $(D \in \mathfrak{X})$). In addition we have $\phi_{w\nu} = \phi_\nu$ (all $w \in W$). On the other hand a famous result of Harish-Chandra asserts that the ϕ_ν exhaust the class of zonal spherical functions on G. In outline, the proof runs as follows. Suppose that ϕ is an arbitrary zonal spherical function on G – then for each $D \in \mathfrak{X}$ we have $D\phi = c_D\phi$ ($c_D \in \mathbf{C}$). The mapping $D \mapsto c_D$ ($D \in \mathfrak{X}$) is a homomorphism of $\mathfrak{X}$ into $\mathbf{C}$ which vanishes on $\mathfrak{X} \cap \mathfrak{k}_c\mathfrak{G}$ (for $c_D = \phi(1; D)$ and ϕ is biinvariant under K); since $\mathfrak{X}/\mathfrak{X} \cap \mathfrak{k}_c\mathfrak{G} \sim I(\mathfrak{a}_{\mathfrak{p}_c})$, this mapping corresponds to a homomorphism of $I(\mathfrak{a}_{\mathfrak{p}_c})$ into $\mathbf{C}$; the latter homomorphism admits an extension to a homomorphism of $S(\mathfrak{a}_{\mathfrak{p}_c})$ into $\mathbf{C}$ which is clearly the extension of a linear function μ on $\mathfrak{a}_\mathfrak{p}$ (cf. Proposition 2.1.3.8(i)). Because a zonal spherical function is determined by its eigenvalues, the function $x \mapsto \int_K \exp\{(\mu - \rho)(H(xk))\}dk$ ($x \in G$) must coincide with ϕ . . . .

Proposition 6.2.2.1 Let ν be a (complex valued) linear function on $\mathfrak{a}_\mathfrak{p}$ – then $\phi_\nu(x) = \phi_{-\nu}(x^{-1})$ (all $x \in G$).

Proof For any $k \in K$, let $\kappa(xk)$ denote the unique element in K such that $xk \in \kappa(xk)A_\mathfrak{p}N^+$ ($x \in G$); we recall that (see Vol. I, number 5.5.1)

$$\int_K f(k)dk = \int_K f(\kappa(xk))\exp\{-2\rho(H(xk))\}dk \qquad (f \in C(K)).$$

Therefore, employing an obvious abuse of notation, we have

$$\phi_\nu(x) = \int_K \exp\{\sqrt{-1}\nu(H(xk)) + \rho\,(H(xk))\}d(\kappa(xk)) \qquad (x \in G).$$

Replacing k by $\kappa(x^{-1}k)$ in this latter integral and making use of the formula $H(x\kappa(x^{-1}k)) = -H(x^{-1}k)$ then leads to the relation

$$\phi_\nu(x) = \int_K \exp\{-\sqrt{-1}\nu(H(x^{-1}k)) - \rho(H(x^{-1}k))\}dk = \phi_{-\nu}(x^{-1}) \qquad (x \in G),$$

as desired. □

We shall now return to the problem mentioned earlier, namely that of showing that the ϕ_ν are actually quasi-bounded. So let $\mu : \mathfrak{a}_\mathfrak{p} \to \mathbf{C}$ be a linear function on $\mathfrak{a}_\mathfrak{p}$ and consider the standard representation U^μ of G (cf. Vol. I, number 5.5.1); we recall that U^μ is realized on the Hilbert space $L^2(K)$ (equipped with the usual inner product) via the prescription

$$U^\mu(x)f(k) = \exp\{-(\mu + 2\rho)(H(x^{-1}k))\}f(\kappa(x^{-1}k)) \qquad (k \in K,\ x \in G;\ f \in L^2(K)).$$

The representation U^μ is not necessarily irreducible; however it is clear that U^μ is 'class one' relative to K, that is there exists a non-zero vector f in $L^2(K)$ such that $U^\mu(k)f = f$ (all $k \in K$) – thus take $f = 1$ . . . . This being the case, consider now the function ψ_μ defined by the rule $\psi_\mu(x) = (U^\mu(x)1, 1)$ ($x \in G$) – then ψ_μ is a zonal spherical function on G. To see this, let us recall that the Hilbert space adjoint of $U^{\mu}(x)$ is $U^{-\bar{\mu}-2\rho}(x^{-1})$ $(x \in G)$ (cf. Vol. I, number 5.5.1) – therefore

$$\begin{aligned}\psi_{\mu}(x) &= (U^{\mu}(x)1, 1) \\ &= (1, U^{-\bar{\mu}-2\rho}(x^{-1})1) = \int_K e^{\mu(H(xk))}\,dk \qquad (x \in G).\end{aligned}$$

Define the linear function ν by the relation $\mu = \sqrt{-1}\nu - \rho$ – then it follows that $\psi_{\mu} = \phi_{\nu}$, whence ψ_{μ} is a zonal spherical function on G, as claimed. Of course ψ_{μ} is quasi-bounded. Conversely, if we are given any zonal spherical function ϕ_{ν} whatsoever, define μ by the relation $(\mu + \rho)/\sqrt{-1} = \nu$ and consider the corresponding standard representation U^{μ} of G on $L^2(K)$; reading the preceding argument backwards, we immediately see that ϕ_{ν} occurs as a 'coefficient' in U^{μ} and hence is quasi-bounded.

Summary Every zonal spherical function ϕ_{ν} on G occurs as a coefficient in some (not necessarily irreducible) continuous representation of G on a Hilbert space (which will, in general, depend on ϕ_{ν}).

One may ask: What is the necessary and sufficient condition on ν to ensure that the corresponding zonal spherical function ϕ_{ν} is bounded or, more particularly, to ensure that ϕ_{ν} is positive definite? [Consider the commutative Banach algebra $I^1(G)$ obtained by completing $I_c(G)$ with respect to the L^1-norm – then, for instance, the answer to the first question tells us what the maximal ideals in $I^1(G)$ are (see the discussion preceding Theorem 6.1.1.11).] These questions will be dealt with systematically in a later chapter – however at this point it will be convenient to note the following fact: If ν is real valued on $\mathfrak{a}_{\mathfrak{p}}$, then the corresponding zonal spherical function ϕ_{ν} is positive definite. Thus consider the unitary representations $U^{\sigma,\mu}$ of the principal P-series for G unitarily induced by unitary representations $mhn \mapsto h^{\mu+\rho}$ of the minimal parabolic subgroup $P = MA_{\mathfrak{p}}N^+$ (hence σ itself is the class of the trivial one dimensional representation of M). If ν is real and if μ is defined by the relation $\mu + \rho = \sqrt{-1}\nu$, then $\mu + \rho$ is pure imaginary and the map $h \mapsto h^{\mu+\rho}$ $(h \in A_{\mathfrak{p}})$ is a unitary character of $A_{\mathfrak{p}}$; moreover all such arise in this way. The representation space $E^{\sigma,\mu}$ for $U^{\sigma,\mu}$ is comprised of those Borel functions f on G such that: (1) $f(xmhn) = h^{-(\mu+2\rho)}f(x)$ $(x \in G, mhn \in P = MA_{\mathfrak{p}}N^+)$; (2) $\int_K |f(k)|^2 dk < \infty$. Plainly the function f, $x \mapsto e^{-(\mu+2\rho)(H(x))}$ $(x \in G)$, is in $E^{\sigma,\mu}$ and is left fixed by the operators $U^{\sigma,\mu}(k)$ $(k \in K)$. Finally

$$\begin{aligned}(U^{\sigma,\mu}(x)f, f) &= \int_K f(x^{-1}k)\,dk \\ &= \int_K e^{-(\mu+2\rho)(H(x^{-1}k))}\,dk = \phi_{-\nu}(x^{-1}) = \phi_{\nu}(x) \qquad (x \in G).\end{aligned}$$

Since the representation $U^{\sigma,\mu}$ is unitary, it is now clear that ϕ_{ν} is indeed positive definite whenever ν is real valued on $\mathfrak{a}_{\mathfrak{p}}$.

As above, let G be a connected semi-simple Lie group with finite center, K a maximal compact subgroup; let $G = KA_{\mathfrak{p}}N^+$ be an Iwasawa decomposition for G; normalize the various Haar measures involved so that

$$d_G(x) = h^{2\rho}\, dk d_{A_{\mathfrak{p}}}(h) d_{N^+}(n) \qquad (x \in G;\, x = khn).$$

Fix a class $\delta \in \hat{K}$ together with an irreducible unitary representation μ_δ of K on a Hilbert space E_δ. Consider the integral transform defined for functions f in $I_{c,\delta}(G)$ by the rule

$$F_f^\delta(h) = h^\rho \int_K \int_{N^+} f(khn)\mu_\delta(k^{-1}) d_{N^+}(n) dk \qquad (h \in A_{\mathfrak{p}}).$$

We shall call the map $f \mapsto F_f^\delta$ $(f \in I_{c,\delta}(G))$ the *Abel transformation* on G (of type δ). [The reason for the adopted terminology is simply this. It will be shown that the map $f \mapsto F_f^\delta$ $(f \in I_{c,\delta}(G))$ is one-to-one; this being so, take $G = \mathbf{SL}(2, \mathbf{R})$ and δ the class of the trivial one dimensional representation of K – then, in this case, the problem of inverting the transformation $f \mapsto F_f^\delta$ is essentially equivalent to that of solving the familiar Abel integral equation.]

Note According to an Example in 6.1.1, the algebra $I_{c,\delta}(G)$ is isomorphic, via the map

$$f \mapsto \Psi_f^\delta,\ \Psi_f^\delta(x) = \int_K \mu_\delta(k^{-1}) f(kx) dk \qquad (x \in G),$$

to an algebra $\mathfrak{A}_{c,\delta}(G)$ of (μ_δ, μ_δ)-spherical functions on G. It is clear, therefore, that if $f \in I_{c,\delta}(G)$, then

$$F_f^\delta(h) = h^\rho \int_{N^+} \Psi_f^\delta(hn) d_{N^+}(n) \qquad (h \in A_{\mathfrak{p}}).$$

In what follows we shall frequently make use of the well-known rule which asserts that, for fixed $h \in A_{\mathfrak{p}}$,

$$\int_{N^+} f(n) d_{N^+}(n) = h^{2\rho} \int_{N^+} f(hnh^{-1}) d_{N^+}(n) \qquad (\text{all } f \in C_c(N^+)).$$

The Abel transformation $f \mapsto F_f^\delta$ is evidently a linear mapping of $I_{c,\delta}(G)$ into the space of all compactly supported continuous functions $F: A_{\mathfrak{p}} \to \mathrm{Hom}_{\mathbf{C}}(E_\delta, E_\delta)$. If $f \mapsto f^*$ is the usual involution in $I_{c,\delta}(G)$, then plainly $F_{f^*}^\delta = (F_f^\delta)^*$ where $(F_f^\delta)^*$ denotes the function $h \mapsto \overline{{}^tF_f^\delta(h^{-1})}$ on $A_{\mathfrak{p}}$. Let $f, g \in I_{c,\delta}(G)$ – then we claim that $F_{f*g}^\delta = F_f^\delta * F_g^\delta$. In fact, for $h \in A_{\mathfrak{p}}$, we have

$$\begin{aligned} F_{f*g}^\delta(h) &= h^\rho \int_K \int_{N^+} \int_G f(khnx^{-1}) g(x) \mu_\delta(k^{-1}) d_G(x) d_{N^+}(n) dk \\ &= h^\rho \int_K \int_{N^+} \int_K \int_{A_{\mathfrak{p}}} \int_{N^+} f(khn(\tilde{k}\tilde{h}\tilde{n})^{-1}) g(\tilde{k}\tilde{h}\tilde{n}) \mu_\delta(k^{-1}) \\ &\quad \cdot \tilde{h}^{2\rho} d_{N^+}(\tilde{n}) d_{A_{\mathfrak{p}}}(\tilde{h}) d\tilde{k}\, d_{N^+}(n) dk \end{aligned}$$

$$= h^{\rho}\int_K\int_{N^+}\int_K\int_{A_\mathfrak{p}}\int_{N^+} f(khn\tilde{h}^{-1})g(\tilde{k}\tilde{h}\tilde{n})\mu_{\check{\delta}}(k^{-1}\tilde{k}^{-1})$$
$$\cdot\ \tilde{h}^{2\rho}d_{N^+}(\tilde{n})d_{A_\mathfrak{p}}(\tilde{h})d\tilde{k}d_{N^+}(n)dk$$
$$= h^{\rho}\int_{A_\mathfrak{p}}\left\{\int_K\int_{N^+} f(kh\tilde{h}^{-1}n)\mu_{\check{\delta}}(k^{-1})d_{N^+}(n)dk\right.$$
$$\left.\times\int_K\int_{N^+} g(\tilde{k}\tilde{h}\tilde{n})\mu_{\check{\delta}}(\tilde{k}^{-1})d_{N^+}(\tilde{n})d\tilde{k}\right\}d_{A_\mathfrak{p}}(\tilde{h})$$
$$= \int_{A_\mathfrak{p}} F_f^{\delta}(h\tilde{h}^{-1})F_g^{\delta}(\tilde{h})d_{A_\mathfrak{p}}(\tilde{h})$$

which serves to establish our contention. Finally we claim that the Abel transformation $f \mapsto F_f^{\delta}$ ($f \in I_{c,\delta}(G)$) is one-to-one. Thus suppose that $F_f^{\delta} = 0$ ($f \in I_{c,\delta}(G)$) – then we are to show that $f = 0$. This will be done as follows. Consider the standard representations U^{μ} of G on $L^2(K)$ per the discussion at the beginning of Vol. I, number 5.5.1; according to Corollary 5.5.1.6, the U^{μ} form a *complete* system of representations of G (i.e. of $C_c(G)$). Therefore it will be enough to prove: $F_f^{\delta} = 0$ ($f \in I_{c,\delta}(G)$) $\Rightarrow$ $U^{\mu}(f) = 0$ for all linear functions $\mu\colon \mathfrak{a}_\mathfrak{p} \to \mathbf{C}$. Choose an orthonormal basis $a_1, \ldots, a_{d(\delta)}$ for $E_{\check{\delta}}$; write $\mu_{\check{\delta}}(k)a_i = \sum_j a_{ji}(k)a_j$ ($k \in K$). Given $\phi, \psi \in L^2(K)$, we shall now calculate $(U^{\mu}(f)\phi, \psi)$ ($f \in I_{c,\delta}(G)$). Since $f = \bar{\chi}_{\delta}*f*\bar{\chi}_{\delta}$, one has

$$(U^{\mu}(f)\phi, \psi) = (U^{\mu}(\bar{\chi}_{\delta}*f*\bar{\chi}_{\delta})\phi, \psi) = (U^{\mu}(f)(\bar{\chi}_{\delta}*\phi), \bar{\chi}_{\delta}*\psi).$$

Evidently

$$\bar{\chi}_{\delta}*\phi = \sum_{i,j}\phi_{ij}a_{ij},\ \bar{\chi}_{\delta}*\psi = \sum_{r,s}\psi_{rs}a_{rs}$$

where

$$\phi_{ij} = d(\delta)\int_K a_{ji}(k)\phi(k^{-1})dk,\ \psi_{rs} = d(\delta)\int_K a_{sr}(k)\psi(k^{-1})dk.$$

Hence

$$(U^{\mu}(f)\phi, \psi) = \sum_{i,j}\sum_{r,s}\phi_{ij}\bar{\psi}_{rs}(U^{\mu}(f)a_{ij}, a_{rs}).$$

On the other hand, for $\Phi, \Psi \in L^2(K)$, we have

$$(U^{\mu}(f)\Phi, \Psi)$$
$$= \int_G\int_K f(x)e^{-(\mu+2\rho)(H(x^{-1}k))}\Phi(\kappa(x^{-1}k))\overline{\Psi(k)}dk\,d_G(x)$$
$$= \int_K\int_G f(kx^{-1})e^{-(\mu+2\rho)(H(x))}\Phi(\kappa(x))\overline{\Psi(k)}d_G(x)dk$$
$$= \int_K\left\{\int_K\int_{A_\mathfrak{p}}\int_{N^+} f(kn^{-1}h^{-1}\tilde{k}^{-1})h^{-\mu}\Phi(\tilde{k})\overline{\Psi(k)}d_{N^+}(n)d_{A_\mathfrak{p}}(h)d\tilde{k}\right\}dk$$
$$= \int_K\left\{\int_K\int_{A_\mathfrak{p}}\int_{N^+} f(kn^{-1}h^{-1})h^{-\mu}\Phi(\tilde{k})\overline{\Psi(\tilde{k}k)}d_{N^+}(n)d_{A_\mathfrak{p}}(h)d\tilde{k}\right\}dk\ \ (f = f_K)$$
$$= \int_K\int_{A_\mathfrak{p}}\int_{N^+} f(khn)h^{\mu+2\rho}\Psi^{*}*\Phi(k^{-1})d_{N^+}(n)d_{A_\mathfrak{p}}(h)dk.$$

Let us apply this last relation to the matrix coefficients $\Phi = a_{ij}$, $\Psi = a_{rs}$ – thus

$$
\begin{aligned}
&(U^{\mu}(f)a_{ij}, a_{rs})\\
&= \int_K \int_{A_{\mathfrak{p}}} \int_{N^+} f(khn)h^{\mu+2\rho} a_{sr} * a_{ij}(k^{-1}) d_{N^+}(n) d_{A_{\mathfrak{p}}}(h) dk\\
&= \frac{\delta_{ir}}{d(\delta)} \int_K \int_{A_{\mathfrak{p}}} \int_{N^+} f(khn)h^{\mu+2\rho} a_{sj}(k^{-1}) d_{N^+}(n) d_{A_{\mathfrak{p}}}(h) dk\\
&= \frac{\delta_{ir}}{d(\delta)} \int_{A_{\mathfrak{p}}} h^{\mu+\rho} F_f^{\delta}(h)_{sj} d_{A_{\mathfrak{p}}}(h)
\end{aligned}
$$

and so

$$(U^{\mu}(f)\phi, \psi) = d(\delta)^{-1} \sum_{i,r,s} \phi_{ir}\bar{\psi}_{is} \int_{A_{\mathfrak{p}}} h^{\mu+\rho} F_f^{\delta}(h)_{sr} d_{A_{\mathfrak{p}}}(h).$$

Therefore, if $F_f^{\delta} = 0$ ($f \in I_{c,\delta}(G)$), then $(U^{\mu}(f)\phi, \psi) = 0$ for all $\phi, \psi \in L^2(K)$; this implies that $U^{\mu}(f) = 0$, whence $f = 0$, μ being arbitrary.

Summary The Abel transformation $f \mapsto F_f^{\delta}$ is a *-linear isomorphism of the algebra $I_{c,\delta}(G)$ onto a certain subalgebra of $C_c(A_{\mathfrak{p}}; \mathrm{Hom}_{\mathbf{C}}(E_{\check{\delta}}, E_{\check{\delta}}))$.

[The continuity properties of the map $f \mapsto F_f^{\delta}$ will not be needed for the present discussion.]

Remark A direct characterization of the range of the map $f \mapsto F_f^{\delta}$ appears to be unknown except in the case when δ is the class of the trivial one dimensional representation of K (and even there it is necessary to invoke Plancherel's Theorem for G/K . . .). This is one of the reasons why the theory for semisimple groups is so much more difficult than that for motion groups.

Let ϕ_ν be a zonal spherical function on G (cf. supra) – then, as a simple calculation shows,

$$\int_G f(x)\phi_\nu(x) d_G(x) = \int_{A_{\mathfrak{p}}} F_f(h) h^{\sqrt{-1}\nu} d_{A_{\mathfrak{p}}}(h) \qquad (\text{all } f \in I_c(G);$$

we shall agree to drop the 'δ' from the notation in this case). This relation suggests that it might very well be possible to construct representations of the algebra $I_{c,\delta}(G)$ ($\delta \in \hat{K}$) by employing the Abel transform; following H. Glover, we shall now show that this is actually the case.

Let M denote the centralizer of $A_{\mathfrak{p}}$ in K – then, for any $m \in \mathsf{M}$, we have

$$
\begin{aligned}
F_f^{\delta}(h) &= h^{\rho} \int_K \int_{N^+} f((mkm^{-1})hn) \mu_{\check{\delta}}(mk^{-1}m^{-1}) d_{N^+}(n) dk\\
&= h^{\rho} \int_K \int_{N^+} f(khn) \left\{ \int_{\mathsf{M}} \mu_{\check{\delta}}(mk^{-1}m^{-1}) dm \right\} d_{N^+}(n) dk
\end{aligned}
$$

$$(h \in A_{\mathfrak{p}};\ \text{all } f \in I_{c,\delta}(G)).$$

In order to make further progress it will be necessary to investigate the integral $\int_{\mathsf{M}} \mu_{\check{\delta}}(mkm^{-1}) dm$ ($k \in K$). To this end fix a class $\sigma \in \hat{\mathsf{M}}$; we shall suppose that σ occurs in $\delta | \mathsf{M}$ exactly $[\delta : \sigma] \geqslant 1$ times; we may assume that

$$\mu_\delta(m) = \begin{pmatrix} \mu_{\sigma_1}(m) & & & & & 0 \\ & \mu_{\sigma_2}(m) & & & & \\ & & \cdot & & & \\ & & & \cdot & & \\ & & & & \cdot & \\ 0 & & & & & \mu_{\sigma_d}(m) \end{pmatrix} \qquad (m \in \mathsf{M})$$

in block diagram form with $\mu_{\sigma_i} \sim \mu_{\check{\sigma}}$ iff $1 \leqslant i \leqslant [\delta : \sigma] (\leqslant d)$ (the other irreducible unitary representations of M which appear in $\check{\delta} | \mathsf{M}$ being grouped consecutively along the diagonal according to their equivalence class). The following notations will facilitate our computations: For $1 \leqslant p, q \leqslant d$, write $\epsilon_{pq} = 1$ or 0 according to whether $\sigma_p = \sigma_q$ or not; set $u(i) = p$ if (i, i) is on the $(p, p)^{th}$ block and write $v(i) = i - \sum_{j=1}^{u(i)-1} d(\sigma_j)$ (the 'local' index); put $I(i) = \{r : u(i) = u(r)\}$ (the indices for the $u(i)^{th}$ block). Now let $k \in K$ – then for i, j lying between 1 and $d(\delta)$ we have

$$\begin{aligned} \left(\int_{\mathsf{M}} \mu_\delta(mkm^{-1}) dm \right)_{ij} &= \int_{\mathsf{M}} \sum_{r,s} a_{ir}(m) a_{rs}(k) a_{sj}(m^{-1}) dm \\ &= \sum_{r,s} a_{rs}(k) \int_{\mathsf{M}} a_{ir}(m) \overline{a_{js}(m)} dm. \end{aligned}$$

In view of our assumptions the integral $\int_{\mathsf{M}} a_{ir}(m) \overline{a_{js}(m)} dm$ vanishes if (i, r) is not on a diagonal block or if (j, s) is not on a diagonal block; accordingly we may suppose that (i, r) is on the $(p, p)^{th}$ block (say) while (j, s) is on the $(q, q)^{th}$ block (say) – but then, thanks to the Schur Orthogonality Relations, we have

$$\int_{\mathsf{M}} a_{ir}(m) \overline{a_{js}(m)} dm = d(\sigma_{u(i)})^{-1} \epsilon_{u(i)u(j)} \delta_{v(i)v(j)} \delta_{v(r)v(s)}$$

$$\left\{ \text{if} \quad \begin{matrix} u(i) = u(r) = p \\ u(j) = u(s) = q \end{matrix} \right. .$$

Hence

$$\left(\int_{\mathsf{M}} \mu_\delta(mkm^{-1}) dm \right)_{ij} = d(\sigma_{u(i)})^{-1} \epsilon_{u(i)u(j)} \delta_{v(i)v(j)} \sum_{r \in I(i)} \sum_{s \in I(j)} a_{rs}(k) \delta_{v(r)v(s)}$$

$$(k \in K).$$

The presence of the factor $\delta_{v(i)v(j)}$ shows that the only possible non-zero elements must lie on the 'local diagonal' of the $(u(i), u(j))^{th}$ block; the presence of the factor $\delta_{v(r)v(s)}$ implies that the summation is the 'partial trace' of the $(u(i), u(j))^{th}$ block. For $1 \leqslant p, q \leqslant d$ with $\epsilon_{pq} = 1$ let I_{pq} denote the matrix which is the identity in the $(p, q)^{th}$ block and zero elsewhere – then, because of the presence of the factor $\epsilon_{u(i)u(j)}$, we can write

$$\left(\int_{\mathsf{M}} \mu_\delta(mkm^{-1}) dm \right)_{ij} = d(\sigma_{u(i)})^{-1} \epsilon_{u(i)u(j)} \delta_{v(i)v(j)} \mathrm{tr}(\mu_\delta(k) I_{u(j)u(i)}) \quad (k \in K).$$

Now fix an $f \in I_{c,\delta}(G)$ – then, in view of what was said above, the ij^{th} term of $F_f^\delta(h)$ $(h \in A_\mathfrak{p})$ is given by the formula

$$(F_f^\delta(h))_{ij} = d(\sigma_{u(i)})^{-1}\epsilon_{u(i)u(j)}\delta_{v(i)v(j)} \times h^\rho \int_K \int_{N^+} f(khn)\mathrm{tr}(I_{u(i)u(j)}\mu_\delta(k))d_{N^+}(n)dk.$$

We see, therefore, that the matrix representing $F_f^\delta(h)$ $(h \in A_\mathfrak{p})$ is a scalar multiple of the 'identity' in each of those blocks for which $\epsilon_{u(i)u(j)} = 1$; furthermore the presence of the factor $\epsilon_{u(i)u(j)}$ implies that the $(u(i), u(j))^{th}$ block of $F_f^\delta(h)$ $(h \in A_\mathfrak{p})$ vanishes identically unless $\sigma_{u(i)} = \sigma_{u(j)}$. To illustrate consider the case when $\sigma_1 = \sigma_2 = \sigma_3 = \check{\sigma}$ and $\sigma_4 \neq \check{\sigma}$; then μ_δ is partitioned into 16 blocks and the matrix representing F_f^δ assumes the form

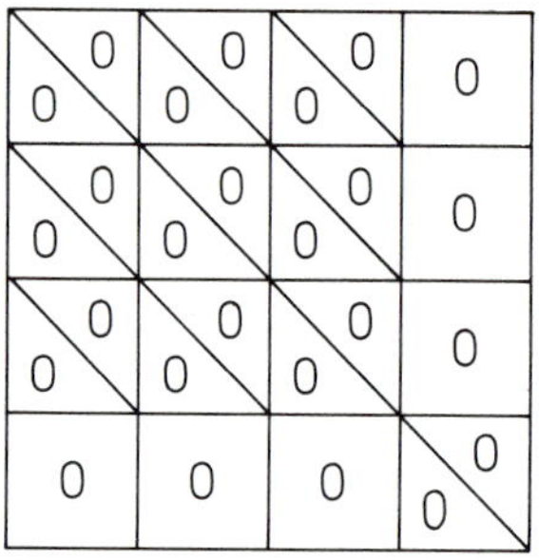

Since σ occurs $[\delta : \sigma]$ $(\geqslant 1)$ times in $\delta\,|\mathrm{M}$, the preceding considerations suggest that we define, for each $f \in I_{c,\delta}(G)$, a $[\delta : \sigma] \times [\delta : \sigma]$ matrix by the rule

$$(F_f^{\delta,\sigma}(h))_{pq} = d(\sigma)^{-1}h^\rho \int_K \int_{N^+} f(khn)\mathrm{tr}(I_{pq}\mu_\delta(k))d_{N^+}(n)dk \qquad (h \in A_\mathfrak{p}).$$

Here the indices p and q run between 1 and $[\delta : \sigma]$ (hence stand for an element of a matrix rather than a block). The assignment $f \mapsto F_f^{\delta,\sigma}$ $(f \in I_{c,\delta}(G))$ is obviously linear in f. If $f \mapsto f^*$ is the usual involution in $I_{c,\delta}(G)$, then $F_{f^*}^{\delta,\sigma} = (F_f^{\delta,\sigma})^*$ – thus, for $h \in A_\mathfrak{p}$, we have

$$\begin{aligned}(F_{f^*}^{\delta,\sigma}(h))_{pq} &= d(\sigma)^{-1}h^\rho \int_K \int_{N^+} \overline{f(n^{-1}h^{-1}k^{-1})}\mathrm{tr}(I_{pq}\mu_\delta(k))d_{N^+}(n)dk \\ &= d(\sigma)^{-1}h^{-\rho}\overline{\left\{\int_K \int_{N^+} f(kh^{-1}n)\mathrm{tr}(I_{qp}\mu_\delta(k))d_{N^+}(n)dk\right\}} \\ &= \overline{(F_f^{\delta,\sigma}(h^{-1}))_{qp}} = \overline{({}^tF_f^{\delta,\sigma}(h^{-1}))}_{pq}\end{aligned}$$

which serves to establish our contention. Now let $f, g \in I_{c,\delta}(G)$ – then we claim that $F_{f*g}^{\delta,\sigma} = F_f^{\delta,\sigma} * F_g^{\delta,\sigma}$. In fact

$$\begin{aligned}(F_{f*g}^{\delta,\sigma}(h))_{pq} &= (F_{f*g}^\delta(h))_{pd(\sigma),qd(\sigma)} = (F_f^\delta * F_g^\delta(h))_{pd(\sigma),qd(\sigma)} \\ &= \int_{A_\mathfrak{p}} \sum_{i=1}^{d(\sigma)} (F_f^\delta(h\tilde{h}^{-1}))_{pd(\sigma),i}(F_g^\delta(\tilde{h}))_{i,qd(\sigma)}d_{A_\mathfrak{p}}(\tilde{h})\end{aligned}$$

$$= \int_{A_\mathfrak{p}} \sum_{i=1}^{[\delta:\sigma]} (F_f^\delta(h\tilde{h}^{-1}))_{pd(\sigma),id(\sigma)} (F_g^\delta(\tilde{h}))_{id(\sigma),qd(\sigma)} d_{A_\mathfrak{p}}(\tilde{h})$$

$$= \int_{A_\mathfrak{p}} \sum_{i=1}^{[\delta:\sigma]} (F_f^{\delta,\sigma}(h\tilde{h}^{-1}))_{pi} (F_g^{\delta,\sigma}(\tilde{h}))_{iq} d_{A_\mathfrak{p}}(\tilde{h})$$

$$= (F_f^{\delta,\sigma} * F_g^{\delta,\sigma}(h))_{pq} \qquad (\text{all } h \in A_\mathfrak{p}),$$

as desired.

Now let $\mu : \mathfrak{a}_\mathfrak{p} \to \mathbf{C}$ be an arbitrary linear function on $\mathfrak{a}_\mathfrak{p}$; define a matrix function $\Phi_\delta^{\sigma,\mu}$ on G by the prescription

$$(\Phi_\delta^{\sigma,\mu}(x))_{pq} = d(\sigma)^{-1} \int_K \operatorname{tr}(I_{pq}\mu_\delta(\kappa(k^{-1}xk)))e^{\mu(H(xk))}dk \qquad (x \in G)$$

where $1 \leqslant p, q \leqslant [\delta : \sigma]$ – then $\Phi_\delta^{\sigma,\mu}$ is 'essentially' a spherical function on G of type δ. [In passing let us observe that

$$(\Phi_\delta^{\sigma,\mu}(f))_{pq} = \int_G f(x)(\Phi_\delta^{\sigma,\mu}(x))_{pq} d_G(x)$$

$$= \int_{A_\mathfrak{p}} (F_f^{\delta,\sigma}(h))_{pq} h^{\mu+\rho} d_{A_\mathfrak{p}}(h) \qquad (\text{all } f \in I_{c,\delta}(G)).]$$

Thus, in the first place, it is clear that $\Phi_\delta^{\sigma,\mu}$ is K-central. Secondly

$$\chi_\delta * (\Phi_\delta^{\sigma,\mu}(x))_{pq} = \int_K \chi_\delta(k)(\Phi_\delta^{\sigma,\mu}(k^{-1}x))_{pq} dk$$

$$= d(\sigma)^{-1} \int_K \int_K \chi_\delta(k) \operatorname{tr}(I_{pq}\mu_\delta(\kappa(\tilde{k}k^{-1}x\tilde{k}^{-1})))e^{\mu(H(x\tilde{k}^{-1}))} d\tilde{k}dk$$

$$= d(\sigma)^{-1} \int_K \int_K \chi_\delta(k) \operatorname{tr}(\mu_\delta(k^{-1})\mu_\delta(\kappa(x\tilde{k}^{-1}))I_{pq}\mu_\delta(\tilde{k}))$$

$$\times\, e^{\mu(H(x\tilde{k}^{-1}))} dk d\tilde{k}$$

$$= d(\sigma)^{-1} \int_K \operatorname{tr}(\mu_\delta(\kappa(xk^{-1}))I_{pq}\mu_\delta(k))e^{\mu(H(xk^{-1}))} dk$$

$$= (\Phi_\delta^{\sigma,\mu}(x))_{pq} \qquad (\text{all } x \in G)$$

where, in going from the third equality to the fourth, we used the familiar rule which asserts that

$$T = d(\delta) \int_K \mu_\delta(k^{-1}) \operatorname{tr}(\mu_\delta(k)T) dk \qquad (\text{all } T \in \operatorname{Hom}_\mathbf{C}(E_\delta, E_\delta)).$$

In order to show that $\Phi_\delta^{\sigma,\mu}$ is quasi-bounded we shall proceed as follows. For p (respectively q) as above, let $I(p) = \{i : u(i) = p\}$ (respectively $I(q) = \{i : u(i) = q\}$) – then

$$(\Phi_\delta^{\sigma,\mu}(x))_{pq} = d(\sigma)^{-1} \int_K \operatorname{tr}(I_{pq}\mu_\delta(\kappa(k^{-1}xk)))e^{\mu(H(xk))} dk$$

$$= d(\sigma)^{-1} \int_K e^{\mu(H(xk))} \sum_{i,j,t=1}^{d(\delta)} (I_{pq})_{ij} \overline{a_{jt}(k^{-1})}\, \overline{a_{ti}(\kappa(xk))} dk$$

$$= d(\sigma)^{-1} \sum_{i \in I(p)} \sum_{j \in I(q)} \delta_{v(i)v(j)} \sum_{t=1}^{d(\delta)} \int_K a_{tj}(k) \overline{a_{ti}(\kappa(xk))}\, e^{\mu(H(xk))} dk$$

$$= d(\sigma)^{-1} \sum_{i \in I(p)} \sum_{j \in I(q)} \delta_{v(i)v(j)} \sum_{t=1}^{d(\delta)} (a_{tj}, U^{-\bar{\mu}-2\rho}(x^{-1})a_{ti})$$

$$= d(\sigma)^{-1} \sum_{i \in I(p)} \sum_{j \in I(q)} \delta_{v(i)v(j)} \sum_{t=1}^{d(\delta)} (U^{\mu}(x)a_{tj}, a_{ti}) \qquad (\text{all } x \in G)$$

since the Hilbert space adjoint of $U^{\mu}(x)$ is $U^{-\bar{\mu}-2\rho}(x^{-1})$ $(x \in G)$ (cf. Vol. I, number 5.5.1). The quasi-boundedness of $\Phi_{\delta}^{\sigma,\mu}$ is thereby established. Finally we claim that the assignment $f \mapsto \int_G f(x)\Phi_{\delta}^{\sigma,\mu}(x)d_G(x)$ $(f \in I_{c,\delta}(G))$ is a representation of the algebra $I_{c,\delta}(G)$ which is even a *-representation if $\mathscr{R}(\mu) = -\rho$. In fact

$$\begin{aligned}
\Phi_{\delta}^{\sigma,\mu}(f*g) &= \int_{A_{\mathfrak{p}}} h^{\mu+\rho} F_{f*g}^{\delta,\sigma}(h) d_{A_{\mathfrak{p}}}(h) \\
&= \int_{A_{\mathfrak{p}}} h^{\mu+\rho} F_f^{\delta,\sigma} * F_g^{\delta,\sigma}(h) d_{A_{\mathfrak{p}}}(h) \\
&= \int_{A_{\mathfrak{p}}} \int_{A_{\mathfrak{p}}} h^{\mu+\rho} F_f^{\delta,\sigma}(h\bar{h}^{-1}) F_g^{\delta,\sigma}(\bar{h}) d_{A_{\mathfrak{p}}}(\bar{h}) d_{A_{\mathfrak{p}}}(h) \\
&= \int_{A_{\mathfrak{p}}} \int_{A_{\mathfrak{p}}} (h\bar{h}^{-1})^{\mu+\rho} F_f^{\delta,\sigma}(h\bar{h}^{-1}) \bar{h}^{\mu+\rho} F_g^{\delta,\sigma}(\bar{h}) d_{A_{\mathfrak{p}}}(h) d_{A_{\mathfrak{p}}}(\bar{h}) \\
&= \Phi_{\delta}^{\sigma,\mu}(f)\Phi_{\delta}^{\sigma,\mu}(g) \qquad (\text{all } f, g \in I_{c,\delta}(G))
\end{aligned}$$

which proves that $\Phi_{\delta}^{\sigma,\mu}$ is a representation of $I_{c,\delta}(G)$. An easy computation tells us that

$$(\Phi_{\delta}^{\sigma,\mu}(f))^* = \Phi_{\delta}^{\sigma,-\bar{\mu}-2\rho}(f^*) \qquad (\text{all } f \in I_{c,\delta}(G)).$$

Therefore if $\mathscr{R}(\mu) = -\rho$, then the map $f \mapsto \int_G f(x)\Phi_{\delta}^{\sigma,\mu}(x)d_G(x)$ actually defines a *-representation of the *-algebra $I_{c,\delta}(G)$. Our observations show, therefore, that $\Phi_{\delta}^{\sigma,\mu}$ is a spherical function on G of type δ provided the representation $f \mapsto \Phi_{\delta}^{\sigma,\mu}(f)$ $(f \in I_{c,\delta}(G))$ is irreducible.

By an *elementary representation* of the algebra $I_{c,\delta}(G)$ we shall understand any representation of the form $f \mapsto \int_G f(x)\Phi_{\delta}^{\sigma,\mu}(x)d_G(x)$ $(f \in I_{c,\delta}(G))$. The basis for this terminology is simply this. Fix a linear function μ: $\mathfrak{a}_{\mathfrak{p}} \to \mathbf{C}$; fix a class σ which occurs in $\delta | M$ and consider the elementary representation $U^{\sigma,\mu}$ of G – then δ occurs exactly $[\delta : \sigma]$ $(\geqslant 1)$ times in $U^{\sigma,\mu} | K$. Moreover there is defined on a certain $[\delta : \sigma]$-dimensional subspace of the δth isotypic component of $U^{\sigma,\mu}$ a representation of the algebra $I_{c,\delta}(G)$; a little reflection then leads without difficulty to the conclusion that this representation of $I_{c,\delta}(G)$ is equivalent to the elementary representation $\Phi_{\delta}^{\sigma,\mu}$ defined above.

Here is the main result of the present number.

Theorem 6.2.2.2 (Glover) Let G be a connected semi-simple Lie group with finite center, K a maximal compact subgroup of G; fix a class $\delta \in \hat{K}$; let ρ be a semi-norm on G – then any ρ-continuous finite dimensional irreducible representation of the algebra $I_{c,\delta}(G)$ is equivalent to an irreducible component of one of its elementary representations.

[In view of Theorem 5.5.1.5 and the discussion supra, our contention is clear (modulo Theorem 6.1.1.11, of course).]

Corollary 6.2.2.3 Let G be a connected semi-simple Lie group with finite center, K a maximal compact subgroup of G; let $U^{\sigma,\mu}$ be an elementary representation of G; for $\delta \in \hat{K}$, let $\psi_{\delta}^{\sigma,\mu}$ denote the function $x \mapsto \mathrm{tr}(\mathsf{P}^{\sigma,\mu}(\delta)U^{\sigma,\mu}(x)\mathsf{P}^{\sigma,\mu}(\delta))$ $(x \in G)$ – then

$$\psi_{\delta}^{\sigma,\mu}(x) = d(\sigma)^{-1}\int_K \chi_{\delta}*\chi_{\sigma}(\kappa(k^{-1}xk))e^{\mu(H(xk))}dk \qquad (x \in G).$$

Proof Fix $x \in G$ – then we have

$$\begin{aligned}\psi_{\delta}^{\sigma,\mu}(x) &= d(\delta)\mathrm{tr}(\Phi_{\delta}^{\sigma,\mu}(x))\\ &= \frac{d(\delta)}{d(\sigma)}\sum_{p=1}^{[\delta:\sigma]}\int_K \mathrm{tr}(I_{pp}\mu_{\delta}(\kappa(k^{-1}xk)))e^{\mu(H(xk))}dk\\ &= d(\sigma)^{-1}\int_K \chi_{\delta}*\chi_{\sigma}(\kappa(k^{-1}xk))e^{\mu(H(xk))}dk,\end{aligned}$$

as desired. □

Remark A comment on our procedure is in order. Thus, in the case of motion groups, we first found all 'reasonable' irreducible representations of a given $I_{c,\delta}(G)$ and then utilized this result to 'classify' $\bar{G}$ (cf. Theorems 6.2.1.3 and 6.2.1.4); for semi-simple groups this procedure was reversed in that we first 'classified' $\bar{G}$ (essentially via certain deep results on differential equations) and then found the 'reasonable' irreducible representations of $I_{c,\delta}(G)$. Recently Rader [1] has obtained a direct 'elementary' proof of Theorem 6.2.2.2.

Example In order to illustrate, in a convenient way, some of the general theory developed above we shall now place a restrictive assumption on G. Let M denote the centralizer of $A_{\mathfrak{p}}$ in K; in what follows it will be supposed that the pair (K, M) has the following property: Fix a class $\delta \in \hat{K}$ – then a given element $\sigma \in \hat{\mathsf{M}}$ occurs at most once in $\delta \mid \mathsf{M}$. [Here is an instance of this phenomenon. Let n be an integer $\geqslant 2$; let G be the identity component of the orthogonal group associated with the indefinite quadratic form $-x_0^2 + x_1^2 + \cdots + x_n^2$ (i.e. let G be the generalized Lorentz group) – then G admits an Iwasawa decomposition, $G = KA_{\mathfrak{p}}N^+$, where K is the maximal compact subgroup of rotations around the x_0-axis (thus $K \sim \mathbf{SO}(n)$), $A_{\mathfrak{p}}$ is a one parameter subgroup of matrices of the form

$$A_{\mathfrak{p}} = \left\{\begin{pmatrix} \cosh(t) & \sinh(t) & 0 \\ \sinh(t) & \cosh(t) & \\ 0 & & I_{n-1}\end{pmatrix} : \ t \in \mathbf{R}\right\}$$

(I_{n-1} denoting the unit matrix of order $n-1$), and N^+ is a nilpotent group homeomorphic to $\mathbf{R}^{n-1}$. The centralizer M of $A_{\mathfrak{p}}$ in K consists of the rotations in the subspace $(x_2, x_3, \ldots, x_n)$ which leave fixed x_0 and x_1 – hence $\mathsf{M} \sim \mathbf{SO}(n-1)$. It is a classical fact that the restriction to M of any irreducible representation of K contains a given irreducible representation of M at most once. [For a simple proof, see Dixmier, 'Sur les représentations de certains groupes orthogonaux', C.R. Acad. Sc. Paris, vol. 89 (1960), pp. 3263–3265.] Another instance of the phenomenon in question is discussed in Stein [1, p. 576].] Under these conditions it is clear that the matrix representing F_f^δ ($f \in I_{c,\delta}(G)$) assumes diagonal form relative to the basis for $E_{\mathfrak{z}}$ which we are using. There is an important consequence which may be drawn from this observation, namely: For every $\delta \in \hat{K}$, the algebras $I_{c,\delta}(G)$ are commutative. Thus if $f, g \in I_{c,\delta}(G)$, then $F_{f*g}^\delta = F_f^\delta * F_g^\delta = F_g^\delta * F_f^\delta = F_{g*f}^\delta \Rightarrow f*g = g*f$, the Abel transformation being injective. The commutativity of the algebra $I_{c,\delta}(G)$ then implies that δ occurs at most once in every TCI Banach representation of G (Proposition 6.1.1.6), that every TCI Banach representation of $I_{c,\delta}(G)$ is one dimensional (Proposition 4.5.1.10) and that every spherical function on G satisfies a convenient functional equation (cf. Theorem 6.1.1.7). In the present situation we have an explicit formula for the spherical functions on G of type δ; thus letting σ run through those classes in $\hat{M}$ for which $[\delta : \sigma] \geqslant 1$, every spherical function on G of type δ is then seen to be given by the prescription

$$x \mapsto \int_K \frac{\chi_\delta * \chi_\sigma(\kappa(k^{-1}xk))}{\chi_\delta * \chi_\sigma(1)} \exp\{(\sqrt{-1}\nu - \rho)(H(xk))\}\, dk \qquad (x \in G)$$

for a suitable choice of the linear function ν ($\nu : \mathfrak{a}_\mathfrak{p} \to \mathbf{C}$) (our function being even positive definite for real ν).

For completeness let us record the following result.

Theorem 6.2.2.4 (Harish-Chandra) Let G be a connected semi-simple Lie group with finite center, K a maximal compact subgroup of G; let U be a TCI Banach representation of G on E; fix a class $\delta_0 \in \hat{K}$ which occurs in $U|K$ – then there exists a linear function $\mu : \mathfrak{a}_\mathfrak{p} \to \mathbf{C}$ with the property that $\omega(D : \mu + \rho) \in \operatorname{Ker}(\mu_{\delta_0}) \Rightarrow D \in \operatorname{Ker}(U_{\delta_0})$ ($D \in \mathfrak{X}$). This being so, fix one such μ and let M denote the centralizer of $A_\mathfrak{p}$ in K – then we can choose an element $T_\mu \in \operatorname{Hom}_{\mathbf{C}}(E(\delta_0), E(\delta_0))$ such that:

(i) $U_{\delta_0}(m)T_\mu = T_\mu U_{\delta_0}(m)$ (all $m \in M$);

(ii) $\Psi_{\delta_0}^U(x)(=\mathsf{P}(\delta_0)U(x)\mathsf{P}(\delta_0)) = \int_K U_{\delta_0}(\kappa(xk))T_\mu U_{\delta_0}(k^{-1})e^{\mu(H(xk))}dk$ (all $x \in G$).

In view of Lemma 5.5.1.9, we need only deal with statements (i) and (ii); in turn their proofs are contained in the following discussion.

Consider the standard representation U^μ of G on $L^2(K)$ (cf. Vol. I, number 5.5.1); let T_0 denote the linear transformation in $L^2(K; \delta_0)$ defined by the rule $T_0(f) = (f, \bar{\chi}_{\delta_0})\bar{\chi}_{\delta_0}$ ($f \in L^2(K; \delta_0)$) – then we claim that

$$\mathsf{P}^\mu(\delta_0)U^\mu(x)\mathsf{P}^\mu(\delta_0) = \int_K U_{\delta_0}^\mu(\kappa(xk))T_0 U_{\delta_0}^\mu(k^{-1})e^{\mu(H(xk))}dk \quad (\text{all } x \in G).$$

This is seen as follows. To begin with, note that

$$(U_{\delta_0}^\mu(k)\bar{\chi}_{\delta_0}, \bar{\chi}_{\delta_0}) = \bar{\chi}_{\delta_0} * \bar{\chi}_{\delta_0}(k^{-1}) = (U_{\delta_0}^\mu(k)\bar{\chi}_{\delta_0})(1) \qquad (\text{all } k \in K);$$

thus, since $L^2(K; \delta_0)$ is spanned by vectors of the form $U_{\delta_0}^\mu(k)\bar{\chi}_{\delta_0}$ ($k \in K$), we deduce that $f(1) = (f, \bar{\chi}_{\delta_0})$ (all $f \in L^2(K; \delta_0)$) and so, for any $f \in L^2(K; \delta_0)$, we have

$$f(k) = (f, U_{\delta_0}^\mu(k)\bar{\chi}_{\delta_0}) \qquad (\text{all } k \in K).$$

Now suppose that $f, g \in L^2(K; \delta_0)$ – then, since $d(\kappa(xk)) = e^{-2\rho(H(xk))}dk$ $(x \in G)$, we have

$$(U^\mu(x)f, g) = \int_K f(\kappa(x^{-1}k))\overline{g(k)} \exp\{-(\mu + 2\rho)(H(x^{-1}k))\}\, dk$$

$$= \int_K f(k)\overline{g(\kappa(xk))}e^{\mu(H(xk))}dk \qquad (\text{all } x \in G).$$

On the other hand, if $f, g \in L^2(K; \delta_0)$, then also

$$\begin{aligned}(U^\mu_{\delta_0}(\kappa(xk))T_0(U^\mu_{\delta_0}(k^{-1})f), g) &= (U^\mu_{\delta_0}(k^{-1})f, \bar{\chi}_{\delta_0})(U^\mu_{\delta_0}(\kappa(xk))\bar{\chi}_{\delta_0}, g)\\ &= (f, U^\mu_{\delta_0}(k)\bar{\chi}_{\delta_0})\overline{(g, U^\mu_{\delta_0}(\kappa(xk))\bar{\chi}_{\delta_0})}\\ &= f(k)\overline{g(\kappa(xk))} \qquad (\text{all } x \in G,\ k \in K).\end{aligned}$$

Consequently, for all $f, g \in L^2(K; \delta_0)$, we have

$$(U^\mu(x)f, g) = \int_K (U^\mu_{\delta_0}(\kappa(xk))T_0(U^\mu_{\delta_0}(k^{-1})f), g)e^{\mu(H(xk))}dk \quad (\text{all } x \in G)$$

from which the asserted formula for $\mathsf{P}^\mu(\delta_0)U^\mu(x)\mathsf{P}^\mu(\delta_0)$ $(x \in G)$ then follows at once.

Because $\kappa(xkm) = \kappa(xk)m$ and $H(xkm) = H(xk)$ $(m \in \mathsf{M})$, we can write

$$\begin{aligned}\mathsf{P}^\mu(\delta_0)U^\mu(x)\mathsf{P}^\mu(\delta_0) &= \int_K U^\mu_{\delta_0}(\kappa(xk))T_0U^\mu_{\delta_0}(k^{-1})e^{\mu(H(xk))}dk\\ &= \int_K U^\mu_{\delta_0}(\kappa(xk))\left[\int_{\mathsf{M}} U^\mu_{\delta_0}(m)T_0U^\mu_{\delta_0}(m^{-1})dm\right]\\ &\quad \times U^\mu_{\delta_0}(k^{-1})e^{\mu(H(xk))}dk \qquad (\text{all } x \in G)\end{aligned}$$

which shows that T_0 can be replaced by $\int_{\mathsf{M}} U^\mu_{\delta_0}(m)T_0U^\mu_{\delta_0}(m^{-1})dm$ in the expression for $\mathsf{P}^\mu(\delta_0)U^\mu(x)\mathsf{P}^\mu(\delta_0)$ $(x \in G)$; as this latter transformation commutes with the $U^\mu_{\delta_0}(m)$ $(m \in \mathsf{M})$, contentions (i) and (ii) of Theorem 6.2.2.4 then follow without difficulty from Theorem 5.5.1.5.

Note It is clear that Theorem 6.2.2.4 also provides one, in principle at least, with an 'explicit formula' for the spherical trace functions on G.

Chapter 7

Topology on the Dual Plancherel Measure

Introduction

The purpose of the present chapter is to set down in a precise fashion the main facts as regards the structure of the unitary dual $\hat{G}$ of a locally compact group G (which satisfies the second axiom of countability, say). Since these topics are dealt with more than adequately by Dixmier [15], we shall give very few proofs; instead we shall be content to look at a number of examples which serve to illustrate the abstract theory, full details being provided in the case of a semi-simple G.

7.1 Topology on the Dual

7.1.1 Generalities

Let G be a locally compact group which satisfies the second axiom of countability; let $\hat{G}$ be the unitary dual of G (i.e. the set of unitary equivalence classes of irreducible unitary representations of G). In the present number we shall recall the definition and properties of the topology with which $\hat{G}$ is customarily equipped; the reader who is unfamiliar with this circle of ideas can find complete details in Dixmier [15].

Let $L^1(G)$ be the space of integrable functions on G with respect to left invariant Haar measure – then, as we know, $L^1(G)$ is a Banach *-algebra. Moreover there is a natural one-to-one correspondence between the unitary equivalence classes of unitary representations of G and those of the non-degenerate *-representations of $L^1(G)$; under this correspondence, irreducible unitary representations of G give rise to irreducible *-representations of $L^1(G)$ and vice versa. Given $f \in L^1(G)$, put

$$\|f\| = \sup \|U(f)\| \qquad (\leqslant \|f\|_1)$$

where U runs through the set of non-degenerate *-representations of $L^1(G)$ – then the assignment $f \mapsto \|f\|$ defines a norm on $L^1(G)$, the *minimal regular norm*. The completion of $L^1(G)$ with respect to the minimal regular norm is a C^*-algebra $C^*(G)$ (say), the so-called *group C^*-algebra* of G. The correspondence between representations of G and of $L^1(G)$ carries over to an analogous correspondence between unitary representations of G and *-representations of $C^*(G)$, irreducible representations of one corresponding to irreducible representations of the other. The unitary dual

$\hat{C}^*(G)$ of $C^*(G)$ (i.e. the set of unitary equivalence classes of (non-zero) irreducible *-representations of $C^*(G)$) carries a topology, the hull-kernel topology: Let $\hat{\mathscr{V}}$ be a subset of $\hat{C}^*(G)$ – then, by definition, the closure $\hat{\mathscr{V}}^{cl}$ of $\hat{\mathscr{V}}$ is the set of all $\hat{U}$ in $\hat{C}^*(G)$ which are weakly contained in $\hat{\mathscr{V}}$, that is for which

$$\bigcap_{\hat{V} \in \hat{\mathscr{V}}} \operatorname{Ker}(\hat{V}) \subset \operatorname{Ker}(\hat{U}).$$

The assignment $\hat{\mathscr{V}} \mapsto \hat{\mathscr{V}}^{cl}$ determines a closure operation in $\hat{C}^*(G)$, the topology thus defined being the *hull-kernel topology;* unless the contrary is stated, it will be assumed that $\hat{C}^*(G)$ is equipped with this topology.

By definition, then, the *dual space* $\hat{G}$ of G is the set $\hat{G}$ supplied with the topology arising from the canonical correspondence $\hat{G} \leftrightarrow \hat{C}^*(G)$. The topological space $\hat{G}$ is a locally compact Baire space which is even second countable (since, by assumption, G is); however $\hat{G}$ will not, in general, be Hausdorff. . . . [In this connection, let us recall that if G is postliminaire, then there exists in $\hat{G}$ an open, everywhere dense, locally compact subset which, in the relative topology, is Hausdorff.] If G is postliminaire, then $\hat{G}$ is a T_0-space; if G is liminaire, then $\hat{G}$ is a T_1-space (i.e. 'points are closed').

The topology on $\hat{G}$ can be described in somewhat different terms – thus:

Proposition 7.1.1.1 (Fell) Fix a subset $\hat{\mathscr{V}}$ of $\hat{G}$; let $\hat{U} \in \hat{G}$ – then the following conditions are equivalent:

(i) $\hat{U} \in \hat{\mathscr{V}}^{cl}$;

(ii) There exists a (non-zero) function of positive type associated with $\hat{U}$ which is the uniform limit on compacta of functions of positive type associated with elements of $\hat{\mathscr{V}}$;

(iii) Every function of positive type associated with $\hat{U}$ is the uniform limit on compacta of functions of positive type associated with $\hat{\mathscr{V}}$.

Note Suppose that G is abelian – then the preceding result serves to ensure us that the usual topology on the dual (group) $\hat{G}$ coincides with the hull-kernel topology.

Examples (1) The dual of a compact group is discrete; the dual of a discrete group is compact.

(2) Suppose that G is the infinite dihedral group – then $\hat{G}$ can be identified with $(0, 1) \cup \{a\} \cup \{b\} \cup \{c\} \cup \{d\}$. A point t in $(0, 1)$ approaches both a and b when t tends to zero in the usual sense; a point t in $(0, 1)$ approaches both c and d when t tends to one in the usual sense – schematically (cf. Dixmier [19]):

(3) Let us take for G the Heisenberg group – then, as we have seen in Vol. I, number 5.4.2 (Example 3), the unitary dual $\hat{G}$ of G can be identified with ($\mathbf{R}$ –

$\{0\}) \cup \mathbf{R}^2$. The topology on $\hat{G}$ is the ordinary topology in the parameters with the exception that when a point t in $\mathbf{R} - \{0\}$ tends to 0 in the usual sense it approaches (in $\hat{G}$) every point of $\mathbf{R}^2$ – schematically (cf. Dixmier [9]):

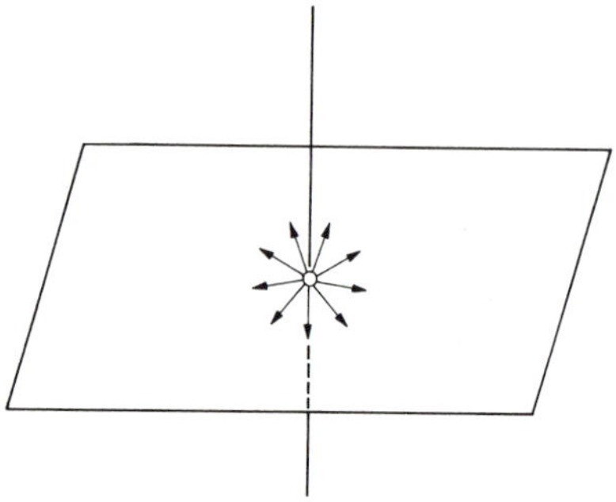

(4) Suppose that G is the '$\alpha t + \beta$' group – then, as we have seen in Vol. I, number 5.4.2 (Example 1), $\hat{G}$ can be identified with the subset

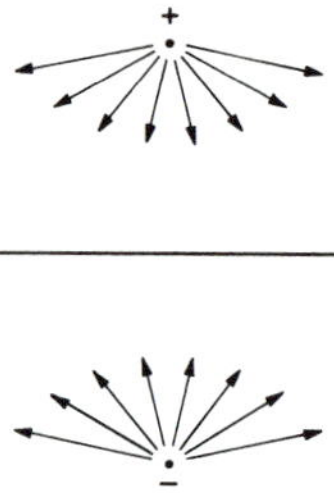

of the ordinary Euclidean plane. The closure of $+$ (respectively $-$) is $\hat{G} - \{-\}$ (respectively $\hat{G} - \{+\}$); the points $\{+\}, \{-\}$ are both open but not closed .... [For the details, consult Fell [2].]

(5) Suppose that $G = \mathbf{SL}(2, \mathbf{C})$ – then, as is well-known, the irreducible unitary representations of G fall into three distinct classes, namely the trivial one dimensional representation I (say), the representations in the principal P-series (P a minimal parabolic subgroup of G), and the representations in the complementary series. The representations $U^{m,r}$ of the principal P-series are indexed by a pair (m, r) with m an integer and r a real number; two representations in the principal P-series corresponding to distinct parameter pairs (m_1, r_1), (m_2, r_2) are unitarily equivalent iff $m_1 = -m_2, r_1 = -r_2$; in view of this, let us agree to index the principal P-series by the pairs (m, r) for $m \geqslant 0$ (with $r \geqslant 0$ when $m = 0$). The representations U^s in the complementary series are indexed by a real number s with $-1 < s < 0$; distinct values of s $(-1 < s < 0)$ give rise to inequivalent representations; no representation in the complementary series is (unitarily) equivalent to a member of the principal P-series; finally the point $s = -1$ corresponds to the trivial one dimensional representation of G. The topology of $\hat{G}$ is the ordinary topology which it acquires when regarded as a subset of the plane with the exception that if t approaches -1 in the ordinary sense then t approaches both I and $U^{2,0}$ in $\hat{G}$ – schematically (cf. Fell [1]):

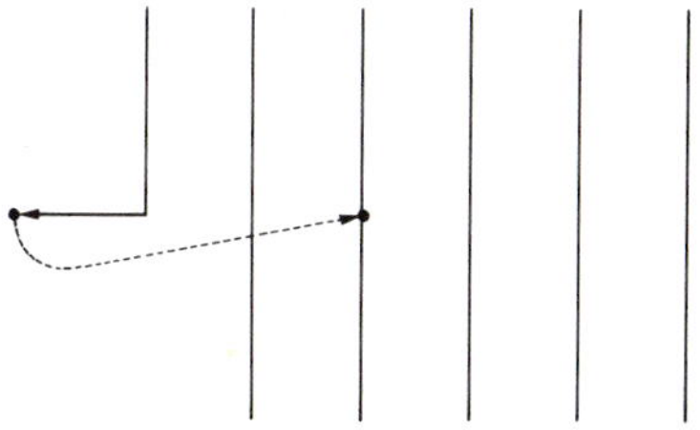

In passing we remark that Fell has determined the structure of $C^*(\mathbf{SL}(2, \mathbf{C}))$ (see his paper 'The structure of algebras of operator fields,' Acta Math., vol. 106 (1961), pp. 233–280).

Let U be a *-representation of $C^*(G)$, $\mathscr{V}$ a family of *-representations of $C^*(G)$ – then U is said to be *weakly contained* in $\mathscr{V}$ provided

$$\bigcap_{V \in \mathscr{V}} \mathrm{Ker}(V) \subset \mathrm{Ker}(U).$$

Let U be a unitary representation of G, $\mathscr{V}$ a family of unitary representations of G – then U is said to be weakly contained in $\mathscr{V}$ if this is the case when U and $\mathscr{V}$ are considered as representations of $C^*(G)$. [Naturally these agreements are invariant under unitary equivalence; accordingly they may be applied to an equivalence class of unitary representations rather than just the representation itself.]

Definition Let U be a unitary representation of G – then by the *support of* U, written spt (U), we shall understand the subset of $\hat{G}$ comprised of those classes which are weakly contained in U.

Evidently the support of a given unitary representation of G is a closed subset of $\hat{G}$.

Definition The *reduced dual* $\hat{G}_r$ of G is the support of the left regular representation.

Suppose that G is unimodular – then every (irreducible) square integrable unitary representation of G is contained in the left regular representation, therefore belongs to $\hat{G}_r$ (hence $\hat{G} = \hat{G}_r$ if G is compact). Evidently $\hat{G} = \hat{G}_r$ whenever G is abelian . . . .

For orientation let us recall the following result.

Proposition 7.1.1.2 (Takenouchi) Let G be a locally compact group which satisfies the second axiom of countability – then the following conditions are equivalent:

(i) The dual of G is equal to the reduced dual;

(ii) The trivial one dimensional representation of G belongs to the reduced dual;

(iii) Every continuous function on G of positive type is the uniform limit on compacta of functions of the form $f*\bar{f}'$ ($f \in C_c(G)$);

(iv) The function 1 is the uniform limit on compacta of functions of the form $f*\bar{f}'$ ($f \in C_c(G)$).

[For any $f \in C_c(G)$, $\bar{f}'$ denotes the function $x \mapsto \overline{f(x^{-1})}$ ($x \in G$).]

Examples (1) A classical lemma due to Takenouchi [1] (cf. Dixmier [15, p. 319]) asserts that if G is the semi-direct product of two groups H and K, each verifying the conditions of Proposition 7.1.1.2, then G itself verifies these conditions. To illustrate, suppose that G is a motion group – then G admits no 'complementary series', i.e. $\hat{G} = \hat{G}_r$.

(2) (Takenouchi) Let G be a (non-compact) connected semi-simple Lie group with finite center – then it is always the case that $\hat{G} \neq \hat{G}_r$ (therefore, in a certain sense, G always admits 'complementary series'). Thus let $G = KA_{\mathfrak{p}}N^+$ be an Iwasawa decomposition for G; normalize the relevant Haar measures in such a way as to ensure that $d_G(x) = f_{2\rho}(s)dkd_S(s)$ whenever $x = ks$ ($k \in K$, $s \in S(= A_{\mathfrak{p}}N^+)$). [Here $f_{2\rho}(s) = h^{2\rho}$ ($s = hn\ldots$); given any $x \in G$, let us agree to denote by $\kappa(x)$ (respectively $\sigma(x)$) the unique element in K (respectively S) such that $x = \kappa(x)\sigma(x)$.] Fix an element x_0 in G having the property that $f_{2\rho}(\sigma(x_0k_0)) < 4^{-1}$ for some k_0 in K; let ω be the subset of K consisting of those k with $f_{2\rho}(\sigma(x_0k)) \leqslant 4^{-1}$ – then certainly $vol(\omega) = \int_\omega dk > 0$. Choose $\delta > 0$ and $\epsilon > 0$ in such a way that $4\epsilon/(1-\epsilon) \leqslant \delta^2 < vol(\omega)$ – then there is no $f \in C_c(G)$ such that

$$|f*f^*(x) - 1| \leqslant \epsilon$$

for all x in the compact set $\Omega = K \cup Kx_0K$. In fact suppose to the contrary that such an f actually does exist; put $f_0(x) = \int_K f(kx)dk$ ($x \in G$) – then, in view of the relation $|f*f^*(k_1xk_2) - 1| \leqslant \epsilon$ ($k_1, k_2 \in K$; $x \in \Omega$), we infer by integration that $|f_0*f_0^*(x) - 1| \leqslant \epsilon$ ($x \in \Omega$). Put $F(x) = \|f_0\|_2^{-1}f_0(x)$ ($x \in G$); because $|\,\|f_0\|_2^2 - 1| \leqslant \epsilon$, for any $x \in \Omega$, we have

$$\begin{aligned}|F*F^*(x) - 1| &\leqslant \|f_0\|_2^{-2}|f_0*f_0^*(x) - 1| + \|f_0\|_2^{-2}|\,\|f_0\|_2^2 - 1| \\ &\leqslant \frac{2\epsilon}{(1-\epsilon)} \leqslant \frac{\delta^2}{2},\end{aligned}$$

whence

$$\int_G |F(xy) - F(y)|^2 d_G(y) \leqslant \delta^2 \qquad (\text{all } x \in \Omega).$$

Keeping in mind the left invariance of F under K, we then find that

$$\begin{aligned}&\int_\omega \Big(\int_S |F(\sigma(xk)s) - F(s)|^2 f_{2\rho}(s)d_S(s)\Big)\,dk \\ &\leqslant \int_K \Big(\int_S |F(\sigma(xk)s) - F(s)|^2 f_{2\rho}(s)d_S(s)\Big)\,dk \\ &= \int_G |F(xy) - F(y)|^2 d_G(y) \leqslant \delta^2 \qquad (\text{all } x \in \Omega).\end{aligned}$$

Now let us take $x = x_0$; thus we have

$$\begin{aligned}\int_S |F(\sigma(x_0k)s)|^2 f_{2\rho}(s)d_S(s) &= \left\{\frac{1}{f_{2\rho}(\sigma(x_0k))}\right\}\int_S |F(s)|^2 f_{2\rho}(s)d_S(s) \\ &\geqslant 4 \qquad (\text{all } k \in \omega)\end{aligned}$$

and so

$$\delta \geqslant \left\{\int_\omega \left(\int_S |F(\sigma(x_0k)s) - F(s)|^2 f_{2\rho}(s) d_S(s)\right) dk\right\}^{1/2}$$

$$\geqslant \left\{\int_\omega \left(\int_S |F(\sigma(x_0k)s)|^2 f_{2\rho}(s) d_S(s)\right) dk\right\}^{1/2}$$

$$-\left\{\int_\omega \left(\int_S |F(s)|^2 f_{2\rho}(s) d_S(s)\right) dk\right\}^{1/2}$$

$$\geqslant 2\sqrt{vol(\omega)} - \sqrt{vol(\omega)} = \sqrt{vol(\omega)},$$

which contradicts the assumption that $\delta^2 < vol\,(\omega)$. It therefore follows that $\hat{G} \neq \hat{G}_r$, as contended. [In passing we remark that the preceding result constitutes one of the main steps in the proof of the following theorem (cf. Takenouchi [1]): Let G be a connected Lie group – then $\hat{G} = \hat{G}_r$ iff the quotient of G by its radical is compact.]

7.1.2 Applications to Semi-Simple Lie Groups

Let G be an acceptable connected semi-simple Lie group with finite center, $G = KA_{\mathfrak{p}}N^+$ an Iwasawa decomposition for G; let M be the centralizer of $A_{\mathfrak{p}}$ in K and put $P = \mathsf{M}A_{\mathfrak{p}}N^+$ – then P is a minimal parabolic subgroup of G. The primary objective of the present number is to study the position of the principal P-series in the dual space $\hat{G}$ of G; accordingly the notations introduced in Vol. I, section 5.5 are now assumed to be in force.

Let W denote the Weyl group of the pair $(\mathfrak{g}, \mathfrak{a}_{\mathfrak{p}})$ – then, under the quotient topology, the orbit space $\mathcal{Q}_P = W\backslash(\hat{\mathsf{M}} \times \hat{A}_{\mathfrak{p}})$ is locally compact, Hausdorff, and second countable. Owing to Theorem 5.5.3.3, the elements of the principal P-series for G, modulo unitary equivalence, are in a one-to-one correspondence with the points of $\mathcal{Q}_P$; this being so, let $\hat{G}_P$ denote the subset of $\mathcal{Q}_P$ comprised of those representations in the principal P-series, modulo unitary equivalence, which are irreducible – then, thanks to Theorem 5.5.2.1, $\hat{G}_P$ is an everywhere dense subset of $\mathcal{Q}_P$ (per the topology supra). [We remind the reader that the inclusion $\hat{G}_P \subset \mathcal{Q}_P$ is, in general, proper . . . .] On the other hand, we can, in the obvious way, also view $\hat{G}_P$ qua a subset of $\hat{G}$. Therefore $\hat{G}_P$ has two topologies: The *natural topology* (inherited from $\mathcal{Q}_P$) and the *hull-kernel topology* (inherited from $\hat{G}$). We intend to show that these topologies are the same.

Suppose that we are given a point q in $\mathcal{Q}_P$ which does not belong to $\hat{G}_P$– then we may associate with q, in the obvious fashion, a finite number ($\leqslant [W]$) of elements in $\hat{G}$ (cf. Corollary 5.5.2.2). We shall agree to denote by $\check{G}_P$ the subset of $\hat{G}$ thus arising upon letting q range through the complement of $\hat{G}_P$ in $\mathcal{Q}_P$ – then $\hat{G}_P \cap \check{G}_P = \varnothing$ (cf. the discussion following Corollary 5.5.2.2).

Theorem 7.1.2.1 (Lipsman) Fix a subset $\mathcal{S} \subset \hat{G}_P$ – then the hull-kernel closure of $\mathcal{S}$ in $\hat{G}$ consists precisely of those points in $\hat{G}_P \cup \check{G}_P$ which are associated with the points in the natural closure of $\mathcal{S}$ in $\mathcal{Q}_P$.

[Needless to say, with any given point in $\mathcal{Q}_P$ there is associated one or more points in $\hat{G}_P \cup \check{G}_P$ (cf. supra).]

Corollary 7.1.2.2 The hull-kernel topology on $\hat{G}_P$ coincides with the natural topology (hence is Hausdorff).

The proof of the present theorem depends on a classical lemma from the theory of C^*-algebras.

Let $C^*(G)$ be the group C^*-algebra of G – then $C^*(G)$ contains a dense self-adjoint subalgebra $\mathfrak{A}$ having the property that each of its elements is *boundedly represented* in $\hat{C}^*(G)$. [To say that an element f in $C^*(G)$ is boundedly represented in $\hat{C}^*(G)$ means that there exists an integer $n\ (= n(f))$ such that rank $(U(f)) \leqslant n$ for all $U\ (\hat{U} \in \hat{C}^*(G))$.] In fact, due to the density of $C_c^\infty(G)$ in $C^*(G)$, it is clear that we may take for $\mathfrak{A}$ the algebra $C_*^\infty(G)$ comprised of all finite linear combinations of elements $\bar{\chi}_{\delta_1} * f * \bar{\chi}_{\delta_2}$ $(\delta_1, \delta_2 \in \hat{K}; f \in C_c^\infty(G))$ (here, of course, we have to keep in mind that K is uniformly large in $G \ldots$). [In passing we remark that it is trivial that the convolution on $L^1(G) \times L^1(K)$ can be extended to a convolution on $C^*(G) \times L^1(K)$ so as to be continuous in both variables; similarly for the convolution on $L^1(K) \times L^1(G) \ldots$.]

Lemma 7.1.2.3 Retain the above notations and assumptions; let $\{\hat{U}_n\}$ be a net of elements in $\hat{C}^*(G)$ and $\hat{V}_1, \ldots, \hat{V}_r$ a finite sequence of (not necessarily distinct) elements in $\hat{C}^*(G)$ such that

$$\lim_n \operatorname{tr}(\hat{U}_n(f)) = \sum_{i=1}^r \operatorname{tr}(\hat{V}_i(f)) \qquad (\text{all } f \in C_*^\infty(G)).$$

Then, for any $\hat{U}$ in $\hat{C}^*(G)$, $\hat{U}_n \to \hat{U}$ (in the hull-kernel topology) iff $\hat{U} = \hat{V}_i$ for some i lying between 1 and r.

[For the proof of this result, the reader is referred to Fell [1, p. 391].]

Notation Let $q \in \mathcal{Q}_P$ – then, by F_q, we shall understand the character of q (in the obvious sense). Let $q \in \hat{G}_P$ – then, by $\hat{U}_q$, we shall understand that element in $\hat{G}$ canonically associated with q.

Proof of Theorem 7.1.2.1 (1) Let q be a point in the natural closure of $\mathcal{S}$ in $\mathcal{Q}_P$ – then there exists a sequence $\{q_n\}$ in $\mathcal{S}$ such that $q_n \to q$ in the natural topology. Put $\hat{U}_n = \hat{U}_{q_n}$ – then, in view of Theorem 5.5.3.1 and the Lebesgue Dominated Convergence Theorem, we have

$$\begin{aligned}\lim_n \operatorname{tr}(\hat{U}_n(f)) &= \lim_n \int_G f(x)F_n(x)d_G(x)\\ &= \int_G f(x)F_q(x)d_G(x) \qquad (\text{all } f \in C_*^\infty(G))\end{aligned}$$

and so

$$\lim_n \operatorname{tr}(\hat{U}_n(f)) = \sum_i \operatorname{tr}(\hat{V}_i(f)) \qquad (\text{all } f \in C_*^\infty(G))$$

where $\hat{V}_i$ runs through those elements in $\hat{G}_P \cup \check{G}_P$ which are associated with q. According to Lemma 7.1.2.3 supra, each $\hat{V}_i$ belongs to the hull-kernel closure of $\mathscr{S}$ in $\hat{G}$ . . . .

(2) Let $\hat{U}$ be a point in the hull-kernel closure of $\mathscr{S}$; choose a net $\{\hat{U}_n\}$ in $\mathscr{S}$ such that $\hat{U}_n \to \hat{U}$ in the hull-kernel topology – then no subnet $\{\hat{V}_n\}$ of $\{\hat{U}_n\}$ can converge in the natural topology to the point at infinity in $\mathscr{Q}_P$. [For if so, then, thanks to the Riemann-Lebesgue Lemma (cf. Theorem 5.5.4.1), it would necessarily be the case that

$$\operatorname{tr}(\hat{V}_n(f)) \to 0 \qquad (\text{all } f \in C_*^\infty(G)).$$

However, in view of Lemma 7.1.2.3, this would mean that $\{\hat{V}_n\}$ converges hull-kernelwise to no limit at all, an impossibility . . . .] Consequently, all the natural cluster points of $\{\hat{U}_n\}$ lie in the finite part of $\mathscr{Q}_P$. Let q be such a cluster point; without changing the notation, let us suppose that $\{\hat{U}_n\}$ itself converges to q – then

$$\lim_n \operatorname{tr}(\hat{U}_n(f)) = \sum_i \operatorname{tr}(\hat{V}_i(f)) \qquad (\text{all } f \in C_*^\infty(G))$$

where $\hat{V}_i$ runs over those elements in $\hat{G}_P \cup \check{G}_P$ which are associated with q. This serves to show, then, that the only possible hull-kernel limits of the net $\{\hat{U}_n\}$ must be the points which are associated with q; in particular $\hat{U}$ is associated with q . . . .

Hence the theorem. □

Problem At the time of this writing the following question had not been resolved: What are the closures of the subsets of $\check{G}_P$?

Remark Because the elements in the class one principal P-series are always irreducible (cf. Theorem 5.5.2.3), it follows from the above discussion that the natural topology on this set is the same as the hull-kernel topology.

Note As has been shown by Kazdan [1], the study of the dual space of G can indeed be very fruitful. Thus suppose that G is a connected simple Lie group which, moreover, admits a faithful finite dimensional representation and has the property that its split-rank is not less than 2; let Γ be a discrete subgroup of G such that G/Γ is of finite volume – then Γ is finitely generated and $\Gamma/[\Gamma, \Gamma]$ is, itself, finite. This result verifies a conjecture of Selberg. Its proof depends on relating the topological structure of $\hat{\Gamma}$ to the algebraic structure of Γ; in turn, there exist close connections between $\hat{\Gamma}$ and $\hat{G}$ . . . . [For an exposition of Kazdan's work see C. Delaroche and A. Kirillov, 'Sur les relations entre l'espace dual d'un groupe et la structure de ses sous-groupes fermés,' Seminaire Bourbaki, n° 343 (1968), Paris; cf., too, the paper of Wang [1].]

7.2 Plancherel Measure

7.2.1 Generalities

Let G be a locally compact unimodular group satisfying the second axiom of countability and which, moreover, is postliminaire; let $\hat{G}$ be the dual space of G (cf. 7.1.1) – then the abstract Plancherel Theorem for G reads as follows (cf. Dixmier [15, p. 328]).

Theorem 7.2.1.1 (Segal) Retain the above notations and assumptions; fix a Haar measure on G – then there exists a unique positive measure μ on $\hat{G}$ such that

$$\int_G |f(x)|^2 d_G(x) = \int_{\hat{G}} \operatorname{tr}(\hat{U}(f)\hat{U}(f)^*) d\mu(\hat{U})$$

for all $f \in L^1(G) \cap L^2(G)$.

Remark Implicit in the statement of the Plancherel Theorem is the fact that $\hat{U}(f)$ $(f \in L^1(G) \cap L^2(G))$ is of the Hilbert-Schmidt class for μ-almost all $\hat{U}$ $(\hat{U} \in \hat{G})$.

The measure μ supra is called the *Plancherel measure* for $\hat{G}$ (associated with the given Haar measure on G). If the Haar measure on G is multiplied by a positive scalar k, then the Plancherel measure on $\hat{G}$ is multiplied by k^{-1}; the support of the Plancherel measure μ is equal to the support of the left regular representation of G, i.e. the support of μ is equal to the reduced dual $\hat{G}_r$ of G.

Note The set $\hat{G}$ is, of course, a topological space; consequently $\hat{G}$ may also be regarded as a Borel space, the (standard . . .) Borel structure on $\hat{G}$ being that canonically determined by its topology; this being so, the Plancherel measure μ on $\hat{G}$ is then to be construed as a positive measure per the ambient underlying Borel structure on $\hat{G}$.

Theorem 7.2.1.2 (Harish-Chandra) Retain the above notations and assumptions; fix a Haar measure on G and let μ be the corresponding Plancherel measure on $\hat{G}$; let $\hat{U} \in \hat{G}$ – then $\hat{U}$ is square integrable iff $\mu(\{\hat{U}\}) > 0$ and, in this event, $\mu(\{\hat{U}\})$ is equal to the formal dimension $d_{\hat{U}}$ of $\hat{U}$.

[For the proof of this result, the reader is referred to Dixmier [15, p. 329].]

Remark Let U be an (irreducible) square integrable unitary representation of G, viewed as a subrepresentation of the left regular representation of G on $L^2(G)$; let P_U be the central projection associated with U – then, for every $f \in L^1(G) \cap L^2(G)$, one has $\int_G |\mathsf{P}_U(f)(x)|^2 d_G(x) = d_U \operatorname{tr}(U(f)U(f)^*)$ (cf. Dixmier [15, p. 281]). Consequently the operator $U(f)$ $(f \in L^1(G) \cap L^2(G))$ is of the Hilbert-Schmidt class (cf. Theorem 4.5.7.4).

In practice, it is usually a difficult problem to explicitly compute the Plancherel measure for a given group; here are some examples where the calculation has actually been carried out.

Examples (1) If G is abelian, then the classical Plancherel Theorem shows that the Plancherel measure is nothing more than the Haar measure (conveniently normalized) on the dual group $\hat{G}$.

(2) If G is compact, then the Peter-Weyl Theorem implies that the Plancherel measure assigns to each point $\hat{U}$ in $\hat{G}$ the mass $\dim(\hat{U})$ (provided $vol(G) = \int_G d_G(x) = 1 \ldots$).

(3) Suppose that G is a connected, simply connected nilpotent Lie group of dimension n, say (thus G is liminaire) – then, as is known (cf. Vol. I, number 5.1.1, Example 3), the points of $\hat{G}$ are in a one-to-one correspondence $\hat{U}_\lambda \leftrightarrow \lambda$ with the points of the orbit space $\Lambda = G\backslash\mathfrak{g}^\vee$ (G operating on $\mathfrak{g}^\vee$ via the coadjoint representation). Let d denote the dimension of a generic orbit in Λ (i.e. the dimension of an orbit in general position); choose a system $\lambda_1, \ldots, \lambda_{n-d}$ of generators for the field of rational G-invariants on $\mathfrak{g}^\vee$ – then there exists a differential form

$$(*) \quad R(\lambda_1, \ldots, \lambda_{n-d})d\lambda_1 \cdots d\lambda_{n-d},$$

where R is a certain (unique) rational function on $\mathfrak{g}^\vee$, such that

$$\int_G |f(x)|^2 d_G(x) = \int_\Lambda \operatorname{tr}(\hat{U}_\lambda(f)\hat{U}_\lambda(f)^*)R(\lambda)d\lambda \qquad (\text{all } f \in L^1(G) \cap L^2(G)).$$

[Here $R(\lambda)d\lambda$ is the measure on Λ canonically deduced from the differential form (*).] This is substantially the Plancherel Theorem for G; for clarification and amplification, see Pukanszky [2], [4]; see also Dixmier [3], [5] and Kirillov [1], [2], [3], [4]. [The following point should be mentioned. It is tacitly supposed that the orbit space Λ is equipped with the quotient topology; now the points of Λ are in a natural one-to-one correspondence $\lambda \leftrightarrow \hat{U}_\lambda$ with the points of $\hat{G}$ and so it is only natural to ask: Is this correspondence a homeomorphism? There is evidence to indicate that the answer is 'yes'; however, the problem has not yet been settled (cf. Dixmier [20], Pukanszky [5]).]

(4) Let $G = \mathbf{SL}(2, \mathbf{R})$ – then the irreducible unitary representations of G fall into four distinct classes, namely the trivial one dimensional representation I (say), the representations in the principal P-series (with one exception . . .), the discrete series, and the complementary series. For well-known reasons, it is permissible to index the representations $U^{\pm,r}$ in the principal P-series by the disjoint union $[0, \infty) \cup [0, \infty)$; on the other hand, the representations U_n in the discrete series are indexed by the set $\{\pm 1, \pm 2, \ldots\}$; this being so, the Plancherel formula reads as follows (relative to a suitable normalization of the pertinent Haar measures):

$$\begin{aligned} 8\pi \int_G |f(x)|^2 d_G(x) = &\sum_{\substack{-\infty<n<\infty \\ n\neq 0}} |n| \operatorname{tr}(U_n(f)U_n(f)^*) \\ &+ \int_0^\infty \frac{r}{2} \tanh\left(\frac{\pi r}{2}\right) \operatorname{tr}(U^{+,r}(f)U^{+,r}(f)^*)dr \\ &+ \int_0^\infty \frac{r}{2} \coth\left(\frac{\pi r}{2}\right) \operatorname{tr}(U^{-,r}(f)U^{-,r}(f)^*)dr \end{aligned}$$

$$(f \in L^1(G) \cap L^2(G)).$$

We shall derive this result in due course (cf. the Epilogue); in passing observe that neither $\{I\}$ nor the representations in the complementary series contribute to the Plancherel formula for G. [The situation as regards $G = \mathbf{SL}(2, \mathbf{C})$ will be explicated in 7.2.2 infra.]

(5) Let $\mathbf{k}$ be a (non-discrete) locally compact, totally disconnected field, the characteristic of whose residual class field is odd – then Sally and Shalika [2] have written down an explicit Plancherel formula for $G = \mathbf{SL}(2, \boldsymbol{k})$.

Problem Let G be a connected semi-simple Lie group with finite center, $\mathfrak{g}$ its Lie algebra – then G operates on $\mathfrak{g}$ via the adjoint action, and one can ask: How does $L^2(\mathfrak{g})$ decompose relative to the canonical unitary representation of G?

7.2.2 The Plancherel Theorem for Complex Connected Semi-Simple Lie Groups

Let G be a complex connected semi-simple Lie group which is, moreover, acceptable; let $G = KA_{\mathfrak{p}}N^+$ be an Iwasawa decomposition for G, M the centralizer of $A_{\mathfrak{p}}$ in K, $P = MA_{\mathfrak{p}}N^+$ the associated minimal parabolic subgroup of G – then, since G is complex, M is a maximal torus in K. It will now be convenient to make a change in our notation; thus let us agree to write J_- for M and J_+ for $A_{\mathfrak{p}}$ – then $J = J_-J_+$ is a Cartan subgroup of G. The unitary dual $\hat{J}_-$ of J_- can be identified with a lattice in the dual space of the Lie algebra of J_-; on the other hand, the unitary dual $\hat{J}_+$ of J_+ can be identified with the Lie algebra of J_+ itself. A given pair $(\lambda_-, \lambda_+) \in \hat{J}_- \times \hat{J}_+$ $(\sim \hat{J})$ determines a unitary character $\xi_{\lambda_-, \lambda_+}$ (say) of J via the prescription

$$(j_-, j_+) \mapsto \xi_{\lambda_-}(j_-)\xi_{\lambda_+}(j_+) = e^{\Lambda_{-,+}(H_- + H_+)},$$

$\Lambda_{-,+}$ being the linear function on $\mathfrak{j}_c$ which coincides with $\sqrt{-1}\,\lambda_-$ on the Lie algebra of J_- and with $\sqrt{-1}\lambda_+$ on the Lie algebra of J_+ (here $j_- = \exp(H_-)$ $(H_- \in \mathfrak{j}_-)$, $j_+ = \exp(H_+)$ $(H_+ \in \mathfrak{j}_+) \ldots$). The representations U^{λ_-, λ_+} in the principal P-series for G can be indexed by the pairs (λ_-, λ_+); because G is complex, every member of the principal P-series for G is irreducible (cf. Theorem 5.5.2.4).

We come now to the Plancherel Theorem for G.

Theorem 7.2.2.1 (Harish-Chandra) Let G be an acceptable complex connected semi-simple Lie group, $P = J_-J_+R_u(P)$ a minimal parabolic subgroup of G – then, for a certain normalization of the pertinent Haar measures,

$$\int_G |f(x)|^2 d_G(x) = (2\pi)^{-\dim(G)/2} \iint_{\hat{J}_- \times \hat{J}_+} \operatorname{tr}(U^{\lambda_-, \lambda_+}(f) U^{\lambda_-, \lambda_+}(f)^*) \times \prod_{\alpha > 0} (\sqrt{-1}\Lambda_{-,+}, \alpha)\, d\lambda_- d\lambda_+$$

for all $f \in L^1(G) \cap L^2(G)$.

[In order to avoid any confusion, let us keep in mind that G is, in particular, liminaire (cf. Vol. I, number 4.5.7).]

Note Let W denote the Weyl group of the pair $(\mathfrak{g}, \mathfrak{j}_+)$ – then, thanks to Theorem 5.5.2.4, the orbit space $\mathcal{Q}_P = W \backslash (\hat{J}_- \times \hat{J}_+)$ may be regarded as a subset of G which is, in fact, *closed* (cf. Theorem 7.1.2.1). Taking into account the W-invariance of the (non-negative...) factor $\prod_{\alpha>0}(\sqrt{-1}\Lambda_{-,+}, \alpha)$, we thus see that the integral on the right supra can actually be taken over $\mathscr{Q}_P = \hat{G}_P$ (as is to be expected...). These considerations serve to show, therefore, that the reduced dual $\hat{G}_r$ of G is equal to $\hat{G}_P$, hence is a Hausdorff space. [We emphasize, however, that the full strength of Theorem 5.5.2.4 is not needed for the deduction of the Plancherel Theorem; rather Theorem 5.5.2.1 is quite sufficient.]

Remark The Plancherel Theorem supra admits an obvious interpretation in the language of direct integral theory....

Turning to the proof of our theorem, first note that it will be enough to establish the asserted relation for functions f in $C_c^\infty(G)$ (this follows from a familiar completeness argument – see the Appendix to the present number); because $\|f\|_2^2 = f * f^*(1)$ $(f \in C_c^\infty(G))$, it is clear that we need only show that

$$f(1) = (2\pi)^{-\dim(G)/2} \iint_{\hat{J}_- \times \hat{J}_+} T^{\lambda_-,\lambda_+}(f) \prod_{\alpha>0} (\sqrt{-1}\Lambda_{-,+}, \alpha) d\lambda_- d\lambda_+$$

for all $f \in C_c^\infty(G)$ (T^{λ_-,λ_+} the character of U^{λ_-,λ_+}) – this will now be done.

Fix an $f \in C_c^\infty(G)$; let Φ_f denote the invariant integral of f calculated relative to the Cartan subgroup J – then, since J is fundamental, Theorem 8.5.1.6 tells us that

$$f(1) = \mathbf{M}_G^{-1}\Phi_f(1; \Pi).$$

Because the pair $(\mathfrak{g}_c, \mathfrak{j}_c)$ possesses no singular imaginary roots, Φ_f is, in the case at hand, a C^∞ function on J (cf. Theorem 8.5.1.4); in fact, Φ_f actually belongs to the Schwartz space on J, i.e. $\Phi_f \in \mathscr{C}(J)$ (cf. Corollary 8.5.1.2). This being so, it then follows immediately that

$$\Phi_f(1; \Pi) = (2\pi)^{-l/4} \iint_{\hat{J}_- \times \hat{J}_+} \hat{\Phi}_f(\lambda_-, \lambda_+) \prod_{\alpha>0} (-\Lambda_{-,+}, \alpha) d\lambda_- d\lambda_+ \quad (l = \operatorname{rank}(G)),$$

$\hat{\Phi}_f$ being, of course, the Fourier transform of Φ_f. On the other hand, modulo the notational changes agreed to above, the discussion in Vol. I, number 5.5.3 provides us with the relation

$$\begin{aligned} T^{\lambda_-,\lambda_+}(f) &= \iint_{J_- \times J_+} \Phi_f(j_-, j_+)\xi_{\lambda_-,\lambda_+}(j_-, j_+) d_{J_-}(j_-) d_{J_+}(j_+) \\ &= (2\pi)^{l/4}\hat{\Phi}_f(\lambda_-, \lambda_+) \qquad ((\lambda_-, \lambda_+) \in \hat{J}_- \times \hat{J}_+). \end{aligned}$$

Since G is complex, the number r of positive roots of the pair $(\mathfrak{g}_c, \mathfrak{j}_c)$ is even; so, putting everything together, we then find that

$$f(1) = \mathbf{M}_G^{-1}(2\pi)^{-l/2} \iint_{\hat{J}_- \times \hat{J}_+} T^{\lambda_-,\lambda_+}(f) \prod_{\alpha>0} (\Lambda_{-,+}, \alpha) d\lambda_- d\lambda_+,$$

which is equivalent to the desired result, $\mathbf{M}_G$ being equal to

$$(\sqrt{-1})^r(2\pi)^{\dim(G)/2}(2\pi)^{-l/2}.$$

Remark It is not difficult to extend the preceding considerations to those (acceptable) connected semi-simple Lie groups with finite center which have the property that they possess but *one* conjugacy class of Cartan subgroups.

Appendix Here we intend to establish the following result.

Lemma (Dixmier) Let G be a unimodular Lie group, Λ a locally compact Hausdorff space countable at infinity, μ a positive measure on Λ; suppose that with each point $\lambda \in \Lambda$ there is attached a unitary representation U_λ(say) of G; suppose further that the function $\lambda \mapsto \operatorname{tr}(U_\lambda(f)U_\lambda(f)^*)$ is μ-integrable with

$$\int_G |f(x)|^2 d_G(x) = \int_\Lambda \operatorname{tr}(U_\lambda(f)U_\lambda(f)^*)d\mu(\lambda)$$

for every $f \in C_c^\infty(G)$–then the function $\lambda \mapsto \operatorname{tr}(U_\lambda(f)U_\lambda(f)^*)$ is μ-integrable with

$$\int_G |f(x)|^2 d_G(x) = \int_\Lambda \operatorname{tr}(U_\lambda(f)U_\lambda(f)^*)d\mu(\lambda)$$

for every $f \in L^1(G) \cap L^2(G)$.

Proof Fix $f \in L^1(G) \cap L^2(G)$; choose a sequence $(f_1, f_2, \ldots)$ in $C_c^\infty(G)$ such that $g_n = f_1 + \ldots + f_n$ approaches f both in $L^1(G)$ and $L^2(G)$ with $\|f_n\|_2 \leqslant 2^{-n}$ $(n = 1, 2, \ldots)$. Set

$$\Phi(\lambda) = \sum_{n=1}^{\infty} (\operatorname{tr}(U_\lambda(f_n)U_\lambda(f_n)^*))^{1/2} \qquad (\lambda \in \Lambda);$$

since

$$\left(\int_\Lambda^* |\Phi(\lambda)|^2 d\mu(\lambda)\right)^{1/2} \leqslant \sum_{n=1}^{\infty} \left(\int_\Lambda^* \operatorname{tr}(U_\lambda(f_n)U_\lambda(f_n)^*)d\mu(\lambda)\right)^{1/2}$$
$$\leqslant \sum_{n=1}^{\infty} \|f_n\|_2 < \infty,$$

it follows that $\Phi(\lambda) < \infty$ except (possibly) on a μ-null set N_Φ. For every $\lambda \notin N_\Phi$, the sequence $\{U_\lambda(g_n)\}$ converges in the sense of the Hilbert-Schmidt norm; because $\|U_\lambda(f) - U_\lambda(g_n)\| \leqslant \|f - g_n\|_1 \to 0$, it must therefore be the case that $U_\lambda(f)$ is of the Hilbert-Schmidt class (all $\lambda \notin N_\Phi$). Owing to the relation

$$(\operatorname{tr}(U_\lambda(f)U_\lambda(f)^*))^{1/2} \leqslant \sum_{n=1}^{\infty} (\operatorname{tr}(U_\lambda(f_n)U_\lambda(f_n)^*))^{1/2} = \Phi(\lambda) \qquad (\lambda \in \Lambda),$$

we have $\int_\Lambda^* \operatorname{tr}(U_\lambda(f)U_\lambda(f)^*)d\mu(\lambda) < \infty$; moreover, due to the fact that

$$\operatorname{tr}(U_\lambda(f)U_\lambda(f)^*) = \lim_{n\to\infty} \operatorname{tr}(U_\lambda(g_n)U_\lambda(g_n)^*) \qquad (\lambda \in \Lambda - N_\Phi),$$

the function $\lambda \mapsto \operatorname{tr}(U_\lambda(f)U_\lambda(f)^*)$ $(\lambda \in \Lambda)$ is μ-measurable, hence μ-integrable. Finally, in view of the estimate

$$(\operatorname{tr}((U_\lambda(f) - U_\lambda(g_n))(U_\lambda(f) - U_\lambda(g_n))^*))^{1/2} \leqslant \sum_{i=n+1}^{\infty} (\operatorname{tr}(U_\lambda(f_i)U_\lambda(f_i)^*))^{1/2},$$

we have

$$\int_{\Lambda} (\mathrm{tr}((U_\lambda(f) - U_\lambda(g_n))(U_\lambda(f) - U_\lambda(g_n))^*))^{1/2} d\mu(\lambda) \leqslant \sum_{i=n+1}^{\infty} 2^{-i} = 2^{-n},$$

which shows that $\int_{\Lambda} \mathrm{tr}(U_\lambda(f)U_\lambda(f)^*)d\mu(\lambda)$ is the limit of the

$$\int_{\Lambda} \mathrm{tr}(U_\lambda(g_n)U_\lambda(g_n)^*)d\mu(\lambda) = \|g_n\|_2^2,$$

i.e. $\int_{\Lambda} \mathrm{tr}(U_\lambda(f)U_\lambda(f)^*)d\mu(\lambda) = \|f\|_2^2$. Hence the lemma. □

Chapter 8

Analysis on a Semi-Simple Lie Group

Introduction

Let G be a connected semi-simple Lie group with finite center – then the primary objective of the present chapter is to develop machinery which, by itself, is more or less independent of representation theory but which, as will become apparent in later chapters, is of decisive importance in the Harmonic Analysis on G. The basic idea which underlies most of the discussion is that of transferring a given analytical question on G, via some device or other, to an analogous question on one of its Cartan subgroups ('non-abelian' $\rightarrow$ 'abelian'. . .), or, at least to some lower dimensional subgroup (where, presumably, induction may be applied. . .). There is little to be gained by giving a detailed analysis of the main results at this point; instead we shall refer the reader to the introductory remarks which (usually) preface a given number in a given section.

Notational Conventions Let (G, K) be a reductive pair satisfying the assumptions set down in Vol. I, number 1.1.5. Let $\mathfrak{j}$ be a θ-stable Cartan subalgebra of $\mathfrak{g}$ such that $\bar{\mathfrak{j}}_\mathfrak{p}$ $(= \mathfrak{j}_\mathfrak{p} \cap \bar{\mathfrak{g}}) \neq \{0\}$.

(1) Fix compatible orders on the spaces of real valued linear functions on $\mathfrak{j}_\mathfrak{p}$ and $\mathfrak{j}_\mathfrak{p} + \sqrt{-1}\mathfrak{j}_\mathfrak{k}$; let Φ^+ be the corresponding set of positive roots of the pair $(\mathfrak{g}_c, \mathfrak{j}_c)$, Φ_I^+ the subset of positive imaginary roots – then Φ_I^+ can also be regarded as the set of positive roots of the pair $(\mathfrak{l}_c, \mathfrak{j}_c)$, $\mathfrak{l}$ the centralizer of $\mathfrak{j}_\mathfrak{p}$ in $\mathfrak{g}$. Let L be the centralizer of $\mathfrak{j}_\mathfrak{p}$ in G, L the identity component of L; let $\mathfrak{m} = \mathfrak{l} \cap \mathfrak{k} + [\mathfrak{l}, \mathfrak{l}] \cap \mathfrak{p}$ – then $\mathfrak{l} = \mathfrak{m} + \mathfrak{j}_\mathfrak{p}$ (direct sum) with $\mathsf{L} = \mathsf{M}J_\mathfrak{p}$, $L = MJ_\mathfrak{p}$ where the symbols M and M are to be assigned the evident interpretations (cf. Vol. I, number 1.2.4). Given a real linear function σ on $\mathfrak{j}_\mathfrak{p}$, let $\mathfrak{g}^\sigma$ denote the set of all $X \in \mathfrak{g}$ such that $[H, X] = \sigma(H)X$ (all $H \in \mathfrak{j}_\mathfrak{p}$); put $\mathfrak{n}^+ = \sum_{\sigma>0} \mathfrak{g}^\sigma$, $\mathfrak{n}^- = \theta(\mathfrak{n}^+)$; let $N^\pm$ be the analytic subgroup of G corresponding to $\mathfrak{n}^\pm$ – then L normalizes $N^\pm$ and $P = \mathsf{M}J_\mathfrak{p}N^+$ is a 'parabolic' subgroup of G (which is even cuspidal. . .).

(2) Extend $\mathfrak{j}_\mathfrak{p}$ to a maximal abelian subspace $\mathfrak{a}_\mathfrak{p}$ of $\mathfrak{p}$; fix an order on the dual of $\mathfrak{a}_\mathfrak{p}$ compatible with the one already placed on the dual of $\mathfrak{j}_\mathfrak{p}$; let Σ^+ denote the set of positive roots of the pair $(\mathfrak{g}, \mathfrak{a}_\mathfrak{p})$; let $\mathfrak{g} = \mathfrak{k} + \mathfrak{a}_\mathfrak{p} + \mathfrak{n}^+$ be the corresponding Iwasawa decomposition for $\mathfrak{g}$ – then, of course, the $\mathfrak{n}^+$ appearing in this decomposition will differ, in general, from the $\mathfrak{n}^+$ introduced in (1) above. Accordingly, on the occasions when it is necessary to consider both $\mathfrak{a}_\mathfrak{p}$ and $\mathfrak{j}_\mathfrak{p}$ simultaneously, we shall adopt the following conventions. Let Σ_1^+ denote the subset of Σ^+ consisting of those λ which vanish identically on $\mathfrak{j}_\mathfrak{p}$; let $\Sigma_2^+ = \Sigma^+ - \Sigma_1^+$ and put

$$\mathfrak{n}_1^+ = \sum_{\lambda \in \Sigma^+_1} \mathfrak{g}^\lambda, \; \mathfrak{n}_2^+ = \sum_{\lambda \in \Sigma^+_2} \mathfrak{g}^\lambda$$

so that $\mathfrak{n}_1^+$, $\mathfrak{n}_2^+$ are nilpotent subalgebras of $\mathfrak{g}$ with $\mathfrak{n}^+ = \mathfrak{n}_1^+ + \mathfrak{n}_2^+$ (direct sum); evidently $\mathfrak{n}_2^+$ is an ideal in $\mathfrak{n}^+$ and, moreover, coincides with the $\mathfrak{n}^+$ appearing in (1) above. Finally put $\mathfrak{a}_\mathfrak{p}^1 = \mathfrak{a}_\mathfrak{p} \cap \mathfrak{m}$, $\mathfrak{a}_\mathfrak{p}^2 = \mathfrak{j}_\mathfrak{p}$ – then clearly $\mathfrak{a}_\mathfrak{p} = \mathfrak{a}_\mathfrak{p}^1 + \mathfrak{a}_\mathfrak{p}^2$ (direct sum). Assign to the symbols $A_\mathfrak{p}$, $N^\pm$, A_1, A_2, $N_1^\pm$, $N_2^\pm$ the obvious meanings – then

$$G = KA_{\mathfrak{p}}N^{\pm}, L = MA_2, M = M_K A_1 N_1^{\pm} \ (M_K = M \cap K),$$
$$A_{\mathfrak{p}} = A_1 A_2, N^{\pm} = N_1^{\pm} N_2^{\pm}.$$

8.1 Preliminaries

8.1.1 Acceptable Groups

Fix a reductive pair (G, K) per Vol. I, number 1.1.5; unexplained notations and conventions are as in Vol. I, section 1.1.

In order to avoid certain technical difficulties (cf. 8.5), it will be convenient to introduce the notion of an acceptable reductive Lie group.

Let $\mathfrak{j}$ be a Cartan subalgebra of $\mathfrak{g}$, J the Cartan subgroup of G associated with $\mathfrak{j}$. Let G_c be a complexification of G (cf. Vol. I, number 1.1.5) and let J_c denote the Cartan subgroup of G_c associated with $\mathfrak{j}_c$ – then J_c is abelian and connected. It is clear that $i_c(\mathrm{J}) \subset J_c$ (i_c the homomorphism of G into G_c). Let μ be a linear function on $\mathfrak{j}_c$ – then there exists at most one holomorphic homomorphism ξ_μ of J_c into $\mathbf{C}^*$ such that

$$\xi_\mu(\exp H) = e^{\mu(H)} \qquad (H \in \mathfrak{j}_c).$$

Furthermore $\xi_\mu \circ i_c$ is a homomorphism of J into $\mathbf{C}^*$ which is independent of the choice of the complexification G_c (so long as it can be defined by means of G_c at all). For brevity, we shall write ξ_μ instead of $\xi_\mu \circ i_c$; it will sometimes be convenient to denote the value of the character ξ_μ at j by the symbol j^μ ($j \in \mathrm{J}$).

Let Φ^+ denote the set of positive roots of the pair $(\mathfrak{g}_c, \mathfrak{j}_c)$ with respect to some ordering; set $\rho = 2^{-1} \sum_{\alpha > 0} \alpha$.

Definition G is said to be *acceptable* if G admits a complexification G_c with the property that ξ_ρ can be defined on J_c.

It should be noted that the condition of acceptability is independent of: (1) The choice of the ordering for the roots of the pair $(\mathfrak{g}_c, \mathfrak{j}_c)$; (2) The choice of the Cartan subalgebra $\mathfrak{j}$ of $\mathfrak{g}$. [Let W denote the Weyl group of the pair $(\mathfrak{g}_c, \mathfrak{j}_c)$ – then (1) is a consequence of the fact that $\rho - w(\rho)$ ($w \in W$) is an integral linear combination of positive roots (cf. Proposition 1.1.2.5). On the other hand, (2) is clear since all Cartan subalgebras of $\mathfrak{g}_c$ are conjugate.]

A complexification G_c of G will be termed acceptable if ξ_ρ can be defined on J_c. [Recall that in the semi-simple case, ξ_ρ is always defined on J_c if G_c is simply connected – cf. Propositions 1.2.2.1, 1.2.2.2.]

Remark The philosophy behind the condition of acceptability is this: G always admits a finite covering group which is acceptable – for our purposes this will be quite sufficient. [In fact, since the center of a connected complex semi-simple Lie group is finite, it is clear that there exists a finite covering group G^* of G which itself admits a complexification G_c^* with the property that $\bar{G}_c^* \cap C_c^* = \{1\}$ and $\bar{G}_c^*$ simply connected ($\bar{G}_c^*$, C_c^* the complex analytic subgroups of G_c^* corresponding to $\bar{\mathfrak{g}}_c$ and $\mathfrak{c}_c$, respectively). Because G^* is a *finite* covering group of G, one sees without difficulty that the pair (G^*, K^*) (K^* the analytic subgroup of G^* with Lie algebra $\mathfrak{k}$) also verifies the general assumptions which were set down in Vol. I, number 1.1.5.]

Example The group $\mathbf{SL}(2, \mathbf{R})/\{\pm 1\}$ is not acceptable. . . .

To our standing hypotheses on the pair (G, K) let us add the additional assumption that G is acceptable.

Fix a θ-stable Cartan subalgebra $\mathfrak{j}$ of $\mathfrak{g}$ and let J be the Cartan subgroup of G associated with $\mathfrak{j}$; let Φ denote the set of roots of the pair $(\mathfrak{g}_c, \mathfrak{j}_c)$. Introduce compatible orderings into the duals of $\mathfrak{j}_\mathfrak{p}$ and $\mathfrak{j}_\mathfrak{p} + \sqrt{-1}\,\mathfrak{j}_\mathfrak{k}$, respectively; let Φ^+ denote the corresponding set of positive roots – then, utilizing the notations which were introduced in Vol. I, number 1.3.2, it is clear that $\Phi_R^+ \cup \Phi_C^+$ is the set of positive roots which do not vanish on $\mathfrak{j}_\mathfrak{p}$ while on the other hand all the elements of Φ_I^+, when restricted to $\mathfrak{j}_\mathfrak{p}$, vanish identically there. If $\alpha \in \Phi$, then $\alpha^\theta \in \Phi$ ($\alpha^\theta(H) = \alpha(\theta H)$, $H \in \mathfrak{j}_c$); in particular note that

$$\alpha \in \Phi_R^+ \cup \Phi_C^+ \Rightarrow -\alpha^\theta \in \Phi_R^+ \cup \Phi_C^+$$

and $\alpha \in \Phi_I^+ \Rightarrow \alpha = \alpha^\theta$. In the same vein, if $\alpha \in \Phi$, then $\alpha^\sigma \in \Phi$ ($\alpha^\sigma(H) = \overline{\alpha(\sigma H)}$, $H \in \mathfrak{j}_c$); in particular note that

$$\alpha \in \Phi_R^+ \cup \Phi_C^+ \Rightarrow \alpha^\sigma \in \Phi_R^+ \cup \Phi_C^+$$

and $\alpha \in \Phi_I^+ \Rightarrow \alpha = -\alpha^\sigma$.

If α is an element of Φ, then it is clear that the character ξ_α is defined on J (in fact we have $\mathrm{Ad}\,(j)X_\alpha = \xi_\alpha(j)X_\alpha$ ($j \in \mathsf{J}$, $X_\alpha \in \mathfrak{g}_c^\alpha$) . . .).

Proposition 8.1.1.1 Retain the above notations and assumptions; fix a root α in Φ – then, for all $j \in \mathsf{J}$, we have:

(i) $\xi_\alpha(j^{-1}) = \xi_{-\alpha}(j)$; (ii) $\overline{\xi_\alpha(j^{-1})} = \xi_{\alpha^\theta}(j)$; (iii) $\overline{\xi_\alpha(j)} = \xi_{\alpha^\sigma}(j)$.

Proof If $X_\alpha \in \mathfrak{g}_c^\alpha$, $X_{-\alpha} \in \mathfrak{g}_c^{-\alpha}$, then $[X_\alpha, X_{-\alpha}] \in \mathfrak{j}_c$ – hence

$$[X_\alpha, X_{-\alpha}] = \mathrm{Ad}(j)[X_\alpha, X_{-\alpha}] = \xi_\alpha(j)\xi_{-\alpha}(j)[X_\alpha, X_{-\alpha}] \qquad (j \in \mathsf{J})$$

which gives (i). Let us turn now to (ii). In the first place, since $\mathrm{Ad}\,(\theta(j)) = \theta \circ \mathrm{Ad}\,(j) \circ \theta$ ($j \in \mathsf{J}$), it is evident that $\xi_{\alpha^\theta}(j) = \xi_\alpha(\theta j)$ ($j \in \mathsf{J}$); secondly, in view of the fact that J_K is compact, the restriction of ξ_α to J_K is a unitary character of J_K (i.e., a homomorphism of J_K into $\mathbf{T}$); finally, for arbitrary $j \in \mathsf{J}$, write $j = j_K j_\mathfrak{p}$ ($j_K \in \mathsf{J}_K$, $j_\mathfrak{p} \in J_\mathfrak{p}$) – then

$$\overline{\xi_{\alpha^\theta}(j)} = \overline{\xi_\alpha(j_K j_\mathfrak{p}^{-1})} = \overline{\xi_\alpha(j_K)}\xi_\alpha(j_\mathfrak{p}^{-1}) = \xi_\alpha(j^{-1})$$

which is (ii). The proof of (iii) is left to the reader. ☐

Keeping to the above notations, let us put

$$\diamondsuit(j)(= \diamondsuit_\mathsf{J}(j)) = \prod_{\alpha > 0} (1 - \xi_\alpha(j^{-1})) \qquad (j \in \mathsf{J}).$$

Evidently the set of points in J where $\diamondsuit(j) \neq 0$ is precisely J', J' the regular elements in J. Taking into account Proposition 8.1.1.1, we have

$$\overline{\diamondsuit(j)} = \prod_{\alpha>0} (1 - \xi_{\alpha^\theta}(j)) = (-1)^{r_I} \prod_{\alpha \in \Phi_I^+} \xi_\alpha(j) \cdot \diamondsuit(j) \qquad (j \in \mathrm{J}),$$

where r_I is the number of roots in Φ_I^+. Given $j \in \mathrm{J}$, write

$$\Delta(j)(=\Delta_{\mathrm{J}}(j)) = \xi_\rho(j) \prod_{\alpha>0} (1 - \xi_\alpha(j^{-1})).$$

It is clear that $\det(\mathrm{Ad}(j^{-1}) - 1)_{\mathfrak{g}_c/\mathfrak{j}_c} = (-1)^r \Delta(j)^2$ $(j \in \mathrm{J})$, r the number of positive roots of the pair $(\mathfrak{g}_c, \mathfrak{j}_c)$.

Definition Fix $j \in \mathrm{J}$ and let α be a root of the pair $(\mathfrak{g}_c, \mathfrak{j}_c)$ – then we say that j and α *commute* if $\xi_\alpha(j) = 1$.

For any $j \in \mathrm{J}$, define an integer $r(R:j) \geqslant 0$ as follows. Write

$$j = j_K j_\mathfrak{p} \qquad (j_K \in \mathrm{J}_K, j_\mathfrak{p} \in J_\mathfrak{p})$$

– then $r(R:j)$ is, by definition, the number of positive real roots which commute with j_K. If α is a real root of the pair $(\mathfrak{g}_c, \mathfrak{j}_c)$, then $\alpha(H) = 0$ for all $H \in \mathfrak{j}_\mathfrak{k}$ – hence $r(R:j)$ depends only on the component of j_K in J_K and so the function $j \mapsto r(R:j)$ $(j \in \mathrm{J})$ is locally constant, i.e. constant on each component of J.

Proposition 8.1.1.2 Let r denote the number of positive roots of the pair $(\mathfrak{g}_c, \mathfrak{j}_c)$ – then $\overline{\Delta(j)} = (-1)^{r+r(R:j)}\Delta(j)$ (all $j \in \mathrm{J}$).

Proof Set

$$\diamondsuit_I(j) = \prod_{\alpha \in \Phi_I^+} (1 - \xi_\alpha(j^{-1})), \quad \diamondsuit_{R,C}(j) = \prod_{\alpha \in \Phi_R^+ \cup \Phi_C^+} (1 - \xi_\alpha(j^{-1})) \qquad (j \in \mathrm{J})$$

where, for brevity, we omit the sub 'J' from the notation; plainly

$$\Delta(j) = \xi_\rho(j) \diamondsuit_I(j) \diamondsuit_{R,C}(j) \qquad (\text{all } j \in \mathrm{J}).$$

It was pointed out above that the set $\Phi_R^+ \cup \Phi_C^+$ is σ-stable; therefore, on the basis of Proposition 8.1.1.1(iii), we find that

$$\overline{\diamondsuit_{R,C}(j)} = \diamondsuit_{R,C}(j) \qquad (\text{all } j \in \mathrm{J}).$$

Otherwise said, $\diamondsuit_{R,C}$ is a real valued function on J. On the other hand, since $\alpha \in \Phi_I^+ \Rightarrow \alpha^\sigma = -\alpha$, it must be the case that

$$\overline{\diamondsuit_I(j)} = (-1)^{r_I} \xi_{2\rho_I}(j) \diamondsuit_I(j) \qquad (j \in \mathrm{J}),$$

r_I the number of roots in Φ_I^+, ρ_I one-half the sum of the roots in Φ_I^+. If

$$j = j_K j_\mathfrak{p} \quad (j_K \in \mathrm{J}_K, j_\mathfrak{p} \in J_\mathfrak{p}),$$

then $\overline{\xi_\rho(j)} = \xi_\rho(j_K^{-1} j_\mathfrak{p})$ while $\xi_{2\rho_I}(j_\mathfrak{p}) = 1$ – hence

$$\begin{aligned}\overline{\Delta(j)} &= (-1)^{r_I} \xi_\rho(j_K^{-1} j_\mathfrak{p}) \xi_{2\rho_I}(j_K) \diamondsuit_I(j) \diamondsuit_{R,C}(j) \\ &= (-1)^{r_I} \xi_{2\rho}(j_K^{-1}) \xi_{2\rho_I}(j_K) \Delta(j) = (-1)^{r_I} \xi_{2\rho_{R,C}}(j_K^{-1}) \Delta(j),\end{aligned}$$

where $\rho_{R,C}$ is one-half the sum of the roots in $\Phi_R^+ \cup \Phi_C^+$. The positive complex roots occur in pairs $(\alpha, \alpha^\sigma)(\alpha \in \Phi_C^+)$; moreover,

$$\xi_\alpha(j_K)\xi_{\alpha^\sigma}(j_K) = |\xi_\alpha(j_K)|^2 = 1 \qquad (\alpha \in \Phi_C^+),$$

whence $\xi_{2\rho_{R,C}}(j_K) = \xi_{2\rho_R}(j_K)$, ρ_R one-half the sum of the roots in Φ_R^+. But for any α in Φ_R^+, $\xi_\alpha(j_K)$ is both real and unimodular, hence is ± 1 – therefore

$$\xi_{2\rho_R}(j_K) = \prod_{\alpha \in \Phi_R^+} \xi_\alpha(j_K) = (-1)^{s(R:j)},$$

$s(R:j)$ the number of roots $\alpha \in \Phi_R^+$ such that $\xi_\alpha(j_K) = -1$. [For α in Φ_R^+, it is clear that $\xi_\alpha(j_K)$ is unimodular; to verify that $\xi_\alpha(j_K)$ is also real, one need only check that $\xi_\alpha | \mathsf{Z}(J_\mathfrak{p})$ takes on real values alone ($\mathsf{Z}(J_\mathfrak{p})$ as defined in Proposition 1.4.1.3). So suppose that $h \in \mathsf{Z}(J_\mathfrak{p})$; let $i_c : G \to G_c$ be a complexification of G; choose $H \in \mathfrak{j}_\mathfrak{p}$ such that $i_c(h) = \exp\sqrt{-1}H$ – then $\xi_\alpha(h) = e^{\sqrt{-1}\alpha(H)} = \xi_{\alpha^\theta}(h) = e^{-\sqrt{-1}\alpha(H)} \Rightarrow 2\sqrt{-1}\alpha(H) = 2\pi\sqrt{-1}n$ $(n \in \mathbf{Z}) \Rightarrow \alpha(H) = \pi n. \ldots$] Since $r(R:j) + s(R:j)$ is the total number of roots in Φ_R^+ and since the roots in Φ_C^+ occur in pairs, we have

$$r(R:j) + s(R:j) + r_I \equiv r \bmod 2.$$

Thus $s(R:j) + r_I \equiv r + r(R:j) \bmod 2$ and so

$$\overline{\Delta(j)} = (-1)^{s(R:j)+r_I}\Delta(j) = (-1)^{r+r(R:j)}\Delta(j),$$

as we wished to prove. □

Given $j \in \mathsf{J}$, put

$$\diamondsuit_R(j) = \prod_{\alpha \in \Phi_R^+} (1 - \xi_\alpha(j^{-1}))$$

and define a function ϵ_R on $\mathsf{J}'(=\mathsf{J} \cap G')$ by the rule $\epsilon_R(j) = \operatorname{sign}\{\diamondsuit_R(j)\}$ $(j \in \mathsf{J}')$. [Let $t \in \mathbf{R}$ – then sign $(t) = 1, -1$, or 0 according as $t > 0$, $t < 0$, or $t = 0$.] As we know, the group $W(G, \mathsf{J}_0)$ operates on J and hence on $\mathfrak{j}$; a consideration of eigenvalues shows that $\mathfrak{j}_\mathfrak{p}$ is $W(G, \mathsf{J}_0)$-stable – hence $\mathsf{Z}(J_\mathfrak{p})$ and J_K are too. For any $w \in W(G, \mathsf{J}_0)$, set

$$\Phi_R^+(w) = \{\alpha \in \Phi_R^+ : \alpha^{w^{-1}} > 0\}, \ \Phi_R^-(w) = \{\alpha \in \Phi_R^+ : \alpha^{w^{-1}} < 0\}$$

and then define a homomorphism ϵ_w of J_K by the rule

$$\epsilon_w(j_K) = \operatorname{sign}\{\prod_{\alpha \in \Phi_R^+(w)} \xi_\alpha(w \cdot j_K)\} \qquad (j_K \in \mathsf{J}_K).$$

Finally, writing $\Pi_R = \prod_{\alpha \in \Phi_R^+} H_\alpha$ (cf. Vol. I, number 1.3.2 – the product lies in the symmetric algebra of $\mathfrak{j}_c$), define $\epsilon_R(w)$ by the equation $w\Pi_R = \epsilon_R(w)\Pi_R$ $(w \in W(G, \mathsf{J}_0))$.

The next result tells us how the function $j \mapsto \epsilon_R(j)$ $(j \in \mathsf{J}')$ transforms under the action of $W(G, \mathsf{J}_0)$.

Proposition 8.1.1.3 Fix $j \in \mathsf{J}'$ and $w \in W(G, \mathsf{J}_0)$; write $j = j_K j_\mathfrak{p}$ $(j_K \in \mathsf{J}_K, j_\mathfrak{p} \in J_\mathfrak{p})$ – then

$$\epsilon_R(w \cdot j) = \epsilon_R(j)\epsilon_R(w)\epsilon_w(j_K).$$

Proof First note that $\xi_\alpha(w\cdot j) = \xi_{\alpha w^{-1}}(j)$ for any $j \in \mathrm{J}$ – hence, for $j \in \mathrm{J}'$, we have

$$\begin{aligned}\epsilon_R(w\cdot j) &= \operatorname{sign}\{\prod_{\alpha\in\Phi_R^+}(1-\xi_{\alpha w^{-1}}(j^{-1}))\}\\ &= \operatorname{sign}\{\prod_{\alpha\in\Phi_R^+(w)}(1-\xi_{\alpha w^{-1}}(j^{-1}))\prod_{\alpha\in\Phi_R^-(w)}(1-\xi_{\alpha w^{-1}}(j^{-1}))\}\\ &= (-1)^{r_R^-(w)}\operatorname{sign}\{\prod_{\alpha\in\Phi_R^-(w)}\xi_{-\alpha w^{-1}}(j)\}\cdot\operatorname{sign}\{\Diamond_R(j)\}\end{aligned}$$

where $r_R^-(w)$ is the number of roots in $\Phi_R^-(w)$. Since $(-1)^{r_R^-(w)} = \epsilon_R(w)$ and $\operatorname{sign}\{\xi_{-\alpha w^{-1}}(j)\} = \operatorname{sign}\{\xi_\alpha(w\cdot j_K)\}$, the contention of the proposition is now clear. □

Corollary 8.1.1.4 Retain the above notations and assumptions; fix $w_1, w_2 \in W(G, \mathrm{J}_0)$ – then $\epsilon_{w_1w_2}(j_K) = \epsilon_{w_1}(w_2\cdot j_K)\epsilon_{w_2}(j_K)$ (all $j_K \in \mathrm{J}_K$).

Example The preceding considerations may be conveniently illustrated in the case when $G = \mathrm{U}(p, q)$; thus, agreeing to use the notations which were introduced in Vol. I, number 1.4.2, fix a $j \in \mathrm{J}^i (0 \leqslant i \leqslant q)$ – then we saw that the eigenvalues of j were the numbers $e^{\Phi_1}, \ldots, e^{\Phi_{p-i}}, e^{z_i}, \ldots, e^{z_1}, e^{z_{-1}}, \ldots, e^{z_{-i}}, e^{\Psi_{q-i}}, \ldots, e^{\Psi_1}$. Let us denote these eigenvalues by $\Lambda_1(j), \Lambda_2(j), \ldots, \Lambda_n(j)$ $(j \in \mathrm{J}^i)$ according to the above order; we may take for the positive 'roots' of the pair (G, J^i) the maps $j \mapsto \Lambda_\mu(j)/\Lambda_\nu(j) (j \in \mathrm{J}^i)$ where $1 \leqslant \mu < \nu \leqslant n$ – then, for i lying between 0 and q, we have

$$\Phi_{CP}^+ = \left\{\frac{e^{\Phi_\mu}}{e^{\Phi_\nu}}: 1 \leqslant \mu < \nu \leqslant p - i\right\} \cup \left\{\frac{e^{\Psi_\mu}}{e^{\Psi_\nu}}: 1 \leqslant \nu < \mu \leqslant q - i\right\},$$

$$\Phi_{SI}^+ = \left\{\frac{e^{\Phi_\mu}}{e^{\Psi_\nu}}: 1 \leqslant \mu \leqslant p - i,\ 1 \leqslant \nu \leqslant q - i\right\},$$

$$\Phi_R^+ = \left\{\frac{e^{z_k}}{e^{z_{-k}}}: 1 \leqslant k \leqslant i\right\},$$

the remaining positive roots being complex. The character ξ_ρ is given on J^i by

$$\xi_\rho(j) = (\Lambda_1^{n-1}(j)\ldots\Lambda_n^{-(n-1)}(j))^{1/2} \qquad (j \in \mathrm{J}^i)$$

and so ξ_ρ is a one-valued or two-valued function on J^i according to whether n is odd or even, whence $\mathrm{U}(p, q)$ is not acceptable if n is even. As we clearly have

$$\Diamond_R^i(j) = \prod_{k=1}^{i}(1 - e^{-2t_k}) \qquad (j \in \mathrm{J}^i),$$

it follows that $\epsilon_R^i(j) = \operatorname{sign}\left\{\prod_{k=1}^{i} t_k\right\}$ (j regular in J^i).

8.1.2 Normalization of Invariant Measures

Fix a reductive pair (G, K) per Vol. I, number 1.1.5. In this number we shall proceed to establish our conventions as regards the normalization of certain invariant measures on $\mathfrak{g}$, G etc. – these conventions will be assumed to be in force in the sequel unless specific mention is made to the contrary. [In the present number and the following one, we shall constantly be concerned with how invariant measures are transformed under certain mappings; the reader who is unfamiliar with the techniques involved here is advised to first consult Helgason [2, Chapter 10].] By way of notation, $\mathfrak{j}$ will denote a generic θ-stable Cartan subalgebra of $\mathfrak{g}$, J the Cartan subgroup of G associated with $\mathfrak{j}$. It will be convenient to fix once and for all a maximal set $\mathfrak{j}^i$ $(1 \leqslant i \leqslant r)$ of θ-stable Cartan subalgebras of $\mathfrak{g}$, no two of which are conjugate; J^i will denote the Cartan subgroup of G associated with $\mathfrak{j}^i$ $(1 \leqslant i \leqslant r)$.

For the moment, let us suppose that G is semi-simple (so that, according to our conventions, G has finite center). In this situation the Lie algebra $\mathfrak{g}$ is canonically a real Hilbert space relative to the scalar product $(X, Y)_\theta = -B(X, \theta Y)(X, Y \in \mathfrak{g})$; there results, then, a privileged Haar measure $d_\mathfrak{g}$ on $\mathfrak{g}$, namely that in terms of which the hypercube determined by an orthonormal basis has unit measure. Notice that this Haar measure is independent of the particular orthonormal basis chosen since the matrix of an orthonormal change of basis is orthogonal and so the absolute value of its determinant is one. In the same manner we may select a canonical Haar measure $d_\mathfrak{j}$ on the Cartan subalgebra $\mathfrak{j}$ (or, for that matter, on any linear subspace of $\mathfrak{g}$).

Remark The measure $d_\mathfrak{g}$ is independent of the choice of the Cartan involution θ. Thus, if $\tilde{\theta}$ is another Cartan involution of $\mathfrak{g}$, then there exists an x in G such that $\tilde{\theta} = \mathrm{Ad}(x)\circ\theta\circ\mathrm{Ad}(x^{-1})$ and so, for all $X, Y \in \mathfrak{g}$, we have

$$(X, Y)_\theta = -B(X, \theta Y) = -B(\mathrm{Ad}(x)X, \tilde{\theta}(\mathrm{Ad}(x)Y)) = (\mathrm{Ad}(x)X, \mathrm{Ad}(x)Y)_{\tilde{\theta}}.$$

Since $|\det(\mathrm{Ad}(x))| = 1$, the contention is clear. It is equally evident that the canonical Euclidean measure on $\mathfrak{j}$ is independent of the Cartan involution relative to which $\mathfrak{j}$ is stable.

Turning now to the general case, fix once and for all a Euclidean structure on $\mathfrak{g}$ which renders $\mathfrak{c}_\mathfrak{k}$, $\bar{\mathfrak{g}}$, and $\mathfrak{c}_\mathfrak{p}$ mutually orthogonal and has the property that its restriction to $\bar{\mathfrak{g}}$ agrees with $(.,.)_\theta$ there. The Haar measure on $\bar{\mathfrak{g}}$ is determined per the previous agreement; as for the Haar measure on $\mathfrak{c}_\mathfrak{k}$ (respectively $\mathfrak{c}_\mathfrak{p}$), we shall take the canonical element of volume determined by the fixed Euclidean structure with which the space has been equipped. The Haar measure on $\mathfrak{c}$ is then fixed when we require that $d_\mathfrak{c}(C) = d_{\mathfrak{c}_\mathfrak{k}}(C_\mathfrak{k})d_{\mathfrak{c}_\mathfrak{p}}(C_\mathfrak{p})$ $(C = C_\mathfrak{k} + C_\mathfrak{p} \ldots)$; therefore the Haar measure on $\mathfrak{g}$ (respectively $\hat{\mathfrak{g}}$) is fixed when we stipulate that $d_\mathfrak{g}(X) = d_{\bar{\mathfrak{g}}}(\bar{X})d_\mathfrak{c}(C)$ $(X = \bar{X} + C \ldots)$ (respectively $d_{\hat{\mathfrak{g}}}(\hat{X}) = d_{\bar{\mathfrak{g}}}(\bar{X})d_{\mathfrak{c}_\mathfrak{k}}(C_\mathfrak{k})$ $(\hat{X} = \bar{X} + C_\mathfrak{k} \ldots)$). Entirely analogous agreements are understood to be in force per the Haar measures on $\mathfrak{j}$, $\bar{\mathfrak{j}}$, and $\hat{\mathfrak{j}}$.

As usual, set $\mathfrak{g}(\mathfrak{j}) = \bigcup_{x \in G} x(\mathfrak{j}')$ (cf. Vol. I, number 1.3.4) – then, relative to our determination of $d_\mathfrak{g}$ and $d_\mathfrak{j}$, Proposition 1.4.2.2 tells us that the G-invariant measure $d_{G/J}$ on G/J can be normalized in such a way that

$$\int_{\mathfrak{g}(\mathfrak{j})} f(X)d_\mathfrak{g}(X) = \iint_{G/J \times \mathfrak{j}'} |\pi(H)|^2 f(\dot{x}H)d_{G/J}(\dot{x})d_\mathfrak{j}(H) \quad (f \in C_c(\mathfrak{g}(\mathfrak{j}))).$$

Because $\mathfrak{g}' = \bigcup_{i=1}^{r} \mathfrak{g}(\mathfrak{j}^i)$ (cf. Proposition 1.3.4.1) and, since this is a disjoint union whose complement in $\mathfrak{g}$ is of measure zero, we have

$$\text{(I)} \int_{\mathfrak{g}} f(X)d_{\mathfrak{g}}(X) = \sum_{i=1}^{r} \iint_{G/\mathrm{J}^i \times \mathfrak{j}^i} |\pi^i(H)|^2 f(\dot{x}H)d_{G/\mathrm{J}^i}(\dot{x})d_{\mathfrak{j}^i}(H)$$

$$(f \in C_c(\mathfrak{g})).$$

Here d_{G/J^i} is the G-invariant measure on G/J^i, $d_{\mathfrak{j}^i}$ the canonical Haar measure on $\mathfrak{j}^i$ ($i = 1, \ldots, r$).

Let J_0 denote the center of J – then, as we know, J/J_0 is finite. Assign to each point in J/J_0 mass one; this determines a J-invariant measure on J/J_0 – thus the total volume of J/J_0 is $[\mathrm{J} : \mathrm{J}_0]$ ($[\mathrm{J} : \mathrm{J}_0]$ the finite index of J_0 in J). Because the invariant measures on G/J and J/J_0 have now been fixed, there exists a unique determination of the G-invariant measure d_{G/J_0} on G/J_0 such that for all $f \in C_c(G/\mathrm{J}_0)$, we have

$$\int_{G/\mathrm{J}_0} f(x\mathrm{J}_0)d_{G/\mathrm{J}_0}(\dot{x}) = \int_{G/\mathrm{J}} \left\{\int_{\mathrm{J}/\mathrm{J}_0} f(xj\mathrm{J}_0)d_{\mathrm{J}/\mathrm{J}_0}(j)\right\} d_{G/\mathrm{J}}(\dot{x}).$$

Of the groups $\mathrm{J}^1, \ldots, \mathrm{J}^r$, assume that J^1 is fundamental – then $\mathrm{J}^1 = \mathrm{J}^1_0$ (cf. Proposition 1.4.1.4). Write $\mathrm{J}^1 = \mathrm{J}^1_K J^1_{\mathfrak{p}}$ where $\mathrm{J}^1_K = \mathrm{J}^1 \cap K$, $J^1_{\mathfrak{p}} = \exp(\mathfrak{j}^1_{\mathfrak{p}})$. Normalize the Haar measure on J^1 in the following way: Write $d_{\mathrm{J}^1}(j) = d^1_K(j_K)d^1_{\mathfrak{p}}(j_{\mathfrak{p}})$ where $j = j_K j_{\mathfrak{p}}$ ($j_K \in \mathrm{J}^1_K, j_{\mathfrak{p}} \in J^1_{\mathfrak{p}}$). Here d^1_K is the Haar measure on J^1_K which assigns to J^1_K total mass 1 while $d^1_{\mathfrak{p}}$ is the Haar measure on $J^1_{\mathfrak{p}}$ which is the transport via the exponential map of the privileged Haar measure on $\mathfrak{j}^1_{\mathfrak{p}}$. Now normalize the Haar measure d_G on G by the requirement

$$\int_G f(x)d_G(x) = \int_{G/\mathrm{J}^1} \left\{\int_{\mathrm{J}^1} f(xj)d_{\mathrm{J}^1}(j)\right\} d_{G/\mathrm{J}^1}(\dot{x}) \qquad (f \in C_c(G)).$$

The G-invariant measure d_{G/J^1} on G/J^1 is, of course, to be taken in the normalization agreed to earlier. Having fixed the Haar measure d_G on G and the G-invariant measure d_{G/J_0} on G/J_0, we shall normalize the Haar measure d_{J_0} on J_0 via the stipulation

$$\int_G f(x)d_G(x) = \int_{G/\mathrm{J}_0} \left\{\int_{\mathrm{J}_0} f(xj)d_{\mathrm{J}_0}(j)\right\} d_{G/\mathrm{J}_0}(\dot{x}) \qquad (f \in C_c(G)).$$

The Haar measure d_{J} on J is then fixed when we require that $d_{\mathrm{J}} = d_{\mathrm{J}_0}$ on J_0; notice that the corresponding uniquely determined J-invariant measure on J/J_0 is precisely the measure $d_{\mathrm{J}/\mathrm{J}_0}$ supra.

As is usual, set $G(\mathrm{J}) = \bigcup_{x \in G} x\mathrm{J}'x^{-1}$ (cf. Vol. I, number 1.4.1); owing to Proposition 1.4.2.3 and the above agreements, for all $f \in C_c(G(\mathrm{J}))$, we have

$$\int_{G(\mathrm{J})} f(x)d_G(x) = [W(G, \mathrm{J}_0)]^{-1} \int_{\mathrm{J}'} |\det(\mathrm{Ad}(j^{-1}) - 1)_{\mathfrak{g}/\mathfrak{j}}| \, d_{\mathrm{J}}(j)$$

$$\times \int_{G/\mathrm{J}_0} f(\overset{*}{j})d_{G/\mathrm{J}_0}(\dot{x}).$$

Because $G' = \bigcup_{i=1}^{r} G(\mathrm{J}^i)$ (cf. Theorem 1.4.1.7) and, since this is a disjoint union whose complement in G is of measure zero, we have

$$(\mathrm{II})\ \int_G f(x)d_G(x) = \sum_{i=1}^{r} [W(G, \mathrm{J}_0^i)]^{-1} \int_{\mathrm{J}^i} |\det(\mathrm{Ad}(j^{-1}) - 1)_{\mathfrak{g}/\mathfrak{j}^i}|\, d_{\mathrm{J}^i}(j) \times \int_{G/\mathrm{J}_0{}^i} f({}^{\dot{x}}j)d_{G/\mathrm{J}_0{}^i}(\dot{x})$$

for all $f \in C_c(G)$. In particular, if G is acceptable, then the present generalization of Weyl's integral formula reads

$$(\mathrm{II}_\mathrm{A})\ \int_G f(x)d_G(x) = \sum_{i=1}^{r} [W(G, \mathrm{J}_0^i)]^{-1} \int_{\mathrm{J}^i} |\Delta^i(j)|^2 d_{\mathrm{J}^i}(j) \times \int_{G/\mathrm{J}_0{}^i} f({}^{\dot{x}}j)d_{G/\mathrm{J}_0{}^i}(\dot{x}) \qquad (f \in C_c(G))$$

where we have set $\Delta^i = \Delta_{\mathrm{J}^i}$ $(i = 1, \ldots, r)$.

Remarks (1) For some purposes it will be best to use a normalization of the invariant measures on G/J_0 and J_0 which differs from the one agreed to above (cf. 8.1.3 infra); fortunately, however, it will always be possible to arrange things in such a manner as to ensure the validity of formula II (although formula I may very well be altered by certant constant factors).

(2) The preceding conventions are applicable, of course, to $\hat{G}$ (or $\bar{G}$). This being so, fix an $f \in C_c(\mathfrak{g})$ and, for $C \in \mathfrak{c}$, write

$$C = C_{\mathfrak{k}} + C_{\mathfrak{p}} \quad (C_{\mathfrak{k}} \in \mathfrak{c}_{\mathfrak{k}},\ C_{\mathfrak{p}} \in \mathfrak{c}_{\mathfrak{p}});$$

let $\hat{f}$ denote the function on $\hat{\mathfrak{g}}$ given by $\hat{f}(\hat{X}) = f(\hat{X} + C_{\mathfrak{p}})\,(\hat{X} \in \hat{\mathfrak{g}})$; let $\bar{f}$ denote the function on $\bar{\mathfrak{g}}$ given by $\bar{f}(\bar{X}) = \hat{f}(\bar{X} + C_{\mathfrak{k}})\,(\bar{X} \in \bar{\mathfrak{g}})$. Now fix an $H \in \mathfrak{j}'$; write

$$H = \hat{H} + C_{\mathfrak{p}}\ (\hat{H} \in \hat{\mathfrak{j}}', C_{\mathfrak{p}} \in \mathfrak{c}_{\mathfrak{p}}),\ \hat{H} = \bar{H} + C_{\mathfrak{k}}\ (\bar{H} \in \bar{\mathfrak{j}}', C_{\mathfrak{k}} \in \mathfrak{c}_{\mathfrak{k}})$$

– then it is clear that

$$\int_{G/\mathrm{J}} f(\dot{x}H)d_{G/\mathrm{J}}(\dot{x}) = \int_{\hat{G}/\hat{\mathrm{J}}} \hat{f}(\dot{\hat{x}}\hat{H})d_{\hat{G}/\hat{\mathrm{J}}}(\dot{\hat{x}}) = \int_{\bar{G}/\bar{\mathrm{J}}} \bar{f}(\dot{\bar{x}}\bar{H})d_{\bar{G}/\bar{\mathrm{J}}}(\dot{\bar{x}}) \qquad (f \in C_c(\mathfrak{g})).$$

Keeping to the above notations and assumptions per our choice of Euclidean structures on $\mathfrak{c}_{\mathfrak{k}}$, $\bar{\mathfrak{g}}$, and $\mathfrak{c}_{\mathfrak{p}}$ it is clear the corresponding norm $\|\,.\,\|$ on $\mathfrak{p}$ is K-invariant. Given $x \in G$, write (uniquely)

$$x = k \exp X \quad (k \in K,\ X \in \mathfrak{p})$$

and put $\sigma(x) = \|X\|$; agreeing to denote the canonical extension to G of the 'Cartan involution' θ by the same symbol, we have $\sigma(x) = \sigma(\theta(x)) = \sigma(x^{-1})$ (all $x \in G$) – moreover σ is biinvariant under K, that is

$$\sigma(k_1 x k_2) = \sigma(x) \quad (k_1, k_2 \in K;\ x \in G).$$

[In passing observe that σ is not C^∞ although σ^2 is . . . .]

Proposition 8.1.2.1 The function σ is subadditive, i.e. $\sigma(xy) \leqslant \sigma(x) + \sigma(y)$ (all $x, y \in G$).

Proof We may obviously assume that $x = \exp X$, $y = \exp Y$ $(X, Y \in \mathfrak{p})$; this being so, write $xy = k \exp Z$ $(k \in K, Z \in \mathfrak{p})$ – then one has

$$\exp 2Z = \theta(xy)^{-1}xy = \exp Y \cdot \exp 2X \cdot \exp Y.$$

Define now a function $t \mapsto Z(t)$ of a real variable t by the rule $\exp(2Z(t)) = \exp(tY)\exp(2X)\exp(tY)$. It is clear that $Z(t)$ is analytic in t; a differentiation with respect to t then gives

$$\left\{\frac{(1 - e^{-2\operatorname{ad}(Z(t))})}{2\operatorname{ad}(Z(t))}\right\} Z'(t) = 2^{-1}(1 + e^{-2\operatorname{ad}(Z(t))})Y$$

where $Z'(t) = dZ(t)/dt$ $(t \in \mathbf{R})$. Hence $(Z(t), Z'(t)) = (Z(t), Y)$ (all $t \in \mathbf{R}$). Suppose first that $Z(t) = 0$ for some $t \in \mathbf{R}$ – then $1 = \exp(tY) \exp(2X)\exp(tY)$, whence $X = -tY$ and so

$$\|Z\| \leqslant \|tY\| + \|Y\| = \|X\| + \|Y\|$$

which yields the assertion of the lemma in this case. On the other hand if $Z(t) \neq 0$ for all $t \in \mathbf{R}$, then $\|Z(t)\|$ is analytic in t and

$$(Z(t), Z'(t)) = \|Z(t)\|\frac{d\|Z(t)\|}{dt} = (Z(t), Y) \qquad (t \in \mathbf{R});$$

thus $|d\|Z(t)\|/dt| \leqslant \|Y\|$ $(t \in \mathbf{R})$ and so by integration we find that $\|Z(1)\| - \|Z(0)\| \leqslant \|Y\|$, i.e. that $\|Z\| \leqslant \|X\| + \|Y\|$, as desired. □

Corollary 8.1.2.2 Let ω be a compact subset of G – then there exist positive constants d_1 and d_2 such that

$$d_1(1 + \sigma(xy)) \leqslant 1 + \sigma(x) \leqslant d_2(1 + \sigma(xy)) \qquad (x \in G, y \in \omega).$$

[This is a straightforward consequence of the preceding lemma.]

8.1.3 Integration Formulas

Let (G, K) be a reductive pair verifying the assumptions set forth in Vol. I, number 1.1.5, $G = KA_{\mathfrak{p}}N^+$ an Iwasawa decomposition for G; let $d_{\mathfrak{g}}$, $d_{\mathfrak{a}_{\mathfrak{p}}}$, and $d_{\mathfrak{n}^+}$ denote the Haar measures on $\mathfrak{g}$, $\mathfrak{a}_{\mathfrak{p}}$, and $\mathfrak{n}^+$, respectively, which are derived from the fixed Euclidean structure on $\mathfrak{g}$ (cf. 8.1.2); let $d_{A_{\mathfrak{p}}}$ and d_{N^+} be the Haar measures on $A_{\mathfrak{p}}$ and N^+, respectively, which correspond to $d_{\mathfrak{a}_{\mathfrak{p}}}$ and $d_{\mathfrak{n}^+}$ under the exponential mapping – then, as we know, the Haar measure on G can be normalized in such a way that for all $f \in C_c(G)$

$$\int_G f(x)d_G(x) = \iiint_{K \times A_{\mathfrak{p}} \times N^+} f(khn)h^{2\rho}dk\,d_{A_{\mathfrak{p}}}(h)d_{N^+}(n) \qquad (\rho = 2^{-1}\sum_{\lambda > 0} m(\lambda)\lambda)$$

where dk is normalized Haar measure on K. On the other hand, let $\mathscr{C}$ denote the positive Weyl chamber in $\mathfrak{a}_{\mathfrak{p}}$ – then there exists a unique determination of the Haar measure on G such that

$$\int_G f(x)d_G(x) = \iiint_{K\times\mathscr{C}\times K} f(k_1 \exp H\, k_2) \triangle (\exp H)\, dk_1 d_{\mathfrak{a}_\mathfrak{p}}(H) dk_2 \qquad (f \in C_c(G))$$

where we have set

$$\triangle(\exp H) = \prod_{\lambda>0} (e^{\lambda(H)} - e^{-\lambda(H)})^{m(\lambda)} \qquad (H \in \mathfrak{a}_\mathfrak{p}).$$

Now suppose that d_G is the Haar measure on G per our agreements in 8.1.2 – then, of course, there exists a unique positive constant $\mathbf{c}_G$ (say) such that

$$\int_G f(x)d_G(x) = \mathbf{c}_G \iiint_{K\times A_\mathfrak{p}\times N^+} f(khn)h^{2\rho}dk d_{A_\mathfrak{p}}(h) d_{N^+}(n) \qquad (f \in C_c(G)).$$

Remark Using the methods and results of 8.4.5 infra, one can actually compute $\mathbf{c}_G$. Thus let $\mathfrak{j}$ be a θ-stable fundamental Cartan subalgebra of $\mathfrak{g}$; let W_K denote the subgroup of the Weyl group W of the pair $(\mathfrak{g}_c, \mathfrak{j}_c)$ which is generated by the compact roots; let $\rho_K = 2^{-1}\sum_\alpha \alpha$, α a positive compact root of the pair $(\mathfrak{g}_c, \mathfrak{j}_c)$ – then

$$\mathbf{c}_G^{-1} = [W_K] \prod_{\alpha\in\Phi^+_{CP}} (\alpha, \rho_K)(2\pi)^{\mathbf{m}_G - \mathbf{r}_G}(2)^{\mathbf{n}_G}$$

where

$$\mathbf{m}_G = 2^{-1}(\dim(G/K) - \operatorname{rank}(G) + \operatorname{rank}(K)),$$
$$\mathbf{n}_G = 2^{-1}(\dim(G/K) - \operatorname{rank}(G/K)),$$
$$\mathbf{r}_G = 2^{-1}(\dim(G) - \operatorname{rank}(G)).$$

Let $\mathfrak{j}$ be a θ-stable Cartan subalgebra of $\mathfrak{g}$, J the Cartan subgroup of G associated with $\mathfrak{j}$; assuming that the pair $(\mathfrak{g}_c, \mathfrak{j}_c)$ admits at least one positive root which is not imaginary (so that $\bar{\mathfrak{j}}_\mathfrak{p}(= \mathfrak{j}_\mathfrak{p} \cap \bar{\mathfrak{g}}) \neq \{0\}$), assign to the symbols $\mathfrak{l}, L, \mathfrak{m}, M, \mathfrak{n}^\pm, N^\pm$ their usual meanings relative to the present context (see the Introduction supra for notational conventions); write $P = MJ_\mathfrak{p}N^+$ and set $\rho_P(H) = 2^{-1}\operatorname{tr}(\operatorname{ad}(H)|\mathfrak{n}^+)$ $(H \in \mathfrak{j}_\mathfrak{p})$.

Proposition 8.1.3.1 Normalize the Haar measure on N^+ so that it corresponds, under the exponential map, to the canonical Euclidean measure on $\mathfrak{n}^+$ – then there exists a unique determination of the Haar measure on M such that

$$\int_G f(x)d_G(x) = \mathbf{c}_G \iiiint_{K\times M\times J_\mathfrak{p}\times N^+} f(kmjn)j^{2\rho_P}dk d_M(m) d_{J_\mathfrak{p}}(j) d_{N^+}(n) \qquad (f \in C_c(G)).$$

[Here the Haar measure on $J_\mathfrak{p}$ is the one which corresponds, under the exponential map, to the canonical Euclidean measure on $\mathfrak{j}_\mathfrak{p}$.]

Proof For the proof it will be best to alter our notation temporarily; thus, as in the Introduction, write $G = KA_{\mathfrak{p}}N^+$, $M = M_K A_1 N_1^+$,

$$A_{\mathfrak{p}} = A_1 A_2 (A_2 = J_{\mathfrak{p}} \ldots),$$

$N^+ = N_1^+ N_2^+$ (so that the N_2^+ at hand plays the role of the N^+ in the statement of our proposition) etc. This being done, normalize the Haar measure on M by the requirement

$$d_M(m) = h_1^{2\rho_1}\, dm_K d_{A_1}(h_1) d_{N_1^+}(n_1)$$

where $m = m_K h_1 n_1$ (the Haar measures on A_1 and N_1^+ being the 'obvious' ones) – then, for any $f \in C_c(G)$, we have

$$\begin{aligned}
\int_G f(x) d_G(x) &= \mathbf{c}_G \iiint_{K\times A_{\mathfrak{p}}\times N^+} f(khn) h^{2\rho} dk d_{A_{\mathfrak{p}}}(h) d_{N^+}(n) \\
&= \mathbf{c}_G \iiiint_{K\times A_{\mathfrak{p}}\times N_1^+\times N_2^+} f(khn_1n_2) h^{2\rho} dk d_{A_{\mathfrak{p}}}(h) d_{N_1^+}(n_1) d_{N_2^+}(n_2) \\
&= \mathbf{c}_G \int\!\!\int\!\!\int\!\!\int\!\!\int\!\!\int_{K\times M_K\times A_1\times N_1^+\times A_2\times N_2^+} f(km_K h_1 n_1 h_2 n_2) h_1^{2\rho_1} h_2^{2\rho_2} \\
&\qquad \times dk dm_K d_{A_1}(h_1) d_{N_1^+}(n_1) d_{A_2}(h_2) d_{N_2^+}(n_2) \\
&= \mathbf{c}_G \iiiint_{K\times M\times A_2\times N_2^+} f(kmh_2n_2) h_2^{2\rho_2} dk d_M(m) d_{A_2}(h_2) d_{N_2^+}(n_2)
\end{aligned}$$

which serves to establish our contention. □

Keeping to the above notations, let J_0 denote the center of J – then we recall that the Haar measure on J_0 has been normalized in such a way as to ensure the validity of the formula $\int_G = \int_{G/\mathsf{J}_0} \int_{\mathsf{J}_0}$. Suppose that we alter our point of view by first normalizing the Haar measure on J_0 via the requirement that $\mathsf{J}_0 \cap K$ be assigned mass 1 (the Haar measure on $J_{\mathfrak{p}}$ being as above) and then forcing the normalization of d_{G/J_0} through the relation $\int_G = \int_{G/\mathsf{J}_0} \int_{\mathsf{J}_0}$; upon demanding that d_J agree with d_{J_0} on J_0, it then follows that the formula

$$\int_{G(\mathsf{J})} f(x) d_G(x) = [W(G, \mathsf{J}_0)]^{-1} \int_{\mathsf{J}'} |\det(\mathrm{Ad}(j^{-1}) - 1)_{\mathfrak{g}/\mathfrak{j}}| d_\mathsf{J}(j) \times \int_{G/\mathsf{J}_0} f({}^{\dot{x}} j) d_{G/\mathsf{J}_0}(\dot{x})$$

is still valid for all $f \in C_c(G(\mathsf{J}))$ but the infinitesimal analogue of this relation (cf. 8.1.2) is only true up to a constant factor. Unless otherwise specified, in the remainder of the present number it will be supposed that the aforestated agreements are in force. [Let us observe that if $\mathfrak{j}$ is fundamental, then the current assumptions as regards the normalization of the relevant invariant measures on J_0 and G/J_0 are not at all new (cf. 8.1.2).]

Here are two simple corollaries of the preceding proposition.

Corollary 8.1.3.2 Retain the above notations and assumptions – then

$$\int_{G/J_0} f(\dot{x})d_{G/J_0}(\dot{x}) = \mathbf{c}_G \iiint_{K\times M\times N^+} f(\dot{\widehat{kmn}})dkd_M(m)d_{N^+}(n)$$

for all $f \in C_c(G/J_0)$.

[One need only remark that when the orders of integration over $J_\mathfrak{p}$ and N^+ are switched, the factor $j^{2\rho_P}$ ($j \in J_\mathfrak{p}$) is wiped out.]

Corollary 8.1.3.3 Retain the above notations and assumptions – then

$$\int_{G/J_0} f(\dot{x})d_{G/J_0}(\dot{x}) = \mathbf{c}_G \iiint_{K\times N^+\times M} f(\dot{\widehat{knm}})dkd_{N^+}(n)d_M(m)$$

for all $f \in C_c(G/J_0)$.

[Bear in mind that M admits no non-trivial homomorphisms into the multiplicative group of positive real numbers (cf. Vol. I, number 1.2.4).

Note Suppose in particular that $\mathfrak{j}_\mathfrak{p} = \mathfrak{a}_\mathfrak{p}$ (= maximal abelian subalgebra of $\mathfrak{p}$) – then M, the identity component of the centralizer in K of $\mathfrak{a}_\mathfrak{p}$, is compact and the unique determination of the Haar measure on M per Proposition 8.1.3.1 is that which assigns to M mass 1. Consequently, for any $f \in C_c(G/J_0)$, we have

$$\int_{G/J_0} f(\dot{x})d_{G/J_0}(\dot{x}) = \mathbf{c}_G \iint_{K\times N^+} f(\dot{\widehat{kn}})dkd_{N^+}(n).$$

Let $\nabla_\mathfrak{l}$ denote the polynomial function on $\mathfrak{l}_c$ which is defined by the rule $\nabla_\mathfrak{l}(X) = \det(\mathrm{ad}(X)|\mathfrak{n}^+)$ ($X \in \mathfrak{l}_c$) – then, in particular,

$$\nabla_\mathfrak{l}(H) = \pi_{R,C}(H) \qquad (H \in \mathfrak{j})$$

where

$$\pi_{R,C} = \prod_{\alpha\in\Phi_R^+\cup\Phi_C^+}\alpha .$$

Suppose that X is an element of $\mathfrak{l}$ with the property that $\nabla_\mathfrak{l}(X) \neq 0$ – then it follows from a standard argument that the assignment

$$Z \mapsto e^{\mathrm{ad}(Z)}X - X \qquad (Z \in \mathfrak{n}^+)$$

is an analytic diffeomorphism of $\mathfrak{n}^+$ onto itself (with Jacobian $|\nabla_\mathfrak{l}(X)| \neq 0 \ldots$), whence, for such an X,

$$\int_{N^+} f(nX)d_{N^+}(n) = |\nabla_\mathfrak{l}(X)|^{-1}\int_{\mathfrak{n}^+} f(X+Z)d_{\mathfrak{n}^+}(Z) \qquad (f \in C_c(\mathfrak{l}+\mathfrak{n}^+)).$$

Proposition 8.1.3.4 Retain the above notations and assumptions; let H be an element in $\mathfrak{j}$ such that $\pi_{R,C}(H) \neq 0$ – then

$$\int_{G/J_0} f(\dot{x}H)\, d_{G/J_0}(\dot{x}) = \mathbf{c}_G \,|\, \pi_{R,C}(H)\,|^{-1} \times \iiint_{K \times M \times \mathfrak{n}^+} f(k(mH + Z))\, dk d_M(m) d_{\mathfrak{n}^+}(Z)$$

for all $f \in C_c(\mathfrak{g})$.

Proof Since it is clear that $\nabla_{\mathfrak{l}}(mH) = \nabla_{\mathfrak{l}}(H) = \pi_{R,C}(H)$ $(m \in M)$, the present result is seen to be an immediate consequence of the preceding remarks and Corollary 8.1.3.3. □

We shall now assume that G is, moreover, *acceptable* – then, as is easily seen, the same must be true of L. Let Φ_I^+ denote the set of positive imaginary roots of the pair $(\mathfrak{g}_c, \mathfrak{j}_c)$ (or, equivalently, the set of positive roots of the pair $(\mathfrak{l}_c, \mathfrak{j}_c) \ldots$); assign to the symbols Φ_R^+ and Φ_C^+ similar connotations. Given $j \in J'$, put

$$\Delta_I(j) = \xi_\rho(j_K) \prod_{\alpha \in \Phi_I^+} (1 - \xi_\alpha(j^{-1})),$$

$$\Delta_{R,C}(j) = \xi_\rho(j_{\mathfrak{p}}) \prod_{\alpha \in \Phi_R^+ \cup \Phi_C^+} (1 - \xi_\alpha(j^{-1}))$$

so that $\Delta = \Delta_I \cdot \Delta_{R,C} \cdot$ [Here j_K is the component of j in J_K, $j_{\mathfrak{p}}$ the component of j in $J_{\mathfrak{p}}$.]

Proposition 8.1.3.5 Retain the above notations and assumptions – then the invariant measure on L/J can be normalized in such a way that

$$\epsilon_R(j)\Delta(j) \int_{G/J_0} f(^{\dot{x}}j) d_{G/J_0}(\dot{x})$$

$$= \mathbf{c}_G \Delta_I(j) \xi_\rho(j_{\mathfrak{p}}) \iint_{L/J \times N^+} f_K(^{\dot{l}}j \cdot n) d_{L/J}(\dot{l}) d_{N^+}(n) \qquad (j \in J')$$

for all $f \in C_c(G)$.

[Given any f in $C_c(G)$, we agree to write $f_K(x) = \int_K f(kxk^{-1})dk$ $(x \in G)$.]

The integrals appearing in the statement of our proposition exist (cf. 8.1.4); this being so, let us preface the proof with a couple of lemmas.

Lemma 8.1.3.6 Let L be the centralizer of $\mathfrak{j}_{\mathfrak{p}}$ in G; fix an $l \in \mathsf{L}$ such that $\det(\mathrm{Ad}(l^{-1}) - 1)_{\mathfrak{n}^+} \neq 0$ – then the mapping ξ defined by the rule

$$n \mapsto l^{-1} n l n^{-1} \qquad (n \in N^+)$$

is an analytic diffeomorphism of N^+ onto itself.

Proof It is clear that ξ is analytic; this being the case, fix $n \in N^+$, $Z \in \mathfrak{n}^+$ – then

$$\xi(n \exp(tZ)) = l^{-1}nln^{-1} \exp(t \operatorname{Ad}(nl^{-1})Z) \exp(-t \operatorname{Ad}(n)Z) \quad (t \in \mathbf{R}),$$

whence

$$\det((d\xi)_n) = \det(\operatorname{Ad}(l^{-1}) - 1)_{\mathfrak{n}^+} \neq 0$$

which implies that ξ is everywhere regular. Now let $\sigma_0 < \ldots < \sigma_r$ be the positive roots of the pair $(\mathfrak{g}, \mathfrak{j}_\mathfrak{p})$ arranged in increasing order (bear in mind that $\mathfrak{n}^+ = \sum_{\sigma>0} \mathfrak{g}^\sigma \ldots$); put $\mathfrak{n}_i^+ = \sum_{j \geqslant i} \mathfrak{g}^{\sigma_j}$ where $0 \leqslant i \leqslant r$ and set $\mathfrak{n}_{r+1}^+ = \{0\}$; let

$$A_1 = \operatorname{Ad}(l^{-1}) | \mathfrak{n}^+,$$

$A_2 = -1$ and $A = A_1 + A_2$ – then all the conditions of Lemma 1.1.4.2 are met in the obvious way. Consequently the map

$$Z \mapsto \log(\exp(\operatorname{Ad}(l^{-1})Z) \cdot \exp(-Z)) \qquad (Z \in \mathfrak{n}^+)$$

defines a bijective polynomial map of $\mathfrak{n}^+$ onto itself. Hence the lemma. □

Corollary 8.1.3.7 Fix an element $l \in \mathsf{L}$ such that $\det(\operatorname{Ad}(l^{-1}) - 1)_{\mathfrak{n}^+} \neq 0$ – then

$$\int_{N^+} f(nln^{-1}) d_{N^+}(n) = |\det(\operatorname{Ad}(l^{-1}) - 1)_{\mathfrak{n}^+}|^{-1} \int_{N^+} f(ln) d_{N^+}(n)$$

for all $f \in C_c(G)$.

Lemma 8.1.3.8 Fix an element $j \in \mathsf{J}'$ – then

$$\epsilon_R(j)\Delta(j) = \Delta_I(j)\xi_\rho(j_\mathfrak{p}) |\det(\operatorname{Ad}(j^{-1}) - 1)_{\mathfrak{n}^+}|$$

where $j_\mathfrak{p}$ is the component of j in $J_\mathfrak{p}$.

Proof If α belongs to $\Phi_R^+ \cup \Phi_C^+$, then the same is true of $-\alpha^\theta$; therefore, in view of the fact that $\xi_{-\alpha^\theta}(j^{-1}) = \overline{\xi_\alpha(j^{-1})}$ (cf. Proposition 8.1.1.1), we have

$$\begin{aligned} |\det(\operatorname{Ad}(j^{-1}) - 1)_{\mathfrak{n}^+}| &= \prod_{\alpha \in \Phi_R^+ \cup \Phi_C^+} |\xi_\alpha(j^{-1}) - 1| \\ &= \prod_{\alpha \in \Phi_R^+} |\xi_\alpha(j^{-1}) - 1| \prod_{\substack{\alpha \in \Phi_C^+ \\ \alpha > -\alpha^\theta}} |\xi_\alpha(j^{-1}) - 1|^2 \\ &= \epsilon_R(j) \prod_{\alpha \in \Phi_R^+ \cup \Phi_C^+} (1 - \xi_\alpha(j^{-1})) \end{aligned}$$

from which our assertion follows at once. □

Proof of Proposition 8.1.3.5 Since $\mathfrak{j}$ is a fundamental Cartan subalgebra of $\mathfrak{l}$, the Cartan subgroup of L associated with $\mathfrak{j}$ is J (= the identity component of J); similarly the Cartan subgroup of M associated with $\mathfrak{j}_{\mathfrak{k}}$ is J_K (= the identity component of J_K). The Haar measure on M is fixed via Proposition 8.1.3.1; agreeing to assign to J_K total mass 1 (which will not conflict with the fact that $\mathsf{J}_0 \cap K$ has also been assigned mass 1. . .), the invariant measure on $M/J_K(\sim L/J)$ is determined when we require that $\int_M = \int_{M/J_K} \int_{J_K}$. Now fix an $f \in C_c(G)$ – then, for all $j \in \mathsf{J}'$, we have

$$\epsilon_R(j)\Delta(j) \int_{G/\mathsf{J}_0} f({}^{\dot{x}}j)d_{G/\mathsf{J}_0}(\dot{x})$$

$$= \mathbf{c}_G\epsilon_R(j)\Delta(j) \iint_{N^+\times M} f_K(n({}^m j)n^{-1})d_{N^+}(n)d_M(m)$$

$$= \mathbf{c}_G\Delta_I(j)\xi_\rho(j_\mathfrak{p})\,|\det(\mathrm{Ad}\,(j^{-1}) - 1)_{\mathfrak{n}^+}| \iint_{N^+\times L/J} f_K(n({}^{\dot{l}} j)n^{-1})d_{L/J}(\dot{l})d_{N^+}(n)$$

$$= \mathbf{c}_G\Delta_I(j)\xi_\rho(j_\mathfrak{p}) \iint_{L/J\times N^+} f_K({}^{\dot{l}} j\cdot n)d_{L/J}(\dot{l})d_{N^+}(n),$$

as we wished to prove. □

Note Suppose in particular that $\mathfrak{j}_\mathfrak{p} = \mathfrak{a}_\mathfrak{p}$ (= maximal abelian subalgebra of $\mathfrak{p}$) – then it is clear that for any $f \in C_c(G)$ we have

$$\epsilon_R(j)\Delta(j) \int_{G/\mathsf{J}_0} f({}^{\dot{x}}j)d_{G/\mathsf{J}_0}(\dot{x}) = \mathbf{c}_G\Delta_I(j)\xi_\rho(j_\mathfrak{p}) \int_{N^+} f_K(jn)d_{N^+}(n) \qquad (j \in \mathsf{J}').$$

Appendix Let G be a connected semi-simple Lie group with finite center, $G = KA_\mathfrak{p}N^+$ an Iwasawa decomposition for G; let M be the centralizer of $A_\mathfrak{p}$ in K.

Lemma The Haar measure on $N^-(= (\theta N^+))$ can be normalized in such a way that

$$\int_K f(k)dk = \iint_{\mathsf{M}\times N^-} f(m\kappa(n))e^{-2\rho(H(n))}\,dmd_{N^-}(n) \qquad (\rho = 2^{-1} \sum_{\lambda>0} m(\lambda)\lambda).$$

This normalization of d_{N^-} is characterized by the condition that

$$\int_{N^-} e^{-2\rho(H(n))}d_{N^-}(n) = 1.$$

Proof Let $S = A_\mathfrak{p}N^+, P = \mathsf{M}A_\mathfrak{p}N^+$; let d_S be the left invariant Haar measure on S, d_P the left invariant Haar measure on P – then it can be assumed that

$$d_G(x) = \delta_S(s^{-1})dkd_S(s)\ (x = ks),\ d_P(p) = dmd_S(s) \qquad (p = ms).$$

Because N^-P is a dense open submanifold of G whose complement has zero Haar measure (cf. Proposition 1.2.3.5), it follows from Lemma 5.5.1.4 that the Haar measure on N^- can be so normalized that

$$\int_G f(x)d_G(x) = \iint_{N^-\times P} f(np)\delta_P(p^{-1})d_{N^-}(n)d_P(p)$$
$$= \iint_{N^-\times P} f(\kappa(n)p)\delta_P(p^{-1})e^{-2\rho(H(n))}d_{N^-}(n)d_P(p)$$
$$= \iiint_{N^-\times M\times S} f(\kappa(n)ms)\delta_S(s^{-1})e^{-2\rho(H(n))}d_{N^-}(n)dm d_S(s)$$

for all $f \in C_c(G)$. Fix a function $g \in C_c(S)$ such that $\int_S g(s)\delta_S(s^{-1})d_S(s) = 1$; define a function F on G by the rule $F(ks) = f(k)g(s)$ $(k \in K, s \in S)$, f a given function in $C(K)$ – then, on the one hand we have

$$\int_G F(x)d_G(x) = \int_K f(k)dk,$$

while on the other

$$\int_G F(x)d_G(x) = \iint_{N^-\times M} f(\kappa(n)m)e^{-2\rho(H(n))}d_{N^-}(n)dm.$$

Since the invariant measure on N^- remains unchanged under the action of M by inner automorphisms, we deduce that

$$\int_K f(k)dk = \iint_{M\times N^-} f(m\kappa(n))e^{-2\rho(H(n))}dm d_{N^-}(n) \qquad (\text{all } f \in C(K)),$$

as desired. Finally, taking $f = 1$ leads to the relation

$$\int_{N^-} e^{-2\rho(H(n))}d_{N^-}(n) = 1. \quad \square$$

Corollary Let $\mu : \mathfrak{a}_\mathfrak{p} \to \mathbf{C}$ be a linear function on $\mathfrak{a}_\mathfrak{p}$; put $\mu_+ = \mu + \rho$, $\mu_- = \mu - \rho$ – then

$$\int_K \exp\{\mu(H(hk)) - \rho(H(hk))\}\,dk$$
$$= h^{\mu_-}\int_{N^-} \exp\{\mu_-(H(hnh^{-1})) - \mu_+(H(n))\}\,d_{N^-}(n)$$

for all $h \in A_\mathfrak{p}$.

[The reader will observe that this result allows one to express zonal spherical functions on G by means of an integral over N^-.]

Taking $\mu = 0$ in the preceding corollary yields the relation

$$h^\rho \int_K e^{-\rho(H(hk))}dk = \int_{N^-} \exp\{-\rho(H(hnh^{-1})) - \rho(H(n))\}\,d_{N^-}(n),$$

valid for all $h \in A_\mathfrak{p}$.

8.1.4 A Theorem of Compacity

Let (G, K) be a reductive pair per Vol. I, number 1.1.5 – then the main result of the present number can be formulated as follows.

Theorem 8.1.4.1 (Harish-Chandra) Let γ be a semi-simple element in G, J a Cartan subgroup of G containing γ; let G_γ be the centralizer of γ in G, $x \mapsto \dot{x}$ the natural mapping of G onto the factor space G/G_γ – then there exists a neighborhood $\mathcal{O}(\gamma)$ of γ in J with the following property: Given any compact set ω in G, there exists a compact set Ω in G/G_γ such that $xjx^{-1} \in \omega$ $(x \in G, j \in \mathcal{O}(\gamma)) \Rightarrow \dot{x} \in \Omega$.

[Since γ is semi-simple, it actually does lie in some Cartan subgroup of G; cf. Vol. I, number 1.4.3.]

The importance of this theorem for subsequent developments can not be overemphasized; indeed the fundamental properties of the invariant integral (cf. infra) rest essentially upon it.

Before proceeding to the details, let us make a few remarks of a general nature. In the first place, it can be assumed that γ lies in a θ-stable Cartan subgroup J (say) of G. Secondly, an easy argument shows that it is sufficient to consider just the case when G is semi-simple (use the isomorphism theorems). Moreover, since $Z \subset \mathsf{G}_\gamma$ (Z the center of G), it is clear that one may replace G by any other connected group locally isomorphic to it. This being so, it may then be assumed that G is the real analytic subgroup of G_c corresponding to $\mathfrak{g}$, G_c the simply connected complex analytic Lie group corresponding to $\mathfrak{g}_c$ (clearly G is closed in G_c). [Note that this assumption implies that the Cartan subgroups of G are abelian; cf. Theorem 1.4.1.5.] If J_c (respectively $\mathsf{G}_{\gamma,c}$) is the Cartan subgroup of G_c associated with $\mathfrak{j}_c$ (respectively the centralizer of γ in G_c), then $\mathsf{J} = J_c \cap G$, $\mathsf{G}_\gamma = \mathsf{G}_{\gamma,c} \cap G$ – therefore G/G_γ may be regarded as a subset of $G_c/\mathsf{G}_{\gamma,c}$.

Lemma 8.1.4.2 There exists a neighborhood $\mathcal{O}_c(\gamma)$ of γ in J_c with the following property: Given any compact set ω_c in G_c, there exists a compact set Ω_c in $G_c/\mathsf{G}_{\gamma,c}$ such that $xjx^{-1} \in \omega_c$ $(x \in G_c, j \in \mathcal{O}_c(\gamma)) \Rightarrow \dot{x} \in \Omega_c$.

Accepting this lemma, Theorem 8.1.4.1 would follow if it could be shown that G/G_γ is closed in $G_c/\mathsf{G}_{\gamma,c}$ and that the canonical injection of G/G_γ into $G_c/\mathsf{G}_{\gamma,c}$ is a homeomorphism. Plainly this injection is continuous – hence the theorem is a consequence of:

Lemma 8.1.4.3 Let $\{\dot{x}_n\}$ be a sequence in G/G_γ which converges in $G_c/\mathsf{G}_{\gamma,c}$ to some point $\dot{x}$ – then $\dot{x} \in G/\mathsf{G}_\gamma$ and $\dot{x}_n \to \dot{x}$ in G/G_γ.

The discussion below will be devoted to giving the proofs of the two lemmas supra.

We shall keep to the notations and conventions which were introduced above; in particular, then, G denotes the real analytic subgroup of G_c corresponding to $\mathfrak{g}$, G_c the simply connected complex analytic Lie group with Lie algebra $\mathfrak{g}_c$, while γ denotes some fixed element in the Cartan subgroup J $(= J_c \cap G)$.

Adopting the usual notations, assume that the roots Φ of the pair $(\mathfrak{g}_c, \mathfrak{j}_c)$ have been ordered in some way and set $\mathfrak{n}_c^+ = \sum_{\alpha>0} \mathfrak{g}_c^\alpha$, $\mathfrak{n}_c^- = \sum_{\alpha<0} \mathfrak{g}_c^\alpha$; let N_c^+, N_c^- denote the corresponding complex analytic subgroups of G_c. Divide the positive roots Φ^+ into two disjoint subsets Φ_1^+, Φ_2^+ as follows: Put α in Φ_1^+ if α commutes with γ (i.e. if $\xi_\alpha(\gamma) = 1 \ldots$); otherwise put α in Φ_2^+. Set $\mathfrak{n}_c^1 = \sum_\alpha \mathfrak{g}_c^\alpha$, $\alpha \in \Phi_1^+$; $\mathfrak{n}_c^2 = \sum_\alpha \mathfrak{g}_c^\alpha$, $\alpha \in \Phi_2^+$; let N_c^1 be the (complex) analytic subgroup of G_c corresponding to $\mathfrak{n}_c^1$ and let N_c^2 denote the set of all elements in N_c^+ of the form $\exp X (X \in \mathfrak{n}_c^2)$. Evidently N_c^2 is closed in N_c^+. Finally, if $\alpha_1 < \alpha_2 < \ldots < \alpha_r$ are the roots in Φ^+ arranged in increasing order, let $\mathfrak{n}_i = \sum_{j=i}^r \mathfrak{g}_c^{\alpha_j}$ – then the assumptions of Lemma 1.1.4.1 are all fulfilled in the obvious way.

Choose a compact neighborhood $\mathcal{O}_c(\gamma)$ of γ in J_c such that if $\alpha \in \Phi_2^+$, then ξ_α never takes the value 1 on $\mathcal{O}_c(\gamma)$; as will be seen shortly, such a neighborhood verifies the conditions of Lemma 8.1.4.2 above (per any compact set ω_c in G_c).

Lemma 8.1.4.4 There exist compact sets ω^1, ω^2 in N_c^1, N_c^2, respectively, having the following property. Suppose that $n_2 j n_1 n_2^{-1} \in \omega_c$ for some $j \in \mathcal{O}_c(\gamma)$, $n_1 \in N_c^1$, and $n_2 \in N_c^2$ – then $n_1 \in \omega^1$ and $n_2 \in \omega^2$.

We shall defer the proof of this lemma momentarily and see first how Lemma 8.1.4.2 follows from it.

Let U denote the real analytic subgroup of G_c corresponding to the compact real form $\mathfrak{u} = \mathfrak{k} + \sqrt{-1}\mathfrak{p}$ – then, in view of Iwasawa's Theorem, we have $G_c = U N_c^2 N_c^1 J_c$. Note too that $N_c^1 J_c \subset \mathsf{G}_{\gamma, c}$.

Proof of Lemma 8.1.4.2 Fix a compactum ω_c in G_c and suppose that

$$xjx^{-1} \in \omega_c \; (x \in G_c, \; j \in \mathcal{O}_c(\gamma));$$

write

$$x = u n_2 n_1 j_x \; (u \in U, \; n_2 \in N_c^2, \; n_1 \in N_c^1, \; j_x \in J_c).$$

If $n_{1,x} = j^{-1} n_1 j n_1^{-1}$, then

$$xjx^{-1} = u(n_2 j n_{1,x} n_2^{-1})u^{-1} \in \omega_c$$

and hence $n_2 j n_{1,x} n_2^{-1} \in U \omega_c U$. But $U \omega_c U$ is compact and so, choosing compact sets ω^1, ω^2 in accordance with Lemma 8.1.4.4, we have that $n_2 \in \omega^2$ while $x \in U\omega^2 N_c^1 J_c \subset U\omega^2 \mathsf{G}_{\gamma,c}$ – therefore $\dot{x} \in \Omega_c$, $\Omega_c = \widehat{U\dot{\omega}^2}$. □

Proof of Lemma 8.1.4.4 Recall that $J_c N_c^+$ is closed in G_c and that the mapping $(j, n) \mapsto jn$ of $J_c \times N_c^+$ into G_c is, in particular, a homeomorphism. Define $\log n \; (n \in N_c^+)$ in the usual way and denote by $t_\alpha \; (\alpha \in \Phi^+)$ the Cartesian coordinates in the complex Euclidean space $\mathfrak{n}_c^+$ corresponding to the basis $\{X_\alpha : \alpha \in \Phi^+\}$ (X_α a fixed non-zero element in $\mathfrak{g}_c^\alpha$). Evidently, in order to prove the lemma, it is enough to show that $t_\alpha(\log n_1)$ and $t_\alpha(\log n_2)$ remain bounded for every $\alpha \in \Phi^+$.

(1) Put $\mathfrak{n}_c^+(\alpha) = \sum_{0<\beta<\alpha} \mathbf{C}X_\beta$ and let P_α denote the projection of $\mathfrak{n}_c^+$ onto $\mathfrak{n}_c^+(\alpha)$ given by $\mathsf{P}_\alpha(X_\beta) = X_\beta$ $(0 < \beta < \alpha)$, $\mathsf{P}_\alpha(X_\beta) = 0$ $(\beta \geqslant \alpha)$. In view of the lemma in the Appendix to the present number, there exists a polynomial function p_α on $\mathfrak{n}_c^+(\alpha) \times \mathfrak{n}_c^+(\alpha) \times \mathfrak{n}_c^+(\alpha)$ such that

$$\begin{aligned} &t_\alpha(\log(\exp X_1 \exp X_2 \exp X_3)) \\ &\qquad = t_\alpha(X_1 + X_2 + X_3) + p_\alpha(\mathsf{P}_\alpha(X_1), \mathsf{P}_\alpha(X_2), \mathsf{P}_\alpha(X_3)) \end{aligned}$$

for all $X_1, X_2, X_3 \in \mathfrak{n}_c^+$ – we shall use this fact in a moment.

(2) Observe that $n_2 j n_1 n_2^{-1} = j(j^{-1}n_2 j)n_1 n_2^{-1}$; therefore

$$t_\alpha(\log(j^{-1}n_2 j n_1 n_2^{-1}))$$

must remain bounded for all $\alpha \in \Phi^+$, ω_c being compact.

(3) Now suppose that the assertion of the lemma were false; let β be the lowest root in Φ^+ such that $|t_\beta(\log n_1)| + |t_\beta(\log n_2)|$ does not remain bounded when n_1, n_2 and j vary in such a way as to fulfill our assumptions. Put $X_i = \log n_i$ $(i = 1, 2)$ – then $\log(j^{-1}n_2 j) = \operatorname{Ad}(j^{-1})X_2$ and from (1) we have

$$\begin{aligned} &t_\beta(\log(j^{-1}n_2 j n_1 n_2^{-1})) \\ &= t_\beta((\operatorname{Ad}(j^{-1}) - 1)X_2 + X_1) + p_\beta(\mathsf{P}_\beta(\operatorname{Ad}(j^{-1})X_2), \mathsf{P}_\beta(X_1), \mathsf{P}_\beta(-X_2)). \end{aligned}$$

Therefore

$$\begin{aligned} &(\xi_\beta(j^{-1}) - 1)t_\beta(X_2) + t_\beta(X_1) \\ &\qquad = t_\beta(\log(j^{-1}n_2 j n_1 n_2^{-1})) - p_\beta(\mathsf{P}_\beta(\operatorname{Ad}(j^{-1})X_2), \mathsf{P}_\beta(X_1), \mathsf{P}_\beta(-X_2)). \end{aligned}$$

The right hand side of this equality remains bounded as n_1, n_2 and j vary. So look at the left hand side. If $\beta \in \Phi_1^+$, then $t_\beta(X_2) = 0$; if $\beta \in \Phi_2^+$, then $t_\beta(X_1) = 0$ and, since $(\xi_\beta(j^{-1}) - 1)^{-1} = \xi_\beta(j)(1 - \xi_\beta(j))^{-1}$ is bounded for $j \in \mathcal{O}_c(\gamma)$, we see that in either case $|t_\beta(X_1)| + |t_\beta(X_2)|$ remains bounded. But this contradicts the definition of β . . . .

The proof of the lemma is now complete. □

It remains to give the proof of Lemma 8.1.4.3. As there, let $\{\dot{x}_n\}$ be a sequence in G/G_γ which converges in $G_c/\mathsf{G}_{\gamma,c}$ to some point $\dot{x}$; select elements $x_n \in G$, $x \in G_c$ lying in the cosets $\dot{x}_n$ and $\dot{x}$, respectively – then obviously $x_n \gamma x_n^{-1} \to x\gamma x^{-1}$ in G_c and therefore the $x_n\gamma x_n^{-1}$ stay within a compact subset of G (G is closed in G_c). However γ is a semi-simple element of G and so the orbit $G\cdot\gamma$ is a *closed* subset of G (cf. Proposition 1.4.3.4); consequently the $\dot{x}_n$ are all contained in a compact subset of G/G_γ and from this the assertion of Lemma 8.1.4.3 follows immediately. [The fact that γ is semi-simple is crucial here; for a well-known theorem of Arens then ensures us that the natural map of G/G_γ, endowed with the quotient topology, onto $G\cdot\gamma$, endowed with the induced topology, is a homeomorphism . . . .]

Harish-Chandra's 'Theorem of Compacity' is thereby completely proved.

Appendix Let N be a connected, simply connected, real or complex nilpotent Lie group with Lie algebra $\mathfrak{n}$. Let $X_1, \ldots, X_r$ be a basis for $\mathfrak{n}$ over $\mathbf{k}$ ($\mathbf{k} = \mathbf{R}$ or $\mathbf{C}$) such that

$$[X_{i_1}, X_{i_2}] \in \sum_{j>i_2} \mathbf{k}X_j \qquad (1 \leqslant i_1 < i_2 \leqslant r).$$

Put $\mathfrak{n}_0 = \{0\}$, $\mathfrak{n}_i = \sum_{j=1}^{i} \mathbf{k}X_j$ $(1 \leqslant i \leqslant r)$, and let P_i denote the projection of $\mathfrak{n}$ onto $\mathfrak{n}_i$ given by $\mathsf{P}_i(X_j) = X_j$ $(1 \leqslant j \leqslant i)$, $\mathsf{P}_i(X_j) = 0$ $(i < j \leqslant r)$; let $\mathfrak{n}_i^m$ denote the Cartesian product of $\mathfrak{n}_i$ with itself m times.

Lemma Let m be a positive integer – then, for each i $(1 \leqslant i \leqslant r)$, there exists a polynomial mapping p_i of $\mathfrak{n}_{i-1}^m$ into $\mathfrak{n}_i$ such that

$$\begin{aligned} &\mathsf{P}_i(\log(\exp Y_1 \cdots \exp Y_m)) \\ &= \mathsf{P}_i(Y_1 + \cdots + Y_m) + p_i(\mathsf{P}_{i-1}(Y_1), \ldots, \mathsf{P}_{i-1}(Y_m)) \end{aligned}$$

for all $Y_1, \ldots, Y_m \in \mathfrak{n}$.

Proof We proceed by induction on $r = \dim(\mathfrak{n})$. If $r = 1$, then $\mathfrak{n}$ is abelian and the assertion is clear. So suppose that $r \geqslant 2$; put $\mathfrak{n}^* = \mathfrak{n}/\mathbf{k}X_r$ and consider N/N_r, N_r the one parameter subgroup of N corresponding to X_r. Because X_r lies in the center of $\mathfrak{n}$, $\log(\exp X \exp Y) = X + Y$ $(X \in \mathfrak{n}, Y \in \mathbf{k}X_r)$ and thus the exponential map of $\mathfrak{n}^*$ onto N/N_r is still injective. Let $X \mapsto X^*$ $(X \in \mathfrak{n})$ denote the natural mapping of $\mathfrak{n}$ onto $\mathfrak{n}^*$ and let $\mathfrak{n}_i^*$ be the image of $\mathfrak{n}_i$ $(1 \leqslant i \leqslant r)$ under this mapping. Define the projection P_i^* of $\mathfrak{n}^*$ onto $\mathfrak{n}_i^*$ by the prescription $\mathsf{P}_i^*(X^*) = (\mathsf{P}_i(X))^*$ $(X \in \mathfrak{n}, 1 \leqslant i \leqslant r)$. Since $\dim(\mathfrak{n}^*) < \dim(\mathfrak{n})$, the induction hypothesis is applicable to $\mathfrak{n}^*$. Hence for each i there exists a polynomial mapping q_i of $\mathfrak{n}_{i-1}^m$ into $\mathfrak{n}_i$ such that

$$\begin{aligned} &\mathsf{P}_i(\log(\exp Y_1 \cdots \exp Y_m)) \\ &\equiv \mathsf{P}_i(Y_1 + \cdots + Y_m) + q_i(\mathsf{P}_{i-1}(Y_1), \ldots, \mathsf{P}_{i-1}(Y_m)) \bmod \mathbf{k}X_r \end{aligned}$$

for all $Y_1, \ldots, Y_m \in \mathfrak{n}$. But if $i < r$, then $\mathfrak{n}_i \cap (\mathbf{k}X_r) = \{0\}$ which implies that the congruence is an equality in this case. If $i = r$, then P_r reduces to the identity mapping of $\mathfrak{n}$. Moreover, as is well-known, the map $(Y_1, \ldots, Y_m) \mapsto \log(\exp Y_1 \ldots \exp Y_m)$ is a polynomial mapping of $\mathfrak{n}^m = \mathfrak{n}_r^m$ into $\mathfrak{n}$ (cf. Helgason [2, p. 227]). This being so, put

$$p_r(Y_1, \ldots, Y_m) = \log(\exp Y_1 \cdots \exp Y_m) - (Y_1 + \cdots + Y_m).$$

Since X_r lies in the center of $\mathfrak{n}$, it is clear that

$$p_r(Y_1, \ldots, Y_m) = p_r(\mathsf{P}_{r-1}(Y_1), \ldots, \mathsf{P}_{r-1}(Y_m)),$$

whence

$$\log(\exp Y_1 \cdots \exp Y_m) = (Y_1 + \cdots + Y_m) + p_r(\mathsf{P}_{r-1}(Y_1), \ldots, \mathsf{P}_{r-1}(Y_m))$$

which proves the assertion for $i = r$ too. □

8.1.5 The Standard Semi-Norm on a Semi-Simple Lie Group

Let G be a connected semi-simple Lie group with finite center, K a maximal compact subgroup of G; let $G = KA_{\mathfrak{p}}N^+$ be an Iwasawa decomposition for

G – then, of course, $G = KA_{\mathfrak{p}}K$. Let θ be the ambient Cartan involution in the Lie algebra $\mathfrak{g}$, $(.\,,.)_\theta$ the associated Euclidean structure on $\mathfrak{g}$; for any element $T \in \mathrm{Hom}_{\mathbf{R}}(\mathfrak{g}, \mathfrak{g})$, let T^* denote its adjoint relative to $(.\,,.)_\theta$.

Definition Given x in G, put $\|x\|_\theta^2 = \mathrm{tr}(\mathrm{Ad}(x)\,\mathrm{Ad}(x)^*)$ – then the assignment $x \mapsto \|x\|_\theta$ is called a *standard semi-norm* on G (relative to θ).
[It is clear, among other things, that $\|xy\|_\theta \leqslant \|x\|_\theta \|y\|_\theta$ $(x, y \in G) \ldots .]$

Observe that the adjoint $\mathrm{Ad}(x)^*$ of $\mathrm{Ad}(x)$ is $\mathrm{Ad}(\theta(x))^{-1}$ (all $x \in G$); this implies in particular that (1) $\mathrm{Ad}(k)$ $(k \in K)$ is unitary, (2) $\mathrm{Ad}(h)$ $(h \in A_{\mathfrak{p}})$ is self-adjoint with eigenvalues 1, $h^{\pm\lambda}$ $(\lambda \in \Sigma^+)$ (Σ the set of roots of the pair $(\mathfrak{g}, \mathfrak{a}_{\mathfrak{p}})$). It follows, therefore, that $\|k_1hk_2\|_\theta = \|h\|_\theta$ $(k_1, k_2 \in K;\ h \in A_{\mathfrak{p}})$.

Remark If θ_1 and θ_2 are two Cartan involutions on $\mathfrak{g}$ giving rise to standard semi-norms $\|\,.\,\|_{\theta_1}$ and $\|\,.\,\|_{\theta_2}$, respectively, then it is easy to see that $\|\,.\,\|_{\theta_1} \asymp \|\,.\,\|_{\theta_2}$.

In what follows, if no confusion can arise, the subscript 'θ' will be dropped from the notation.

Lemma 8.1.5.1 Let $x \in G$ – then $\|x\| = \|x^{-1}\|$.

Proof Fix $x \in G$ and write $x = k_1hk_2$ $(k_1, k_2 \in K;\ h \in A_{\mathfrak{p}})$ – then $\|x\| = \|h\|$. Since θ is a unitary transformation on $\mathfrak{g}$ and since $\mathrm{Ad}(h^{-1}) = \theta \circ \mathrm{Ad}(h) \circ \theta^{-1}$, we see that $\|x\| = \|h\| = \|h^{-1}\| = \|x^{-1}\|$, as desired. □

Let $\mathscr{C}$ be the positive Weyl chamber implicit in our choice of an Iwasawa decomposition for G.

Lemma 8.1.5.2 Let μ be a real valued linear function on $\mathfrak{a}_{\mathfrak{p}}$ – then there exists a non-negative integer n such that

$$e^{\mu(H)} \leqslant \|\exp H\|^n$$

for all $H \in \mathscr{C}^{cl}$.

Proof Let $\{\lambda_1, \ldots, \lambda_l\}$ be the fundamental system of roots for Σ associated with $\mathscr{C}$; set $r = \sum_{\lambda > 0} m(\lambda)$ and choose a positive integer s such that $s(\sum_i \lambda_i(H)) \geqslant \mu(H)$ (all $H \in \mathscr{C}^{cl}$) – then

$$\|\exp H\|^2 \geqslant \sum_{\lambda>0} m(\lambda)e^{2\lambda(H)} \geqslant r \exp\Big(2r^{-1} \sum_{\lambda>0} m(\lambda)\lambda(H)\Big) \qquad (H \in \mathscr{C}^{cl})$$

and so

$$\|\exp H\|^{rs} \geqslant \exp\Big(s \sum_{\lambda>0} m(\lambda)\lambda(H)\Big) \geqslant e^{\mu(H)} \qquad (H \in \mathscr{C}^{cl})$$

which proves that the contention of the lemma is met with $n = rs$. □

Lemma 8.1.5.3 Let U be a representation of G on a finite dimensional complex Hilbert space E – then there exists a non-negative integer n such that $\|U(x)\|_{HS} \leq \|x\|^n$ for all x in G.

Proof By passing to an equivalent representation if necessary, it can be supposed that $U \mid K$ is unitary while $U \mid A_{\mathfrak{p}}$ is self-adjoint; since $G = K\exp(\mathscr{C}^{cl})K$, we can then assume that $x = h \in \exp(\mathscr{C}^{cl})$. This being the case, let $\Lambda_1, \ldots, \Lambda_r$ be the distinct weights of U with respect to $\mathfrak{a}_c$; in view of Lemma 8.1.5.2, there exists an integer $n \geqslant 0$ such that $e^{\Lambda_i(H)} \leqslant \|\exp H\|^n$ $(1 \leqslant i \leqslant r)$ for all $H \in \mathscr{C}^{cl}$; therefore, if $d = \dim(E)$, then

$$\|U(h)\|_{HS}^2 \leqslant d \max_i e^{2\Lambda_i(H)} \leqslant d\|h\|^{2n}$$

where $h = \exp H$ $(H \in \mathscr{C}^{cl})$. Hence the lemma. □

Lemma 8.1.5.4 There exists a positive integer n such that

$$\int_G \|x\|^{-n} \, d_G(x) < \infty.$$

Proof Choose a linear function μ on $\mathfrak{a}_{\mathfrak{p}}$ such that the integral of $e^{-\mu}$ taken over $\mathscr{C}$ is finite; owing to Lemma 8.1.5.2, there exists a positive integer n such that $\exp(\mu(H) + 2\rho(H)) \leqslant \|\exp H\|^n$ for all H in $\mathscr{C}$ (as usual, $\rho = 2^{-1} \sum_{\lambda > 0} m(\lambda)\lambda$) – therefore, using a familiar integration formula, we find that

$$\int_G \|x\|^{-n} \, d_G(x) = \int_{A_{\mathfrak{p}}} \|h\|^{-n} \bigcirc(h) \, d_{A_{\mathfrak{p}}}(h) \leqslant \int_{\mathscr{C}} e^{-\mu(H)} \, d_{\mathfrak{a}_{\mathfrak{p}}}(H) < \infty$$

relative to a suitable normalization of the various Haar measures. □

8.1.6 Completely Invariant Sets

Let $\mathfrak{g}$ be a reductive Lie algebra over **R**, G a connected Lie group with Lie algebra $\mathfrak{g}$ – then G operates on $\mathfrak{g}$ in the customary way.

Notation As usual, $\mathscr{N}$ will denote the set of nilpotent elements in $\mathfrak{g}$, $\mathscr{S}$ the set of semi-simple elements in $\mathfrak{g}$.

Let Ω be a subset of $\mathfrak{g}$. We shall say that Ω is *completely invariant* provided it has the following property: If $\omega \subset \Omega$ and ω is compact, then $(G\cdot\omega)^{cl} \subset \Omega$. Completely invariant sets play an important role in the theory of central distributions on $\mathfrak{g}$. Accordingly, at this juncture, it will be convenient to set down a few of their elementary properties.

The following lemma is clear in view of Proposition 1.3.5.7.

Lemma 8.1.6.1 Let Ω be a completely invariant subset of $\mathfrak{g}$, X an element of Ω. If X_s is the semi-simple component of X, then $X_s \in \Omega$.

Let $\emptyset$ denote the empty set, Ω a completely invariant subset of $\mathfrak{g}$. If Ω_0 is a G-invariant closed subset of Ω with $\Omega_0 \cap \mathscr{S} = \emptyset$, then $\Omega_0 = \emptyset$; likewise if $\Omega_0 \cap \mathscr{S} = \{0\}$, then $\Omega_0 \subset \Omega \cap \mathscr{N}$. On the other hand, suppose that Ω_0 is an open G-invariant subset of Ω with $\Omega \cap \mathscr{S} \subset \Omega_0$ – then $\Omega_0 = \Omega$.

Lemma 8.1.6.2 Let Ω be an open completely invariant subset of $\mathfrak{g}$ – then the following conditions are equivalent:
(i) $\Omega \cap \mathscr{N} \neq \emptyset$; (ii) $0 \in \Omega$; (iii) $\mathscr{N} \subset \Omega$.

Proof (i) ⇒ (ii) by Lemma 8.1.6.1; (ii) ⇒ (iii) by Lemma 1.3.5.6; (iii) ⇒ (i) is obvious. □

Existence of Completely Invariant Sets in the Algebra To begin with, let us suppose that $\mathfrak{g}$ is semi-simple. For any $r > 0$, let $\mathfrak{g}(r)$ denote the set of all $X \in \mathfrak{g}$ such that $|\mathscr{I}(\lambda)| < r$ for every eigenvalue λ of ad (X) ($\mathscr{I}(\lambda)$ the imaginary part of λ). Evidently $\mathfrak{g}(r)$ is an open, completely invariant neighborhood of zero in $\mathfrak{g}$; moreover it is clear that $\mathscr{N} \subset \mathfrak{g}(r)$. Note too that $\mathfrak{g}(r)$ is connected (for $X \in \mathfrak{g}(r) \Rightarrow tX \in \mathfrak{g}(r)$ $(0 \leqslant t \leqslant 1)$). Here is an important property of $\mathfrak{g}(r)$. Suppose that $r \leqslant \pi$ ($\pi = 3.14\ldots$) – then the restriction to $\mathfrak{g}(r)$ of the exponential mapping of $\mathfrak{g}$ into G is regular and injective. [If $X \in \mathfrak{g}(r)$, $r \leqslant \pi$, then ad (X) can never have an eigenvalue of the form $2\pi\sqrt{-1}n$, n a non-zero integer – therefore

$$(1 - e^{-\mathrm{ad}(X)})/\mathrm{ad}\,(X) = \sum_{m=0}^{\infty} (-1)^m(\mathrm{ad}\,(X))^m/(m+1)!$$

is a non-singular transformation of $\mathfrak{g}$, whence exp is regular at X. . . . Next assume that $\exp X_1 = \exp X_2$ ($X_1, X_2 \in \mathfrak{g}(r)$). – then X_1, X_2 commute and so $\exp(X_1 - X_2) = \exp(X_1)\exp(-X_2) = 1$. Let λ_{ji} $(1 \leqslant j \leqslant s_i)$ denote all the distinct eigenvalues of ad (X_i) $(i = 1, 2)$ – then every eigenvalue of ad $(X_1 - X_2)$ must be of the form $\lambda_{j1} - \lambda_{k2}$ for some j and k (Lie's Theorem) and, of course, $|\mathscr{I}(\lambda_{j1} - \lambda_{k2})| < 2\pi$. Therefore $\exp(\lambda_{j1} - \lambda_{k2}) \neq 1$ unless $\lambda_{j1} = \lambda_{k2}$; but since $e^{\mathrm{ad}(X_1 - X_2)} = 1$, zero is the only possible eigenvalue for ad $(X_1 - X_2)$. This means that ad $(X_1 - X_2)$ is nilpotent, hence ad $(X_1 - X_2) = 0$ (the map $X \mapsto e^{\mathrm{ad}(X)}$ is one-to-one on $\mathscr{N}$ . . .) and so it follows that $X_1 = X_2$, $\mathfrak{g}$ being semi-simple.] Now assume that $\mathfrak{g}$ is reductive; select an open, relatively compact neighborhood $\mathscr{O}$ of zero in $\mathfrak{c}$ (the center of $\mathfrak{g}$) on which the exponential mapping is injective; fix a number r $(0 < r \leqslant \pi)$ and put $\Omega = \mathscr{O} + \bar{\mathfrak{g}}(r)$ – then it is clear, in view of what was said above, that Ω is an open, completely invariant neighborhood of zero in $\mathfrak{g}$ on which the exponential mapping is regular and injective.

Lemma 8.1.6.3 Let ω be a relatively compact subset of $\mathfrak{g}$, $\mathfrak{j}$ a Cartan subalgebra of $\mathfrak{g}$ – then $\mathfrak{j} \cap (G\cdot\omega)$ is a relatively compact subset of $\mathfrak{j}$.

Proof Choose relatively compact sets $\omega_{\mathfrak{c}}$ and $\omega_{\bar{\mathfrak{g}}}$ in $\mathfrak{c}$ and $\bar{\mathfrak{g}}$, respectively, such that $\omega \subset \omega_{\mathfrak{c}} + \omega_{\bar{\mathfrak{g}}}$ – then $G\cdot\omega \subset \omega_{\mathfrak{c}} + G\cdot\omega_{\bar{\mathfrak{g}}}$, whence $\mathfrak{j} \cap (G\cdot\omega) \subset \omega_{\mathfrak{c}} + \mathfrak{j} \cap (G\cdot\omega_{\bar{\mathfrak{g}}})$ which shows that there is no loss of generality in assuming that $\mathfrak{g}$ is semi-simple. This being the case, let r be the number of positive roots of the pair $(\mathfrak{g}_c, \mathfrak{j}_c)$; write

$$\det(t - \mathrm{ad}\,(X)) = t^l\left\{t^{2r} + \sum_{1}^{2r} t^{2r-i}p_i(X)\right\} \qquad (X \in \mathfrak{g}_c),$$

t an indeterminate – then the p_i are G-invariant polynomial functions on $\mathfrak{g}_c$ and therefore remain bounded on $G\cdot\omega$. If $H \in \mathfrak{j}$, then we have

$$\det(t - \mathrm{ad}\,(H)) = t^l \prod_{\alpha>0} (t^2 - \alpha(H)^2).$$

Therefore the elementary symmetric functions in $\alpha(H)^2$ $(\alpha > 0)$ remain bounded as H varies in $\mathfrak{j} \cap (G\cdot\omega)$ – thus the $\alpha(H)$ remain bounded and so $\mathfrak{j} \cap (G\cdot\omega)$ is relatively compact, $\mathfrak{g}$ being semi-simple. □

As above, let $\mathfrak{g}$ be a reductive Lie algebra over **R**, G a connected Lie group with Lie algebra $\mathfrak{g}$ – then G operates on itself via inner automorphisms.

Let Ω be a subset of G. We shall say that Ω is *completely invariant* provided it has the following property: If $\omega \subset \Omega$ and ω is compact, then $(G\cdot\omega)^{cl} \subset \Omega$.

Owing to Proposition 1.4.3.5, the following result is clear.

Lemma 8.1.6.4 Let Ω be a completely invariant subset of G, Ω_0 a G-invariant subset of Ω which is closed in Ω. If Ω_0 contains no semi-simple element of Ω, then Ω_0 is empty.

Existence of Completely Invariant Sets in the Group Let us suppose that G is semi-simple; fix r, $0 < r \leqslant \pi$, and set $\Omega = \exp(\mathfrak{g}(r))$ – then Ω is a completely invariant open subset of G. Thus fix a compact subset $\omega \subset \Omega$ and let $\{X_n\}$, $\{x_n\}$ be sequences in $\mathfrak{g}(r)$ and G, respectively, such that $\exp X_n \in \omega$ and $x_n \cdot (\exp X_n)$ converges to some point y in G. Since $\log(\omega) \cap \mathfrak{g}(r)$ is compact, by passing to subsequences, if necessary, we may assume that $X_n \to X$ and $x_n X_n \to Y$ where $X \in \log(\omega)$, $Y \in \mathfrak{g}$ (cf. the lemma infra). But $\operatorname{ad}(X)$ and $\operatorname{ad}(Y)$ have the same eigenvalues – thus $Y \in \mathfrak{g}(r)$ and so $y = \exp Y \in \Omega$, as desired.
The immediately preceding discussion is completed by the following result.

Lemma 8.1.6.5 Let $\mathfrak{g}$ be a semi-simple Lie algebra over **R**, G a connected Lie group with Lie algebra $\mathfrak{g}$; fix r, $0 < r \leqslant \pi$, and suppose that $\{X_n\}$ is a sequence in $\mathfrak{g}(r)$ such that the $\exp X_n$ lie in a compact subset of G – then the X_n lie in a compact subset of $\mathfrak{g}$.

Proof Fix a Cartan decomposition $\mathfrak{g} = \mathfrak{k} + \mathfrak{p}$; put $\mathfrak{u} = \mathfrak{k} + \sqrt{-1}\mathfrak{p}$ and let τ denote the conjugation of $\mathfrak{g}_c$ with respect to $\mathfrak{u}$ – then the positive definite Hermitian form $(.\,,.)_\tau$ equips $\mathfrak{g}_c$ with the structure of a Hilbert space. If G_c is a connected complex analytic group with Lie algebra $\mathfrak{g}_c$, then we shall agree to write $\|x\|_\tau$ for $\|\operatorname{Ad}(x)\|_{HS}$ $(x \in G_c)$. Fix a τ-stable Cartan subalgebra $\mathfrak{j}_c$ of $\mathfrak{g}_c$; as usual Φ (respectively Φ^+) will denote the set of roots (respectively positive roots) of the pair $(\mathfrak{g}_c, \mathfrak{j}_c)$.
(1) With no loss of generality, we may assume that G is the real analytic subgroup of G_c corresponding to $\mathfrak{g}$, G_c the simply connected complex analytic Lie group corresponding to $\mathfrak{g}_c$. If U denotes the analytic subgroup of G_c corresponding to $\mathfrak{u}$, then it is clear that

$$\mathfrak{g}_c = \bigcup_{u \in U} \operatorname{Ad}(u)(\mathfrak{j}_c + \mathfrak{n}_c^+).$$

This being so, choose

$$u_n \in U,\ Y_n \in \mathfrak{j}_c + \mathfrak{n}_c^+$$

such that $X_n = \operatorname{Ad}(u_n) Y_n$; since $\operatorname{Ad}(u_n)$ is unitary, $\|X_n\|_\tau = \|Y_n\|_\tau$ and $\|\exp X_n\|_\tau = \|\exp Y_n\|_\tau$. Write $Y_n = H_n + Z_n$ $(H_n \in \mathfrak{j}_c, Z_n \in \mathfrak{n}_c^+)$; the eigenvalues of $\operatorname{ad}(Y_n)$ and $\operatorname{ad}(H_n)$ are the same, whence $|\mathscr{I}(\alpha(H_n))| < r \leqslant \pi$ for all $\alpha \in \Phi^+$. However $|e^{\alpha(H_n)}|$ also remains bounded $(\alpha \in \Phi^+)$. Therefore $\|H_n\|_\tau$ is bounded. [We write $\|X\|_\tau$ for $\|\operatorname{ad}(X)\|_{HS}$ $(X \in \mathfrak{g}_c)$.]
(2) Let $F(t) = \sum_0^\infty (-1)^m t^m/(m+1)! = (1 - e^{-t})/t$ $(t \in \mathbf{C})$. Fix a non-zero element $H_0 \in \mathfrak{j}_c$ and let $t_\alpha (\alpha \in \Phi^+)$ denote the complex coordinates in $\mathfrak{n}_c^+$ corresponding to the basis X_α $(\alpha \in \Phi^+)$. If $H \in \mathfrak{j}_c$, $Z \in \mathfrak{n}_c^+$, then it is clear that $\operatorname{Ad}(\exp(H+Z))H_0 - H_0 \in \mathfrak{n}_c^+$. For any $\beta \in \Phi^+$, consider the (holomorphic) function

$$f_\beta(H, Z) = t_\beta(\operatorname{Ad}(\exp(H+Z))H_0 - H_0) + e^{\beta(H)}F(\beta(H))\beta(H_0)t_\beta(Z)$$

on $\mathfrak{j}_c + \mathfrak{n}_c^+$. It is easy to see that if $\beta < \gamma$ $(\gamma \in \Phi^+)$, then $f_\beta(H, Z + tX_\gamma) = f_\beta(H, Z)$ $(t \in \mathbf{C})$. Hence f_β depends only on H and $t_\alpha(Z)$ $(0 < \alpha \leqslant \beta)$. Let us now show that f_β depends only on H and $t_\alpha(Z)$ $(0 < \alpha < \beta)$. For this purpose, it is enough to show that $(d/dt) f_\beta(H, Z + tX_\beta)|_{t=0} = 0$. Now

$$(d/dt)\operatorname{Ad}(\exp(H + Z + tX_\beta))H_0|_{t=0} = \operatorname{Ad}(\exp(H+Z))(\operatorname{ad}(Y)H_0),$$

where

$$Y = \{(1 - e^{-\mathrm{ad}(H+Z)})/\mathrm{ad}(H+Z)\}X_\beta \equiv F(\beta(H))X_\beta \bmod \sum_{\gamma>\beta} \mathbf{C}X_\gamma.$$

Therefore

$$\begin{aligned}(d/dt)f_\beta(H, Z + tX_\beta)|_{t=0} \\ = F(\beta(H))t_\beta(\mathrm{Ad}(\exp(H+Z))[X_\beta, H_0]) + e^{\beta(H)}F(\beta(H))\beta(H_0).\end{aligned}$$

Since $[X_\beta, H_0] = -\beta(H_0)X_\beta$ and

$$\mathrm{Ad}(\exp(H+Z))X_\beta \equiv e^{\beta(H)}X_\beta \bmod \sum_{\gamma>\beta} \mathbf{C}X_\gamma,$$

we see that $(d/dt)f_\beta(H, Z + tX_\beta)|_{t=0} = 0$, as contended.

(3) Suppose the assertion of the lemma is false – then there exists a root $\beta \in \Phi^+$ such that $t_\beta(Z_n)$ does not remain bounded. Select the least such root β – then $\|H_n\|_\tau$ and $t_\alpha(Z_n)$ $(0 < \alpha < \beta)$ all remain bounded and therefore, in view of (2), $f_\beta(H_n, Z_n)$ remains bounded as $n \to \infty$. By assumption $\|\exp Y_n\|_\tau$ and so, also, $\|\mathrm{Ad}\,(\exp\,(H_n + Z_n))\|_\tau$ stays bounded. Thus the same holds for

$$e^{\beta(H_n)}F(\beta(H_n))\beta(H_0)t_\beta(Z_n) = f_\beta(H_n, Z_n) - t_\beta(\mathrm{Ad}\,(\exp\,(H_n + Z_n))H_0 - H_0).$$

However $e^{-\beta(H_n)}F(\beta(H_n))^{-1}$ remains bounded; choosing H_0 such that $\beta(H_0) \neq 0$, we find that $t_\beta(Z_n)$ is bounded too which contradicts the choice of β.

This completes the proof of the lemma. □

We return now to the case of a general, reductive G.

Lemma 8.1.1.6 Let ω be a relatively compact subset of G, J a Cartan subgroup of G – then $J \cap (G\cdot\omega)$ is a relatively compact subset of J.

[We shall leave the proof to the reader.]

8.2 Differential Operators on Reductive Lie Groups and Algebras

8.2.1 Radial Components of Differential Operators on a Manifold

In this number we shall establish some general results concerning differential operators on a manifold. The remaining numbers of this section are essentially applications of these generalities to various concrete situations.

Notation Let M_1, M_2 be two C^∞ manifolds, f a C^∞ function on $M_1 \times M_2$. If $p_i \in M_i$ and D_i is a differential operator on M_i $(i = 1, 2)$, then we shall denote the value of $(D_1 \times D_2)f$ at (p_1, p_2) by $f(p_1; D_1 : p_2; D_2)$; D_i will be suppressed in this notation whenever it is 1. All other unexplained notations and conventions are as in Appendix 2.

Let M_1 and M_2 be two C^∞ manifolds of dimensions m_1 and m_2 respectively; let μ_1 (respectively μ_2) be a positive measure on M_1 (respectively M_2) which on any coordinate neighborhood is a non-zero C^∞ multiple of Lebesgue measure. Suppose that $\pi : M_1 \to M_2$ is a surjective C^∞ map such that rank $(d\pi) = m_2$ on all of M_1.

Theorem 8.2.1.1 (Harish-Chandra) For every $v \in C_c^\infty(M_1)$ there exists a unique function $f_v \in C_c^\infty(M_2)$ such that

$$\int_{M_1} (F \circ \pi)(p_1)v(p_1)d\mu_1(p_1) = \int_{M_2} F(p_2)f_v(p_2)d\mu_2(p_2)$$

for all $F \in C_c^\infty(M_2)$. Moreover spt $(f_v) \subset \pi(\text{spt}(v))$ and the map $v \mapsto f_v$ is a continuous surjection of $C_c^\infty(M_1)$ onto $C_c^\infty(M_2)$.

Corollary 8.2.1.2 Let F be a continuous function on M_2 – then

$$\int_{M_1} (F \circ \pi)(p_1)v(p_1)d\mu_1(p_1) = \int_{M_2} F(p_2)f_v(p_2)d\mu_2(p_2) \qquad (v \in C_c^\infty(M_1)).$$

Corollary 8.2.1.3 Let F be a measurable function on M_2 – then F is locally summable (with respect to μ_2) iff $F \circ \pi$ is locally summable (with respect to μ_1) and, in case this is so, we have

$$\int_{M_1} (F \circ \pi)(p_1)v(p_1)d\mu_1(p_1) = \int_{M_2} F(p_2)f_v(p_2)d\mu_2(p_2) \qquad (v \in C_c^\infty(M_1)).$$

The uniqueness of f_v is clear – our main task, therefore, is to establish its existence. We shall begin by proving the theorem and its corollaries in a special case.

For any $r > 0$, let I_r denote the open interval $(-r, r)$ in $\mathbf{R}$ and J_r the closed interval $[-r, r]$. Let π denote the projection $(x_1, \ldots, x_{m_1}) \mapsto (x_1, \ldots, x_{m_2})$ of $\mathbf{R}^{m_1}$ onto $\mathbf{R}^{m_2}$ $(m_1 \geqslant m_2)$.

Lemma 8.2.1.4 For any $v \in C_c^\infty(I_r^{m_1})$ define $g_v \in C_c^\infty(I_r^{m_2})$ by the prescription

$$g_v(x_1, \ldots, x_{m_2}) = \int v(x_1, \ldots, x_{m_1})dx_{m_2+1} \ldots dx_{m_1}.$$

Then the map $v \mapsto g_v$ is a continuous mapping of $C_c^\infty(I_r^{m_1})$ onto $C_c^\infty(I_r^{m_2})$ and spt $(g_v) \subset \pi(\text{spt}(v))$.

Proof Plainly $g_v \in C_c^\infty(I_r^{m_2})$ and spt $(g_v) \subset \pi(\text{spt}(v))$. For any s, $0 < s < r$, put $C_s^\infty(I_r^{m_1}) = C_\omega^\infty(I_r^{m_1})$ where $\omega = J_s^{m_1}$; it is not difficult to see that our mapping is continuous on $C_s^\infty(I_r^{m_1})$ – therefore, in view of the definition of the Schwartz topology, it is also continuous on $C_c^\infty(I_r^{m_1})$. In order to show that the map $v \mapsto g_v$ is surjective, we may assume that $m_1 > m_2$. Fix a ϕ in $C_c^\infty(I_r^{m_1 - m_2})$ with the property that

$$\int \phi(x_{m_2+1}, \ldots, x_{m_1})\, dx_{m_2+1} \ldots dx_{m_1} = 1.$$

Given $g \in C_c^\infty(I_r^{m_2})$, put $v = g \times \phi$ – then $v \in C_c^\infty(I_r^{m_1})$ and $g = g_v$. □

Let us now prove Theorem 8.2.1.1 in the special case when $M_1 = I_r^{m_1}$, $M_2 = I_r^{m_2}$ and π is the projection used in Lemma 8.2.1.4. Let $d_1x = dx_1 \ldots dx_{m_1}$, $d_2x = dx_1 \ldots dx_{m_2}$ – then $d\mu_1 = \phi_1 d_1x$, $d\mu_2 = \phi_2 d_2x$ where

ϕ_1 and ϕ_2 are positive C^∞ functions on M_1 and M_2, respectively. For any $v \in C_c^\infty(M_1)$, define f_v on M_2 as follows: $f_v = \phi_2^{-1} g_{\phi_1 v}$ – obviously f_v fulfills the condition of Theorem 8.2.1.1. Moreover Lemma 8.2.1.4 implies that the map $v \mapsto f_v$ is a continuous mapping of $C_c^\infty(M_1)$ into $C_c^\infty(M_2)$ with $\mathrm{spt}\,(f_v) \subset \pi(\mathrm{spt}\,(v))$. Since the endomorphisms $v \mapsto \phi_1 v$ and $g \mapsto \phi_2^{-1} g$ of $C_c^\infty(M_1)$ and $C_c^\infty(M_2)$, respectively, are surjective, another appeal to Lemma 8.2.1.4 tells us that the mapping $v \mapsto f_v$ is surjective. Finally Corollaries 8.2.1.2, 8.2.1.3 are, in this case, direct consequences of Fubini's Theorem.

We turn now to the general situation.

Proof of Theorem 8.2.1.1 (1) Fix a point $p_1 \in M_1$ and choose a small open connected neighborhood $\mathcal{O}_1^*$ of p_1 in M_1; let $\mathcal{O}_2^* = \pi(\mathcal{O}_1^*)$. Since rank $(d\pi) = m_2$, π is an open mapping – thus $\mathcal{O}_2^*$ is open in M_2 and so if $\mathcal{O}_1^*$ is sufficiently small, we can choose a coordinate system $\{x_1^1, \ldots, x_{m_1}^1\}$ on $\mathcal{O}_1^*$ such that $x_i^1(p_1) = 0$ $(1 \leqslant i \leqslant m_1)$ and $x_j^1 = x_j^2 \circ \pi$ $(1 \leqslant j \leqslant m_2)$ where $\{x_1^2, \ldots, x_{m_2}^2\}$ is a coordinate system on $\mathcal{O}_2^*$ (cf. Chevalley [1, p. 80]). Fix a small number $r > 0$ and let $\mathcal{O}_1$ be the set of all $p_1^* \in \mathcal{O}_1^*$ where $|x_i^1(p_1^*)| < r$ $(1 \leqslant i \leqslant m_1)$. If r is small enough, the map

$$p_1^* \mapsto (x_1^1(p_1^*), \ldots, x_{m_1}^1(p_1^*))$$

is a diffeomorphism of $\mathcal{O}_1$ with $I_r^{m_1}$; put $\mathcal{O}_2 = \pi(\mathcal{O}_1)$ – then it is clear from the above discussion that Theorem 8.2.1.1 and its corollaries hold for the triple $(\mathcal{O}_1, \mathcal{O}_2, \pi)$. Conclusion: For every point $p_1 \in M_1$, there exists an open neighborhood $\mathcal{O}_1(p_1)$ of p_1 in M_1 such that Theorem 8.2.1.1 and its corollaries hold for the triple $(\mathcal{O}_1(p_1), \pi(\mathcal{O}_1(p_1)), \pi)$.

(2) Fix a compact subset ω_1 of M_1 – then the collection

$$\{\mathcal{O}_1(p_1) : p_1 \in \omega_1\}$$

is an open covering of ω_1 in M_1. From this covering of ω_1, choose a finite subcovering $\{\mathcal{O}_1^i\}$ $(1 \leqslant i \leqslant t)$; select $\psi_1^i \in C_c^\infty(\mathcal{O}_1^i)$ such that $\psi_1^1 + \cdots + \psi_1^t = 1$ on ω_1; put $\mathcal{O}_2^i = \pi(\mathcal{O}_1^i)$ and for any $v \in C_{\omega_1}^\infty(M_1)$, set $v_i = \psi_1^i v$ – then $v = v_1 + \cdots + v_t$, where $v_i \in C_c^\infty(\mathcal{O}_1^i)$. Let $f_{v_i} \in C_c^\infty(\mathcal{O}_2^i)$ be the function which corresponds to v_i upon application of Theorem 8.2.1.1 to the triple $(\mathcal{O}_1^i, \mathcal{O}_2^i, \pi)$; put $f_v = \sum_1^t f_{v_i}$. If F is a continuous function on M_2, then

$$\int_{M_1} (F \circ \pi) v \, d\mu_1 = \sum_1^t \int_{M_1} (F \circ \pi) v_i \, d\mu_1.$$

But

$$\int_{M_1} (F \circ \pi) v_i \, d\mu_1 = \int_{\mathcal{O}^i_1} (F \circ \pi) v_i \, d\mu_1 = \int_{\mathcal{O}^i_2} F f_{v_i} \, d\mu_2 = \int_{M_2} F f_{v_i} \, d\mu_2$$

from Corollary 8.2.1.2 applied to $(\mathcal{O}_1^i, \mathcal{O}_2^i, \pi)$ – therefore

$$\int_{M_1} (F \circ \pi) v \, d\mu_1 = \sum_1^t \int_{M_2} F f_{v_i} \, d\mu_2 = \int_{M_2} F f_v \, d\mu_2.$$

In addition $\operatorname{spt}(f_v) \subset \bigcup_i \operatorname{spt}(f_{v_i})$; however $\operatorname{spt}(f_{v_i}) \subset \pi(\operatorname{spt}(v_i)) \subset \pi(\operatorname{spt}(v))$, whence $\operatorname{spt}(f_v) \subset \pi(\operatorname{spt}(v))$. Clearly the mapping $v \mapsto f_v$ is continuous on $C^\infty_{\omega_1}(M_1)$ – therefore the mapping $v \mapsto f_v$ is continuous on $C_c^\infty(M_1)$. It remains to verify that this map is surjective. Given $g \in C_c^\infty(M_2)$, choose the compact set ω_1 above in such a way that $\operatorname{spt}(g) \subset \omega_2 = \pi(\omega_1)$. Evidently $\{\mathcal{O}_2^i\}$ $(1 \leqslant i \leqslant t)$ is an open covering for ω_2 in M_2; select $\psi_2^i \in C_c^\infty(\mathcal{O}_2^i)$ such that $\psi_2^1 + \cdots + \psi_2^t = 1$ on ω_2; put $g_i = \psi_2^i g$ so that $g = g_1 + \cdots + g_t$. Theorem 8.2.1.1 is applicable to the triple $(\mathcal{O}_1^i, \mathcal{O}_2^i, \pi)$ and so there exists $v_i \in C_c^\infty(\mathcal{O}_1^i)$ such that $g_i = f_{v_i}$. Let $v = v_1 + \cdots + v_t$ – then it is clear that $g = f_v$ which serves to establish the surjectivity of our mapping.

This proves Theorem 8.2.1.1 and Corollary 8.2.1.2. □

Proof of Corollary 8.2.1.3 Fix a point $p_2 \in M_2$; let $p_1 \in M_1$ be such that $\pi(p_1) = p_2$; choose $\mathcal{O}_1(p_1) = \mathcal{O}_1$ as above and put $\mathcal{O}_2 = \pi(\mathcal{O}_1)$ – then Corollary 8.2.1.3 holds for $(\mathcal{O}_1, \mathcal{O}_2, \pi)$, whence $F \circ \pi$ is locally summable on $\mathcal{O}_1$ iff F is locally summable on $\mathcal{O}_2$. Since p_2 is an arbitrary point in M_2, the first statement of the corollary follows. The validity of the second is likewise clear. □

Retain the preceding notations.

Proposition 8.2.1.5 Let T be a distribution on M_2 – then the mapping τ_T, $v \mapsto T(f_v)$ $(v \in C_c^\infty(M_1))$, is a distribution on M_1. Moreover T is completely determined by τ_T.

Proof This is plain, the map $v \mapsto f_v$ being a continuous surjection of $C_c^\infty(M_1)$ onto $C_c^\infty(M_2)$. □

Let G be a Lie transformation group operating on a C^∞ manifold M – then, to each X in the Lie algebra $\mathfrak{g}$ of G there corresponds a vector field $\bar{X}$ on M (A 2.4). On the other hand, the elements of $\mathfrak{g}$ may also be viewed as left invariant vector fields on G; this will be done in what follows unless explicitly stated to the contrary. Given $f \in C^\infty(M)$, define a function $(x, p) \mapsto f(x : p)$ of two variables on $G \times M$ by the rule

$$f(x : p) = f(x \cdot p) \qquad (x \in G, p \in M).$$

The next statement is obvious.

Lemma 8.2.1.6 Let D be a differential operator on M, $X \in \mathfrak{g}$ – then $f(x; X : p; D) = f(x : p; D(-\bar{X}))$ (all $f \in C^\infty(M)$).

We are now in a position to take up the main theme in this number, namely radial components of differential operators on a manifold. The chief results, which were very kindly communicated to the author by Sigurdur Helgason, may be formulated as follows.

Let M be a C^∞ manifold, p a point in M and M_p the tangent space to M at p. Let G be a Lie transformation group of M. We recall (A 2.4) that a C^∞ function f on an open subset of M is said to be *locally invariant* if $\bar{X}f = 0$ for each vector field $\bar{X}$ on M induced by the action of G ($X \in \mathfrak{g}$, $\mathfrak{g}$ the Lie algebra of G).

Let N be a submanifold of M satisfying the following *transversality condition:* For each $q \in N$, $M_q = N_q + (G\cdot q)_q$ (direct sum).

Theorem 8.2.1.7 (Helgason) Let D be a differential operator on M – then there exists a unique differential operator $\mathfrak{r}(D)$ on N such that

$$(Df)^{\sim} = \mathfrak{r}(D)\tilde{f}$$

for each C^∞ locally invariant f on M (the tilde denoting restriction to N).

Definition The operator $\mathfrak{r}(D)$ is called the *radial component* of D.

Suppose now that μ (respectively ν) is a positive measure on M (respectively N) which on any coordinate neighborhood is a non-zero C^∞ multiple of Lebesgue measure. Assume that G is unimodular. Given $v \in C_c^\infty(G \times N)$ there exists, according to Theorem 8.2.1.1 (with π, $(x, q) \mapsto x\cdot q$ $(x \in G, q \in N)$), a unique function $f_v \in C_c^\infty(G\cdot N)$ such that

$$\int_{G\times N} F(x\cdot q)v(x:q)d_G(x)d\nu(q) = \int_M F(p)f_v(p)d\mu(p) \qquad (F \in C_c^\infty(G\cdot N)).$$

Let $V_v \in C_c^\infty(N)$ denote the function $q \mapsto \int_G v(x:q)d_G(x)$ $(q \in N)$.

Theorem 8.2.1.8 (Helgason) Suppose that G leaves μ invariant; let T be a G-invariant distribution on $G\cdot N$ – then there exists a unique distribution $\tilde{T}$ on N such that

$$\tilde{T}(V_v) = T(f_v) \qquad (\text{all } v \in C_c^\infty(G \times N)).$$

If D is a G-invariant differential operator on M, then

$$(D\cdot T)^{\sim} = \mathfrak{r}(D)\cdot\tilde{T}.$$

[We recall that $D\cdot T$ denotes the distribution on M given by the rule $f \mapsto T(D^\natural f)$ $(f \in C_c^\infty(M))$, $D^\natural$ the adjoint of D with respect to the measure μ; similarly for $\mathfrak{r}(D)\cdot\tilde{T}$.]

Remark The notation $\tilde{T}$ is appropriate. For if T is a G-invariant locally summable function F_T on $G\cdot N$, then the restriction $\tilde{F}_T$ to N verifies $F_T(f_v) = \tilde{F}_T(V_v)$ $(v \in C_c^\infty(G \times N))$ – thus

$$\begin{aligned} F_T(f_v) &= \int_M F_T(p)f_v(p)\,d\mu(p) = \int_{G\times N} F_T(x\cdot q)v(x:q)\,d_G(x)\,d\nu(q) \\ &= \int_{G\times N} \tilde{F}_T(q)v(x:q)\,d_G(x)\,d\nu(q) = \int_N \tilde{F}_T(q)V_v(q)\,d\nu(q) = \tilde{F}_T(V_v). \end{aligned}$$

Roughly speaking, then, one may view the operation $T \mapsto \tilde{T}$ as 'restriction' of invariant distributions.

Example Here is a simple realization of the above set up. Take for G the orthogonal group of degree n and let G operate in the usual way on $M = \mathbf{R}^n$. Identify $\mathbf{R}$ with a subspace of $\mathbf{R}^n$ under the mapping $t \mapsto (t, 0, \ldots, 0)$ ($t \in \mathbf{R}$); let $\mathbf{R}^+$ be the set of positive real numbers; put $N = \mathbf{R}^+$ – then $G \cdot N$ is the complement of the origin in $\mathbf{R}^n$. All the preceding conditions are now obviously met. In particular, if $D = \sum_i (\partial/\partial x_i)^2$ is the Laplacian on M, then $\natural(D) = \partial^2/\partial r^2 + (n-1)/r\,\partial/\partial r$, as is to be expected.

Other examples will be considered later.

Let us turn now to the proofs of the preceding statements. For any point $p \in M$, let G_p denote the stabilizer of p in G. Fix a point $q_0 \in N$; let us agree to view G in the usual way as a principal fiber bundle over G/H (here we have written H for G_{q_0}).

Lemma 8.2.1.9 There exists an open relatively compact neighborhood $\mathcal{O}_N(q_0)$ of q_0 in N and a relatively compact local cross section S through 1 in G over an open neighborhood of $1H$ in G/H such that the mapping ϕ, $(s, q) \to s \cdot q$, is a diffeomorphism of $S \times \mathcal{O}_N(q_0)$ onto an open neighborhood $\mathcal{O}_M(q_0)$ of q_0 in M.

Proof Let $\mathfrak{h}$ denote the Lie algebra of H; select a subspace $\mathfrak{s} \subset \mathfrak{g}$ such that $\mathfrak{g} = \mathfrak{h} + \mathfrak{s}$ (direct sum). The map $\exp : \mathfrak{s} \to G$ is regular at the origin; hence our lemma will follow when it is shown that the mapping ψ

$$(X, q) \mapsto \exp X \cdot q$$

of $\mathfrak{s} \times N$ into M is regular at $(0, q_0)$. Since $d\psi_{(0,q_0)}(0 \times N_{q_0}) = N_{q_0}$, it is enough to verify that $d\psi_{(0,q_0)}(\mathfrak{s} \times 0) = (G \cdot q_0)_{q_0}$. But let $X \in \mathfrak{s}$, $Y \in \mathfrak{h}$ – then $\exp(tX)\exp(tY) = \exp(t(X + Y) + O(t^2))$ ($t \in \mathbf{R}$) and so the curves $t \mapsto \exp(tX) \cdot q_0$, $t \mapsto \exp(t(X + Y)) \cdot q_0$ have the same tangent vector at q_0, which implies that $d\psi_{(0,q_0)}(\mathfrak{s} \times 0) = (G \cdot q_0)_{q_0}$. □

Remark Fix a point $s \cdot q \in \mathcal{O}_M(q_0)$ ($s \in S, q \in \mathcal{O}_N(q_0)$); the transversality assumption implies that the tangent space $(\phi(S \times q))_{s \cdot q}$ equals $(G \cdot q)_{s \cdot q}$ – hence, since $\phi(S \times q) \subset G \cdot s \cdot q$, $\phi(S \times q)$ is an open submanifold of $G \cdot s \cdot q$ and so there exists a neighborhood $\mathcal{O}_G(1)$ of 1 in G such that $\mathcal{O}_G(1) \cdot s \cdot q \subset \phi(S \times q)$.

Proof of Theorem 8.2.1.7 (1) Let $q_0 \in N$; choose $\mathcal{O}_N(q_0)$ and S as in Lemma 8.2.1.9. For $u \in C^\infty(\mathcal{O}_N(q_0))$ define a C^∞ function f_u on $\mathcal{O}_M(q_0)$ by the rule: $f_u(s \cdot q) = u(q)$ ($s \in S$, $q \in \mathcal{O}_N(q_0)$). It is obvious that f_u is locally invariant on $\mathcal{O}_M(q_0)$. Consider the mapping

$$\natural(D)_{q_0; S, \mathcal{O}_N(q_0)} : C^\infty(\mathcal{O}_N(q_0)) \to C^\infty(\mathcal{O}_N(q_0))$$

given by $u \mapsto (Df_u)^\sim$(the tilde denoting restriction to $\mathcal{O}_N(q_0)$) – this map is an endomorphism of $C^\infty(\mathcal{O}_N(q_0))$ which decreases supports and hence is a differential operator on $\mathcal{O}_N(q_0)$ (cf. A2.1).

(2) Keep fixed $\mathcal{O}_N(q_0)$ but let S^* be another local cross section with the property of Lemma 8.2.1.9. Define ϕ^*, $\mathcal{O}_M^*(q_0)$ and f_u^* by means of this S^*. Claim:

$$\natural(D)_{q_0;\,S,\,\mathcal{O}_N(q_0)} = \natural(D)_{q_0;S^*,\,\mathcal{O}_N(q_0)}.$$

To see this, fix a point $\bar{q}_0 \in \mathcal{O}_N(q_0)$ and a compact neighborhood $\bar{\mathcal{O}}$ of $\bar{q}_0$ in $\mathcal{O}_N(q_0)$ – then there exists an open connected neighborhood $\bar{G}$ of 1 in G such that $\bar{G}\cdot\bar{\mathcal{O}} \subset \mathcal{O}_M(q_0) \cap \mathcal{O}_M^*(q_0)$. As will be seen below (cf. (3)), the claim will follow provided we can show that $f_u = f_u^*$ on $\bar{G}\cdot\bar{\mathcal{O}}$. To verify that $f_u = f_u^*$ on $\bar{G}\cdot\bar{\mathcal{O}}$, choose $\bar{q} \in \bar{\mathcal{O}}$ – then $\bar{G}\cdot\bar{q} \subset \bigcup_{q\in\mathcal{O}_N(q_0)} S\cdot q$. Let q be an arbitrary element in $\mathcal{O}_N(q_0)$ and consider a point $\bar{p} \in \bar{G}\cdot\bar{q} \cap S\cdot q$; write $\bar{p} = \bar{x}\cdot\bar{q} = s\cdot q$ ($\bar{x} \in \bar{G}$, $s \in S$). In view of the Remark following Lemma 8.2.1.9, there exists an open connected neighborhood $\mathcal{O}_G(1)$ of 1 in G such that $\mathcal{O}_G(1)\cdot s\cdot q \subset S\cdot q$. We may, shrinking $\mathcal{O}_G(1)$ if necessary, assume that $\mathcal{O}_G(1)\bar{x} \subset \bar{G}$. But then $\mathcal{O}_G(1)\bar{x}\cdot\bar{q} \subset \bar{G}\cdot\bar{q} \cap S\cdot q$; by dimension, $\mathcal{O}_G(1)\bar{x}\cdot\bar{q}$ is open in $\bar{G}\cdot\bar{q}$. Conclusion: Each $\bar{p} \in \bar{G}\cdot\bar{q} \cap S\cdot q$ is in an open subset of $\bar{G}\cdot\bar{q}$ which is itself contained in $\bar{G}\cdot\bar{q} \cap S\cdot q$, whence $\bar{G}\cdot\bar{q} \cap S\cdot q$ is open in $\bar{G}\cdot\bar{q}$. Therefore, by connectedness, $\bar{G}\cdot\bar{q} \subset S\cdot q$ for exactly one $q \in \mathcal{O}_N(q_0)$; since $\bar{q} \in S\cdot q$, it must be the case that $q = \bar{q}$. Now let $\bar{p} \in \bar{G}\cdot\bar{\mathcal{O}}$ – we shall prove that $f_u(\bar{p}) = f_u^*(\bar{p})$. Thus write

$$\bar{p} = \bar{x}\cdot\bar{q} = s\cdot\bar{q} = s^*\cdot\bar{q} \qquad (\bar{x} \in \bar{G};\ s \in S,\ s^* \in S^*,\ \bar{q} \in \bar{\mathcal{O}})$$

– then $f_u(\bar{p}) = f_u(s\cdot\bar{q}) = u(\bar{q}) = f_u^*(s^*\cdot\bar{q}) = f_u^*(\bar{p})$, as desired.

(3) The considerations in (2) imply that $Df_u = Df_u^*$ on the interior of $\bar{G}\cdot\bar{\mathcal{O}}$ – therefore

$$\natural(D)_{q_0;S,\,\mathcal{O}_N(q_0)}(u) = \natural(D)_{q_0;S^*,\,\mathcal{O}_N(q_0)}(u)$$

on the interior of $\bar{\mathcal{O}}$. Since u is an arbitrary element of $C^\infty(\mathcal{O}_N(q_0))$ and $\bar{q}_0$ is an arbitrary point in $\mathcal{O}_N(q_0)$, it follows that

$$\natural(D)_{q_0;S,\,\mathcal{O}_N(q_0)} = \natural(D)_{q_0;S^*,\,\mathcal{O}_N(q_0)}$$

if S and S^* both have the property of Lemma 8.2.1.9.

(4) In view of (3), we can, for each relatively compact open neighborhood $\mathcal{O}_N(q_0)$ of q_0 for which an S per Lemma 8.2.1.9 exists, define

$$\natural(D)_{q_0;\mathcal{O}_N(q_0)} = \natural(D)_{q_0;S,\,\mathcal{O}_N(q_0)}.$$

Moreover it is clear that if $\mathcal{O}_N^1(q_0)$, $\mathcal{O}_N^2(q_0)$ are open relatively compact neighborhoods of q_0 such that S_1, S_2 with the property of Lemma 8.2.1.9 exist, then $\natural(D)_{q_0;\mathcal{O}_{N^1}(q_0)} = \natural(D)_{q_0;\mathcal{O}_{N^2}(q_0)}$ on $C^\infty(\mathcal{O}_N^1(q_0) \cap \mathcal{O}_N^2(q_0))$.

(5) Thanks to (4), for each $q \in N$, $u \in C^\infty(q)$, we can define the linear form $\natural(D)_q$ by $\natural(D)_q(u) = \natural(D)_{q;\mathcal{O}_N(q)}(u)$, $\mathcal{O}_N(q)$ an open relatively compact neighborhood of q such that $u \in C^\infty(\mathcal{O}_N(q))$. Finally, for $u \in C^\infty(N)$, let $\natural(D)u$ denote the function $q \mapsto \natural(D)_q(u)$ $(q \in N)$ – then $\natural(D)$ is the differential operator on N satisfying the condition of our theorem. Thus let f be a locally invariant C^∞ function on M; fix a point $q_0 \in N$ – we must show that $(Df)\tilde{}(q_0) = (\natural(D)\tilde{f})(q_0)$. For this purpose, choose $\mathcal{O}_N(q_0)$, S

as above and consider $f_{\tilde{f}} \in C^\infty(\mathcal{O}_M(q_0))$ – then f and $f_{\tilde{f}}$ are both locally invariant and coincide on $\mathcal{O}_N(q_0)$. It suffices to verify, therefore, that $f = f_{\tilde{f}}$ on $\mathcal{O}_M(q_0)$. But this is clear. . . .

As $\natural(D)$ is evidently unique, the proof of the theorem is now complete. ☐

Remark It is easy to see that the following extension of Theorem 8.2.1.7 is valid. Let D be a differential operator on an open subset $\mathcal{O}(M)$ of M – then there exists a unique differential operator $\natural(D)$ on $\mathcal{O}(N) = N \cap \mathcal{O}(M)$ with the property that if f is a locally invariant C^∞ function on an open subset $\mathscr{P}(M) \subset \mathcal{O}(M)$, then $(Df)^\sim = \natural(D)\tilde{f}$ on $\mathscr{P}(N) = N \cap \mathscr{P}(M)$ (the tilde denoting restriction to $\mathscr{P}(N)$).

In order to prove Theorem 8.2.1.8 some preparation is necessary. Fix a point $q_0 \in N$; let $\mathcal{O}_N(q_0)$ and $S\cdot q_0$ be as in Lemma 8.2.1.9 – assume moreover that they are coordinate neighborhoods of q_0 in N and $G\cdot q_0$, respectively. Let D be a differential operator on $\mathcal{O}_M(q_0)$; by shrinking $\mathcal{O}_N(q_0)$ and S, if necessary, it may be supposed that D is of finite order on $\mathcal{O}_M(q_0)$. We shall now define what is meant by the transversal part D_T of D.

Let $p \in \mathcal{O}_M(q_0)$; write, uniquely, $p = s\cdot q$ $(s \in S,\ q \in \mathcal{O}_N(q_0))$. Given $f \in C^\infty(\mathcal{O}_M(q_0))$, let f_p denote the restriction of f to $s\cdot\mathcal{O}_N(q_0)$ and extend f_p to a locally invariant C^∞ function F_p in a neighborhood of p. The mapping which assigns to f the function $p \mapsto DF_p(p)$ is an endomorphism of $C^\infty(\mathcal{O}_M(q_0))$ which decreases supports, hence is a differential operator on $\mathcal{O}_M(q_0)$ which will be called the *transversal part* of D and denoted by D_T.

Lemma 8.2.1.10 Retain the above notations – then $D - D_T \in \mathfrak{D}(\mathcal{O}_M(q_0))\bar{\mathfrak{g}}(\mathcal{O}_M(q_0))$ where $\mathfrak{D}(\mathcal{O}_M(q_0))$ is the algebra of all differential operators on $\mathcal{O}_M(q_0)$ and $\bar{\mathfrak{g}}(\mathcal{O}_M(q_0))$ is the set of restrictions of the elements in $\bar{\mathfrak{g}}$ to $\mathcal{O}_M(q_0)$.

[Here we have written $\bar{\mathfrak{g}}$ for $\{\bar{X}: X \in \mathfrak{g}\}$.]

Proof Let $\{x_1, \ldots, x_n\}$ be a coordinate system on $\mathcal{O}_N(q_0)$ and $\{x_{n+1}, \ldots, x_m\}$ a coordinate system on $S\cdot q_0$ – then the map

$$p \mapsto \{x_1(q), \ldots, x_n(q), x_{n+1}(s\cdot q_0), \ldots, x_m(s\cdot q_0)\}$$

defines a coordinate system on $\mathcal{O}_M(q_0)$ $(p = s\cdot q;\ s \in S,\ q \in \mathcal{O}_N(q_0))$. It is clear that $D - D_T$ annihilates all C^∞ functions on $\mathcal{O}_M(q_0)$ which are constant in the last $m - n$ variables $x_{n+1}, \ldots, x_m$. Write, in the usual notations (cf. A2.1),

$$D - D_T = \sum_I a_I(x_1, \ldots, x_m)\partial^{|I|}/\partial x_1^{i_1} \ldots \partial x_m^{i_m}.$$

Then we have $D - D_T = \sum_{n+1}^m D_i \circ \partial/\partial x_i$ $(D_i \in \mathfrak{D}(\mathcal{O}_M(q_0)))$. Because each of the $\partial/\partial x_i$ $(n+1 \leqslant i \leqslant m)$ can be expressed as a C^∞ linear combination of certain $\tilde{X}$'s $(X \in \mathfrak{g})$ on $\mathcal{O}_M(q_0)$, the contention of the lemma is now evident. □

Lemma 8.2.1.11 Let $D \in \mathfrak{D}(\mathcal{O}_M(q_0))$ – then

$$(D_T f)^\sim = \mathfrak{f}(D)\tilde{f} \qquad (\text{all } f \in C^\infty(\mathcal{O}_M(q_0))).$$

Proof Let $q \in \mathcal{O}_N(q_0)$ – then it is clear that

$$(D_T f)^\sim(q) = DF_q(q) = (\mathfrak{f}(D)\tilde{f})(q). \quad \square$$

Proof of Theorem 8.2.1.8 (1) We shall first show that there exists a unique distribution $\tilde{T}$ on N such that

$$\tilde{T}(V_v) = T(f_v) \text{ for all } v \in C_c^\infty(G \times N).$$

(a) As in Proposition 8.2.1.5, let τ_T denote the distribution on $G \times N$ defined by $\tau_T(v) = T(f_v)$ $(v \in C_c^\infty(G \times N))$. Claim: τ_T is invariant under the action $(y, q) \mapsto x\cdot(y, q) = (xy, q)$ of G on $G \times N$. Thus, suppose we knew that $f_{v^x} = f_v^x$ (all $x \in G$) – then

$$\tau_T^x(v) = \tau_T(v^{x^{-1}}) = T(f_{v^{x^{-1}}}) = T(f_v^{x^{-1}}) = T(f_v) = \tau_T(v),$$

the second from the last equality being true in view of the G-invariance of T. To show that $f_{v^x} = f_v^x$, observe that for all $F \in C^\infty(G\cdot N)$, we have

$$\begin{aligned}
\int_M F(p) f_{v^x}(p)\,d\mu(p) &= \int_{G\times N} F\circ\pi(y:q)v^x(y:q)\,d_G(y)\,d\nu(q) \\
&= \int_{G\times N} (F\circ\pi)^{x^{-1}}(y:q)v(y:q)\,d_G(y)\,d\nu(q) \\
&= \int_M F^{x^{-1}}(p) f_v(p)\,d\mu(p) \\
&= \int_M F(p) f_v^x(p)\,d\mu(p) \quad (\mu \text{ is } G\text{-invariant}) \\
&\Rightarrow f_{v^x} = f_v^x.
\end{aligned}$$

(b) Fix a $u \in C_c^\infty(N)$ and consider the map S_u given by the rule $w \mapsto \tau_T(w \times u)$ $(w \in C_c^\infty(G))$ – then S_u is a distribution on G. Owing to the invariance of τ_T, S_u is left invariant and so (Proposition 5.2.1.2) there exists a constant c_u with the property that

$$S_u(w) = c_u \int_G w(x)\,d_G(x) \qquad (\text{all } w \in C_c^\infty(G)).$$

Fix $w_0 \in C_c^\infty(G)$ such that $\int_G w_0(x)\,d_G(x) = 1$. The mapping $u \mapsto \tau_T(w_0 \times u) = c_u$ is a distribution on N, call it $\tilde{T}$; plainly

$$\tau_T(w \times u) = \tilde{T}(u) \int_G w(x)\,d_G(x).$$

Let us check that $\tilde{T}(V_v) = T(f_v)$ $(v \in C_c^\infty(G \times N))$. To this end, note that the map σ_T, $v \mapsto \tilde{T}(V_v)$, is a distribution on $C_c^\infty(G \times N)$ (the map $v \mapsto V_v$ being continuous); but $\tau_T(w \times u) = \tilde{T}(V_{w\times u}) = \sigma_T(w \times u)$, whence $\tau_T = \sigma_T$ (the set $\{w \times u : w \in C_c^\infty(G), u \in C_c^\infty(N)\}$ is total in $C_c^\infty(G \times N)$) – therefore $\tilde{T}(V_v) = \tau_T(v) = T(f_v)$ (all $v \in C_c^\infty(G \times N)$), as desired. The uniqueness of $\tilde{T}$ is evident. Finally note that $\tilde{T} = 0 \Rightarrow T = 0$ (the map $v \mapsto f_v$ is surjective). The proof of the first assertion of Theorem 8.2.1.8 is now complete.

(2) The second assertion of the theorem is a strictly local affair, that is, it is enough to prove that for each $q_0 \in N$ there exists a neighborhood $\mathcal{O}_N^*(q_0)$ of q_0 in N such that $(D\cdot T)^{\sim}(u) = (\text{♀}(D)\cdot\tilde{T})(u)$ for all $u \in C_c^\infty(\mathcal{O}_N^*(q_0))$. So choose $\mathcal{O}_N(q_0)$, S as before the statement of Lemma 8.2.1.10; pick neighborhoods $\mathcal{O}_N^*(q_0)$ of q_0 in $\mathcal{O}_N(q_0)$ and G^* of 1 in G such that $G^*\cdot\mathcal{O}_N^*(q_0) \subset \mathcal{O}_M(q_0)$. Select $w_0 \in C_c^\infty(G^*)$ with $\int_G w_0 = 1$; fix $u \in C_c^\infty(\mathcal{O}_N^*(q_0))$ and put $v = w_0 \times u$ (thus $V_v = u$) – then if we show that $D\cdot T(f_v) = \text{♀}(D)\cdot\tilde{T}(V_v)$, the assertion will follow, u being arbitrary.

(a) Let us consider D as an element of $\mathfrak{D}(\mathcal{O}_M(q_0))$ – then D_T lies in $\mathfrak{D}(\mathcal{O}_M(q_0))$, and, according to Lemma 8.2.1.10, we can write $D - D_T = \sum_i D_i \circ \bar{X}_i$ ($D_i \in \mathfrak{D}(\mathcal{O}_M(q_0))$), whence, by iteration, we find that

$$D = D_T + \Sigma\, D_{i_1\ldots i_r}\circ\bar{X}_{i_1}\circ\cdots\circ\bar{X}_{i_r},$$

the $D_{i_1\ldots i_r}$ being transversal (i.e. coincide with their transversal part).

(b) Given $F \in C_c^\infty(G\cdot N)$, define a function $(x, p) \mapsto F(x:p)$ on $G \times G\cdot N$ by the rule $F(x:p) = F(x\cdot p)$ ($x \in G$, $p \in G\cdot N$) (cf. Lemma 8.2.1.6) – then $F(x:p;\,D) = DF(x\cdot p)$ by the invariance of D.

(c) At this point a computation will be made, the significance of which will become apparent shortly. Thus, for any $F \in C_c^\infty(G\cdot N)$, we have

$$\begin{aligned}
\int_M F(p)D^\sharp f_v(p)d\mu(p) &= \int_M DF(p)f_v(p)d\mu(p)\\
&= \int_{G\times N} DF(x\cdot q)v(x:q)d_G(x)d\nu(q)\\
&= \int_{G\times N} F(x:q;D)v(x:q)d_G(x)d\nu(q)\\
&= \int_{G\times N} F(x:q;D_T)v(x:q)d_G(x)\,d\nu(q)\\
&\quad+\Sigma\int_{G\times N} F(x:q;D_{i_1\ldots i_r}\circ\bar{X}_{i_1}\circ\ldots\circ\bar{X}_{i_r})\\
&\quad\times v(x:q)d_G(x)d\nu(q)\\
&= \int_{G\times N} F(x:q;\text{♀}(D))v(x:q)d_G(x)d\nu(q)\\
&\quad+\Sigma(-1)^r\int_{G\times N} F(x;X_{i_r}\ldots X_{i_1}:q;\text{♀}(D_{i_1\ldots i_r}))\\
&\quad\times v(x:q)d_G(x)d\nu(q) \text{ (Lemmas 8.2.1.11, 8.2.1.6)}
\end{aligned}$$

$$
\begin{aligned}
&= \int_{G\times N} F(x\cdot q)v(x:q;\mathfrak{r}(D)^{\sharp})d_G(x)d\nu(q) \\
&\quad +\Sigma(-1)^r \int_{G\times N} F(x\cdot q) \\
&\quad \times v(x;X_{i_1}^{\sharp}\ldots X_{i_r}^{\sharp}:q;\mathfrak{r}(D_{i_1\ldots i_r})^{\sharp})d_G(x)d\nu(q) \\
&= \int_{G\times N} F(x\cdot q)v_0(x:q)d_G(x)d\nu(q) \\
&\quad +\Sigma \int_{G\times N} F(x\cdot q)v_{i_1\ldots i_r}(x:q)d_G(x)d\nu(q) \\
&\qquad\qquad\qquad (G \text{ is unimodular}) \\
&= \int_M F(p)f_{v_0}(p)d\mu(p) \\
&\quad +\Sigma \int_M F(p)f_{v_{i_1\ldots i_r}}(p)d\mu(p),
\end{aligned}
$$

where we have set

$$
\begin{aligned}
v_0(x:q) &= v(x:q;\mathfrak{r}(D)^{\sharp}), \\
v_{i_1\ldots i_r}(x:q) &= v(x;X_{i_1}\ldots X_{i_r}:q;\mathfrak{r}(D_{i_1\ldots i_r})^{\sharp}),
\end{aligned}
$$

$(x \in G^*, q \in \mathcal{O}_N^*(q_0))$. Since F is arbitrary, we conclude that

$$D^{\sharp}f_v = f_{v_0} + \Sigma f_{v_{i_1\ldots i_r}}.$$

(d) It is now easy to show that $D\cdot T(f_v) = \mathfrak{r}(D)\cdot\tilde{T}(V_v)$. For we have

$$
\begin{aligned}
D\cdot T(f_v) = T(D^{\sharp}f_v) &= T(f_{v_0}) + \Sigma\, T(f_{v_{i_1\ldots i_r}}) \\
&= \tilde{T}(V_{v_0}) + \Sigma\, \tilde{T}(V_{v_{i_1\ldots i_r}}) = \mathfrak{r}(D)\cdot\tilde{T}(V_v),
\end{aligned}
$$

it being obvious that $V_{v_{i_1\ldots i_r}} = 0$.

This completes the proof of the theorem. □

8.2.2 Radial Components of Polynomial Differential Operators on a Reductive Lie Algebra

Let $\mathfrak{g}$ be a reductive Lie algebra over **R**, $\mathfrak{j}$ a Cartan subalgebra of $\mathfrak{g}$ – then Int $(\mathfrak{g})$ operates on $\mathfrak{g}$ in the usual way, viz. $x\cdot X = xX$ $(x \in \text{Int}(\mathfrak{g})$, $X \in \mathfrak{g})$. The set up in 8.2.1 may be realized by taking $M = \mathfrak{g}$, $N = \mathfrak{j}'$, $G = \text{Int}(\mathfrak{g})$ (plainly the transversality condition is satisfied here). In particular, to each differential operator D on $\mathfrak{g}$, there is assigned its radial component $\mathfrak{r}(D)$ on $\mathfrak{j}'$. Now, for arbitrary $D \in \mathfrak{D}(\mathfrak{g})$, the general theory does not tell us what $\mathfrak{r}(D)$ 'looks like'; however, if D is suitably specialized, then the exact form of $\mathfrak{r}(D)$ can be determined – this is the case, for instance, if D is an invariant (relative to the action of Int $(\mathfrak{g})$) polynomial differential operator with constant coefficients (Theorem 8.2.2.4). Results of this nature play a basic role in what follows.

Let E be a finite dimensional vector space over $\mathbf{R}$, D a differential operator on E. Fix a point $X \in E$ – then it is clear that there exists a unique element $p \in S(E_c)$ such that $f(X; D) = f(X; \partial(p))$ for all $f \in C^\infty(E)$ (cf. Vol. I, number 2.1.4). We shall call $\partial(p)$ the *local expression* of D at X and denote it by D_X.

The usual notations will be employed – thus Φ will denote the set of roots of the pair $(\mathfrak{g}_c, \mathfrak{j}_c)$, Φ^+ the positive roots with respect to some ordering of Φ. By B we shall understand the non-degenerate symmetric Ad-invariant bilinear form on $\mathfrak{g}$ obtained by taking the product of the Killing form on the derived algebra of $\mathfrak{g}$ with a non-degenerate symmetric bilinear form on the center of $\mathfrak{g}$; extend B to $\mathfrak{g}_c$ (by linearity over $\mathbf{C}$) and identify $\mathfrak{g}_c$ with its dual under B; choose non-zero elements $X_\alpha \in \mathfrak{g}_c^\alpha$, $X_{-\alpha} \in \mathfrak{g}_c^{-\alpha}$ $(\alpha \in \Phi^+)$ such that $B(X_\alpha, X_{-\alpha}) = 1$ – then $H_\alpha = [X_\alpha, X_{-\alpha}]$ is that element of $\mathfrak{j}_c$ which corresponds to α under the identification of $\mathfrak{j}_c$ with its dual (via the restriction of B to $\mathfrak{j}_c$). With these agreements, $S(\mathfrak{g}_c)$ is to be identified with $P(\mathfrak{g}_c)$, $S(\mathfrak{j}_c)$ with $P(\mathfrak{j}_c)$ – hence, in particular, the elements of the algebra $\mathfrak{P}(\mathfrak{g})$ of polynomial differential operators on $\mathfrak{g}$ are now finite linear combinations of terms of the form $p\partial(q)$ $(p, q \in S(\mathfrak{g}_c))$. For any $X \in \mathfrak{g}_c$, let L_X denote the endomorphism of $S(\mathfrak{g}_c)$ corresponding to multiplication by X in $S(\mathfrak{g}_c)$ and, as usual, let $d(X)$ denote the unique derivation of $S(\mathfrak{g}_c)$ which coincides on $\mathfrak{g}_c$ with ad (X). We shall write, in this number, G for any connected Lie group with Lie algebra $\mathfrak{g}$ (G operating on $\mathfrak{g}$ via the adjoint action); the elements of the universal enveloping algebra $\mathfrak{G}$ of $\mathfrak{g}_c$ are then to be construed as left invariant differential operators on G.

Given an $f \in C^\infty(\mathfrak{g})$, we may define a function $(x, X) \mapsto f(x : X)$ of two variables on $G \times \mathfrak{g}$ via the prescription $f(x : X) = f(x \cdot X)$ $(x \in G, X \in \mathfrak{g})$. Using the convention of writing introduced after Theorem 2.3.1.2, let us begin with:

Proposition 8.2.2.1 Let $f \in C^\infty(\mathfrak{g})$; let $X_1, \ldots, X_r \in \mathfrak{g}_c$ and $p \in S(\mathfrak{g}_c)$ – then

$$f(x\,; X_1 \cdot \ldots \cdot X_r : X\,; \partial(p)) = f(x : X\,; \partial(q)) \qquad (x \in G, X \in \mathfrak{g})$$

where

$$q = (L_{[X_1, X]} + d(X_1)) \ldots (L_{[X_r, X]} + d(X_r))p.$$

Proof We shall use induction on r.

(1) The case $r = 1$. Plainly $f(x : X; \partial(p)) = f(xX; \partial(xp))$ so that for $X_1 \in \mathfrak{g}$,

$$f(x\,; X_1 : X\,; \partial(p)) = \frac{d}{dt} f(x_t^1 X\,; \partial(x_t^1 p))\big|_{t=0} \qquad (t \in \mathbf{R})$$

where $x_t^1 = x \exp(tX_1)$ $(t \in \mathbf{R})$. Now

$$\frac{d}{dt} f(x_t^1 X; \partial(x_t^1 p))\,|_{t=0} = \frac{d}{dt} f(x_t^1 X; \partial(xp))\,|_{t=0} + \frac{d}{dt} f(xX; \partial(x_t^1 p))\,|_{t=0}$$

and since

$$f(x_t^1 X; \partial(xp)) = f(x : \exp(tX_1)X; \partial(p)),$$
$$f(xX; \partial(x_t^1 p)) = f(x : X; \partial(\exp(tX_1)p)),$$

we get

$$f(x; X_1 : X; \partial(p)) = f(x : X; \partial((L_{[X_1, X]} + d(X_1))p)),$$

as desired.

(2) The case $r > 1$. Set

$$q_1 = (L_{[X_2, X]} + d(X_2)) \ldots (L_{[X_r, X]} + d(X_r))p;$$

owing to the induction hypothesis, then, we have

$$f(x; X_2 \cdot \ldots \cdot X_r : X; \partial(p)) = f(x : X; \partial(q_1)).$$

Consequently, for $X_1 \in \mathfrak{g}$,

$$\begin{aligned} &f(x; X_1 \cdot \ldots \cdot X_r : X; \partial(p)) \\ &= \frac{d}{dt} f(x \exp(tX_1); X_2 \cdot \ldots \cdot X_r : X; \partial(p))\,|_{t=0} \\ &= \frac{d}{dt} f(x \exp(tX_1) : X; \partial(q_1))\,|_{t=0} \\ &= f(x; X_1 : X; \partial(q_1)) = f(x : X; \partial(q)) \qquad \text{(by (1))}. \end{aligned}$$

Hence the proposition. □

Fix $X_0 \in \mathfrak{g}$; put

$$U_{X_0}(X) = L_{[X, X_0]} + d(X) \qquad (X \in \mathfrak{g}_c)$$

– then the mapping $X \mapsto U_{X_0}(X)$ is a representation of $\mathfrak{g}_c$ on $S(\mathfrak{g}_c)$ and, as always, the extension of U_{X_0} to $\mathfrak{G}$ will be denoted by the same symbol. Given X in $\mathfrak{g}$, define a linear map $\boldsymbol{\Gamma}_X$ of $\mathfrak{G} \otimes_{\mathbf{C}} S(\mathfrak{g}_c)$ into $S(\mathfrak{g}_c)$ by the rule

$$\boldsymbol{\Gamma}_X(D \otimes p) = U_X(D)p \qquad (D \in \mathfrak{G},\ p \in S(\mathfrak{g}_c)).$$

In these notations, then, Proposition 8.2.2.1 says that for all $f \in C^\infty(\mathfrak{g})$,

$$\begin{aligned} f(x; D : X; \partial(p)) &= f(x : X; \partial(U_X(D)p)) \\ &= f(x : X; \partial(\boldsymbol{\Gamma}_X(D \otimes p))) \qquad (x \in G,\ X \in \mathfrak{g}). \end{aligned}$$

The mapping $\boldsymbol{\Gamma}_X (X \in \mathfrak{g})$ possesses two important properties which we shall now proceed to explicate. Let $\lambda : S(\mathfrak{g}_c) \to \mathfrak{G}$ denote the canonical space bijection; put

$$\mathfrak{s}_c = \sum_{\alpha > 0} (\mathfrak{g}_c^\alpha + \mathfrak{g}_c^{-\alpha}),\ \mathfrak{S} = \lambda(S(\mathfrak{s}_c)),\ \mathfrak{S}^m = \lambda(S^m(\mathfrak{s}_c)),\ \mathfrak{S}^+ = \sum_{m \geqslant 1} \mathfrak{S}^m.$$

Lemma 8.2.2.2 Fix an element H in $\mathfrak{j}'$ – then $\mathbf{\Gamma}_H$ defines a bijective mapping between $\mathfrak{S} \otimes_{\mathbf{C}} S(\mathfrak{j}_c)$ and $S(\mathfrak{g}_c)$.

Proof We shall prove by induction on m that

$$\sum_{e_1+e_2\leqslant m} \mathbf{\Gamma}_H(\mathfrak{S}^{e_1} \otimes_{\mathbf{C}} S^{e_2}(\mathfrak{j}_c)) = \sum_{e\leqslant m} S^e(\mathfrak{g}_c).$$

Since the spaces

$$\sum_{e_1+e_2\leqslant m} \mathfrak{S}^{e_1} \otimes_{\mathbf{C}} S^{e_2}(\mathfrak{j}_c), \quad \sum_{e\leqslant m} S^e(\mathfrak{g}_c)$$

have the same dimension, this will suffice to establish our lemma. The assertion being clear for $m = 0$, assume that $m \geqslant 1$; since the left hand side is contained in the right hand side, in view of the induction hypothesis it is enough to prove that

$$\sum_{e_1+e_2\leqslant m} \mathbf{\Gamma}_H(\mathfrak{S}^{e_1} \otimes_{\mathbf{C}} S^{e_2}(\mathfrak{j}_c)) + \sum_{e\leqslant m-1} S^e(\mathfrak{g}_c) \supset S^m(\mathfrak{g}_c).$$

The space $S^m(\mathfrak{g}_c)$ is spanned by elements of the form $X_1 \ldots X_{e_1}H_1 \ldots H_{e_2}$ $(X_i \in \mathfrak{s}_c, H_j \in \mathfrak{j}_c)$ where $e_1 + e_2 = m$; because H is regular, we can choose $\tilde{X}_i \in \mathfrak{s}_c$ such that $[H, \tilde{X}_i] = -X_i$ $(1 \leqslant i \leqslant e_1)$ – thus

$$\mathbf{\Gamma}_H(\tilde{X}_1 \ldots \cdot \tilde{X}_{e_1} \otimes H_1 \ldots H_{e_2}) \equiv X_1 \ldots X_{e_1}H_1 \ldots H_{e_2} \bmod \sum_{e\leqslant m-1} S^e(\mathfrak{g}_c)$$

and so, thanks to familiar properties of λ, we have

$$\mathbf{\Gamma}_H(\lambda(\tilde{X}_1 \ldots \tilde{X}_{e_1}) \otimes H_1 \ldots H_{e_2}) \equiv X_1 \ldots X_{e_1}H_1 \ldots H_{e_2} \bmod \sum_{e\leqslant m-1} S^e(\mathfrak{g}_c)$$

which serves to prove our assertion. □

Lemma 8.2.2.3 Fix an element p in $S(\mathfrak{g}_c)$ – then there exists an integer $d \geqslant 0$ and a polynomial mapping γ_p of $\mathfrak{j}$ into $\mathfrak{S} \otimes_{\mathbf{C}} S(\mathfrak{j}_c)$ such that $\mathbf{\Gamma}_H(\gamma_p(H)) = \pi(H)^d p$ (all $H \in \mathfrak{j}$).

[Here, as is customary, $\pi = \prod_{\alpha>0} \alpha$.]

Proof Fix an integer $m \geqslant 0$ and let

$$E_m = \sum_{e_1+e_2\leqslant m} \mathfrak{S}^{e_1} \otimes_{\mathbf{C}} S^{e_2}(\mathfrak{j}_c);$$

fix a bijective linear mapping C carrying $\sum_{e\leqslant m} S^e(\mathfrak{g}_c)$ onto E_m and put $A_H = C\mathbf{\Gamma}_H$ $(H \in \mathfrak{j})$ – then it is obvious that the assignment $H \mapsto A_H$ defines a polynomial mapping of $\mathfrak{j}$ into the space of endomorphisms of E_m, and, in view of what has been said above,

$$\delta_H = \det(A_H) = 0 \Rightarrow \pi(H) = 0 \qquad (H \in \mathfrak{j}),$$

whence δ_H divides π^d for some integer $d \geqslant 0$. Let t be an indeterminate – then

$$\det(t - A_H) = (-1)^M \delta_H + t \sum_{i=1}^{M} p_i(H) t^{i-1} \qquad (H \in \mathfrak{j}),$$

the p_i being polynomial functions on $\mathfrak{j}$ ($M = \dim(E_m)$). Agreeing to write B_H for

$$(-1)^{M+1}\sum_{i=1}^{M} p_i(H)A_H^{i-1} \qquad (H \in \mathfrak{j}),$$

we see without difficulty that $A_H B_H = B_H A_H = \delta_H I$ ($H \in \mathfrak{j}$), I the identity mapping in E_m. It can be supposed that $p \in \sum_{e \leqslant m} S^e(\mathfrak{g}_c)$ – then

$$\Gamma_H(B_H Cp) = C^{-1}A_H B_H Cp = \delta_H p \qquad (H \in \mathfrak{j}).$$

Now choose a polynomial function q on $\mathfrak{j}$ such that $\pi^d = \delta_H q$ (cf. supra); put $\gamma_p(H) = q(H)B_H Cp$ ($H \in \mathfrak{j}$) – then $\Gamma_H(\gamma_p(H)) = \pi(H)^d p$ (all $H \in \mathfrak{j}$), as we wished to prove. □

Given any element H in $\mathfrak{j}'$ and $p \in S(\mathfrak{g}_c)$, it follows that there exists a unique element $\wp_H(p) \in S(\mathfrak{j}_c)$ such that $p - \wp_H(p) \in \Gamma_H(\mathfrak{S}^+ \otimes_{\mathbf{C}} S(\mathfrak{j}_c))$; write $\wp_H(\partial(p)) = \partial(\wp_H(p))$ ($p \in S(\mathfrak{g}_c)$) and then extend $\wp_H$ to all of $\mathfrak{P}(\mathfrak{g})$ by setting $\wp_H(D)(= \wp(D)_H) = \wp_H(D_H)$ ($D \in \mathfrak{P}(\mathfrak{g})$) where D_H is the local expression of D at H. It is easy to see that the local expression at any point $H \in \mathfrak{j}'$ of the radial component $\wp(D)$ of D ($D \in \mathfrak{P}(\mathfrak{g})$) is in fact $\wp(D)_H$. [Regarding $\mathfrak{j}'$ as an open submanifold of $\mathfrak{j}$, Lemma 8.2.2.3 implies that for each polynomial differential operator D on $\mathfrak{g}$ there exists an analytic differential operator on $\mathfrak{j}'$ whose local expression at $H \in \mathfrak{j}'$ is $\wp(D)_H$. To verify that this differential operator is $\wp(D)$ ($D \in \mathfrak{P}(\mathfrak{g})$), fix a locally invariant C^∞ function f on $\mathfrak{g}$; given a $D \in \mathfrak{P}(\mathfrak{g})$, choose elements $S_i \in \mathfrak{S}^+$, $p_i \in S(\mathfrak{j}_c)$ and analytic functions a_i on $\mathfrak{j}'$ such that

$$f(H; D) = f(H; \wp(D)_H) + \sum_{i=1}^{r} a_i(H) f(1; S_i : H; \partial(p_i)) \qquad (H \in \mathfrak{j}').$$

Owing to the local invariance of f, $f(1; S_i : H; \partial(p_i)) = 0$. Therefore the local expression at H ($H \in \mathfrak{j}'$) of $\wp(D)$ ($D \in \mathfrak{P}(\mathfrak{g})$) is indeed $\wp(D)_H$.]

Let $\mathfrak{P}_I(\mathfrak{g})$ denote the subalgebra of $\mathfrak{P}(\mathfrak{g})$ consisting of those elements $D \in \mathfrak{P}(\mathfrak{g})$ which are invariant under G, $\mathfrak{Z}(\mathfrak{g})$ the subalgebra of $\mathfrak{P}_I(\mathfrak{g})$ generated by $I(\mathfrak{g}_c) \cup \partial(I(\mathfrak{g}_c))$ ($I(\mathfrak{g}_c)$ the G-invariants in $S(\mathfrak{g}_c)$). Our main task now is to obtain an *explicit formula* for $\wp(D)$, $D \in \mathfrak{Z}(\mathfrak{g})$. Obviously, for $p \in I(\mathfrak{g}_c)$, $\wp(p) = \bar{p}\,|\,\mathfrak{j}'$, $\bar{p}$ the restriction of p to $\mathfrak{j}_c$. On the other hand, it is a simple matter to show that the mapping $D \mapsto \wp(D)$ is a homomorphism of $\mathfrak{P}_I(\mathfrak{g})$ into the algebra $\mathfrak{D}(\mathfrak{j}')$ of differential operators on $\mathfrak{j}'$ – therefore it suffices to determine $\wp(D)$ for $D \in \partial(I(\mathfrak{g}_c))$.

The result in question can be formulated as follows.

Theorem 8.2.2.4 (Harish-Chandra) If $p \in I(\mathfrak{g}_c)$, then $\wp(\partial(p)) = \pi^{-1}\partial(\bar{p})\circ\pi$.

Let ω denote the polynomial function $X \mapsto B(X, X)$ ($X \in \mathfrak{g}_c$) (the '*Casimir polynomial*' of $\mathfrak{g}$). The first thing that will be done is to verify the assertion of Theorem 8.2.2.4 for ω (clearly $\omega \in I(\mathfrak{g}_c)$) – the general case will then follow by an algebraic device.

It is clear that $\omega = \tilde{\omega} + 2\sum_{\alpha>0} X_\alpha X_{-\alpha}$ ($\tilde{\omega}$ the restriction of ω to $\mathfrak{j}_c$). Claim:

$$\mathfrak{f}(\partial(\omega)) = \partial(\tilde{\omega}) + 2\sum_{\alpha>0} \alpha^{-1}\partial(H_\alpha).$$

Thus consider any locally invariant C^∞ function f on $\mathfrak{g}$ – then

$$f(H;\mathfrak{f}(\partial(\omega))) = f(H;\partial(\omega)) = f(H;\partial(\tilde{\omega})) + 2\sum_{\alpha>0} f(H;X_\alpha X_{-\alpha})$$

$$(\text{all } H \in \mathfrak{j}').$$

According to Proposition 8.2.2.1,

$$f(1;X_\alpha\cdot X_{-\alpha}:H) = f(1:H;\partial(\alpha(H)H_\alpha - \alpha(H)^2 X_\alpha X_{-\alpha})) \qquad (H \in \mathfrak{j}').$$

But $f(1;X_\alpha\cdot X_{-\alpha}:H) = f(H:\bar{X}_{-\alpha}\circ\bar{X}_\alpha)$ (Lemma 8.2.1.6) $= 0$, f being locally invariant. Our claim is now evident. On the other hand, let us compute $\pi^{-1}\partial(\tilde{\omega})\circ\pi$. Let $\alpha_1, \ldots, \alpha_r$ denote the distinct positive roots of the pair $(\mathfrak{g}_c, \mathfrak{j}_c)$ – then

$$\{\partial(\tilde{\omega}), \pi\} = \sum_{i=1}^{r} \alpha_1\alpha_2\ldots\alpha_{i-1}\{\partial(\tilde{\omega}), \alpha_i\}\circ(\alpha_{i+1}\ldots\alpha_r)$$

$$= 2\sum_{i=1}^{r} \alpha_1\alpha_2\ldots\alpha_{i-1}\partial(H_{\alpha_i})\circ(\alpha_{i+1}\ldots\alpha_r).$$

One has $\{\partial(H_\alpha), \beta\} = \beta(H_\alpha)$ for any two roots α and β and so

$$\partial(\tilde{\omega})\circ\pi = \pi\partial(\tilde{\omega}) + 2\pi\sum_{i=1}^{r} \alpha_i^{-1}\partial(H_{\alpha_i}) + q,$$

q a certain element in $S(\mathfrak{j}_c)$ which is in fact equal to zero. [To see this, apply both sides of the last equality to the constant function 1 and get $\partial(\tilde{\omega})\pi = q$; but $\partial(\tilde{\omega})\pi = 0$ (Corollary 2.1.5.12(i)), whence $q = 0$.] It is now obvious that $\mathfrak{f}(\partial(\omega))$ and $\pi^{-1}\partial(\tilde{\omega})\circ\pi$ are equal when viewed as differential operators on $\mathfrak{j}'$.

Remark The discussion above shows that the radial component $\mathfrak{f}(\partial(\omega))$ of $\partial(\omega)$ becomes singular when considered on all of $\mathfrak{j}$. But this is not unusual: Consider the radial component $\partial^2/\partial r^2 + (n-1)/r\,\partial/\partial r$ of the Laplacian on $\mathbf{R}^n$.

To prove Theorem 8.2.2.4 in complete generality, a couple of technical lemmas will be needed.

Lemma 8.2.2.5 Let $d_{\partial(\omega)}$ denote the derivation of $\mathfrak{P}(\mathfrak{g})$ given by

$$d_{\partial(\omega)}(D) = 2^{-1}\{\partial(\omega), D\} \qquad (D \in \mathfrak{P}(\mathfrak{g})).$$

Let p be a homogeneous element in $S(\mathfrak{g}_c)$ of degree m – then $d_{\partial(\omega)}^m(p) = m!\,\partial(p)$.

Proof It is clear that $d_{\partial(\omega)}(X) = \partial(X)$ ($X \in \mathfrak{g}_c$); proceeding by induction on m, let us show that

$$d_{\partial(\omega)}^m(X_1X_2\ldots X_m) = m!\,\partial(X_1X_2\ldots X_m)$$

for $X_1, \ldots, X_m \in \mathfrak{g}_c$ – this will suffice to establish the lemma. Since $d^2_{\partial(\omega)}(X) = 0$ ($X \in \mathfrak{g}_c$), Leibnitz' rule for derivations implies that

$$\begin{aligned} d^m_{\partial(\omega)}(X_1X_2 \ldots X_m) &= (d^m_{\partial(\omega)}(X_1 \ldots X_{m-1}))\circ X_m \\ &\quad + md^{m-1}_{\partial(\omega)}(X_1 \ldots X_{m-1})\circ d_{\partial(\omega)}(X_m) \\ &= m!\,\partial(X_1 \ldots X_{m-1})\circ d_{\partial(\omega)}(X_m) \\ &= m!\,\partial(X_1 \ldots X_{m-1}X_m). \quad \square \end{aligned}$$

Lemma 8.2.2.6 Let $d_{\mathfrak{l}(\partial(\omega))}$ denote the derivation of $\mathfrak{D}(\mathfrak{j}')$ given by

$$d_{\mathfrak{l}(\partial(\omega))}(D) = 2^{-1}\{\mathfrak{l}(\partial(\omega)), D\} \qquad (D \in \mathfrak{D}(\mathfrak{j}')).$$

Let p be a homogeneous element in $I(\mathfrak{g}_c)$ of degree m – then $d^m_{\mathfrak{l}(\partial(\omega))}(\tilde{p}) = m!\,\mathfrak{l}(\partial(p))$.

Proof Since the map $D \mapsto \mathfrak{l}(D)$ is a homomorphism of $\mathfrak{P}_I(\mathfrak{g})$ into $\mathfrak{D}(\mathfrak{j}')$,

$$d_{\mathfrak{l}(\partial(\omega))}\circ\mathfrak{l}(D) = \mathfrak{l}\circ d_{\partial(\omega)}(D)$$

for $D \in \mathfrak{P}_I(\mathfrak{g})$ – hence

$$m!\,\mathfrak{l}(\partial(p)) = \mathfrak{l}(d^m_{\partial(\omega)}(p)) = d^m_{\mathfrak{l}(\partial(\omega))}(\mathfrak{l}(p)) = d^m_{\mathfrak{l}(\partial(\omega))}(\tilde{p}). \quad \square$$

Proof of Theorem 8.2.2.4 It may be supposed that $p \in I(\mathfrak{g}_c)$ is homogeneous of degree m.

(1) Let A be an associative algebra over **C** with a unit 1. For any $x \in A$ let d_x denote the derivation of A given by $d_x(y) = 2^{-1}(xy - yx)$ $(y \in A)$. By induction on k ($k \geqslant 0$) one finds that

$$d^k_x(y) = 2^{-k}\sum_{i=0}^{k}\binom{k}{i}(-1)^i x^{k-i}yx^i.$$

In particular, suppose that z is an element of A with an inverse z^{-1}; set $x_z = z^{-1}xz$ – then, if y is an element of A commuting with z, we have

$$d^k_{x_z}(y) = z^{-1}d^k_x(y)z.$$

(2) Specialize the set up in (1) by taking $A = \mathfrak{D}(\mathfrak{j}')$, $x = \partial(\tilde{\omega})$, $y = \tilde{p}$, $z = \pi$ – then $x_z = z^{-1}xz = \pi^{-1}\partial(\tilde{\omega})\circ\pi = \mathfrak{l}(\partial(\omega))$, as was noted earlier. Hence, using Lemmas 8.2.2.5, 8.2.2.6, we get

$$m!\,\mathfrak{l}(\partial(p)) = d^m_{\mathfrak{l}(\partial(\omega))}(\tilde{p}) = \pi^{-1}d^m_{\partial(\tilde{\omega})}(\tilde{p})\circ\pi = m!\,\pi^{-1}\partial(\tilde{p})\circ\pi.$$

The contention of the theorem is now clear. $\square$

Let W denote the Weyl group of the pair $(\mathfrak{g}_c, \mathfrak{j}_c)$. Evidently W operates on $\mathfrak{P}(\mathfrak{j})$; let $\mathfrak{P}_I(\mathfrak{j})$ denote the corresponding set of W-invariants – then there exists a unique homomorphism $\tilde{\mathfrak{l}}$ of $\mathfrak{Z}(\mathfrak{g})$ into $\mathfrak{P}_I(\mathfrak{j})$ such that $\tilde{\mathfrak{l}}(p) = \tilde{p}$ and $\tilde{\mathfrak{l}}(\partial(p)) = \partial(\tilde{p})$ $(p \in I(\mathfrak{g}_c))$. [The uniqueness of $\tilde{\mathfrak{l}}$ is obvious, $\mathfrak{Z}(\mathfrak{g})$ being generated by $I(\mathfrak{g}_c) \cup \partial(I(\mathfrak{g}_c))$; as for the existence, simply set $\tilde{\mathfrak{l}}(D) = \pi\mathfrak{l}(D)\circ\pi^{-1}$ $(D \in \mathfrak{Z}(\mathfrak{g}))$.]

For certain purposes it will be necessary to have at hand a generalization of Theorem 8.2.2.4 and supporting discussion. Because the transversality condition need not be satisfied in this more general setting, the results in question are not really special cases of those obtained in 8.2.1 – on the other hand the proofs, for the most part, are simply variants of those already presented and so we shall not stop to give them (cf. Harish-Chandra [24]).

Let $\mathfrak{g}$ be a reductive Lie algebra over $\mathbf{R}$, G a connected Lie group with Lie algebra $\mathfrak{g}$ – then G operates on $\mathfrak{g}$ in the usual way. Let $\mathfrak{h}$ be a subalgebra of $\mathfrak{g}$ which is reductive in $\mathfrak{g}$ and of the same rank as $\mathfrak{g}$; let H be the analytic subgroup of G corresponding to $\mathfrak{h}$. Put $\delta_{\mathfrak{h}}(X) = \det(\mathrm{ad}(X)_{\mathfrak{g}/\mathfrak{h}})$ $(X \in \mathfrak{h})$ and let $'\mathfrak{h}$ be the subset of $\mathfrak{h}$ comprised of those X where $\delta_{\mathfrak{h}}(X) \neq 0$; let ϕ denote the mapping $(x, X) \mapsto xX$ of $G \times \mathfrak{h}$ into $\mathfrak{g}$ – then ϕ is everywhere of rank n on $G \times {'\mathfrak{h}}$ ($n = \dim(\mathfrak{g})$). [The proof of this fact offers no difficulty (cf. Lemma 1.4.3.1); the point, of course, is that Theorem 8.2.1.1 is now at our disposal. Needless to say, when $'\mathfrak{h}$ is empty, one must make the obvious convention. . . .]

Let $\Omega_{\mathfrak{h}}$ be an open subset of $'\mathfrak{h}$ which is completely invariant in $\mathfrak{h}$. We have:

(1) Given $v \in C_c^\infty(G \times \Omega_{\mathfrak{h}})$, there exists a unique function $f_v \in C_c^\infty(G \cdot \Omega_{\mathfrak{h}})$ such that

$$\int_{G \times \Omega_{\mathfrak{h}}} F(xX) v(x : X) d_G(x) d_{\mathfrak{h}}(X) = \int_{\Omega_{\mathfrak{g}}} F(X) f_v(X) d_{\mathfrak{g}}(X)$$

for all $F \in C_c^\infty(\Omega_{\mathfrak{g}})$ $(\Omega_{\mathfrak{g}} = G \cdot \Omega_{\mathfrak{h}})$.

(2) Let $D \in \mathfrak{P}(\mathfrak{g})$ – then there exists a differential operator $\natural_{\mathfrak{h}}(D)$ on $'\mathfrak{h}$ such that

$$f(X; D) = f(X; \natural_{\mathfrak{h}}(D)) \qquad (X \in {'\mathfrak{h}})$$

for each locally invariant C^∞ function f on $\Omega_{\mathfrak{g}}$.

(3) Let T be a G-central distribution on $\Omega_{\mathfrak{g}}$ (cf. 8.3.1) – then there exists a unique H-central distribution $\tilde{T}$ on $\Omega_{\mathfrak{h}}$ such that $T(f_v) = \tilde{T}(V_v)$ (all $v \in C_c^\infty(G \times \Omega_{\mathfrak{h}})$) where

$$V_v(X) = \int_G v(x : X) d_G(x) \qquad (X \in \Omega_{\mathfrak{h}}).$$

In addition, if $D \in \mathfrak{P}_I(\mathfrak{g})$, then

$$(D \cdot T)^\sim = \natural_{\mathfrak{h}}(D) \cdot \tilde{T}.$$

[Actually Helgason has proved a theorem (unpublished) which contains the above results as a corollary.]

In the present more general set up Theorem 8.2.2.4 reads as follows.

Theorem 8.2.2.4 (bis) (Harish-Chandra) Retain the above notations; let T be a G-central distribution on $\Omega_{\mathfrak{g}}$ – then

$$\natural_{\mathfrak{h}}(\partial(p)) \cdot \tilde{T} = (\delta_{\mathfrak{h}}^{-1/2} \partial(\tilde{p})) \cdot (\delta_{\mathfrak{h}}^{1/2} \tilde{T}) \qquad (\tilde{p} = p \mid \mathfrak{h}_c; p \in I(\mathfrak{g}_c)).$$

The reader will find the proof of this theorem in Harish-Chandra [24, p. 556]. [As in the proof of Theorem 8.2.2.4 itself, one first deals with the case when p is the Casimir polynomial of $\mathfrak{g}$ by direct computation; the general case is then reduced to this one via the previously employed algebraic device.]

We shall conclude this number with a few remarks on the nature of the solutions to certain systems of differential equations; for the present discussion, it can be safely assumed that $\mathfrak{g}$ is semi-simple.

Let $I(\mathfrak{j}_c)$ denote the set of invariants in $S(\mathfrak{j}_c)$ for the Weyl group W of the pair $(\mathfrak{g}_c, \mathfrak{j}_c)$.

Proposition 8.2.2.7 Let T be a distribution on an open subset $\mathcal{O}$ of $\mathfrak{j}$; suppose that T is an eigendistribution of $I(\mathfrak{j}_c)$, i.e. suppose that for each $p \in I(\mathfrak{j}_c)$ there exists a complex number c_p such that $\partial(p)\cdot T = c_p T$ – then T coincides, in the sense of distributions, with an analytic function on $\mathcal{O}$.

Proof Let $H_1, \ldots, H_l$ be a basis for $\mathfrak{j}$ over $\mathbf{R}$ and set $\Delta = \sum_1^l H_i^2 \in S(\mathfrak{j}_c)$; let t be an indeterminate and consider the polynomial

$$\prod_{w \in W} (t - \Delta^w) = t^{[W]} + p_1 t^{[W]-1} + \cdots + p_{[W]}$$

with coefficients $p_i \in I(\mathfrak{j}_c)$ ($[W]$ the order of W) – then

$$0 = \prod_{w \in W} (\Delta - \Delta^w) = \Delta^{[W]} + p_1 \Delta^{[W]-1} + \cdots + p_{[W]}$$

$\Rightarrow$

$$\{\partial(\Delta)^{[W]} + c_{p_1}\partial(\Delta)^{[W]-1} + \cdots + c_{p_{[W]}}\}\cdot T = 0.$$

Hence T is a distribution solution to an elliptic differential equation, therefore, by a well-known Regularity Theorem, T coincides with an analytic function on $\mathcal{O}$. □

The symmetric algebra $S(\mathfrak{j}_c)$ is a free abelian module of rank $[W]$ over $I(\mathfrak{j}_c)$; accordingly we may choose homogeneous elements $q_1 = 1, q_2, \ldots, q_{[W]}$ such that $S(\mathfrak{j}_c) = \sum_i I(\mathfrak{j}_c) q_i$ (cf. Corollary 2.1.3.7).

Proposition 8.2.2.8 Let ϕ be an analytic function on a non-empty open connected subset $\mathcal{O}$ of $\mathfrak{j}$ which is, moreover, an eigenfunction of $I(\mathfrak{j}_c)$. If the derivatives $\partial(q_i)\phi$ $(1 \leqslant i \leqslant [W])$ all vanish simultaneously at some point $H_0 \in \mathcal{O}$, then ϕ must be identically zero.

Proof Given $p \in S(\mathfrak{j}_c)$, there exist elements $p_1, \ldots, p_{[W]} \in I(\mathfrak{j}_c)$ such that $p = \sum_i p_i q_i$; by assumption there exists $c_i \in \mathbf{C}$ such that $\partial(p_i)\phi = c_i \phi$ $(i = 1, \ldots, [W])$ – therefore

$$\phi(H_0; \partial(p)) = \sum_i c_i \phi(H_0; \partial(q_i)) = 0.$$

This shows that all the derivatives of ϕ vanish at H_0 and so, in view of the connectedness of $\mathcal{O}$, ϕ must be identically zero. □

Corollary 8.2.2.9 Suppose that $\tilde{H}$ is a point in $\mathfrak{j}_c$ such that $\pi(\tilde{H}) \neq 0$ and that ϕ is an analytic function on a non-empty open connected subset $\mathcal{O}$ of $\mathfrak{j}$ satisfying the system of differential equations

$$(*) \quad \partial(p)\phi = p(\tilde{H})\phi \qquad (p \in I(\mathfrak{j}_c)).$$

Then there exist unique constants $c_w (w \in W)$ such that

$$\phi(H) = \sum_{w \in W} c_w \exp(B(H, w\tilde{H})) \qquad (H \in \mathcal{O}).$$

Proof Since $\tilde{H}$ is regular, the $[W]$ points $w\tilde{H}$ $(w \in W)$ are all distinct. Put $\phi_w(H) = \exp(B(H, w\tilde{H}))$ $(H \in \mathcal{O})$ – then the ϕ_w $(w \in W)$ are linearly independent analytic functions on $\mathcal{O}$ and since $\partial(H)\phi_w = B(H, w\tilde{H})\phi_w$, it is clear that $\partial(p)\phi_w = p(\tilde{H})\phi_w$ for all $p \in I(\mathfrak{j}_c)$. Let E denote the vector space over **C** consisting of all analytic solutions to the system $(*)$ – then our corollary will follow when it is shown that $\dim(E) \leqslant [W]$. Choose a point $H_0 \in \mathcal{O}$; if $\dim(E) > [W]$, then there exists a non-zero $\phi \in E$ satisfying the $[W]$ linear conditions $\phi(H_0; \partial(q_i)) = 0$ $(1 \leqslant i \leqslant [W])$ – but according to Proposition 8.2.2.8 this is impossible. □

Proposition 8.2.2.10 Let ϕ be an analytic function on a non-empty open connected subset $\mathcal{O}$ of $\mathfrak{j}$. If ϕ is annihilated by all homogeneous elements in $I(\mathfrak{j}_c)$ of positive degree, then there exists a polynomial function $p \in S(\mathfrak{j}_c)$ such that $\phi = p$ on $\mathcal{O}$.

Proof For any $H \in \mathfrak{j}_c$, consider the polynomial

$$\prod_{w \in W} (t - wH) = t^{[W]} + p_1 t^{[W]-1} + \cdots + p_{[W]}$$

in the indeterminate t with coefficients $p_i \in I(\mathfrak{j}_c)$ – then

$$H^{[W]} + H^{[W]-1}p_1 + \cdots + p_{[W]} = 0$$

and so

$$\partial(H^{[W]} + H^{[W]-1}p_1 + \cdots + p_{[W]})\phi = 0.$$

Because the p_i are homogeneous of positive degree, our hypothesis on ϕ tells us that $\partial(p_i)\phi = 0$ $(1 \leqslant i \leqslant [W])$ – hence $\partial(H^{[W]})\phi = 0$. Consequently, if $H_1, \ldots, H_l$ is a basis for $\mathfrak{j}$ over **R**, then

$$\partial(H_i^{[W]})\phi = 0 \qquad (1 \leqslant i \leqslant l)$$

and this implies that ϕ is a polynomial of degree $\leqslant l[W]$. □

Corollary 8.2.2.11 Let $\mathcal{O}$ and $\tilde{\mathcal{O}}$ be non-empty open connected subsets of $\mathfrak{j}'$; suppose that

$$\sigma(H : \tilde{H}) = \sum_{w \in W} c_w(H) \exp(\sqrt{-1}B(\tilde{H}, wH)) \qquad (H \in \mathcal{O}, \tilde{H} \in \tilde{\mathcal{O}})$$

where the c_w $(w \in W)$ are C^∞ functions on $\mathcal{O}$. If

$$\sigma(H;\partial(\tilde{p}):\tilde{H}) = p(\sqrt{-1}\tilde{H})\sigma(H:\tilde{H}) \qquad (\text{all } p \in I(\mathfrak{g}_c);\ \tilde{p} = p\,|\,\mathfrak{j}_c),$$

then each c_w is constant on $\mathcal{O}$.

Proof For the moment let's fix an $\tilde{H} \in \tilde{\mathcal{O}}$ and let λ_w denote the linear function $H \mapsto \sqrt{-1}B(\tilde{H}, wH)$ on $\mathfrak{j}_c$; since $\pi(\tilde{H}) \neq 0$, the λ_w $(w \in W)$ are all distinct and so the exponentials e^{λ_w} are linearly independent over $S(\mathfrak{j}_c)$; in turn this means that the coefficients of e^{λ_w} in

$$\sigma(H;\partial(\tilde{p}):\tilde{H}) - p(\sqrt{-1}\tilde{H})\sigma(H:\tilde{H})$$

must all be zero. Now, for a fixed w, such a coefficient is of the form

$$c_w(H;\partial(\tilde{p})) + \sum_i p_i(\tilde{H})f_i(H) - p(\sqrt{-1}\tilde{H})c_w(H),$$

where the p_i are certain homogeneous elements in $S(\mathfrak{j}_c)$ of positive degree and the f_i are certain C^∞ functions on $\mathcal{O}$. Owing to the arbitrariness of $\tilde{H}$ in $\tilde{\mathcal{O}}$, then, it follows that $c_w(H;\partial(\tilde{p})) = 0$ $(H \in \mathcal{O})$ for all homogeneous elements $p \in I(\mathfrak{g}_c)$ of positive degree – therefore, thanks to Propositions 8.2.2.7 and 8.2.2.10, c_w is a polynomial function on $\mathcal{O}$. Now fix $\tilde{H} \in \tilde{\mathcal{O}}$ again; in view of the equation $\sigma(H;\partial(\tilde{p}):\tilde{H}) = p(\sqrt{-1}\tilde{H})\sigma(H:\tilde{H})$, we conclude from Corollary 8.2.2.9 that

$$\sigma(H:\tilde{H}) = \sum_{w\in W} \tilde{c}_w e^{\lambda_w(H)} \qquad (H \in \mathcal{O}),$$

where the $\tilde{c}_w$ are certain complex numbers which are independent of H. Because the e^{λ_w} $(w \in W)$ are linearly independent over $S(\mathfrak{j}_c)$, we see that actually $c_w = \tilde{c}_w$ (all $w \in W$). Hence the corollary. □

8.2.3 Radial Components of Left Invariant Differential Operators on a Reductive Lie Group

Let $\mathfrak{g}$ be a reductive Lie algebra over **R**, G a connected Lie group with Lie algebra $\mathfrak{g}$. [Thus, strictly speaking, we are now in a setting somewhat more general than that envisioned in Vol. I, number 1.1.5; for the purposes at hand, however, this will cause no difficulty.] Let $\mathfrak{j}$ be a Cartan subalgebra of $\mathfrak{g}$, J the Cartan subgroup of G associated with $\mathfrak{j}$. The group G operates on itself by inner automorphisms – the set up in 8.2.1 may then be realized by taking $M = G$, $N = \mathrm{J}'$, $G = G$ (it is obvious that the transversality condition is satisfied here). Our main objective in this number is: Viewing the elements of $\mathfrak{G}$ (the universal enveloping algebra of $\mathfrak{g}_c$, $\mathfrak{g}$ the Lie algebra of G) as *left* invariant differential operators on G, find an explicit formula for the radial component $\mathfrak{r}(Z)$ $(Z \in \mathfrak{Z}$, $\mathfrak{Z}$ the center of $\mathfrak{G})$ (cf. Theorem 8.2.3.9 infra). In order to get at this result it will be necessary to develop some preliminary material (which will be useful in other connections as well).

Let G be a Lie group, $\mathfrak{g}$ its Lie algebra, $\mathfrak{G}$ the universal enveloping algebra of $\mathfrak{g}_c$.

Let $\mathcal{O}_G$ be a non-empty open subset of G, D a differential operator on $\mathcal{O}_G$; fix a point $x_0 \in \mathcal{O}_G$.

Definition The unique element $D_{x_0} \in \mathfrak{G}$ with the property that $f(x_0; D) = f(x_0; D_{x_0})$ for all $f \in C^\infty(\mathcal{O}_G)$ will be called the *local expression* of D at x_0.

[For the existence and uniqueness of D_{x_0}, see Helgason [2, p. 98].]

Fix $X \in \mathfrak{g}$ – then the prescription

$$f(x; X) = df(x \exp(tX))/dt\,|_{t=0} \qquad (x \in G, f \in C^\infty(G))$$

defines a left invariant differential operator on G, and, in the well-known way, extends to a representation of $\mathfrak{G}$ on the algebra of all left invariant differential operators on G. On the other hand, the prescription

$$f(X; x) = df(\exp(tX)x)/dt\,|_{t=0} \qquad (x \in G, f \in C^\infty(G))$$

defines a right invariant differential operator on G, and, as is easily seen, extends to a representation of $\tilde{\mathfrak{G}}$ (the reverse algebra of $\mathfrak{G}$) on the algebra of all right invariant differential operators on G. This being so, fix an element $f \in C^\infty(G)$ – then a simple computation tells us that $f(D; x) = f(x; x^{-1}D)$ ($D \in \mathfrak{g}$; here $xD = \mathrm{Ad}(x)D$ $(x \in G)$). Now put $\boldsymbol{\mathfrak{G}} = \mathfrak{G} \otimes_C \mathfrak{G}$ and define a multiplication in $\boldsymbol{\mathfrak{G}}$ by the rule

$$(D_1 \otimes D_2)(D_3 \otimes D_4) = (D_3 D_1 \otimes D_2 D_4) \qquad (D_i \in \mathfrak{G}, i = 1, \ldots, 4).$$

Then the space $C^\infty(G)$ acquires the structure of a left $\boldsymbol{\mathfrak{G}}$-module when we write

$$(D_1 \otimes D_2)f(x) = f(D_1; x; D_2) \qquad (x \in G, f \in C^\infty(G)).$$

Note Needless to say, one can interpret the preceding remarks in the language of convolution (cf. A2.4); for various reasons, which will become apparent in due course, we choose not to do so.

Keeping to the above notations, for any $X \in \mathfrak{g}_c$, let L_X and R_X, respectively, denote the endomorphisms $D \mapsto XD$ and $D \mapsto DX$ $(D \in \mathfrak{G})$ of $\mathfrak{G}$. Fix an $x \in G$ and put

$$U_x(X) = L_{x^{-1}X} - R_X \qquad (X \in \mathfrak{g}_c)$$

– then the mapping $X \mapsto U_x(X)$ is a representation of $\mathfrak{g}_c$ on $\mathfrak{G}$ and, as always, the extension of U_x to $\mathfrak{G}$ will be denoted by the same symbol. Let $\boldsymbol{\Gamma}_x$ denote the linear mapping of $\mathfrak{G} \otimes_C \mathfrak{G}$ into $\mathfrak{G}$ defined by the rule

$$\boldsymbol{\Gamma}_x(D_1 \otimes D_2) = U_x(D_1)D_2 \qquad (D_1, D_2 \in \mathfrak{G}).$$

Employing the notations of Vol. I, number 2.3.1, we see that $\boldsymbol{\Gamma}_x$ defines a linear mapping of $\mathfrak{G}^{m_1} \otimes_C \mathfrak{G}^{m_2}$ into $\mathfrak{G}_{m_1+m_2}$.

Now the group G operates on itself by inner automorphisms and so, given an $f \in C^\infty(G)$, we may define a function $(x, y) \mapsto f(x:y)$ of two variables on $G \times G$ by the prescription $f(x:y) = f(xyx^{-1})$ $(x, y \in G)$; it is clear that $f(x:y;D) = f(xyx^{-1}; xD)$ $(x, y \in G;\ D \in \mathfrak{G})$. This being the case, the following result (which is the reason for introducing the map $\boldsymbol{\Gamma}_x$ $(x \in G)$ in the first place) should be compared with the observations made immediately after Proposition 8.2.2.1.

Proposition 8.2.3.1 Let $f \in C^\infty(G)$; let $D_1, D_2 \in \mathfrak{G}$ – then

$$f(x;D_1:y;D_2) = f(x:y;\boldsymbol{\Gamma}_y(D_1 \otimes D_2))$$

for all $x, y \in G$.

Proof Let $X \in \mathfrak{g}$; plainly

$${}^{\exp(tX)}y = y \exp(ty^{-1}X)\exp(-tX) \qquad (y \in G, t \in \mathbf{R});$$

on the other hand, it is evident that

$$f(x \exp(tX):y) = f(x:{}^{\exp(tX)}y) \qquad (x \in G, t \in \mathbf{R}).$$

Hence

$$f(x;X:y) = f(x:y;y^{-1}X - X) = f(x:X;y) - f(x:y;X) \quad (x, y \in G).$$

By differentiating with respect to y, we thus find that

$$\begin{aligned} f(x;X:y;D_2) &= f(x:y;(y^{-1}X)D_2 - D_2X) \\ &= f(x:y;U_y(X)D_2). \end{aligned}$$

Finally, if $X_1, \ldots, X_r \in \mathfrak{g}_c$ and $D_1 = X_1X_2\ldots X_r$, then an induction on r shows that $f(x;D_1:y;D_2) = f(x:y;\boldsymbol{\Gamma}_y(D_1 \otimes D_2))$, as desired. □

Let $x \mapsto \phi(x)$ $(x \in G)$ be an automorphism of G, $D \mapsto \phi(D)$ the corresponding automorphism of $\mathfrak{G}$.

Lemma 8.2.3.2 Fix $x \in G$ and $D_1, D_2 \in \mathfrak{G}$ – then

$$\boldsymbol{\Gamma}_{\phi(x)}(\phi(D_1) \otimes \phi(D_2)) = \phi(\boldsymbol{\Gamma}_x(D_1 \otimes D_2)).$$

Proof In fact if ϕ denotes the automorphism $D \mapsto \phi(D)$ of $\mathfrak{G}$, then one finds from the definitions that

$$U_{\phi(x)}(\phi(X)) = \phi U_x(X)\phi^{-1} \qquad (X \in \mathfrak{g}).$$

The lemma is an immediate consequence of this observation. □

Suppose that $X_1, \ldots, X_m$ are elements of $\mathfrak{g}_c$ and $(i_1, \ldots, i_m)$ is a permutation of $(1, \ldots, m)$; recall that

$$X_1X_2\ldots X_m - X_{i_1}X_{i_2}\ldots X_{i_m} \in \mathfrak{G}_{m-1}.$$

Lemma 8.2.3.3 Let X_i and Y_j $(1 \leqslant i \leqslant r, 1 \leqslant j \leqslant s)$ be elements in $\mathfrak{g}_c$; fix an $x \in G$ and put $X'_i = x^{-1}X_i - X_i$ $(1 \leqslant i \leqslant r)$ – then

$$\Gamma_x(\lambda(X_1X_2 \ldots X_r) \otimes \lambda(Y_1Y_2 \ldots Y_s)) \equiv \lambda(X'_1X'_2 \ldots X'_rY_1Y_2 \ldots Y_s) \mod \mathfrak{G}_{(r+s-1)}.$$

[Here $\lambda : S(\mathfrak{g}_c) \rightarrow \mathfrak{G}$ is the canonical space bijection between the symmetric algebra $S(\mathfrak{g}_c)$ of $\mathfrak{g}_c$ and $\mathfrak{G}$.]

Proof By induction on r we find that

$$\Gamma_x(X_1X_2 \ldots X_r \otimes Y_1Y_2 \ldots Y_s) \equiv X'_1X'_2 \ldots X'_rY_1Y_2 \ldots Y_s \mod \mathfrak{G}_{(r+s-1)}$$

where all the products are in $\mathfrak{G}$. Our assertion thus follows from the comment above. □

Henceforth it will be supposed that $\mathfrak{g}$ is reductive; in this situation it is possible to say a good deal more about Γ_x $(x \in G)$. Thus fix a Cartan subalgebra $\mathfrak{j}$ of $\mathfrak{g}$ and let J denote the Cartan subgroup of G associated with $\mathfrak{j}$; put

$$\mathfrak{z}_c = \sum_{\alpha>0} (\mathfrak{g}_c^{\alpha} + \mathfrak{g}_c^{-\alpha}),\ \mathfrak{S} = \lambda(S(\mathfrak{z}_c)),\ \mathfrak{S}^m = \lambda(S^m(\mathfrak{z}_c)),\ \mathfrak{S}^+ = \sum_{m\geqslant 1} \mathfrak{S}^m$$

and $\mathfrak{J} = \lambda(S(\mathfrak{j}_c))$, $\mathfrak{J}^m = \lambda(S^m(\mathfrak{j}_c))$ (m a non-negative integer).

Lemma 8.2.3.4 Fix $j \in J'$ – then Γ_j defines a bijective mapping between $\mathfrak{S} \otimes_{\mathbf{C}} \mathfrak{J}$ and $\mathfrak{G}$. In addition

$$\sum_{e_1+e_2\leqslant m} \Gamma_j(\mathfrak{S}^{e_1} \otimes_{\mathbf{C}} \mathfrak{J}^{e_2}) = \mathfrak{G}_m \qquad (m \geqslant 0).$$

Proof (1) We shall first prove by induction on m that

$$\sum_{e_1+e_2\leqslant m} \Gamma_j(\mathfrak{S}^{e_1} \otimes_{\mathbf{C}} \mathfrak{J}^{e_2}) = \mathfrak{G}_m.$$

This being clear for $m = 0$, assume that $m \geqslant 1$; since the left hand side is contained in the right hand side, in view of the induction hypothesis, it is enough to prove that

$$\sum_{e_1+e_2\leqslant m} \Gamma_j(\mathfrak{S}^{e_1} \otimes_{\mathbf{C}} \mathfrak{J}^{e_2}) + \mathfrak{G}_{m-1} \supset \mathfrak{G}^m.$$

Fix two integers $e_1, e_2 \geqslant 0$ such that $e_1 + e_2 = m$ and suppose that $X_i \in \mathfrak{z}_c$ $(1 \leqslant i \leqslant e_1)$, $H_j \in \mathfrak{j}_c$ $(1 \leqslant j \leqslant e_2)$; let $X = X_1X_2 \ldots X_{e_1} \in S(\mathfrak{z}_c)$ and $H = H_1H_2 \ldots H_{e_2} \in S(\mathfrak{j}_c)$ (take $X = 1$ if $e_1 = 0$ and $H = 1$ if $e_2 = 0$) – then it suffices to verify that

$$\lambda(XH) \in \sum_{e_1+e_2\leqslant m} \Gamma_j(\mathfrak{S}^{e_1} \otimes_{\mathbf{C}} \mathfrak{J}^{e_2}) + \mathfrak{G}_{m-1}.$$

If $e_1 = 0$ this is clear since $\Gamma_j(1 \otimes \lambda(H)) = \lambda(H)$. Hence we may assume that $e_1 > 0$. Because $\mathfrak{z}_c$ is J-stable and j is regular, we can choose $\tilde{X}_i \in \mathfrak{z}_c$ such that $(\operatorname{Ad}(j)^{-1} - 1)\tilde{X}_i = X_i$ $(1 \leqslant i \leqslant e_1)$; set $\tilde{X} = \tilde{X}_1\tilde{X}_2 \ldots \tilde{X}_r \in$

$S(\mathfrak{z}_c)$ – then Lemma 8.2.3.3 implies that

$$\Gamma_j(\lambda(\tilde{X}) \otimes \lambda(H)) \equiv \lambda(XH) \bmod \mathfrak{G}_{m-1}$$

and this gives us what we want.

(2) The spaces $\sum_{e_1+e_2 \leqslant m} (\mathfrak{S}^{e_1} \otimes_{\mathbf{C}} \mathfrak{J}^{e_2})$, $\mathfrak{G}_m$ have the same dimension by Proposition 2.3.1.3 (ii). Therefore Γ_j is injective.

This completes the proof of the lemma. □

Corollary 8.2.3.5 Fix $D \in \mathfrak{G}$ – then, for any $j \in \mathsf{J}'$, there exists a unique element $\text{♀}(D)_j \in \mathfrak{J}$ such that $D - \text{♀}(D)_j \in \Gamma_j(\mathfrak{S}^+ \otimes_{\mathbf{C}} \mathfrak{J})$. Moreover, if $D \in \mathfrak{G}_m$, then $\text{♀}(D)_j \in \mathfrak{J}_m$ and

$$\Gamma_j^{-1}(D - \text{♀}(D)_j) \in \sum_{e_2 \geqslant 0} \sum_{1 \leqslant e_1 \leqslant m-e_2} \mathfrak{S}^{e_1} \otimes_{\mathbf{C}} \mathfrak{J}^{e_2}.$$

[As will be seen below, $\text{♀}(D)_j$ ($j \in \mathsf{J}'$) is the local expression at j of the radial component $\text{♀}(D)$ of D ($D \in \mathfrak{G}$).]

Lemma 8.2.3.6 Let $D \in \mathfrak{G}$ – then there exists an integer $d \geqslant 0$ such that the mapping

$$j \mapsto [\det(\mathrm{Ad}(j^{-1}) - 1)_{\mathfrak{z}_c}]^d \text{♀}(D)_j \qquad (j \in \mathsf{J}')$$

can be extended to an analytic mapping of J into $\mathfrak{J}$.

Proof Employing the notations which were introduced during the proof of Lemma 8.2.3.4, it is clear that we may assume that $D = \lambda(XH) \in \mathfrak{G}_m$ ($m \geqslant 1$) and use induction on m. Let $A(j)$ denote the restriction of $\mathrm{Ad}(j^{-1}) - 1$ to $\mathfrak{z}_c$ ($j \in \mathsf{J}$). If t is an indeterminate, then $\det(t - A(j)) = \sum_{k=0}^{s} D_k(j)t^k$ – here $s = \dim(\mathfrak{z}_c)$, the D_k ($0 \leqslant k \leqslant s$) are analytic functions on J, $D_s = 1$ and $D_0(j) = (-1)^s \det(A(j))$. Agreeing to write $B(j)$ for

$$(-1)^{s+1} \sum_{k=0}^{s-1} D_{k+1}(j)A(j)^k \qquad (j \in \mathsf{J}),$$

we see without difficulty that $\det(A(j))I = A(j)B(j) = B(j)A(j)$ (all $j \in \mathsf{J}$). This being so, put

$$X_i(j) = B(j)X_i \qquad (1 \leqslant i \leqslant e_1),$$

$X(j) = \prod_i X_i(j)$ ($j \in \mathsf{J}$) – then, thanks to Lemma 8.2.3.3, it is clear that

$$\Gamma_j(\lambda(X(j)) \otimes H) - \det(A(j))^{e_1}\lambda(XH) \in \mathfrak{G}_{m-1}.$$

The assertion of the lemma now follows from the induction hypothesis. □

Note Let Φ be the set of roots of the pair $(\mathfrak{g}_c, \mathfrak{j}_c)$. Each $\alpha \in \Phi$ determines, in a natural way, a character ξ_α of J; cf. 8.1.1. Let Ξ be the space (over $\mathbf{C}$) comprised of the finite linear combinations of characters in the multiplicative group generated by the ξ_α ($\alpha \in \Phi$). Fix a $D \in \mathfrak{G}$ – then, as can be seen from the proof of the preceding lemma, it is possible to choose elements $H_1, \ldots, H_p \in \mathfrak{J}$, $\xi_1, \ldots, \xi_p \in \Xi$ and an integer $d \geqslant 0$ such that

$$\mathfrak{f}(D)_j = [\det(\mathrm{Ad}(j^{-1}) - 1)_{\mathfrak{g}_c}]^{-d} \sum_{i=1}^{p} \xi_i(j) H_i$$

for all $j \in \mathsf{J}'$.

Regarding J' as an open submanifold of J, Lemma 8.2.3.6 implies that for each $D \in \mathfrak{G}$ there exists an analytic differential operator on J' whose local expression at $j \in \mathsf{J}'$ is $\mathfrak{f}(D)_j$. Let us verify that this differential operator is $\mathfrak{f}(D)$. Fix a locally invariant C^∞ function f on G. There exist $S_i \in \mathfrak{S}^+$, $H_i \in \mathfrak{J}$ $(1 \leqslant i \leqslant p$, say) such that

$$D = \mathfrak{f}(D)_j + \sum_{i=1}^{p} a_i(j) \Gamma_j(S_i \otimes H_i) \qquad (j \in \mathsf{J}', a_i \in C^\infty(\mathsf{J}')).$$

Hence $f(j; D) = f(j; \mathfrak{f}(D)_j) + \sum_{i=1}^{p} a_i(j) f(1; S_i : j; H_i)$. But f is locally invariant and so $f(1; S_i : j; H_i) = 0$. Therefore the local expression at j of $\mathfrak{f}(D)$ is indeed $\mathfrak{f}(D)_j$ $(j \in \mathsf{J}')$.

Here is a variant of the preceding considerations which is occasionally useful.

Keeping to the above notations, let η_α denote the function $(1 - \xi_\alpha^{-1})^{-1}$ on J' and let $\mathfrak{R}$ be the ring of analytic functions on J' generated over $\mathbf{C}$ by 1 and the η_α $(\alpha \in \Phi)$. An easy differentiation shows that

$$H\eta_\alpha = \alpha(H)\eta_\alpha(1 - \eta_\alpha) \qquad (H \in \mathfrak{j}),$$

whence the ring $\mathfrak{R}$ is stable under the differential operators in $\mathfrak{J}$.

Lemma 8.2.3.7 Let $D \in \mathfrak{G}$ – then there exist $S_i \in \mathfrak{S}$, $H_i \in \mathfrak{J}$ and $\eta_i \in \mathfrak{R}$ $(1 \leqslant i \leqslant p)$ such that

$$\Gamma_j^{-1}(D) = \sum_{i=1}^{p} \eta_i(j)(S_i \otimes H_i) \qquad (j \in \mathsf{J}').$$

Proof Set $d(X) = L_X - R_X$ $(X \in \mathfrak{g}_c)$ – then $d(X)$ is a derivation of $\mathfrak{G}$. Clearly

$$U_j(X_\alpha) = \xi_\alpha(j)^{-1} L_{X_\alpha} - R_{X_\alpha} = -(1 - \xi_\alpha(j)^{-1}) L_{X_\alpha} + d(X_\alpha) \qquad (j \in \mathsf{J})$$

and therefore

$$\eta_\alpha(j) U_j(X_\alpha) = -(L_{X_\alpha} - \eta_\alpha(j) d(X_\alpha)) \qquad (j \in \mathsf{J}').$$

This shows that

$$\begin{aligned}(-1)^r \eta_{\alpha_1}(j) \ldots \eta_{\alpha_r}(j) \Gamma_j(X_{\alpha_1} \ldots X_{\alpha_r} \otimes H) \\ = (L_{X_{\alpha_1}} - \eta_{\alpha_1}(j) d(X_{\alpha_1})) \ldots (L_{X_{\alpha_r}} - \eta_{\alpha_r}(j) d(X_{\alpha_r}))H\end{aligned}$$

for $j \in \mathsf{J}'$, $H \in \mathfrak{J}$ and $\alpha_1, \ldots, \alpha_r \in \Phi$. The assertion of the lemma now follows by induction on the degree of D. □

Corollary 8.2.3.8 Retain the above notations; let $f \in C^\infty(G)$ – then

$$f(xjx^{-1}; xD) = \sum_{i=1}^{p} \eta_i(j) f(x; S_i : j; H_i) \qquad (x \in G, j \in \mathrm{J}').$$

Proof In fact

$$\begin{aligned} f(xjx^{-1}; xD) &= f(x:j; D) = f(x:j; \Gamma_j(\sum_i \eta_i(j)(S_i \otimes H_i))) \\ &= \sum_i \eta_i(j) f(x; S_i : j; H_i), \end{aligned}$$

the last equality being a consequence of Proposition 8.2.3.1. □

As above, let $\mathfrak{g}$ be a reductive Lie algebra over $\mathbf{R}$, G a connected Lie group with Lie algebra $\mathfrak{g}$; fix a Cartan subalgebra $\mathfrak{j}$ of $\mathfrak{g}$ and let J be the Cartan subgroup of G associated with $\mathfrak{j}$; let $\gamma : \mathfrak{Z} \to I(\mathfrak{j}_c)$ denote the canonical isomorphism between the center $\mathfrak{Z}$ of $\mathfrak{G}$ and the set $I(\mathfrak{j}_c)$ of Weyl group invariants in $S(\mathfrak{j}_c) \sim \mathfrak{J}$ (see Vol. I, number 2.3.3).

The following result is the group theoretic analogue to Theorem 8.2.2.4.

Theorem 8.2.3.9 (Harish-Chandra) If $Z \in \mathfrak{Z}$, then

$$\mathfrak{f}(Z) = |D_l|^{-1/2} \gamma(Z) \circ |D_l|^{1/2}$$

as differential operators on J'.

[Here D_l is the analytic function on G which was introduced in Vol. I, number 1.4.1 (l the rank of G); notice, in particular, that $D_l(j) = \det(\mathrm{Ad}(j^{-1}) - 1)_{\mathfrak{g}_c}$ $(j \in \mathrm{J})$.]

Proof of Theorem 8.2.3.9 Let

$$\Omega_Z = \mathfrak{f}(Z) - |D_l|^{-1/2} \gamma(Z) \circ |D_l|^{1/2} \qquad (Z \in \mathfrak{Z}).$$

We are to show that $\Omega_Z = 0$ (all $Z \in \mathfrak{Z}$). To this end it may be supposed, for rather obvious reasons, that G is semi-simple; in fact, if G_c is the simply connected complex analytic group with Lie algebra $\mathfrak{g}_c$, then, without any loss of generality, we may assume that G is the real analytic subgroup of G_c corresponding to $\mathfrak{g}_c$. If J_c is the Cartan subgroup of G_c associated with $\mathfrak{j}_c$, then certainly $\mathrm{J} \subset J_c$; let $\check{\mathfrak{j}}$ be the complete inverse image of J in $\mathfrak{j}_c$ under the exponential mapping – then $\check{\mathfrak{j}}$ is a closed real Lie subgroup of $\mathfrak{j}_c$ and its Lie algebra is $\mathfrak{j}$. Put

$$\Delta(H) = \prod_{\alpha > 0} (e^{\alpha(H)/2} - e^{-\alpha(H)/2}) \qquad (H \in \mathfrak{j}_c);$$

plainly $D_l(\exp H) = (-1)^r \Delta^2(H)$ (r the number of positive roots of the pair $(\mathfrak{g}_c, \mathfrak{j}_c)$). Let $\mathfrak{j}_c(\Delta)$ denote the open submanifold of $\mathfrak{j}_c$ consisting of those points H where $\Delta(H) \neq 0$. Note that $\check{\mathfrak{j}}(\Delta) = \check{\mathfrak{j}} \cap \mathfrak{j}_c(\Delta)$ is the complete inverse image of J' in $\mathfrak{j}_c$ under the exponential mapping. Consequently, to any differential operator D on J' there corresponds exactly one differential operator $\check{D}$ on $\check{\mathfrak{j}}(\Delta)$ with the following property: If $f \in C^\infty(\mathrm{J}')$ and if $\check{f}(H) = f(\exp H)$ $(H \in \check{\mathfrak{j}}(\Delta))$, then $\check{f}(H; \check{D}) = f(\exp H; D)$. Evidently $\check{D} = 0$ iff $D = 0$; this being the case, we see that our problem of proving that $\Omega_Z = 0$ is reduced to showing that $\check{\Omega}_Z = 0$. Let us describe how this is going to be done. It is clear that

$$\check{\Omega}_Z = \check{\mathfrak{f}}(Z) - \Delta^{-1}\gamma(Z)\circ\Delta;$$

so, if $\mathfrak{w}$ denotes the (ordered) additive group generated by the roots α in Φ, then it follows from Lemma 8.2.3.6 above that there exist elements $\sigma_1, \ldots, \sigma_p$ in $\mathfrak{w}$, $H_1, \ldots, H_p$ in $\mathfrak{J}$ and an integer $d \geqslant 0$ such that

$$\check{\Omega}_Z = \Delta^{-d} \sum_1^p e^{\sigma_i} H_i.$$

Proceeding by contradiction, assume that $\check{\Omega}_Z \neq 0$. We may suppose that $\sigma_1, \ldots, \sigma_p$ are all distinct, that σ_1 is the highest element (relative to the ordering of $\mathfrak{w}$) among $\sigma_1, \ldots, \sigma_p$, and that none of the H_i are zero $(i = 1, \ldots, p)$. With these agreements, it will be seen below that necessarily $\langle H_1, \mathbf{e}^\Lambda \rangle = 0$ for all dominant integral Λ, a contradiction since $H_1 \neq 0$. So fix a $\Lambda \in \mathfrak{W}^+$; let E_Λ be a finite dimensional simple $\mathfrak{g}_c$-module with highest weight Λ, U_Λ the corresponding representation of $\mathfrak{g}_c$ on E_Λ and extend U_Λ to both $\mathfrak{G}$ and G_c. The character ξ_Λ of the irreducible holomorphic representation U_Λ of G_c on E_Λ is given, on J_c, by Weyl's formula:

$$\xi_\Lambda(\exp H)\Delta(H) = \sum_{w \in W} \det(w) e^{\langle H, w(\Lambda+\rho)\rangle} \qquad (H \in \mathfrak{j}_c).$$

Because ξ_Λ is a central function on G_c (i.e. $\xi_\Lambda(xyx^{-1}) = \xi_\Lambda(y); x, y \in G_c$), the very definition of radial component tells us that $\xi_\Lambda(j; Z) = \xi_\Lambda(j; \mathfrak{f}(Z))$ on J'. Consequently, if $\check{\mathfrak{f}}(Z)_H$ denotes the local expression of $\check{\mathfrak{f}}(Z)$ at $H \in \check{\mathfrak{j}}(\Delta)$, we have that $\check{\xi}_\Lambda(H; \check{\mathfrak{f}}(Z)_H) = \xi_\Lambda(\exp H; Z)$. But, for any $a \in E_\Lambda$, $U_\Lambda(Z)a = \kappa_\Lambda(Z)a = \langle \gamma(Z), \mathbf{e}^{\Lambda+\rho}\rangle a$ (Vol. I, number 2.4.3) – thus $Z\xi_\Lambda = \langle \gamma(Z), \mathbf{e}^{\Lambda+\rho}\rangle \xi_\Lambda$, whence

$$\check{\xi}_\Lambda(H; \check{\mathfrak{f}}(Z)_H) = \langle \gamma(Z), \mathbf{e}^{\Lambda+\rho}\rangle \check{\xi}_\Lambda(H) \qquad (H \in \check{\mathfrak{j}}(\Delta)).$$

Put $\check{\xi}'_\Lambda = \check{\xi}_\Lambda \Delta$ – then, since $\gamma(Z) \in I(\mathfrak{j}_c)$,

$$\gamma(Z)\check{\xi}'_\Lambda = \langle \gamma(Z), \mathbf{e}^{\Lambda+\rho}\rangle \check{\xi}'_\Lambda$$

and so we find that

$$\begin{aligned} \check{\xi}_\Lambda(H; \check{\Omega}_Z) &= \check{\xi}_\Lambda(H; \check{\mathfrak{f}}(Z) - \Delta^{-1}\gamma(Z)\circ\Delta) \\ &= \check{\xi}_\Lambda(H; \check{\mathfrak{f}}(Z)) - \Delta(H)^{-1}\check{\xi}'_\Lambda(H; \gamma(Z)) = 0 \\ &\qquad\qquad (\text{all } H \in \check{\mathfrak{j}}(\Delta)). \end{aligned}$$

On the other hand, let $\Lambda = \lambda_1, \ldots, \lambda_q$ be the distinct weights of U_Λ, $m(\lambda_k)$ the multiplicity of λ_k $(k = 1, \ldots, q)$. Evidently

$$\check{\xi}_\Lambda(H) = \sum_{k=1}^q m(\lambda_k) e^{\lambda_k(H)} \qquad (H \in \mathfrak{j}_c).$$

Therefore

$$0 = \xi_{\Lambda}(H;\check{\Omega}_Z) = \Delta(H)^{-d}\sum_{i=1}^{p}\sum_{k=1}^{q} m(\lambda_k)\langle H_i, \mathbf{e}^{\lambda_k}\rangle e^{\sigma_i(H)+\lambda_k(H)}$$

$$(\text{all } H \in \check{\mathfrak{j}}(\Delta)).$$

From this relation we deduce without difficulty that

$$\sum_{i=1}^{p}\sum_{k=1}^{q} m(\lambda_k)\langle H_i, \mathbf{e}^{\lambda_k}\rangle e^{\sigma_i+\lambda_k}$$

vanishes identically on $\mathfrak{j}_c$. However $\sigma_1 + \Lambda = \sigma_1 + \lambda_1 > \sigma_i + \lambda_k$ unless $i = k = 1$; moreover the exponentials of distinct linear functions are linearly independent – hence $\langle H_1, \mathbf{e}^{\Lambda}\rangle = 0$. Since Λ is an arbitrary element in $\mathfrak{W}^+$, H_1 is necessarily zero and this gives the sought for contradiction.

The proof of the theorem is now complete. □

Keeping to the assumption that G is reductive, we shall conclude this number with the presentation of a certain collection of results which, strictly speaking, can not be regarded as special cases of the generalities obtained in 8.2.1 (the transversality condition may be absent . . .); on the other hand the proofs proceed, in the main, along entirely familiar lines and therefore will be omitted (cf. Harish-Chandra [27]). [Actually Helgason has proved a general theorem (unpublished) which contains the facts in question as a corollary.]

Let $\mathfrak{g}$ be a reductive Lie algebra over $\mathbf{R}$, G a connected Lie group with Lie algebra $\mathfrak{g}$. Let γ be a semi-simple element in G, $\mathfrak{g}_\gamma$ the centralizer of γ in $\mathfrak{g}$; let G_γ be the analytic subgroup of G corresponding to $\mathfrak{g}_\gamma$; given x_γ in G_γ, put

$$\delta_\gamma(x_\gamma) = \begin{cases} 1 \text{ if } \gamma \text{ is central,} \\ \det(\mathrm{Ad}(\gamma x_\gamma)^{-1} - 1)_{\mathfrak{g}/\mathfrak{g}_\gamma} \quad \text{if } \gamma \text{ is non-central.} \end{cases}$$

Clearly δ_γ is an analytic function on G_γ and $\delta_\gamma(1) \neq 0$. Let $'G_\gamma$ be the set of all points $x_\gamma \in G_\gamma$ where $\delta_\gamma(x_\gamma) \neq 0$ – then $'G_\gamma$ is an open neighborhood of 1 in G_γ.

Lemma 8.2.3.10 Let ϕ denote the mapping $(x, x_\gamma) \mapsto x(\gamma x_\gamma)x^{-1}$ of $G \times G_\gamma$ into G; let $n = \dim(G)$ – then ϕ is everywhere of rank n on $G \times {}'G_\gamma$.

[For the proof, see Lemma 1.4.3.1; the point, of course, is that Theorem 8.2.1.1 is now at our disposal.]

Suppose that our semi-simple element γ belongs to Ω, Ω an open and completely invariant subset of G; put $\Omega_\gamma = \gamma^{-1}\Omega \cap {}'G_\gamma$ – then Ω_γ is an open and completely invariant neighborhood of 1 in G_γ. Set $\mathcal{O}_G(\gamma) = G\cdot(\gamma\Omega_\gamma)$; according to Lemma 8.2.3.10, $\mathcal{O}_G(\gamma)$ is an open neighborhood of γ in Ω.

We have:

(1) Given $v \in C_c^\infty(G \times \Omega_\gamma)$, there exists a unique function $f_v \in C_c^\infty(\mathcal{O}_G(\gamma))$ such that

$$\int_{G\times\Omega_\gamma} F(x(\gamma x_\gamma)x^{-1})v(x : x_\gamma)d_G(x)d_{G_\gamma}(x_\gamma) = \int_{\mathcal{O}_G(\gamma)} F(x)f_v(x)d_G(x)$$

for all $F \in C_c^\infty(\mathcal{O}_G(\gamma))$.

(2) Let D be a differential operator on $\mathcal{O}_G(\gamma)$ – then there exists a differential operator $\wp_\gamma(D)$ on Ω_γ such that

$$f(\gamma x_\gamma ; D) = f(\gamma x_\gamma ; \wp_\gamma(D)) \qquad (x_\gamma \in \Omega_\gamma)$$

for each locally invariant C^∞ function f on $\mathcal{O}_G(\gamma)$.

(3) Let T be a central distribution on $\mathcal{O}_G(\gamma)$ – then there exists a unique distribution $\tilde{T}$ on Ω_γ, central relative to G_γ, such that $T(f_v) = \tilde{T}(V_v)$ $(v \in C_c^\infty(G \times \Omega_\gamma))$ where

$$V_v(x_\gamma) = \int_G v(x : x_\gamma)d_G(x) \qquad (x_\gamma \in \Omega_\gamma).$$

Moreover, if D is a G-invariant differential operator on $\mathcal{O}_G(\gamma)$, then

$$(D \cdot T)^\sim = \wp_\gamma(D)\cdot\tilde{T}.$$

Let $\mathfrak{G}_\gamma$ denote the universal enveloping algebra of $\mathfrak{g}_{\gamma,c}$; invoking the notations of Example 3 following Proposition 2.3.3.6, let $\mu_{\mathfrak{g}/\mathfrak{g}_\gamma}$ denote the canonical isomorphism of $\mathfrak{Z}$ into $\mathfrak{Z}_\gamma$ (the center of $\mathfrak{G}_\gamma$).

Theorem 8.2.3.9 (bis) (Harish-Chandra) Let T be a central distribution on $\mathcal{O}_G(\gamma)$ – then

$$\wp_\gamma(Z)\cdot\tilde{T} = |\delta_\gamma|^{-1/2}\mu_{\mathfrak{g}/\mathfrak{g}_\gamma}(Z)\cdot(|\delta_\gamma|^{1/2}\tilde{T})$$

for all $Z \in \mathfrak{Z}$.

[We shall not stop to give the proof of this result; the details may be found in Harish-Chandra [27, p. 474].]

The statement of Theorem 8.2.3.9 (bis) should be compared with that of Theorem 8.2.3.9. Suppose, in particular, that γ is regular; in this situation the centralizer $\mathfrak{g}_\gamma$ of γ in $\mathfrak{g}$ is a Cartan subalgebra $\mathfrak{j}_\gamma$ (say) of $\mathfrak{g}$ while G_γ becomes the analytic subgroup J_γ of G corresponding to $\mathfrak{j}_\gamma$. One may then proceed just as in the proof of Theorem 8.2.3.9 and show that

$$\wp_\gamma(Z) = |\delta_\gamma|^{-1/2}\gamma(Z)\circ|\delta_\gamma|^{1/2} \qquad (Z \in \mathfrak{Z})$$

as differential operators on $J_\gamma \cap (\gamma^{-1}G')$.

8.2.4 The Connection between Differential Operators in the Algebra and on the Group

Let $\mathfrak{g}$ be a reductive Lie algebra over $\mathbf{R}$, G a connected Lie group with Lie algebra $\mathfrak{g}$; let $\mathfrak{G}$ be the universal enveloping algebra of the complexification $\mathfrak{g}_c$ of $\mathfrak{g}$.

Many problems on G can be transformed, by means of the exponential mapping, into analogous questions on the Lie algebra $\mathfrak{g}$ – we shall now indicate one of the pertinent techniques used for effecting such reductions.

Put

$$\Upsilon(X) = |\det\{(e^{\mathrm{ad}(X/2)} - e^{-\mathrm{ad}(X/2)})/\mathrm{ad}(X)\}|^{1/2} \qquad (X \in \mathfrak{g}).$$

Then Υ is analytic around every point $X_0 \in \mathfrak{g}$ where $\Upsilon(X_0) \neq 0$ and of course the exponential mapping of $\mathfrak{g}$ into G is regular at X_0 iff $\Upsilon(X_0) \neq 0$.

Let $\mathcal{O}_{\mathfrak{g}}$ be an open subset of $\mathfrak{g}$ on which exp is regular and injective; set $\mathcal{O}_G = \exp(\mathcal{O}_{\mathfrak{g}})$. For any function ϕ on $\mathcal{O}_{\mathfrak{g}}$, let f_ϕ denote the function on $\mathcal{O}_G$ given by $f_\phi(\exp X) = \Upsilon(X)^{-1}\phi(X)$ $(X \in \mathcal{O}_{\mathfrak{g}})$ – then f_ϕ is C^∞ or analytic iff the same holds for ϕ (in particular the map $\phi \mapsto f_\phi$ establishes an isomorphism (of topological vector spaces) between $C_c^\infty(\mathcal{O}_{\mathfrak{g}})$ and $C_c^\infty(\mathcal{O}_G)$). It is clear that for any differential operator D on $\mathcal{O}_G$ there exists a unique differential operator $\nabla(D)$ on $\mathcal{O}_{\mathfrak{g}}$ such that $Df_\phi = f_{\nabla(D)\phi}$ for $\phi \in C^\infty(\mathcal{O}_{\mathfrak{g}})$. Moreover D is analytic iff $\nabla(D)$ is analytic.

If the Euclidean measure $d_{\mathfrak{g}}(X)$ on $\mathfrak{g}$ is normalized in a suitable way, then $d_G(x) = \Upsilon(X)^2 d_{\mathfrak{g}}(X)$ $(x = \exp X,\ X \in \mathcal{O}_{\mathfrak{g}})$ – hence

$$\int_{\mathfrak{g}} \phi_1(X)\phi_2(X)d_{\mathfrak{g}}(X) = \int_G f_{\phi_1}(x)f_{\phi_2}(x)d_G(x)$$

for $\phi_1 \in C^\infty(\mathcal{O}_{\mathfrak{g}})$, $\phi_2 \in C_c^\infty(\mathcal{O}_{\mathfrak{g}})$.

Lemma 8.2.4.1 Let D be a differential operator on $\mathcal{O}_G$ – then $\nabla(D^\sharp) = \nabla(D)^\sharp$.

Proof If $\phi_1, \phi_2 \in C_c^\infty(\mathcal{O}_{\mathfrak{g}})$, then

$$\int_G (D^\sharp f_{\phi_1})f_{\phi_2}d_G(x) = \int_G f_{\phi_1}f_{\nabla(D)\phi_2}d_G(x) = \int_{\mathfrak{g}} \phi_1(\nabla(D)\phi_2)d_{\mathfrak{g}}(X)$$
$$= \int_{\mathfrak{g}} (\nabla(D)^\sharp\phi_1)\phi_2 d_{\mathfrak{g}}(X) = \int_G f_{\nabla(D)^\sharp\phi_1}f_{\phi_2}d_G(x)$$

$\Rightarrow$

$$f_{\nabla(D)^\sharp\phi_1} = D^\sharp f_{\phi_1} = f_{\nabla(D^\sharp)\phi_1} \Rightarrow \nabla(D)^\sharp = \nabla(D^\sharp). \quad \square$$

Let $\mathfrak{j}$ be a Cartan subalgebra of $\mathfrak{g}$, $I(\mathfrak{j}_c)$ the set of Weyl group invariants in $S(\mathfrak{j}_c)$. The center $\mathfrak{Z}$ of $\mathfrak{G}$ and $I(\mathfrak{j}_c)$ are isomorphic under the map γ; composing γ with the inverse of the Chevalley restriction map gives an isomorphism $Z \mapsto p_Z$ of $\mathfrak{Z}$ onto the set $I(\mathfrak{g}_c)$ of G-invariants in $S(\mathfrak{g}_c)$: Call this isomorphism (which does not depend on $\mathfrak{j}_c$) the *canonical isomorphism* of $\mathfrak{Z}$ with $I(\mathfrak{g}_c)$.

Here is the relationship which exists between $\mathfrak{Z}$ and $\partial(I(\mathfrak{g}_c))$.

Proposition 8.2.4.2 Suppose that $\mathcal{O}_{\mathfrak{g}}$ is an open subset of $\mathfrak{g}$ on which exp is regular and injective. Let ϕ be a locally invariant C^∞ function on $\mathcal{O}_{\mathfrak{g}}$ – then $Zf_\phi = f_{\partial(p_Z)\phi}$ for all $Z \in \mathfrak{Z}$.

Proof Fix $Z \in \mathfrak{Z}$ and let $\mathcal{O}'_{\mathfrak{g}} = \mathcal{O}_{\mathfrak{g}} \cap \mathfrak{g}'$ – then it suffices to prove that $\nabla(Z)\phi = \partial(p_Z)\phi$ on $\mathcal{O}'_{\mathfrak{g}}$. Fix a point $H_0 \in \mathcal{O}'_{\mathfrak{g}}$ and let $\mathfrak{j}$ denote the centralizer of H_0 in $\mathfrak{g}$ – then $\mathfrak{j}$ is a Cartan subalgebra of $\mathfrak{g}$. Let $\mathcal{O}(H_0)$ be an open, connected neighborhood of H_0 in $\mathfrak{j} \cap \mathcal{O}'_{\mathfrak{g}}$ – then it is enough to show that $\phi(H; \nabla(Z)) = \phi(H; \partial(p_Z))$ for $H \in \mathcal{O}(H_0)$. By hypothesis ϕ is locally invariant and so it follows from Theorem 8.2.2.4 that

$$\phi(H;\partial(p_Z)) = \phi(H;\mathfrak{r}(\partial(p_Z))) = \pi(H)^{-1}\phi(H;\partial(\tilde{p}_Z)\circ\pi) \qquad (H \in \mathcal{O}(H_0)).$$

[Here π denotes the product of the positive roots of the pair $(\mathfrak{g}_c, \mathfrak{j}_c)$, $\tilde{p}_Z$ the restriction of p_Z to $\mathfrak{j}_c$.] Let J be the analytic subgroup of G corresponding to $\mathfrak{j}$. Since $\mathcal{O}(H_0) \subset \mathcal{O}'_{\mathfrak{g}}$, $\Upsilon(H)$ is non-zero and therefore $\exp H \in J'$ for $H \in \mathcal{O}(H_0)$. But f_ϕ is obviously locally invariant with respect to G – Theorem 8.2.3.9 thus tells us that

$$f_\phi(\exp H;Z) = |D_l(\exp H)|^{-1/2} f_\phi(\exp H;\gamma(Z)\circ|D_l|^{1/2}) \qquad (H \in \mathcal{O}(H_0)).$$

Plainly $|D_l(\exp H)|^{1/2} = \Upsilon(H)|\pi(H)|$ – therefore

$$|D_l(\exp H)|^{1/2} f_\phi(\exp H) = |\pi(H)|\phi(H) \qquad (H \in \mathcal{O}(H_0)).$$

Because $\mathcal{O}(H_0)$ is connected and π is nowhere zero there, we have that $|\pi(H)| = \epsilon\pi(H)$ $(H \in \mathcal{O}(H_0))$ where $\epsilon = |\pi(H_0)|/\pi(H_0)$. Hence

$$\begin{aligned} f_\phi(\exp H;Z) &= \pi(H)^{-1}\Upsilon(H)^{-1}\phi(H;\partial(\tilde{p}_Z)\circ\pi) \\ &= \Upsilon(H)^{-1}\phi(H;\partial(p_Z)) \qquad (H \in \mathcal{O}(H_0)). \end{aligned}$$

On the other hand, from the definition of $\nabla(Z)$, it is clear that $\phi(X; \nabla(Z)) = \Upsilon(X)f_\phi(\exp X;Z)$ (all $X \in \mathcal{O}_G$), whence $\phi(H;\nabla(Z)) = \phi(H;\partial(p_Z))$ on $\mathcal{O}(H_0)$, as contended. □

The preceding proposition admits an important consequence; in order to formulate and prove it, some preparation is necessary.

For any distribution T on $\mathcal{O}_G$, let σ_T denote the distribution on $\mathcal{O}_{\mathfrak{g}}$ given by $\sigma_T(\phi) = T(f_\phi)$ $(\phi \in C_c^\infty(\mathcal{O}_{\mathfrak{g}}))$ – then it follows from Lemma 8.2.4.1 that $\sigma_{D\cdot T} = \nabla(D)\cdot\sigma_T$.

Suppose, moreover, that $\mathcal{O}_{\mathfrak{g}}$ is G-stable – then obviously $\mathcal{O}_G$ is also G-stable. Since Υ is invariant under G, it is clear that $(f_\phi)^x = f_{\phi^x}$, $\nabla(xD) = x(\nabla(D))$ and $\sigma_{T^x} = (\sigma_T)^x$ $(x \in G)$. [The action, here, of G on itself and its Lie algebra is the obvious one.]

Proposition 8.2.4.3 Retain the above notations but suppose in addition that $\mathcal{O}_{\mathfrak{g}}$ is completely invariant. Let T be a central distribution on $\mathcal{O}_G$ (cf. 8.3.3) – then $\sigma_{Z\cdot T} = \partial(p_Z)\cdot\sigma_T$ (all $Z \in \mathfrak{Z}$).

The proof depends upon the following technical lemma.

Lemma 8.2.4.4 (Harish-Chandra) Let $\mathfrak{g}$ be a reductive Lie algebra over **R**, Ω an open and completely invariant subset of $\mathfrak{g}$; let D be an analytic central differential operator of finite order on Ω such that D

annihilates every central C^∞ function on Ω – then D also annihilates every central distribution T on Ω (cf. 8.3.1).

[The proof of this result may be found in Harish-Chandra [26, p. 36].]

Proof of Proposition 8.2.4.3 It has been noted above that $\sigma_{Z\cdot T} = \nabla(Z)\cdot\sigma_T$; on the other hand, Proposition 8.2.4.2 tells us that $\nabla(Z)\phi = \partial(p_Z)\phi$ for every central C^∞ function ϕ on $\mathcal{O}_{\mathfrak{g}}$, whence, on the basis of the preceding lemma, $\sigma_{Z\cdot T} = \partial(p_Z)\cdot\sigma_T$ (all Z in $\mathfrak{Z}$) (σ_T is clearly central). □

8.3 Central Eigendistributions on Reductive Lie Algebras and Groups

8.3.1 The Main Theorem in the Algebra

In this number and the next we shall present the main facts as regards central eigendistributions on a reductive Lie algebra. [It is a question here of 'hard analysis'; the methods of representation theory hardly enter in at all.]

Let $\mathfrak{g}$ be a reductive Lie algebra over $\mathbf{R}$, G a connected Lie group with Lie algebra $\mathfrak{g}$; let $I(\mathfrak{g}_c)$ be the set of invariants in $S(\mathfrak{g}_c)$ for the usual action of G on $S(\mathfrak{g}_c)$. Let Ω be an open, completely invariant subset of $\mathfrak{g}$. A distribution T on Ω is said to be *central* if it is invariant under G; a distribution T on Ω is said to be an *eigendistribution* (of $I(\mathfrak{g}_c)$) if there exists a homomorphism $\kappa : I(\mathfrak{g}_c) \to \mathbf{C}$ with the property that $\partial(p)\cdot T = \kappa(p)T$ (all $p \in I(\mathfrak{g}_c)$).

Theorem 8.3.1.1 (Harish-Chandra) Let $\mathfrak{g}$ be a reductive Lie algebra over $\mathbf{R}$, Ω an open and completely invariant subset of $\mathfrak{g}$; let T be a central distribution on Ω which is annihilated by an ideal I in $I(\mathfrak{g}_c)$ of finite codimension (thus $\partial(p)\cdot T = 0$ for all $p \in I$) – then, in the sense of distributions, T agrees with a locally summable function F_T (say) on Ω which is actually analytic on $\Omega' = \Omega \cap \mathfrak{g}'$.

[This result applies, of course, to central eigendistributions T on Ω.]

The proof of Theorem 8.3.1.1 (which proceeds by induction on the dimension of $\mathfrak{g}$), when taken in all its details, is difficult and lengthy (cf. Harish-Chandra [26, p. 11]). [A discussion of what's involved here is given below.]

Corollary 8.3.1.2 Retain the above notations – then, for any $D \in \mathfrak{P}_I(\mathfrak{g})$ (cf. 8.2.2), the distribution $D\cdot T$ fulfills the conditions of Theorem 8.3.1.1 and hence DF_T is locally summable on Ω with $D\cdot T = D\cdot F_T = DF_T$.

[This is not quite obvious (although, of course, it is if $D = \partial(p)$ (some $p \in I(\mathfrak{g}_c)$)); we shall defer the details for the time being.]

For the sake of simplicity, let us suppose in what follows that $\Omega = \mathfrak{g}$.

The first step in the proof of Theorem 8.3.1.1 is to show that T agrees with an analytic function F_T (say) on $\mathfrak{g}'$. This is done as follows. Fix a point $H_0 \in \mathfrak{g}'$ – then it will be enough to show that T coincides with an analytic function around H_0. For this purpose, let $\mathfrak{j}$ and J be the centralizers of H_0 in $\mathfrak{g}$ and G respectively; as we know, $\mathfrak{j}$ is a Cartan subalgebra of $\mathfrak{g}$ while J is the Cartan subgroup of G associated with $\mathfrak{j}$. [The fact that G is an 'arbitrary' connected Lie group with Lie algebra $\mathfrak{g}$ does not cause any difficulty . . . .] Choose a small open connected neighborhood $\mathcal{O}(H_0)$ of H_0 in $\mathfrak{j}'$ with the property that the map ϕ,

$$(\dot{x}, H) \mapsto \dot{x}H \quad (\dot{x} \in G/J,\ H \in \mathcal{O}(H_0)),$$

is one-to-one. Let $\tilde{T}$ denote the distribution on $\mathcal{O}(H_0)$ which corresponds to $T|G\cdot\mathcal{O}(H_0)$ under the process described in 8.2.1 and put $\sigma = \pi\tilde{T}$ (π as always). Agreeing to denote the restriction to $\mathfrak{j}_c$ of an element $p \in I(\mathfrak{g}_c)$ by $\bar{p}$, Theorems 8.2.1.8 and 8.2.2.4 tell us that $\partial(\bar{p})\cdot\sigma = (\partial(\bar{p})\circ\pi)\cdot\tilde{T} = (\pi\circ\mathfrak{f}(\partial(p))\cdot\tilde{T} = 0$ for all $p \in I$; let $\bar{I} = \{\bar{p} : p \in I(\mathfrak{g}_c)\}$ – then $\bar{I}$ is an ideal of finite codimension in $I(\mathfrak{j}_c)$ and so, since $S(\mathfrak{j}_c)$ is a finite module over $I(\mathfrak{j}_c)$, it follows that σ is actually annihilated by an ideal of finite codimension I_S (say) in $S(\mathfrak{j}_c)$. In turn this latter fact implies that σ is analytic on $\mathcal{O}(H_0)$. To see this, let $H_1, \ldots, H_l$ be a basis for $\mathfrak{j}$ over $\mathbf{R}$; set $\Delta = \sum_{i=1}^{l} H_i^2$ and choose complex numbers $c_i \in \mathbf{C}$ ($1 \leqslant i \leqslant m$, say) such that

$$\square = \Delta^m + \sum_{i=1}^{m} c_i \Delta^{m-i} \in I_S.$$

Then $\square$ is an analytic, elliptic differential operator on $\mathfrak{j}$. But $\square\cdot\sigma = 0$, whence, by the Regularity Theorem, σ coincides, in the sense of distributions, with an analytic function F_σ (say) on $\mathcal{O}(H_0)$. Thanks to Lemma 2 in the Appendix to this number, then, we see that there exist distinct linear functions λ_i and polynomial functions p_i on $\mathfrak{j}_c$ ($1 \leqslant i \leqslant r$) such that

$$F_\sigma(H) = \sum_{i=1}^{r} p_i(H) e^{\lambda_i(H)} \qquad (H \in \mathcal{O}(H_0)).$$

We can now define an analytic function F_T on $G\cdot\mathcal{O}(H_0)$ by the prescription

$$F_T(\dot{x}H) = F_\sigma(H)/\pi(H) \qquad (\dot{x} \in G/J,\ H \in \mathcal{O}(H_0));$$

as the notation suggests, $T = F_T$ on $G\cdot\mathcal{O}(H_0)$. Indeed, if $v \in C_c^\infty(G \times \mathcal{O}(H_0))$, then

$$\begin{aligned} T(f_v) = \tilde{T}(V_v) &= \int_{G\times\mathcal{O}(H_0)} v(x:H)F_T(xH)d_G(x)d_{\mathfrak{j}}(H) \\ &= \int_{G\cdot\mathcal{O}(H_0)} f_v(X)F_T(X)d_{\mathfrak{g}}(X), \end{aligned}$$

whence $T = F_T$ on $G\cdot\mathcal{O}(H_0)$ (the map $v \mapsto f_v$ of $C_c^\infty(G \times \mathcal{O}(H_0))$ into $C_c^\infty(G\cdot\mathcal{O}(H_0))$ being surjective).

At this point, then, we know that T agrees with an analytic function

F_T on $\mathfrak{g}'$; locally F_T itself can be written as the quotient of two analytic functions, the denominator being π.

Lemma 8.3.1.3 Retain the above notations – then F_T is locally summable on $\mathfrak{g}$.

Proof Let $l = \text{rank}(\mathfrak{g})$, t an indeterminate; let $d_l(X)$ denote the coefficient of t^l in $\det(t - \text{ad}(X))$ $(X \in \mathfrak{g})$ – then $|d_l|^{-1/2}$ is locally summable on $\mathfrak{g}$ (cf. 8.4.1 infra). Because the singular set is of measure zero, our lemma will follow when it is shown that a given point X_0 in $\mathfrak{g}$ necessarily admits a neighborhood $\mathcal{O}(X_0)$ in $\mathfrak{g}$ with the property that $|d_l|^{1/2}|F_T|$ remains bounded on $\mathcal{O}'(X_0)(= \mathcal{O}(X_0) \cap \mathfrak{g}')$. For this purpose fix a positive definite quadratic form Q on $\mathfrak{g}$; let $\mathcal{O}$ denote the set of all X in $\mathfrak{g}$ such that $Q(X - X_0) < 1$; define the polynomial function p on $\mathfrak{g}$ by the rule

$$p(X) = (Q(X - X_0) - 1)d_l(X) \qquad (X \in \mathfrak{g})$$

and let $\mathfrak{g}_p = \{X \in \mathfrak{g} : p(X) \neq 0\}$ – then, thanks to a well-known theorem of Whitney [1, p. 547], $\mathfrak{g}_p$ has but a finite number of connected components. [Theorem (Whitney): A real algebraic variety V has at most a finite number of connected components; furthermore if $\tilde{V}$ is a subvariety of V, then $V - \tilde{V}$ has but a finite number of connected components too.] It is clear that any connected component of $\mathcal{O}'$ $(= \mathcal{O} \cap \mathfrak{g}')$ is also a connected component of $\mathfrak{g}_p$ – thus $\mathcal{O}'$ has only a finite number of connected components and so it will be enough to show that $|d_l|^{1/2}|F_T|$ remains bounded on a connected component $\mathcal{O}_0$ of $\mathcal{O}'$. To this end, fix an element $H_0 \in \mathcal{O}_0$ and use the notation which was introduced above. Let $\mathscr{C}_0$ denote the connected component of $(\dot{1}, H_0)$ in $\phi^{-1}(\mathcal{O}_0)$ – then $\phi(\mathscr{C}_0) = \mathcal{O}_0$. [Since ϕ is regular, $\phi(\mathscr{C}_0)$ is open in $\mathcal{O}_0$; therefore, due to the connectedness of $\mathcal{O}_0$, we need only verify that $\phi(\mathscr{C}_0)$ is closed in $\mathcal{O}_0$. So let $(\dot{x}_k, H_k)$ $(k \geqslant 1)$ be a sequence in $\mathscr{C}_0$ such that $X_k = \dot{x}_k H_k$ converges to some point X in $\mathcal{O}_0$. The X_k lie in a bounded subset of $\mathfrak{g}$; therefore, in view of Lemma 8.1.6.3 supra, the H_k themselves remain bounded in $\mathfrak{j}$ – hence, by passing to a subsequence if necessary, we may assume that the H_k converge to a point H (say) in $\mathfrak{j}$. Now $d_l(H_k) = d_l(\dot{x}_k H_k) \to d_l(X) \neq 0$, whence H is regular; this implies that J is the centralizer of H in G and therefore that the $\dot{x}_k$ remain within a compact subset of G/J (cf. Lemma 8.4.1.1); again, without loss of generality, it can be assumed that the $\dot{x}_k$ converge to $\dot{x}$ (say) in G/J – therefore

$$(\dot{x}_k, H_k) \to (\dot{x}, H) \Rightarrow \dot{x}H = X \in \mathcal{O}_0 \Rightarrow (\dot{x}, H) \in \phi^{-1}(\mathcal{O}_0) \Rightarrow$$
$$(\dot{x}, H) \in \mathscr{C}_0 \Rightarrow X = \dot{x}H \in \phi(\mathscr{C}_0),$$

as desired.] Choose a small open connected neighborhood $\mathcal{O}(\dot{1})$ of $\dot{1}$ in G/J; it can be assumed that $\mathcal{O}(\dot{1}) \times \mathcal{O}(H_0) \subset \mathscr{C}_0$. Consider the function

$$(\dot{x}, H) \mapsto F_T(\dot{x}H) - \pi(H)^{-1} \sum_i p_i(H) e^{\lambda_i(H)}$$

on $\mathscr{C}_0$; it is obviously analytic and, moreover, vanishes identically on

$\mathcal{O}(\dot{1}) \times \mathcal{O}(H_0)$, hence vanishes identically on $\mathscr{C}_0, \mathscr{C}_0$ being connected. Consequently

$$|d_l(\dot{x}H)|^{1/2}\,|F_T(\dot{x}H)| = |\sum_i p_i(H)e^{\lambda_i(H)}| \qquad (\text{all } (\dot{x}, H) \in \mathscr{C}_0).$$

Now we have seen above that $\phi(\mathscr{C}_0) = \mathcal{O}_0$ is contained in the bounded set $\mathcal{O}$; therefore, on the basis of reasoning which was used once before (Lemma 8.1.6.3), we then deduce that the projection of $\mathscr{C}_0$ onto $\mathfrak{j}$ is a bounded subset of $\mathfrak{j}$ – but this implies that $|d_l|^{1/2}\,|F_T|$ is bounded on $\phi(\mathscr{C}_0) = \mathcal{O}_0$, thus finishing the proof of the lemma. □

It remains to establish that $T = F_T$ on *all* of $\mathfrak{g}$ (and not just $\mathfrak{g}'$). [In view of the possible singularities of F_T, this fact is not at all obvious.] The proof proceeds by induction on the dimension of $\mathfrak{g}$.

To begin with let us suppose that the center $\mathfrak{c}$ of $\mathfrak{g}$ is non-trivial – then $\dim(\bar{\mathfrak{g}}) < \dim(\mathfrak{g})$ and the induction hypothesis is applicable to $\bar{\mathfrak{g}}$ (the derived algebra of $\mathfrak{g}$). Fix Euclidean measures $d_{\mathfrak{c}}(C)$ and $d_{\bar{\mathfrak{g}}}(\bar{X})$ on $\mathfrak{c}$ and $\bar{\mathfrak{g}}$, respectively, such that $d_{\mathfrak{g}}(X) = d_{\mathfrak{c}}(C)d_{\bar{\mathfrak{g}}}(\bar{X})$ if $X = C + \bar{X}$. Given any $g \in C_c^\infty(\mathfrak{c})$, let T_g denote the distribution on $\bar{\mathfrak{g}}$ defined by

$$T_g(h) = T(g \times h) \qquad (h \in C_c^\infty(\bar{\mathfrak{g}}));$$

if $\bar{G}$ is the analytic subgroup of G corresponding to $\bar{\mathfrak{g}}$, then it is clear that T_g is $\bar{G}$-central – furthermore it is obvious that T_g is annihilated by an ideal of finite codimension in $I(\bar{\mathfrak{g}}_c)$ (since $I(\mathfrak{g}_c) = S(\mathfrak{c}_c)I(\bar{\mathfrak{g}}_c)$). Consequently, in view of the induction hypothesis, T_g coincides with a locally summable function F_g (say) on $\bar{\mathfrak{g}}$. Now, as has been noted earlier, $T = F_T$ on $\mathfrak{g}'$; it follows, therefore, that

$$T_g(h) = T(g \times h) = \int g(C)h(\bar{X})F_T(C + \bar{X})d_{\mathfrak{c}}(C)d_{\bar{\mathfrak{g}}}(\bar{X})$$
$$(\text{all } h \in C_c^\infty(\bar{\mathfrak{g}}')),$$

whence

$$F_g(\bar{X}) = \int g(C)F_T(C + \bar{X})d_{\mathfrak{c}}(C) \qquad (\bar{X} \in \bar{\mathfrak{g}}');$$

but F_g is locally summable on $\bar{\mathfrak{g}}$ while F_T is locally summable on $\mathfrak{g}$, so

$$T(g \times h) = T_g(h) = \int g(C)h(\bar{X})F_T(C + \bar{X})d_{\mathfrak{c}}(C)d_{\bar{\mathfrak{g}}}(\bar{X}) = F_T(g \times h)$$
$$(\text{all } h \in C_c^\infty(\bar{\mathfrak{g}})),$$

which shows that $T = F_T$ on $\mathfrak{g}$ (the $g \times h$'s being total in $C_c^\infty(\mathfrak{g})$).

The above discussion implies that $\mathfrak{g}$ can be taken semi-simple. Fix a non-zero semi-simple element Γ in $\mathfrak{g}$ – we shall now prove that $T = F_T$ around Γ. To this end, let $\mathfrak{g}_\Gamma$ denote the centralizer of Γ in $\mathfrak{g}$, G_Γ the analytic subgroup of G corresponding to $\mathfrak{g}_\Gamma$; we recall that $\mathfrak{g}_\Gamma$ is reductive in $\mathfrak{g}$ with rank $(\mathfrak{g})$ = rank $(\mathfrak{g}_\Gamma)$. Put $\delta_\Gamma(X_\Gamma) = \det(\operatorname{ad}(X_\Gamma)_{\mathfrak{g}/\mathfrak{g}_\Gamma})$ $(X_\Gamma \in \mathfrak{g}_\Gamma)$ and let $'\mathfrak{g}_\Gamma = \{X_\Gamma \in \mathfrak{g}_\Gamma : \delta_\Gamma(X_\Gamma) \neq 0\}$ – then $\delta_\Gamma(\Gamma) \neq 0$. Let Ω_Γ be the set of all $X_\Gamma \in \mathfrak{g}_\Gamma$ such that $|\delta_\Gamma(X_\Gamma)| > |\delta_\Gamma(\Gamma)|/2$ – then Ω_Γ is an open neighborhood of Γ in $'\mathfrak{g}_\Gamma$ which is, in fact, completely invariant in $\mathfrak{g}_\Gamma$. Let $\tilde{T}$ be the distribution on Ω_Γ corresponding to T per Theorem 8.2.2.4 (bis) and supporting discussion – then $\tilde{T}$ is G_Γ-central. Put $\sigma = |\delta_\Gamma|^{1/2}\tilde{T}$; since $\delta_\Gamma^2 > 0$ on Ω_Γ, σ is also a G_Γ-central distribution on Ω_Γ which is, moreover, annihilated by an ideal I_Γ in $I(\mathfrak{g}_{\Gamma,c})$ of finite codimension. [To see this, we need only bear in mind that $I(\mathfrak{g}_{\Gamma,c})$ is a finite module over $I(\mathfrak{g}_c; \mathfrak{g}_{\Gamma,c})$ (cf. Proposition 2.1.5.10) . . . .] Because $\mathfrak{g}$ is semi-simple and Γ is non-zero, $\dim(\mathfrak{g}_\Gamma) < \dim(\mathfrak{g})$ and the induction hypothesis is applicable to the triple $(\sigma, \Omega_\Gamma, I_\Gamma)$. [This is one of the reasons why it is necessary to work with triples (T, Ω, I) in the first place.] Consequently, if Ω'_Γ denotes the set of all points in Ω_Γ which are regular in $\mathfrak{g}_\Gamma$, then σ coincides with a locally summable function F_σ (say) on Ω_Γ which is even analytic on Ω'_Γ. Since $\Omega_\Gamma \subset {}'\mathfrak{g}_\Gamma$, it is clear that $\Omega'_\Gamma \subset \mathfrak{g}'$. If $v \in C_c^\infty(G \times \Omega'_\Gamma)$, then

$$T(f_v) = \int v(x : X_\Gamma)F_T(xX_\Gamma)d_G(x)d_{\mathfrak{g}_\Gamma}(X_\Gamma)$$

$$= \int V_v(X_\Gamma)F_T(X_\Gamma)d_{\mathfrak{g}_\Gamma}(X_\Gamma) = \tilde{T}(V_v),$$

whence

$$F_\sigma(X_\Gamma) = |\delta_\Gamma(X_\Gamma)|^{1/2}F_T(X_\Gamma) \qquad (\text{all } X_\Gamma \in \Omega'_\Gamma).$$

Therefore, keeping in mind Corollary 8.2.1.3, we deduce, by the obvious argument, that $T(f_v) = F_T(f_v)$ (all $v \in C_c^\infty(G \times \Omega_\Gamma)$) which serves to show that $T = F_T$ around Γ (the map $v \mapsto f_v$ being surjective . . .).

Let $\mathcal{N}$ be the set of all nilpotent elements in $\mathfrak{g}$ – then the support of $T - F_T$ is contained in $\mathcal{N}$ (on the basis of the above discussion and Lemma 8.1.6.1, this is clear). In order to exploit this fact one appeals to the following lemma.

Lemma 8.3.1.4 Let $\mathfrak{g}$ be a semi-simple Lie algebra over **R**; let T be a central distribution on $\mathfrak{g}$ with support contained in $\mathcal{N}$ and such that $\partial(p(\omega))\cdot T = 0$ for some non-zero polynomial p (ω the Casimir polynomial of $\mathfrak{g}$) – then $T = 0$.

[The proof of this result, which relies heavily on the fact that $\mathcal{N}$ is the union of *finitely* many G-orbits (Kostant), may be found in Harish-Chandra [23, p. 306].]

This being so, suppose now that there exists an integer $m \geqslant 0$ and a complex number c such that $(\partial(\omega) - c)^m\cdot T = 0$ – then we claim that $T = F_T$. To verify the claim let us proceed by induction on m. If $m = 0$, then $T = 0$ and our contention is trivially true. If $m \geqslant 1$, put $T_0 = (\partial(\omega) - c)\cdot T$ – then T_0 satisfies the same general conditions as T and $(\partial(\omega) - c)^{m-1}\cdot T_0 = 0$. Because $T = F_T$ on $\mathfrak{g}'$ and F_T is analytic there, it must be the case that $T_0 = (\partial(\omega) - c)F_T$ on $\mathfrak{g}'$; therefore, by the induction hypothesis, $T_0 = F_0$ where $F_0 = (\partial(\omega) - c)F_T$ – thus

$$(\partial(\omega) - c)\cdot(T - F_T + F_T) = F_0$$

and so $(\partial(\omega) - c)\cdot(T - F_T) = \partial(\omega)F_T - \partial(\omega)\cdot F_T$. It can be shown that $\partial(\omega)F_T = \partial(\omega)\cdot F_T$ – therefore $(\partial(\omega) - c)\cdot(T - F_T) = 0$, whence $T = F_T$ on $\mathfrak{g}$ (Lemma 8.3.1.4). [Due to the possible singularities of F_T, it is not obvious that $\partial(\omega)F_T = \partial(\omega)\cdot F_T$ (cf. Harish-Chandra [26, p. 14]); we shall not stop to prove this here since entirely similar techniques of proof will be encountered later on when we take up the difficult question of 'regularizing' integrals on $\mathfrak{g}$ (cf. Volume III).]

It is now a simple matter to finish the proof of Theorem 8.3.1.1 in the case of a semi-simple $\mathfrak{g}$. Let $\mathbf{T}$ be the vector space comprised of all distributions on $\mathfrak{g}$ of the form $\partial(p)\cdot T$ ($p \in I(\mathfrak{g}_c)$) – then certainly $\dim(\mathbf{T}) < \infty$ and, of course, every element of $\mathbf{T}$ satisfies the same conditions as T itself. The mapping $S \mapsto \partial(\omega)\cdot S$ is an endomorphism of $\mathbf{T}$; therefore, by linear algebra, it is possible to pick a basis $\{T_i : 1 \leqslant i \leqslant n\}$ for $\mathbf{T}$ over $\mathbf{C}$ with the following property: There exist complex numbers c_i and integers $m_i \geqslant 0$ such that $(\partial(\omega) - c_i)^{m_i}\cdot T_i = 0$ $(1 \leqslant i \leqslant n)$. Owing to what was said above, there exist locally summable functions F_i on $\mathfrak{g}$ which are analytic on $\mathfrak{g}'$ and such that $T_i = F_i$ on $\mathfrak{g}$ $(1 \leqslant i \leqslant n)$. Because the T_i constitute a basis for $\mathbf{T}$, our given T can be expressed as a linear combination of them; as the same must be true of F_T and the F_i, it then follows that $T = F_T$ on $\mathfrak{g}$, thus completing the proof of Theorem 8.3.1.1 (modulo the two omissions supra).

We have yet to prove Corollary 8.3.1.2. Since the distribution $D\cdot T$ is G-central, it is enough to verify that the dimension of the space of all distributions of the form $\partial(p)\cdot(D\cdot T)$ ($p \in I(\mathfrak{g}_c)$) is finite. For this purpose we need a technical lemma.

Given $p \in S(\mathfrak{g}_c)$, let R_p and ∇_p denote the endomorphisms $D \mapsto D \circ \partial(p)$ and $D \mapsto \{\partial(p), D\}$ ($D \in \mathfrak{P}(\mathfrak{g})$), respectively, of $\mathfrak{P}(\mathfrak{g})$.

Lemma 8.3.1.5 Fix $p \in S(\mathfrak{g}_c)$ – then, for every $D \in \mathfrak{P}(\mathfrak{g})$, there exists an integer $d \geqslant 0$ such that $\nabla_p^d(D) = 0$.

Proof Let A be the set of all $p \in S(\mathfrak{g}_c)$ for which the statement of the lemma is valid – then A is a subalgebra of $S(\mathfrak{g}_c)$. [To see this, first note that $\nabla_p, R_p, \nabla_q, R_q$ ($p, q \in S(\mathfrak{g}_c)$) all commute with each other and satisfy the relation $\nabla_{pq} = \nabla_p\nabla_q + R_p\nabla_q + \nabla_p R_q$. Now fix p, q in A and $D \in \mathfrak{P}(\mathfrak{g})$ and choose an integer $d \geqslant 0$ such that $\nabla_p^d(D) = \nabla_q^d(D) = 0$ – then it is obvious that $(\nabla_p + \nabla_q)^{2d}(D) = 0$ and

$$\nabla_{pq}^{3d}(D) = (\nabla_p\nabla_q + R_p\nabla_q + \nabla_p R_q)^{3d}(D) = 0$$

which shows that $p + q$ and pq are both in A . . . .] On the other hand, if $p \in P(\mathfrak{g}_c)$, $q \in S(\mathfrak{g}_c)$ and $X \in \mathfrak{g}_c$, then clearly

$$\nabla_X^d(p\partial(q)) = (\nabla_X^d p)\partial(q) = 0$$

if $d > \deg(p)$ – thus $\mathfrak{g} \subset A$ and so $A = S(\mathfrak{g}_c)$. $\square$

Proof of Corollary 8.3.1.2 Let **T** denote the space of all distributions on $\mathfrak{g}$ of the form $\partial(p)\cdot T(p \in I(\mathfrak{g}_c))$ – then $\dim(\mathbf{T}) < \infty$. Because the algebra $I(\mathfrak{g}_c)$ is abelian, there exists a basis $T_1, \ldots, T_n$ for **T** over **C** and homomorphisms $\chi_1, \ldots, \chi_n$ of $I(\mathfrak{g}_c)$ into **C** such that

$$(\partial(p) - \chi_i(p))^n \cdot T_i = 0 \qquad (1 \leqslant i \leqslant n;\ \text{all } p \in I(\mathfrak{g}_c));$$

therefore, since T is a linear combination of the T_i, there is no loss of generality in assuming in addition that

$$(\partial(p) - \chi(p))^n \cdot T = 0 \qquad (\text{all } p \in I(\mathfrak{g}_c))$$

for some integer $n \geqslant 0$ and some homomorphism χ of $I(\mathfrak{g}_c)$ into **C**. This being so, fix $p \in I(\mathfrak{g}_c)$ and choose an integer $d \geqslant 0$ with the property that $\nabla_p^d(D) = 0$ – then

$$(\partial(p) - \chi(p))^{d+n} \circ D = (\nabla_p + R_p - \chi(p))^{d+n} D$$
$$= \sum_{i=0}^{d+n} \binom{d+n}{i} (R_p - \chi(p))^{d+n-i} \nabla_p^i(D)$$

and hence $(\partial(p) - \chi(p))^{d+n} \cdot (D \cdot T) = 0$. [For if $i \geqslant d$, then $\nabla_p^i(D) = 0$ while if $i \leqslant d$, then $(\partial(p) - \chi(p))^{d+n-i} \cdot T = 0$.] Choose $p_1, \ldots, p_l$ in $I(\mathfrak{g}_c)$ with the property that $I(\mathfrak{g}_c) = \mathbf{C}[p_1, \ldots, p_l]$ – then there clearly exists an integer $N \geqslant 0$ such that $(\partial(p_i) - \chi(p_i))^N \cdot T = 0$ $(1 \leqslant i \leqslant l)$. But this means that the space of all distributions of the form $\partial(p) \cdot (D \cdot T)$ $(p \in I(\mathfrak{g}_c))$ has dimension at most N^l. Therefore Theorem 8.3.1.1 is applicable to $D \cdot T$. □

Appendix We shall record here two elementary, well-known results from the theory of ordinary differential equations.

Lemma 1 Let I be an open interval of the real line **R**. Let q be a polynomial of degree d (with complex coefficients). If $\lambda_1, \ldots, \lambda_r$ are the distinct roots of q with multiplicities $m_1, \ldots, m_r$, then the analytic solutions ϕ on I to

$$q(d/dt)\phi = 0$$

are of the form

$$\phi(t) = p_1(t)e^{\lambda_1 t} + \cdots + p_r(t)e^{\lambda_r t} \qquad (t \in I)$$

where the p_i $(i = 1, \ldots, r)$ are polynomials (with complex coefficients) of degree less than m_i.

The proof of this fact is elementary and may be left to the reader; moreover, an easy induction yields without difficulty:

Lemma 2 Let $\mathcal{O}$ be an open connected subset of $\mathbf{R}^l$. Let $q_1, \ldots, q_l$ be polynomials (in one variable) of degrees $d_1, \ldots, d_l$, respectively – then the analytic solutions $\phi = \phi(t_1, \ldots, t_l)$ on $\mathcal{O}$ of the system:

$$q_i(\partial/\partial t_i)\phi = 0 \qquad (i = 1, \ldots, l)$$

are linear combinations with complex coefficients of functions of the form

$$t_1^{n_1} \cdots t_l^{n_l} e^{\lambda_1 t_1 + \cdots + \lambda_l t_l}$$

where the n_i are integers, λ_i is a root of q_i with multiplicity m_i and $0 \leqslant n_i < m_i$ $(i = 1, \ldots, l)$.

8.3.2 Properties of F_T — I

Let $\mathfrak{g}$ be a reductive Lie algebra over **R**, G a connected Lie group with Lie algebra $\mathfrak{g}$.

Let Ω be an open, completely invariant subset of $\mathfrak{g}$; let T be a central distribution on Ω which is annihilated by an ideal I in $I(\mathfrak{g}_c)$ of finite codimension – then, according to Theorem 8.3.1.1, T agrees, in the sense of distributions, with a locally summable function F_T which is even analytic on $\Omega' = \Omega \cap \mathfrak{g}'$. The objective of the present number is to set down the basic properties which the function F_T possesses; we shall be particularly interested in the behavior of F_T and its derivatives on the singular set.

Let $\mathfrak{j}$ be a Cartan subalgebra of $\mathfrak{g}$; let $\mathfrak{j}'(R)$ denote the complement in $\mathfrak{j}$ of the set of zeros of the real roots of the pair $(\mathfrak{g}_c, \mathfrak{j}_c)$. Let $F_{T,\mathfrak{j}}$ be the analytic function on $\Omega \cap \mathfrak{j}'$ defined by the rule

$$F_{T,\mathfrak{j}}(H) = \pi(H)F_T(H) \qquad (H \in \Omega \cap \mathfrak{j}'),$$

π as always. [The motivation lying behind the introduction of $F_{T,\mathfrak{j}}$ is clear in view of the discussion in 8.3.1.]

Proposition 8.3.2.1 Retain the above notations – then $F_{T,\mathfrak{j}}$ can be extended to an analytic function on $\Omega \cap \mathfrak{j}'(R)$.

[For the proof, which is rather long and technical, see the Appendix to the present number.]

In what follows we shall denote the (analytic) extension of $F_{T,\mathfrak{j}}$ to $\Omega \cap \mathfrak{j}'(R)$ by the same symbol.

Proposition 8.3.2.2 Retain the above notations; let H_0 be a point in $\Omega \cap \mathfrak{j}$ and D an element in $\mathfrak{P}(\mathfrak{j})$ such that $w_\alpha(D) = -D$ for every real root α of the pair $(\mathfrak{g}_c, \mathfrak{j}_c)$ which vanishes at H_0 – then $DF_{T,\mathfrak{j}}$ can be extended to a continuous function around H_0 (i.e. there exists a neighborhood $\mathcal{O}(H_0)$ of H_0 in $\Omega \cap \mathfrak{j}$ and a continuous function F on $\mathcal{O}(H_0)$ such that $F = DF_{T,\mathfrak{j}}$ on $\mathcal{O}(H_0) \cap \mathfrak{j}'(R)$).

Proof The proof is entirely analogous to that of Theorem 8.4.4.1 infra. Thus fix a point H_0 in $\Omega \cap \mathfrak{j}$ per the statement of our proposition. Choose an open, convex, relatively compact neighborhood $\mathcal{O}(H_0)$ of H_0 in $\Omega \cap \mathfrak{j}$; things can clearly be arranged so as to ensure that no real root α of the pair $(\mathfrak{g}_c, \mathfrak{j}_c)$ vanishes anywhere on $\mathcal{O}(H_0)$ unless $\alpha(H_0) = 0$. Let $\mathcal{O}^{\ddagger}(H_0)$ be the set of all points in $\mathcal{O}(H_0)$ which are either regular or semi-regular – then $\mathcal{O}^{\ddagger}(H_0)$ is open and connected (cf. Lemma 8.4.4.5 below). Furthermore it follows from Proposition 8.3.2.1 and the lemma in the Appendix to this number that there exists a continuous function F on

$\mathcal{O}'(H_0)$ such that $DF_{T,\mathfrak{j}} = F$ on $\mathcal{O}'(H_0) \cap \mathfrak{j}'(R)$. Now the set $\mathcal{O}(H_0) \cap \mathfrak{j}'$ has only a finite number of connected components, say $\mathscr{C}_1, \ldots, \mathscr{C}_r$; in view of what we have seen in 8.3.1, there exists an analytic function F_i on $\mathfrak{j}$ such that $F_{T,\mathfrak{j}} = F_i$ on $\mathscr{C}_i$ $(1 \leqslant i \leqslant r)$ – therefore $DF_{T,\mathfrak{j}}$ is of class C^∞ on $\mathscr{C}_i^{cl}$ (cf. 8.4.4). But $\mathscr{C}_i^{cl} = (\mathscr{C}_i \cap \mathcal{O}'(H_0))^{cl}$ and $DF_{T,\mathfrak{j}} = F$ on $\mathscr{C}_i \cap \mathcal{O}'(H_0)$ – hence F is also of class C^∞ on $\mathscr{C}_i^{cl}$ $(1 \leqslant i \leqslant r)$. Fix a Euclidean norm $\|\,.\,\|$ on $\mathfrak{j}$ and put $\mu(F) = \sup |F(H_1; \partial(H_2))|$ where H_1 and H_2 vary in $\mathcal{O}'(H_0)$ and $\mathfrak{j}$, respectively, under the sole restriction that $\|H_2\| \leqslant 1$ – then certainly $\mu(F) < \infty$. Arguing now just as in the proof of Theorem 8.4.4.1 below, we deduce without difficulty that

$$|F(H_2) - F(H_1)| \leqslant \mu(F)\,\|H_2 - H_1\| \qquad (\text{all } H_1, H_2 \text{ in } \mathcal{O}'(H_0)).$$

This, however, obviously means that $DF_{T,\mathfrak{j}}$ can be extended to a continuous function on $\mathcal{O}(H_0)$. □

Corollary 8.3.2.3 Let D be an element in $\mathfrak{P}(\mathfrak{j})$ such that $w_\alpha(D) = -D$ for every real root of the pair $(\mathfrak{g}_c, \mathfrak{j}_c)$ – then $DF_{T,\mathfrak{j}}$ admits a continuous extension to $\Omega \cap \mathfrak{j}$.

On the basis of Proposition 8.3.2.2, this statement is obvious. The extended function will again be denoted by $DF_{T,\mathfrak{j}}$; the symbol $F_{T,\mathfrak{j}}(H; D)$ will stand for the value of $DF_{T,\mathfrak{j}}$ at H $(H \in \Omega \cap \mathfrak{j})$.

Making the customary identifications, define the differential operator $\partial(\pi)$ in the usual way (cf. 8.2.2) – then, in view of what has been said above, it is clear that $\partial(\pi)F_{T,\mathfrak{j}}$ is a continuous function on $\Omega \cap \mathfrak{j}$. [In passing, let us note that since the differential operator $\partial(\pi)\circ\pi$ does not depend on the choice of an ordering for the roots of the pair $(\mathfrak{g}_c, \mathfrak{j}_c)$, the same must be true of the function $\partial(\pi)F_{T,\mathfrak{j}}$.]

Proposition 8.3.2.4 Retain the above notations; let $\mathfrak{j}^1$ and $\mathfrak{j}^2$ be two Cartan subalgebras of $\mathfrak{g}$ – then

$$\partial(\pi^1)F_{T,\mathfrak{j}^1} = \partial(\pi^2)F_{T,\mathfrak{j}^2}$$

on $\Omega \cap \mathfrak{j}^1 \cap \mathfrak{j}^2$.

Here is an easy consequence of this result.

Let $D \in \mathfrak{Z}(\mathfrak{g})$ and define $\tilde{\gamma}(D)$ as in 8.2.2; bearing in mind Corollary 8.3.1.2 and the definition of $\tilde{\gamma}(D)$, it is clear that $F_{D\cdot T,\mathfrak{j}} = \tilde{\gamma}(D)F_{T,\mathfrak{j}}$ – therefore

$$\partial(\pi)F_{D\cdot T,\mathfrak{j}} = (\partial(\pi)\circ\tilde{\gamma}(D))F_{T,\mathfrak{j}}$$

admits a continuous extension to $\Omega \cap \mathfrak{j}$.

Corollary 8.3.2.5 Retain the above notations; let $\mathfrak{j}^1$ and $\mathfrak{j}^2$ be two Cartan subalgebras of $\mathfrak{g}$ – then

$$(\partial(\pi^1)\circ\tilde{\gamma}_1(D))F_{T,\mathfrak{j}^1} = (\partial(\pi^2)\circ\tilde{\gamma}_2(D))F_{T,\mathfrak{j}^2}$$

on $\Omega \cap \mathfrak{j}^1 \cap \mathfrak{j}^2$.

[This follows upon applying Proposition 8.3.2.4 to $D \cdot T$ instead of T.]

The proof of Proposition 8.3.2.4, which proceeds by induction on the dimension of $\mathfrak{g}$, is contained in the following discussion.

Fix a point $H_0 \in \Omega \cap \mathfrak{j}^1 \cap \mathfrak{j}^2$; we are to show that

$$F_{T,\mathfrak{j}^1}(H_0\,;\partial(\pi^1)) = F_{T,\mathfrak{j}^2}(H_0\,;\partial(\pi^2)).$$

First let us suppose that the center $\mathfrak{c}$ of $\mathfrak{g}$ is non-trivial – then $\dim(\bar{\mathfrak{g}}) < \dim(\mathfrak{g})$ and the induction hypothesis is applicable to $\bar{\mathfrak{g}}$. Write $H_0 = C_0 + \bar{H}_0$ ($C_0 \in \mathfrak{c}$, $\bar{H}_0 \in \bar{\mathfrak{g}}$); choose an open relatively compact neighborhood $\mathcal{O}(C_0)$ of C_0 in $\mathfrak{c}$; let $\bar{\Omega}$ denote the subset of $\bar{\mathfrak{g}}$ comprised of those $\bar{X}$ such that $[\mathcal{O}(C_0)]^{cl} + \bar{X} \subset \Omega$ – then $\bar{\Omega}$ is an open, completely invariant neighborhood of $\bar{H}_0$ in $\bar{\mathfrak{g}}$ provided $\mathcal{O}(C_0)$ is sufficiently small. Fix $g \in C_c^\infty(\mathcal{O}(C_0))$ and consider the distribution T_g on $\bar{\Omega}$ defined by the rule $T_g(h) = T(g \times h)$ ($h \in C_c^\infty(\bar{\Omega})$); it is obvious that T_g is $\bar{G}$-central ($\bar{G}$ the analytic subgroup of G corresponding to $\bar{\mathfrak{g}}$) and is annihilated by an ideal of finite codimension in $I(\bar{\mathfrak{g}}_c)$) – therefore Theorem 8.3.1.1 is applicable to the triple ($\bar{\mathfrak{g}}$, $\bar{\Omega}$, T_g). Furthermore, since $\dim(\bar{\mathfrak{g}}) < \dim(\mathfrak{g})$, Proposition 8.3.2.4 is applicable to T_g via the induction hypothesis. This being so, put

$$F_{g,\mathfrak{j}}(H) = \int_{\mathfrak{c}} g(C) F_{T,\mathfrak{j}}(C + H) d_{\mathfrak{c}}(C) \qquad (H \in \bar{\Omega} \cap \mathfrak{j}')$$

(here we have to bear in mind the relationship which exists between F_g and F_T (cf. 8.3.1)) – then it must be the case that

$$F_{g,\mathfrak{j}^1}(H\,;\partial(\pi^1)) = F_{g,\mathfrak{j}^2}(H\,;\partial(\pi^2)) \qquad (\text{all } H \in \bar{\Omega} \cap \mathfrak{j}^1 \cap \mathfrak{j}^2),$$

whence $\partial(\pi^1)F_{T,\mathfrak{j}^1}$ and $\partial(\pi^2)F_{T,\mathfrak{j}^2}$ coincide around H_0, g being an arbitrary element in $C_c^\infty(\mathcal{O}(C_0))$.

It can now be assumed that $\mathfrak{g}$ is semi-simple. To begin with, let us suppose that $H_0 = \Gamma$ is non-zero – then the centralizer $\mathfrak{g}_\Gamma$ of Γ in $\mathfrak{g}$ is reductive in $\mathfrak{g}$, of the same rank as $\mathfrak{g}$ and, moreover, $\dim(\mathfrak{g}_\Gamma) < \dim(\mathfrak{g})$. [Since Γ lies in a Cartan subalgebra of $\mathfrak{g}$, it is semi-simple . . . .] In what follows we shall identify $\mathfrak{g}_\Gamma$ with its dual by means of the restriction to $\mathfrak{g}_\Gamma$ of the Killing form of $\mathfrak{g}$. Introduce δ_Γ, $'\mathfrak{g}_\Gamma$ in the usual way; let $\mathfrak{j} = \mathfrak{j}^1$ or $\mathfrak{j}^2$ – then $\mathfrak{j} \subset \mathfrak{g}_\Gamma$. Let Φ denote the set of roots of the pair $(\mathfrak{g}_c, \mathfrak{j}_c)$, Φ_Γ the set of roots of the pair $(\mathfrak{g}_{\Gamma,c}, \mathfrak{j}_c)$, $\Phi_\daleth$ the complement of Φ_Γ in Φ – then, in the obvious notations, we have $\pi = \pi_\Gamma \pi_\daleth$ and it is clear that $w_\alpha(\pi_\daleth) = \pi_\daleth$ for all $\alpha \in \Phi_\Gamma^+$. Consequently, in view of Chevalley's Restriction Theorem, there exists an element $p \in I(\mathfrak{g}_{\Gamma,c})$ which, when restricted to $\mathfrak{j}^1$, gives $\pi_\daleth^1$; but $\delta_\Gamma | \mathfrak{j}^1 = (-1)^{r_\daleth}(\pi_\daleth^1)^2$ ($r_\daleth$ the cardinality of $\Phi_\daleth^+$) – thus, by Chevalley's Theorem again, $\delta_\Gamma = (-1)^{r_\daleth} p^2$ and so p, when restricted to $\mathfrak{j}^2$, gives $\epsilon\pi_\daleth^2$ ($\epsilon = \pm 1$). Now put $\Omega_\Gamma = \Omega \cap {'\mathfrak{g}_\Gamma}$ – then Ω_Γ is an open neighborhood of Γ in $\mathfrak{g}_\Gamma$ which is completely invariant (relative to $\mathfrak{g}_\Gamma$). Let $\tilde{T}$ denote the distribution on Ω_Γ corresponding to T via the customary procedure; let $T_\Gamma = p\tilde{T}$ – then, thanks to Theorem 8.2.2.4(bis), Theorem

8.3.1.1 is applicable to the triple $(\mathfrak{g}_\Gamma, \Omega_\Gamma, T_\Gamma)$. Because $\dim(\mathfrak{g}_\Gamma) < \dim(\mathfrak{g})$, Proposition 8.3.2.4 and its corollary are also applicable to the triple $(\mathfrak{g}_\Gamma, \Omega_\Gamma, T_\Gamma)$ (through the induction hypothesis); thus, in view of the definition of $\tilde{\mathfrak{I}}$, we may infer that

$$\partial(\pi_\Gamma^1 \tilde{p}_1) F_{T_\Gamma, \mathfrak{j}^1} = \partial(\pi_\Gamma^2 \tilde{p}_2) F_{T_\Gamma, \mathfrak{j}^2} \qquad (\tilde{p}_i = p \mid \mathfrak{j}_c^i, i = 1, 2)$$

on $\Omega_\Gamma \cap \mathfrak{j}^1 \cap \mathfrak{j}^2$. [Here, of course,

$$F_{T_\Gamma, \mathfrak{j}}(H) = \pi_\Gamma(H) p(H) F_T(H) \qquad (\text{all } H \in \Omega_\Gamma \cap \mathfrak{j}').]$$

However $\pi_\Gamma^1 \tilde{p}_1 = \pi^1$, $\pi_\Gamma^2 \tilde{p}_2 = \epsilon \pi^2$, whence $F_{T_\Gamma, \mathfrak{j}^1} = F_{T, \mathfrak{j}^1}$ and $F_{T_\Gamma, \mathfrak{j}^2} = F_{T, \mathfrak{j}^2}$ on $\Omega_\Gamma \cap (\mathfrak{j}^1)'$ and $\Omega_\Gamma \cap (\mathfrak{j}^2)'$, respectively; this shows that

$$\partial(\pi^1) F_{T, \mathfrak{j}^1} = \partial(\pi^2) F_{T, \mathfrak{j}^2}$$

on $\Omega_\Gamma \cap \mathfrak{j}^1 \cap \mathfrak{j}^2$ which, since $\Gamma \in \Omega_\Gamma \cap \mathfrak{j}^1 \cap \mathfrak{j}^2$, serves to establish our contention in this case.

It remains to consider the case when $H_0 = 0$ (with, of course, $\mathfrak{g}$ semi-simple); we shall deal with it by means of the following lemma.

Lemma 8.3.2.6 Retain the immediately preceding notations but assume in addition that $0 \in \Omega$; given any Cartan subalgebra $\mathfrak{j}$ of $\mathfrak{g}$, put $c(\mathfrak{j}) = F_{T, \mathfrak{j}}(0; \partial(\pi))$ – then the value of $c(\mathfrak{j})$ is the same for all Cartan subalgebras $\mathfrak{j}$ of $\mathfrak{g}$.

Proof We shall break the proof of this lemma up into several steps.

(1) Let $\mathfrak{j}^1$ and $\mathfrak{j}^2$ be two Cartan subalgebras of $\mathfrak{g}$: If $\mathfrak{j}^1 \cap \mathfrak{j}^2 \neq \{0\}$, then $c(\mathfrak{j}^1) = c(\mathfrak{j}^2)$. [Because Ω is an open neighborhood of zero in $\mathfrak{g}$, there exists a non-zero H in $\mathfrak{j}^1 \cap \mathfrak{j}^2$ such that

$$tH \in \Omega \cap \mathfrak{j}^1 \cap \mathfrak{j}^2 \qquad (0 \leqslant t \leqslant 1);$$

in view of what was proved above, it is clear that

$$F_{T, \mathfrak{j}^1}(tH; \partial(\pi^1)) = F_{T, \mathfrak{j}^2}(tH; \partial(\pi^2)) \qquad (0 < t \leqslant 1).$$

Our assertion now follows upon letting t approach zero.]

(2) Let $\mathfrak{j}$ be a Cartan subalgebra of $\mathfrak{g}$ – then $c(\mathfrak{j}) = c(x(\mathfrak{j}))$ (all $x \in G$). [One may assume, without loss of generality, that $F_{T, x(\mathfrak{j})}$ is defined by $x(\pi)$; thus

$$F_{T, x(\mathfrak{j})}(xH) = F_{T, \mathfrak{j}}(H) \qquad (H \in \Omega \cap \mathfrak{j}')$$

and so, as in (1), we need only differentiate and let H approach zero.]

(3) Fix a Cartan involution θ of $\mathfrak{g}$ and let $\mathfrak{g} = \mathfrak{k} + \mathfrak{p}$ be the corresponding Cartan decomposition (evidently we can now assume that $\mathfrak{g}$ is not compact). If $\mathfrak{j}$ is a θ-stable Cartan subalgebra of $\mathfrak{g}$, put $l_+(\mathfrak{j}) = \dim(\mathfrak{j} \cap \mathfrak{p})$, $l_-(\mathfrak{j}) = \dim(\mathfrak{j} \cap \mathfrak{k})$ (thus $l_+(\mathfrak{j}) + l_-(\mathfrak{j}) = l$, l the rank of $\mathfrak{g}$); write $l_+ = \sup_\mathfrak{j} l_+(\mathfrak{j})$, $l_- = \sup_\mathfrak{j} l_-(\mathfrak{j})$ ($\mathfrak{j}$ running through the θ-stable Cartan subalgebras of $\mathfrak{g}$) and fix two θ-stable Cartan subalgebras $\mathfrak{j}^+$ and $\mathfrak{j}^-$ (say) such that $l_+ = l_+(\mathfrak{j}^+)$, $l_- = l_-(\mathfrak{j}^-)$. Let $\mathfrak{j}$ be an arbitrary θ-stable

Cartan subalgebra of $\mathfrak{g}$ – then $c(\mathfrak{j}) = c(\mathfrak{j}^+)$ if $l_+(\mathfrak{j}) > 0$ while $c(\mathfrak{j}) = c(\mathfrak{j}^-)$ if $l_-(\mathfrak{j}) > 0$. [To see this, let K be the analytic subgroup of G corresponding to $\mathfrak{k}$ – then $\mathfrak{j}^+ \cap \mathfrak{p}$ (respectively $\mathfrak{j}^- \cap \mathfrak{k}$) is a maximal abelian subspace of $\mathfrak{p}$ (respectively $\mathfrak{k}$) and so, for well-known reasons, we can choose elements k^+, k^- in K with the property that $k^+(\mathfrak{j} \cap \mathfrak{p}) \subset \mathfrak{j}^+ \cap \mathfrak{p}$, $k^-(\mathfrak{j} \cap \mathfrak{k}) \subset \mathfrak{j}^- \cap \mathfrak{k}$. It follows, therefore, that $\dim (k^+(\mathfrak{j}) \cap \mathfrak{j}^+) \geqslant l_+(\mathfrak{j})$, $\dim (k^-(\mathfrak{j}) \cap \mathfrak{j}^-) \geqslant l_-(\mathfrak{j})$. Our contention is then seen to be a consequence of (1) and (2).]

(4) Let $\mathfrak{j}^+$ and $\mathfrak{j}^-$ be as in (3) – then $c(\mathfrak{j}^+) = c(\mathfrak{j}^-)$. [Our statement follows from (3) if $l_-(\mathfrak{j}^+) + l_+(\mathfrak{j}^-) \geqslant 1$; hence we may assume that $\mathfrak{j}^+ \subset \mathfrak{p}$ and $\mathfrak{j}^- \subset \mathfrak{k}$. Because $\mathfrak{j}^+$ is contained in $\mathfrak{p}$, the pair $(\mathfrak{g}_c, \mathfrak{j}_c^+)$ admits at least one positive real root α_0 (say); using this α_0, construct H^*, X^*, Y^* as in the discussion near the end of Vol. I, number 1.3.4 and let ν be the automorphism of $\mathfrak{g}_c$ per Lemma 1.3.4.4 – then

$$\mathfrak{h}^- = \nu(\mathfrak{j}_c^+) \cap \mathfrak{g} = \sigma_{\alpha_0} + \mathbf{R}(X^* - Y^*)$$

is a θ-stable Cartan subalgebra of $\mathfrak{g}$. [Here σ_{α_0} is the hyperplane consisting of those H in $\mathfrak{j}^+$ where $\alpha_0(H) = 0$.] Since $\mathfrak{h}^- \cap \mathfrak{p} = \sigma_{\alpha_0}$ and $\mathfrak{h}^- \cap \mathfrak{k} = \mathbf{R}(X^* - Y^*)$, it is obvious that $l_+(\mathfrak{h}^-) = l - 1$ and $l_-(\mathfrak{h}^-) = 1$; therefore, if $l \geqslant 2$, then, on the basis of (3), it must be the case that $c(\mathfrak{j}^+) = c(\mathfrak{j}^-)$; on the other hand, if $l = 1$, then zero is a semi-regular element of non-compact type and so our assertion follows from the lemma in the Appendix to this number.]

(5) Let $\mathfrak{j}^1$ and $\mathfrak{j}^2$ be any two Cartan subalgebras of $\mathfrak{g}$ – then $c(\mathfrak{j}^1) = c(\mathfrak{j}^2)$. [Without loss of generality it can be assumed that both $\mathfrak{j}^1$ and $\mathfrak{j}^2$ are θ-stable (cf. Proposition 1.3.1.1 and bear in mind (2) above); but then the results obtained in (3) and (4) lead at once to the sought for conclusion.]

This completes the proof of the lemma. □

Thanks to Lemma 8.3.2.6 and the discussion preceding it, Proposition 8.3.2.4 is now established.

In order to describe the next result on F_T, we shall need an elementary lemma.

Lemma 8.3.2.7 There exists a unique G-invariant analytic differential operator $\nabla_{\mathfrak{g}}$ on $\mathfrak{g}'$ with the following property: Let $\mathfrak{j}$ be a Cartan subalgebra of $\mathfrak{g}$ – then $f(H; \nabla_{\mathfrak{g}}) = f(H; \partial(\pi)\circ\pi)$ for $f \in C^\infty(\mathfrak{g})$ (all $H \in \mathfrak{j}'$).

Proof Because two distinct Cartan subalgebras cannot have a regular element in common, the uniqueness of $\nabla_{\mathfrak{g}}$ is plain. As for the existence, one may proceed as follows. Let J be the Cartan subgroup of G associated with $\mathfrak{j}$ (thus, by definition, J is the centralizer of $\mathfrak{j}$ in G); let $x \mapsto \dot{x}$ denote the natural projection of G onto G/J – then, in view of Proposition 1.4.2.2, the mapping ϕ, $(\dot{x}, H) \mapsto \dot{x}H = xH$, of $G/J \times \mathfrak{j}'$ onto $\mathfrak{g}(\mathfrak{j})$ $(= \bigcup_{x \in G} x(\mathfrak{j}'))$ is an everywhere regular, $[W(G, J)]$-to-one map. Since ϕ is locally an analytic diffeomorphism and since $\partial(\pi)\circ\pi$ is invariant under $W(G, J)$, it is clear that there exists a G-invariant analytic differential operator ∇ on $\mathfrak{g}(\mathfrak{j})$ such that

$$f(\dot{x}H;\nabla)=f(\dot{x}:H;\partial(\pi)\circ\pi) \qquad (\dot{x}\in G/J,\ H\in\mathfrak{j}')$$

for $f\in C^{\infty}(\mathfrak{g}(\mathfrak{j}))$ (obvious notation). This being so, choose now a maximal set $\mathfrak{j}^1,\ldots,\mathfrak{j}^r$ of Cartan subalgebras of $\mathfrak{g}$, no two of which are conjugate under G; define the differential operator ∇_i on $\mathfrak{g}(\mathfrak{j}^i)$ as above ($i=1,\ldots,r$); bearing in mind that $\mathfrak{g}'$ is the disjoint union of the open sets $\mathfrak{g}(\mathfrak{j}^i)$ (cf. Proposition 1.3.4.1), we see that $\nabla_{\mathfrak{g}}$ can be defined by setting $\nabla_{\mathfrak{g}}=\nabla_i$ on $\mathfrak{g}(\mathfrak{j}^i)$ ($i=1,\ldots,r$). □

Proposition 8.3.2.8 Retain the above notations; let $D\in\mathfrak{Z}(\mathfrak{g})$ – then $(\nabla_{\mathfrak{g}}\circ D)F_T$ admits a continuous extension to all of Ω.

Proof As might be expected, the proof proceeds by induction on the dimension of $\mathfrak{g}$. In view of Corollary 8.3.1.2, we need only deal explicitly with the case $D=1$.

(1) Let us assume that the center $\mathfrak{c}$ of $\mathfrak{g}$ is not trivial; fix a point $X_0=C_0+\bar{X}_0$ in Ω ($C_0\in\mathfrak{c}$, $\bar{X}_0\in\bar{\mathfrak{g}}$) – then we have to prove that $\nabla_{\mathfrak{g}}F_T$ can be extended to a continuous function around X_0. Choose an open, relatively compact, connected neighborhood $\mathcal{O}(C_0)$ of C_0 in $\mathfrak{c}$ such that $[\mathcal{O}(C_0)]^{cl}+\bar{X}_0\subset\Omega$; let $\bar{\Omega}$ be the set of all $\bar{X}\in\bar{\mathfrak{g}}$ such that

$$[\mathcal{O}(C_0)]^{cl}+\bar{X}\subset\Omega.$$

Then $\bar{\Omega}$ is an open, completely invariant subset of $\bar{\mathfrak{g}}$. Because $I\cap S(\mathfrak{c}_c)$ is an ideal of finite codimension in $S(\mathfrak{c}_c)$, the space E_I comprised of those analytic functions ϕ on $\mathcal{O}(C_0)$ such that $\partial(u)\phi=0$ (all $u\in I\cap S(\mathfrak{c}_c)$) is finite dimensional (cf. the Appendix to 8.3.1); let ϕ_i ($1\leqslant i\leqslant n$) be a basis for E_I over $\mathbf{C}$. Fix an $\bar{X}\in\bar{\Omega}'=\bar{\Omega}\cap\mathfrak{g}'$ – then it is obvious that $F_T(C+\bar{X};\partial(u))=0$ (all $u\in I\cap S(\mathfrak{c}_c)$) ($C\in\mathcal{O}(C_0)$) and so

$$F_T(C+\bar{X})=\sum_{i=1}^{n}\phi_i(C)F_i(\bar{X}) \qquad (C\in\mathcal{O}(C_0))$$

for certain complex numbers $F_i(\bar{X})$. Thanks to the analyticity of F_T on Ω', the F_i ($1\leqslant i\leqslant n$) may be regarded as analytic functions on $\bar{\Omega}'$. Given $g\in C_c^{\infty}(\mathcal{O}(C_0))$, let T_g denote the distribution on $\bar{\Omega}$ defined by the rule $T_g(h)=T(g\times h)$ ($h\in C_c^{\infty}(\bar{\Omega})$) – then, as we know, the induction hypothesis is applicable to the triple $(\bar{\mathfrak{g}},\bar{\Omega},T_g)$, whence, in particular, $\nabla_{\bar{\mathfrak{g}}}F_g$ admits a continuous extension to $\bar{\Omega}$. But

$$F_g(\bar{X})=\sum_{i=1}^{n}F_i(\bar{X})\int_{\mathfrak{c}}g(C)\phi_i(C)d_{\mathfrak{c}}(C) \qquad (\bar{X}\in\bar{\Omega}')$$

and since this is true for every $g\in C_c^{\infty}(\mathcal{O}(C_0))$, it follows from the linear independence of the ϕ_i that $\nabla_{\bar{\mathfrak{g}}}F_i$ can also be extended continuously to $\bar{\Omega}$ ($1\leqslant i\leqslant n$). [We have tacitly employed Lemma 8.3.5.4 infra.] The relation

$$F_T(C+\bar{X};\nabla_{\mathfrak{g}})=\sum_{i}\phi_i(C)F_i(\bar{X};\nabla_{\bar{\mathfrak{g}}}) \qquad (C\in\mathcal{O}(C_0),\ \bar{X}\in\bar{\Omega})$$

now implies that $\nabla_{\mathfrak{g}} F_T$ extends to a continuous function on $\mathcal{O}(C_0) + \bar{\Omega}$, as desired.

(2) The discussion in (1) implies that $\mathfrak{g}$ can be taken semi-simple. This being so, let Ω_0 be the set of all points $X_0 \in \Omega$ such that $\nabla_{\mathfrak{g}} F_T$ can be extended to a continuous function around X_0; clearly Ω_0 is an open, G-stable subset of Ω – therefore, to conclude that $\Omega_0 = \Omega$, we need only verify that every semi-simple element of Ω lies in Ω_0 (cf. Lemma 8.1.6.1 and subsequent observations).

(a) Fix a semi-simple element $\Gamma \in \Omega$ and assume first that Γ is non-zero – then the notations and results which were obtained during the proof of Proposition 8.3.2.4 are available. Hence, in particular, the induction hypothesis is applicable to the triple $(\mathfrak{g}_\Gamma, \Omega_\Gamma, T_\Gamma)$, whence $(\nabla_\Gamma \circ \partial(p))(p\tilde{F}_T)$ extends to a continuous function f_Γ (say) on Ω_Γ (bear in mind that $T_\Gamma = p\tilde{T} = p\tilde{F}_T$, $\tilde{F}_T$ the restriction of F_T to Ω_Γ). Let G_Γ denote the analytic subgroup of G corresponding to $\mathfrak{g}_\Gamma$, $x \mapsto \dot{x}$ the natural projection of G onto G/G_Γ; choose a local analytic cross section s for G/G_Γ over G_Γ and a small neighborhood $\mathcal{O}(\Gamma)$ of Γ in Ω_Γ such that the mapping ϕ,

$$(\dot{x}, X_\Gamma) \mapsto s(\dot{x})X_\Gamma \qquad (X_\Gamma \in \mathcal{O}(\Gamma)),$$

is an analytic diffeomorphism onto its image $N_{\mathfrak{g}}(\Gamma)$ (say) which we may assume to be connected and contained in Ω – this is clearly possible to arrange. Extend f_Γ to $N_{\mathfrak{g}}(\Gamma)$ by writing $f_\Gamma(\phi(\dot{x}, X_\Gamma)) = f_\Gamma(X_\Gamma)$ – then f_Γ is certainly continuous on $N_{\mathfrak{g}}(\Gamma)$ and, moreover, $f_\Gamma = \nabla_{\mathfrak{g}} F_T$ on $N_{\mathfrak{g}}(\Gamma) \cap \mathfrak{g}'$ which serves to establish our contention in this case. [Let $X \in N_{\mathfrak{g}}(\Gamma) \cap \mathfrak{g}'$; write $X = \phi(\dot{x}, X_\Gamma)$ – then, since $\nabla_{\mathfrak{g}} F_T$ is G-central, $F_T(X; \nabla_{\mathfrak{g}}) = F_T(X_\Gamma; \nabla_{\mathfrak{g}})$ and so it will be enough to verify that $f_\Gamma(X_\Gamma) = F_T(X_\Gamma; \nabla_{\mathfrak{g}})$. Because X_Γ is regular in both $\mathfrak{g}$ and $\mathfrak{g}_\Gamma$, its centralizer $\mathfrak{j}$ in $\mathfrak{g}_\Gamma$ is a Cartan subalgebra of $\mathfrak{g}_\Gamma$ and $X_\Gamma \in \Omega \cap \mathfrak{j}'$ – therefore

$$F_T(X_\Gamma; \nabla_{\mathfrak{g}}) = F_T(X_\Gamma; \partial(\pi) \circ \pi) = F_T(X_\Gamma; \partial(\pi_\Gamma) \circ \pi_\Gamma \circ \mathfrak{P}_\Gamma(\partial(p) \circ p)),$$

where, of course, $\mathfrak{P}_\Gamma$ is computed relative to the pair $(\mathfrak{g}_\Gamma, \mathfrak{j})$ (plainly $\mathfrak{P}_\Gamma(\partial(p) \circ p) = \pi_\Gamma^{-1} \partial(\pi_1) \circ \pi)$). It is clear that $\tilde{F}_T$ is G_Γ-central – thus

$$\tilde{F}_T(H; \partial(p) \circ p) = \tilde{F}_T(H; \mathfrak{P}_\Gamma(\partial(p) \circ p)) \qquad (\text{all } H \in \Omega_\Gamma \cap \mathfrak{j}')$$

and so

$$f_\Gamma(X_\Gamma) = \tilde{F}_T(X_\Gamma; \nabla_\Gamma \circ (\partial(p) \circ p)) = \tilde{F}_T(X_\Gamma; \partial(\pi_\Gamma) \circ \pi_\Gamma \circ \mathfrak{P}_\Gamma(\partial(p) \circ p))$$
$$= F_T(X_\Gamma; \nabla_{\mathfrak{g}}).]$$

(b) It remains to consider the possibility that $0 \in \Omega$. In this eventuality, the set $\mathcal{N}$ of nilpotent elements in $\mathfrak{g}$ is contained in Ω (cf. Lemma 8.1.6.2); moreover, we know that $\nabla_{\mathfrak{g}} F_T$ can be extended to a continuous function $\mathfrak{F}_T$ (say) on $\Omega \cap \complement \mathcal{N}$ – therefore it will be sufficient to establish the following result: There exists a number c such that if $\{X_k\}$ is a sequence in Ω' which converges to some element X in $\mathcal{N}$, then $\mathfrak{F}_T(X_k) \to c$. [Define $\mathfrak{j}^i$ and $\mathfrak{g}(\mathfrak{j}^i)$ as during the proof of Lemma 8.3.2.7 $(1 \leqslant i \leqslant r)$; define the numbers $c(\mathfrak{j}^i)$ per Lemma 8.3.2.6 ($1 \leqslant i \leqslant r$) – then $c(\mathfrak{j}^1) = \ldots = c(\mathfrak{j}^r) = c$ (say). Because $\mathfrak{g}'$ is the union of the $\mathfrak{g}(\mathfrak{j}^i)$, for each index k we can select an index i_k and elements $x_k \in G$, $H_k \in (\mathfrak{j}^{i_k})'$ with the property that $X_k = x_k H_k$; since $X_k \to X$, $H_k \to 0$ (cf. Lemma 8.1.6.3), whence

$$\mathfrak{F}_T(H_k) = F_T(H_k; \nabla_{\mathfrak{g}}) \to c \Rightarrow \mathfrak{F}_T(X_k) \to c \ldots .]$$

This completes the proof of the proposition. □

The preceding proposition tells us that $\mathfrak{F}_T = \nabla_{\mathfrak{g}} F_T$ extends to a continuous function on Ω. For later work it will be important to know that certain derivatives of $\mathfrak{F}_T$ are locally summable on Ω. So let $D \in \mathfrak{Z}(\mathfrak{g})$ – then we claim that $D\mathfrak{F}_T$ is locally summable on Ω. Thus, to begin with, suppose that $\mathfrak{j}$ is a Cartan subalgebra of $\mathfrak{g}$, D a polynomial differential operator on $\mathfrak{j}$ – then $D\tilde{\mathfrak{F}}_T$ is locally bounded on $\Omega \cap \mathfrak{j}$, that is, $D\tilde{\mathfrak{F}}_T$ is bounded on $\omega \cap \mathfrak{j}'$ for every compact subset ω of $\Omega \cap \mathfrak{j}$ (tilde being restriction to $\mathfrak{j}$). [Fix a point $H_0 \in \Omega \cap \mathfrak{j}$; fix a positive definite quadratic form Q on $\mathfrak{j}$ and for any $\epsilon > 0$, let $\mathcal{O}(\epsilon)$ denote the set of all $H \in \mathfrak{j}$ such that $Q(H - H_0) < \epsilon^2$ – then $\mathcal{O}(\epsilon) \subset \Omega$ provided ϵ is chosen sufficiently small. Moreover $\mathcal{O}'(\epsilon)(= \mathcal{O}(\epsilon) \cap \mathfrak{g}')$ has but finitely many connected components (cf. the proof of Lemma 8.3.1.3). Because $D\tilde{\mathfrak{F}}_T$ remains bounded on each component of $\mathcal{O}'(\epsilon)$, it must therefore remain bounded on $\mathcal{O}'(\epsilon) \ldots .]$ This being the case, suppose now that $D \in \mathfrak{Z}(\mathfrak{g})$ – then

$$(D\mathfrak{F}_T)^{\sim} = \mathring{\gamma}(D)\tilde{F}_T = \pi^{-1}(\mathring{\gamma}(D) \circ \pi)\tilde{\mathfrak{F}}_T$$

(for the definition of $\mathring{\gamma}$, see the discussion following the proof of Theorem 8.2.2.4). But $\mathring{\gamma}(D) \circ \pi \in \mathfrak{P}(\mathfrak{j})$; therefore, in view of the earlier observation, $\pi(D\mathfrak{F}_T)^{\sim}$ is locally bounded on $\Omega \cap \mathfrak{j}$, whence $|d_l|^{1/2} \, |D\mathfrak{F}_T|$ is locally bounded on Ω (cf. the proof of Lemma 8.3.1.3). Because $|d_l|^{-1/2}$ is locally summable (cf. 8.4.1 infra), the local summability of $D\mathfrak{F}_T$ ($D \in \mathfrak{Z}(\mathfrak{g})$) is clear.

Here is an important consequence of Proposition 8.3.2.8.

Proposition 8.3.2.9 Retain the above notations; suppose that F_T is locally constant on Ω' (that is constant on each component of Ω') – then T is locally constant on Ω.

Proof It is not difficult to see that there exists a positive number M such that $\partial(\pi)\pi = M$ for every Cartan subalgebra $\mathfrak{j}$ of $\mathfrak{g}$. [One can reason directly or else apply Proposition 8.3.2.4 with $T = 1$ and $\Omega = \mathfrak{g}$.] Since F_T is locally constant on Ω', it must therefore be the case that $\nabla_{\mathfrak{g}} F_T = MF_T$; on the basis of Proposition 8.3.2.8, we may then conclude that F_T can be extended to a continuous function on Ω. Hence T is locally constant on Ω. □

Appendix Agreeing to keep to the above notations, let us prove that $F_{T,\mathfrak{j}}$ can be extended to an analytic function on $\Omega \cap \mathfrak{j}'(R)$. For this purpose, fix a point H_0 in $\Omega \cap \mathfrak{j}'(R)$ – then it will be enough to prove that there exists an analytic function on an open neighborhood $\mathcal{O}(H_0)$ of H_0 in $\Omega \cap \mathfrak{j}'(R)$ which has the property that its restriction to $\mathcal{O}(H_0) \cap \mathfrak{j}'$ agrees with $F_{T,\mathfrak{j}}$ there.

Step I To begin with, let us suppose that $H_0 = \Gamma$ is a semi-regular element of *compact* type. Let $\mathfrak{g}_\Gamma$ denote the centralizer of Γ in $\mathfrak{g}$ and introduce δ_Γ and $'\mathfrak{g}_\Gamma$ in the usual way; let G_Γ be the analytic subgroup of G corresponding to $\mathfrak{g}_\Gamma$, $x \mapsto \dot{x}$ the natural mapping of G onto G/G_Γ; choose a local analytic cross section s for G/G_Γ over G_Γ and a small G_Γ-stable neighborhood Ω_Γ of Γ in $'\mathfrak{g}_\Gamma$ such that the mapping ϕ, $(\dot{x}, X_\Gamma) \mapsto s(\dot{x})X_\Gamma$ ($X_\Gamma \in \Omega_\Gamma$), is an analytic diffeomorphism onto its image $N_\mathfrak{g}(\Gamma)$ (say) which we may assume to be connected and contained in Ω. [The fact that the derived algebra of $\mathfrak{g}_\Gamma$ is compact serves to ensure us that Ω_Γ can be taken to be G_Γ-invariant.] Let $\tilde{T}$ be the distribution on Ω_Γ which corresponds to T via the customary procedure; put $\sigma = |\delta_\Gamma|^{1/2}\tilde{T}$ – then, of course, σ is annihilated by an ideal I_Γ in $I(\mathfrak{g}_{\Gamma,c})$ of finite codimension. However, since the derived algebra of $\mathfrak{g}_\Gamma$ is, in the present case, compact, it follows readily that σ actually satisfies an elliptic differential equation with constant coefficients, hence coincides with an analytic function F_σ (say) on Ω_Γ which is in fact G_Γ-central (Ω_Γ being G_Γ-stable). Consider now the analytic function F on $N_\mathfrak{g}(\Gamma)$ defined by the rule

$$F(\phi(\dot{x}, X_\Gamma)) = |\delta_\Gamma(X_\Gamma)|^{-1/2}F_\sigma(X_\Gamma) \qquad (X_\Gamma \in \Omega_\Gamma).$$

Then the usual argument leads at once to the conclusion that $F_T = F$ on $N_\mathfrak{g}(\Gamma)$ which proves, of course, that $F_{T,\mathfrak{j}}$ extends to an analytic function around Γ in $\Omega \cap \mathfrak{j}'(R)$ (and even somewhat more . . .).

Step II Let us now suppose that $H_0 = \Gamma$ is a semi-regular element of *non-compact* type – then the notations and results revolving around Lemmas 1.3.4.3 and 1.3.4.4 are available (in particular it can be assumed that $\mathfrak{j} = \mathfrak{h}_\Gamma^- \ldots$). Introduce G_Γ, $\mathfrak{g}_\Gamma$ $(= \mathfrak{c}_\Gamma + \mathfrak{l}_\Gamma \ldots)$, δ_Γ, and $'\mathfrak{g}_\Gamma$ per usual; let Ω_Γ be the set comprised of those X_Γ in $\Omega \cap \mathfrak{g}_\Gamma$ such that $|\delta_\Gamma(X_\Gamma)| > |\delta_\Gamma(\Gamma)|/2$ – then Ω_Γ is an open neighborhood of Γ in $'\mathfrak{g}_\Gamma$ which is completely invariant (with respect to $\mathfrak{g}_\Gamma$). Fix an open, convex neighborhood $\Omega_\mathfrak{c}$ $(\mathfrak{c} = \mathfrak{c}_\Gamma \ldots)$ of Γ in $\mathfrak{c}$ such that $[\Omega_\mathfrak{c}]^{cl}$ is compact and contained in Ω_Γ; let $\Omega_\mathfrak{l}$ $(\mathfrak{l} = \mathfrak{l}_\Gamma \ldots)$ denote the set of all Z in $\mathfrak{l}$ such that $[\Omega_\mathfrak{c}]^{cl} + Z \subset \Omega_\Gamma$ – then it is clear that $\Omega_\mathfrak{l}$ is a completely invariant neighborhood of zero in $\mathfrak{l}$. Proceeding in the familiar fashion, put $\sigma = |\delta_\Gamma|^{1/2}\tilde{T}$ – then σ is annihilated by an ideal of finite codimension in $I(\mathfrak{g}_{\Gamma,c})$. This being the case, let ω be the Casimir polynomial for $\mathfrak{l}$ – then there exist complex numbers $c_0 = 1, c_1, \ldots, c_m$ such that

$$\partial(\omega)^m \cdot \sigma + \sum_{i=1}^{m} c_i \partial(\omega)^{m-i} \cdot \sigma = 0.$$

Now fix a g in $C_c^\infty(\Omega_\mathfrak{c})$ and let σ_g denote the distribution on $\Omega_\mathfrak{l}$ defined by the rule $\sigma_g(h) = \sigma(g \times h)$ ($h \in C_c^\infty(\Omega_\mathfrak{l})$) – then σ_g is obviously L-central (L the analytic subgroup of G corresponding to $\mathfrak{l}$) and, moreover, we have the relation

$$\partial(\omega)^m \cdot \sigma_g + \sum_{i=1}^{m} c_i \partial(\omega)^{m-i} \cdot \sigma_g = 0.$$

Since $I(\mathfrak{l}_c) = \mathbf{C}[\omega]$, Theorem 8.3.1.1 and its corollary are applicable to the triple $(\mathfrak{l}, \Omega_\mathfrak{l}, \sigma_g)$; thus there exists an analytic function σ_g (say) on $\Omega_\mathfrak{l}'$ (the points in $\Omega_\mathfrak{l}$ which are regular in $\mathfrak{l}$) which is locally summable on $\Omega_\mathfrak{l}$ and such that

$$\int_\mathfrak{l} \{\partial(\omega)^n h \cdot \sigma_g - h \cdot \partial(\omega)^n \sigma_g\}\, d_\mathfrak{l}(Z) = 0 \qquad (\text{all } h \in C_c^\infty(\Omega_\mathfrak{l}))$$

for $n \geqslant 0$. Given $\epsilon > 0$, let $\Omega_{\mathfrak{l}}(\epsilon)$ denote the subset of $\mathfrak{l}$ consisting of those Z such that $|\omega(Z)| < 8\epsilon^2$; by taking ϵ sufficiently small, we can arrange that tH^* and $t(X^* - Y^*)$ both lie in $\Omega_{\mathfrak{l}}$ whenever $|t| \leqslant \epsilon$ $(t \in \mathbf{R})$, hence $[\Omega_{\mathfrak{l}}(\epsilon)]^{cl} \subset \Omega_{\mathfrak{l}}$. Evidently, then, there exist three analytic functions w_g, u_g, v_g on $\mathbf{R}$ such that

$$\begin{aligned} w_g(\xi) &= \xi\sigma_g(\xi H^*) && (0 < \xi \leqslant \epsilon), \\ u_g(\xi) &= \xi\sigma_g(\xi(X^* - Y^*)) && (0 < \xi \leqslant \epsilon), \\ v_g(\xi) &= \xi\sigma_g(\xi(X^* - Y^*)) && (-\epsilon \leqslant \xi < 0). \end{aligned}$$

Define the distributions T_n $(n \geqslant 0)$ on $\mathfrak{l}$ as in Corollary 8.4.2.8 infra with (w, u, v) replaced by (w_g, u_g, v_g) – then it follows easily from the explicit computations on $\mathfrak{l}$ which are given in 8.4.2 below that necessarily

$$T_n(h) = M \int_{\mathfrak{l}} \{\partial(\omega)^n h \cdot \sigma_g - h \cdot \partial(\omega)^n \sigma_g\} d_{\mathfrak{l}}(Z) = 0$$

for all $h \in C_c^\infty(\Omega_{\mathfrak{l}}(\epsilon/2))$ (M a certain positive constant). We therefore may conclude from Corollary 8.4.2.8 that $u_g = v_g$. Let I_ϵ denote the open interval $(-\epsilon, \epsilon)$ in $\mathbf{R}$; set $\mathfrak{j}(\epsilon) = \Omega_{\mathfrak{c}} + I_\epsilon(X^* - Y^*)$ $(\mathfrak{j} = \mathfrak{h}^- (= \mathfrak{h}_\Gamma^-) \ldots)$ – then it is clear that for all C in $\Omega_{\mathfrak{c}}$, we have

$$\xi |\delta_\Gamma(C + \xi(X^* - Y^*))|^{1/2} = 2^{-1}\sqrt{-1}\gamma\pi(C + \xi(X^* - Y^*)) \qquad (|\xi| < \epsilon)$$

where $\gamma = |\pi(\Gamma)\alpha^{-1}(\Gamma)|/\pi(\Gamma)\alpha^{-1}(\Gamma)$ (α the unique positive singular imaginary root of the pair $(\mathfrak{g}_c, \mathfrak{j}_c)$ which vanishes at Γ, $\alpha(X^* - Y^*) = -2\sqrt{-1} \ldots$). Because

$$\sigma_g(Z) = \int_{\mathfrak{c}} g(C) |\delta_\Gamma(C + Z)|^{1/2} F_T(C + Z)\, d_{\mathfrak{c}}(C) \qquad (Z \in \Omega_{\mathfrak{l}}'),$$

we then deduce that

$$u_g(\xi) = 2^{-1}\sqrt{-1}\gamma \int_{\mathfrak{c}} F_{T,\mathfrak{j}}(C + \xi(X^* - Y^*)) g(C)\, d_{\mathfrak{c}}(C) \qquad (0 < \xi < \epsilon),$$

$$v_g(\xi) = 2^{-1}\sqrt{-1}\gamma \int_{\mathfrak{c}} F_{T,\mathfrak{j}}(C + \xi(X^* - Y^*)) g(C)\, d_{\mathfrak{c}}(C) \qquad (-\epsilon < \xi < 0).$$

Let J_ϵ denote the open interval $(0, \epsilon)$ in $\mathbf{R}$; let $\mathscr{C}^\pm$ be the connected component of $\Omega \cap \mathfrak{j}'$ containing $\Omega_{\mathfrak{c}} \pm J_\epsilon(X^* - Y^*)$ – then, in view of what has been said in 8.3.1, there exist analytic functions F^+ and F^- on $\mathfrak{j}$ such that $F_{T,\mathfrak{j}} = F^+$ on $\mathscr{C}^+$ while $F_{T,\mathfrak{j}} = F^-$ on $\mathscr{C}^-$, whence

$$u_g(\xi) = 2^{-1}\sqrt{-1}\gamma \int_{\mathfrak{c}} F^+(C + \xi(X^* - Y^*)) g(C)\, d_{\mathfrak{c}}(C) \qquad (\xi \in \mathbf{R})$$

$$v_g(\xi) = 2^{-1}\sqrt{-1}\gamma \int_{\mathfrak{c}} F^-(C + \xi(X^* - Y^*)) g(C)\, d_{\mathfrak{c}}(C) \qquad (\xi \in \mathbf{R}).$$

But we have seen above that $u_g = v_g$ for every $g \in C_c^\infty(\Omega_{\mathfrak{c}})$ – thus $F^+ = F^-$ and so $F_{T,\mathfrak{j}} = F^+ = F^-$ on $\mathfrak{j}' \cap \mathfrak{j}(\epsilon)$. Our assertion in this case now follows from the fact that $\mathfrak{j}(\epsilon)$ is a neighborhood of Γ in $\mathfrak{j}$.

Step III Fix an arbitrary point H_0 in $\Omega \cap \mathfrak{j}'(R)$; let $\mathscr{O}(H_0)$ be an open convex neighborhood of H_0 in $\Omega \cap \mathfrak{j}'(R)$; let $\mathscr{O}^\sharp(H_0)$ be the set consisting of all regular and semi-regular elements in $\mathscr{O}(H_0)$ – then $\mathscr{O}^\sharp(H_0)$ is open and connected (cf. Lemma 8.4.4.5 infra). Furthermore, the above discussion shows that there exists an analytic function on $\mathscr{O}^\sharp(H_0)$ which has the property that its restriction to $\mathscr{O}^\sharp(H_0) \cap \mathfrak{j}'$ coincides with $F_{T,\mathfrak{j}}$ there; on the other hand, if

$\mathscr{C}$ is any component of $\mathcal{O}^{\sharp}(H_0) \cap \mathfrak{j}' = \mathcal{O}(H_0) \cap \mathfrak{j}'$, then, as we know, there exists an analytic function on $\mathfrak{j}$ which agrees with $F_{T,\mathfrak{j}}$ on $\mathscr{C}$; of course this implies that $F_{T,\mathfrak{j}}$ can be extended to an analytic function on $\mathcal{O}(H_0)$.

Proposition 8.3.2.1 is therefore completely proved. □

It is convenient to isolate part of the preceding discussion in a separate result. So fix a semi-regular element Γ in $\Omega \cap \mathfrak{j}'(R)$ of the non-compact type; introduce $\mathfrak{g}_\Gamma$ $(= \mathfrak{c}_\Gamma + \mathfrak{l}_\Gamma)$, $\delta_\Gamma, \Omega_\Gamma, \Omega_\mathfrak{c}$ $(\mathfrak{c} = \mathfrak{c}_\Gamma)$, and $\Omega_\mathfrak{l}$ $(\mathfrak{l} = \mathfrak{l}_\Gamma)$ as above; relative to Γ, assign to the symbols $\mathfrak{h}^+$ $(= \mathfrak{h}_\Gamma^+)$ and $\mathfrak{h}^-$ $(= \mathfrak{h}_\Gamma^-)$ their usual meanings – then there is no loss of generality in taking $\mathfrak{h}^- = \mathfrak{j}$ and, moreover, we may assume that the positive roots of the pair $(\mathfrak{g}_c, \mathfrak{h}_c^+)$ go into those of the pair $(\mathfrak{g}_c, \mathfrak{h}_c^-)$ under the automorphism ν (cf. Lemma 1.3.4.4).

Lemma If D is a polynomial differential operator on $\mathfrak{h}^+$ such that $w_+(D) = -D$ (w_+ the Weyl reflection determined by α_+), then $DF_{T,\mathfrak{h}^+}$ can be extended to a continuous function on $\Omega_\mathfrak{c} + (-\epsilon, \epsilon)H^*$ for sufficiently small positive ϵ. Furthermore

$$F_{T,\mathfrak{h}^+}(H; D) = F_{T,\mathfrak{h}^-}(H; \nu D) \qquad (\text{all } H \in \Omega_\mathfrak{c}).$$

[Adopting the notations which were used above, observe that Corollary 8.4.2.8 provides us with the relations

$$(-1)^n(D_\xi^{2n+1}w_g)_0 = (D_\xi^{2n+1}u_g)_0 = (D_\xi^{2n+1}v_g)_0 \qquad (n \geqslant 0),$$

the subscript 0 denoting the value at zero $(g \in C_c^\infty(\Omega_\mathfrak{c}))$. This being so, let $\mathscr{C}_+^+$ be the connected component of $\Omega' \cap \mathfrak{h}^+$ containing $\Omega_\mathfrak{c} + J_\epsilon H^*$, $\mathscr{C}_-^+$ the connected component of $\Omega' \cap \mathfrak{h}^-$ containing $\Omega_\mathfrak{c} + J_\epsilon(X^* - Y^*)$ $(J_\epsilon = (0, \epsilon))$ – then there exist analytic functions F_+^+ and F_-^+ on $\mathfrak{h}^+$ and $\mathfrak{h}^-$, respectively, such that $F_{T,\mathfrak{h}^+} = F_+^+$ on $\mathscr{C}_+^+$ and $F_{T,\mathfrak{h}^-} = F_-^+$ on $\mathscr{C}_-^+$. Since w_g admits an obvious integral representation in terms of F_+^+ (in fact $w_g(\xi) = 2^{-1}\gamma \int_\mathfrak{c} F_+^+(C + \xi H^*)g(C)d_\mathfrak{c}(C)$ $(\xi \in \mathbf{R}) \ldots$) and since $\nu(H^*) = \sqrt{-1}(X^* - Y^*)$, it follows that

$$F_+^+(C; \partial(H^*)^{2n+1}) = F_-^+(C; \partial(\nu(H^*))^{2n+1}) \qquad (\text{all } C \in \mathfrak{c}),$$

g being an arbitrary element of $C_c^\infty(\Omega_\mathfrak{c})$. The remaining details of the argument are more or less obvious and may be left to the reader to fill in.]

8.3.3 The Main Theorem on the Group

Let $\mathfrak{g}$ be a reductive Lie algebra over $\mathbf{R}$, G a connected Lie group with Lie algebra $\mathfrak{g}$; let $\mathfrak{G}$ be the universal enveloping algebra of $\mathfrak{g}_c$, $\mathfrak{Z}$ the center of $\mathfrak{G}$.

Let Ω be an open, completely invariant subset of G. A distribution T on Ω is said to be *central* if it is invariant under the canonical action of G on Ω through inner automorphisms; a distribution T on Ω is said to be $\mathfrak{Z}$-*finite* if the space spanned by the $Z \cdot T$ $(Z \in \mathfrak{Z})$ is finite dimensional (in particular T is said to be an *eigendistribution* of $\mathfrak{Z}$ if there exists a homomorphism $\kappa : \mathfrak{Z} \to \mathbf{C}$ such that $Z \cdot T = \kappa(Z)T$ (all $Z \in \mathfrak{Z}$)).

Theorem 8.3.3.1 (Harish-Chandra) Let $\mathfrak{g}$ be a reductive Lie algebra over $\mathbf{R}$, G a connected Lie group with Lie algebra $\mathfrak{g}$; let Ω be an open, completely invariant subset of G; let T be a central distribution on Ω

which is annihilated by an ideal $\mathfrak{J}$ in $\mathfrak{Z}$ of finite codimension (thus $Z \cdot T = 0$ for all $Z \in \mathfrak{J}$) – then, in the sense of distributions, T agrees with a locally summable function F_T (say) on Ω which is actually analytic on $\Omega' = \Omega \cap G'$.

[Suppose, in particular, that G is semi-simple with finite center – then, as has been noted in Vol. I, number 4.5.8, the significance of this result for Harmonic Analysis on G is simply this: The characters of the irreducible unitary representations of G are central eigendistributions on G and each character determines the representation to within unitary equivalence.]

Proof of Theorem 8.3.3.1 The proof proceeds by induction on $\dim(G)$. Let Ω_0 be the set of all points $\gamma \in \Omega$ with the following property: There exists an open neighborhood $N_G(\gamma)$ of γ in Ω and a locally summable function F_T on $N_G(\gamma)$ such that F_T is analytic on $N_G(\gamma) \cap G'$ and $T = F_T$ on $N_G(\gamma)$. Clearly Ω_0 is an open and invariant subset of Ω and, of course, the theorem will follow if we can prove that $\Omega = \Omega_0$; in turn, for this purpose, it will be enough to verify that Ω_0 contains all the semi-simple points of Ω (cf. Lemma 8.1.6.4). With this in mind, relative to a given semi-simple element γ in Ω we shall agree to employ, for the most part without comment, the notations and results which were outlined near the end of 8.2.3 supra.

(1) We shall first show that there exists an analytic function F_T on Ω' such that $T = F_T$ on Ω' ($\Omega' = \Omega \cap G'$). So fix a γ in Ω'; let $\mathfrak{j}_\gamma$ be the centralizer of γ in $\mathfrak{g}$, J_γ the analytic subgroup of G with Lie algebra $\mathfrak{j}_\gamma$; let $\tilde{T}$ denote the distribution on Ω_γ ($= \gamma^{-1}\Omega \cap {}'J_\gamma$) corresponding to T via the results outlined earlier. Put $\Sigma = |\delta_\gamma|^{1/2}\tilde{T}$ – then, on the basis of Theorem 8.2.3.9 (bis), we find that

$$\gamma(Z)\cdot\Sigma = (\gamma(Z)\circ|\delta_\gamma|^{1/2})\cdot\tilde{T} = (|\delta_\gamma|^{1/2}\mathfrak{x}_\gamma(Z))\cdot\tilde{T} = 0$$

for $Z \in \mathfrak{J}$. Now $S(\mathfrak{j}_{\gamma,c})$ is a finite module over $\gamma(\mathfrak{Z}) = I(\mathfrak{j}_{\gamma,c})$ and thus it is clear that $I = S(\mathfrak{j}_{\gamma,c})\gamma(\mathfrak{J})$ has finite codimension, m say, in $S(\mathfrak{j}_{\gamma,c})$. Let $H_1, \ldots, H_l$ be a basis for $\mathfrak{j}_\gamma$ over $\mathbf{R}$; set $\Delta = \sum_1^l H_i^2 \in S(\mathfrak{j}_{\gamma,c})$ and then choose $c_i \in \mathbf{C}$ such that

$$\Box = \Delta^m + \sum_1^m c_i\Delta^{m-i} \in I.$$

Evidently, $\Box$ is an analytic elliptic differential operator on J_γ. Hence, since $\Box\cdot\Sigma = 0$, we may conclude that Σ coincides, in the sense of distributions, with an analytic function F_Σ on Ω_γ. Let $x \mapsto \dot{x}$ denote the projection of G onto G/J_γ; choose open neighborhoods $\mathcal{O}$ and $\mathcal{O}_\gamma$ of 1 in G and Ω_γ, respectively, such that the map $(\dot{x}, j_\gamma) \mapsto x(\gamma j_\gamma)x^{-1} = {}^{\dot{x}}(\gamma j_\gamma)$ is an analytic diffeomorphism of $\dot{\mathcal{O}} \times \mathcal{O}_\gamma$ onto its image $N_G(\gamma)$ (say). On the open subset $N_G(\gamma)$ define an analytic function F_T by the rule

$$F_T({}^{\dot{x}}(\gamma j_\gamma)) = |\delta_\gamma(j_\gamma)|^{-1/2}F_\Sigma(j_\gamma) \qquad (\dot{x} \in \dot{\mathcal{O}},\ j_\gamma \in \mathcal{O}_\gamma).$$

Claim: $T = F_T$ on $N_G(\gamma)$. To see this, note that for all $v \in C_c^\infty(\mathcal{O} \times \mathcal{O}_\gamma)$, we have

$$\begin{aligned}\int_{N_G(\gamma)} F_\Gamma(x) f_v(x) d_G(x) &= \int_{\mathcal{O}\times\mathcal{O}_\gamma} F_T({}^x(\gamma j_\gamma)) v(x : j_\gamma) d_G(x) d_{J_\gamma}(j_\gamma)\\ &= \int_{\mathcal{O}_\gamma} V_v(j_\gamma) \,|\, \delta_\gamma(j_\gamma) \,|^{-1/2} F_\Sigma(j_\gamma) d_{J_\gamma}(j_\gamma)\\ &= \tilde{T}(V_v) = T(f_v)\end{aligned}$$

which establishes the claim, the map $v \mapsto f_v$ being a surjection of $C_c^\infty(\mathcal{O} \times \mathcal{O}_\gamma)$ onto $C_c^\infty(N_G(\gamma))$. Conclusion: There exists an analytic function F_T on Ω' such that $T = F_T$ on Ω'.

(2) Fix a semi-simple element $\gamma \in \Omega$; we shall assume that γ does not lie in the center of G – therefore $\dim(\mathfrak{g}_\gamma) < \dim(\mathfrak{g})$ ($\mathfrak{g}_\gamma$ the centralizer of γ in $\mathfrak{g}$). Let $\tilde{T}$ denote the distribution on Ω_γ corresponding to T per the discussion at the end of 8.2.3 – then $\tilde{T}$ is central (relative to G_γ) and $\mathfrak{f}_\gamma(Z)\cdot\tilde{T} = (Z\cdot T)^\sim = 0$ for all $Z \in \mathfrak{J}$, whence $\mu_{\mathfrak{g}/\mathfrak{g}_\gamma}(Z)\cdot\Sigma = 0$ (all $Z \in \mathfrak{J}$), $\Sigma = |\,\delta_\gamma\,|^{1/2}\tilde{T}$. Now $\mathfrak{Z}_\gamma$ is a finite module over $\mu_{\mathfrak{g}/\mathfrak{g}_\gamma}(\mathfrak{Z})$ and thus it is clear that $\mathfrak{J}_\gamma = \mathfrak{Z}_\gamma \mu_{\mathfrak{g}/\mathfrak{g}_\gamma}(\mathfrak{J})$ has finite codimension in $\mathfrak{Z}_\gamma$. Denoting by Ω'_γ the set of those elements in Ω_γ which are regular in G_γ, it follows from the induction hypothesis that Σ coincides, in the sense of distributions, with a locally summable function F_Σ on Ω_γ which is analytic on Ω'_γ. Let ϕ denote the mapping $(x, x_\gamma) \mapsto x(\gamma x_\gamma)x^{-1}$ of $G \times \Omega_\gamma$ into Ω; put

$$N_G(\gamma)(= \mathcal{O}_G(\gamma)) = \phi(G \times \Omega_\gamma).$$

Clearly $\phi(G \times \Omega'_\gamma) = N'_G(\gamma)$, $N'_G(\gamma) = N_G(\gamma) \cap \Omega'$ (for the detailed verification, see the Appendix to this number). Since $T = F_T$ on Ω', we have that

$$\begin{aligned}T(f_v) &= \int_{G\times\Omega'_\gamma} F_T({}^x(\gamma x_\gamma)) v(x : x_\gamma) d_G(x) d_{G_\gamma}(x_\gamma)\\ &= \int_{G\times\Omega'_\gamma} |\delta_\gamma(x_\gamma)|^{-1/2} F_\Sigma(x_\gamma) v(x : x_\gamma) d_G(x) d_{G_\gamma}(x_\gamma) \quad (v \in C_c^\infty(G \times \Omega'_\gamma))\end{aligned}$$

which shows that the analytic function

$$(x, x_\gamma) \mapsto F_T({}^x(\gamma x_\gamma)) - |\,\delta_\gamma(x_\gamma)\,|^{-1/2} F_\Sigma(x_\gamma)$$

vanishes on $G \times \Omega'_\gamma$. Therefore $F_T \circ \phi$ is locally summable on $G \times \Omega_\gamma$ and so F_T is locally summable on $N_G(\gamma)$ (Corollary 8.2.1.3); moreover, by the usual computation, we find that

$$\int_{N_G(\gamma)} F_T(x) f_v(x) d_G(x) = T(f_v) \qquad (v \in C_c^\infty(G \times \Omega_\gamma)),$$

whence $T = F_T$ on $N_G(\gamma)$ from which we conclude that $\gamma \in \Omega_0$.

(3) It remains to consider the case when γ is central; without loss of generality it may be supposed that $\gamma = 1$ (translation by γ^{-1}). In order to prove that $1 \in \Omega_0$, we shall employ the reduction technique discussed in 8.2.4 and 'lift' our problem to a corresponding question in $\mathfrak{g}$. Fix an open and completely invariant neighborhood $\mathcal{O}_\mathfrak{g}$ of zero in $\mathfrak{g}$ on which exp is

regular and injective and such that $\mathcal{O}_G = \exp(\mathcal{O}_{\mathfrak{g}}) \subset \Omega$. [This is possible; cf. the discussion in 8.1.6.] As in Proposition 8.2.4.3, let σ_T denote the distribution on $\mathcal{O}_{\mathfrak{g}}$ corresponding to T; by that proposition, then,

$$\partial(p_Z)\cdot\sigma_T = 0 \qquad (\text{all } Z \in \mathfrak{Z}).$$

Now it is plain that the image of $\mathfrak{Z}$ in $I(\mathfrak{g}_c)$ under the canonical isomorphism $Z \mapsto p_Z$ is itself of finite codimension in $I(\mathfrak{g}_c)$. Owing to Theorem 8.3.1.1, σ_T coincides, in the sense of distributions, with a locally summable function $F = F_{\sigma_T}$ which is analytic on $\mathcal{O}_{\mathfrak{g}}' = \mathcal{O}_{\mathfrak{g}} \cap \mathfrak{g}'$. Define the function f_F as in the discussion preceding Lemma 8.2.4.1; obviously f_F is locally summable on $\mathcal{O}_G$ and $T = f_F$ on $\mathcal{O}_G$. However $T = F_T$ on $\mathcal{O}_G \cap \Omega'$ and so it follows that $f_F = F_T$ almost everywhere on $\mathcal{O}_G$. Consequently F_T is locally summable on $\mathcal{O}_G$ and $T = F_T$ on $\mathcal{O}_G$. Therefore $1 \in \Omega_0$, as desired.

The proof of the theorem is now complete. □

Let F_T be the locally summable function on Ω corresponding to T. Fix $Z \in \mathfrak{Z}$ – then the distribution $Z\cdot T$ also satisfies the condition of Theorem 8.3.3.1 and it is clear that $Z\cdot T = ZF_T$ on Ω'. Therefore ZF_T is also locally summable on Ω and $Z\cdot T = ZF_T$ on Ω – hence:

Corollary 8.3.3.2 Retain the above notations and assumptions – then, for any $Z \in \mathfrak{Z}$, the function ZF_T on Ω' is locally summable on Ω and $Z\cdot T = ZF_T$. Consequently

$$\int_\Omega f(x)(ZF_T(x))d_G(x) = \int_\Omega (Z^\sharp f(x))F_T(x)d_G(x) \qquad (f \in C_c^\infty(\Omega)).$$

Let us assume now that $\mathfrak{g}$ is semi-simple and that G has finite center; fix a maximal compact subgroup K of G – the usual notations are then assumed to be in force. [As will become apparent, the ensuing discussion is applicable without essential change to the reductive pairs (G, K) of Vol. I, number 1.1.5.] Let T be a non-zero central eigendistribution (of $\mathfrak{Z}$) on G – thus there exists a homomorphism $\kappa: \mathfrak{Z} \to \mathbf{C}$ such that $Z\cdot T = \kappa(Z)T$ (all $Z \in \mathfrak{Z}$). In view of Theorem 8.3.3.1, T coincides on G with a locally summable function F_T which is even analytic on G'. Problem: Find a 'formula' for F_T on G'. Just what, then, does this entail? In the first place G' will, in general, not be connected – however, in the present situation G' has only a finite number of connected components $G_1, \ldots, G_s$, say (cf. Corollary 1.4.1.10) and so it is enough to determine F_T on each of the G_k $(1 \leqslant k \leqslant s)$. Choose the Cartan subgroups $\mathsf{J}^1, \ldots, \mathsf{J}^r$ as usual; again, since K is compact here, each of the $(\mathsf{J}^i)'$ has but a finite number of connected components (see Proposition 1.4.1.9). Moreover, given a G_k, it is clear that there exists an i $(1 \leqslant i \leqslant r)$ and a component C^i of $(\mathsf{J}^i)'$ such that $G_k = \bigcup_{x\in G} x\mathsf{C}^i x^{-1}$. But F_T is central – therefore F_T is known on G_k as soon as it is known on C^i and, in turn, for this latter purpose, it will be sufficient to determine F_T on some non-empty open subset of C^i.

Remarks (1) Observe that if G is either compact or complex, then F_T is completely determined by its restriction to any non-empty open subset of G' (for in either case G' is connected).

(2) It is possible to give examples where F_T vanishes identically on one of the components G_k without vanishing identically on G'. [Consider, for instance, the characters of the principal P-series, P a minimal parabolic subgroup of G.]

With all this in mind, let us proceed to the formulation of the result.

Theorem 8.3.3.3 (Harish-Chandra) Let G be a connected semi-simple Lie group with finite center, K a maximal compact subgroup of G; let T be a central eigendistribution (of $\mathfrak{Z}$) on G, F_T the locally summable function on G which agrees with T (in the sense of distributions); let $\mathfrak{j}$ be a θ-stable Cartan subalgebra of $\mathfrak{g}$, J the Cartan subgroup of G associated with $\mathfrak{j}$ – then there exists a linear function Λ on $\mathfrak{j}_c$ with the following property. Given $j_0 \in J'$ we can find polynomial functions p_w $(w \in W)$ on $\mathfrak{j}_c$ such that

$$F_T(j_0 \exp H) = |D_l(j_0 \exp H)|^{-1/2} \sum_{w \in W} p_w(H) e^{w(\Lambda+\rho)(H)}$$

for all H lying sufficiently near zero in $\mathfrak{j}$. The linear function $\Lambda + \rho$ is unique to within an operation of W – moreover, if $[W_\Lambda]$ is the order of the subgroup W_Λ consisting of those $w \in W$ for which $w(\Lambda + \rho) = \Lambda + \rho$, then the degree of every p_w is necessarily less than $[W_\Lambda]$.

[The usual notations are in force here – thus W denotes the Weyl group of the pair $(\mathfrak{g}_c, \mathfrak{j}_c)$, ρ one-half the sum of the positive roots in some ordering etc.]

Plainly the preceding formula for F_T bears a striking resemblance to the character formulas of Weyl.

Proof of Theorem 8.3.3.3 Let $\tilde{T}$ denote the distribution on J' corresponding to T in the by now familiar manner; the Remark following the statement of Theorem 8.2.1.8 tells us that $\tilde{T} = \tilde{F}_T$, $\tilde{F}_T$ the restriction of F_T to J'. Put $\Sigma = |D_l|^{1/2}\tilde{F}_T$.

(1) Claim: There exists a linear function Λ on $\mathfrak{j}_c$ with the property that Σ satisfies the following system of differential equations:

$$(*) \quad u\Sigma = \langle u, e^{\Lambda+\rho}\rangle \Sigma \qquad (\text{all } u \in I(\mathfrak{j}_c)).$$

To see this, first note that Theorem 2.4.3.2 implies the existence of a $\Lambda \in \mathfrak{j}_c^\vee$ such that $\kappa = \kappa_\Lambda$, i.e. such that

$$\kappa(Z) = \langle \gamma^*(Z), e^\Lambda\rangle = \langle \gamma(Z), e^{\Lambda+\rho}\rangle \qquad (\text{all } Z \in \mathfrak{Z}).$$

On the other hand, Theorem 8.2.1.8, combined with Theorem 8.2.3.9, tells us that $\gamma(Z)\Sigma = \kappa(Z)\Sigma = \kappa_\Lambda(Z)\Sigma$ (all $Z \in \mathfrak{Z}$); since $\gamma(\mathfrak{Z}) = I(\mathfrak{j}_c)$, our claim is now clear. [In passing, let us bear in mind that the linear function $\Lambda + \rho$ is determined to within an operation of W; of course this too is contained in the statement of Theorem 2.4.3.2.]

(2) Fix $u \in S(\mathfrak{j}_c)$; let t be an indeterminate and consider the polynomial

$$\prod_{w \in W} (t - u^w) = t^{[W]} + u_1 t^{[W]-1} + \cdots + u_{[W]}$$

with coefficients $u_i \in I(\mathfrak{j}_c)$ ($[W]$ the order of W). Setting $t = u$, we find that

$$u^{[W]}\Sigma + \langle u_1, \mathbf{e}^{\Lambda+\rho}\rangle u^{[W]-1}\Sigma + \cdots + \langle u_{[W]}, \mathbf{e}^{\Lambda+\rho}\rangle \Sigma = 0;$$

but $\mathbf{e}^{\Lambda+\rho}$ is a homomorphism of $S(\mathfrak{j}_c)$ and so we conclude that

$$(**) \quad \prod_{w \in W} (u - \langle u^w, \mathbf{e}^{\Lambda+\rho}\rangle)\,\Sigma = 0 \qquad (\text{all } u \in S(\mathfrak{j}_c)).$$

(3) Fix a point $j_0 \in \mathrm{J}'$; using the systems $(*)$, $(**)$, we shall now determine Σ (and hence F_T) on a neighborhood in J' of j_0. Choose a small open connected neighborhood $\mathcal{O}$ of zero in $\mathfrak{j}$ on which exp is regular and injective and such that $j_0 \exp H \in \mathrm{J}'$ for all $H \in \mathcal{O}$ – then $\check{\Sigma}(H) = \Sigma(j_0 \exp H)$ is an analytic function on $\mathcal{O}$. Let $H_1, \ldots, H_l$ be a basis for $\mathfrak{j}$ over $\mathbf{R}$ and let $t_1, \ldots, t_l$ denote the corresponding coordinates in $\mathfrak{j}$. Define polynomials q_i in an indeterminate t by the prescription

$$q_i(t) = \prod_{w \in W} (t - w(\Lambda + \rho)(H_i)) \qquad (i = 1, \ldots, l).$$

Owing to $(**)$, then, $q_i(\partial/\partial t_i)\check{\Sigma} = q_i(H_i)\check{\Sigma} = 0$ and so Lemma 2 in the Appendix to 8.3.1 gives us distinct linear functions $\lambda_1, \ldots, \lambda_r$ on $\mathfrak{j}_c$ and non-zero polynomials $p_1, \ldots, p_r$ such that

$$\check{\Sigma} = p_1 e^{\lambda_1} + \cdots + p_r e^{\lambda_r}$$

on $\mathcal{O}$. Claim: For each λ_i, there exists a $w_i \in W$ such that $\lambda_i = w_i(\Lambda + \rho)$ $(i = 1, \ldots, r)$. To establish this, it suffices, in view of Proposition 2.1.3.8 (ii), to verify that

$$\langle u, \mathbf{e}^{\lambda_i}\rangle = \langle u\ \mathbf{e},^{\Lambda+\rho}\rangle \qquad (\text{all } u \in I(\mathfrak{j}_c)).$$

For this purpose, we use $(*)$ to derive that

$$0 = (\partial(u) - \langle u, \mathbf{e}^{\Lambda+\rho}\rangle)\,\check{\Sigma} = p_1' e^{\lambda_1} + \cdots + p_r' e^{\lambda_r} \qquad (u \in I(\mathfrak{j}_c))$$

where $p_i' = (\langle u, \mathbf{e}^{\lambda_i} - \mathbf{e}^{\Lambda+\rho}\rangle) p_i + p_i''$ and $\deg(p_i'') < \deg(p_i)$. But the exponentials of distinct linear functions are linearly independent over the ring of polynomial functions on $\mathfrak{j}_c$ – thus $p_i' = 0$ and so

$$\langle u, \mathbf{e}^{\lambda_i}\rangle = \langle u, \mathbf{e}^{\Lambda+\rho}\rangle \qquad (\text{all } u \in I(\mathfrak{j}_c)),$$

as desired.

(4) Let W_Λ be the subgroup of W consisting of those $w \in W$ which fix $\Lambda + \rho$. We shall now prove that the degree of each p_i is less than $[W_\Lambda]$. Suppose that the degree of p_1 (say) is d_1. Choose an element $H_0 \in \mathfrak{j}$ such that $(\Lambda + \rho)(H_0) \neq w(\Lambda + \rho)(H_0)$ $(w \in W)$ unless $w \in W_\Lambda$ and the degree of the polynomial $p(t) = p_1(tH_0)$ in t $(t \in \mathbf{R})$ is d_1 – this is clearly possible. Put $\sigma(t) = \check{\Sigma}(tH_0)$ $(t \in \mathbf{R})$ – then σ is an analytic function defined on some open interval I in $\mathbf{R}$ containing the origin and it is

obvious that

$$\prod_{w\in W} (d/dt - w(\Lambda + \rho)(H_0))\sigma = 0 \qquad (t \in I).$$

Set $\tilde{p}_i = p_i(tH_0)$ and $\tilde{\lambda}_i = \lambda_i(H_0)$ – then $\sigma(t) = \sum_1^r \tilde{p}_i(t)e^{\tilde{\lambda}_i t}$ $(t \in I)$. The numbers $\tilde{\lambda}_1, \ldots, \tilde{\lambda}_r$ are all distinct and, in fact, are the roots of the polynomial $\prod_{w\in W} (t - w(\Lambda + \rho)(H_0))$; because these roots occur with multiplicity $[W_\Lambda]$, Lemma 1 in the Appendix to 8.3.1 then implies that the degree of p cannot exceed $[W_\Lambda] - 1$.

All the assertions of Theorem 8.3.3.3 are now evident. □

Observe that if the group W_Λ above reduces to $\{1\}$, then the p_w must be constants.

Example Keeping to the above notations, consider the following problems:

(**P**$_1$) Find all central eigendistributions on G;

(**P**$_2$) Find a criterion which would ensure whether or not a central eigendistribution on G is the character of some TCI Hilbert representation of G;

(**P**$_3$) Decide whether or not every central eigendistribution on G is expressible as a finite linear combination of characters corresponding to TCI Hilbert representations of G.

Needless to say, when G is compact problems (**P**$_1$), (**P**$_2$), and (**P**$_3$) have been completely solved (we recall, in particular, that any central eigendistribution on a compact G is proportional to a character of $G \ldots$). On the other hand, at the time of this writing, the answers to these questions for non-compact G are not known; Hirai [7], however, has made progress in the case of certain classical groups – thus, for example, problem (**P**$_3$) admits an affirmative answer for $G = \mathbf{SU}(p, 1)$ $(p \geqslant 1)$.

Appendix Let Ω be a completely invariant open subset of G, γ an element of Ω which is semi-simple in G; relative to γ, introduce $\mathfrak{g}_\gamma$, G_γ and Ω_γ in the usual way. Let x_γ be an element of Ω_γ which is regular in G_γ and let $\mathfrak{j}_\gamma$ denote the centralizer of x_γ in $\mathfrak{g}_\gamma$ – then $\mathfrak{j}_\gamma$ is a Cartan subalgebra of $\mathfrak{g}_\gamma$ and therefore also of $\mathfrak{g}$, the ranks of $\mathfrak{g}$ and $\mathfrak{g}_\gamma$ being the same.

Lemma Let J_γ be the analytic subgroup of G with Lie algebra $\mathfrak{j}_\gamma$ – then the following two conditions on an element j_γ in J_γ are equivalent:

(i) $j_\gamma \in x_\gamma^{-1}\Omega_\gamma$ and $\det(\mathrm{Ad}(x_\gamma j_\gamma)^{-1} - 1)_{\mathfrak{g}_\gamma/\mathfrak{j}_\gamma} \neq 0$;

(ii) $j_\gamma \in (\gamma x_\gamma)^{-1}\Omega$ and $\det(\mathrm{Ad}(\gamma x_\gamma j_\gamma)^{-1} - 1)_{\mathfrak{g}/\mathfrak{j}_\gamma} \neq 0$.

Proof Since $\mathrm{Ad}(\gamma) = 1$ on $\mathfrak{g}_\gamma$, it is clear that

$$\begin{aligned} &\det(\mathrm{Ad}(\gamma x_\gamma j_\gamma)^{-1} - 1)_{\mathfrak{g}/\mathfrak{j}_\gamma} \\ &= \det(\mathrm{Ad}(\gamma x_\gamma j_\gamma)^{-1} - 1)_{\mathfrak{g}/\mathfrak{g}_\gamma} \det(\mathrm{Ad}(x_\gamma j_\gamma)^{-1} - 1)_{\mathfrak{g}_\gamma/\mathfrak{j}_\gamma} \end{aligned}$$

for all $j_\gamma \in J_\gamma$. Now suppose that (i) holds – then $x_\gamma j_\gamma \in \Omega_\gamma$ and therefore $\gamma x_\gamma j_\gamma \in \Omega$ with $\det(\mathrm{Ad}(\gamma x_\gamma j_\gamma)^{-1} - 1)_{\mathfrak{g}/\mathfrak{j}_\gamma} \neq 0$, whence (i) ⇒ (ii). In a similar way we verify that (ii) ⇒ (i). □

Corollary Retain the above notations – then γx_γ is regular in G and $\mathfrak{j}_\gamma$ is the centralizer of γx_γ in $\mathfrak{g}$.

[This follows upon taking $j_\gamma = 1$ in the above lemma.]

8.3.4 Properties of F_T – II

Let $\mathfrak{g}$ be a reductive Lie algebra over **R**, G a connected Lie group with Lie algebra $\mathfrak{g}$ – in what follows it will be assumed, in addition, that G is acceptable. [Just as in earlier numbers we shall be taking the liberty of using some of the definitions and notations of Vol. I, section 1.4 even though we are not necessarily in the setting of Vol. I, number 1.1.5; for the present discussion, it will be clear that no difficulties can arise from this practice.]

Let Ω be an open, completely invariant subset of G; let T be a central distribution on Ω which is annihilated by an ideal $\mathfrak{I}$ in $\mathfrak{Z}$ of finite codimension – then, according to Theorem 8.3.3.1, T agrees, in the sense of distributions, with a locally summable function F_T which is even analytic on $\Omega' = \Omega \cap \mathfrak{g}'$. In this number we shall study the behavior of F_T and its derivatives around the singular points in Ω (cf. 8.3.2). For the most part this study proceeds along the following lines: Reduce the problem at hand to its analogue in the Lie algebra and then quote the appropriate result from 8.3.2. With this in mind, we shall first develop the tools which are needed to effect such reductions and then after that take up the properties of F_T itself.

Let $\mathfrak{j}$ be a Cartan subalgebra of $\mathfrak{g}$, Φ the set of roots of the pair $(\mathfrak{g}_c, \mathfrak{j}_c)$, Φ^+ the set of positive roots relative to some ordering of Φ; let J be the Cartan subgroup of G associated with $\mathfrak{j}$. Fix a point γ in $\Omega \cap J$ – then γ is semi-simple. Let $\mathfrak{g}_\gamma$ be the centralizer of γ in $\mathfrak{g}$, G_γ the analytic subgroup of G with Lie algebra $\mathfrak{g}_\gamma$. Invoking the usual notations, put $\Omega_\gamma = \gamma^{-1}\Omega \cap {}'G_\gamma$ and let $\tilde{T}$ denote the distribution on Ω_γ which corresponds to T by the now familiar process (cf. part (2) of the proof of Theorem 8.3.3.1 – the fact that γ may be central causes no problems). If Ω_γ' denotes the set of those elements in Ω_γ which are regular in G_γ, then, as has been seen during the proof of Theorem 8.3.3.1 $\Sigma\,(= |\delta_\gamma|^{1/2}\tilde{T})$ coincides, in the sense of distributions, with a locally summable function F_Σ on Ω_γ which is analytic on Ω_γ' – moreover

$$F_T(\gamma x_\gamma) = |\delta_\gamma(x_\gamma)|^{-1/2} F_\Sigma(x_\gamma) \qquad (\text{all } x_\gamma \in \Omega_\gamma').$$

Because Ω_γ is an open, completely invariant neighborhood of 1 in G_γ, it is possible to choose an open, completely invariant neighborhood $\mathcal{O}_{\mathfrak{g}_\gamma}$ of zero in $\mathfrak{g}_\gamma$ on which the exponential mapping defines an analytic diffeomorphism of $\mathcal{O}_{\mathfrak{g}_\gamma}$ onto an open subset $\mathcal{O}_{G_\gamma}$ of Ω_γ. Consider the function Υ_γ on $\mathfrak{g}_\gamma$ per 8.2.4; as there, for any $\phi \in C_c^\infty(\mathcal{O}_{\mathfrak{g}_\gamma})$, define $f_\phi \in C_c^\infty(\mathcal{O}_{G_\gamma})$ by the prescription

$$f_\phi(\exp X_\gamma) = \Upsilon_\gamma(X_\gamma)^{-1}\phi(X_\gamma) \qquad (X_\gamma \in \mathcal{O}_{\mathfrak{g}_\gamma}).$$

Let σ_Σ be the distribution on $\mathcal{O}_{\mathfrak{g}_\gamma}$ given by

$$\sigma_\Sigma(\phi) = \Sigma(f_\phi) \qquad (\phi \in C_c^\infty(\mathcal{O}_{\mathfrak{g}_\gamma}));$$

since Σ is annihilated by an ideal of finite codimension in $\mathfrak{Z}_\gamma$, it follows from Proposition 8.2.4.3 that σ_Σ is annihilated by an ideal of finite codimension in $I(\mathfrak{g}_{\gamma,c})$ – therefore, thanks to Theorem 8.3.1.1, σ_Σ coincides on $\mathcal{O}_{\mathfrak{g}_\gamma}$ with a locally summable function F_σ (say) which is, itself, analytic on $\mathcal{O}'_{\mathfrak{g}_\gamma} = \mathcal{O}_{\mathfrak{g}_\gamma} \cap \mathfrak{g}'_\gamma$ ($\mathfrak{g}'_\gamma$ the set of elements in $\mathfrak{g}_\gamma$ which are regular in $\mathfrak{g}_\gamma$). [In what follows we shall assume (as we may) that $\exp(\mathcal{O}'_{\mathfrak{g}_\gamma}) \subset \Omega'_\gamma$.]

Lemma 8.3.4.1 Retain the above notations – then

$$F_\sigma(X_\gamma) = \Upsilon_\gamma(X_\gamma)\,|\,\delta_\gamma(\exp X_\gamma)\,|^{1/2} F_T(\gamma \exp X_\gamma) \qquad (\text{all } X_\gamma \in \mathcal{O}'_{\mathfrak{g}_\gamma}).$$

Proof Fix $\phi \in C_c^\infty(\mathcal{O}_{\mathfrak{g}_\gamma})$ – then

$$\int \phi F_\sigma d_{\mathfrak{g}_\gamma}(X_\gamma) = \sigma_\Sigma(\phi) = \Sigma(f_\phi) = \int \phi \Upsilon_\gamma F_\Sigma d_{\mathfrak{g}_\gamma}(X_\gamma).$$

Because $\exp(\mathcal{O}'_{\mathfrak{g}_\gamma}) \subset \Omega'_\gamma$,

$$F_\Sigma(\exp X_\gamma) = |\,\delta_\gamma(\exp X_\gamma)\,|^{1/2} F_T(\gamma \exp X_\gamma) \qquad (\text{all } X_\gamma \in \mathcal{O}'_{\mathfrak{g}_\gamma})$$

and so our assertion is obvious. □

Define Φ_γ and Φ_γ^+ in the obvious way; set $\pi_\gamma = \prod_{\alpha \in \Phi_\gamma^+} \alpha$.

Lemma 8.3.4.2 Let $\mathcal{O}$ be an open, connected neighborhood of zero in $\mathcal{O}_{\mathfrak{g}_\gamma} \cap \mathfrak{j}$ – then

$$\pi_\gamma(H)\Upsilon_\gamma(H)\,|\,\delta_\gamma(\exp H)\,|^{1/2} = c_\gamma \Delta(\gamma \exp H) \qquad (H \in \mathcal{O})$$

where c_γ is a constant. Let $r - r_\gamma$ denote the cardinality of $\Phi^+ - \Phi_\gamma^+$ – then $c_\gamma^2 = (-1)^{r-r_\gamma}\operatorname{sign}\{\delta_\gamma(1)\}$ while

$$c_\gamma = |\,\delta_\gamma(1)\,|^{1/2}\{\xi_\rho(\gamma) \prod_{\alpha \in \Phi^+ - \Phi_\gamma^+} (1 - \xi_\alpha(\gamma)^{-1})\}^{-1}.$$

Proof Put $\rho_\gamma = 2^{-1}\sum_\alpha \alpha$, $\alpha \in \Phi_\gamma^+$ – then it is clear that for all $H \in \mathfrak{j}$, we have

$$\begin{aligned}\delta_\gamma(\exp H) &= \prod_{\alpha \in \Phi^+ - \Phi_\gamma^+} \{(\xi_\alpha(\gamma \exp H)^{-1} - 1)(\xi_\alpha(\gamma \exp H) - 1)\}\\ &= (-1)^{r-r_\gamma}\xi_{2(\rho-\rho_\gamma)}(\gamma)\{e^{\rho-\rho_\gamma(H)} \prod_{\alpha \in \Phi^+ - \Phi_\gamma^+} (1 - \xi_\alpha(\gamma \exp H)^{-1})\}^2.\end{aligned}$$

Because $\delta_\gamma(\exp H)$ is real and non-zero (all $H \in \mathcal{O}$), the function

$$H \mapsto |\,\delta_\gamma(\exp H)\,|^{-1/2} e^{\rho-\rho_\gamma(H)} \prod_{\alpha \in \Phi^+ - \Phi_\gamma^+} (1 - \xi_\alpha(\gamma \exp H)^{-1})$$

is analytic on $\mathcal{O}$ and its fourth power is a constant; therefore, in view of the connectedness of $\mathcal{O}$, it must be the case that

$$|\,\delta_\gamma(\exp H)\,|^{1/2} = \tilde{c}_\gamma e^{\rho-\rho_\gamma(H)} \prod_{\alpha \in \Phi^+ - \Phi_\gamma^+} (1 - \xi_\alpha(\gamma \exp H)^{-1}) \qquad (H \in \mathcal{O})$$

where

$$\tilde{c}_\gamma = |\,\delta_\gamma(1)\,|^{1/2} \prod_{\alpha \in \Phi^+ - \Phi_\gamma^+} (1 - \xi_\alpha(\gamma)^{-1})^{-1}.$$

On the other hand, it is certainly true that

$$\pi_\gamma(H)\Upsilon_\gamma(H) = \prod_{\alpha \in \Phi_\gamma^+} (e^{\alpha(H)/2} - e^{-\alpha(H)/2}) \qquad (\text{all } H \in \mathcal{O}).$$

Hence

$$\pi_\gamma(H)\Upsilon_\gamma(H)|\delta_\gamma(\exp H)|^{1/2} = c_\gamma \Delta(\gamma \exp H) \qquad (\text{all } H \in \mathcal{O})$$

where $c_\gamma = \tilde{c}_\gamma \xi_\rho(\gamma)^{-1}$. That the constant c_γ has the value stated in the assertion of our lemma is clear from its definition; on the other hand,

$$c_\gamma^2 \prod_{\alpha \in \Phi^+ - \Phi_\gamma^+} (1 - \xi_\alpha(\gamma^{-1}))^2 = |\delta_\gamma(1)| = \delta_\gamma(1) \operatorname{sign}\{\delta_\gamma(1)\}$$
$$= (-1)^{r - r_\gamma} \operatorname{sign}\{\delta_\gamma(1)\}\xi_{2(\rho - \rho_\gamma)}(\gamma) \prod_{\alpha \in \Phi^+ - \Phi_\gamma^+} (1 - \xi_\alpha(\gamma^{-1}))^2$$

and so $c_\gamma^2 = (-1)^{r - r_\gamma} \operatorname{sign}\{\delta_\gamma(1)\}$, as contended. □

Let Φ_R^+ be the set of positive real roots of the pair $(\mathfrak{g}_c, \mathfrak{j}_c)$; for $j \in \mathrm{J}$, put $\diamondsuit_R(j) = \prod_{\alpha \in \Phi^+_R} (1 - \xi_\alpha(j^{-1}))$ and set $\mathrm{J}'(R) = \{j \in \mathrm{J} : \diamondsuit_R(j) \neq 0\}$. Let $F_{T,\mathrm{J}}$ be the analytic function on $\Omega \cap \mathrm{J}'$ defined by the rule

$$F_{T,\mathrm{J}}(j) = \Delta(j) F_T(j) \qquad (j \in \Omega \cap \mathrm{J}'),$$

Δ as always.

Proposition 8.3.4.3 Retain the above notations – then $F_{T,\mathrm{J}}$ can be extended to an analytic function on $\Omega \cap \mathrm{J}'(R)$.

Proof Fix an element $j_0 = \gamma$ (say) in $\Omega \cap \mathrm{J}'(R)$ and introduce $\mathfrak{g}_\gamma, G_\gamma, \Omega_\gamma$ etc. as above. It will be enough to show that there exists an analytic function on an open neighborhood $\mathcal{O}(\gamma)$ of γ in $\Omega \cap \mathrm{J}'(R)$ with the property that its restriction to $\mathcal{O}(\gamma) \cap \mathrm{J}'$ agrees with $F_{T,\mathrm{J}}$ there. This being so, choose $\mathcal{O}$ per Lemma 8.3.4.2 – then $\gamma \exp(\mathcal{O})$ is an open neighborhood of γ in $\Omega \cap \mathrm{J}$, whence $\mathcal{O}(\gamma) = \gamma \exp(\mathcal{O}) \cap \mathrm{J}'(R)$ is an open neighborhood of γ in $\Omega \cap \mathrm{J}'(R)$. The preceding lemmas tell us that

$$\pi_\gamma(H) F_\sigma(H) = c_\gamma F_{T,\mathrm{J}}(\gamma \exp H)$$

(all $H \in \mathcal{O}'$ (= the $\mathfrak{g}_\gamma$-regular elements in $\mathcal{O}$)); on the other hand, according to Proposition 8.3.2.1, the function

$$H \mapsto \pi_\gamma(H) F_\sigma(H) \qquad (H \in \mathcal{O}_{\mathfrak{g}_\gamma} \cap \mathfrak{j}'_\gamma)$$

admits an analytic extension to $\mathcal{O}_{\mathfrak{g}_\gamma} \cap \mathfrak{j}'_\gamma(R)$. [Here

$$\mathfrak{j}'_\gamma = \{H \in \mathfrak{j} : \pi_\gamma(H) \neq 0\};$$

similarly $\mathfrak{j}'_\gamma(R)$ consists of those H in $\mathfrak{j}$ such that $\prod_{\alpha \in \Phi_{\gamma, R}^+} \alpha(H) \neq 0$ $(\Phi^+_{\gamma, R} = \Phi^+_\gamma \cap \Phi^+_R)$.] Since a given element j in the set $\mathcal{O}(\gamma)$ can be written uniquely in the form $\gamma \exp H$ with $H \in \mathcal{O} \cap \mathfrak{j}'_\gamma(R)$, the contention of the present proposition is now clear. □

In what follows we shall denote the (analytic) extension of $F_{T,\mathrm{J}}$ to $\Omega \cap \mathrm{J}'(R)$ by the same symbol.

Proposition 8.3.4.4 Retain the above notations; let j_0 be a point in $\Omega \cap \mathrm{J}$ and D an element in the universal enveloping algebra of $\mathfrak{j}_c$

such that $w_\alpha(D) = -D$ for every real root α of the pair $(\mathfrak{g}_c, \mathfrak{j}_c)$ which commutes with j_0 – then $DF_{T,J}$ can be extended to a continuous function around j_0.

[In view of what has been said above and Proposition 8.3.2.2, this is obvious.]

For any root α of the pair $(\mathfrak{g}_c, \mathfrak{j}_c)$ define the element H_α as usual (cf. 8.2.2) and put $\Pi = \prod_{\alpha>0} H_\alpha$ – then Π is a differential operator on J and, thanks to Proposition 8.3.4.4, $\Pi F_{T,J}$ can be extended to a continuous function on $\Omega \cap J$.

Proposition 8.3.4.5 Retain the above notations; let J^1 and J^2 be two Cartan subgroups of G – then

$$\Pi^1 F_{T,J^1} = \Pi^2 F_{T,J^2}$$

on $\Omega \cap J^1 \cap J^2$.

Proof Let $\mathfrak{j}^1$ and $\mathfrak{j}^2$ be the Lie algebras of J^1 and J^2, respectively; fix a point $j_0 = \gamma$ (say) in $\Omega \cap J^1 \cap J^2$ and introduce $\mathfrak{g}_\gamma$, G_γ, Ω_γ etc. as above relative to γ – then $\mathfrak{j}^1$ and $\mathfrak{j}^2$ are, of course, Cartan subalgebras of $\mathfrak{g}_\gamma$. Put $\mathfrak{j} = \mathfrak{j}^1$ or $\mathfrak{j}^2$ and define

$$\Pi_\gamma = \prod_{\alpha \in \Phi_{\gamma}^+} H_\alpha, \quad \Pi/\Pi_\gamma = \prod_{\alpha \in \Phi^+ - \Phi_{\gamma}^+} H_\alpha$$

in the notation employed earlier – then $\Pi = \Pi_\gamma \cdot \Pi/\Pi_\gamma$. Since

$$\Pi^{w_\alpha} = -\Pi, \; \Pi_\gamma^{w_\alpha} = -\Pi_\gamma \qquad (\text{all } \alpha \in \Phi_\gamma^+),$$

it is clear that Π/Π_γ is invariant under the Weyl group of the pair $(\mathfrak{g}_{\gamma,c}, \mathfrak{j}_c)$. This being so, choose an element p in $I(\mathfrak{g}_{\gamma,c})$ which, when restricted to $\mathfrak{j}_c^1$, gives Π^1/Π_γ^1 (the terminology being justified upon making the customary identifications); it is easy to see that things can be arranged so as to ensure that the restriction of p to $\mathfrak{j}_c^2$ is Π^2/Π_γ^2 – this can be accomplished without altering the other conditions. [Let G_c be an acceptable complexification of G, $G_{\gamma,c}$ the analytic subgroup of G_c corresponding to $\mathfrak{g}_{\gamma,c}$ – then we can choose an $x_\gamma \in G_{\gamma,c}$ with the property that $x_\gamma(\mathfrak{j}_c^1) = \mathfrak{j}_c^2$. Because the definition of $\Pi^i F_{T,J^i}$ $(i = 1, 2)$ is independent of the ordering of the roots, we may suppose that the positive roots of the pair $(\mathfrak{g}_c, \mathfrak{j}_c^1)$ are mapped by x_γ to the positive roots of the pair $(\mathfrak{g}_c, \mathfrak{j}_c^2)$ . . . . In passing note that this argument tells us that it is permissible to also assume that $c_\gamma^1 = c_\gamma^2$ (c_γ^1 and c_γ^2 being as in Lemma 8.3.4.2 relative to $\mathfrak{j}^1$ and $\mathfrak{j}^2$, respectively).] Using Corollary 8.3.2.5, therefore, we deduce that

$$\partial(\pi_\gamma^1 \cdot \tilde{p}_1)(\pi_\gamma^1 F_\sigma) = \partial(\pi_\gamma^2 \cdot \tilde{p}_2)(\pi_\gamma^2 F_\sigma) \qquad (\tilde{p}_i = p \,|\, \mathfrak{j}_c^i,\ i = 1, 2)$$

on $\mathcal{O}_{\mathfrak{g}_\gamma} \cap \mathfrak{j}^1 \cap \mathfrak{j}^2$. [Strictly speaking, the 'H_α' appearing in the definition of Π_γ^i $(i = 1, 2)$ are not the same as the 'H_α' which are defined relative to the Killing form of the derived algebra of $\mathfrak{g}_\gamma$; however this causes no difficulty. . . .] Now let $\mathscr{C}_\gamma$ be an open convex neighborhood of zero in

$\mathcal{O}_{\mathfrak{g}_\gamma}$ – then, as has been seen above,

$$\pi_\gamma^i(H)F_\sigma(H) = c_\gamma^i\Delta^i(\gamma \exp H)F_T(\gamma \exp H) \qquad (H \in \mathscr{C}_\gamma' \cap \mathfrak{j}^i,\ i = 1, 2)$$

and so, since $c_\gamma^1 = c_\gamma^2 \neq 0$, it follows that

$$\Pi^1 F_{T,\mathrm{J}^1}(\gamma) = \Pi^2 F_{T,\mathrm{J}^2}(\gamma),$$

which concludes the proof, $\gamma = j_0$ being an arbitrary element of $\Omega \cap \mathrm{J}^1 \cap \mathrm{J}^2$. □

Our next task will be to formulate the group theoretic analogue of Proposition 8.3.2.8; as there, we shall have need of a preliminary lemma.

Lemma 8.3.4.6 There exists a unique G-invariant analytic differential operator ∇_G on G' with the following property: Let J be a Cartan subgroup of G – then $f(j; \nabla_G) = f(j; \Pi \circ \Delta_{\mathrm{J}})$ for $f \in C^\infty(G)$ (all $j \in \mathrm{J}'$).

[The proof of this fact is substantially the same as that of Lemma 8.3.2.7 and will be left to the reader.]

Proposition 8.3.4.7 Retain the above notations; let $Z \in \mathfrak{Z}$ – then $(\nabla_G \circ Z)F_T$ admits a continuous extension to all of Ω.

Proof Since the distribution $Z \cdot T$ satisfies the same conditions as T itself, it will be sufficient to consider just the case when $Z = 1$. Let Ω_0 be the set of all points $\gamma \in \Omega$ with the following property: There exists an open neighborhood $N_G(\gamma)$ of γ in Ω and a continuous function F on $N_G(\gamma)$ such that $F = \nabla_G F_T$ on $N_G(\gamma) \cap G'$. Clearly Ω_0 is an open and invariant subset of Ω and, of course, the proposition will follow when it is shown that $\Omega = \Omega_0$; in turn, for this purpose we need only verify that Ω_0 contains all the semi-simple points of Ω (cf. the discussion in 8.1.6). So fix a semi-simple element γ in Ω; relative to γ, introduce $\mathfrak{g}_\gamma$, G_γ, Ω_γ etc. as above and fix an open convex neighborhood $\mathscr{C}_\gamma$ of zero in $\mathcal{O}_{\mathfrak{g}_\gamma}$. Let $\mathfrak{j}$ be a Cartan subalgebra of $\mathfrak{g}_\gamma$ (and hence of $\mathfrak{g}$), J the Cartan subgroup of G associated with $\mathfrak{j}$ – then $\gamma \in \Omega \cap \mathrm{J}$. Let p be that element in $I(\mathfrak{g}_{\gamma,c})$ which, when restricted to $\mathfrak{j}_c$, gives Π/Π_γ; let c_γ be the constant of Lemma 8.3.4.2 corresponding to $\mathfrak{j}$ – then

$$F_\sigma(X_\gamma; \nabla_{\mathfrak{g}_\gamma} \circ \partial(p)) = c_\gamma F_T(\gamma \exp X_\gamma; \nabla_G)$$

for $X_\gamma \in \mathscr{C}_\gamma'$. This being so, it now follows from Proposition 8.3.2.8 that there exists an open neighborhood $\tilde{\Omega}_\gamma$ of 1 in Ω_γ and a continuous function F on $\tilde{\Omega}_\gamma$ such that $F(x_\gamma) = F_T(\gamma x_\gamma; \nabla_G)$ for all $x_\gamma \in \tilde{\Omega}_\gamma'$. To finish the proof, choose a local analytic cross section s of G/G_γ over G_γ and a small neighborhood $\mathcal{O}$ of 1 in $\tilde{\Omega}_\gamma$ such that the mapping ϕ,

$$(\dot{x}, x_\gamma) \mapsto {}^{s(\dot{x})}(\gamma x_\gamma) \qquad (x_\gamma \in \mathcal{O}),$$

is an analytic diffeomorphism onto its image $N_G(\gamma) \subset \Omega$; extend F to

$N_G(\gamma)$ by writing $F(\phi(\dot{x}, x_y)) = F(x_y)$ – then F is continuous (on $N_G(\gamma)$) and since $\nabla_G F_T$ is G-central, it is clear that $F = \nabla_G F_T$ on $N_G(\gamma) \cap G'$. □

8.3.5 Rapidly Decreasing Functions on a Euclidean Space

In this number we shall review some well-known facts which will be used constantly in the sequel; at the same time, it will be convenient to present a few results on the fundamental solutions to certain partial differential equations.

Let E be a finite dimensional vector space over $\mathbf{R}$ – adopt the notations of Vol. I, number 2.1.4. Let D be a polynomial differential operator on E, i.e. let $D \in \mathfrak{P}(E)$; put

$$|f|_D = \sup_{X \in E} |f(X; D)| \qquad (f \in C^\infty(E)).$$

The space $\mathscr{C}(E)$ consists of all $f \in C^\infty(E)$ for which $|f|_D < \infty$ for all $D \in \mathfrak{P}(E)$; the elements of $\mathscr{C}(E)$ are called the *rapidly decreasing* functions on E. It is well-known (Schwartz [2]) that the semi-norms $\{|\,.\,|_D : D \in \mathfrak{P}(E)\}$ equip $\mathscr{C}(E)$ with the structure of a locally convex, complete, Hausdorff, topological vector space – moreover $\mathscr{C}(E)$ is a topological algebra under convolution.

Definition A distribution T on E, continuous for the topology on $C_c^\infty(E)$ induced by the $\{|\,.\,|_D : D \in \mathfrak{P}(E)\}$, is called a *tempered distribution*.

If T is tempered, then T admits a unique continuous extension to the space $\mathscr{C}(E)$.

[The preceding considerations generalize in the usual way to open subsets of E.]

Let $Q : E \times E \to \mathbf{R}$ denote a positive definite bilinear form on E. Fix an orthonormal basis in E relative to the Euclidean norm $\|\,.\,\|$ determined by Q; let $x_1, \ldots, x_n$ denote the corresponding set of coordinates ($n = \dim(E)$); set $dX = dx_1 \ldots dx_n$ – then dX is the usual canonical element of volume on $E \sim \mathbf{R}^n$.

Let F be a locally summable function on E – then F is said to be *tempered* if the distribution $f \mapsto \int_E fFdX$ is tempered ($f \in C_c^\infty(E)$).

Examples (1) Let F be a measurable function on E – then the mapping $f \mapsto \int_E fFdX$ is a tempered distribution provided there exists an integer $m > 0$ such that $r = \sup_{X \in E} |F(X)|(1 + \|X\|)^{-m} < \infty$. [Thus select an integer $M > 0$ such that $R = \int_E (1 + \|X\|)^{-M}\, dX < \infty$ – then

$$\left|\int_E f(X)F(X)dX\right| \leqslant rR \sup_{X \in E} (1 + \|X\|)^{m+M} |f(X)| < \infty.]$$

(2) Let $F(X) = \sum_1^m p_i(X)e^{\lambda_i(X)}$ ($X \in E$) where the p_i are (complex valued) polynomial functions on E and the λ_i are distinct complex valued linear functions on E. Let $\mathcal{O}$ be a non-empty open subset of E which contains tX whenever $X \in \mathcal{O}, t \geqslant 1$ (in which case $\mathcal{O}$ is termed *full*) – then the rule $f \mapsto \int_{\mathcal{O}} f(X)F(X)dX$ ($f \in C_c^\infty(\mathcal{O})$) defines a tempered distribution on $\mathcal{O}$ iff $\mathscr{R}(\lambda_i(X)) \leqslant 0$ for all $X \in \mathcal{O}$ ($i = 1, \ldots, m$). [Use induction on m and (1).]

Given $f \in \mathscr{C}(E)$, put

$$\hat{f}(X) = \frac{1}{(2\pi)^{n/2}} \int_E f(Y) e^{\sqrt{-1}Q(X,Y)} dY \qquad (X \in E).$$

Then $\hat{f}$ is just the usual Fourier transform of f. We recall that $\hat{f} \in \mathscr{C}(E)$ and, in fact, that the map $f \mapsto \hat{f}$ is a homeomorphism of $\mathscr{C}(E)$. Of course Fourier inversion holds here:

$$f(X) = \frac{1}{(2\pi)^{n/2}} \int_E \hat{f}(Y) e^{-\sqrt{-1}Q(X,Y)} dY \qquad (X \in E).$$

Let T be a tempered distribution – then the Fourier transform $\hat{T}$ of T is defined by $\hat{T}(f) = T(\hat{f})$ $(f \in \mathscr{C}(E))$. The distribution $\hat{T}$ is also tempered.

Given a polynomial function p on E, let $p^\sharp$ be defined by the prescription $p^\sharp(X) = p(-\sqrt{-1}\,X)$ $(X \in E)$ – then: (1) $(pf)^\wedge = \partial(p^\sharp)\hat{f}$; (2) $(\partial(p)f)^\wedge = (p^\sharp)\hat{f}$ $(f \in \mathscr{C}(E))$. [Statements (1), (2) are, of course, nothing more than the usual rules of 'operational calculus'.]

Remark Note that the requirement of positive definiteness placed on Q is not really necessary – one need only request that Q be non-degenerate. Thus, let $X_1, \ldots, X_n$ be a basis in E, $x_1, \ldots, x_n$ the corresponding coordinates of a generic element $X \in E$; let dX denote the Lebesgue measure in E determined by requiring that the measure of the hypercube $\{X: 0 \leqslant x_i \leqslant 1, i = 1, \ldots, n\}$ be one. On the other hand, let $\tilde{X}_1, \ldots, \tilde{X}_n$ be the basis in E dual to $\{X_i\}$ with respect to Q, i.e. such that $Q(X_i, \tilde{X}_j) = \delta_{ij}$. Denote the coordinates of a generic element $\tilde{X} \in E$ with respect to $\{\tilde{X}_i\}$ by $\tilde{x}_1, \ldots, \tilde{x}_n$ and let $d\tilde{X}$ denote the Lebesgue measure in E determined by requiring that the measure of the hypercube $\{\tilde{X}: 0 \leqslant \tilde{x}_i \leqslant 1, i = 1, \ldots, n\}$ be one. The Fourier transform of $f \in \mathscr{C}(E)$ is then defined by

$$\hat{f}(\tilde{X}) = \frac{1}{(2\pi)^{n/2}} \int_E f(Y) e^{\sqrt{-1}Q(\tilde{X},Y)}\, dY \qquad (\tilde{X} \in E)$$

whereas the Fourier inversion now reads

$$f(X) = \frac{1}{(2\pi)^{n/2}} \int_E \hat{f}(\tilde{Y}) e^{-\sqrt{-1}Q(X,\tilde{Y})}\, d\tilde{Y} \qquad (X \in E).$$

Here is a realization of this set up. Let $\mathfrak{g}$ be a semi-simple Lie algebra over $\mathbf{R}$, $\mathfrak{g} = \mathfrak{k} + \mathfrak{p}$ a Cartan decomposition of $\mathfrak{g}$, θ the corresponding Cartan involution; let $X_1, \ldots, X_k, Y_1, \ldots, Y_p$ be an orthonormal basis for $\mathfrak{g}$ relative to the Euclidean structure $(\,.\,,\,.\,)_\theta$ $(X_i \in \mathfrak{k}, Y_j \in \mathfrak{p})$ – then the canonically determined Lebesgue measure is $d_\mathfrak{g}$. Notice that $-X_1, \ldots, -X_k, Y_1, \ldots, Y_p$ is the corresponding dual basis for the Killing form B (B playing the role of Q) and, moreover, that $d_\mathfrak{g}$ is also the canonical Lebesgue measure associated with this dual basis.

For later use, let us insert here a few simple observations on divisibility. Let $\lambda \neq 0$ be a real linear function on E, r a real number, $\mathscr{H}$ the hyperplane in E given by $\lambda = r$. Let $\mathscr{C}_{\lambda,r}(E)$ denote the (necessarily closed) subspace of $\mathscr{C}(E)$ consisting of those $f \in \mathscr{C}(E)$ which vanish identically on $\mathscr{H}$.

Lemma 8.3.5.1 For every $f \in \mathscr{C}_{\lambda,r}(E)$ there exists a unique element $g_f \in \mathscr{C}(E)$ such that $f = (\lambda - r)g_f$; denote g_f by $(\lambda - r)^{-1}f$ – then the mapping $f \mapsto (\lambda - r)^{-1}f$ of $\mathscr{C}_{\lambda,r}(E)$ into $\mathscr{C}(E)$ is continuous.

Proof We may assume that $r = 0$. Let X_λ denote the unique element in E such that $Q(X, X_\lambda) = \lambda(X)$ for all $X \in E$; let $X_1, \ldots, X_n$ be a basis for E such that $X_1 = X_\lambda / Q(X_\lambda, X_\lambda)$ and $X_2, \ldots, X_n$ are in the hyperplane $\lambda = 0$. If $x_1, \ldots, x_n$ are the corresponding coordinates, then $f(0, x_2, \ldots, x_n) = 0 \Rightarrow f(x_1, \ldots, x_n) = x_1 g(x_1, \ldots, x_n)$ where

$$g(x_1, \ldots, x_n) = \int_0^1 \frac{d}{dt} f(tx_1, x_2, \ldots, x_n)\, dt. \quad \square$$

Lemma 8.3.5.2 Let λ_i be non-zero real linear functions on E, r_i real numbers $(1 \leqslant i \leqslant s)$; assume that the hyperplanes $\mathscr{H}_i : \lambda_i = r_i$ are all distinct; put $\pi = \prod_1^s (\lambda_i - r_i)$ and let $\mathscr{C}_{\mathscr{H}}(E)$ denote the subspace of all $f \in \mathscr{C}(E)$ which vanish identically on $\bigcup_{i=1}^s \mathscr{H}_i$ – then, for every $f \in \mathscr{C}_{\mathscr{H}}(E)$, there exists a unique element $g_f = \pi^{-1} f$ in $\mathscr{C}(E)$ such that $f = \pi g_f$. In addition the map $f \mapsto \pi^{-1} f$ is a continuous mapping of $\mathscr{C}_{\mathscr{H}}(E)$ into $\mathscr{C}(E)$.

[This statement follows from Lemma 8.3.5.1 by induction on s.]

Corollary 8.3.5.3 Consider $C_c^\infty(E) \cap \mathscr{C}_{\mathscr{H}}(E)$ as a topological subspace of $C_c^\infty(E)$ – then $\pi^{-1} f \in C_c^\infty(E)$ (all $f \in C_c^\infty(E) \cap \mathscr{C}_{\mathscr{H}}(E)$) and the mapping $f \mapsto \pi^{-1} f$ of $C_c^\infty(E) \cap \mathscr{C}_{\mathscr{H}}(E)$ into $C_c^\infty(E)$ is continuous.

The following elementary fact is occasionally useful.

Lemma 8.3.5.4 Let $\mathscr{O}$ be a non-empty open subset of E, $f_1, \ldots, f_r$ a finite number of continuous functions on $\mathscr{O}$; suppose that the f_i are linearly independent over $\mathbf{C}$ – then there exists an $f \in C_c^\infty(\mathscr{O})$ such that

$$\int_{\mathscr{O}} f(X) f_i(X)\, dX = \begin{cases} 1 & \text{if } \; i = 1 \\ 0 & \text{if } \; i \neq 1 \end{cases} \qquad (1 \leqslant i \leqslant r).$$

Proof Let T_i denote the distribution on $\mathscr{O}$ given by

$$T_i(f) = \int_{\mathscr{O}} f(X) f_i(X)\, dX \qquad (f \in C_c^\infty(\mathscr{O})).$$

Let $\mathscr{E}$ be a vector space over $\mathbf{C}$ of dimension r; let $a_1, \ldots, a_r$ be a basis for $\mathscr{E}$, and, for any $f \in C_c^\infty(\mathscr{O})$, set $T(f) = \sum_1^r T_i(f) a_i$; let $\tilde{\mathscr{E}}$ be the subspace of $\mathscr{E}$ consisting of the $T(f)$ ($f \in C_c^\infty(\mathscr{O})$) – then $\tilde{\mathscr{E}} = \mathscr{E}$. [Suppose not; choose a non-zero linear function λ on $\mathscr{E}$ such that $\lambda | \tilde{\mathscr{E}}$ vanishes. Put $c_i = \lambda(a_i)$ $(1 \leqslant i \leqslant r)$; since λ is non-zero, not all the c_i are zero. Let $f_\lambda = \sum_1^r c_i f_i$; because the f_i are linearly independent, $f_\lambda \neq 0$ and so there exists $f \in C_c^\infty(\mathscr{O})$ for which

$$\int_{\mathscr{O}} f(X) f_\lambda(X)\, dX \neq 0.$$

However

$$\int_{\mathscr{O}} f(X) f_\lambda(X)\, dX = \sum_1^r c_i T_i(f) = \lambda(T(f)) = 0,$$

a contradiction. Hence $\tilde{\mathscr{E}} = \mathscr{E}$.] This being the case, select f such that $T(f) = a_1$ – then this f meets the above requirements. $\square$

Retain the above notations and assumptions. Let n_1, n_2 be two non-negative integers such that $n_1 + n_2 = n \geqslant 2$; put

$$u = \sum_1^{n_1} x_i^2 - \sum_{n_1+1}^{n} x_i^2, \quad \square = \sum_1^{n_1} (\partial/\partial x_i)^2 - \sum_{n_1+1}^{n} (\partial/\partial x_i)^2.$$

DeRham [1] has shown that there exists a function ξ on $\mathbf{R}$ with the following properties. Let $\Xi(X) = \xi(u(X))$ $(X \in E)$ – then Ξ is a locally summable function on E and

$$\square^{[n/2]}.\Xi = \delta$$

in the sense of distributions. [Here δ denotes the Dirac measure concentrated at the origin while $[n/2]$ is the greatest integer in $n/2$.] It is actually possible to give explicit expressions for ξ in terms of the Heaviside function $t \mapsto Y(t)$ ($Y(t)$ being 0 or 1 according as $t \leqslant 0$ or $t > 0$, $t \in \mathbf{R}$). There are four cases:

(1) $\xi(t) = (-1)^{(n_1-1)/2}a_n^{-1}Y(t)|t|^{-1/2}$ if $n \equiv n_1 \equiv 1 \bmod 2$;

(2) $\xi(t) = (-1)^{n_1/2}a_n^{-1}Y(-t)|t|^{-1/2}$ if $n \equiv n_2 \equiv 1 \bmod 2$;

(3) $\xi(t) = -(-1)^{n_1/2}b_n^{-1}\log|t|$ if $n \equiv n_1 \equiv 0 \bmod 2$;

(4) $\xi(t) = (-1)^{(n_1-1)/2}c_n^{-1}Y(t)$ if $n_1 \equiv n_2 \equiv 1 \bmod 2$.

Here

$$a_n = 2^{n-1}\pi^{(n-1)/2}\Gamma((n-1)/2), \quad b_n = \pi c_n = 2^n\pi^{n/2}\Gamma(n/2),$$

Γ the classical Gamma function.

Remark (Harish-Chandra) The preceding considerations are valid for the space $\mathscr{C}(E)$ too. Thus set $u_1 = \sum_1^{n_1} x_i^2$, $u_2 = \sum_{n_1+1}^{n} x_i^2$ – then, for m a sufficiently large positive integer, it is clear that

$$r = \int_E |\Xi|(1+u_1)^{-m}(1+u_2)^{-m}\,dX < \infty$$

and so

$$\int_E |\Xi f|\,dX \leqslant r \sup_{X \in E} |(1+u_1(X))^m(1+u_2(X))^m f(X)|$$

for $f \in \mathscr{C}(E)$. Hence Ξ is tempered. Since $\square$ is self-adjoint, we have

$$f(0) = \int_E \Xi\square^{[n/2]}f\,dX \qquad (\text{all } f \in \mathscr{C}(E)).$$

These results of deRham will play an important role later on in this chapter.

We shall conclude this number by presenting some elementary facts dealing with the nature of fundamental solutions to certain elliptic Laplacians – they will be used during the course of the proof of one of the main theorems in 8.5.

Set $d(X) = (x_1^2 + \cdots + x_n^2)^{1/2}$ for $X \in E \sim \mathbf{R}^n$.

Lemma 8.3.5.5 Let r be a real number, $D = \partial^k/\partial x_{i_1} \cdots \partial x_{i_k}$ – then

$$Dd^r = \sum_{i=0}^{k} p_i d^{r-i-k},$$

$$D(d^r \log d) = \sum_{i=0}^{k} P_i d^{r-i-k} + (\log d)\sum_{i=0}^{k} Q_i d^{r-i-k}$$

where p_i, P_i, Q_i are homogeneous polynomials in $(x_1, \ldots, x_n)$ of degree i.

[This follows by induction on the integer k.]

Corollary 8.3.5.6 If $r > k$, then d^r and $d^r \log d$ are functions of class C^k on $\mathbf{R}^n$.

Corollary 8.3.5.7 Retain the above notations – then $d^{k-1}Dd$ remains bounded on $\mathbf{R}^n$.

Lemma 8.3.5.8 Let $h \in C^\infty(\mathbf{R})$ be such that it is constant on the intervals $(-\infty, 0]$ and $[1, +\infty)$. Choose two numbers r_1, r_2 with $0 < r_2 \leqslant r_1 \leqslant 1$ and put $H_{r_1,r_2}(X) = h(r_2^{-1}(d(X) - r_1))$ $(X \in E)$ – then, for each integer $k \geqslant 0$, there exists a number $c_k \geqslant 0$, independent of r_1 and r_2, such that

$$|DH_{r_1,r_2}| \leqslant c_k r_2^{-k}$$

for $D = \partial^k/\partial x_{i_1} \cdots \partial x_{i_k}$ $(1 \leqslant i_1, \ldots, i_k \leqslant n)$.

Proof We shall proceed by induction on k. If $k = 0$, we may take $c_0 = \sup|h|$. So assume that $k \geqslant 1$ and put $h'(t) = dh/dt$ $(t \in \mathbf{R})$ – then h' satisfies the conditions of the lemma too and

$$\partial H_{r_1,r_2}/\partial x_i = r_2^{-1} H'_{r_1,r_2} \cdot \partial d/\partial x_i \qquad (1 \leqslant i \leqslant n),$$

where

$$H'_{r_1,r_2}(X) = h'(r_2^{-1}(d(X) - r_1)) \qquad (X \in E).$$

Since $H'_{r_1,r_2}(X) = 0$ unless $r_1 \leqslant d(X) \leqslant r_1 + r_2$, we have

$$\sup_X |DH_{r_1,r_2}| \leqslant r_2^{-1} \sup_{d(X) \geqslant r_2} |D'(H'_{r_1,r_2} \cdot \partial d/\partial x_{i_k})|,$$

where $D' = \partial^{k-1}/\partial x_{i_1} \cdots \partial x_{i_{k-1}}$. Expanding $D'(H'_{r_1,r_2} \cdot \partial d/\partial x_{i_k})$ by the Leibnitz rule, using Corollary 8.3.5.7, and the induction hypothesis applied to H'_{r_1,r_2}, the result follows. □

Put $\Delta = \sum_{i=1}^n (\partial/\partial x_i)^2$ and let δ denote the Dirac measure concentrated at the origin.

The proof of the following result may be found in Schwartz [1, p. 47].

Lemma 8.3.5.9 If n is odd

$$\Delta^{k+(n-1)/2} \cdot d^{2k-1} = c_{k,n}\delta \qquad (k \geqslant 1)$$

and if n is even

$$\Delta^{k+n/2} \cdot (d^{2k} \log d) = c'_{k,n}\delta \qquad (k \geqslant 0).$$

Here $c_{k,n}$ and $c'_{k,n}$ are non-zero numbers and the above relations are taken in the sense of the theory of distributions.

Remark For the record, the constants $c_{k,n}$, $c'_{k,n}$ are:

$$c_{k,n} = (2(k + (n-1)/2) - n)(2(k + (n-1)/2) - 2 - n) \cdots (4 - n)(2 - n) \times 2^{k+(n-1)/2-1}(k - 1 + (n-1)/2)! \frac{2(\sqrt{\pi})^n}{\Gamma(n/2)};$$

$$c'_{k,n} = [[(2(k + n/2) - n)(2(k + n/2) - 2 - n) \cdots (4 - n)(2 - n)]] \times 2^{k-1+n/2}(k - 1 + n/2)! \frac{2(\sqrt{\pi})^n}{\Gamma(n/2)}$$

where in the formula [[. . .]] the factor zero is omitted.

Lemma 8.3.5.10 Fix integers $m_1 \geqslant 0$, $m_2 \geqslant 1$ – then there exists an integer $m_3 \geqslant 1$ and a function f_0 on $\mathbf{R}^n$ of class $C^{2m_3(m_2-1)+m_1}$ such that

$$\Delta^{m_2 m_3} \cdot f_0 = \delta.$$

Proof Choose m_3 so that $2m_3 > m_1 + n$; consider the two cases: n odd or n even. If n is odd, then $k = m_3m_2 - (n-1)/2$ is an integer and $2k = 2m_3m_2 - n + 1 > m_1 + 1 \Rightarrow k \geqslant 1$. Put $f_0 = c_{k,n}^{-1}d^{2k-1}$ – then

$$\Delta^{m_2m_3}\cdot f_0 = \Delta^{k+(n-1)/2}\cdot f_0 = \delta$$

and $2k - 1 = 2m_3m_2 - n = 2m_3(m_2 - 1) + 2m_3 - n > 2m_3(m_2 - 1) + m_1$. Thus f_0 is of class $C^{2m_3(m_2-1)+m_1}$ by Corollary 8.3.5.6. On the other hand, if n is even, put

$$k = m_3m_2 - n/2 \Rightarrow 2k = 2m_3m_2 - n > m_1 \Rightarrow k \geqslant 0.$$

Put $f_0 = (c'_{k,n})^{-1}d^{2k}\log d$ – then $2k = 2m_3m_2 - n > 2m_3(m_2 - 1) + m_1$ and so f_0 again is of class $C^{2m_3(m_2-1)+m_1}$. □

8.3.6 Tempered Distributions on a Reductive Lie Algebra

Let $\mathfrak{g}$ be a reductive Lie algebra over $\mathbf{R}$, $\theta : \mathfrak{g} \to \mathfrak{g}$ an involutive automorphism of $\mathfrak{g}$ whose restriction to the derived algebra $\bar{\mathfrak{g}} = [\mathfrak{g}, \mathfrak{g}]$ is a Cartan involution (of $\bar{\mathfrak{g}}$); write, as usual, $\mathfrak{g} = \mathfrak{k} + \mathfrak{p}$ and let (G, K) be a pair corresponding to $\mathfrak{g}$ per Vol. I, number 1.1.5.

Let T be a central distribution on $\mathfrak{g}$ which is annihilated by an ideal I of finite codimension in $I(\mathfrak{g}_c)$ – then, as we know, T agrees, in the sense of distributions, with a locally summable function F_T which is actually analytic on $\mathfrak{g}'$. The primary objective of this number is to give a growth condition on F_T which is both necessary and sufficient for T to be tempered.

Select a maximal set $\mathfrak{j}^1, \ldots, \mathfrak{j}^r$ of mutually non-conjugate θ-stable Cartan subalgebras of $\mathfrak{g}$; fix a Euclidean norm $\|\,.\,\|$ on $\mathfrak{g}$.

Theorem 8.3.6.1 (Harish-Chandra) Let T be a central distribution on $\mathfrak{g}$ which is annihilated by an ideal I in $I(\mathfrak{g}_c)$ of finite codimension – then T is tempered iff there exists an integer $m \geqslant 0$ such that

$$\sup_{H \in (\mathfrak{j}^i)'} (1 + \|H\|)^{-m} |\pi^i(H)| |F_T(H)| < \infty \qquad (1 \leqslant i \leqslant r).$$

The sufficiency of the condition is an easy consequence of a familiar integration formula and Theorem 8.4.1.2 below (the notations of which we shall agree to adopt).

Proof of Theorem 8.3.6.1 *Sufficiency* Let $f \in C_c^\infty(\mathfrak{g})$ – then

$$T(f) = \int_{\mathfrak{g}} f(X)F_T(X)d_{\mathfrak{g}}(X)$$

$$= \sum_{i=1}^{r} (-1)^{r_I^i} \int_{(\mathfrak{j}^i)'} \epsilon_R^i(H)\pi^i(H)F_T(H)\phi_f^i(H)d_{\mathfrak{j}^i}(H)$$

and so, for a suitable non-negative integer M, we have

$$|T(f)| \leq \sum_{i=1}^{r} \sup_{(\mathfrak{j}^i)'} ((1 + \|H\|)^M |\phi_f^i(H)|) \qquad (\text{all } f \in C_c^\infty(\mathfrak{g})).$$

On the basis of Theorem 8.4.1.2, it is now clear that T is tempered. □

The proof of the necessity rests on the following lemma.

Lemma 8.3.6.2 Let T be a central tempered distribution on $\mathfrak{g}$; let $\mathfrak{j}$ be a Cartan subalgebra of $\mathfrak{g}$, $\mathscr{C}$ a Weyl chamber in $\mathfrak{j}$ – then there exists an integer $n \geqslant 0$ such that $\pi^n F_{T,\mathfrak{j}}$ is tempered on $\mathscr{C}$.

[Let us recall that $F_{T,\mathfrak{j}}$ is the analytic function on $\mathfrak{j}'$ defined by the rule $F_{T,\mathfrak{j}}(H) = \pi(H)F_T(H)$ $(H \in \mathfrak{j}')$ (cf. 8.3.2).]

Proof of Theorem 8.3.6.1 *Necessity* We shall agree to write $\mathfrak{j}$ for $\mathfrak{j}^i$ $(i = 1, \ldots, r)$; fix a Weyl chamber $\mathscr{C}$ in $\mathfrak{j}$ – then it is clear that we need only produce an integer $m \geqslant 0$ such that

$$|F_{T,\mathfrak{j}}(H)| \lesssim (1 + \|H\|)^m \qquad (\text{all } H \in \mathscr{C}).$$

To this end, we recall that it was shown in 8.3.1 that there exist distinct linear functions λ_i and polynomial functions p_i on $\mathfrak{j}_c$ such that

$$F_{T,\mathfrak{j}}(H) = \sum_i p_i(H)e^{\lambda_i(H)} \qquad (\text{all } H \in \mathscr{C}).$$

[In reality this result was established only on an open subset of $\mathscr{C}$; but then, by analytic continuation, it holds throughout $\mathscr{C}$.] On the other hand, Lemma 8.3.6.2 tells us that $\pi^n F_{T,\mathfrak{j}}$ is tempered on $\mathscr{C}$ (for some integer $n \geqslant 0$), i.e. that $\sum_i q_i e^{\lambda_i}$ is tempered on $\mathscr{C}$ (where $q_i = \pi^n p_i$) – therefore, in view of Example 2 in 8.3.5, it must be the case that $\mathscr{R}(\lambda_i) \leqslant 0$ on $\mathscr{C}$ (it being clear that $\mathscr{C}$ is full) and so the desired estimate is obvious. □

Before giving the proof of Lemma 8.3.6.2, a preliminary remark is needed.

Fix a Cartan subalgebra $\mathfrak{j}$ of $\mathfrak{g}$. Given $f \in C^\infty(\mathfrak{g})$, define as in 8.2.2, a function of two variables by the prescription

$$f(x:X) = f(xX) \qquad (x \in G, X \in \mathfrak{g}).$$

Fix an element D in $\mathfrak{P}(\mathfrak{g})$ – then $f(xH; D) = f(x:H; x^{-1}D)$ $(H \in \mathfrak{j})$. There exist linearly independent elements D_i in $\mathfrak{P}(\mathfrak{g})$ and analytic functions a_i on G $(1 \leqslant i \leqslant p)$ such that $x^{-1}D = \sum_i a_i(x)D_i$ (all $x \in G$), whence

$$f(xH; D) = \sum_i a_i(x)f(x:H; D_i).$$

Furthermore it is possible to find an integer $d \geqslant 0$ and elements $D_{ij} \in \mathfrak{P}(\mathfrak{j})$, $S_{ij} \in \mathfrak{S}^+$ $(1 \leqslant j \leqslant q)$ such that

$$f(x:H; D_i) = \pi(H)^{-d} \sum_j f(x; S_{ij}: H; D_{ij}) \qquad (H \in \mathfrak{j}').$$

[To see this, it is necessary to make two observations: (1) Consider the map $\boldsymbol{\Gamma}_X$ $(X \in \mathfrak{g})$ as defined in the discussion following Proposition 8.2.2.1 – then, for every $H \in \mathfrak{j}'$, $\boldsymbol{\Gamma}_H$ defines a bijective mapping between $\mathfrak{S} \otimes_{\mathbf{C}} S(\mathfrak{j}_c)$ and $S(\mathfrak{g}_c)$ (for the proof of this assertion, see Lemma 8.2.2.2); (2) Let $p \in S(\mathfrak{g}_c)$ – then there exists an integer $d \geqslant 0$ and a polynomial mapping

γ_p of $\mathfrak{j}$ into $\mathfrak{S} \otimes_{\mathbf{C}} S(\mathfrak{j}_c)$ such that

$$\Gamma_H(\gamma_p(H)) = \pi(H)^d \cdot p \qquad (\text{all } H \in \mathfrak{j})$$

(cf. Lemma 8.2.2.3).]

Now let $\mathfrak{j}$ be a θ-stable Cartan subalgebra of $\mathfrak{g}$; let J be the Cartan subgroup of G associated with $\mathfrak{j}$; introduce, in the usual way, the group $W(G, J) = J^*/J$ (cf. Vol. I, number 1.4.2) – we recall that $W(G, J)$ operates on G/J as well as $\mathfrak{j}'$. Choose a function α in $C_c^\infty(G/J)$ such that

$$\int_{G/J} \alpha(\dot{x}) d_{G/J}(\dot{x}) = 1$$

and put

$$\alpha_0 = [W(G, J)]^{-1} \sum_{w \in W(G, J)} \alpha^w .$$

Select a compact set ω in G such that spt $(\alpha_0) \subset \dot{\omega}$ where $w(\dot{\omega}) = \dot{\omega}$ (all $w \in W(G, J)$); for any $g \in C_c^\infty(\mathfrak{j}')$ define a function $f_g \in C_c^\infty(\mathfrak{g})$ by the prescription

$$f_g(\dot{x}H) = [W(G, J)]^{-1} \alpha_0(\dot{x}) \sum_{w \in W(G, J)} g^w(H) \qquad (\dot{x} \in \dot{\omega},\ H \in \operatorname{spt}(g)).$$

Fix $D \in \mathfrak{P}(\mathfrak{g})$; if $x \in \omega$ and $H \in \mathfrak{j}'$, then, in view of what was said above, we have

$$f_g(xH; D) = \pi(H)^{-d} \sum_{i,j} a_i(x) \alpha_0(x; S_{ij}) g_0(H; D_{ij})$$

where

$$g_0 = [W(G, J)]^{-1} \sum_{w \in W(G, J)} g^w .$$

Because ω is compact, it is clear that

$$\sup |Df_g| \leqslant c_D \sum_{i,j} \sup |\pi^{-d} D_{ij} g_0| \qquad (\text{all } g \in C_c^\infty(\mathfrak{j}'))$$

where c_D is a constant which depends only on D.

This discussion may be summarized as follows.

Lemma 8.3.6.3 Retain the above notations and assumptions; fix a $D \in \mathfrak{P}(\mathfrak{g})$ – then there exists an integer $d \geqslant 0$ and a finite number of polynomial differential operators D_i on $\mathfrak{j}$ such that

$$\sup |Df_g| \leqslant \sum_i \sup |\pi^{-d} D_i g| \qquad (\text{all } g \in C_c^\infty(\mathfrak{j}')).$$

Proof of Lemma 8.3.6.2 It can be supposed that $\mathfrak{j}$ is θ-stable. Since T is tempered, there exist a finite number of polynomial differential operators D_i on $\mathfrak{g}$ such that $|T(f)| \leqslant \sum_i \sup |D_i f|$ (all $f \in C_c^\infty(\mathfrak{g})$). As will become apparent, this means that we might just as well assume that $|T(f)| \leqslant \sup |Df|$ (all $f \in C_c^\infty(\mathfrak{g})$) for some $D \in \mathfrak{P}(\mathfrak{g})$ and then use the notation of Lemma 8.3.6.3; thus, by that lemma, there exists an integer

$d \geqslant 0$ and a finite number of polynomial differential operators D_i on $\mathfrak{j}$ such that

$$|T(f_g)| \leqslant \sup|Df_g| \leqslant \sum_i \sup|\pi^{-d}D_i g| \qquad (\text{all } g \in C_c^\infty(\mathfrak{j}')).$$

On the other hand

$$T(f_g) = \int f_g F_T d_\mathfrak{g}(X) = (-1)^{r_I} \int \epsilon_R \phi_{f_g} F_{T,\mathfrak{j}} d_\mathfrak{j}(H);$$

moreover

$$\phi_{f_g}(H) = \epsilon_R(H)\pi(H) \int_{G/J} f_g(\dot{x}H) d_{G/J}(\dot{x}) = \epsilon_R(H)\pi(H) g_0(H) \qquad (H \in \mathfrak{j}')$$

and so we see that

$$T(f_g) = (-1)^{r_I} \int \pi g_0 F_{T,\mathfrak{j}} d_\mathfrak{j}(H) = (-1)^{r_I} \int \pi g F_{T,\mathfrak{j}} d_\mathfrak{j}(H)$$

provided we take into account the fact that

$$F_{T,\mathfrak{j}}^w = \det(w) F_{T,\mathfrak{j}} \qquad (\text{all } w \in W(G, J)).$$

[As above, we have taken the liberty of introducing the invariant integral of 8.4.1. . . .] Now put $h = \pi^{n-1} g \, (n \geqslant 1)$ – then

$$\left| \int g\pi^n F_{T,\mathfrak{j}} d_\mathfrak{j}(H) \right| = |T(f_h)| \leqslant \sum_i \sup|\pi^{-d} D_i(\pi^{n-1}g)| \qquad (\text{all } g \in C_c^\infty(\mathfrak{j}')).$$

By taking n sufficiently large, we are ensured that $\pi^{-1} D_i \circ \pi^{n-1}$ is a polynomial differential operator on $\mathfrak{j}$. Therefore $\pi^n F_{T,\mathfrak{j}}$ is tempered on $\mathfrak{j}'$, hence is tempered on any Weyl chamber $\mathscr{C}$. □

8.3.7 Rapidly Decreasing Functions on a Reductive Lie Group

Let (G, K) be a reductive pair per Vol. I, number 1.1.5. Following Harish-Chandra [30], we shall, in this number, consider a certain space $\mathscr{C}(G)$ of C^∞ functions which plays, on G, a role similar to that played by the rapidly decreasing functions on $\mathbf{R}^n$ (8.3.5). However the complete justification for the introduction of $\mathscr{C}(G)$ will become apparent only later on.

Before proceeding to the definition of the space of rapidly decreasing functions and the exploration of its basic properties, it will first be necessary to consider the function $\boldsymbol{\vdash\!\circ\!\dashv}$ (definition below).

So let $\mathfrak{g}$ be a semi-simple Lie algebra over $\mathbf{R}$, $\mathfrak{g} = \mathfrak{k} + \mathfrak{a}_\mathfrak{p} + \mathfrak{n}^+$ an Iwasawa decomposition for $\mathfrak{g}$; let G be a connected Lie group with Lie algebra $\mathfrak{g}$, $G = KA_\mathfrak{p}N^+$ the corresponding Iwasawa decomposition of G – we shall assume that G has finite center (whence K is compact).

Definition Given x in G, write $x = \kappa(x)\exp(H(x))n(x)$ $(H(x) \in \mathfrak{a}_\mathfrak{p})$. Put

$$\vdash\!\circ\!\dashv(x) = \int_K e^{-\rho(H(xk))} dk \qquad (\rho = 2^{-1} \sum_{\lambda > 0} m(\lambda)\lambda).$$

Thus $\vdash\!\circ\!\dashv$ is nothing more than the zonal spherical function on G corresponding to the trivial linear function on $\mathfrak{a}_\mathfrak{p}$. [Since any two maximal abelian subspaces of $\mathfrak{p}$ are K-conjugate, $\vdash\!\circ\!\dashv$ is actually independent of the choice of $\mathfrak{a}_\mathfrak{p}$.]

Proposition 8.3.7.1 For all $x \in G$, we have

$$\vdash\!\circ\!\dashv(x) = \vdash\!\circ\!\dashv(x^{-1}).$$

[This is an immediate consequence of Proposition 6.2.2.1.]

Proposition 8.3.7.2 Let ω be a compact subset of G – then

$$\vdash\!\circ\!\dashv(y_1 x y_2) \lesssim \vdash\!\circ\!\dashv(x) \qquad (x \in G;\ y_1, y_2 \in \omega).$$

Proof For the purposes of the present proof, there clearly is no loss of generality in assuming that G is the real analytic subgroup of G_c corresponding to $\mathfrak{g}$, G_c the simply connected complex analytic group with Lie algebra $\mathfrak{g}_c$. This being so, extend $\mathfrak{a}_\mathfrak{p}$ to a Cartan subalgebra $\mathfrak{a}$ of $\mathfrak{g}$ and then by complexification to a Cartan subalgebra $\mathfrak{a}_c$ of $\mathfrak{g}_c$; let Φ^+ denote the positive roots of the pair $(\mathfrak{g}_c, \mathfrak{a}_c)$, $\{\alpha_1, \ldots, \alpha_l\}$ the corresponding fundamental system; set $H_i = 2H_{\alpha_i}/(\alpha_i, \alpha_i)$ $(i = 1, \ldots, l)$ and select linear functions $\Lambda_1, \ldots, \Lambda_l$ on $\mathfrak{a}_c$ such that $\Lambda_i(H_j) = \delta_{ij}$ $(1 \leqslant i, j \leqslant l)$ – then, as we know, there exists an irreducible representation $U_i = U_{\Lambda_i}$ of $\mathfrak{g}_c$ on a finite dimensional space $E_i = E_{\Lambda_i}$ with highest weight Λ_i. Write $\rho = \sum_i r_i \Lambda_i$ $(r_i \in \mathbf{R})$; let a_i be a unit vector in E_i belonging to the weight Λ_i (we assume, as we may, that E_i carries the structure of a Hilbert space relative to which $U_i \,|\, K$ is unitary) – then it is clear that

$$\vdash\!\circ\!\dashv(x) = \vdash\!\circ\!\dashv(x^{-1}) = \int_K \prod_{i=1}^{l} \| U_i(x^{-1}k)a_i \|^{-r_i}\, dk \qquad (x \in G).$$

Now suppose that y ranges through a compact subset ω of G – then we claim that $\vdash\!\circ\!\dashv(xy) \lesssim \vdash\!\circ\!\dashv(x)$ $(x \in G, y \in \omega)$. Thus, for $x, y \in G$, we have

$$e^{-\rho(H(y^{-1}x^{-1}k))} = \prod_{i=1}^{l} \| U_i(y^{-1}x^{-1}k)a_i \|^{-r_i} \qquad (k \in K)$$

and

$$\begin{aligned} \| U_i(y) \|^{-1} \| U_i(x^{-1}k)a_i \| &\leqslant \| U_i(y^{-1}x^{-1}k)a_i \| \\ &\leqslant \| U_i(y^{-1}) \| \cdot \| U_i(x^{-1}k)a_i \| \qquad (k \in K). \end{aligned}$$

Therefore, if we put

$$M = \sup_{y \in \omega} \prod_{i=1}^{l} \{\| U_i(y) \|^{r_i} + \| U_i(y^{-1}) \|^{-r_i}\},$$

then it follows that

$$e^{-\rho(H(y^{-1}x^{-1}k))} \leqslant M e^{-\rho(H(x^{-1}k))} \qquad (k \in K)$$

for all $x \in G$, $y \in \omega$. Of course this gives our claim and, in so doing, establishes the proposition. □

Relative to the ordering of the set Σ of roots of the pair $(\mathfrak{g}, \mathfrak{a}_{\mathfrak{p}})$ which is implicit in the choice of an Iwasawa decomposition for $\mathfrak{g}$, let $\mathscr{C}$ be the corresponding positive Weyl chamber.

Proposition 8.3.7.3 For all $h \in \exp(\mathscr{C}^{cl})$, we have

$$1 \leqslant h^{\rho}\,\Xi(h).$$

Proof Altering the notation a little, fix $h \in A_{\mathfrak{p}}, h^{+} \in \exp(\mathscr{C}^{cl})$ – then we intend to prove that $h^{\rho}\,\Xi(h) \leqslant (hh^{+})^{\rho}\,\Xi(hh^{+})$. [Needless to say, the contention of the proposition follows upon taking $h = 1$. . . .] Thus, employing the usual notations, we see without difficulty that

$$H(hh^{+}k) = H(h\kappa(h^{+}k)) + H(h^{+}k) \qquad (k \in K)$$

and so, in view of a standard integration formula, we have

$$\int_K e^{-\rho(H(hk))}dk = \int_K \exp\{-\rho(H(hh^{+}k)) - \rho(H(h^{+}k))\}dk$$
$$\leqslant (h^{+})^{\rho} \int_K e^{-\rho(H(hh^{+}k))}dk$$

from Lemma 3.3.2.4 (which is surely applicable . . .). Multiplying this inequality through by h^{ρ} then leads to the conclusion that $h^{\rho}\,\Xi(h) \leqslant (hh^{+})^{\rho}\,\Xi(hh^{+})$, as desired. $\square$

The function Ξ, being biinvariant under K, is completely determined by its restriction to $\exp(\mathscr{C}^{cl})$ (since $G = K\exp(\mathscr{C}^{cl})K$). The following theorem provides us with an estimate on the growth of Ξ at 'infinity'.

Theorem 8.3.7.4 (Harish-Chandra) There exists a non-negative integer d such that

$$\Xi(h) \preceq h^{-\rho}(1 + \sigma(h))^{d} \qquad (\text{all } h \in \exp(\mathscr{C}^{cl})).$$

[In other words $\Xi(h)$ approaches zero very rapidly as h tends to infinity in $\exp(\mathscr{C}^{cl})$.]

Proof The proof depends upon the considerations which may be found in section 9.1 infra; we shall feel free to use the notations and results thereof. Set $\tilde{\Xi}(h) = h^{\rho}\,\Xi(h)$ $(h \in A_{\mathfrak{p}})$; fix an H_0 in $\mathscr{C}$ such that $\lambda_i(H_0) = 1$ $(1 \leqslant i \leqslant l)$ and put $\Psi(t) = \tilde{\Xi}(\exp(tH_0))$ $(t \in \mathbf{R})$ – then, since the principal exponents of $\tilde{\Xi}$ are all equal to zero, we have an expansion

$$\Psi(t) = \sum_{\lambda \in \mathsf{L}} p_{\lambda}(t)e^{-t\lambda(H_0)} \qquad (t > 0),$$

the p_{λ} $(\lambda \in \mathsf{L})$ being polynomials which are uniformly bounded in degree by d (say). Because the series

$$\sum_{\lambda \in \mathsf{L}} |p_{\lambda}(t)e^{-t\lambda(H_0)}|$$

converges uniformly for $t > T$ (T being chosen sufficiently large), it follows that there exists a finite number of distinct elements $\lambda_0 = 0$, $\lambda_1, \ldots, \lambda_r \in \mathsf{L}$ such that

$$\left| \Psi(t) - \sum_{i=0}^{r} p_{\lambda_i}(t) e^{-t\lambda_i(H_0)} \right| \leqslant 1$$

for all $t > T$; but since $\lambda(H_0) = m_\lambda$ ($\lambda \in \mathsf{L}$), it is obvious that

$$\lim_{t\to\infty} \sum_{i=1}^{r} p_{\lambda_i}(t) e^{-t\lambda_i(H_0)} = 0,$$

whence $\Psi(t) \leqslant 2 + p_0(t)$ for all sufficiently large positive values of t. As the degree of p_0 cannot exceed d, this means that we can select a positive constant M such that $\Psi(t) \leqslant M(1 + t)^d$ (all $t \geqslant 0$). Now put

$$\Lambda(H) = \max_{1\leqslant i \leqslant l} \lambda_i(H) \qquad (H \in \mathfrak{a}_\mathfrak{p})$$

– then it is obvious that $\Lambda(H)H_0 - H \in \mathscr{C}^{cl}$ ($H \in \mathfrak{a}_\mathfrak{p}$). Therefore, on the basis of the *proof* of Proposition 8.3.7.3, we have

$$\widetilde{\vdash\!\circ\!\dashv}(\exp H) \leqslant \Psi(\Lambda(H)) \leqslant M(1 + \Lambda(H))^d$$

for all $H \in \mathscr{C}^{cl}$. However $\| H \| \geqslant \Lambda(H)$ (all $H \in \mathscr{C}^{cl}$) and so

$$\widetilde{\vdash\!\circ\!\dashv}(\exp H) \leqslant M(1 + \| H \|)^d \qquad (\text{all } H \in \mathscr{C}^{cl})$$

which proves the theorem. □

Remark Let d_Σ denote the degree of the polynomial function π_Σ (here, as in Vol. I, number 2.1.5, π_Σ denotes the product of the positive indivisible roots of the pair $(\mathfrak{g}, \mathfrak{a}_\mathfrak{p})$) – then it can be shown that one may actually assume that the integer d which figures in Theorem 8.3.7.4 is $\leqslant d_\Sigma$ (cf. Harish-Chandra [18, p. 282]).

Finally, we remind the reader that $\vdash\!\circ\!\dashv$ verifies the functional equation

$$\vdash\!\circ\!\dashv(x) \vdash\!\circ\!\dashv(y) = \int_K \vdash\!\circ\!\dashv(xky)\, dk \qquad (\text{all } x, y \in G).$$

Let us pass now from the case of a connected semi-simple Lie group (with finite center) to the reductive pairs (G, K) of Vol. I, number 1.1.5. We can write $G = \widehat{G} \times C_\mathfrak{p}$, $\widehat{G} = \bar{G}C_\mathfrak{k}$; the function $\vdash\!\circ\!\dashv$ is defined initially only relative to the pair $(\bar{G}, \bar{K})$ but may be extended to all of G by the prescription $\vdash\!\circ\!\dashv(x) = \vdash\!\circ\!\dashv((\hat{x}, c_\mathfrak{p})) = \vdash\!\circ\!\dashv(\hat{x}) = \vdash\!\circ\!\dashv(\bar{x}c_\mathfrak{k}) = \vdash\!\circ\!\dashv(\bar{x})$ ($x \in G$ – we are employing the obvious notations). [Extending $\vdash\!\circ\!\dashv$ from $\bar{G}$ to $\widehat{G}$ via the rule $\vdash\!\circ\!\dashv(\bar{x}c_\mathfrak{k}) = \vdash\!\circ\!\dashv(\bar{x})$ is, of course, legitimate (since $\bar{G} \cap C_\mathfrak{k} \subset \bar{K}$).] We note that $\vdash\!\circ\!\dashv$ is then biinvariant under K and takes the constant value 1 on the center of G; furthermore $\vdash\!\circ\!\dashv$ is continuous.

Introduce the subadditive function σ per 8.1.2.

Proposition 8.3.7.5 There exists a number $r \geqslant 0$ such that

$$\int_G \Xi^2(x)(1+\sigma(x))^{-r} d_G(x) < \infty.$$

Proof Plainly $\Xi(\hat{x}c) = \Xi(\hat{x})$ and $\sigma(\hat{x}c) \geqslant \max(\sigma(\hat{x}), \sigma(c))$ $(\hat{x} \in \widehat{G}, c \in C_{\mathfrak{p}})$ – hence our integral is majorized by

$$\left(\int_{\widehat{G}} \Xi^2(\hat{x})(1+\sigma(\hat{x}))^{-r/2} d_{\widehat{G}}(\hat{x})\right)\left(\int_{C_{\mathfrak{p}}} (1+\sigma(c))^{-r/2} d_{C_{\mathfrak{p}}}(c)\right).$$

The integral on the right is obviously convergent if r is chosen large enough – therefore it suffices to consider the case when $\mathfrak{c}_{\mathfrak{p}} = \{0\}$. But then it is evident that we can replace G by G/Z and thus assume that G is semi-simple. By a standard integration formula we have

$$\begin{aligned} &\int_G \Xi^2(x)(1+\sigma(x))^{-r} d_G(x) \\ &\leq \int_{\mathscr{C}} \nabla(\exp H)\, \Xi^2(\exp H)(1+\|H\|)^{-r} d_{\mathfrak{a}_{\mathfrak{p}}}(H). \end{aligned}$$

However

$$\Xi(h) \leq h^{-\rho}(1+\sigma(h))^d, \quad \nabla(h) \leqslant h^{2\rho} \qquad (h \in \exp(\mathscr{C}))$$

and so

$$\int_G \Xi^2(x)(1+\sigma(x))^{-r} d_G(x) \leq \int_{\mathscr{C}} (1+\|H\|)^{2d-r} d_{\mathfrak{a}_{\mathfrak{p}}}(H) < \infty$$

if r is chosen sufficiently large. □

Remark Suppose that $\mathfrak{c}_{\mathfrak{p}} = \{0\}$ – then one proves, as above, that $\Xi(1+\sigma)^r \in L^p(G)$ for $p > 2$, $r \in \mathbf{R}$.

Define the analytic function D_l on G as usual ($l = \operatorname{rank}(G)$).

Proposition 8.3.7.6 There exists a number $r \geqslant 0$ such that

$$\int_G |D_l(x)|^{-1/2}\, \Xi(x)(1+\sigma(x))^{-r} d_G(x) < \infty.$$

[The proof of this result will be deferred until later on in the present chapter; cf. Proposition 8.5.7.3 infra.]

As above, let (G, K) be a reductive pair fulfilling the general assumptions which were laid down in Vol. I, number 1.1.5. In what follows, the notations and conventions of 8.2.3 (as regards differential operators on G) will be employed without comment.

Let us turn now to the definition of the space $\mathscr{C}(G)$.

Fix an open set $\mathscr{O}$ in G; given an $f \in C^\infty(\mathscr{O})$, put

$$_{D_1}|f|_{r, D_2} = \sup_{x \in \mathscr{O}} (1+\sigma(x))^r\, \Xi^{-1}(x) |f(D_1; x; D_2)|$$

$$(D_1, D_2 \in \mathfrak{G};\ r \in \mathbf{R}).$$

Let $\mathscr{C}(\mathcal{O})$ denote the subspace of $C^\infty(\mathcal{O})$ consisting of those f such that ${}_{D_1}|f|_{r,D_2} < \infty$ for all $r \in \mathbf{R}$ and pairs $(D_1, D_2) \in \mathfrak{G} \times \mathfrak{G}$. Topologize $\mathscr{C}(\mathcal{O})$ by means of the semi-norms ${}_{D_1}|\,.\,|_{r,D_2}$: In this way $\mathscr{C}(\mathcal{O})$ acquires the structure of a locally convex, complete, Hausdorff, topological vector space.

Bearing in mind the properties of $\vdash\!\circ\!\dashv$ which were noted above, one easily establishes the validity of the following result.

Lemma 8.3.7.7 Let $\mathcal{O}$ be an open subset of G; fix $x_1, x_2 \in G$ and, for any f on $\mathcal{O}$, let ${}_{x_1}f_{x_2}$ denote the function on $x_1\mathcal{O}x_2$ given by

$$ {}_{x_1}f_{x_2}(x) = f(x_1^{-1}xx_2^{-1}) \qquad (x \in x_1\mathcal{O}x_2). $$

Then the mapping $f \mapsto {}_{x_1}f_{x_2}$ is an isomorphism (for topological vector spaces) of $\mathscr{C}(\mathcal{O})$ onto $\mathscr{C}(x_1\mathcal{O}x_2)$.

Of course we obtain the space $\mathscr{C}(G)$ by taking $\mathcal{O} = G$ in the above set up. In view of Proposition 8.3.7.5, it is clear that $\mathscr{C}(G) \subset L^2(G)$ (however the elements of $\mathscr{C}(G)$ need not belong to $L^1(G)$ – cf. Chapter 10); indeed there are continuous inclusions $C_c^\infty(G) \to \mathscr{C}(G) \to L^2(G)$. [Later on we shall discuss the important question: When is a square summable C^∞ function actually 'rapidly decreasing', i.e. actually in $\mathscr{C}(G)$?] The primary objective of the present number is to show that $C_c^\infty(G)$ is dense in $\mathscr{C}(G)$; the non-trivial proof of this result hinges essentially on the fact that the left and right 'regular' representations of G on $\mathscr{C}(G)$ are differentiable (cf. infra).

For the applications, it will be necessary to consider a somewhat more general situation. Thus let G be a Lie group whose identity component is G. Let $\mathcal{O}$ be an open subset of G which meets but a finite number of the components of G; choose $x_i \in \mathsf{G}$ and open sets $\mathcal{O}_i$ in G such that $\mathcal{O}$ is the disjoint union of the $x_i\mathcal{O}_i$ $(1 \leqslant i \leqslant r)$. For any $f \in C^\infty(\mathcal{O})$, let f_i denote the function on $\mathcal{O}_i$ given by $f_i(x) = f(x_ix)$ $(x \in \mathcal{O}_i)$. By definition the space $\mathscr{C}(\mathcal{O})$ will consist of those C^∞ functions f on $\mathcal{O}$ with the property that $f_i \in \mathscr{C}(\mathcal{O}_i)$ $(1 \leqslant i \leqslant r)$. Assign to the Cartesian product of the $\mathscr{C}(\mathcal{O}_i)$ the natural topology and then topologize $\mathscr{C}(\mathcal{O})$ in such a way that the mapping $f \mapsto (f_1, \ldots, f_r)$ of $\mathscr{C}(\mathcal{O})$ into $\prod_1^r \mathscr{C}(\mathcal{O}_i)$ is an isomorphism. Owing to Lemma 8.3.7.7, the structure of $\mathscr{C}(\mathcal{O})$, as a locally convex space, is independent of the choice of x_i and $\mathcal{O}_i$. Note too that the injection of $C_c^\infty(\mathcal{O})$ into $\mathscr{C}(\mathcal{O})$ is continuous.

Suppose, in particular, that G/G is finite – then the preceding considerations apply to any open subset $\mathcal{O}$ of G. Moreover, as will be seen below, the finiteness of G/G implies that $C_c^\infty(\mathsf{G})$ is dense in $\mathscr{C}(\mathsf{G})$ (Theorem 8.3.7.13).

Remark Let E be a locally convex, complete, Hausdorff, topological vector space (over $\mathbf{C}$) – then there is no difficulty in defining the space $\mathscr{C}(\mathsf{G}; E)$ .... Furthermore, the considerations infra will show that $C_c^\infty(\mathsf{G}; E)$ is dense in $\mathscr{C}(\mathsf{G}; E)$, the canonical inclusion being continuous.

For any $f \in \mathscr{C}(G)$, $x \in G$, define $L(x)f$, $R(x)f$ as in the Example following Proposition 4.1.1.1.

Proposition 8.3.7.8 Both L and R are differentiable representations of G on $\mathscr{C}(G)$.

Of course we shall refer to L (respectively R) as the left (respectively right) regular representation of G on $\mathscr{C}(G)$.

To prove the proposition a couple of lemmas will be needed.

Let $\{|\,.\,|_\alpha : \alpha \in \mathfrak{A}\}$ denote the set of all continuous semi-norms on $\mathscr{C}(G)$.

Lemma 8.3.7.9 Let $x \in G$: If $f \in \mathscr{C}(G)$, then so do $L(x)f$ and $R(x)f$. In addition, given any compact set ω in G and $\alpha \in \mathfrak{A}$, we can select $\beta \in \mathfrak{A}$ such that

$$|L(x)f|_\alpha + |R(x)f|_\alpha \leqslant |f|_\beta$$

for all $x \in \omega$ and $f \in \mathscr{C}(G)$.

Proof Fix $D, \tilde{D} \in \mathfrak{G}$ – then

$$R(x)f(\tilde{D}\,;\tilde{x}\,;D) = f(\tilde{D}\,;\tilde{x}x\,;x^{-1}D) \qquad (x, \tilde{x} \in G).$$

There exist linearly independent elements D_i $(1 \leqslant i \leqslant p)$ in $\mathfrak{G}$ and analytic functions a_i on G such that $x^{-1}D = \sum_1^p a_i(x)D_i$ $(x \in G)$. Suppose now that x ranges through our compact set ω. Choose $M_1 > 0$, $M_2 > 0$ such that $\sup_{x\in\omega}|a_i(x)| < M_1 (1 \leqslant i \leqslant p)$, $\Xi(\tilde{x}x) \leqslant M_2\,\Xi(\tilde{x})$ $(x \in \omega, \tilde{x} \in G$; cf. Proposition 8.3.7.2) – then, for $r \in \mathbf{R}$, $x \in \omega$,

$$\sup_{\tilde{x}\in G} \frac{(1+\sigma(\tilde{x}))^r}{\Xi(\tilde{x})} |R(x)f(\tilde{D}\,;\tilde{x}\,;D)|$$

$$\leqslant M_1 \sum_i \sup_{\tilde{x}\in G} \frac{(1+\sigma(\tilde{x}))^r}{\Xi(\tilde{x})} |f(\tilde{D}\,;\tilde{x}x\,;D_i)|.$$

Write

$$\sup_{\tilde{x}\in G} \left(\frac{1+\sigma(\tilde{x})}{\Xi(\tilde{x})}\right)^r |f(\tilde{D}\,;\tilde{x}x;D_i)|$$

$$= \sup_{\tilde{x}\in G} \left(\frac{1+\sigma(\tilde{x})}{1+\sigma(\tilde{x}x)}\right)^r \frac{\Xi(\tilde{x}x)}{\Xi(\tilde{x})} \frac{(1+\sigma(\tilde{x}x))^r}{\Xi(\tilde{x}x)} |f(\tilde{D}\,;\tilde{x}x\,;D_i)| \qquad (x \in \omega).$$

Since $\Xi(\tilde{x}x) \leqslant M_2\,\Xi(\tilde{x})$ and since there exist $d_1, d_2 > 0$ such that

$$d_1 \leqslant (1+\sigma(\tilde{x}))/(1+\sigma(\tilde{x}x)) \leqslant d_2 \qquad (x \in \omega, \tilde{x} \in G)$$

(cf. Corollary 8.1.2.2), it is obvious that the sup on the right hand side of the immediately preceding equality is finite. It follows, therefore, that $R(x)f \in \mathscr{C}(G)$ (all $x \in G$). In a similar way we find that $L(x)f \in \mathscr{C}(G)$ (all $x \in G$). Moreover, the above discussion clearly establishes the second assertion of the lemma too. □

Notice that Lemma 8.3.7.9 implies that the second condition of Proposition 4.1.1.1 is met by L and R; to get at the first, we need another lemma.

Lemma 8.3.7.10 Fix an $r \geqslant 0$ – then, for any $\epsilon > 0$, there exists a neighborhood $\mathcal{O}$ of 1 in G and an element $\beta \in \mathfrak{A}$ such that

$$\sup_G (1+\sigma)^r \Xi^{-1} |R(x)f - f| \leqslant \epsilon |f|_\beta$$

for $x \in \mathcal{O}$ and $f \in \mathscr{C}(G)$. Furthermore, the semi-norm $|\,.\,|_\beta$ is independent of ϵ.

[A similar statement is, of course, valid for L.]

Proof Choose a basis $X_1, \ldots, X_n$ for $\mathfrak{g}$ over $\mathbf{R}$; place a norm on $\mathfrak{g}$ and then pick a positive constant t_0 such that

$$\max_{1 \leqslant i \leqslant n} |t_i| \leqslant t_0 \left\| \sum_i t_i X_i \right\| \qquad (t_i \in \mathbf{R}).$$

Let $f \in \mathscr{C}(G)$ and $X \in \mathfrak{g}$ – then

$$f(x \exp X) - f(x) = \int_0^1 f(x \exp(tX); X)dt \qquad (x \in G)$$

and so

$$|f(x \exp X) - f(x)| \leqslant t_0 \|X\| \sum_i \int_0^1 |f(x \exp(tX); X_i)|\, dt.$$

Therefore there exists a constant $M > 0$ such that

$$\sup_G (1+\sigma)^r \Xi^{-1} |R(\exp X)f - f|$$
$$\leqslant M \|X\| \sum_i \sup_G (1+\sigma)^r \Xi^{-1} |X_i f|$$

provided $\|X\| \leqslant 1$ $(X \in \mathfrak{g})$. The statement of the lemma is now clear. □

Corollary 8.3.7.11 Fix $\alpha \in \mathfrak{A}$ and $\epsilon > 0$ – then there exists a neighborhood $\mathcal{O}$ of 1 in G and a $\beta \in \mathfrak{A}$ such that

$$|L(x)f - f|_\alpha + |R(x)f - f|_\alpha \leqslant \epsilon |f|_\beta$$

for $x \in \mathcal{O}$ and $f \in \mathscr{C}(G)$. Moreover, the semi-norm $|\,.\,|_\beta$ is independent of ϵ.

Proof Using the notations introduced during the proof of Lemma 8.3.7.9, write

$$R(x)f(\tilde{D}; \tilde{x}; D) - f(\tilde{D}; \tilde{x}; D)$$
$$= \sum_i (a_i(x) - 1) f(\tilde{D}; \tilde{x}x; D_i) + \{f(\tilde{D}; \tilde{x}x; D) - f(\tilde{D}; \tilde{x}; D)\}.$$

Put $\mathbf{D} = \tilde{D} \otimes D$, $\mathbf{D}_i = \tilde{D} \otimes D_i$ – then $\mathbf{D}$ and $\mathbf{D}_i$ define continuous endomorphisms of $\mathscr{C}(G)$ and the above relation may be written as

$$\mathbf{D}(R(x)f - f) = \sum_i (a_i(x) - 1)R(x)\mathbf{D}_i f + (R(x) - 1)\mathbf{D}f.$$

In view of Lemmas 8.3.7.9, 8.3.7.10, it is obvious that, given $r \geqslant 0$, there exists a neighborhood $\mathscr{O}$ of 1 in G and $\beta \in \mathfrak{A}$ (independent of ϵ) such that

$$\sup_G (1 + \sigma)^r \, \Xi^{-1} \, |\mathbf{D}(R(x)f - f)| \leqslant \epsilon \, | f |_\beta$$

for $x \in \mathscr{O}$ and $f \in \mathscr{C}(G)$. A similar argument applies to L. Hence the corollary. □

Proof of Proposition 8.3.7.8 To verify that L and R are continuous representations of G on $\mathscr{C}(G)$ we shall appeal to Proposition 4.1.1.1. As has been remarked above, the second condition of this proposition is satisfied by L and R; on the other hand, the first is a simple consequence of Corollary 8.3.7.11. We still have to show that L and R are actually differentiable representations of G on $\mathscr{C}(G)$. It suffices to do this for R, the proof for L being similar. So let $f \in \mathscr{C}(G)$, $X \in \mathfrak{g}$ and put

$$\phi_t = t^{-1}(R(x_t)f - f) - Xf \qquad (t \in \mathbf{R},\ t \neq 0)$$

where $x_t = \exp(tX)$ – then we must prove that $\phi_t \to 0$ in $\mathscr{C}(G)$ as $t \to 0$. Let $D, \tilde{D} \in \mathfrak{G}$, r some non-negative real number; evidently, on any compact subset Ω of G,

$$\sup_{\tilde{x} \in \Omega} | \phi_t(\tilde{D} ; \tilde{x} ; D) | \, \Xi(\tilde{x})^{-1}(1 + \sigma(\tilde{x}))^r \to 0$$

as $t \to 0$. On the other hand, it is clear that

$$\phi_t(\tilde{x}) = \int_0^1 \{ f(\tilde{x}x_{ts} ; X) - f(\tilde{x} ; X) \} \, ds$$

$$\Rightarrow$$

$$\phi_t(\tilde{D} ; \tilde{x} ; D) = \sum_i \int_0^1 a_i(x_{ts}) g(\tilde{x}x_{ts} ; D_i X) \, ds - g(\tilde{x} ; DX),$$

where $g = (\tilde{D} \otimes 1)f$ and a_i, D_i $(1 \leqslant i \leqslant p)$ have the same meaning as in the proof of Lemma 8.3.7.9. Fix a compact neighborhood $\omega = \omega^{-1}$ of 1 in G and choose $M > 1$ such that $1 + \sigma(x) \leqslant M$, $|a_i(x)| \leqslant M (1 \leqslant i \leqslant p)$ and $\Xi(\tilde{x}x) \leqslant M \, \Xi(\tilde{x})$ for $x \in \omega$, $\tilde{x} \in G$. Select a $\delta > 0$ with the property that $x_t \in \omega$ for $|t| \leqslant \delta$ – then

$$\begin{aligned} &| \phi_t(\tilde{D} ; \tilde{x} ; D) | \, \Xi(\tilde{x})^{-1}(1 + \sigma(\tilde{x}))^r \\ &\leqslant M^{r+2} \sum_i \sup_{x \in \omega} | g(\tilde{x}x ; D_i X) | \, \Xi(\tilde{x}x)^{-1}(1 + \sigma(\tilde{x}x))^r \\ &\quad + | g(\tilde{x} ; DX) | \, \Xi(\tilde{x})^{-1}(1 + \sigma(\tilde{x}))^r \end{aligned}$$

provided $|t| \leqslant \delta$. Finally, fix an $\epsilon > 0$. Since $g \in \mathscr{C}(G)$, there exists a compact set $\tilde{\omega} \subset G$ outside of which

$$| g(\tilde{x} ; D_i X) | \, \Xi(\tilde{x})^{-1}(1 + \sigma(\tilde{x}))^r \leqslant \epsilon \qquad (\tilde{x} \in G - \tilde{\omega})$$

for $0 \leqslant i \leqslant p$ $(D_0 = D)$. Set $\boldsymbol{\Omega} = \tilde{\omega}\omega$; plainly, if $\tilde{x}$ does not belong to the compact set $\boldsymbol{\Omega}$, then

$$|\phi_t(\tilde{D}\,;\tilde{x}\,;D)| \vdash\!\circ\!\dashv (\tilde{x})^{-1}(1+\sigma(\tilde{x}))^r \leqslant (p+1)M^{r+2}\epsilon \qquad (|t| \leqslant \delta).$$

This, combined with our earlier observation, serves to show that $\phi_t \to 0$ in $\mathscr{C}(G)$ as $t \to 0$. Hence f is differentiable under R – in fact $R_\infty(X)f = Xf$ (all $X \in \mathfrak{g}$). $\square$

Remark Adopt the notations of Proposition 4.4.3.5 – then, by an easy modification of the argument given there, we deduce that for any $f \in \mathscr{C}(G)$ the series

$$\sum_{\delta_1, \delta_2 \in \hat{K}} \chi_{\delta_1} * f * \chi_{\delta_2}$$

converges absolutely to f in $\mathscr{C}(G)$.

Let $\hat{K}$ denote the set of all equivalence classes of finite dimensional irreducible representations of K; for any finite subset F of $\hat{K}$, put $\bar{\chi}_F = \sum_{\delta \in F} \bar{\chi}_\delta$ and let $\mathscr{C}_F(G)$ denote the (closed) subspace of $\mathscr{C}(G)$ comprised of those elements f in $\mathscr{C}(G)$ which verify the relation $f = \bar{\chi}_F * f * \bar{\chi}_F$.

Let $\mathfrak{A}_R^+ = \{{}_D|\,.\,|_r : r \geqslant 0,\ D \in \mathfrak{G}\}$, $\mathfrak{A}_L^+ = \{|\,.\,|_{r,D} : r \geqslant 0,\ D \in \mathfrak{G}\}$; the proof of the following lemma can be found in the Appendix to the present number.

Lemma 8.3.7.12 Let F be a finite subset of $\hat{K}$ – then each of the three sets of semi-norms $\mathfrak{A}_R^+$, $\mathfrak{A}_L^+$, and $\mathfrak{A}$ define the same topology on $\mathscr{C}_F(G)$.

Let G be a Lie group whose identity component is G.

Here is the main result of the present number.

Theorem 8.3.7.13 (Harish-Chandra) Suppose that G/G is finite – then $C_c^\infty(\mathsf{G})$ is dense in $\mathscr{C}(\mathsf{G})$.

Proof It is obviously enough to deal with the case $\mathsf{G} = G$. Moreover, in view of the Remark above, we need only show that $C_c^\infty(G) \cap \mathscr{C}_F(G)$ is dense in $\mathscr{C}_F(G)$ for every finite subset F of $\hat{K}$ – this will now be done.

(1) Given $t > 0$, let G_t denote the open set consisting of those $x \in G$ with $\sigma(x) < t$. Fix $s > 0$; choose $\phi \in C_c^\infty(G_s)$ such that $\phi(k_1 x k_2) = \phi(x)$ $(k_1, k_2 \in K;\ x \in G)$ and $\int_G \phi(x) d_G(x) = 1$. Put $g_t = (1 - \psi_t) * \phi$, ψ_t denoting the characteristic function of G_t; plainly $g_t \in C_c^\infty(G)$. One easily checks that

$$g_t(x) = \begin{cases} 0 \text{ if } \sigma(x) \leqslant t - s, \\ 1 \text{ if } \sigma(x) \geqslant t + s, \end{cases};$$

$$|g_t(x\,;D)| \leqslant \int_G |\phi(y\,;D)|\, d_G(y) \qquad (x \in G\,; D \in \mathfrak{G}).$$

(2) Fix $r \geqslant 0$, $f \in \mathscr{C}_F(G)$ and set $f_t = (1 - g_t)f = (\psi_t * \phi)f$. Since ϕ

and ψ_t are invariant to the right and to the left under K, we see that $f_t \in C_c^\infty(G) \cap \mathscr{C}_F(G)$. Claim: $\sup_G (1+\sigma)^r {\vdash\!\circ\!\dashv}^{-1} |D(f-f_t)| \to 0$ as $t \to \infty$ (all $D \in \mathfrak{G}$). Thus fix $D \in \mathfrak{G}$ – then $D(f-f_t) = \sum_1^p (D_i' g_t)\cdot(D_i f)$ where the D_i, D_i' are certain elements in $\mathfrak{G}$ $(1 \leqslant i \leqslant p)$. Consider the three possibilities.

(a) $\sigma(x) \geqslant t + s$; in this situation, in view of (1), $f(x) - f_t(x) = g_t(x)f(x) = f(x)$, so that, for $r \geqslant 0$,

$$\begin{aligned}&(1+\sigma(x))^r {\vdash\!\circ\!\dashv}(x)^{-1} |f(x;D) - f_t(x;D)| \\ &\leqslant (1+t)^{-1} \sup_G (1+\sigma)^{r+1} {\vdash\!\circ\!\dashv}^{-1} |Df|.\end{aligned}$$

(b) $\sigma(x) \leqslant t - s$; as was noted in (1), $g_t(x)$ necessarily vanishes when $\sigma(x) \leqslant t - s$, whence $f(x) - f_t(x) = 0$ in this case.

(c) $t - s < \sigma(x) < t + s$; let us assume that $t > s$ – then we have that

$$\begin{aligned}&(1+\sigma(x))^r {\vdash\!\circ\!\dashv}(x)^{-1} |f(x;D) - f_t(x;D)| \\ &\leqslant \sum_1^p r_i (1+\sigma(x))^r {\vdash\!\circ\!\dashv}(x)^{-1} |f(x;D_i)| \\ &\leqslant \sum_1^p r_i (1+t-s)^{-1} \sup_G (1+\sigma)^{r+1} {\vdash\!\circ\!\dashv}^{-1} |D_i f|\end{aligned}$$

where $r_i = \int_G |\phi(y;D_i')|\, d_G(y) (1 \leqslant i \leqslant p)$.

Conclusion: $\sup_G (1+\sigma)^r {\vdash\!\circ\!\dashv}^{-1} |D(f-f_t)| \to 0$ as $t \to \infty$, as claimed. Therefore, by Lemma 8.3.7.12, f_t converges to f in $\mathscr{C}_F(G)$.

The proof of the theorem is now complete. □

Since the left and right regular representations of G on $\mathscr{C}(G)$ are differentiable, it follows, in particular, that

$$C_c^\infty(G) * \mathscr{C}(G) \subset \mathscr{C}(G) \quad \text{and} \quad \mathscr{C}(G) * C_c^\infty(G) \subset \mathscr{C}(G).$$

This result will now be strengthened; namely, we shall show that $\mathscr{C}(G)$ itself is closed under convolution – in fact that convolution, $(f, g) \mapsto f*g$, defines a jointly continuous multiplication on $\mathscr{C}(G)$ (so that $\mathscr{C}(G)$ is a convolution algebra).

Let $f, g \in \mathscr{C}(G)$ – then $f*g$ exists as a C^∞ function on G. Claim: If $X, Y \in \mathfrak{g}$, then $(X \otimes Y)(f*g) = (X \otimes 1)f*(1 \otimes Y)g$. To see this, fix $x \in G$, $X \in \mathfrak{g}$ – then, for variable $y \in G$, we know that

$$|t^{-1}[f(\exp(tX)xy) - f(xy)] - f(X;xy)| \qquad (t \in \mathbf{R}, t \neq 0)$$

converges pointwise to zero (as $t \to 0$) and, moreover, converges to zero in the topology of $\mathscr{C}(G)$ (as $t \to 0$) – in particular, there exists a constant $M_x > 0$ such that for sufficiently small t

$$|t^{-1}[f(\exp(tX)xy) - f(xy)] - f(X;xy)| \leqslant M_x {\vdash\!\circ\!\dashv}(y) \qquad (\text{all } y \in G).$$

But then, in view of Proposition 8.3.7.5, it is possible to choose a number $r_x \geqslant 0$ such that for sufficiently small t

$$\begin{aligned}&|\{t^{-1}[f(\exp(tX)xy) - f(xy)] - f(X;xy)\}g(y^{-1})| \\ &\leqslant M_x {\vdash\!\circ\!\dashv}(y) \cdot {\vdash\!\circ\!\dashv}(y)(1+\sigma(y))^{-r_x} \qquad (\text{all } y \in G)\end{aligned}$$

with $\Xi^2 (1 + \sigma)^{-r_2}$ in $L^1(G)$. Therefore, by a Theorem of Lebesgue, it follows that

$$
\begin{aligned}
f*g(X\,;x) &= \lim_{t\to 0} t^{-1}\,[f*g(\exp(tX)x) - f*g(x)] \\
&= \lim_{t\to 0}\int_G t^{-1}[f(\exp(tX)xy) - f(xy)]g(y^{-1})d_G(y) \\
&= \int_G \lim_{t\to 0} t^{-1}[f(\exp(tX)xy) - f(xy)]\,g(y^{-1})d_G(y) \\
&= \int_G f(X\,;xy)g(y^{-1})d_G(y) = (X \otimes 1)f*g(x) \\
&\qquad\qquad\qquad\qquad (\text{all } x \in G,\ X \in \mathfrak{g}).
\end{aligned}
$$

In a similar way we find that

$$
f*g(y;\ Y) = f*(1 \otimes Y)g(y) \qquad (\text{all } y \in G,\ Y \in \mathfrak{g}).
$$

Hence the claim.

Let f_i and g_j be variable elements in $\mathscr{C}(G)$ converging to zero; let $D_1, D_2 \in \mathfrak{G}$, $s \geqslant 0$ – then we intend to show that $f_i*g_j \in \mathscr{C}(G)$ and that

$$
{}_{D_1}|\,f_i*g_j\,|_{s,D_2} \to 0.
$$

Of course this will prove that $\mathscr{C}(G)$ is a convolution algebra. So choose $r \geqslant s$ such that

(1) $\displaystyle\int_{C_\mathfrak{p}} (1 + \sigma(c))^{(s-r)/2} d_{C_\mathfrak{p}}(c) < \infty,$

(2) $\displaystyle\int_{\hat{G}} \frac{\Xi^2(\hat{x})}{(1 + \sigma(\hat{x}))^{(r-s)/2}} d_{\hat{G}}(\hat{x}) < \infty\,;$

put

$$
R_{r,i} = {}_{D_1}|\,f_i\,|_r,\quad S_{s,j} = |\,g_j\,|_{s,D_2}.
$$

Then, on the basis of what has been said above, we have

$$
\begin{aligned}
&|\,f_i*g_j(D_1\,;x\,;D_2)\,| \\
&= |\{(D_1 \otimes 1)f_i*(1 \otimes D_2)g_j\}(x)\,| \\
&\leqslant \int_G |f_i(D_1\,;y)|\,|\,g_j(y^{-1}x\,;D_2)\,|\,d_G(y) \\
&\leqslant R_{r,i}S_{s,j}\int_G \frac{\Xi(y)}{(1 + \sigma(y))^r}\cdot\frac{\Xi(y^{-1}x)}{(1 + \sigma(y^{-1}x))^s} d_G(y) \\
&\leqslant R_{r,i}S_{s,j}\int_G \frac{\Xi(y)}{(1 + \sigma(y))^r}\cdot\frac{\Xi(y^{-1}x)}{(1 + \sigma(x))^s}(1 + \sigma(y))^s d_G(y) \\
&\leqslant R_{r,i}S_{s,j}(1 + \sigma(x))^{-s}\Bigl(\int_{\hat{G}} \frac{\Xi(\hat{y})\,\Xi(\hat{y}^{-1}\hat{x})}{(1 + \sigma(\hat{y}))^{(r-s)/2}} d_{\hat{G}}(\hat{y})\Bigr) \\
&\quad\times \Bigl(\int_{C_\mathfrak{p}} (1 + \sigma(c))^{(s-r)/2} d_{C_\mathfrak{p}}(c)\Bigr).
\end{aligned}
$$

As will become apparent shortly, this estimate allows us to assume that

G is semi-simple. Revising the notation slightly, choose $r \geqslant s$ such that

$$M = \int_G \frac{\Xi^2(x)}{(1+\sigma(x))^{r-s}} d_G(x) < \infty.$$

Using a well-known integration formula, we then find that

$$\begin{aligned}
&|f_i * g_j(D_1 ; x ; D_2)| \\
&\quad \leqslant R_{r,i} S_{s,j} (1+\sigma(x))^{-s} \iiint_{K \times \mathscr{C} \times K} \Xi(k_1 \exp H\, k_2)(1+\sigma(k_1 \exp H\, k_2))^{s-r} \\
&\qquad \times \Xi(k_2^{-1} \exp(-H) k_1^{-1} x) \frown (\exp H) dk_1 d_{\mathfrak{a}_\mathfrak{p}}(H) dk_2 \\
&\quad = R_{r,i} S_{s,j} (1+\sigma(x))^{-s} \int_{\mathscr{C}} \Xi(\exp H)(1+\sigma(\exp H))^{s-r} \\
&\qquad \times \left(\int_K \Xi(\exp(-H) k^{-1} x) dk \right) \frown (\exp H) d_{\mathfrak{a}_\mathfrak{p}}(H) \\
&\quad = R_{r,i} S_{s,j} \Xi(x)(1+\sigma(x))^{-s} \int_G \Xi^2(x)(1+\sigma(x))^{s-r} d_G(x) \\
&\quad = M R_{r,i} S_{s,j} \Xi(x)(1+\sigma(x))^{-s} \qquad (\text{all } x \in G),
\end{aligned}$$

whence $f_i * g_j \in \mathscr{C}(G)$. In addition, holding one of i and j fixed while allowing the other to vary, we deduce that convolution on $\mathscr{C}(G)$ is separately continuous; that it is jointly continuous follows from the fact that $\mathscr{C}(G)$ is a Fréchet space (it is easy to see that there exists a countable basis for the semi-norms on $\mathscr{C}(G)$).

The following statement serves to summarize the preceding observations.

Proposition 8.3.7.14 (Rader) The space $\mathscr{C}(G)$ is a topological algebra under convolution.

Appendix Let us agree to use the notations revolving around the statement of Lemma 8.3.7.12; our goal here is to establish the following lemma.

Lemma 1 Let F be a finite subset of $\hat{K}$ – then each of the three sets of semi-norms $\mathfrak{A}_R^+$, $\mathfrak{A}_L^+$, and $\mathfrak{A}$ define the same topology on $\mathscr{C}_F(G)$.

The proof of this lemma depends upon the following result which we shall accept for the moment.

Lemma 2 Given two elements $D, \tilde{D} \in \mathfrak{G}$, there exists a finite number of $D_i \in \mathfrak{G}$ $(1 \leqslant i \leqslant p)$ with the following property: If Ψ is any C^∞ μ-spherical function on G with values in E (say), then

$$\|\Psi(\tilde{D}; x; D)\| \leqslant \sum_{i=1}^{p} \|\Psi(x; D_i)\|, \quad \|\Psi(\tilde{D}; x; D)\| \leqslant \sum_{i=1}^{p} \|\Psi(D_i; x)\|$$

for all $x \in G$ ($\|\,.\,\|$ a norm on E).

Proof of Lemma 1 Consider $C(K \times K)$ as a Banach space under the norm:

$$\|f\| = \sup_{k_1, k_2 \in K} |f(k_1, k_2)| \qquad (f \in C(K \times K)).$$

Let $\mu_1(k)f, f\mu_2(k)$ $(k \in K)$, respectively, denote the functions

$$(k_1, k_2) \mapsto f(k^{-1}k_1, k_2) \quad \text{and} \quad (k_1, k_2) \mapsto f(k_1, k_2k^{-1}) \qquad (k_1, k_2 \in K).$$

Then $\mu = (\mu_1, \mu_2)$ is a double (Banach) representation of K on $C(K \times K)$. Let C_F be the subspace of $C(K \times K)$ consisting of those f such that

$$f = \int_K \bar{\chi}_F(k)\mu_1(k)f\,dk = \int_K \bar{\chi}_F(k)f\mu_2(k)\,dk;$$

clearly C_F is a finite dimensional space which is, moreover, invariant under μ – write μ_F for $\mu | C_F$. Now take an arbitrary $f \in \mathscr{C}_F(G)$. Define a C^∞ μ_F-spherical function Ψ_f on G with values in C_F as follows: For $x \in G$, $\Psi_f(x)$ is to be the function $(k_1, k_2) \mapsto f(k_1^{-1}xk_2^{-1})$ $(k_1, k_2 \in K)$ in C_F. Evidently, for $D_1, D_2 \in \mathfrak{G}$, $x \in G$,

$$\|\Psi_f(D_1; x; D_2)\| = \sup_{k_1, k_2 \in K} |f(k_1D_1; k_1xk_2^{-1}; k_2D_2)|.$$

The present lemma is therefore a consequence of Lemma 2. □

Before giving the proof of Lemma 2 we shall set up some notation. Let $\mathfrak{a}_\mathfrak{p}$ be a maximal abelian subspace of $\mathfrak{p}$; introduce an order in the space of real linear functions λ on $\mathfrak{a}_\mathfrak{p}$ and, for any such λ, let $\mathfrak{g}^\lambda$ denote the subspace of $\mathfrak{g}$ comprised of those $X \in \mathfrak{g}$ such that $[H, X] = \lambda(H)X$ for all $H \in \mathfrak{a}_\mathfrak{p}$; let Σ^+ be the set of positive roots of the pair $(\mathfrak{g}, \mathfrak{a}_\mathfrak{p})$, $\{\lambda_1, \ldots, \lambda_l\}$ the corresponding fundamental system of roots in Σ^+, $\mathscr{C}$ the associated positive Weyl chamber; put $\mathfrak{n}^+ = \sum_{\lambda > 0} \mathfrak{g}^\lambda$ – then $\mathfrak{g} = \mathfrak{k} + \mathfrak{a}_\mathfrak{p} + \mathfrak{n}^+$ (direct sum) and, in the obvious notations, $\mathfrak{G} = \mathfrak{K}\mathfrak{A}_\mathfrak{p}\mathfrak{N}^+$.

Proof of Lemma 2 (1) Fix an integer $m \geqslant 0$ such that $D, \tilde{D} \in \mathfrak{G}_m$. Choose a basis B for $\mathfrak{G}_m$ with the property that every element $D_b \in B$ has the form $D_b = D_\mathfrak{K} D_{\mathfrak{A}_\mathfrak{p}} D_{\mathfrak{N}^+}$ where $D_\mathfrak{K} \in \mathfrak{K}$, $D_{\mathfrak{A}_\mathfrak{p}} \in \mathfrak{A}_\mathfrak{p}$, $D_{\mathfrak{N}^+} \in \mathfrak{N}^+$ and

$$hD_{\mathfrak{N}^+} = (\prod_i h^{m_i\lambda_i})D_{\mathfrak{N}^+} \qquad (h \in A_\mathfrak{p} = \exp(\mathfrak{a}_\mathfrak{p})),$$

the m_i being non-negative integers – then

$$kD = \sum_{D_b \in B} a_b(k)D_b, \quad k\tilde{D} = \sum_{D_b \in B} \tilde{a}_b(k)D_b \qquad (k \in K),$$

the a_b and $\tilde{a}_b$ being continuous functions on K.

(2) Since any two norms on E are equivalent, we may assume that

$$\|\mu_1(k_1)a\mu_2(k_2)\| = \|a\|$$

for $k_1, k_2 \in K$ and $a \in E$. Let

$$M_0 = \sup_{k \in K} \max_{D_b \in B} (|a_b(k)|, |\tilde{a}_b(k)|).$$

Given $x \in G$, write $x = k_1hk_2$ $(k_1, k_2 \in K;\ h \in \exp(\mathscr{C}^{cl}))$ – then clearly

$$\|\Psi(\tilde{D}; x; D)\| \leqslant \|\Psi(k_1^{-1}\tilde{D}; h; k_2D)\| \leqslant M_0^2 \sum_{\tilde{D}_b, D_b \in B} \|\Psi(\tilde{D}_b; h; D_b)\|.$$

Write $\tilde{D}_b = \tilde{D}_\mathfrak{K}\tilde{D}_{\mathfrak{A}_\mathfrak{p}}\tilde{D}_{\mathfrak{N}^+}$; denoting the representation of $\mathfrak{K}$ corresponding to μ_i again by μ_i $(i = 1, 2)$, we then have that

$$\|\Psi(\tilde{D}_b; h; D_b)\| = \|\mu_1(\tilde{D}_\mathfrak{K})\Psi(h; \tilde{D}_{\mathfrak{A}_\mathfrak{p}}(h^{-1}\tilde{D}_{\mathfrak{N}^+})D_b)\| \leqslant M_1\|\Psi(h; \tilde{D}_{\mathfrak{A}_\mathfrak{p}}\tilde{D}_{\mathfrak{N}^+}D_b)\|$$

where $M_1 = \sup_{\tilde{D}_b \in B}\|\mu_1(\tilde{D}_\mathfrak{K})\|$ (bear in mind that $h^{\lambda_i} \geqslant 1$ $(1 \leqslant i \leqslant l)$). Let $D_i (1 \leqslant i \leqslant p)$ be a basis for the subspace of $\mathfrak{G}$ spanned by the $k(\tilde{D}_{\mathfrak{A}_\mathfrak{p}}\tilde{D}_{\mathfrak{N}^+}D_b)$

($\tilde{D}_b$, $D_b \in B$; $k \in K$). Evidently we can choose a number $M_2 \geqslant 0$ with the following property: If $k(\tilde{D}_{\mathfrak{A}_\mathfrak{p}} \tilde{D}_{\mathfrak{N}^+} D_b) = \sum_1^p c_i D_i$ ($c_i \in \mathbf{C}$), then $|c_i| \leqslant M_2$. This shows that

$$\|\Psi(\tilde{D}; x; D)\| \leqslant M \sum_{i=1}^{p} \|\Psi(x; D_i)\|$$

where $M = M_0^2 M_1 M_2$.

Since our hypotheses are symmetrical with respect to the left and to the right, the assertion of the lemma is now clear. □

8.3.8 Tempered Distributions on a Reductive Lie Group

Let (G, K) be a reductive pair per Vol. I, number 1.1.5; we shall assume that G is acceptable. Let G be a Lie group whose identity component is G and such that the factor space G/G is finite.

Definition A distribution T on G, continuous for the topology which $C_c^\infty(\mathsf{G})$ inherits as a subspace of $\mathscr{C}(\mathsf{G})$, is said to be a *tempered distribution.*

[The notion of 'tempered distribution' on the Lie algebra $\mathfrak{g}$ of G is, of course, well-known – cf. 8.3.5.]

Let T be a distribution on G; if T is tempered, then, thanks to Theorem 8.3.7.13, T admits a unique continuous extension to the space $\mathscr{C}(\mathsf{G})$.

Assign to the symbols $\mathfrak{A}$, $\mathfrak{A}_L^+$, $\mathfrak{A}_R^+$ the usual meanings (cf. Lemma 8.3.7.12).

Proposition 8.3.8.1 Let T be a central, tempered distribution on G – then T is continuous in the topology on $\mathscr{C}(G)$ defined by $\mathfrak{A}_L^+$.

Proof (1) Fix a function $\phi \in C_c^\infty(G)$ such that $\int_G \phi(y) d_G(y) = 1$ and, for any $f \in C_c^\infty(G)$, put $f_\phi(x) = \int_G \phi(y) f({}^y x) d_G(y)$ ($x \in G$). Suppose we could show the following: Given $|\,.\,|_\alpha$ ($\alpha \in \mathfrak{A}$), there exists $M > 0$ and a finite set $|\,.\,|_{r_1, D_1}, \ldots, |\,.\,|_{r_p, D_p}$ of elements in $\mathfrak{A}_L^+$ such that

$$|f_\phi|_\alpha \leqslant M \sum_1^p |f|_{r_i, D_i} \qquad (\text{all } f \in C_c^\infty(G)).$$

Then we claim that the assertion of the proposition is a simple consequence of this statement. In fact, on the one hand, since T is tempered, there exists an $\alpha \in \mathfrak{A}$ such that $|T(f)| \leqslant |f|_\alpha$ (all $f \in C_c^\infty(G)$) while, on the other, $T(f) = T(f_\phi)$, T being central. Conclusion:

$$|T(f)| = |T(f_\phi)| \leqslant |f_\phi|_\alpha \leqslant M \sum_1^p |f|_{r_i, D_i} \qquad (\text{all } f \in C_c^\infty(G))$$

which is what we wanted to prove.

(2) To verify the unproved contention in (1), first note that

$$f_\phi(z_1 x z_2) = \int_G \phi(y z_1^{-1}) f(y x z_2 z_1 y^{-1}) d_G(y) \qquad (z_1, z_2 \in G).$$

Hence, for fixed D, $\tilde{D} \in \mathfrak{G}$, one can choose $D_1, \ldots, D_p \in \mathfrak{G}$ such that

$$|f_\phi(\tilde{D}; x; D)| \leqslant \sum_1^p \sup_{y \in \omega} |f(yxy^{-1}; D_i)| \qquad (x \in G)$$

for all $f \in C_c^\infty(G)$ (here $\omega = \operatorname{spt}(\phi)$). Fix $r \geqslant 0$ – then, due to Lemma 8.1.2.1 and Proposition 8.3.7.2, there exists $M > 0$ with the property that

$$\sup_G (1 + \sigma)^r \, \Xi^{-1} |(\tilde{D} \otimes D) f_\phi| \leqslant M \sum_1^p \sup_G (1 + \sigma)^r \, \Xi^{-1} |D_i f|$$

and so the statement in (1) is now evident.

The proof of the proposition is therefore complete. □

Let T be a distribution on G; we recall (cf. 8.3.3) that T is said to be $\mathfrak{Z}$-finite if the space spanned by the distributions of the form $Z \cdot T \; (Z \in \mathfrak{Z})$ is finite dimensional. According to Theorem 8.3.3.1, a central $\mathfrak{Z}$-finite distribution T is actually a locally summable function F_T which is itself analytic on the regular set G'.

Select a maximal set $\mathfrak{j}^1, \ldots, \mathfrak{j}^r$ of mutually non-conjugate θ-stable Cartan subalgebras of $\mathfrak{g}$; let J^i be the Cartan subgroup of G associated with $\mathfrak{j}^i$ $(1 \leqslant i \leqslant r)$.

The first objective of the present number is to establish:

Theorem 8.3.8.2 (Harish-Chandra) Let T be a central, $\mathfrak{Z}$-finite distribution on G – then T is tempered iff there exists an integer $m \geqslant 0$ such that

$$\sup_{j \in (J^i)'} (1 + \sigma(j))^{-m} |D_l(j)|^{1/2} |F_T(j)| < \infty \qquad (1 \leqslant i \leqslant r).$$

[Thus, if T is a central, tempered, $\mathfrak{Z}$-finite distribution on G, then, of necessity, there exists a non-negative integer m such that

$$\sup_{x \in G'} (1 + \sigma(x))^{-m} |D_l(x)|^{1/2} |F_T(x)| < \infty.$$

Consequently $T(f) = \int_G f(x) F_T(x) d_G(x)$ for all $f \in \mathscr{C}(G)$ (here we have to bear in mind Proposition 8.3.7.6).]

In broad outline, the proof of this result is similar to its analogue in the algebra (cf. Theorem 8.3.6.1); as there, it will be necessary to employ invariant integral theory (cf. 8.5). [We remark, in passing, that the origin for the hypotheses of our theorem is the 'local expression' for F_T which was obtained in Theorem 8.3.3.3.]

Proof of Theorem 8.3.8.2 *Sufficiency* Choose $M > 0$ such that

$$|D_l(j)|^{1/2} |F_T(j)| \leqslant M(1 + \sigma(j))^m \qquad (j \in (J^i)', \; 1 \leqslant i \leqslant r);$$

bearing in mind that $|D_l(j)| = |\Delta_i(j)|^2 \; (j \in J^i)$, use of a familiar integration formula gives

$$|T(f)| = \left| \int_G f(x) F_T(x) d_G(x) \right|$$
$$\leqslant M \sum_{i=1}^{r} [W(G, \mathrm{J}_0^i)]^{-1} \int_{\mathrm{J}^i} (1 + \sigma(j))^m \, |\Phi_f^i(j)| \, d_{\mathrm{J}^i}(j)$$

for all $f \in C_c^\infty(G)$. [Here Φ_f^i denotes the invariant integral of f calculated relative to J^i $(i = 1, \ldots, r)$.] In view of Lemma 8.5.8.3 infra, it is now clear that T is tempered. □

The proof of the necessity is, however, not so simple and requires some preparation.

Let $\mathfrak{j}$ be a θ-stable Cartan subalgebra of $\mathfrak{g}$, J the Cartan subgroup of G associated with $\mathfrak{j}$. We recall that the group $W(G, \mathrm{J}_0)$ operates on the factor space G/J_0 (as well as J') (cf. Vol. I, number 1.4.2). This being so, fix a function $\alpha_0 \in C_c^\infty(G/\mathrm{J}_0)$ such that $\alpha_0(w \cdot \dot{x}) = \alpha(\dot{x})$ (all $w \in W(G, \mathrm{J}_0)$) with $\int_{G/\mathrm{J}_0} \alpha_0 = 1$; select a compact subset $\omega = \omega^{-1}$ of G which has the property that spt $(\alpha_0) \subset \dot{\omega}$. For any $g \in C_c^\infty(\mathrm{J}')$, define a function $f_g \in C_c^\infty(G(\mathrm{J}))$ by the prescription

$$f_g({}^x j) = \alpha_0(\dot{x}) \Delta(j)^{-1} \sum_{w \in W(G, \mathrm{J}_0)} \det(w) g^w(j) \qquad (j \in \mathrm{J}', x \in G).$$

Let us agree to use the notations revolving around Lemma 8.2.3.7 and its corollary – then the following estimate obtains.

Lemma 8.3.8.3 Given $D \in \mathfrak{G}$, there exist $H_i \in \mathfrak{J}$ and $\eta_i \in \mathfrak{R}$ $(1 \leqslant i \leqslant p)$ such that

$$|f_g({}^x j; D)| \leqslant |\xi_\rho(j)|^{-1} \sum_{i=1}^{p} \sum_w |\eta_i(j) g^w(j; H_i)| \qquad (j \in \mathrm{J}', x \in \omega)$$

for all $g \in C_c^\infty(\mathrm{J}')$.

[Here $\sum_w$ denotes a sum over the elements w of $W(G, \mathrm{J}_0)$.]

Proof Let $D_1, \ldots, D_n$ be a basis for the vector space spanned over $\mathbf{C}$ be the xD $(x \in G)$ – then $x^{-1}D = \sum_1^n a_i(x) D_i$ $(x \in G)$ where the a_i are analytic functions on G. Owing to Corollary 8.2.3.8, it is thus possible to choose analytic functions ϕ_i on G and elements $S_i \in \mathfrak{S}$, $H_i \in \mathfrak{J}$, $\eta_i \in \mathfrak{R}$ $(1 \leqslant i \leqslant p)$ such that

$$f({}^x j; D) = \sum_{i=1}^{p} \phi_i(x) \eta_i(j) f(x; S_i : j; H_i) \qquad (j \in \mathrm{J}', x \in G)$$

for all $f \in C^\infty(G)$. Put $\psi(x) = \alpha_0(\dot{x})$, $\psi_i(x) = \phi_i(x) \psi(x; S_i)$ $(x \in G)$ – then it is clear that

$$f_g({}^x j; D) = \sum_{i=1}^{p} \psi_i(x) \eta_i(j) \sum_w \det(w) g^w(j; H_i \circ \Delta^{-1}) \qquad (j \in \mathrm{J}', x \in \omega)$$

for all $g \in C_c^\infty(\mathsf{J}')$. We can write

$$H_i \circ \Delta^{-1} = \xi_\rho^{-1} \circ {}_iH \circ \eta \qquad ({}_iH \in \mathfrak{J})$$

where $\eta = \prod_{\alpha>0} \eta_\alpha$ belongs to the ring $\mathfrak{R}$. But $\mathfrak{R}$ is stable under both the operations of $\mathfrak{J}$ and $W(G, \mathsf{J}_0)$. Therefore the assertion of the lemma is a consequence of the compactness of ω. □

Let us keep to the above notations; thus, in particular, $\mathfrak{j}$ will denote a fixed θ-stable Cartan subalgebra of $\mathfrak{g}$, J the Cartan subgroup of G associated with $\mathfrak{j}$. Fix a component C of J'; because K is compact, J' has but a finite number of connected components (cf. Proposition 1.4.1.9). Therefore, assuming that T is tempered, the proof of our theorem will be complete when it is shown that there exists a number $M \geqslant 0$ and a non-negative integer m such that

$$|\overline{\Delta(j)F_T(j)}| \leqslant M(1 + \sigma(j))^m \qquad (\text{all } j \in \mathsf{C}).$$

[The reason for using $\bar{\Delta}$ instead of Δ will become apparent shortly.]

Let $\mathfrak{j}_\mathfrak{p}'$ denote the set of all $H \in \mathfrak{j}_\mathfrak{p}$ such that $\alpha(H) \neq 0$ for every root α of the pair $(\mathfrak{g}_c, \mathfrak{j}_c)$ which does not vanish identically on $\mathfrak{j}_\mathfrak{p}$; obviously $\mathfrak{j}_\mathfrak{p}'$ has only a finite number of connected components. Fix a connected component $\mathscr{C}^+$ of $\mathfrak{j}_\mathfrak{p}'$ such that C intersects $\mathsf{J}_K J^+$ ($J^+ = \exp(\mathscr{C}^+)$) – then it will be sufficient to establish the required inequality on $\mathsf{J}^\pm = \mathsf{C} \cap (\mathsf{J}_K J^+)$.

Fix an element $j_0 \exp H_0$ ($j_0 \in \mathsf{J}_K$, $H_0 \in \mathscr{C}^+$) in $\mathsf{J}^\pm$ and let $\mathfrak{z}$ denote the centralizer of $j_0 J_K$ in $\mathfrak{g}$; obviously all the roots $\Phi_\mathfrak{z}$ of the pair $(\mathfrak{z}_c, \mathfrak{j}_c)$ are real. We may arrange our choice of compatible orderings in the duals of $\mathfrak{j}_\mathfrak{p}$ and $\mathfrak{j}_\mathfrak{p} + \sqrt{-1}\mathfrak{j}_\mathfrak{k}$ in such a way that an element $\lambda \in \mathfrak{j}_\mathfrak{p}^\vee$ is positive whenever $\lambda(H_0) > 0$. Denoting by Φ^+ the set of positive roots of the pair $(\mathfrak{g}_c, \mathfrak{j}_c)$ with respect to this order, put $\Phi_\mathfrak{z}^+ = \Phi^+ \cap \Phi_\mathfrak{z}$. We shall then use this order for the definition of Δ and ξ_ρ.

Define a function $F_{T,\mathsf{J}}$ on J' by the rule $F_{T,\mathsf{J}}(j) = \overline{\Delta(j)}F_T(j)$ ($j \in \mathsf{J}'$); notice that $F_{T,\mathsf{J}}$ extends to an analytic function on $\mathsf{J}'(R)$ – this is a consequence of Proposition 8.1.1.2 and Proposition 8.3.4.3. [Of course the present definition of $F_{T,\mathsf{J}}$ is slightly different from the one given in 8.3.4.]

Lemma 8.3.8.4 Let J^0 denote the orbit of $\mathsf{J}^\pm$ under the action of $W(G, \mathsf{J}_0)$ – then there exists a non-negative integer m and elements $H_i \in \mathfrak{J}$, $\eta_i \in \mathfrak{R}$ ($1 \leqslant i \leqslant p$) such that

$$\left| \int_\mathsf{J} g(j)F_{T,\mathsf{J}}(j)d_\mathsf{J}(j) \right| \leqslant \sum_{i=1}^{p} \sup_{j \in \mathsf{J}'} (1 + \sigma(j))^m |\eta_i(j)g(j; H_i)|$$

for $g \in C_c^\infty(\mathsf{J}^0)$.

Proof To see where we are headed, observe that

$$|T(f_g)| = \left| \int_\mathsf{J} g(j)F_{T,\mathsf{J}}(j)d_\mathsf{J}(j) \right| \qquad (g \in C_c^\infty(\mathsf{J}^0)).$$

(1) Define the compact set ω as above; evidently, if $g \in C_c^\infty(\mathsf{J}^0)$, then spt $(f_g) \subset \omega \cdot \mathsf{J}^\pm$. Fix $D \in \mathfrak{G}$ and choose $M \geqslant 1$ such that $\Xi(^x y) \leqslant M \Xi(y)$, $1 + \sigma(^x y) \leqslant M(1 + \sigma(y))$ $(x \in \omega,\ y \in G)$. Since $\omega = \omega^{-1}$, it follows from Lemma 8.3.8.3 that

$$\begin{aligned} &|f_g(^x j; D)|\, \Xi(^x j)^{-1}(1 + \sigma(^x j))^m \\ &\leqslant M^{m+1} |f_g(^x j; D)|\, \Xi(j)^{-1}(1 + \sigma(j))^m \\ &\leqslant M^{m+1} |\xi_\rho(j)|^{-1}\, \Xi(j)^{-1}(1 + \sigma(j))^m \sum_{i=1}^{p} \sum_{w} |\eta_i(j) g^w(j; H_i)| \end{aligned}$$

for $g \in C_c^\infty(\mathsf{J}^0)$, $j \in \mathsf{J}^\pm$, $x \in \omega$, and $m \geqslant 0$, whence

$$\begin{aligned} &\sup_{x \in G} |f_g(x; D)|\, \Xi(x)^{-1}(1 + \sigma(x))^m \\ &\leqslant M^{m+1} \sup_{j \in \mathsf{J}^\pm} |\xi_\rho(j)\, \Xi(j)|^{-1}(1 + \sigma(j))^m \sum_{i=1}^{p} \sum_{w} |\eta_i(j) g^w(j; H_i)| \\ &\hphantom{\leqslant M^{m+1}} \qquad\qquad\qquad\qquad\qquad\qquad\qquad\qquad (\text{all } g \in C_c^\infty(\mathsf{J}^0)). \end{aligned}$$

(2) Extend $\mathfrak{j}_\mathfrak{p}$ to a maximal abelian subspace $\mathfrak{a}_\mathfrak{p}$ of $\mathfrak{p}$ and define an order on the dual of $\mathfrak{a}_\mathfrak{p}$ which is compatible with the one already chosen on $\mathfrak{j}_\mathfrak{p}^\vee$. Now $\mathsf{J}^\pm \subset j_0 J_K J^+$, and so, in view of Proposition 8.3.7.3, it is clear that $|\xi_\rho(j)|\, \Xi(j) \geqslant 1$ $(j \in \mathsf{J}^\pm)$ – therefore

$$\begin{aligned} &\sup_{x \in G} |f_g(x; D)|\, \Xi(x)^{-1}(1 + \sigma(x))^m \\ &\leqslant M^{m+1} \sup_{j \in J'} (1 + \sigma(j))^m \sum_{i=1}^{p} \sum_{w} |\eta_i(j) g^w(j; H_i)| \qquad (\text{all } g \in C_c^\infty(\mathsf{J}^0)). \end{aligned}$$

Since the ring $\mathfrak{R}$ is $W(G, \mathsf{J}_0)$-stable, the present lemma thus follows from Proposition 8.3.8.1. □

Fix a non-empty, open, relatively compact, connected subset J^- of $j_0 J_K$ with $J^- J^+ \subset \mathsf{J}^\pm$ and such that ξ_α $(\alpha \in \Phi^+)$ never takes the value 1 on J^- unless $\alpha \in \Phi_\delta^+$ (clearly this is possible). For $f^- \in C_c^\infty(J^-)$, $f^+ \in C_c^\infty(J^+)$, define the function $f^- \times f^+$ in $C_c^\infty(\mathsf{J}^\pm)$ via the evident requirement; put

$$F_{T,\mathsf{J}}^+(H) = \int_{J_K} F_{T,\mathsf{J}}(j_0 j_K \exp H) f^-(j_0 j_K) d_{J_K}(j_K) \qquad (H \in \mathcal{C}^+).$$

Then, relative to an obvious normalization of the Haar measures on J, J_K and $\mathfrak{j}_\mathfrak{p}$, we have

$$\int_{\mathsf{J}} (f^- \times f^+) F_{T,\mathsf{J}} d_{\mathsf{J}}(j) = \int_{\mathcal{C}^+} F_{T,\mathsf{J}}^+ f^+ d_{\mathfrak{j}_\mathfrak{p}}(H).$$

Now fix an $f^- \in C_c^\infty(J^-)$. Let $\pi_+(H) = \prod_{\alpha \in \Phi_+} \alpha(H)$ $(H \in \mathfrak{j}_c)$ – then the restriction of π_+ to $\mathfrak{j}_\mathfrak{p}$ is a polynomial function on $\mathfrak{j}_\mathfrak{p}$ which is not identically zero (if $\mathfrak{j}_\mathfrak{p} \neq \{0\}$). [Here Φ_+ denotes the complement in Φ^+ of the set Φ_I^+ of positive imaginary roots of the pair $(\mathfrak{g}_c, \mathfrak{j}_c)$.] On the basis of what has been said so far and Lemma 8.3.8.4, it is then seen without difficulty that there exists an integer $n \geqslant 0$ such that the distribution T^+

on $\mathscr{C}^+$ defined by the rule

$$f^+ \mapsto \int_{\mathscr{C}^+} f^+ \pi_+^n F_{T,\mathrm{J}}^+ d_{\mathrm{j}_\mathfrak{p}}(H) \qquad (f^+ \in C_c^\infty(\mathscr{C}^+))$$

is tempered.

Proof of Theorem 8.3.8.2 *Necessity* (1) Since $j_0 J_K J^+ \subset \mathrm{J}'(R)$ and since $F_{T,\mathrm{J}}$ is analytic on $\mathrm{J}'(R)$, using the $\mathfrak{Z}$-finiteness of F_T, we see, by an argument similar to the one which was employed during the proof of Theorem 8.3.3.3, that

$$F_{T,\mathrm{J}}(j_0 j_K \exp H) = \sum_{i=1}^{r} \xi_i(j_K) \sum_{k=1}^{s} p_{ik}(H) e^{\lambda_k(H)} \qquad (j_K \in J_K,\ H \in \mathscr{C}^+)$$

where the ξ_i $(1 \leqslant i \leqslant r)$ are distinct characters of J_K, the λ_k $(1 \leqslant k \leqslant s)$ are distinct linear functions on $\mathrm{j}_{\mathfrak{p}_c}$ and the p_{ik} are polynomial functions on $\mathrm{j}_{\mathfrak{p}_c}$.

(2) Fix k $(1 \leqslant k \leqslant s)$ – then we claim that $p_{ik} = 0$ $(1 \leqslant i \leqslant r)$ unless $\mathscr{R}(\lambda_k) \leqslant 0$ on $\mathscr{C}^+$. Thus suppose that $\mathscr{R}(\lambda_k(H^+)) > 0$ for some $H^+ \in \mathscr{C}^+$; in view of the formula for $F_{T,\mathrm{J}}$ in (1) and the fact that T^+ is tempered on $\mathscr{C}^+$, we deduce from Example 2 in 8.3.5 that necessarily

$$\sum_{i=1}^{r} p_{ik} \int_{J_K} \xi_i(j_K) f^-(j_0 j_K) d_{J_K}(j_K) = 0.$$

The arbitrariness of f^- in $C_c^\infty(J^-)$ together with Lemma 8.3.5.4 supra (or rather a modest variant of it . . .) then implies that $p_{ik} = 0$ $(1 \leqslant i \leqslant r)$. Hence the claim.

Because of (2), it is obvious that we can select a number $M \geqslant 0$ and a non-negative integer m such that

$$|\overline{\Delta(j)}F_T(j)| \leqslant M(1 + \sigma(j))^m \qquad (j \in \mathrm{J}^{\pm}).$$

This, of course, suffices to complete the proof of Theorem 8.3.8.2. □

Here is a useful consequence of Theorem 8.3.8.2.

Proposition 8.3.8.5 Let T be a central, tempered, $\mathfrak{Z}$-finite distribution on G – then

$$T(f) = \sum_{i=1}^{r} [W(G, \mathrm{J}_0^i)]^{-1} \int_{\mathrm{J}^i} \epsilon_R^i(j) F_{T,\mathrm{J}^i}(j) \Phi_f^i(j) d_{\mathrm{J}^i}(j) \qquad (\text{all } f \in \mathscr{C}(G)).$$

[As before, $F_{T,\mathrm{J}^i} = \bar{\Delta}_i F_T$ on $\mathrm{J}^i \cap G'$ $(i = 1, \ldots, r)$.]

Proof Fix an $f \in \mathscr{C}(G)$ and choose $f_n \in C_c^\infty(G)$ such that $f_n \to f$ in $\mathscr{C}(G)$ – then $T(f) = \lim_n T(f_n)$. Now it is clear that

$$T(f_n) = \sum_{i=1}^{r} [W(G, \mathrm{J}_0^i)]^{-1} \int_{\mathrm{J}^i} \epsilon_R^i(j) F_{T,\mathrm{J}^i}(j) \Phi_{f_n}^i(j) d_{\mathrm{J}^i}(j) \qquad (n = 1, 2, \ldots)$$

and so, if we take into account the fact that

$$|F_{T,\mathrm{J}^i}(j)| \leqslant M(1+\sigma(j))^m \qquad (j \in \mathrm{J}^i \cap G', i = 1, \ldots r)$$

for numbers m, $M \geqslant 0$, the present proposition is then seen to follow from Lemma 8.5.8.3 infra. □

Let T be a central, $\mathfrak{Z}$-finite distribution on G, $T_\delta(\delta \in \hat{K})$ its δth Fourier component (cf. Vol. I, number 4.4.3) – then each T_δ is an analytic function on G (see Lemma 2 in the Appendix to 8.3.9 below). If in addition T is tempered, then it is actually possible to estimate the T_δ's.

Theorem 8.3.8.6 (Harish-Chandra) Let T be a central, $\mathfrak{Z}$-finite, tempered distribution on G – then, given $\delta \in \hat{K}$, there exists a number $M_\delta \geqslant 0$ and an integer $m_\delta \geqslant 0$ such that

$$|T_\delta(x)| \leqslant M_\delta \Xi(x)(1+\sigma(x))^{m_\delta} \qquad (\text{all } x \in G).$$

[In the terminology of number 9.3.1, this theorem says that the Fourier components of a central, $\mathfrak{Z}$-finite, tempered distribution satisfy the 'weak inequality'.]

The contention of the present theorem is contained in the following lemma.

Lemma 8.3.8.7 Let T be a K-finite, $\mathfrak{Z}$-finite distribution on G which is tempered – then there exists a number $M \geqslant 0$ and a non-negative integer m such that

$$|T(x)| \leqslant M \Xi(x)(1+\sigma(x))^m \qquad (\text{all } x \in G).$$

[To say that T is K-finite means that the span of its left and right translates under K is finite dimensional; evidently our hypotheses on T entail that T is actually an analytic function on G (see the Appendix to 8.3.9 infra).]

Proof (1) There exists a non-negative integer $m \geqslant 0$ such that

$$\int_G |T(x)| \Xi(x)(1+\sigma(x))^{-m} d_G(x) < \infty.$$

To see this, we proceed as follows. Since T is K-finite both to the left and to the right as well as $\mathfrak{Z}$-finite, on the basis of Corollary 8.3.9.2 below there exist elements $\phi_i \in C_c^\infty(G)$ $(i = 1, 2)$ such that $T = \phi_1 * T = T * \phi_2$; set $\psi_i = \phi_i'$ $(\phi_i'(x) = \phi_i(x^{-1}),\ x \in G)$ – then it is clear that $T(f) = T(\psi_1 * f * \psi_2)$ (all $f \in C_c^\infty(G)$). Because T is tempered, the absolute value of the number $T(f)$ is dominated by a sum of terms of the form

$$_{D_1}|f|_{r,D_2} \qquad (f \in C_c^\infty(G)),$$

i.e. is dominated by a sum of terms of the form

$$ {}_{D_1}|\psi_1 * f * \psi_2|_{r, D_2} \qquad (f \in C_c^\infty(G)). $$

This being so, one then establishes without difficulty the existence of a positive constant R (say) with the property that

$$ {}_{D_1}|\psi_1 * f * \psi_2|_{r, D_2} \leqslant R|f|_r $$

for all f in $C_c^\infty(G)$ (here we have to bear in mind Proposition 8.3.7.2). It therefore follows that

$$ \left| \int_G f(x)T(x)d_G(x) \right| \leqslant M|f|_m \qquad (\text{all } f \in C_c^\infty(G)) $$

for certain positive constants M, m, that is

$$ \left| \int_G f(x)T(x)\,\Xi(x)(1+\sigma(x))^{-m}d_G(x) \right| \leqslant M \sup_G |f| $$
$$ (\text{all } f \in C_c^\infty(G)) $$

from which we can read off our contention.

(2) In view of the relation $T = T*\phi_2$, we have the estimate

$$ |T(x)| \leqslant \int_G |T(xy)|\,|\psi_2(y)|d_G(y) \leq \int_G |T(xy)| \frac{\Xi(y)}{(1+\sigma(y))^m} d_G(y) $$
$$ (\text{all } x \in G). $$

(3) Owing to the K-finiteness of T, there exists a finite subset $F \subset \hat{K}$ such that $T = \bar{\chi}_F * T$ ($\bar{\chi}_F = \sum_{\delta \in F} \bar{\chi}_\delta$); taking into account the inequality

$$ \left(\frac{1}{1+\sigma(xy)}\right)^m \leqslant \left(\frac{1+\sigma(x)}{1+\sigma(y)}\right)^m \qquad (x, y \in G;\ m \geqslant 0), $$

we then find that

$$ \begin{aligned} |T(x)| &\leqslant \|\bar{\chi}_F\|_\infty \int_K |T(k^{-1}x)|\,dk \\ &\leq \|\bar{\chi}_F\|_\infty \int_K\int_G |T(k^{-1}xy)| \frac{\Xi(y)}{(1+\sigma(y))^m} d_G(y)dk \\ &\leq \int_G |T(y)| \left[\int_K \frac{\Xi(x^{-1}ky)}{(1+\sigma(x^{-1}ky))^m} dk \right] d_G(y) \\ &= \Xi(x)(1+\sigma(x))^m \cdot \int_G |T(y)|\,\Xi(y)(1+\sigma(y))^{-m}d_G(y) \end{aligned} $$

for all x in G.

The assertion of our lemma is now clear, the integral on the right being finite (cf. (1)). □

Proposition 8.3.8.8 Let T be a central, $\mathfrak{Z}$-finite, tempered distribution on G; let T_δ be the δth Fourier component of T – then

$$ T_\delta(f) = \int_G f(x)T_\delta(x)d_G(x) \qquad (\text{all } f \in \mathscr{C}(G)). $$

[This is an immediate consequence of the preceding theorem provided we bear in mind Proposition 8.3.7.5]

It is not difficult to see what the significance of Theorem 8.3.8.6 is for Harmonic Analysis on G. Thus let U be a TCI Hilbert representation of G on E, T_U its character; let $T_{U,\delta}$ denote the δth Fourier component of T_U – then as has been pointed out in number 6.1.2, $T_{U,\delta} = \psi_\delta^U$, ψ_δ^U the spherical trace function on G of type δ which is associated with U. Consequently, if T_U is in addition tempered, then Theorem 8.3.8.6 tells us that the growth of the ψ_δ^U's is not entirely arbitrary. . . . [The importance of such estimates will be brought out later on.]

Note In passing let us observe that the K-finite coefficients of the irreducible unitary representations of G with tempered characters necessarily satisfy the 'weak inequality'. . . .

Remark Let T be a central, $\mathfrak{Z}$-finite, tempered distribution on G – then, given $\delta \in \hat{K}$, there exists a number $M \geqslant 0$ and an integer $m \geqslant 0$, *independent* of δ, such that

$$|T_\delta(x)| \leqslant M \Xi(x)(1 + \sigma(x))^m$$

for all $x \in G$ (cf. Harish-Chandra [30, p. 51]). This estimate leads at once to the conclusion that the K-finite coefficients of the irreducible unitary representations of G with tempered characters are bounded by a multiple of $\Xi(1 + \sigma)^n$ (some $n \in \mathbf{N}$); in the applications this fact can sometimes be used to determine whether the character of a given irreducible unitary representation of G is tempered or not.

Examples (1) Let G be a connected semi-simple Lie group with finite center which is, moreover, acceptable; let $G = KA_\mathfrak{p}N^+$ be an Iwasawa decomposition for G, $P = MA_\mathfrak{p}N^+$ a minimal parabolic subgroup of G – then the characters of the (unitary) representations $U^{\sigma,\mu}$ in the principal P-series for G are tempered. [In fact, on the basis of Theorem 8.3.8.2 supra, our assertion is seen to be an immediate consequence of Theorem 5.5.3.1.]

(2) Let (G, K) be a reductive pair fulfilling the usual hypotheses (with G acceptable); fix a semi-simple element γ in G and let G_γ be the centralizer of γ in G – then it can be shown that the integral

$$\int_{G/\mathsf{G}_\gamma} f({}^{\dot{x}}\gamma)\, d_{G/\mathsf{G}_\gamma}(\dot{x})$$

converges for every $f \in \mathscr{C}(G)$ and that the assignment

$$f \mapsto \int_{G/\mathsf{G}_\gamma} f({}^{\dot{x}}\gamma)\, d_{G/\mathsf{G}_\gamma}(\dot{x})$$

defines on G a central, tempered distribution. [These facts will be dealt with in detail when we come to the Selberg Principle for G; cf. Chapter 9.]

(3) Let us take $G = \mathbf{SL}(2, \mathbf{R})$ – then, as is well-known, the characters F^μ of the representations U^μ in the unitary 'complementary series' for G are concentrated on the set of hyperbolic elements and are given there by the prescription

$$F^\mu(x) = \frac{|\lambda_x|^\mu + |\lambda_x|^{-\mu}}{|\lambda_x - \lambda_x^{-1}|} \qquad (0 < \mu < 1),$$

where λ_x is an eigenvalue of the matrix x. Surely, then, such representations do not have tempered characters. . . .

8.3.9 Tools for Harmonic Analysis on G

Let (G, K) be a reductive pair per Vol. I, number 1.1.5. Let $\mathfrak{G}$ be the universal enveloping algebra of $\mathfrak{g}_c$, $\mathfrak{Z}$ the center of $\mathfrak{G}$; let $\mathfrak{K}$ denote the universal enveloping algebra of $\mathfrak{k}_c$, $\mathfrak{Z}_{\mathfrak{k}}$ the center of $\mathfrak{K}$.

Let E be a locally convex, complete, Hausdorff, topological vector space.

Theorem 8.3.9.1 (Harish-Chandra) Let U be a differentiable representation of G on E; let $\mathcal{O}$ be a neighborhood of 1 in G – then, for a given K-finite, $\mathfrak{Z}$-finite element a_0 of E, there exists a K-central function ϕ in $C_c^\infty(G)$ with spt $(\phi) \subset \mathcal{O}$ such that $U(\phi)a_0 = a_0$.

[The terms 'K-finite' and '$\mathfrak{Z}$-finite' are to be assigned the evident interpretation.]

Corollary 8.3.9.2 Let f be an element of $C^\infty(G; E)$ which is both K-finite and $\mathfrak{Z}$-finite for the left (respectively right) regular representation of G on $C^\infty(G; E)$ – then there exists a K-central function ϕ in $C_c^\infty(G)$, with support arbitrarily close to 1, such that $f = \phi * f$ (respectively $f = f * \phi$).

[This corollary, which follows at once from the theorem above, will play an important role at various places in the sequel.]

It will be convenient to break up the proof of Theorem 8.3.9.1 into a number of lemmas.

Given a finite subset F of $\hat{K}$, we shall agree to write E_F in place of $\sum_{\delta \in F} E(\delta)$, P_F in place of $\sum_{\delta \in F} \mathsf{P}(\delta)$.

Lemma 8.3.9.3 Retain the above notations and assumptions; fix a finite subset F of $\hat{K}$; let a_0 be a vector in E_F which is annihilated by an ideal $\mathfrak{J}_0$ of finite codimension in $\mathfrak{Z}$ – then the space $\mathsf{P}_F(\mathfrak{G}a_0)$ is finite dimensional.

Proof The proof of the present lemma rests essentially on a fundamental algebraic result of Harish-Chandra, namely Theorem 2.2.1.1. Thus let $\mathfrak{J}$ be the kernel of the mapping $D \mapsto U(D)a_0$ in $\mathfrak{K}$; since $a_0 \in E_F$, $\mathfrak{K}/\mathfrak{J}$ is finite dimensional and the natural representation of $\mathfrak{k}_c$ on $\mathfrak{K}/\mathfrak{J}$ is semi-simple. According to the cited theorem, $\bar{\mathfrak{G}} = \mathfrak{G}/\mathfrak{G}\mathfrak{J}$ may be regarded as a locally finite K-module, whence $\bar{\mathfrak{G}} = \sum_{\delta \in \hat{K}} \bar{\mathfrak{G}}(\delta)$ where each $\bar{\mathfrak{G}}(\delta)$ $(\delta \in \hat{K})$ is a *finite* module over $\mathfrak{Z}$. The mapping ϕ which assigns to each coset $D + \mathfrak{G}\mathfrak{J}(D \in \mathfrak{G})$ the element $U(D)a_0$ is a surjection of $\mathfrak{G}$-modules; since $\sum_{\delta \in \hat{K}} E(\delta)$ is $\mathfrak{G}$-stable (cf. Proposition 4.4.5.18), it must be the case that $\mathfrak{G}a_0 = \sum_{\delta \in \hat{K}} \mathfrak{G}a_0 \cap E(\delta)$, whence

$$\phi(\bar{\mathfrak{G}}(\delta)) = \mathsf{P}(\delta)(\mathfrak{G}a_0) = \mathfrak{G}a_0 \cap E(\delta) \qquad (\delta \in \hat{K}).$$

Choose elements $\bar{D}_1, \ldots, \bar{D}_n$ in $\bar{\mathfrak{G}}(\delta)$ such that $\bar{\mathfrak{G}}(\delta) = \sum_i \mathfrak{Z}\bar{D}_i$ – then $\mathsf{P}(\delta)(\mathfrak{G}a_0) = \sum_i U(\mathfrak{Z})U(D_i)a_0$ and so our assertion is now obvious, a_0 being $\mathfrak{Z}$-finite. □

Lemma 8.3.9.4 Retain the above notations and assumptions; fix a finite subset F of $\hat{K}$; let a_0 be a vector in E_F which is annihilated by an ideal $\mathfrak{J}_0$ of finite codimension in $\mathfrak{Z}$ – then, for every $\mathbf{z} \in E^*$, the function $\phi_{\mathbf{z}}$, $x \mapsto \langle \mathsf{P}_F(U(x)a_0), \mathbf{z}\rangle$ $(x \in G)$, is analytic.

Proof The hypotheses imply that the span of the left translates of $\phi_{\mathbf{z}}$ by the elements of K is finite dimensional; on the other hand $\phi_{\mathbf{z}}$ is annihilated by $\mathfrak{J}_0$ – our contention is therefore seen to be a consequence of Lemma 1 in the Appendix to the present number. □

Lemma 8.3.9.5 Retain the above notations and assumptions; fix a finite subset F of $\hat{K}$; let a_0 be a vector in E_F which is annihilated by an ideal $\mathfrak{J}_0$ of finite codimension in $\mathfrak{Z}$ – then, for every $x \in G$, the vector $\mathsf{P}_F(U(x)a_0)$ lies in $\mathsf{P}_F(\mathfrak{G}a_0)$.

Proof The space $\mathsf{P}_F(\mathfrak{G}a_0)$, being finite dimensional (Lemma 8.3.9.3 supra), is closed in E; therefore we need only show that if $\mathbf{z}$ is an element of E^* which vanishes on $\mathsf{P}_F(\mathfrak{G}a_0)$, then of necessity the scalar function $\phi_{\mathbf{z}}$ is identically zero. But the vanishing of $\mathbf{z}$ on $\mathsf{P}_F(\mathfrak{G}a_0)$ clearly implies that all the derivatives of $\phi_{\mathbf{z}}$, when evaluated at 1, are zero; however $\phi_{\mathbf{z}}$ is analytic (Lemma 8.3.9.4) . . . . □

Proof of Theorem 8.3.9.1 Select a finite subset F of $\hat{K}$ such that a_0 lies in E_F; choose ϕ_i per the statement of our theorem such that

$$U(\phi_i)a_0 = U(\bar{\chi}_F * \phi_i)a_0 \rightarrow a_0$$

in E – then it is clear that a_0 lies in the closure of the space $\mathscr{S}$ of all vectors of the form $U(\phi)a_0$ where ϕ is a K-central function in $C_c^\infty(G)$ with spt $(\phi) \subset \mathcal{O}$. Because a_0 is $\mathfrak{Z}$-finite, it follows from Lemma 8.3.9.5 that $\mathscr{S}$ is contained in $\mathsf{P}_F(\mathfrak{G}a_0)$, in particular that $\mathscr{S}$ is finite dimensional, hence closed – therefore $a_0 \in \mathscr{S}$, as desired. □

Application Here is a typical application of Theorem 8.3.9.1 (or rather its corollary). Let L denote the left regular representation of G on $L^2(G)$; let f be a non-zero $\mathfrak{Z}$-finite function in $L^2(G)$ such that the span of its left K-translates is finite dimensional (to interpret the phrase $\mathfrak{Z}$-finite here, regard f as a locally summable function on G); let E denote the smallest closed L-stable subspace of $L^2(G)$ containing f – then $E = \sum_{i=1}^{p} E_i$ where the E_i are non-zero, mutually orthogonal, closed, L-stable subspaces of E which are, moreover, irreducible under L. [Let U denote the restriction of L to E; for any finite subset F of $\hat{K}$, define the space E_F as above – then the hypotheses tell us that $f \in E_F$ for some F. In view of Lemma 1 in the Appendix to this number, it is clear that f is analytic; consequently, thanks to Corollary 8.3.9.2 above, there exists a $\phi \in C_c^\infty(G)$ such that $f = \phi * f$, whence $f \in E_\infty$. This being so, consider now the $\mathfrak{G}$-orbit of f – then $(\mathfrak{G}f)^{cl} = E$. For otherwise, by the Hahn-Banach Theorem, we could choose a non-zero continuous linear form $\mathbf{z}$ on E such that $\mathbf{z} \mid \mathfrak{G}f = 0$; the function $x \mapsto \langle L(x)f, \mathbf{z}\rangle$ $(x \in G)$ is, however, analytic and all its derivatives, when evaluated at 1, vanish – therefore $\mathbf{z} = 0$ on E. . . . Furthermore, on the basis of Theorem 2.2.1.1, we see that $\mathfrak{G}f = \sum_{\delta \in \hat{K}} (\mathfrak{G}f)(\delta)$ where $\mathsf{P}_F(\mathfrak{G}f)(= E_F)$ is finite dimensional. Let $\tilde{E}$ be any non-zero closed U-stable subspace of E – then $\mathsf{P}_F(\tilde{E}) \neq \{0\}$. [Suppose, to the contrary, that $\mathsf{P}_F(\tilde{E}) = \{0\}$; let $\tilde{E}^\perp$ denote the orthogonal complement of $\tilde{E}$ in E – then $f \in E_F \subset \tilde{E}^\perp$, whence $E \subset \tilde{E}^\perp$ and therefore $\tilde{E} = \{0\}$, a contradiction.] Let $E_i (1 \leqslant i \leqslant p)$ be a finite set of mutually orthogonal, closed, non-zero subspaces of E which are U-stable – then $E_i \cap E_F \neq \{0\}$ and so $p \leqslant \dim(E_F) < \infty$. The required result thus follows upon assuming that p has the largest possible value.]

Appendix Let (G, K) be a reductive pair per Vol. I, number 1.1.5; let $\mathfrak{G}$ be the universal enveloping algebra of $\mathfrak{g}_c$, $\mathfrak{Z}$ the center of $\mathfrak{G}$; let $\mathfrak{K}$ be the universal enveloping algebra of $\mathfrak{k}_c$, $\mathfrak{Z}_\mathfrak{k}$ the center of $\mathfrak{K}$.

Lemma 1 Let T be a distribution on an open subset $\mathcal{O}$ of G and suppose that the dimension of the space of all distributions of the form $D \cdot T$ $(D \in \mathfrak{Z}_\mathfrak{k}\mathfrak{Z})$ is finite – then T coincides, in the sense of distributions, with an analytic function on $\mathcal{O}$.

Proof The considerations of Example 2 in Vol. I, number 2.3.3 admit the obvious extension to the present setting; thus, adopting the notations which were introduced there, it is clear that $\Delta = \omega + 2\omega_- \in \mathfrak{Z}_\mathfrak{k}\mathfrak{Z}$ and so the distributions $\Delta^i \cdot T (i \geqslant 0)$ cannot all be linearly independent. Consequently there exists an integer $m \geqslant 1$ and complex numbers $c_1, \ldots, c_m$ such that $\Box \cdot T = 0$ where

$$\Box = \Delta^m + \sum_1^m c_i \Delta^{m-i}.$$

But $\Box$ is an elliptic analytic differential operator on G, whence T is an analytic function, as contended. $\Box$

Lemma 2 Let T be a $\mathfrak{Z}$-finite distribution on G – then each of the Fourier components T_δ is an analytic function on G $(\delta \in \hat{K})$.

Proof Each of the irreducible characters $\xi_\delta (\delta \in \hat{K})$ is an eigenfunction of all the operators in $\mathfrak{Z}_\mathfrak{k}$ – hence there exists a homomorphism κ_δ of $\mathfrak{Z}_\mathfrak{k}$ into $\mathbf{C}$ such that $Z\xi_\delta = \kappa_\delta(Z)\xi_\delta$ (all $Z \in \mathfrak{Z}_\mathfrak{k}$). From this we see that $Z \cdot T_\delta = \kappa_\delta(Z) T_\delta$ (all $Z \in \mathfrak{Z}_\mathfrak{k}$) and so the assertion of the present lemma follows from that of Lemma 1 above. $\Box$

8.4 The Invariant Integral on a Reductive Lie Algebra

8.4.1 The Invariant Integral – Definition and Properties

Let $\mathfrak{g}$ be a reductive Lie algebra over **R**, $\theta : \mathfrak{g} \to \mathfrak{g}$ an involutive automorphism of $\mathfrak{g}$ whose restriction to the derived algebra $\bar{\mathfrak{g}} = [\mathfrak{g}, \mathfrak{g}]$ is a Cartan involution (of $\bar{\mathfrak{g}}$); write, as usual, $\mathfrak{g} = \mathfrak{k} + \mathfrak{p}$ and let (G, K) be a pair corresponding to $\mathfrak{g}$ per Vol. I, number 1.1.5.

Let $\mathfrak{j}$ be a θ-stable Cartan subalgebra of $\mathfrak{g}$, J the Cartan subgroup of G associated with $\mathfrak{j}$ (as will become apparent, the θ-stability of $\mathfrak{j}$ is not really necessary for much of our discussion).

Lemma 8.4.1.1 Fix an element $\Gamma \in \mathfrak{j}$; let G_Γ be the centralizer of Γ in G, $x \mapsto \dot{x}$ the natural mapping of G onto the factor space G/G_Γ – then there exists a neighborhood $\mathcal{O}(\Gamma)$ of Γ in $\mathfrak{j}$ with the following property: Given any compact set ω in $\mathfrak{g}$, there exists a compact set Ω in G/G_Γ such that $xH(= \mathrm{Ad}\,(x)H) \in \omega$ $(x \in G,\ H \in \mathcal{O}(\Gamma)) \Rightarrow \dot{x} \in \Omega$.

Proof For the purposes of this lemma, it may be supposed that $\mathfrak{g}$ is semi-simple. Fix r, $0 < r < \pi = 3.14\ldots$, and consider the G-stable, completely invariant set $\mathfrak{g}(r)$ per 8.1.6 – then $\exp|\mathfrak{g}(r)$ is regular and injective. Choose a real number $t \neq 0$ such that $t\Gamma \in \mathfrak{g}(r)$ – then it is obvious that G_Γ is the centralizer of $\gamma = \exp(t\Gamma)$ in G and so there exists a neighborhood $\mathcal{O}(\Gamma)$ of Γ in $\mathfrak{j}$ such that $t\mathcal{O}(\Gamma) \subset \mathfrak{g}(r)$ and $\mathcal{O}(\gamma) = \exp(t\mathcal{O}(\Gamma))$ verifies the conditions of Theorem 8.1.4.1. Given ω, consider $\exp(\mathfrak{g}(r) \cap t\omega)$; plainly $\exp(\mathfrak{g}(r) \cap t\omega)$ is contained in a compact subset ω_G (say) of G. Owing to Theorem 8.1.4.1, there exists a compact set Ω in G/G_Γ such that $xjx^{-1} \in \omega_G$ $(j \in \mathcal{O}(\gamma),\ x \in G) \Rightarrow \dot{x} \in \Omega$. Accordingly, if $xH \in \omega$ for some $x \in G$ and $H \in \mathcal{O}(\Gamma)$, then $t(xH) \in \mathfrak{g}(r) \cap t\omega$ (recall that $\mathfrak{g}(r) = x(\mathfrak{g}(r))$, $x \in G$); thus $\exp(t(xH)) = xjx^{-1} \in \omega_G$ $(j = \exp(tH) \in \mathcal{O}(\gamma))$ and so $\dot{x} \in \Omega$, as desired. □

Ordering the roots of the pair $(\mathfrak{g}_c, \mathfrak{j}_c)$ in some way, let Φ_R^+ denote the set of positive real roots and put $\pi_R = \prod_{\alpha \in \Phi_R^+} \alpha$; let

$$\epsilon_R(H) = \mathrm{sign}\,\{\pi_R(H)\} \qquad (H \in \mathfrak{j}).$$

Definition Let $f \in C_c^\infty(\mathfrak{g})$ – then the *invariant integral* of f relative to the Cartan subalgebra $\mathfrak{j}$ is the function ϕ_f on $\mathfrak{j}'$ defined by

$$\phi_f(H) = \epsilon_R(H)\pi(H) \int_{G/\mathsf{J}} f(\dot{x}H)\,d_{G/\mathsf{J}}(\dot{x}) \qquad (H \in \mathfrak{j}').$$

[In view of the discussion in Vol. I, number 3.2.1, we shall really only be interested in the case when G is non-compact. . . .]

The following observations will help to clarify this definition. In the first place, since H is regular, the centralizer of H in G is the Cartan subgroup J (cf. Theorem 1.4.1.5). Moreover, $\phi_f(H)$ $(H \in \mathfrak{j}')$ is meaningful. Thus, if $\omega = \mathrm{spt}\,(f)$, then, in the notations of Lemma 8.4.1.1,

$$\int_{G/\mathrm{J}} f(\dot{x}H)d_{G/\mathrm{J}}(\dot{x}) = \int_{\Omega} f(\dot{x}H)d_{G/\mathrm{J}}(\dot{x}) \qquad (H \in \mathfrak{j}')$$

and this relation actually holds for all elements in $\mathfrak{j}'$ which lie sufficiently close to H. Therefore it is permissible to differentiate under the integral sign and conclude: ϕ_f is a C^∞ function on $\mathfrak{j}'$ (any $f \in C_c^\infty(\mathfrak{g})$).

Remarks (1) The invariant measure on G/J is, of course, to be taken in the normalization determined in 8.1.2. Choose $\mathfrak{j}^1, \ldots, \mathfrak{j}^r$ in the usual way; assign to $\pi^i, \epsilon_R^i, \phi_f^i$ the obvious meanings ($f \in C_c^\infty(\mathfrak{g})$; $i = 1, \ldots, r$); noting that $\bar{\pi}^i = (-1)^{r_I^i}\pi^i$, r_I^i the number of positive imaginary roots of the pair $(\mathfrak{g}_c, \mathfrak{j}_c^i)$ (the number of positive complex roots is even), we have the important formula

$$\int_{\mathfrak{g}} f(X)\, d_{\mathfrak{g}}(X) = \sum_{i=1}^{r} (-1)^{r_I^i} \int_{\mathfrak{j}^i} \pi^i(H)\epsilon_R^i(H)\phi_f^i(H)\, d_{\mathfrak{j}^i}(H) \qquad (f \in C_c^\infty(\mathfrak{g})).$$

(2) The definition of ϕ_f depends on the choice of an ordering for the roots – this is of little importance however.

(3) The group G operates on $\mathfrak{g}$ via the adjoint action and this action extends to $C_c^\infty(\mathfrak{g})$ (A 2.4); this being so, fix an $f \in C_c^\infty(\mathfrak{g})$ – then it is clear that $\phi_{f^x} = \phi_f$ (all $x \in G$), whence the term 'invariant integral'.

(4) The invariant integral ϕ_f satisfies a simple functional equation. Thus fix a w in $W(G, \mathrm{J})$; agreeing to define $\epsilon_R(w)$ by the prescription $w\pi_R = \epsilon_R(w)\pi_R$, we have

$$\phi_f(wH) = \det(w)\epsilon_R(w)\phi_f(H) \qquad (H \in \mathfrak{j}'),$$

where we have taken into account the fact that the operations of $W(G, \mathrm{J})$ preserve the invariant measure on G/J. In particular, let us observe that ϕ_f is left fixed by the Weyl reflections associated with the real roots of the pair $(\mathfrak{g}_c, \mathfrak{j}_c)$ (these reflections do lie in $W(G, \mathrm{J})$; cf. Lemma 1.3.2.4).

It is natural to wonder why the invariant integral should be an object of study. At the present time, it will not be feasible for us to completely justify its introduction. However, roughly speaking, certain properties of ϕ_f are intimately related to the derivation of the explicit Plancherel formula for G. [Indeed, as was seen in Vol. I, number 3.2.2, the Plancherel formula for compact G is an immediate consequence of the continuity of a certain derivative of ϕ_f.]

We shall now formulate, in a precise fashion, the fundamental properties which the invariant integral possesses; the proofs will be deferred until later numbers.

In the first place it turns out that the invariant integral can actually be defined for all $f \in \mathscr{C}(\mathfrak{g})$ (and not just for $f \in C_c^\infty(\mathfrak{g})$).

Theorem 8.4.1.2 (Harish-Chandra) Fix $f \in \mathscr{C}(\mathfrak{g})$ – then the integral

$$\phi_f(H) = \epsilon_R(H)\pi(H) \int_{G/\mathrm{J}} f(\dot{x}H)d_{G/\mathrm{J}}(\dot{x})$$

is absolutely convergent for all $H \in \mathfrak{j}'$ and the function ϕ_f so defined lies in $\mathscr{C}(\mathfrak{j}')$. Furthermore the mapping $f \mapsto \phi_f$ of $\mathscr{C}(\mathfrak{g})$ into $\mathscr{C}(\mathfrak{j}')$ is continuous and

$$\phi_{\partial(p)f} = \partial(\tilde{p})\phi_f \qquad (\text{all } p \in I(\mathfrak{g}_c);\ \tilde{p} = p \mid \mathfrak{j}_c).$$

Let Φ_{SI}^+ denote the set of positive singular imaginary roots of the pair $(\mathfrak{g}_c, \mathfrak{j}_c)$; write $\pi_{SI} = \prod_{\alpha \in \Phi_{SI}^+} \alpha$ and put

$$\mathfrak{j}'(SI) = \{H \in \mathfrak{j}\colon \pi_{SI}(H) \neq 0\}.$$

The proof of Theorem 8.4.1.2 contains the following fact.

Corollary 8.4.1.3 Let $f \in \mathscr{C}(\mathfrak{g})$ – then ϕ_f admits a unique extension to a rapidly decreasing function on $\mathfrak{j}'(SI)$.

Suppose that $f \in C_c^\infty(\mathfrak{g})$ – then Theorem 8.4.1.2 tells us that $\phi_f \in \mathscr{C}(\mathfrak{j}')$ although in general $\phi_f \notin C_c^\infty(\mathfrak{j}')$. However it is true that the support of ϕ_f in $\mathfrak{j}$ is compact. [Since $\mathfrak{j} \cap G\cdot\text{spt}(f)$ is a relatively compact subset of $\mathfrak{j}$ (cf. 8.1.6), it is enough to verify that the support spt (ϕ_f) of ϕ_f in $\mathfrak{j}$ is contained in $(\mathfrak{j} \cap G\cdot\text{spt}(f))^{cl}$. So fix an element H_0 in spt (ϕ_f) and let $\mathcal{O}(H_0)$ be a neighborhood of H_0 in $\mathfrak{j}$ – then plainly $xH \in$ spt (f) for some $x \in G$ and $H \in \mathcal{O}(H_0) \cap \mathfrak{j}'$ which implies that $\mathcal{O}(H_0) \cap G\cdot\text{spt}(f)$ is not void, whence spt $(\phi_f) \subset (\mathfrak{j} \cap G\cdot\text{spt}(f))^{cl}$, as desired.] Note too that if G has but one conjugacy class of Cartan subgroups, then ϕ_f can be extended in a C^∞ way to all of $\mathfrak{j}$ (cf. the corollary supra). However it is definitely *false* that ϕ_f can, in general, be extended to a continuous function on $\mathfrak{j}$ (let alone C^∞) – this is the case already for $\mathfrak{g} = \mathfrak{sl}(2, \mathbf{R})$ (cf. 8.4.2). On the other hand, it will not be difficult to show that ϕ_f can be extended to a C^∞ function on the closure of each component of $\mathfrak{j}'(SI)$ (i.e. ϕ_f can be extended to each closed 'Weyl chamber' but there may be 'jumps' across each 'chamber' wall).

Although it is in general not true that ϕ_f can be extended to a continuous function on all of $\mathfrak{j}$, nevertheless suitable derivatives of ϕ_f always admit such an extension. Here is the main result along these lines.

Theorem 8.4.1.4 (Harish-Chandra) Fix a point $H_0 \in \mathfrak{j}$ and an element $D \in \mathfrak{P}(\mathfrak{j})$; let $\Phi_{SI}^+(H_0)$ denote the set of all $\alpha \in \Phi_{SI}^+$ such that $\alpha(H_0) = 0$. If $w_\alpha(D) = -D$ for every $\alpha \in \Phi_{SI}^+(H_0)$ (w_α the Weyl reflection corresponding to α), then, for any $f \in C_c^\infty(\mathfrak{g})$, $D\phi_f$ can be extended to a continuous function around H_0 (i.e. there exists a neighborhood $\mathcal{O}(H_0)$ of H_0 in $\mathfrak{j}$ and a continuous function F on $\mathcal{O}(H_0)$ such that $F = D\phi_f$ on $\mathcal{O}(H_0) \cap \mathfrak{j}'$).

The Weyl group W of the pair $(\mathfrak{g}_c, \mathfrak{j}_c)$ operates in the obvious manner on the algebra $\mathfrak{P}(\mathfrak{j})$ of polynomial differential operators on $\mathfrak{j}$; call an element $D \in \mathfrak{P}(\mathfrak{j})$ *skew* if $w_\alpha(D) = -D$ for every root α. Theorem 8.4.1.4

clearly implies that if $D \in \mathfrak{P}(\mathfrak{j})$ is skew, then $D\phi_f$ can be extended to a continuous function on $\mathfrak{j}$ (all $f \in C_c^\infty(\mathfrak{g})$). In such a situation, the value of this extended function at any point $H \in \mathfrak{j}$ will be denoted by $\phi_f(H; D)$ ($f \in C_c^\infty(\mathfrak{g})$).

In the usual way we may agree to view $\partial(\pi)$ as a skew polynomial differential operator on $\mathfrak{j}$ (cf. 8.2.2 – it is, of course, necessary to make here the obvious identifications). Fix an $f \in \mathscr{C}(\mathfrak{g})$ – then, since $C_c^\infty(\mathfrak{g})$ is dense in $\mathscr{C}(\mathfrak{g})$, it follows from Theorems 8.4.1.2 and 8.4.1.4 that $\partial(\pi)\phi_f$ admits a continuous extension to all of $\mathfrak{j}$. With this in mind, we can now formulate a result which is fundamental for subsequent developments (cf. with Theorem 3.2.1.2).

Theorem 8.4.1.5 (Harish-Chandra) There exists a real number $\mathbf{M}_\mathfrak{g}$ such that $\phi_f(0; \partial(\pi)) = \mathbf{M}_\mathfrak{g} f(0)$ (all $f \in \mathscr{C}(\mathfrak{g})$). $\mathbf{M}_\mathfrak{g}$ is non-zero iff $\mathfrak{j}$ is fundamental.

Let us note, right here and now, that $\mathbf{M}_\mathfrak{g}$ is necessarily zero if $\mathfrak{j}$ is not fundamental. [For then the pair $(\mathfrak{g}_c, \mathfrak{j}_c)$ necessarily admits a real root α (say) (cf. Vol. I, number 1.3.3); but it is easy to see that $\phi_f^{w_\alpha} = \phi_f$ for all real α (cf. Remark 4 supra) – thus $(\partial(\pi)\phi_f)^{w_\alpha} = -\partial(\pi)\phi_f$ on $\mathfrak{j}'$ and so $\phi_f(0; \partial(\pi)) = 0$.] However if $\mathfrak{j}$ is fundamental, then $\mathbf{M}_\mathfrak{g}$ is not zero – this will be seen in due course.

Remarks Let us keep to the above notations.

(1) The constant $\mathbf{M}_\mathfrak{g}$ is independent of the choice of the ordering for the roots of the pair $(\mathfrak{g}_c, \mathfrak{j}_c)$.

(2) The constant $\mathbf{M}_\mathfrak{g}$ depends only on the conjugacy class of $\mathfrak{j}$ provided we adhere to the usual conventions regarding the normalization of the various measures involved whenever a conjugate to $\mathfrak{j}$ is under consideration.

We shall conclude this résumé with an application.

Let t be an indeterminate; let $d_l(X)$ be the coefficient of t^l in

$$\det(t - \operatorname{ad}(X)) \qquad (X \in \mathfrak{g}_c).$$

Then d_l is a central polynomial function on $\mathfrak{g}_c$ ($l = \operatorname{rank}(\mathfrak{g}_c)$).

Proposition 8.4.1.6 The function $|d_l|^{-1/2}$ is locally summable on $\mathfrak{g}$.

We shall preface the proof with a simple lemma.

Lemma 8.4.1.7 Let $\mathfrak{j}$ be a θ-stable Cartan subalgebra of $\mathfrak{g}$, $\mathfrak{g}(\mathfrak{j}) = \bigcup_{x \in G} x(\mathfrak{j}')$; let f be a complex valued function on $\mathfrak{j}'$ – then there exists a unique function F_f on $\mathfrak{g}'$ such that:

(i) $F_f(xX) = F_f(X)$ (all $x \in G$, $X \in \mathfrak{g}'$);

(ii) $F_f = [W(G, J)]^{-1}\pi^{-1} \sum_{w \in W(G, J)} \det(w) f^w$ on $\mathfrak{j}'$;

(iii) F_f vanishes outside of $\mathfrak{g}(\mathfrak{j})$.

[Here J denotes the Cartan subgroup of G associated with $\mathfrak{j}$, $W(G, J) =$

J^*/J etc.; the proof of the lemma itself is clear since two points in $\mathfrak{j}'$ are conjugate under G iff they are conjugate under $W(G, J)$.]

Retain the above notations – then, since the mapping $(\dot{x}, H) \mapsto \dot{x}H$ of $G/J \times \mathfrak{j}'$ into $\mathfrak{g}$ is a local homeomorphism, it follows that F_f is measurable iff f is measurable. Furthermore, if f is locally summable on $\mathfrak{j}$, then F_f is locally summable on $\mathfrak{g}$ (evidently f is defined almost everywhere on $\mathfrak{j}$ while F_f is defined almost everywhere on $\mathfrak{g}$, the singular elements being of measure zero). [Given a compact set ω in $\mathfrak{g}$, choose a function f_ω in $C_c^\infty(\mathfrak{g})$ such that $f_\omega \geqslant 0$ and $f_\omega \geqslant 1$ on ω – then

$$\int_\omega |F_f| \, d_{\mathfrak{g}}(X) \leqslant \int_{\mathfrak{g}(\mathfrak{j})} |F_f| f_\omega d_{\mathfrak{g}}(X)$$

and

$$\int_{\mathfrak{g}(\mathfrak{j})} |F_f(X)| f_\omega(X) d_{\mathfrak{g}}(X) \leqslant [W(G, J)]^{-1} \sum_{w \in W(G, J)} \int_{\mathfrak{j}} |\phi_{f_\omega}(H)| \, |f^w(H)| d_{\mathfrak{j}}(H)$$
$$= \int_{\mathfrak{j}} |\phi_{f_\omega}(H)| \, |f(H)| d_{\mathfrak{j}}(H) < \infty,$$

the finiteness of the last integral being clear on the basis of the discussion which follows the statement of Theorem 8.4.1.2.]

Proof of Proposition 8.4.1.6 Let $\mathfrak{j}^i$ $(i = 1, \ldots, r)$ be a maximal set of θ-stable Cartan subalgebras of $\mathfrak{g}$, no two of which are conjugate under G; put $\mathfrak{g}^i = \mathfrak{g}(\mathfrak{j}^i)$ and let ψ_i be the characteristic function of the open set $\mathfrak{g}^i$ $(i = 1, \ldots, r)$ – then $\sum_i \psi_i = 1$ almost everywhere on $\mathfrak{g}$ (since $\mathfrak{g}'$ is the disjoint union of the $\mathfrak{g}^i$). To prove our proposition, it will therefore be enough to verify that $\psi_i |d_l|^{-1/2}$ is locally summable on $\mathfrak{g}^i$ $(i = 1, \ldots, r)$. This being so, fix i and set $\mathfrak{j} = \mathfrak{j}^i$, $\psi = \psi_i$; in Lemma 8.4.1.7, take $f = \pi/|\pi|$ (on $\mathfrak{j}'$) – then it is clear that $F_f = \psi |d_l|^{-1/2}$ and hence, in view of what has been said above, $\psi |d_l|^{-1/2}$ is locally summable on $\mathfrak{g}$. □

8.4.2 Computations in $\mathfrak{sl}(2, \mathbf{R})$

In this number we shall study, by a direct method, the invariant integral on $\mathfrak{sl}(2, \mathbf{R})$; as will be seen in due course, the proofs of certain important properties of the invariant integral on a general reductive $\mathfrak{g}$ can frequently be reduced to this special case.

We shall begin by making a few elementary calculations the significance of which will appear shortly.

Given $f \in C_c^\infty(\mathbf{R}^2)$, put

$$\psi_f(\xi) = \xi \int_0^\infty f(\xi e^t, \xi e^{-t})(e^t - e^{-t}) \, dt \qquad (\xi \in \mathbf{R}' = \mathbf{R} - \{0\}).$$

Clearly ψ_f is a C^∞ function on $\mathbf{R}'$. Let (x_1, x_2) denote the usual coordinates in $\mathbf{R}^2$; write f_1 for $\partial f/\partial x_1$, f_2 for $\partial f/\partial x_2$.

Lemma 8.4.2.1 Retain the above notations – then $\lim_{\xi\to 0} d\psi_f(\xi)/d\xi = -2f(0, 0)$.

Proof By differentiation, one finds that

$$\begin{aligned} d\psi_f(\xi)/d\xi &= \int_0^\infty (f + \xi e^t f_1 + \xi e^{-t} f_2)(e^t - e^{-t})\,dt \\ &= \int_0^\infty (e^t f + \xi e^{2t} f_1)\,dt - \int_0^\infty e^{-t} f\,dt \\ &\quad + \int_0^\infty (\xi f_2 - \xi f_1 - \xi e^{-2t} f_2)\,dt. \end{aligned}$$

Since $|f(x_1, x_2)| \leqslant \sup_{\mathbf{R}^2}|f|$ and $\int_0^\infty e^{-t}\,dt = 1$, it follows from a Theorem of Lebesgue that $\lim_{\xi\to 0}\int_0^\infty e^{-t} f\,dt = f(0, 0)$. Similarly $\lim_{\xi\to 0}\xi\int_0^\infty e^{-2t} f_2\,dt = 0$. Choose a $T > 0$ with the property that $|f(x_1, x_2)| = 0$ unless $|x_1| + |x_2| \leqslant T$ – then, in particular, $f_2(\xi e^t, \xi e^{-t}) = 0$ unless $|\xi| e^t \leqslant T$ and so

$$|\xi| e^t \leqslant T \Rightarrow \int_0^\infty |f_2|\,dt \leqslant \sup_{\mathbf{R}^2}|f_2|\log(T/|\xi|) \Rightarrow \lim_{\xi\to 0}\xi\int_0^\infty f_2\,dt = 0.$$

Likewise $\lim_{\xi\to 0}\xi\int_0^\infty f_1\,dt = 0$. It remains to show, therefore, that

$$\lim_{\xi\to 0}\int_0^\infty (e^t f + \xi e^{2t} f_1)\,dt = -f(0, 0).$$

If $x_1 = \xi e^t$, $x_2 = \xi e^{-t}$, then $d(x_1 f)/dt = (f + x_1 f_1)\xi e^t - \xi^2 f_2$, whence

$$e^t f + \xi e^{2t} f_1 = (f + x_1 f_1)e^t = \xi^{-1}(d(x_1 f)/dt) + \xi f_2.$$

Because $\xi^{-1}\int_0^\infty (d(x_1 f)/dt)\,dt = -\xi^{-1}(x_1 f)_{t=0} = -f(\xi, \xi)$, the assertion of the lemma is now evident. □

Corollary 8.4.2.2 Retain the above notations – then $d\psi_f(\xi)/d\xi$ can be extended to a continuous function on **R**.

Lemma 8.4.2.3 Retain the above notations – then

$$\lim_{\xi\downarrow 0}\psi_f(\xi) = \int_0^\infty f(x_1, 0)\,dx_1, \quad \lim_{\xi\uparrow 0}\psi_f(\xi) = -\int_{-\infty}^0 f(x_1, 0)\,dx_1.$$

Proof It was noted during the proof of Lemma 8.4.2.1 that $\xi\int_0^\infty f e^{-t}\,dt \to 0$ as $\xi \to 0$; thus it suffices to consider the integral $\xi\int_0^\infty f e^t\,dt$. First assume that $0 < \xi \leqslant T$ – then

$$\xi\int_0^\infty f e^t\,dt = \int_\xi^\infty f(x_1, \xi^2 x_1^{-1})\,dx_1 = \int_\xi^T f(x_1, \xi^2 x_1^{-1})\,dx_1,$$

and so from Lebesgue's Theorem, it follows that

$$\lim_{\xi\downarrow 0}\int_\xi^\infty f(x_1, \xi^2 x_1^{-1})\,dx_1 = \int_0^T f(x_1, 0)\,dx_1 = \int_0^\infty f(x_1, 0)\,dx_1.$$

The case $\xi < 0$ is treated analogously. □

[The reader will note that the above lemma shows that ψ_f cannot, in general, be extended to a continuous function on **R**.]

Let $\mathfrak{l}$ be the simple non-compact Lie algebra of dimension 3 spanned over $\mathbf{R}$ by the elements H, X, Y which satisfy the relations $[H, X] = 2X, [H, Y] = -2Y, [X, Y] = H$ – we shall then agree to subscribe to the notations and results of the Example following Corollary 1.3.4.2.

Let L denote a connected Lie group with Lie algebra $\mathfrak{l}$ – then, according to our conventions, L has finite center and thus covers Int $(\mathfrak{l})$ a finite number of times, m say.

The set $\mathfrak{h}^- = \{r(X - Y) : r \in \mathbf{R}\}$ is a fundamental Cartan subalgebra of $\mathfrak{l}$; there are two roots of the pair $(\mathfrak{l}_c, \mathfrak{h}_c^-)$, α_- and $-\alpha_-$, where $\alpha_-(X - Y) = -2\sqrt{-1}$; take α_- to be the positive root – then α_- is singular imaginary and, in the usual notations, $H_{\alpha_-} = (\sqrt{-1}/4)(X - Y)$ (recall that $\omega(X - Y) = -8$, ω the Casimir polynomial of $\mathfrak{l}_c$). The map $\{H, X, Y\} \mapsto \{-H, -Y, -X\}$ is a Cartan involution θ of $\mathfrak{l}$ such that $\theta(\mathfrak{h}^-) = \mathfrak{h}^-$ – in fact here $\mathfrak{h}^- = \mathfrak{h}_{\mathfrak{k}}^- = \mathfrak{k}$.

Proposition 8.4.2.4 Let $f \in C_c^\infty(\mathfrak{l})$; if ϕ_f^- is the invariant integral of f relative to the fundamental Cartan subalgebra $\mathfrak{h}^-$, then $\phi_f^-(0; \partial(\pi)) = -2\pi f(0)$.

[Thus this result verifies Theorem 8.4.1.5 in the special case when $\mathfrak{g} = \mathfrak{sl}(2, \mathbf{R})$.]

The proof of Proposition 8.4.2.4 is contained in the following discussion.

Equip the space $\mathfrak{l}$ with the Euclidean structure $(\,.\,,\,.\,)_\theta$ determined by θ. An orthonormal basis for this structure is given by

$$X_1 = \frac{X - Y}{2\sqrt{2}}, \quad X_2 = \frac{X + Y}{2\sqrt{2}}, \quad X_3 = \frac{H}{2\sqrt{2}};$$

if x_1, x_2, x_3 are the associated Euclidean coordinates, then $dx_1dx_2dx_3$ will denote the corresponding canonical element of volume. On the other hand, it will also be convenient to compute in $\mathfrak{l}$ relative to the measure $dZ = dtdrds$ where $Z = tH + rX + sY$ ($Z \in \mathfrak{l}$; $t, r, s \in \mathbf{R}$). To determine the positive scalar factor by which $dtdrds$ and $dx_1dx_2dx_3$ differ, write

$$H = 2\sqrt{2}X_3, \quad X = \sqrt{2}X_1 + \sqrt{2}X_2, \quad Y = -\sqrt{2}X_1 + \sqrt{2}X_2;$$

the absolute value of the determinant of the matrix

$$\begin{pmatrix} 0 & 0 & 2\sqrt{2} \\ \sqrt{2} & \sqrt{2} & 0 \\ -\sqrt{2} & \sqrt{2} & 0 \end{pmatrix}$$

is $8\sqrt{2}$, whence $dtdrds = (8\sqrt{2})^{-1}dx_1dx_2dx_3$.

Let K denote the analytic subgroup of G corresponding to $\mathfrak{k}(= \mathfrak{h}^-)$; the Haar measure on K is to be normalized so that $\int_K dk = 1$.

Fix an $f \in C_c^\infty(\mathfrak{l})$ – then the invariant integral ϕ_f^- of f relative to $\mathfrak{h}^-$ may be viewed as a function on $\mathbf{R}' = \mathbf{R} - \{0\}$. Thus, if $x_1 \neq 0$, we have

$$\begin{aligned} \phi_f^-(x_1) = \phi_f^-(x_1X_1) &= \pi(x_1X_1)\int_L f(\mathrm{Ad}\,(l)(x_1X_1))\,d_L(l) \\ &= -\frac{\sqrt{-1}}{\sqrt{2}}x_1\int_L f(\mathrm{Ad}\,(l)(x_1X_1))\,d_L(l), \end{aligned}$$

d_L the Haar measure on L normalized as in 8.1.2. Put $f_K(Z) = \int_K f(\mathrm{Ad}\,(k)Z)dk$, $Z \in \mathfrak{l}$. A standard integration formula and a little computation gives

$$\begin{aligned} \phi_f^-(x_1) = -\left(\frac{c^+\sqrt{-1}}{\sqrt{2}}\right)x_1\int_0^\infty f_K\left(\frac{x_1}{2\sqrt{2}}(e^{x_3/\sqrt{2}}X - e^{-x_3/\sqrt{2}}Y)\right) \\ \times (e^{x_3/\sqrt{2}} - e^{-x_3/\sqrt{2}})\,dx_3, \end{aligned}$$

c^+ a certain positive constant which will be determined below. Set $\psi_f^-(\xi) = \xi \int_0^\infty f_K(\xi(e^t X - e^{-t} Y))(e^t - e^{-t})\, dt$ ($\xi \in \mathbf{R}'$, dt the usual Lebesgue measure on $\mathbf{R}$). Replacing x_1 by ξ, $x_3/\sqrt{2}$ by t in the last equality above leads to the relation

$$\phi_f^-(\xi) = (-c^+\sqrt{-1})(2\sqrt{2})\psi_f^-(\xi/2\sqrt{2}).$$

Our next task will be to find the exact value of c^+. For this purpose a preliminary remark is needed. Consider the mapping $(\theta, t, \xi) \mapsto (t', r', s')$ defined by the rule

$$\mathrm{Ad}(\exp(\theta(X - Y)))\,\mathrm{Ad}(\exp(tH))(\xi(X - Y)) = t'H + r'X + s'Y$$

where $0 \leqslant \theta < m\pi$, $t > 0$, $\xi \in \mathbf{R}'$. By computation, one finds that the absolute value of the Jacobian of this mapping is given by

$$|\partial(t', r', s')/\partial(\theta, t, \xi)| = c\xi^2(e^{2t} - e^{-2t}),$$

c a certain positive constant. In order to obtain c, we evaluate the left hand side at $\theta = 0$ – thus $\xi\,\mathrm{Ad}(\exp(tH))(X - Y) = \xi(e^{2t}X - e^{-2t}Y)$ and so

$$\begin{aligned}&\partial/\partial\theta\{\mathrm{Ad}(\exp(\theta(X - Y)))(\xi(e^{2t}X - e^{-2t}Y))\}|_{\theta=0}\\ &= \xi[X - Y, e^{2t}X - e^{-2t}Y] = \xi(e^{2t} - e^{-2t})H.\end{aligned}$$

Therefore

$$\begin{aligned}\partial(t', r', s')/\partial(\theta, t, \xi)|_{\theta=0} &= \det\begin{pmatrix}\xi(e^{2t} - e^{-2t}) & 0 & 0\\ 0 & 2\xi e^{2t} & 2\xi e^{-2t}\\ 0 & e^{2t} & -e^{-2t}\end{pmatrix}\\ &= -4\xi^2(e^{2t} - e^{-2t}) \Rightarrow c = 4.\end{aligned}$$

It was observed during the course of the discussion in the Example referred to above that $\mathfrak{l}^- = \mathfrak{l}(\mathfrak{h}^-) = \mathfrak{l}_2 \cup \mathfrak{l}_3$ (disjoint union); in fact every element in $\mathfrak{l}_2$ (respectively $\mathfrak{l}_3$) may be written in the form

$$\mathrm{Ad}(\exp(\theta(X - Y)))\mathrm{Ad}(\exp(tH))(\xi(X - Y))$$

with $0 \leqslant \theta < m\pi$, $t \geqslant 0$, $\xi > 0$ (respectively $\xi < 0$) and, if $t \neq 0$, this can be done in exactly m ways.

Now on the one hand we have

$$\int_{\mathfrak{l}^-} f(x_1, x_2, x_3)\, dx_1 dx_2 dx_3 = \frac{\sqrt{-1}}{\sqrt{2}} \int_{-\infty}^{\infty} \xi\phi_f^-(\xi)\, d\xi$$

while on the other

$$\begin{aligned}\frac{1}{8\sqrt{2}} \int_{\mathfrak{l}^-} f(x_1, x_2, x_3)\, dx_1 dx_2 dx_3 &= \int_{\mathfrak{l}^-} f(t, r, s)\, dt dr ds\\ &= \int_{\mathfrak{l}_2} f(t, r, s)\, dt dr ds + \int_{\mathfrak{l}_3} f(t, r, s)\, dt dr ds.\end{aligned}$$

Plainly

$$\begin{aligned}&\int_{\mathfrak{l}_2} f(t, r, s)\, dt dr ds\\ &= \frac{4\pi}{m\pi} \int_0^{m\pi} d\theta \int_0^\infty \xi^2 d\xi \int_0^\infty f(\mathrm{Ad}(\exp(\theta(X - Y)))e^{t\,\mathrm{ad}(H)}(\xi(X - Y)))\\ &\quad \times (e^{2t} - e^{-2t})\, dt\\ &= 2\pi \int_0^\infty \xi\psi_f^-(\xi)\, d\xi;\end{aligned}$$

similarly we find that

$$\int_{\mathfrak{l}_3} f(t, r, s)\, dtdrds = 2\pi \int_{-\infty}^{0} \xi \psi_{\bar{f}}(\xi)\, d\xi.$$

Hence

$$\int_{\mathfrak{l}^-} f(x_1, x_2, x_3)\, dx_1 dx_2 dx_3 = 16\pi\sqrt{2} \int_{-\infty}^{\infty} \xi \psi_{\bar{f}}(\xi)\, d\xi$$

$$= 2\pi\sqrt{2} \int_{-\infty}^{\infty} \xi \psi_{\bar{f}}(\xi/2\sqrt{2})\, d\xi = (-\pi/c^+\sqrt{-1}) \int_{-\infty}^{\infty} \xi \phi_{\bar{f}}(\xi)\, d\xi$$

$$= \frac{\sqrt{-1}}{\sqrt{2}} \int_{-\infty}^{\infty} \xi \phi_{\bar{f}}(\xi)\, d\xi \Rightarrow c^+ = \sqrt{2}\pi.$$

We are now in a position to show that

$$\phi_{\bar{f}}(0; \partial(\pi)) = -2\pi f(0)$$

for all $f \in C_c^\infty(\mathfrak{l})$ thus completing the proof of Proposition 8.4.2.4. Owing to Lemma 8.4.2.1 supra, $\lim_{\xi\to 0} d\psi_{\bar{f}}(\xi)/d\xi = -2f(0)$ $(f \in C_c^\infty(\mathfrak{l}))$. But $H_{\alpha_-} = \sqrt{-1}(\sqrt{2}/2)X_1$, whence

$$\phi_{\bar{f}}(0; \partial(\pi)) = \lim_{\xi\to 0} \sqrt{-1}\, \frac{\sqrt{2}}{2} \frac{d}{d\xi} \phi_{\bar{f}}(\xi) = \pi \lim_{\xi\to 0} \frac{d}{d\xi} \psi_{\bar{f}}(\xi) = -2\pi f(0).$$

Let us note in passing, that $\psi_{\bar{f}}$ (or $\phi_{\bar{f}}$) cannot, in general, be extended to a continuous function on all of $\mathbf{R}$ $(f \in C_c^\infty(\mathfrak{l}))$ – this is clear in view of Lemma 8.4.2.3. On the other hand one does have the following result.

Lemma 8.4.2.5 Let $f \in C_c^\infty(\mathfrak{l})$; let n be a non-negative integer – then $d^{2n+1}\psi_{\bar{f}}(\xi)/d\xi^{2n+1}$ can be extended to a continuous function on $\mathbf{R}$.

Proof Put $f_n = \partial(\omega^n)f$ – then

$$d^{2n+1}\psi_{\bar{f}}(\xi)/d\xi^{2n+1} = (-1)^n 2^{3n} d\psi_{\bar{f}_n}(\xi)/d\xi$$

and so our assertion follows from Corollary 8.4.2.2. □

[The following technical observation will be needed in 8.4.5. Let $\mathfrak{g}$ be a semi-simple Lie algebra over $\mathbf{R}$, $B_{\mathfrak{g}}$ the Killing form of $\mathfrak{g}$; let $\theta_{\mathfrak{g}}$ be a Cartan involution of $\mathfrak{g}$, $\mathfrak{j}$ a $\theta_{\mathfrak{g}}$-stable Cartan subalgebra of $\mathfrak{g}$ – we shall assume that the pair $(\mathfrak{g}_c, \mathfrak{j}_c)$ admits a positive singular imaginary root α (say). It will be supposed in addition that there exists a $\theta_{\mathfrak{g}}$-stable non-compact three dimensional Lie algebra $\mathfrak{l}$ such that $\mathfrak{j}_{\mathfrak{k}} \subset \mathfrak{l}$ (hence $\mathfrak{j}_{\mathfrak{k}} = \mathfrak{l}_{\mathfrak{k}}$ and $\mathfrak{j}_{\mathfrak{k}}$ is one dimensional). Pick elements X, Y, H in $\mathfrak{l}$ in the usual way; let $\theta_{\mathfrak{l}}$ be the Cartan involution of $\mathfrak{l}$ determined by the requirement $\{H, X, Y\} \mapsto \{-H, -Y, -X\}$; it will be assumed that $\theta_{\mathfrak{l}} = \theta_{\mathfrak{g}} | \mathfrak{l}$ (cf. the discussion following Lemma 1.3.4.4) – then $\mathfrak{l}$ may be viewed as a Euclidean space in two ways, namely through the structures $(.\,,.)_{\theta_{\mathfrak{g}}} | \mathfrak{l}$ and $(.\,,.)_{\theta_{\mathfrak{l}}}$. Our first goal is to determine the positive scalar factor by which the corresponding Euclidean measures on $\mathfrak{l}$ must differ. To this end, let $H_{\alpha,\mathfrak{g}}$ be that vector in $\mathfrak{j}_{\mathfrak{k}_c}$ such that $\alpha(H) = B_{\mathfrak{g}}(H, H_{\alpha,\mathfrak{g}})$ (all $H \in \mathfrak{j}_{\mathfrak{k}_c}$); similarly let $H_{\alpha,\mathfrak{l}}$ be that vector in $\mathfrak{j}_{\mathfrak{k}_c}$ such that $\alpha(H) = B_{\mathfrak{l}}(H, H_{\alpha,\mathfrak{l}})$ (all $H \in \mathfrak{j}_{\mathfrak{k}_c}$) ($B_{\mathfrak{l}}$ the Killing form of $\mathfrak{l}$) – then it is clear that $H_{\alpha,\mathfrak{g}}/\alpha(H_{\alpha,\mathfrak{g}}) = H_{\alpha,\mathfrak{l}}/\alpha(H_{\alpha,\mathfrak{l}})$ (co-roots . . .). Because the complexification of $\mathfrak{l}$ is simple, there exists a constant r such that $B_{\mathfrak{g}} | \mathfrak{l} = rB_{\mathfrak{l}}$. We claim that $r = \alpha(H_{\alpha,\mathfrak{l}})/\alpha(H_{\alpha,\mathfrak{g}})$ (> 0). Thus let $H \in \mathfrak{j}_{\mathfrak{k}}$ – then $B_{\mathfrak{g}}(H, H_{\alpha,\mathfrak{g}}) = \alpha(H) = B_{\mathfrak{l}}(H, H_{\alpha,\mathfrak{l}})$, whence $rB_{\mathfrak{l}}(H, H_{\alpha,\mathfrak{g}}) = B_{\mathfrak{l}}(H, H_{\alpha,\mathfrak{l}})$ and so the claim is evident. Let $s = r^{-1}$. If X_1, X_2, X_3 is, as above, an orthonormal basis for the structure $(.\,,.)_{\theta_{\mathfrak{l}}}$, then clearly $\tilde{X}_1 = \sqrt{s}\,X_1$, $\tilde{X}_2 = \sqrt{s}\,X_2$, $\tilde{X}_3 = \sqrt{s}\,X_3$ is an orthonormal basis in $\mathfrak{l}$ for the structure $(.\,,.)_{\theta_{\mathfrak{g}}}|\mathfrak{l}$. If $\tilde{x}_1, \tilde{x}_2, \tilde{x}_3$ are the corresponding Euclidean coordinates, $d\tilde{x}_1 d\tilde{x}_2 d\tilde{x}_3$ the associated canonical element of volume, then $d\tilde{x}_1 d\tilde{x}_2 d\tilde{x}_3 = s^{-3/2} dx_1 dx_2 dx_3$. Let $\tilde{d}_L$ be that normalization of the Haar measure on L so that

$$\int_{\mathfrak{l}^-} f(\tilde{x}_1, \tilde{x}_2, \tilde{x}_3)\, d\tilde{x}_1 d\tilde{x}_2 d\tilde{x}_3 = \int_{L\times\mathbf{R}'} |\pi(\tilde{x}_1 \tilde{X}_1)|^2 f(\mathrm{Ad}\,(l)(\tilde{x}_1 \tilde{X}_1))\, \tilde{d}_L(l) d\tilde{x}_1$$

for all $f \in C_c(\mathfrak{l}^-)$. Given $f \in C_c^\infty(\mathfrak{l})$, put

$$\tilde{\phi}_f^-(\tilde{x}_1) = \tilde{\phi}_f^-(\tilde{x}_1 \tilde{X}_1) = \pi(\tilde{x}_1 \tilde{X}_1) \int_L f(\mathrm{Ad}\,(l)(\tilde{x}_1 \tilde{X}_1))\, \tilde{d}_L(l).$$

Then we claim that $\tilde{\phi}_f^-(0; \partial(H_{\alpha,\mathfrak{g}})) = -2\pi f(0)$ (all $f \in C_c^\infty(\mathfrak{l})$). To see this it will first be necessary to calculate the scalar factor by which d_L and $\tilde{d}_L$ differ. Let $f \in C_c(\mathfrak{l}^-)$ be arbitrary; on the one hand we have

$$\int_{\mathfrak{l}^-} f(\tilde{x}_1, \tilde{x}_2, \tilde{x}_3)\, d\tilde{x}_1 d\tilde{x}_2 d\tilde{x}_3 = s^{-1/2} \int_{L\times\mathbf{R}'} |\pi(x_1 X_1)|^2 f(\mathrm{Ad}\,(l)(x_1 X_1))\, \tilde{d}_L(l)\, dx_1$$

while on the other

$$\begin{aligned}\int_{\mathfrak{l}^-} f(\tilde{x}_1, \tilde{x}_2, \tilde{x}_3) d\tilde{x}_1 d\tilde{x}_2 d\tilde{x}_3 &= s^{-3/2} \int_{\mathfrak{l}^-} f(x_1, x_2, x_3) dx_1 dx_2 dx_3 \\ &= s^{-3/2} \int_{L\times\mathbf{R}'} |\pi(x_1 X_1)|^2 f(\mathrm{Ad}\,(l)(x_1 X_1))\, d_L(l)\, dx_1,\end{aligned}$$

whence $\tilde{d}_L = s^{-1} d_L$. Now let $f \in C_c^\infty(\mathfrak{l})$ be arbitrary – then $\tilde{\phi}_f^-(\tilde{x}_1) = s^{-1}\phi_f^-(x_1)$ and so, since $H_{\alpha,\mathfrak{g}} = sH_{\alpha,\mathfrak{l}}$, it must be the case that

$$\tilde{\phi}_f^-(0; \partial(H_{\alpha,\mathfrak{g}})) = -2\pi f(0),$$

as contended.]

The set $\mathfrak{h}^+ = \{rH : r \in \mathbf{R}\}$ is a Cartan subalgebra of $\mathfrak{l}$ which is not conjugate to $\mathfrak{h}^-$; there are two roots of the pair $(\mathfrak{l}_c, \mathfrak{h}_c^+)$, α_+ and $-\alpha_+$, where $\alpha_+(H) = 2$; take α_+ to be the positive root – then α_+ is real and, in the usual notations, $H_{\alpha_+} = 4^{-1}H$ (since $\omega(H) = 8$, ω the Casimir polynomial of $\mathfrak{l}_c$).

Let H^+ be the Cartan subgroup of L associated with $\mathfrak{h}^+$; we shall agree to normalize the invariant measure on L/H^+ according to our customary convention. [In passing observe that H^+ is, in the case at hand, abelian (although not necessarily connected).]

Fix an $f \in C_c^\infty(\mathfrak{l})$ – then the invariant integral ϕ_f^+ of f, calculated relative to $\mathfrak{h}^+$, may be viewed as a function on $\mathbf{R}'$. Thus, if $x_3 \neq 0$, then

$$\begin{aligned}\phi_f^+(x_3) = \phi_f^+(x_3 X_3) &= \pi(x_3 X_3)\epsilon_R(x_3 X_3) \int_{L/\mathsf{H}^+} f(\dot{l}(x_3 X_3))\, d_{L/\mathsf{H}^+}(\dot{l}) \\ &= \frac{|x_3|}{\sqrt{2}} \int_{L/\mathsf{H}^+} f(\dot{l}(x_3 X_3))\, d_{L/\mathsf{H}^+}(\dot{l}).\end{aligned}$$

Put $\psi_f^+(\xi) = \int_{-\infty}^{\infty} f_K(\xi H + rX) dr\, (\xi \in \mathbf{R})$ – then, thanks to what has been said in 8.1.3, it is clear that there exists a positive constant c^+ such that, upon writing ξ for x_3,

$$\phi_f^+(\xi) = c^+ \int_{-\infty}^{\infty} f_K\left(\frac{\xi}{2\sqrt{2}} H + rX\right) dr = c^+ \psi_f^+(\xi/2\sqrt{2}) \qquad (\xi \in \mathbf{R}').$$

In order to determine the exact value of c^+, a preliminary remark is needed.

Consider the mapping $(\theta, \xi, r) \mapsto (t', r', s')$ defined by the rule

$$\mathrm{Ad}(\exp(\theta(X - Y)))(\xi H + rX) = t'H + r'X + s'Y$$

where $0 \leqslant \theta < m\pi$, $\xi > 0$, $r \in \mathbf{R}$. It follows from general considerations (or by direct computation) that

$$|\partial(t', r', s')/\partial(\theta, \xi, r)| = c\xi,$$

c a certain positive constant. To compute c, let us evaluate the Jacobian at $\theta = 0$ – thus

$$\begin{aligned}\partial/\partial\theta\,\{\mathrm{Ad}(\exp(\theta(X - Y)))(\xi H + rX)\}|_{\theta=0} &= [X - Y, \xi H + rX]\\ &= rH - 2\xi X - 2\xi Y\end{aligned}$$

and so

$$\partial(t', r', s')/\partial(\theta, \xi, r)|_{\theta=0} = \det\begin{pmatrix} r & -2\xi & -2\xi \\ 1 & 0 & 0 \\ 0 & 1 & 0\end{pmatrix} = -2\xi \Rightarrow c = 2.$$

Write $L = KN^+A_{\mathfrak{p}}$, $\mathfrak{l}^+ = \mathfrak{l}(\mathfrak{h}^+)$ – then it is clear that every element in $\mathfrak{l}^+$ can be written in the form $t(knH)$ ($k \in K$, $n \in N^+$, $t \in \mathbf{R}^+$), that is, every element in $\mathfrak{l}^+$ can be written in the form

$$\mathrm{Ad}(\exp(\theta(X - Y)))(\xi H + rX)$$

where $0 \leqslant \theta < m\pi$, $\xi > 0$, $r \in \mathbf{R}$ and this can be done in exactly m ways.

Returning to our problem of computing c^+, note first that we have (since $\mathfrak{l}^+ = \mathfrak{l}_1$)

$$\int_{\mathfrak{l}^+} f(x_1, x_2, x_3)\, dx_1 dx_2 dx_3 = \frac{1}{\sqrt{2}}\int_{-\infty}^{\infty} |\xi|\,\phi_f^+(\xi)\, d\xi = \frac{2}{\sqrt{2}}\int_0^{\infty} \xi\phi_f^+(\xi)\, d\xi;$$

on the other hand

$$\begin{aligned}&\frac{1}{8\sqrt{2}}\int_{\mathfrak{l}^+} f(x_1, x_2, x_3)\, dx_1 dx_2 dx_3\\ &= \int_{\mathfrak{l}^+} f(t, r, s)\, dt dr ds\\ &= \frac{2\pi}{m\pi}\int_0^{m\pi} d\theta \int_0^{\infty} \xi\, d\xi \int_{-\infty}^{\infty} f(\mathrm{Ad}(\exp(\theta(X - Y)))(\xi H + rX))\, dr\\ &= 2\pi\int_0^{\infty} \xi\, d\xi \int_{-\infty}^{\infty} f_K(\xi H + rX)\, dr = 2\pi\int_0^{\infty} \xi\psi_f^+(\xi)\, d\xi\end{aligned}$$

from which we deduce at once that $c^+ = 2\pi$.

Given an $f \in C_c^\infty(\mathfrak{l})$, define ψ_f^+ and ψ_f^- as above; put $R(f) = \lim_{\xi \downarrow 0} \psi_f^-(\xi)$, $L(f) = \lim_{\xi \uparrow 0} \psi_f^-(\xi)$ – then, on the basis of Lemma 8.4.2.3, it is clear that $R(f) = \int_0^\infty f_K(rX)\, dr$, $L(f) = -\int_{-\infty}^0 f_K(rX)\, dr$ and so

$$\psi_f^+(0) = \int_{-\infty}^{\infty} f_K(rX)\, dr = R(f) - L(f).$$

Lemma 8.4.2.6 Let $\mathcal{O}$ be an open neighborhood of zero in $\mathfrak{l}$; let c, r, and l be three complex numbers; suppose that

$$cf(0) + rR(f) + lL(f) = 0$$

for all $f \in C_c^\infty(\mathcal{O})$ – then $c = r = l = 0$.

Proof Because $\mathrm{Ad}(\exp(\theta(X - Y)))(X) = (\cos\theta \sin\theta)H + (\cos\theta)^2 X - (\sin\theta)^2 Y$ ($\theta \in \mathbf{R}$), it is clearly possible to choose $f \in C_c^\infty(\mathcal{O})$ such that $f(0) = L(f) = 0$ while $R(f) \neq 0$; consequently $r = 0$. In a similar way one finds that $l = 0$. But then it is obvious that $c = 0$. □

Lemma 8.4.2.7 Let w, u be two C^∞ functions on the closed interval $[0, \infty)$ and v a C^∞ function on the closed interval $(-\infty, 0]$; for $f \in C_c^\infty(\mathfrak{l})$, put

$$T(f) = \int_0^\infty (D_\xi^2 \psi_f^+ \cdot w - \psi_f^+ \cdot D_\xi^2 w)\, d\xi - \int_0^\infty (D_\xi^2 \psi_f^- \cdot u - \psi_f^- \cdot D_\xi^2 u)\, d\xi$$
$$- \int_{-\infty}^0 (D_\xi^2 \psi_f^- \cdot v - \psi_f^- \cdot D_\xi^2 v)\, d\xi$$

where $D_\xi = d/d\xi$ – then T is a central distribution on $\mathfrak{l}$ and the following three conditions on T are mutually equivalent:

(i) $T = 0$;
(ii) $0 \notin \mathrm{spt}(T)$;
(iii) $u(0) = v(0)$ and $(D_\xi w)_0 = (D_\xi u)_0 = (D_\xi v)_0$.
[Here the subscript 0 denotes the value at zero.]

Proof It is obvious that T is a central distribution on $\mathfrak{l}$. This being so, fix an f in $C_c^\infty(\mathfrak{l})$ – then a simple integration leads to the relation

$$T(f) = -(D_\xi \psi_f^+ \cdot w - \psi_f^+ \cdot D_\xi w)_0 + \lim_{\xi \downarrow 0} (D_\xi \psi_f^- \cdot u - \psi_f^- \cdot D_\xi u)$$
$$- \lim_{\xi \uparrow 0} (D_\xi \psi_f^- \cdot v - \psi_f^- \cdot D_\xi v).$$

Since $(D_\xi \psi_f^+)_0 = 0$ and $(D_\xi \psi_f^-)_0 = -2f(0)$, we have

$$T(f) = R(f)\{(D_\xi w)_0 - (D_\xi u)_0\} - L(f)\{(D_\xi w)_0 - (D_\xi v)_0\}$$
$$- 2f(0)\{u(0) - v(0)\}$$

and so our assertion follows immediately from Lemma 8.4.2.6. □

Corollary 8.4.2.8 Retain the notations of Lemma 8.4.2.7; put $w_n = 2^{-3n} D_\xi^{2n} w$, $u_n = (-1)^n 2^{-3n} D_\xi^{2n} u$, $v_n = (-1)^n 2^{-3n} D_\xi^{2n} v$, $f_n = \partial(\omega^n) f$ and

$$T_n(f) = \int_0^\infty (\psi_{f_n}^+ \cdot w - \psi_f^+ \cdot w_n)\, d\xi + \int_0^\infty (\psi_{f_n}^- \cdot u - \psi_f^- \cdot u_n)\, d\xi$$
$$+ \int_{-\infty}^0 (\psi_{f_n}^- \cdot v - \psi_f^- \cdot v_n)\, d\xi$$

for $f \in C_c^\infty(\mathfrak{l})$ $(n = 0, 1, 2, \ldots)$ – then the following three conditions are equivalent:

(i) $T_n = 0$ (all $n \in \mathbf{N}$);
(ii) $0 \notin \mathrm{spt}(T_n)$ (all $n \in \mathbf{N}$);
(iii) $(D_\xi^{2n} u)_0 = (D_\xi^{2n} v)_0$ and $(-1)^n (D_\xi^{2n+1} w)_0 = (D_\xi^{2n+1} u)_0 = (D_\xi^{2n+1} v)_0$ $(n \geqslant 0)$.

Furthermore, if u and v are defined and analytic on $\mathbf{R}$ and $0 \notin \mathrm{spt}(T_n)$ (all $n \in \mathbf{N}$), then $u = v$.

[We shall leave the proof of the corollary for the reader to work out.]

8.4.3 Continuity of the Map $f \mapsto \phi_f$

Let $\mathfrak{g}$ be a reductive Lie algebra over $\mathbf{R}$, $\theta : \mathfrak{g} \to \mathfrak{g}$ an involutive automorphism of $\mathfrak{g}$ whose restriction to the derived algebra $\bar{\mathfrak{g}} = [\mathfrak{g}, \mathfrak{g}]$ is a Cartan involution (of $\bar{\mathfrak{g}}$); write, as usual, $\mathfrak{g} = \mathfrak{k} + \mathfrak{p}$ and let (G, K) be a pair corresponding to $\mathfrak{g}$ per Vol. I, number 1.1.5. Fix a θ-stable Cartan subalgebra $\mathfrak{j}$ of $\mathfrak{g}$ and let J be the Cartan subgroup of G associated with $\mathfrak{j}$.

The objective of the present number will be to establish the following theorem.

Theorem 8.4.3.1 (Harish-Chandra) Fix an $f \in \mathscr{C}(\mathfrak{g})$ – then the integral

$$\phi_f(H) = \epsilon_R(H)\pi(H)\int_{G/\mathrm{J}} f(\dot{x}H)d_{G/\mathrm{J}}(\dot{x})$$

is absolutely convergent for all $H \in \mathfrak{j}'$ and the function ϕ_f so defined lies in $\mathscr{C}(\mathfrak{j}')$. Furthermore the mapping $f \mapsto \phi_f$ of $\mathscr{C}(\mathfrak{g})$ into $\mathscr{C}(\mathfrak{j}')$ is continuous and $\phi_{\partial(p)f} = \partial(\tilde{p})\phi_f$ (all $p \in I(\mathfrak{g}_c); \tilde{p} = p \,|\, \mathfrak{j}_c$).

The proof of this theorem proceeds in two stages. Thus the first thing that shall be done is to prove it for fundamental Cartan subalgebras when rank $(\mathfrak{g})$ = rank $(\mathfrak{k})$ (this is the most difficult case); after that, we shall introduce the 'transformations of descent' which will enable us to reduce the general case to the previously considered special situation.

During the course of our discussion the following result will emerge.

Corollary 8.4.3.2 Retain the preceding notations and assumptions – then, for every f in $\mathscr{C}(\mathfrak{g})$, the function ϕ_f admits a unique extension to $\mathscr{C}(\mathfrak{j}'(SI))$. Furthermore the mapping $f \mapsto \phi_f$ of $\mathscr{C}(\mathfrak{g})$ into $\mathscr{C}(\mathfrak{j}'(SI))$ is continuous.

Notations Let $\|\,.\,\|$ be a norm on $\mathfrak{g}$ which renders $\mathfrak{c}_\mathfrak{k}$, $\mathfrak{c}_\mathfrak{p}$, and $\bar{\mathfrak{g}}$ mutually orthogonal and which agrees on $\bar{\mathfrak{g}}$ with the norm derived from the Euclidean structure $(.\,,\,.)_{\bar{\theta}}$ $(\bar{\theta} = \theta \,|\, \bar{\mathfrak{g}})$; such a norm is obviously θ-invariant.

(1) Given a continuous function f on $\mathfrak{g}$ and a non-negative integer m, write

$$|\,f\,|_m = \sup_{X \in \mathfrak{g}} (1 + \|X\|)^m \,|\,f(X)\,|.$$

(2) Given a continuous function f on $\mathfrak{g}$ and a non-negative integer m, write

$$\|\,f\,\|_m = \sup_{X \in \mathfrak{g}} (1 + \|X + \theta(X)\|)^m \,|\,f(X)\,|.$$

[In passing, let us observe that $\|\,.\,\|_m \leqslant 2^m \,|\,.\,|_m$ since

$$\|X + \theta(X)\| \leqslant 2\,\|X\| \qquad (X \in \mathfrak{g});$$

consequently the rule $\|\,.\,\|_m$ defines a continuous semi-norm on $\mathscr{C}(\mathfrak{g})$ ($m = 0, 1, \ldots$).]

Let us suppose for the time being that rank $(\mathfrak{g})$ = rank $(\mathfrak{k})$ and that $\mathfrak{j}$ is a θ-stable *fundamental* Cartan subalgebra of $\mathfrak{g}$ (hence $\mathfrak{j} \subset \mathfrak{k}$); in this situation, integrals over G/J may be replaced by integrals over G; moreover it is a simple matter to convince oneself that there is no essential loss of generality in assuming that $\mathfrak{g}$ is semi-simple.

Our first objective will be to show that for any $f \in \mathscr{C}(\mathfrak{g})$, the integral

$$\phi_f(H) = \pi(H) \int_G f(xH) d_G(x) \qquad (H \in \mathfrak{j}'(SI))$$

is absolutely convergent and the function ϕ_f thus defined lies in $C^\infty(\mathfrak{j}'(SI))$ with $\phi_{\partial(p)f} = \partial(\tilde{p})\phi_f$ (all $p \in I(\mathfrak{g}_c)$).

Lemma 8.4.3.3 Let m be a fixed non-negative integer – then there exist integers $r, s \geqslant 0$ and a positive number S such that

$$(1 + \|H\|)^n |\pi_{SI}(H)|^r \int_G \|x\|^m |f(xH)| d_G(x) \leqslant S \|f\|_{n+s}$$

for all $H \in \mathfrak{j}'(SI), f \in \mathscr{C}(\mathfrak{g})$ and any non-negative integer n. Furthermore, if ω is a compact subset of $G \times \mathfrak{j}'(SI) \times G$ and if $f \in \mathscr{C}(\mathfrak{g})$, then the integral

$$\int_G \|xyz\|^m |f(xyzH)| d_G(y)$$

converges uniformly with respect to $(x, H, z) \in \omega$.

Proof First choose an integer $t \geqslant 0$ such that $\int_G \|x\|^{-t} d_G(x) < \infty$ (cf. Lemma 8.1.5.4); using Lemma 2 in the Appendix to the present number, select integers $r, s \geqslant 0$ and a positive number T with the property that

$$\|x\|^{m+t} |\pi_{SI}(H)|^r \leqslant T \|xH + \theta(xH)\|^s \qquad (x \in G)$$

for all $H \in \mathfrak{j}$ – then, since $\|H\| \leqslant \|xH + \theta(xH)\|$ (see Lemma 1 in the Appendix to this number), we have

$$(1 + \|H\|)^n |\pi_{SI}(H)|^r \|x\|^m |f(xH)| \leqslant T \|x\|^{-t} \|f\|_{n+s}$$

and so if we put $S = T \int_G \|x\|^{-t} d_G(x)$, our first assertion follows. Turning to the second statement, let us suppose that $f \in \mathscr{C}(\mathfrak{g})$ and $(x, H, z) \in \omega$ – then

$$\begin{aligned} \|xyz\|^m |f(xyzH)| &\leqslant T\, |\pi_{SI}(H)|^{-r} \|xyz\|^{-t} \|f\|_s \\ &\leqslant T |\pi_{SI}(H)|^{-r} \|x^{-1}\|^t \|y\|^{-t} \|z^{-1}\|^t \|f\|_s \\ &\qquad\qquad (y \in G). \end{aligned}$$

It is obvious that $|\pi_{SI}(H)|^{-r} \|x^{-1}\|^t \|z^{-1}\|^t$ remains bounded on the compact set ω and so our assertion concerning uniform convergence follows from the fact that $\int_G \|y\|^{-t} d_G(y) < \infty$. $\square$

Note Let f be a continuous function on $\mathfrak{g}$ such that $\|f\|_m < \infty$ $(m = 0, 1, \ldots)$ – then the proof of Lemma 8.4.3.3 shows that the integral $\int_G |f(xH)| d_G(x)$ is finite for every H in $\mathfrak{j}'(SI)$.

Lemma 8.4.3.4 Fix an $f \in \mathscr{C}(\mathfrak{g})$ and elements $D \in \mathfrak{G}$ and $p \in S(\mathfrak{g}_c)$; let ω be a compact subset of $G \times \mathfrak{j}'(SI)$ – then the integral

$$\int_G |f(xy; D:H; \partial(p))| \, d_G(x)$$

converges uniformly with respect to $(y, H) \in \omega$.

[Here we follow the mode of writing introduced in 8.2.2.]

Corollary 8.4.3.5 Retain the notations of the preceding lemma – then

$$\int_G f(x; D:H; \partial(p)) d_G(x) = 0 \qquad (\text{all } D \in \mathfrak{G}\mathfrak{g}_c).$$

[Fix $H \in \mathfrak{j}'(SI)$; for $y \in G$, put $F(y) = \int_G f(xy : H; \partial(p)) d_G(x)$ – then Lemma 8.4.3.4 implies that F is a C^∞ function on G (with differentiation under the integral sign being permissible). Our contention thus follows from the right invariance of Haar measure on G.]

Proof of Lemma 8.4.3.4 Invoking the notations of 8.2.2, we have

$$f(x; D:X; \partial(p)) = f(x:X; \partial(\Gamma_X(D \otimes p))) \qquad (x \in G,\ X \in \mathfrak{g}).$$

Since D and p are fixed, it follows from the definition of Γ_X that there exist a finite number of linearly independent elements $u_1, \ldots, u_r \in S(\mathfrak{g}_c)$ and polynomials $p_1, \ldots, p_r \in S(\mathfrak{g}_c)$ such that

$$\Gamma_X(D \otimes p) = p_1(X)u_1 + \cdots + p_r(X)u_r \qquad (X \in \mathfrak{g}).$$

Therefore

$$\begin{aligned} f(x; D:X; \partial(p)) &= \sum_{i=1}^{r} p_i(X) f(x:X; \partial(u_i)) \\ &= \sum_{i=1}^{r} p_i(X) f(xX; \partial(xu_i)) \qquad (x \in G,\ X \in \mathfrak{g}). \end{aligned}$$

Now we may obviously select a finite number of linearly independent elements $v_1, \ldots, v_s$ in $S(\mathfrak{g}_c)$ such that for all $x \in G$, $xu_i = \sum_{j=1}^{s} a_{ij}(x)v_j$ $(1 \leqslant i \leqslant r)$; the a_{ij} are analytic functions on G which verify the relation $|a_{ij}(x)| \leqslant M\|x\|^m$ $(x \in G)$ for suitable non-negative integers m and M (cf. Lemma 8.1.5.3) – hence

$$|f(x; D:X; \partial(p))| \leqslant M \|x\|^m \sum_{i,j} |p_i(X)| |f(xX; \partial(v_j))| \qquad (x \in G,\ X \in \mathfrak{g}).$$

The present lemma is thus seen to be a consequence of the preceding one, the integral

$$\int_G \|xy\|^m |f(xyH;\partial(v_j))| \, d_G(x) \qquad (1 \leqslant j \leqslant s)$$

converging uniformly with respect to (y, H) on every compact subset ω of $G \times \mathfrak{j}'(SI)$. □

Fix an f in $\mathscr{C}(\mathfrak{g})$ and put, for $H \in \mathfrak{j}'(SI)$, $\psi_f(H) = \int_G f(xH) d_G(x)$ – then ψ_f is of class C^∞ on $\mathfrak{j}'(SI)$ with

$$\psi_f(H;\partial(p)) = \int_G f(x:H;\partial(p)) d_G(x) \qquad (p \in S(\mathfrak{j}_c)).$$

In fact this is an immediate consequence of the uniform convergence of the integral

$$\int_G |f(x:H;\partial(p))| \, d_G(x) \qquad (p \in S(\mathfrak{j}_c))$$

with respect to H on compact subsets of $\mathfrak{j}'(SI)$ (cf. Lemma 8.4.3.4). It then follows immediately that $\phi_f(f \in \mathscr{C}(\mathfrak{g}))$ is also of class C^∞ on $\mathfrak{j}'(SI)$. As regards the relation $\phi_{\partial(p)f} = \partial(\tilde{p})\phi_f$ (all $p \in I(\mathfrak{g}_c)$), it is apparent that we need only establish equality on $\mathfrak{j}'$. So fix an element H in $\mathfrak{j}'$; put $d = \partial(\tilde{p})\circ\pi$ $(p \in I(\mathfrak{g}_c))$ – then, on the basis of what has been said in 8.2.2, we have

$$f(xH;\partial(p)) = \pi(H)^{-1} f(x:H;d_H) + \sum_{i=1}^{r} f(x;D_i:H;\partial(p_i))$$
$$(x \in G),$$

where $D_i \in \mathfrak{G}\mathfrak{g}_c$, $p_i \in S(\mathfrak{j}_c)$ $(1 \leqslant i \leqslant r)$ and d_H is the local expression of d at H. It therefore follows from Lemma 8.4.3.4 and its corollary that

$$\phi_{\partial(p)f}(H) = \pi(H) \int_G f(xH;\partial(p)) d_G(x)$$
$$= \int_G f(x:H;d_H) d_G(x) = \phi_f(H;\partial(\tilde{p})),$$

as desired.

It remains to prove that $\phi_f \in \mathscr{C}(\mathfrak{j}'(SI))$ (all $f \in \mathscr{C}(\mathfrak{g})$) and that the mapping $f \mapsto \phi_f$ of $\mathscr{C}(\mathfrak{g})$ into $\mathscr{C}(\mathfrak{j}'(SI))$ is continuous; the proof of these two facts is difficult, requiring as it does, Harish-Chandra's 'Monomial 1 Technique'.

Heuristics The main point here is that $D\phi_f$ remains *bounded* on $\mathfrak{j}'(SI)$ for every polynomial differential operator D on $\mathfrak{j}$ (all $f \in \mathscr{C}(\mathfrak{g})$). Since the proof of this fact in the general case is a little complicated, as an illustration let us consider the case when $\mathfrak{g} = \mathfrak{sl}(2, \mathbf{R})$; agreeing to use the notations which were introduced in 8.4.2, our problem is to prove that $d^n\psi_{\bar{f}}/d\xi^n$ remains bounded around $\xi = 0$ for every non-negative integer n $(f \in C_c^\infty(\mathfrak{g}))$ – this is done as follows. Let ω be the Casimir polynomial on $\mathfrak{g}$ – then $d^2\psi_{\bar{f}}/d\xi^2 = -2^3\psi_{\overline{\partial(\omega)f}}$

(cf. Lemma 8.4.2.5). By a crude estimate, one proves easily that there exists an integer $m \geqslant 0$ with the property that $M(f) = \sup_\xi |\xi^m \psi_{\bar{f}}(\xi)| < \infty$ for every $f \in C_c^\infty(\mathfrak{g})$; let m_0 be the least possible such integer – then we claim that $m_0 = 0$. Proceeding by contradiction, let us suppose that $m_0 \geqslant 1$ – then

$$\left| \frac{d^2}{d\xi^2} \psi_{\bar{f}} \right| \leqslant 8 |\psi_{\overline{\partial(\omega)f}}| \leqslant 8 |\xi|^{-m_0} M_0(\partial(\omega)f).$$

If $m_0 \geqslant 2$, then it follows by integration that

$$|\psi_{\bar{f}}| \leqslant |\xi|^{2-m_0} \tilde{M}_0(f)$$

where $\tilde{M}_0(f)$ is a positive constant depending on f; as this contradicts the choice of m_0 we must have $m_0 = 1$; but since $\log|\xi|$ is locally summable around $\xi = 0$, the same argument leads once again to a contradiction – thus $m_0 = 0$ and so $|\psi_{\bar{f}}|$ remains bounded (all $f \in C_c^\infty(\mathfrak{g})$). Since the even derivatives of $\psi_{\bar{f}}$ are given, up to a constant factor, by $\psi_{\overline{\partial(\omega^n)f}}$ $(n = 0, 1, \ldots)$, they too must remain bounded; but then, by integration, the same must hold for the odd derivatives of $\psi_{\bar{f}}$ (all $f \in C_c^\infty(\mathfrak{g})$). . . .

Let us turn now to the proof of the fact that ϕ_f lies in $\mathscr{C}(\mathfrak{j}'(SI))$ for all $f \in \mathscr{C}(\mathfrak{g})$; the continuity of the map $f \mapsto \phi_f$ $(f \in \mathscr{C}(\mathfrak{g}))$ will then become apparent during the course of our discussion.

Let $\mathcal{O}$ be a subset of $\mathfrak{j}'(SI)$. We shall say that $\mathcal{O}$ has the property (P) if for every $u \in S(\mathfrak{j}_c)$ there exists an integer $n \geqslant 0$, a real constant $M \geqslant 1$, and a finite number of elements $p_1, \ldots, p_s$ in $S(\mathfrak{g}_c)$ such that

$$\text{(P)} \quad \sup_{H \in \mathcal{O}} (1 + \|H\|)^m |\phi_f(H; \partial(u))| \leqslant M^m \sum_{j=1}^{s} \|\partial(p_j)f\|_{m+n}$$

for all $f \in \mathscr{C}(\mathfrak{g})$ and every non-negative integer m. Concerning the property (P), it is clear in the first place that the union of a finite number of sets with property (P) again has the same property; likewise if $\mathcal{O}$ possesses property (P), then so does its closure in $\mathfrak{j}'(SI)$. Finally one has the following important observation. Let $[W]$ be the order of the Weyl group W of the pair $(\mathfrak{g}_c, \mathfrak{j}_c)$; choose elements $u_1, \ldots, u_{[W]}$ in $S(\mathfrak{j}_c)$ such that $S(\mathfrak{j}_c) = \sum_i I(\mathfrak{j}_c)u_i$ (cf. Corollary 2.1.3.7); for any $u \in S(\mathfrak{j}_c)$, let $p_i(u)$ denote the (unique) element in $I(\mathfrak{g}_c)$ such that $u = \sum_i \widetilde{p_i(u)} u_i$ and put $D_i(u) = \partial(p_i(u))$ – then, for f in $\mathscr{C}(\mathfrak{g})$, we have seen above that

$$\partial(u)\phi_f = \sum_i \partial(u_i)\phi_{D_i(u)f}$$

and so

$$|\phi_f(H; \partial(u))| \leqslant \sum_i |\phi_{D_i(u)f}(H; \partial(u_i))| \qquad (H \in \mathfrak{j}'(SI))$$

which shows that in order to prove that $\mathcal{O}$ has property (P), one need only check it for $u = u_i$ $(i = 1, \ldots, [W])$.

It is obvious that our theorem will be established when it is shown that $\mathfrak{j}'(SI)$ has the property (P); since $\mathfrak{j}'$ is a dense subset of $\mathfrak{j}'(SI)$ and has but a finite number of connected components, it will be enough to prove that each connected component of $\mathfrak{j}'$ has the property (P). So fix a connected component $\mathscr{C}$ in $\mathfrak{j}'$ and let $\{\alpha_1, \ldots, \alpha_l\}$ be the fundamental system of positive roots for the pair $(\mathfrak{g}_c, \mathfrak{j}_c)$ singled out by this choice of $\mathscr{C}$; choose a basis $\{H_1, \ldots, H_l\}$ for $\mathfrak{j}$ over $\mathbf{R}$ such that $\alpha_i(H_j) = \sqrt{-1}\delta_{ij}$ $(1 \leqslant i, j \leqslant l)$ and let $t_1, \ldots, t_l$ denote the Cartesian coordinates in $\mathfrak{j}$ corresponding to this basis. By a *monomial* T on $\mathfrak{j}$ we shall understand a function of the form $T = t_1^{m_1} \ldots t_l^{m_l}$ where $m_1, \ldots, m_l$ are non-negative integers; it will be convenient to refer to m_i as the *exponent* of t_i in T (thus the monomial whose exponents are all zero is 1). In the set of all monomials a lexicographic order may be introduced as follows. Suppose that $T = t_1^{m_1} \ldots t_l^{m_l}$, $T' = t_1^{m_1'} \ldots t_l^{m_l'}$ are two distinct monomials – then we say that $T < T'$ (or $T' > T$) if $m_i < m_i'$ where i $(1 \leqslant i \leqslant l)$ is the least index such that $m_i \neq m_i'$. It is obvious that every non-empty set of monomials has a lowest element; note too that every monomial takes only positive values on $\mathscr{C}$. A monomial T is said to have the property (P) if there exists an integer $n \geqslant 0$, a real constant $M \geqslant 1$ and a finite set of elements $p_1, \ldots, p_s$ in $S(\mathfrak{g}_c)$ such that

$$\text{(P)} \quad \sup_{H \in \mathscr{C}} T(H)(1 + \|H\|)^m |\phi_f(H; \partial(u_i))| \leqslant M^m \sum_{j=1}^{s} \|\partial(p_j)f\|_{m+n} \qquad (1 \leqslant i \leqslant [W])$$

for all $f \in \mathscr{C}(\mathfrak{g})$ and every non-negative integer m. In order to prove that $\mathscr{C}$ has the property (P), it will clearly be enough to show that the monomial 1 has the property (P).

To begin with, monomials T with the property (P) actually do exist; in fact it is clear that there exists a monomial T (say) such that

$$|\pi_{SI}(H)| \geqslant T(H) \qquad (\text{all } H \in \mathscr{C}).$$

It then follows readily from the *proof* of Lemma 8.4.3.4 and Lemma 8.4.3.3 that a suitable positive integral power of T has the property (P). This being the case, let T be the *lowest* monomial with property (P) – then we have to prove that $T = 1$. Assuming to the contrary that $T \neq 1$, we shall produce a monomial which is lower than T but still has the property (P); this however will contradict our original supposition on T. . . . Let i_0 be the least index $(1 \leqslant i_0 \leqslant l)$ such that the exponent m_0 of t_{i_0} in T is positive; put $t = t_{i_0}$ – then $T = t^{m_0}T_0$ where T_0 is a monomial in which the exponents of t_i are zero for $1 \leqslant i \leqslant i_0$. Write $\partial_t = \partial/\partial t = \partial(H_{i_0})$ and let $H_{i_0}u_i = \sum_j \tilde{q}_{ij}u_j$ where $q_{ij} \in I(\mathfrak{g}_c)$ $(1 \leqslant i, j \leqslant [W])$. Now fix an element f in $\mathscr{C}(\mathfrak{g})$ – then

$$\partial_t(\partial(u_i)\phi_f) = \sum_j \partial(u_j)\phi_{\partial(q_{ij})f} \qquad (1 \leqslant i \leqslant [W]).$$

Set $\phi_i = T_0\partial(u_i)\phi_f$,

$$\psi_i = T_0 \sum_j \partial(u_j)\phi_{\partial(q_{ij})f} \qquad (1 \leqslant i \leqslant [W])$$

– then $\partial_t\phi_i = \psi_i$ (all i). Moreover if $n, M, p_1, \ldots, p_s$ are selected in accordance with (P) for the monomial T, then we have

$$t^{m_0}\mu^m|\phi_i| \leqslant M^m a_m,\ t^{m_0}\mu^m|\psi_i| \leqslant M^m b_m \qquad (1 \leqslant i \leqslant [W],\ m \geqslant 0)$$

on $\mathscr{C}$. Here $\mu(H) = 1 + \|H\|$ $(H \in \mathfrak{j})$ and

$$a_m = \sum_{j=1}^{s} \|\partial(p_j)f\|_{m+n},\ b_m = \sum_{j=1}^{s}\sum_{h,k=1}^{[W]} \|\partial(p_j)\partial(q_{hk})f\|_{m+n}.$$

Let τ_+, τ_- denote the subsets of $\mathscr{C}$ defined by the conditions $t \geqslant 1$ and $0 < t \leqslant 1$, respectively; for any H in $\mathscr{C}$, let H_0 denote the unique point in the intersection of the line $H + rH_{i_0}$ $(r \in \mathbf{R})$ with the 'hyperplane' $\sigma = \tau_+ \cap \tau_-$; put $\nu(H) = \mu(H_0)$ $(H \in \mathscr{C})$ – then, for all $H \in \tau_-$, we have

$$\mu(H) \leqslant N\nu(H) \quad \text{and} \quad \nu(H) \leqslant N\mu(H)$$

where $N = 1 + \|H_{i_0}\|$. Fix an integer $m \geqslant 0$ – then

$$|\psi_i| \leqslant M^m b_m \mu^{-m} t^{-m_0} \leqslant (MN)^m b_m \nu^{-m} t^{-m_0}$$

on τ_-. Suppose that H is a point in τ_- and H_0 is the corresponding point on σ – then, upon integrating the differential equation $\partial_t\phi_i = \psi_i$ along the line segment $t_k = t_k(H)$ $(k \neq i_0,\ 1 \leqslant k \leqslant l)$, $t(H) \leqslant t = t_{i_0} \leqslant 1$, we find that

$$\begin{aligned} |\phi_i(H_0) - \phi_i(H)| &= \left| \int_{t(H)}^{1} \psi_i(H_0 - (1-r)H_{i_0})dr \right| \\ &\leqslant (MN)^m b_m (\nu(H))^{-m} \int_{t(H)}^{1} r^{-m_0} dr, \end{aligned}$$

ν being constant on this segment. For H in τ_-, put

$$\eta(H) = \begin{cases} \{t(H)^{-m_0+1} - 1\}/(m_0 - 1) & \text{if } m_0 \geqslant 2, \\ |\log(t(H))| & \text{if } m_0 = 1. \end{cases}$$

Then we have the estimate

$$|\phi_i(H)| \leqslant |\phi_i(H_0)| + (MN)^m b_m (\nu(H))^{-m}\eta(H) \qquad (\text{all } H \in \tau_-).$$

On the other hand, since $t(H_0) = 1$, $\mu(H_0)^m|\phi_i(H_0)| \leqslant M^m a_m$; moreover $\mu(H_0) = \nu(H)$ and so

$$\begin{aligned} |\phi_i| &\leqslant \{M^m a_m + (MN)^m b_m \eta\}\nu^{-m} \\ &\leqslant \{(MN)^m a_m + (MN^2)^m b_m \eta\}\mu^{-m} \end{aligned}$$

on τ_-. Hence

$$t^{m_0-1}\mu^m|\phi_i| \leqslant (MN^2)^m\{a_m + b_m\eta t^{m_0-1}\}$$

on τ_- while

$$t^{m_0-1}\mu^m|\phi_i| \leqslant t^{m_0}\mu^m|\phi_i| \leqslant M^m a_m \leqslant (MN^2)^m a_m$$

on τ_+.

We shall now distinguish two cases.

(I) The case $m_0 \geqslant 2$: In this situation, $\eta t^{m_0-1} \leqslant 1$ on τ_-, whence

$$t^{m_0-1}\mu^m |\phi_i| \leqslant (MN^2)^m\{a_m + b_m\} \qquad (1 \leqslant i \leqslant [W])$$

on $\mathscr{C}$. Put $T^0 = t^{m_0-1}T_0$ – then it follows that

$$T^0\mu^m |\partial(u_i)\phi_f| \leqslant (MN^2)^m \sum_{j=1}^{r} \{\|\partial(p_j)f\|_{m+n} + \sum_{h,k=1}^{[W]} \|\partial(p_j)\partial(q_{hk})f\|_{m+n}\} \qquad (1 \leqslant i \leqslant [W])$$

on $\mathscr{C}$. Since f and m are arbitrary, it is clear that T^0 has the property (P); however T^0 is lower than T and so we get a contradiction with the definition of T.

(II) The case $m_0 = 1$: In this situation

$$\mu^m |\phi_i| \leqslant (MN^2)^m\{a_m + b_m |\log t|\}$$

on τ_-. Write s for $s([W]^2 + 1)$ and let D_j $(1 \leqslant j \leqslant s)$ denote the s differential operators $\partial(p_j)$ and $\partial(p_j)\partial(q_{hk})$ – then we have

$$T_0\mu^m |\partial(u_i)\phi_f| \leqslant (MN^2)^m \left\{\sum_{j=1}^{s} \|D_j f\|_{m+n}\right\}(1 + |\log t|) \qquad (1 \leqslant i \leqslant [W])$$

on τ_- for any $f \in \mathscr{C}(\mathfrak{g})$ and any $m \geqslant 0$. In particular we may apply this result to $\partial(q_{hk})f$ to conclude that

$$\mu^m |\psi_i| \leqslant (MN^2)^m c_m(1 + |\log t|) \qquad (1 \leqslant i \leqslant [W],\ m \geqslant 0)$$

on τ_-, where

$$c_m = \sum_{j=1}^{s} \sum_{h,k=1}^{[W]} \|D_j\partial(q_{hk})f\|_{m+n}.$$

For H in τ_- we then have the estimate

$$\begin{aligned}|\phi_i(H)| &\leqslant |\phi_i(H_0)| + \int_{t(H)}^{1} |\psi_i(H_0 - (1-r)H_{i_0})|\,dr \\ &\leqslant \{M^m a_m + 2(MN^3)^m c_m\}\nu(H)^{-m}.\end{aligned}$$

Therefore

$$\mu^m |\phi_i| \leqslant 2(MN^4)^m(a_m + c_m)$$

on τ_- while, on the other hand, since $m_0 = 1$,

$$\mu^m |\phi_i| \leqslant t\mu^m |\phi_i| \leqslant M^m a_m \leqslant 2(MN^4)^m a_m$$

on τ_+. It then follows without difficulty that T_0 has the property (P), a contradiction once again.

All the contentions of Theorem 8.4.3.1 and its corollary have therefore been established in the case rank $(\mathfrak{g})$ = rank $(\mathfrak{k})$, $\mathfrak{j}$ fundamental in $\mathfrak{g}$.

In order to prove Theorem 8.4.3.1 in complete generality, it will be necessary to have at hand an artifice which will enable us to reduce the general case to the special one studied above; such a reduction can be carried out through use of the so-called 'transformations of descent'.

Let $\mathfrak{j}$ be a θ-stable Cartan subalgebra of $\mathfrak{g}$; for the purposes of the present discussion, it will be assumed that the pair $(\mathfrak{g}_c, \mathfrak{j}_c)$ admits at least one positive root which is not imaginary (so that $\bar{\mathfrak{j}}_{\mathfrak{p}} = \mathfrak{j}_{\mathfrak{p}} \cap \bar{\mathfrak{g}} \neq \{0\}$). Let $\mathfrak{l}$ be the centralizer of $\mathfrak{j}_{\mathfrak{p}}$ in $\mathfrak{g}$ and introduce L, $\mathfrak{m}$, M, $\mathfrak{n}^{\pm}$, $N^{\pm}$ etc. in the usual way (cf. 8.1.3); write $P = MJ_{\mathfrak{p}}N^{+}$.

Fix an element f in $\mathscr{C}(\mathfrak{g})$; let ω be a compact subset of $\mathfrak{g}$ and, for any $p \in S(\mathfrak{g}_c)$, consider the integral

$$\int_{\mathfrak{n}^+} |f(X + Z; \partial(p))| \, d_{\mathfrak{n}^+}(Z) \qquad (X \in \omega).$$

We claim that this integral converges *uniformly* with respect to X in ω. Thus, to begin with, choose a positive integer n such that

$$\zeta = \int_{\mathfrak{n}^+} (1 + \|Z\|)^{-n} d_{\mathfrak{n}^+}(Z) < \infty.$$

Then, taking into account the relations

$$\begin{aligned} \|Z\| &\leqslant \|Z + \theta(Z)\| \leqslant \|X + Z + \theta(X + Z)\| + \|X + \theta(X)\| \\ &\leqslant \|X + Z + \theta(X + Z)\| + 2N \qquad (X \in \omega,\ Z \in \mathfrak{n}^+) \end{aligned}$$

where $N = \sup_{X \in \omega} \|X\|$, we find that

$$(1 + \|Z\|)^n |f(X + Z; \partial(p))| \leqslant (1 + 2N)^n \|\partial(p)f\|_n \qquad (X \in \omega,\ Z \in \mathfrak{n}^+)$$

which proves that our integral does indeed converge uniformly with respect to $X \in \omega$. Consequently the function $X \mapsto \int_{\mathfrak{n}^+} f(X + Z) d_{\mathfrak{n}^+}(Z)$ $(X \in \mathfrak{g})$ is of class C^∞ on $\mathfrak{g}$, differentiation under the integral sign being permissible. . . .

Given an f in $\mathscr{C}(\mathfrak{g})$, put

$$f_P(X) = \int_{\mathfrak{n}^+} f(X + Z) d_{\mathfrak{n}^+}(Z) \qquad (X \in \mathfrak{l}).$$

Owing to the preceding remarks, $f_P \in C^\infty(\mathfrak{l})$; in fact $f_P \in \mathscr{C}(\mathfrak{l})$ and the mapping $f \mapsto f_P$ of $\mathscr{C}(\mathfrak{g})$ into $\mathscr{C}(\mathfrak{l})$ is continuous. To see this, fix an integer $m \geqslant 0$ and an element $p \in S(\mathfrak{l}_c)$; since

$$\max(\|X\|, \|Z\|) \leqslant \|X + Z\| \qquad (X \in \mathfrak{l},\ Z \in \mathfrak{n}^+),$$

we have

$$(1 + \|X\|)^m (1 + \|Z\|)^n |f(X + Z; \partial(p))| \leqslant |\partial(p)f|_{m+n}$$

and so

$$|\partial(p)f_P|_m \leqslant \zeta |\partial(p)f|_{m+n}$$

from which our assertion follows at once.

For $f \in \mathscr{C}(\mathfrak{g})$, write $f_K(X) = \int_K f(kX)dk$ $(X \in \mathfrak{g})$; it is easy to see that the assignment $f \mapsto f_K$ ($f \in \mathscr{C}(\mathfrak{g})$) is a continuous mapping of $\mathscr{C}(\mathfrak{g})$ into itself. This being the case, given f in $\mathscr{C}(\mathfrak{g})$, put

$$f_{K,P}(X) = \int_{\mathfrak{n}^+} f_K(X + Z) d_{\mathfrak{n}^+}(Z) \qquad (X \in \mathfrak{l}).$$

Then the above discussion implies that the assignment $f \mapsto f_{K,P}$ defines a continuous mapping of $\mathscr{C}(\mathfrak{g})$ into $\mathscr{C}(\mathfrak{l})$; furthermore we have

$$(\partial(p)f)_{K,P} = \partial(p_{\mathfrak{l}}) f_{K,P}$$

for all $p \in I(\mathfrak{g}_c)$ ($p_{\mathfrak{l}}$ the restriction of p to $\mathfrak{l}_c$). . . . [Here we have to observe that $p - p_{\mathfrak{l}} \in S(\mathfrak{g}_c)\mathfrak{n}_c^+$ (all $p \in I(\mathfrak{g}_c)$).]

We shall refer to the mappings $f \mapsto f_P$, $f \mapsto f_{K,P}$ ($f \in \mathscr{C}(\mathfrak{g})$) as the *(infinitesimal) transformations of descent.*

We come now to the proof of Theorem 8.4.3.1.

The case when $\mathfrak{c}_{\mathfrak{p}} \neq \{0\}$ is easily liquidated. For in this situation $\mathfrak{g} = \hat{\mathfrak{g}} + \mathfrak{c}_{\mathfrak{p}}$ with dim $(\hat{\mathfrak{g}}) <$ dim $(\mathfrak{g})$; granting that our theorem holds for $\hat{\mathfrak{g}}$, a simple argument then leads quickly to the conclusion that it also holds for $\mathfrak{g}$.

So it can now be assumed that $\mathfrak{c}_{\mathfrak{p}} = \{0\}$; supposing first that $\mathfrak{j}_{\mathfrak{p}} \neq \{0\}$, we have dim $(\mathfrak{l}) <$ dim $(\mathfrak{g})$ ($\mathfrak{l}$ the centralizer of $\mathfrak{j}_{\mathfrak{p}}$ in $\mathfrak{g}$) and the sought for result is known to be true for the pair $(\mathfrak{l}, \mathfrak{j})$ (since it is true for the pair $(\mathfrak{m}, \mathfrak{j}_{\mathfrak{k}})$ . . .). This being so, fix an f in $\mathscr{C}(\mathfrak{g})$; put $|f|_K(X) = \int_K |f(kX)| dk$ ($X \in \mathfrak{g}$),

$$|f|_{K,P}(X) = \int_{\mathfrak{n}^+} |f|_K(X + Z) d_{\mathfrak{n}^+}(Z) \qquad (X \in \mathfrak{l}).$$

Then, utilizing a standard integration formula modulo constant factors (cf. 8.1.3), we have

$$\begin{aligned}\int_{G/J} |f(\dot{x}H)|\, d_{G/J}(\dot{x}) &\leqslant |\pi_{R,C}(H)|^{-1} \iint_{M \times \mathfrak{n}^+} |f|_K(mH + Z) d_M(m) d_{\mathfrak{n}^+}(Z) \\ &\leqslant |\pi_{R,C}(H)|^{-1} \int_M |f|_{K,P}(mH) d_M(m) < \infty \\ &\qquad\qquad (H \in \mathfrak{j}'),\end{aligned}$$

the latter integral being finite in view of Lemma 8.4.3.3 and subsequent comment. [Choose a positive integer n such that the integral of $Z \mapsto (1 + \|Z\|)^{-n}$ ($Z \in \mathfrak{n}^+$) over $\mathfrak{n}^+$ is finite with value ζ (say) – then, for any integer $m \geqslant 0$, we clearly have

$$\sup_{X \in \mathfrak{l}} (1 + \|X + \theta(X)\|)^m |f|_{K,P}(X) \leqslant \zeta \|f\|_{m+n}$$

which shows that our considerations are in fact applicable to $|f|_{K,P}$. . . .] The integral defining ϕ_f is therefore absolutely convergent; furthermore, in the obvious notations, we have, up to constant factors,

$$\phi_f(H) = \epsilon_R(H)\pi(H)\int_{G/J} f(\dot{x}H)d_{G/J}(\dot{x})$$

$$= \epsilon_R(H)\pi(H)|\pi_R(H)|^{-1}\pi_C(H)^{-1}\iint_{M\times\mathfrak{n}^+} f_K(mH+Z)d_M(m)d_{\mathfrak{n}^+}(Z)$$

$$= \pi_I(H)\int_M f_{K,P}(mH)d_M(m) = \phi^{\mathfrak{l}}_{f_{K,P}}(H) \qquad \text{(say)}, \qquad (H \in \mathfrak{j}')$$

$\phi^{\mathfrak{l}}_{f_{K,P}}$ the invariant integral of $f_{K,P}$ calculated relative to the pair $(\mathfrak{l}, \mathfrak{j})$ (we recall that $f_{K,P} \in \mathscr{C}(\mathfrak{l})$). [Needless to say the invariant integral relative to the pair $(\mathfrak{l}, \mathfrak{j})$ is calculated by means of an integral over L/J, i.e. by an integral over M/J_K (J_K the identity component of J_K), or, what is the same, by an integral over M; of course rank $(\mathfrak{m})$ = rank $(\mathfrak{m}_{\mathfrak{k}})$, $\mathfrak{j}_{\mathfrak{k}}$ being a Cartan subalgebra of $\mathfrak{m}$.] The assignment $f \mapsto f_{K,P} \mapsto \phi^{\mathfrak{l}}_{f_{K,P}} = \phi_f$ defines a continuous mapping of $\mathscr{C}(\mathfrak{g})$ into $\mathscr{C}(\mathfrak{j}')$; in addition $\phi^{\mathfrak{l}}_{f_{K,P}}$ actually lies in $\mathscr{C}(\mathfrak{j}'(SI))$, whence ϕ_f admits a unique extension to $\mathfrak{j}'(SI)$ (all $f \in \mathscr{C}(\mathfrak{g})$); finally, for any p in $I(\mathfrak{g}_c)$, we have

$$\phi_{\partial(p)f} = \phi^{\mathfrak{l}}_{(\partial(p)f)_{K,P}} = \phi^{\mathfrak{l}}_{\partial(p_{\mathfrak{l}})f_{K,P}} = \partial(\tilde{p})\phi^{\mathfrak{l}}_{f_{K,P}} = \partial(\tilde{p})\phi_f \qquad (\text{all } f \in \mathscr{C}(\mathfrak{g})).$$

Theorem 8.4.3.1 and its corollary are thereby established in this case.

Keeping to the assumption that $\mathfrak{c}_{\mathfrak{p}} = \{0\}$, it still remains to consider the possibility that $\mathfrak{j}_{\mathfrak{p}} = \{0\}$; this case was, however, dealt with above. Theorem 8.4.3.1 and its corollary are now completely proved.

Appendix Let $\mathfrak{g}$ be a semi-simple Lie algebra over $\mathbf{R}$, $\mathfrak{g} = \mathfrak{k} + \mathfrak{p}$ a Cartan decomposition of $\mathfrak{g}$ with Cartan involution θ; let G be a connected semi-simple Lie group with Lie algebra $\mathfrak{g}$, K the analytic subgroup of G corresponding to $\mathfrak{k}$ – in what follows it will be assumed that K is compact (so that G has finite center) and that rank (G) = rank (K).

Lemma 1 Let $\mathfrak{j}$ be a θ-stable fundamental Cartan subalgebra of $\mathfrak{g}$ – then, for all H in $\mathfrak{j}$, we have

$$||xH + \theta(xH)||^2 = 2||H||^2 + 2||xH||^2 \qquad (x \in G).$$

[Here, of course, the norm on $\mathfrak{g}$ is that derived from the Euclidean structure $(\cdot\,,\cdot)_\theta$.]

Proof Fix an Iwasawa decomposition $\mathfrak{g} = \mathfrak{k} + \mathfrak{a}_{\mathfrak{p}} + \mathfrak{n}^+$ for $\mathfrak{g}$; agreeing to employ the customary notations, we then have $G = KA_{\mathfrak{p}}K$. Since the elements of K are represented by orthogonal transformations on $\mathfrak{g}$ which commute with θ, it can be supposed that $x = hk$ ($h \in A_{\mathfrak{p}}$, $k \in K$). This being the case, fix an element $H \in \mathfrak{j}$ and set $kH = X$ – then $xH = hX$ and $\theta(xH) = h^{-1}X$, whence

$$||xH + \theta(xH)||^2 = ||hX + h^{-1}X||^2 = ||hX||^2 + ||h^{-1}X||^2 + 2||X||^2$$

(h being self-adjoint). Because $||h^{-1}X|| = ||\theta(hX)|| = ||hX||$, it now follows that

$$||xH + \theta(xH)||^2 = 2||X||^2 + 2||hX||^2 = 2||H||^2 + 2||xH||^2,$$

as we wished to prove. □

Lemma 2 Let $\mathfrak{j}$ be a θ-stable fundamental Cartan subalgebra of $\mathfrak{g}$ – then there exist non-negative integers r and s and a positive number S such that for all $H \in \mathfrak{j}$, we have

$$\|x\| |\pi_{SI}(H)|^r \leqslant S\|xH + \theta(xH)\|^s \qquad (x \in G).$$

[The symbol $\|\,.\,\|$, when applied to elements of G, is the so-called standard semi-norm (on G) relative to the Cartan involution θ (cf. 8.1.5).]

Proof Fix an Iwasawa decomposition $\mathfrak{g} = \mathfrak{k} + \mathfrak{a}_\mathfrak{p} + \mathfrak{n}^+$ for $\mathfrak{g}$; relative to our choice of a positive Weyl chamber $\mathscr{C}$ in $\mathfrak{a}_\mathfrak{p}$, let $\lambda_1, \ldots, \lambda_l$ be the associated fundamental system of positive roots for the pair $(\mathfrak{g}, \mathfrak{a}_\mathfrak{p})$; choose H_i in $\mathfrak{a}_\mathfrak{p}$ with the property that $\lambda_i(H_j) = \delta_{ij}$ $(1 \leqslant i, j \leqslant l)$ and put

$$\Lambda = \max_{i,\lambda} |\lambda(H_i)| \qquad (\lambda \in \Sigma^+, 1 \leqslant i \leqslant l).$$

Let $\mathfrak{m}$ be the centralizer of $\mathfrak{a}_\mathfrak{p}$ in $\mathfrak{k}$. An elementary argument tells us that the mapping ϕ defined by the rule $X \mapsto X + \theta(X)$ $(X \in \mathfrak{m} + \mathfrak{n}^+)$ establishes a linear isomorphism between $\mathfrak{m} + \mathfrak{n}^+$ and $\mathfrak{k}$; let ψ denote its inverse.

(1) We claim that there exists a positive constant R such that

$$\|H\|^{l(r_{SI}-1)} \prod_{i=1}^{l} \|\mathrm{ad}\,(H_i)^2\psi(kH)\| \geqslant R|\pi_{SI}(H)|^l \qquad (\text{all } H \in \mathfrak{j}, k \in K)$$

where r_{SI} is the number of positive singular imaginary roots of the pair $(\mathfrak{g}_c, \mathfrak{j}_c)$. To see this, first observe that

$$\|\mathrm{ad}\,(H_i)^2 X\| \geqslant 2^{-1/2}\|\mathrm{ad}\,(H_i)\phi(X)\| \qquad (X \in \mathfrak{m} + \mathfrak{n}^+, 1 \leqslant i \leqslant l),$$

that is,

$$\|\mathrm{ad}\,(H_i)^2\psi(X)\| \geqslant 2^{-1/2}\|\mathrm{ad}\,(H_i)X\| \qquad (X \in \mathfrak{k}, 1 \leqslant i \leqslant l).$$

Set $\gamma(X) = |\gamma_X|$ where γ_X is an eigenvalue of $\mathrm{ad}\,(X)|\mathfrak{p}_c$ $(X \in \mathfrak{k})$ with the smallest possible absolute value – then we evidently have

$$\prod_{i=1}^{l} \|\mathrm{ad}\,(H_i)^2\psi(X)\| \geqslant 2^{-l/2} \prod_{i=1}^{l} \|\mathrm{ad}\,(H_i)X\| \geqslant 2^{-l/2}\gamma(X)^l \prod_{i=1}^{l} \|H_i\|$$
$$(\text{all } X \in \mathfrak{k}).$$

It is obvious that $\gamma(kX) = \gamma(X)$ $(k \in K, X \in \mathfrak{k})$; moreover the very definition of γ tells us that $\gamma(H) = \min_{\alpha \in \Phi_{SI}^+} |\alpha(H)|$ $(H \in \mathfrak{j})$ (Φ_{SI}^+ the positive singular imaginary roots of the pair $(\mathfrak{g}_c, \mathfrak{j}_c)$) – therefore

$$\prod_{i=1}^{l} \|\mathrm{ad}\,(H_i)^2\psi(kH)\| \geqslant R \min_{\alpha \in \Phi_{SI}^+} |\alpha(H)|^l \qquad (\text{all } H \in \mathfrak{j}, k \in K)$$

where $R = 2^{-l/2} \prod_{i=1}^{l} \|H_i\|$. However $|\alpha(H)| \leqslant \|H\|$ for any $\alpha \in \Phi_{SI}^+$ $(H \in \mathfrak{j})$; thus

$$\|H\|^{r_{SI}-1} \min_{\alpha \in \Phi_{SI}^+} |\alpha(H)| \geqslant \prod_{\alpha \in \Phi_{SI}^+} |\alpha(H)| = |\pi_{SI}(H)| \qquad (H \in \mathfrak{j})$$

and so

$$\|H\|^{l(r_{SI}-1)} \prod_{i=1}^{l} \|\mathrm{ad}\,(H_i)^2\psi(kH)\| \geqslant R|\pi_{SI}(H)|^l \qquad (\text{all } H \in \mathfrak{j}, k \in K),$$

as claimed.

(2) Now fix an element x in G; without loss of generality we can assume that $x = hk$ $(h \in \exp(\mathscr{C}^{cl}), k \in K)$. Let H be an element in $\mathfrak{j}$ such that $\pi_{SI}(H) \neq 0$;

write $\psi(kH) = M + X$ $(M \in \mathfrak{m},\ X \in \mathfrak{n}^+)$ – then $kH = 2M + X + \theta(X)$, whence

$$\|xH + \theta(xH)\|^2 = 2\|H\|^2 + 2\|xH\|^2$$
$$= 16\|M\|^2 + 4\|X\|^2 + 2\|hX\|^2 + 2\|h^{-1}X\|^2,$$

$\mathfrak{m}, \mathfrak{n}^+$, and $\mathfrak{n}^-$ being orthogonal to one another. Let X_λ be the orthogonal projection of X onto $\mathfrak{g}^\lambda$ $(\lambda > 0)$ – then it is clear that

$$\|hX\|^2 \geqslant \Lambda^{-4}\|\mathrm{ad}\,(H_i)^2 hX\|^2 = \Lambda^{-4}\sum_{\lambda>0} h^{2\lambda}\lambda(H_i)^4\|X_\lambda\|^2$$
$$\geqslant \Lambda^{-4}\sum_{\lambda>0} h^{2\lambda_i}\lambda(H_i)^4\|X_\lambda\|^2 = \Lambda^{-4}h^{2\lambda_i}\|\mathrm{ad}\,(H_i)^2X\|^2 \qquad (1 \leqslant i \leqslant l).$$

Put $\mu = \lambda_1 + \cdots + \lambda_l$ – then it follows that

$$\|xH + \theta(xH)\|^l \geqslant \|hX\|^l \geqslant \Lambda^{-2l}h^\mu \prod_{i=1}^{l}\|\mathrm{ad}\,(H_i)^2X\|.$$

In (1) above we saw that there exists a positive constant R such that

$$\|H\|^{l(r_{SI}-1)}\prod_{i=1}^{l}\|\mathrm{ad}\,(H_i)^2\psi(kH)\| \geqslant R|\pi_{SI}(H)|^l \qquad (\text{all } H \in \mathfrak{j}, k \in K)$$

where r_{SI} is the number of positive singular imaginary roots of the pair $(\mathfrak{g}_c, \mathfrak{j}_c)$. Since $\|x\|^2 = \|h\|^2 \leqslant nh^{2\Lambda\mu}$ $(n = \dim(\mathfrak{g}))$ while

$$\mathrm{ad}\,(H_i)^2\psi(kH) = \mathrm{ad}\,(H_i)^2X \qquad (1 \leqslant i \leqslant l),$$

we then find that

$$\|x\| \leqslant \sqrt{n}\,\Lambda^{2\Lambda l}\|xH + \theta(xH)\|^{\Lambda l}\{\prod_{i=1}^{l}\|\mathrm{ad}\,(H_i)^2X\|\}^{-\Lambda}$$
$$\leqslant \sqrt{n}\,\Lambda^{2\Lambda l}\|xH + \theta(xH)\|^{\Lambda l}\{R|\pi_{SI}(H)|^l\|H\|^{-l(r_{SI}-1)}\}^{-\Lambda}$$
$$\leqslant \sqrt{n}\,\Lambda^{2\Lambda l}\|xH + \theta(xH)\|^{\Lambda l}\{R|\pi_{SI}(H)|^l\|xH + \theta(xH)\|^{-l(r_{SI}-1)}\}^{-\Lambda}.$$

Our lemma now follows upon taking $r = \Lambda l$, $s = \Lambda l r_{SI}$, and

$$S = \sqrt{n}\,\Lambda^{2\Lambda l}R^{-\Lambda}. \qquad \square$$

8.4.4 Extension Problems

Let $\mathfrak{g}$ be a reductive Lie algebra over $\mathbf{R}$, $\theta : \mathfrak{g} \to \mathfrak{g}$ an involutive automorphism of $\mathfrak{g}$ whose restriction to the derived algebra $\bar{\mathfrak{g}}$ is a Cartan involution (of $\bar{\mathfrak{g}}$); write, as usual, $\mathfrak{g} = \mathfrak{k} + \mathfrak{p}$ and let (G, K) be a pair corresponding to $\mathfrak{g}$ per Vol. I, number 1.1.5. Let $\mathfrak{j}$ be a θ-stable Cartan subalgebra of $\mathfrak{g}$, J the Cartan subgroup of G associated with $\mathfrak{j}$ – the notations and conventions of 8.4.1 are now assumed to be in force.

Here is the main result of the present number.

Theorem 8.4.4.1 (Harish-Chandra) Fix a point $H_0 \in \mathfrak{j}$ and an element $D \in \mathfrak{P}(\mathfrak{j})$; let $\Phi_{SI}^+(H_0)$ denote the set of all $\alpha \in \Phi_{SI}^+$ such that $\alpha(H_0) = 0$. If $w_\alpha(D) = -D$ for every $\alpha \in \Phi_{SI}^+(H_0)$ (w_α the Weyl reflection corresponding to α), then, for any $f \in C_c^\infty(\mathfrak{g})$, $D\phi_f$ can be extended to a continuous function around H_0 (i.e. there exists a neighborhood $\mathcal{O}(H_0)$

of H_0 in $\mathfrak{j}$ and a continuous function F on $\mathcal{O}(H_0)$ such that $F = D\phi_f$ on $\mathcal{O}(H_0) \cap \mathfrak{j}'$).

Corollary 8.4.4.2 Fix $f \in C_c^\infty(\mathfrak{g})$ – then $\partial(\pi)\phi_f$ admits a continuous extension to all of $\mathfrak{j}$.

The proof of Theorem 8.4.4.1 requires some preparation; in particular, the singular elements in $\mathfrak{j}$ will have to be taken into account – but, as we shall see, it will be enough to confine our attention to the 'best' among these, namely the semi-regular elements. In this latter situation it is possible to relate the invariant integral of a given $f \in C_c^\infty(\mathfrak{g})$ to that of a certain compactly supported C^∞ function on (essentially) $\mathfrak{sl}(2, \mathbf{R})$. The computations in 8.4.2 will then lead us quickly to the sought for result.

Fix a semi-regular element Γ in $\mathfrak{j}$ of the *non-compact* type; let $\mathfrak{g}_\Gamma$ denote the centralizer of Γ in $\mathfrak{g}$ – then we shall agree to use without comment the notations and conventions revolving around Lemmas 1.3.4.3 and 1.3.4.4. [In particular $\mathfrak{h}_\Gamma^+$ and $\mathfrak{h}_\Gamma^-$ will be chosen per the discussion following Lemma 1.3.4.4, thus are θ-stable; note too that $\mathfrak{l}_\Gamma$ can also be taken θ-stable.] Let H_Γ^+ (respectively H_Γ^-) be the Cartan subgroup of G associated with $\mathfrak{h}_\Gamma^+$ (respectively $\mathfrak{h}_\Gamma^-$); let L_Γ be the analytic subgroup of G with Lie algebra $\mathfrak{l}_\Gamma$. Because $L_\Gamma \cap \mathsf{H}_\Gamma^\pm$ is the centralizer of $\mathfrak{h}_\Gamma^\pm$ in L_Γ, $L_\Gamma \cap \mathsf{H}_\Gamma^\pm$ is the Cartan subgroup of L_Γ associated with $\mathfrak{l}_\Gamma \cap \mathfrak{h}_\Gamma^\pm$.

Note In what follows we shall write $\mathfrak{c}$, $\mathfrak{h}^\pm$, $\mathsf{H}^\pm$, and L for $\mathfrak{c}_\Gamma$, $\mathfrak{h}_\Gamma^\pm$, $\mathsf{H}_\Gamma^\pm$, and L_Γ, respectively (Γ being fixed until further notice).

Let L_K be the (compact) one parameter subgroup of G corresponding to $X^* - Y^*$; given $f \in C_c^\infty(\mathfrak{g}_\Gamma)$, write $f_{L_K}(\Xi) = \int_{L_K} f(k\Xi)dk$ $(\Xi \in \mathfrak{g}_\Gamma)$ (dk denoting normalized Haar measure on L_K).

Lemma 8.4.4.3 Fix $f \in C_c^\infty(\mathfrak{g}_\Gamma)$; put

$$\psi_f^+(H) = \int_{-\infty}^{\infty} f_{L_K}(H + tX^*)dt \qquad (H \in \mathfrak{h}^+)$$

and

$$\psi_f^-(H) = \xi \int_0^{\infty} f_{L_K}(H_\mathfrak{c} + \xi(e^t X^* - e^{-t} Y^*))(e^t - e^{-t})dt \qquad (H \in \mathfrak{h}^-)$$

when $\xi = \sqrt{-1}\alpha_-(H)/2 \neq 0$ ($H_\mathfrak{c}$ the component of H in $\mathfrak{c}$) – then there exist positive numbers c^+ and c^- such that

$$|\alpha_+(H)| \int_{L/L\cap \mathsf{H}^+} f(\dot{l}H)d_{L/L\cap\mathsf{H}^+}(\dot{l}) = c^+\psi_f^+(H) \qquad (H \in \mathfrak{h}^+ \cap \mathfrak{g}')$$

and

$$\alpha_-(H) \int_{L/L\cap \mathsf{H}^-} f(\dot{l}H)d_{L/L\cap\mathsf{H}^-}(\dot{l}) = -\sqrt{-1}c^-\psi_f^-(H) \qquad (H \in \mathfrak{h}^- \cap \mathfrak{g}').$$

[This result is an immediate consequence of the computations in 8.4.2.]

Let G_Γ be the centralizer of $\boldsymbol{\Gamma}$ in G, G_Γ the identity component of G_Γ – then $G_\Gamma\mathsf{H}^\pm$ is an open subgroup of G_Γ with $m^\pm = [\mathsf{G}_\Gamma : G_\Gamma\mathsf{H}^\pm] \leqslant 2$ (cf. the first Appendix to the present number). Fix an open convex neighborhood $\mathscr{O}_\Gamma$ of $\boldsymbol{\Gamma}$ in $\mathfrak{g}_\Gamma$ such that: (1) $\mathscr{O}_\Gamma \cap \mathfrak{h}^\pm$ verifies the hypotheses of Lemma 8.4.1.1; (2) There exist but two roots of the pair $(\mathfrak{g}_c, \mathfrak{h}_c^\pm)$ which ever take the value zero on $\mathscr{O}_\Gamma \cap \mathfrak{h}^\pm$. [Since $\boldsymbol{\Gamma}$ is semi-regular, it is possible, of course, to choose $\mathscr{O}_\Gamma$ in this way.] Let $x \mapsto \tilde{x}$ denote the natural mapping of G onto G/G_Γ – then the homogeneous space G/G_Γ carries a G-invariant measure (cf. Proposition 1.4.2.4 . . .). This being so, let us assume that the invariant measures on $G/\mathsf{H}^\pm$ and $\mathsf{G}_\Gamma/\mathsf{H}^\pm$ have been normalized in such a way that

$$\int_{G/\mathsf{H}^\pm} = \int_{G/\mathsf{G}_\Gamma} \int_{\mathsf{G}_\Gamma/\mathsf{H}^\pm}$$

Let $\dot{G}_\Gamma$ denote the image of G_Γ in $\mathsf{G}_\Gamma/\mathsf{H}^\pm$ and select elements $\xi_1 = 1, \ldots, \xi_{m^\pm}$ in G_Γ such that $\mathsf{G}_\Gamma = \bigcup_i \xi_i(G_\Gamma\mathsf{H}^\pm)$ – then $\mathsf{G}_\Gamma/\mathsf{H}^\pm$ is the disjoint union of the $\xi_i\dot{G}_\Gamma$ and $\dot{G}_\Gamma \sim G_\Gamma/G_\Gamma \cap \mathsf{H}^\pm$. [Under the assumption that we are considering $\mathsf{G}_\Gamma/\mathsf{H}^-$ with $m^- = 2$, we can suppose in addition that $\xi_2 H = w_- H$ (all $H \in \mathfrak{h}^-$) where w_- is the Weyl reflection corresponding to the root α_- (see the first Appendix to the present number) – then ξ_2 normalizes $\mathfrak{l}$ (as well as $\mathfrak{h}^-$).]

Fix a relatively compact open set ω in $\mathfrak{g}$ – then, in view of our assumption on $\mathscr{O}_\Gamma$, there exists a compact set $\boldsymbol{\Omega}^\pm$ in G/G_Γ such that $\tilde{x} \in \boldsymbol{\Omega}^\pm$ whenever $xH \in \omega$ ($x \in G$, $H \in \mathscr{O}_\Gamma \cap \mathfrak{h}^\pm$). Normalize the Haar measures on G and G_Γ by the requirement $\int_G = \int_{G/\mathsf{G}_\Gamma} \int_{\mathsf{G}_\Gamma}$; fix a $\psi \in C_c^\infty(G)$ such that $\int_{\mathsf{G}_\Gamma} \psi(x\xi) d_{\mathsf{G}_\Gamma}(\xi) = 1$ if $\tilde{x} \in \boldsymbol{\Omega}^\pm$ ($x \in G$) and, for any $f \in C_c^\infty(\omega)$, put

$$v_f^\pm(\Xi) = \sum_i \int_G \psi(x) f(x\xi_i\Xi) d_G(x) \qquad (\Xi \in \mathfrak{g}_\Gamma; i = 1, \ldots, m^\pm).$$

If $H \in \mathfrak{h}^\pm \cap \mathfrak{g}'$, then it is clear that

$$\int_{G/\mathsf{H}^\pm} f(\dot{x}H) d_{G/\mathsf{H}^\pm}(\dot{x}) = \int_{G/\mathsf{G}_\Gamma} f(H:\tilde{x}) d_{G/\mathsf{G}_\Gamma}(\tilde{x})$$

where

$$f(H:\tilde{x}) = \int_{\mathsf{G}_\Gamma/\mathsf{H}^\pm} f(x(\dot{\xi}H)) d_{\mathsf{G}_\Gamma/\mathsf{H}^\pm}(\dot{\xi}) = \sum_i \int_{\dot{G}_\Gamma} f(x\xi_i(\dot{\xi}H)) d_{\mathsf{G}_\Gamma/\mathsf{H}^\pm}(\dot{\xi}) \qquad (x \in G).$$

Consequently, for $H \in \mathscr{O}_\Gamma \cap \mathfrak{h}^\pm \cap \mathfrak{g}'$, we have

$$\int_{G/\mathsf{H}^\pm} f(\dot{x}H) d_{G/\mathsf{H}^\pm}(\dot{x}) = \int_{\boldsymbol{\Omega}} f(H:\tilde{x}) d_{G/\mathsf{G}_\Gamma}(\tilde{x}) = \int_{\dot{G}_\Gamma} v_f^\pm(\dot{\xi}H) d_{\mathsf{G}_\Gamma/\mathsf{H}^\pm}(\dot{\xi}) \qquad (\text{all } f \in C_c^\infty(\omega)).$$

If we let $\phi_f^{\pm}$ denote the invariant integral of f ($f \in C_c^\infty(\mathfrak{g})$) calculated relative to the θ-stable Cartan subalgebra $\mathfrak{h}^{\pm}$, then, in the evident notations, the above calculations provide us with the relation

$$\phi_f^{\pm}(H) = \epsilon_R^{\pm}(H)\pi^{\pm}(H) \int_{\dot{G}_\Gamma} v_f^{\pm}(\dot{\xi}H) d_{G_\Gamma/\mathsf{H}^{\pm}}(\dot{\xi}) \qquad (H \in \mathcal{O}_\Gamma \cap \mathfrak{h}^{\pm} \cap \mathfrak{g}'),$$

valid for all $f \in C_c^\infty(\omega)$.

In order to exploit the preceding observations, it is necessary to note that $\dot{G}_\Gamma \sim G_\Gamma/G_\Gamma \cap \mathsf{H}^{\pm} \sim L/L \cap \mathsf{H}^{\pm}$ (with preservation of the relevant invariant measures). [This is clear since $G_\Gamma = L_\Gamma \exp(\mathfrak{c}_\Gamma)$.]

Lemma 8.4.4.4 Fix $f \in C_c^\infty(\omega)$ and set $f_\psi(\Xi) = \int_G \psi(x) f(x\Xi) d_G(x)$ ($\Xi \in \mathfrak{g}_\Gamma$); put $\pi_+ = \alpha_+^{-1}\pi^+$, $\pi_- = \alpha_-^{-1}\pi^-$ – then there exist non-zero real numbers r^+ and r^- (independent of f, ψ and ω) such that

$$\phi_f^{+}(H) = r^+\pi_+(H)\psi_{f_\psi}^{+}(H) \qquad (H \in \mathcal{O}_\Gamma \cap \mathfrak{h}^{+} \cap \mathfrak{g}')$$

and

$$\phi_f^{-}(H) = \begin{cases} \sqrt{-1}r^-\pi_-(H)\psi_{f_\psi}^{-}(H)(m^- = 1) \\ \qquad\qquad\qquad\qquad\qquad (H \in \mathcal{O}_\Gamma \cap \mathfrak{h}^{-} \cap \mathfrak{g}'). \\ \sqrt{-1}r^-\pi_-(H)\{\psi_{f_\psi}^{-}(H) - \psi_{f_\psi}^{-}(w_-H)\}(m^- = 2) \end{cases}$$

Proof (1) We shall deal first with $\mathfrak{h}^+$. Since α_+ is a real root, $\pi_+(= \alpha_+^{-1}\pi^+)$ and $\pi_{R,+}(= \alpha_+^{-1}\pi_R^+)$ are both polynomial functions on $\mathfrak{h}_c^+$ which are nowhere zero on the convex set $\mathcal{O}_\Gamma \cap \mathfrak{h}^+$, whence $\epsilon_R^+(H) = \operatorname{sign}\{\pi_{R,+}(\Gamma)\} \cdot \operatorname{sign}\{\alpha_+(H)\}$ ($H \in \mathcal{O}_\Gamma \cap \mathfrak{h}^+$). In the case at hand, $m^+ = 1$ (cf. the first Appendix to the present number) – thus $v_f^+ = f_\psi$ and so, bearing in mind Lemma 8.4.4.3, we find that

$$\begin{aligned} \phi_f^{+}(H) &= \operatorname{sign}\{\pi_{R,+}(\Gamma)\} \cdot \pi_+(H) \,|\alpha_+(H)| \\ &\quad \times \int_{L/L\cap \mathsf{H}^+} d_{L/L\cap\mathsf{H}^+}(\dot{l}) \int_G \psi(x) f(x(\dot{l}H)) d_G(x) \\ &= r^+\pi_+(H)\psi_{f_\psi}^{+}(H) \qquad (H \in \mathcal{O}_\Gamma \cap \mathfrak{h}^{+} \cap \mathfrak{g}'). \end{aligned}$$

Here we have put $r^+ = \operatorname{sign}\{\pi_{R,+}(\Gamma)\} \cdot c^+$ (strictly speaking, this c^+ is not the c^+ of Lemma 8.4.4.3 – the present normalization of the L-invariant measure on $L/L \cap \mathsf{H}^+$ need not be the 'canonical' one).

(2) Let us now consider $\mathfrak{h}^-$. In this case α_- is a singular imaginary root of the pair $(\mathfrak{g}_c, \mathfrak{h}_c^-)$ and so $\epsilon_R^-(H) = \epsilon_R^-(\Gamma)$ ($H \in \mathcal{O}_\Gamma \cap \mathfrak{h}^-$). Hence, for $H \in \mathcal{O}_\Gamma \cap \mathfrak{h}^- \cap \mathfrak{g}'$, we have

$$\begin{aligned} \phi_f^{-}(H) &= \epsilon_R^-(\Gamma)\pi^-(H) \int_{L/L\cap\mathsf{H}^-} d_{L/L\cap\mathsf{H}^-}(\dot{l}) \\ &\quad \times \sum_{i=1}^{m^-} \int_G \psi(x) f(x\xi_i(\dot{l}H)) d_G(x). \end{aligned}$$

It has been pointed out above that each ξ_i normalizes both $\mathfrak{l}$ and $\mathfrak{h}^-$

and so also L and H^- – therefore it is clear that

$$\int_{L/L\cap\mathsf{H}^-} f(x\xi_i(\dot{l}H))d_{L/L\cap\mathsf{H}^-}(\dot{l}) = \int_{L/L\cap\mathsf{H}^-} f(x\dot{l}(\xi_i H))d_{L/L\cap\mathsf{H}^-}(\dot{l}).$$

But $\xi_2 H = w_- H$ if $m^- = 2$ (w_- the Weyl reflection determined by α_-) – thus, in view of Lemma 8.4.4.3, we have

$$\phi_f^-(H) = \begin{cases} \sqrt{-1}r^-\pi_-(H)\psi_{f_\psi}^-(H)(m^-=1) \\ \qquad\qquad\qquad\qquad\qquad (H \in \mathcal{O}_\Gamma \cap \mathfrak{h}^- \cap \mathfrak{g}'). \\ \sqrt{-1}r^-\pi_-(H)\{\psi_{f_\psi}^-(H) - \psi_{f_\psi}^-(w_-H)\}(m^-=2) \end{cases}$$

Here $r^- = -\epsilon_R(\Gamma)c^-$ (again the present c^- need not be the c^- of Lemma 8.4.4.3 . . .).

This completes the proof of the lemma. □

Proof of Theorem 8.4.4.1 Fix a point H_0 in $\mathfrak{j}$ per the statement of the theorem.

(1) To begin with, let us suppose that $\Phi_{SI}^+(H_0)$ is empty – then $H_0 \in \mathfrak{j}'(SI)$ and so the required result is seen to be a consequence of Corollary 8.4.3.2 in this case.

(2) Suppose now that $\Phi_{SI}^+(H_0)$ is not empty but that H_0 is semi-regular – then, of necessity, H_0 is of the non-compact type and, moreover, the unique positive root of the pair $(\mathfrak{g}_c, \mathfrak{j}_c)$ which vanishes at H_0 is singular imaginary ($\Phi_{SI}^+(H_0)$ being non-empty). The present H_0 can play the role of the Γ supra, whence $\mathfrak{j}$ is conjugate to $\mathfrak{h}_\Gamma^-$ (cf. Lemma 1.3.4.3); accordingly, in this situation, our contention is seen to be a consequence of the following statement: Let D be an element in $\mathfrak{P}(\mathfrak{h}_\Gamma^-)$ such that $w_-(D) = -D$ – then, for any $f \in C_c^\infty(\mathfrak{g})$, $D\phi_f^-$ can be extended to a continuous function around Γ $(= H_0)$. [Keeping to the earlier notations, note that $w_-(D\circ\pi_-) = -(D\circ\pi_-)$; therefore, since $\mathfrak{P}(\mathfrak{h}_\Gamma^-) = \mathfrak{P}(\mathfrak{c}_\Gamma)\mathfrak{P}(\mathbf{R}(X^* - Y^*))$, $D\circ\pi_-$ can be written as a finite sum of elements of the form $\Delta\circ\alpha_-^i\partial(X^* - Y^*)^j$ where $\Delta \in \mathfrak{P}(\mathfrak{c}_\Gamma)$ and $i + j$ is odd (note that w_- leaves $\mathfrak{c}_\Gamma$ invariant). Now fix an $f \in C_c^\infty(\mathfrak{g})$ and choose ω large enough to ensure that spt $(f) \subset \omega$ – then, in view of Lemma 8.4.4.4, it will be enough to establish that the function

$$H \mapsto \alpha_-(H)^i\psi_{\Delta f_\psi}^-(H;\partial(X^* - Y^*)^j) \qquad (H \in \mathcal{O}_\Gamma \cap \mathfrak{h}_\Gamma^- \cap \mathfrak{g}')$$

admits a continuous extension to $\mathcal{O}_\Gamma \cap \mathfrak{h}_\Gamma^-$. But, if $i \geqslant 1$, then, since α_- vanishes on $\mathfrak{c}_\Gamma$, this can be done by defining the extended function to be zero on $\mathcal{O}_\Gamma \cap \mathfrak{c}_\Gamma$; if $i = 0$, then the existence of the desired extension follows from Lemma 8.4.2.5 (j being odd).]

(3) Turning to the general case, let H_0 be an arbitrary element in $\mathfrak{j}$. Fix a relatively compact open convex neighborhood $\mathcal{O}(H_0)$ of H_0 in $\mathfrak{j}$ with the following property: If $\alpha \in \Phi_{SI}^+$ and $\alpha(H_0) \neq 0$, then α never takes the value zero on $\mathcal{O}(H_0)$. Let $\mathcal{O}^\sharp(H_0)$ be the set of all points in $\mathcal{O}(H_0)$ which are either regular or semi-regular – then it follows from what

has been said above that $F = D\phi_f$ can be extended to a continuous function on $\mathcal{O}'(H_0)$. We shall now prove that F admits a continuous extension to all of $\mathcal{O}(H_0)$. To see this, fix a Euclidean norm $\| \,.\, \|$ on $\mathfrak{j}$ and set $\mu(F) = \sup |F(H_1; \partial(H_2))|$ where H_1 and H_2 vary in $\mathfrak{j}'$ and $\mathfrak{j}$ under the sole restriction that $\| H_2 \| \leqslant 1$ – then Theorem 8.4.3.1 tells us that $\mu(F) < \infty$. Let H_1 and H_2 be two points in $\mathcal{O}(H_0) \cap \mathfrak{j}'$. If we can show that necessarily $|F(H_1) - F(H_2)| \leqslant \mu(F) \| H_2 - H_1 \|$, then, thanks to the density of $\mathcal{O}(H_0) \cap \mathfrak{j}'$ in $\mathcal{O}(H_0)$, there can be no doubt that F is extendible to a continuous function on $\mathcal{O}(H_0)$. To verify that F satisfies the 'Lipschitz condition' in question, assume first that the straight line segment

$$H_t = (2 - t)H_1 + (t - 1)H_2 \quad (t \in I = [1, 2])$$

lies entirely in $\mathcal{O}'(H_0)$ – then the function $t \mapsto F(H_t)$ is continuous in t ($t \in I$) and, except at the (at most) finite number of $t \in I$ such that $\pi(H_t) = 0$, we have the relation $dF(H_t)/dt = F(H_t; \partial(H_2 - H_1))$. Consequently

$$|F(H_2) - F(H_1)| \leqslant \int_I |F(H_t; \partial(H_2 - H_1))| \, dt \leqslant \mu(F) \|H_2 - H_1\|.$$

On the other hand, if H_1 and H_2 are any two points in $\mathcal{O}(H_0) \cap \mathfrak{j}'$, then we can choose $H_i' \in \mathcal{O}(H_0) \cap \mathfrak{j}'$ arbitrarily near H_i ($i = 1,2$) such that the straight line segment joining H_1' to H_2' lies entirely in $\mathcal{O}'(H_0)$ (cf. Lemma 8.4.4.5 infra); but, in view of the preceding observation (applied to H_1' and H_2'), we then have that

$$|F(H_2') - F(H_1')| \leqslant \mu(F) \| H_2' - H_1' \|;$$

by continuity, the same inequality is valid with H_1' replaced by H_1, H_2' replaced by H_2. . . .

Hence the theorem. □

The proof of the following elementary lemma may be left to the reader.

Lemma 8.4.4.5 Let E be a finite dimensional vector space over **R**, $\mathcal{O}$ a non-empty, open, and convex subset of E. Let $\mathcal{P}$ be the complement in $\mathcal{O}$ of the union of a finite number of affine subspaces of dimension $\leqslant r < n = \dim(E)$. Let $\mathcal{P}'$ be an open and dense subset of $\mathcal{P}$ – then, given any two points X_1, X_2 in $\mathcal{P}$, we can find points X_i' in $\mathcal{P}'$ arbitrarily near X_i ($i = 1, 2$) with the following property: Let I denote the straight line segment joining X_1' to X_2' – then $I \subset \mathcal{P}$ if $r < n - 1$ and the complement of $I \cap \mathcal{P}$ in I is a finite set if $r = n - 1$.

Let E be a finite dimensional Euclidean space over **R**, $\mathcal{O}$ an open subset of E, $\tilde{E}$ a closed subset of E; assume that $\mathcal{O} \cap \tilde{E}$ is open in E and dense in $\tilde{E}$. Let f be a C^∞ function on $\mathcal{O}$ – then f is said to be of class C^∞ on $\tilde{E}$ if, for every $p \in S(E_c)$, there exists a continuous function f_p on $\tilde{E}$ which coincides with $\partial(p)f$ on $\mathcal{O} \cap \tilde{E}$. [It is obvious that f_p, if it exists, is unique.]

Lemma 8.4.4.6 Let $\mathcal{O}$ be a non-empty, open, and convex subset of E; let f be a C^∞ function on $\mathcal{O}$ such that $\sup_{\mathcal{O}} |\partial(p)f| < \infty$ for every $p \in S(E_c)$ – then f is of class C^∞ on $\mathcal{O}^{cl}$.

Proof Fix $p \in S(E_c)$ – then we have to prove that $\partial(p)f$ can be extended to a continuous function on $\mathcal{O}^{cl}$. Because $\partial(p)f$ satisfies the same conditions as f, it will be sufficient to prove that f itself can be so extended. Fix a Euclidean norm $\|\,.\,\|$ on E and let $e_1, \ldots, e_n$ be an orthonormal basis for E over $\mathbf{R}$ (relative to the norm in question); fix a point X_0 in $\mathcal{O}^{cl}$ and, for a given $\epsilon > 0$, let $\mathcal{S}_\epsilon(X_0)$ denote the subset of E comprised of those X with $\|X - X_0\| < \epsilon$ – then, for any two points X_1, X_2 in $\mathcal{O} \cap \mathcal{S}_\epsilon(X_0)$, we have

$$f(X_2) - f(X_1) = \int_0^1 (d/dt)f(X_1 + t(X_2 - X_1))dt,$$

whence

$$|f(X_2) - f(X_1)| \leqslant \sup_{\mathcal{O}} |\partial(X_2 - X_1)f| \leqslant 2\epsilon \max_i \sup_{\mathcal{O}} |\partial(e_i)f|,$$

from which it now follows that f can be extended to a continuous function on $\mathcal{O}^{cl}$. □

Application Returning to the earlier notations and assumptions, fix an f in $\mathscr{C}(\mathfrak{g})$ – then ϕ_f is of class C^∞ on the closure of each connected component of $\mathfrak{j}'(SI)$.

Appendix 1 Fix a semi-regular element Γ in $\mathfrak{g}$ of non-compact type; we shall then agree to use without comment the notations and conventions revolving around Lemmas 1.3.4.3 and 1.3.4.4. As there, introduce the Cartan subalgebras $\mathfrak{h}_\Gamma^+$ and $\mathfrak{h}_\Gamma^-$ of $\mathfrak{g}_\Gamma = \mathfrak{l}_\Gamma + \mathfrak{c}_\Gamma$ (and hence of $\mathfrak{g}$); let H_Γ^+ (respectively H_Γ^-) be the Cartan subgroup of G associated with $\mathfrak{h}_\Gamma^+$ (respectively $\mathfrak{h}_\Gamma^-$); let L_Γ be the analytic subgroup of G corresponding to $\mathfrak{l}_\Gamma$; let G_Γ be the centralizer of Γ in G, G_Γ the identity component of G_Γ – then it is clear that L_Γ is a normal subgroup of G_Γ and, moreover, $G_\Gamma \mathsf{H}_\Gamma^\pm$ is an open subgroup of G_Γ (note that $G_\Gamma = L_\Gamma \exp(\mathfrak{c}_\Gamma)$).

Lemma 1 Retain the above notations – then $[\mathsf{G}_\Gamma : G_\Gamma \mathsf{H}_\Gamma^\pm] \leqslant 2$.

Proof Because $\mathfrak{l}_\Gamma$ is a non-compact simple Lie algebra of dimension 3, the facts outlined in the Example following Corollary 1.3.4.2 are available – we shall employ them without comment during the ensuing discussion. Let ξ be any element in G_Γ – then $\xi(\mathfrak{l}_\Gamma) = \mathfrak{l}_\Gamma$ and so

$$\omega(\xi H^*) = \omega(H^*), \quad \omega(\xi(X^* - Y^*)) = \omega(X^* - Y^*)$$

(ω the Casimir polynomial of $\mathfrak{l}_{\Gamma,c}$). This shows that

$$\xi H^* \in \mathfrak{l}_1, \quad \xi(X^* - Y^*) \in \mathfrak{l}_2 \cup \mathfrak{l}_3.$$

Consequently there exists $l_1, l_2 \in L_\Gamma$ such that

$$l_1^{-1}\xi H^* \in \mathbf{R}^+ H^*, \quad l_2^{-1}\xi(X^* - Y^*) \in \mathbf{R}'(X^* - Y^*),$$

whence

$$l_1^{-1}\xi H^* = H^*, \quad l_2^{-1}\xi(X^* - Y^*) = \pm(X^* - Y^*).$$

If t is sufficiently small and positive, then $\mathbf{\Gamma} + tH^*$ is regular and, thanks to Theorem 1.4.1.5, its centralizer in G is H_Γ^+ – therefore

$$l_1^{-1}\xi \in \mathsf{H}_\Gamma^+ \Rightarrow \xi \in L_\Gamma \mathsf{H}_\Gamma^+ = G_\Gamma \mathsf{H}_\Gamma^+ \Rightarrow [\mathsf{G}_\Gamma : G_\Gamma \mathsf{H}_\Gamma^+] = 1.$$

Turning to $G_\Gamma \mathsf{H}_\Gamma^-$, choose, if possible, an element $\xi_- \in \mathsf{G}_\Gamma$ such that

$$\xi_-(X^* - Y^*) = -(X^* - Y^*);$$

in case $l_2^{-1}\xi(X^* - Y^*) = (X^* - Y^*)$, the immediately preceding argument (applied to $\mathbf{\Gamma} + t(X^* - Y^*)$ $(t \in \mathbf{R}')$) shows that $\xi \in G_\Gamma \mathsf{H}_\Gamma^-$; on the other hand, if $l_2^{-1}\xi(X^* - Y^*) = -(X^* - Y^*)$, then

$$\xi \in L_\Gamma \xi_- \mathsf{H}_\Gamma^- = \xi_- L_\Gamma \mathsf{H}_\Gamma^- = \xi_- G_\Gamma \mathsf{H}_\Gamma^-$$

(since $\xi_-^{-1} l_2^{-1}\xi(X^* - Y^*) = (X^* - Y^*)$), whence $[\mathsf{G}_\Gamma : G_\Gamma \mathsf{H}_\Gamma^-] \leqslant 2$. □

The proof of Lemma 1 actually shows that $[\mathsf{G}_\Gamma : G_\Gamma \mathsf{H}_\Gamma^+] = 1$; moreover it is also clear that $[\mathsf{G}_\Gamma : G_\Gamma \mathsf{H}_\Gamma^-] = 1$ if ξ_- does not exist – however, if ξ_- does exist, then $[\mathsf{G}_\Gamma : G_\Gamma \mathsf{H}_\Gamma^-] = 2$. [Suppose, to the contrary, that $\xi_- \in G_\Gamma \mathsf{H}_\Gamma^- = L_\Gamma \mathsf{H}_\Gamma^-$; write $\xi_- = l_- h_-$ (say) – then

$$l_-(X^* - Y^*) = \xi_-(X^* - Y^*) = -(X^* - Y^*),$$

an impossibility (cf. the Example following Corollary 1.3.4.2).]

Lemma 2 Retain the above notations – then $[\mathsf{G}_\Gamma : G_\Gamma \mathsf{H}_\Gamma^-] = 2$ iff there exists an element $\xi_- \in \mathsf{G}_\Gamma$ which coincides with w_- on $\mathfrak{h}_{\Gamma,c}^-$ (w_- the Weyl reflection corresponding to the root α_-).

Proof It is evident that ξ_- normalizes $\mathfrak{h}_{\Gamma,c}^-$; this being so, choose an element w in the Weyl group of the pair $(\mathfrak{g}_c, \mathfrak{h}_{\Gamma,c}^-)$ which agrees with ξ_- on $\mathfrak{h}_{\Gamma,c}^-$ – then

$$\begin{aligned} w(\mathbf{\Gamma} + t(X^* - Y^*)) &= \xi_-(\mathbf{\Gamma} + t(X^* - Y^*)) = \mathbf{\Gamma} - t(X^* - Y^*) \\ &= w_-(\mathbf{\Gamma} + t(X^* - Y^*)) \qquad (t \in \mathbf{R}). \end{aligned}$$

But, for a suitable choice of t, $\mathbf{\Gamma} + t(X^* - Y^*)$ is regular in $\mathfrak{g}$; since $w_- w$ leaves it fixed, it follows that $w = w_-$, as desired. □

Appendix 2 Let us agree to retain the general assumptions and notations which were introduced earlier in the number.

Fix a semi-regular element $\mathbf{\Gamma}_0$ in $\mathfrak{g}$ of the non-compact type; by conjugating $\mathbf{\Gamma}_0$ if necessary, we can assume that $\mathbf{\Gamma}_0$ belongs to a θ-stable Cartan subalgebra $\mathfrak{j}$ (say) of $\mathfrak{g}$; this being so, relative to $\mathbf{\Gamma}_0$, introduce $\mathfrak{h}^\pm$, $\mathsf{H}^\pm$, L, ν etc. in the usual way; let $\phi_f^\pm$ denote the invariant integral of f $(f \in C_c^\infty(\mathfrak{g}))$ calculated relative to the θ-stable Cartan subalgebra $\mathfrak{h}^\pm$.

Lemma There exists a locally constant function c on the set of semi-regular elements in $\mathfrak{h}^+ \cap \mathfrak{h}^-$ which is nowhere zero and has the property that for every $D \in \mathfrak{P}(\mathfrak{h}^+)$

$$\phi_f^+(\Gamma; D) = c(\Gamma)\{\lim_{t\downarrow 0} \phi_f^-(\Gamma + t(X^* - Y^*); \nu D) - \lim_{t\uparrow 0} \phi_f^-(\Gamma + t(X^* - Y^*); \nu D)\} \qquad (\text{all } f \in C_c^\infty(\mathfrak{g})),$$

Γ semi-regular in $\mathfrak{h}^+ \cap \mathfrak{h}^-$.

[Observe that ϕ_f^+ ($f \in C_c^\infty(\mathfrak{g})$) is of class C^∞ around every semi-regular point in $\mathfrak{h}^+ \cap \mathfrak{h}^-$ (on the basis of Corollary 8.4.3.2, this is clear (α_+ being a real root of the pair $(\mathfrak{g}_c, \mathfrak{h}_c^+)$).]

Proof Fix a semi-regular element Γ in $\mathfrak{h}^+ \cap \mathfrak{h}^-$ ($= \mathfrak{c} =$ the center of $\mathfrak{g}_{\Gamma_0}$) – then certainly $\mathfrak{g}_\Gamma = \mathfrak{g}_{\Gamma_0}$, whence Γ is of the non-compact type. This means that all of the machinery developed in the discussion leading up to the proof of Theorem 8.4.4.1 supra is now applicable in the obvious manner (relative to Γ). To establish our lemma, then, it will be enough to exhibit a non-zero constant c such that the asserted relation holds for all $H \in \mathcal{O}_\Gamma \cap \mathfrak{c}$.

(1) To begin with, let us assume that D is w_+-skew – then we have $(D\phi_f^+)^{w_+} = -(D\phi_f^+)$. Because $D\phi_f^+$ is of class C^∞ on $(\mathfrak{h}^+)'(SI)$ (cf. Corollary 8.4.3.2), it must therefore be the case that $D\phi_f^+ = 0$ on $\mathcal{O}_\Gamma \cap \mathfrak{c}$. On the other hand, $w_-(\nu D) = -\nu D$; so, in view of what was said during part (2) of the proof of Theorem 8.4.4.1, it follows that

$$\lim_{t\downarrow 0} \phi_f^-(H + t(X^* - Y^*); \nu D) - \lim_{t\uparrow 0} \phi_f^-(H + t(X^* - Y^*); \nu D) = 0 \qquad (\text{all } H \in \mathcal{O}_\Gamma \cap \mathfrak{c}).$$

(2) Let us now consider the case when D has the property that it is w_+-invariant. Fix an $f \in C_c^\infty(\mathfrak{g})$ and choose a relatively compact open subset ω of $\mathfrak{g}$ such that spt $(f) \subset \omega$ – then, since $\mathcal{O}_\Gamma \cap \mathfrak{h}^+$ is contained in $(\mathfrak{h}^+)'(SI)$, Lemma 8.4.4.4 tells us that

$$\phi_f^+(H; D) = r^+\psi_{f_\psi}^+(H; D\circ\pi_+) \qquad (H \in \mathcal{O}_\Gamma \cap \mathfrak{h}^+ \cap \mathfrak{g}').$$

In view of the fact that $\mathfrak{P}(\mathfrak{h}^+) = \mathfrak{P}(\mathfrak{c})\mathfrak{P}(\mathbf{R}H^*)$, $D\circ\pi_+$ can be written as a finite linear combination of terms of the form $\tilde{D} = \Delta\circ\alpha_+^i\partial(H^*)^j$ where $\Delta \in \mathfrak{P}(\mathfrak{c})$ and $i + j$ is even (evidently $D\circ\pi_+$ is w_+-invariant); plainly

$$\nu\tilde{D} = (-1)^{j/2}\Delta\circ\alpha_-^i\partial(X^* - Y^*)^j.$$

(a) Suppose that $i \geqslant 1$ – then it is clear that

$$\psi_{f_\psi}^+(H; \tilde{D}) = \alpha_+(H)^i\psi_{\Delta f_\psi}^+(H; \partial(H^*)^j) = 0 \qquad (\text{all } H \in \mathcal{O}_\Gamma \cap \mathfrak{c})$$

and likewise

$$\lim_{t\to 0} \psi_{f_\psi}^-(H + t(X^* - Y^*); \nu\tilde{D}) = 0 \qquad (\text{all } H \in \mathcal{O}_\Gamma \cap \mathfrak{c}).$$

(b) Suppose that $i = 0$ (in which case j is even) – then it follows from the computations in 8.4.2 that

$$\begin{aligned}\psi_{f_\psi}^+(H; \tilde{D}) &= \psi_{\Delta f_\psi}^+(H; \partial(H^*)^j) \\ &= (-1)^{j/2}\{\lim_{t\downarrow 0} \psi_{\Delta f_\psi}^-(H + t(X^* - Y^*); \partial(X^* - Y^*)^j) \\ &\quad - \lim_{t\uparrow 0} \psi_{\Delta f_\psi}^-(H + t(X^* - Y^*); \partial(X^* - Y^*)^j)\} \\ &= \lim_{t\downarrow 0} \psi_{f_\psi}^-(H + t(X^* - Y^*); \nu\tilde{D}) \\ &\quad - \lim_{t\uparrow 0} \psi_{f_\psi}^-(H + t(X^* - Y^*); \nu\tilde{D})\end{aligned}$$

for all $H \in \mathcal{O}_\Gamma \cap \mathfrak{c}$. Put $c = -\sqrt{-1}r^+/m^-r^-$ ($\neq 0$); taking into account Lemma 8.4.4.4, we then have

$$\phi_f^+(H;\tilde{D}) = c\{\lim_{t\downarrow 0}\phi_f^-(H + t(X^* - Y^*);\nu\tilde{D}) - \lim_{t\uparrow 0}\phi_f^-(H + t(X^* - Y^*);\nu\tilde{D})\} \qquad (\text{all } H \in \mathcal{O}_\Gamma \cap \mathfrak{c}).$$

Given an arbitrary $D \in \mathfrak{P}(\mathfrak{h}^+)$, set

$$D' = 2^{-1}(D + w_+D), \quad 'D = 2^{-1}(D - w_+D)$$

– then $D = D' + 'D$ with $w_+(D') = D'$, $w_+('D) = -'D$. This being the case, there can now be no doubt as to the validity of our lemma. ☐

Theorem (Harish-Chandra) Let $(\mathfrak{g}, \mathfrak{k})$ be a reductive pair satisfying the usual conditions; suppose that rank $(\mathfrak{g})$ = rank $(\mathfrak{k})$ and fix a Cartan subalgebra $\mathfrak{t}$ of $\mathfrak{k}$ (hence of $\mathfrak{g}$). Let f be a function in $\mathscr{C}(\mathfrak{g})$ whose invariant integral calculated relative to every non-compact Cartan subalgebra of $\mathfrak{g}$ vanishes identically – then the invariant integral ϕ_f of f calculated relative to the compact Cartan subalgebra $\mathfrak{t}$ extends to a rapidly decreasing function on $\mathfrak{t}$.

Proof Fix an element $H_0 \in \mathfrak{t}$. If $H_0 \in \mathfrak{t}'(SI)$, then, as we know, ϕ_f is C^∞ in a neighborhood of H_0; if H_0 has the property that there exists exactly one positive singular imaginary root of the pair $(\mathfrak{g}_c, \mathfrak{t}_c)$ which annihilates it, then, in view of our hypotheses on f and the preceding lemma (which is surely applicable even though f need only lie in $\mathscr{C}(\mathfrak{g})\ldots$), ϕ_f can again be extended to a C^∞ function in a neighborhood of H_0. It follows, therefore, that ϕ_f extends to a C^∞ function on the set of those points in $\mathfrak{t}$ which are either regular or semi-regular. Since $D\phi_f$ is bounded on $\mathfrak{t}'$ for every $D \in \mathfrak{P}(\mathfrak{t})$, the by now familiar 'Lipschitz condition' argument serves to ensure us that ϕ_f is a C^∞ function on all of $\mathfrak{t}$ (cf. part (3) of the proof of Theorem 8.4.4.1). The fact that ϕ_f belongs to $\mathscr{C}(\mathfrak{t})$ is then clear. ☐

8.4.5 The Main Theorem

Let $\mathfrak{g}$ be a reductive Lie algebra over **R**, $\theta : \mathfrak{g} \to \mathfrak{g}$ an involutive automorphism of $\mathfrak{g}$ whose restriction to the derived algebra $\bar{\mathfrak{g}}$ is a Cartan involution (of $\bar{\mathfrak{g}}$); write, as usual, $\mathfrak{g} = \mathfrak{k} + \mathfrak{p}$ and let (G, K) be a pair corresponding to $\mathfrak{g}$ per Vol. I, number 1.1.5 – the notations and conventions of 8.4.1 are now assumed to be in force.

In what follows it will be supposed that G is *not* compact.

The present number will be devoted to proving the following theorem.

Theorem 8.4.5.1 (Harish-Chandra) Let $\mathfrak{j}$ be a θ-stable fundamental Cartan subalgebra of $\mathfrak{g}$ – then there exists a non-zero real number $\mathbf{M}_\mathfrak{g}$ such that $\phi_f(0; \partial(\pi)) = \mathbf{M}_\mathfrak{g} f(0)$ (all $f \in \mathscr{C}(\mathfrak{g})$). In fact if r denotes the number of positive roots of the pair $(\mathfrak{g}_c, \mathfrak{j}_c)$, then

$$\mathbf{M}_\mathfrak{g} = (-1)^{\mathbf{m}_\mathfrak{g}}(2\pi)^r \qquad (\pi = 3.14\ldots)$$

where

$$\mathbf{m}_\mathfrak{g} = 2^{-1}(\dim(G/K) - \operatorname{rank}(G) + \operatorname{rank}(K)).$$

[In view of the discussion revolving around the statement of Theorem 8.4.1.5, it is clear that we need only consider the case when $\mathfrak{j}$ is fundamental. Let us also bear in mind that the function $H \mapsto \phi_f(H; \partial(\pi))$ $(H \in \mathfrak{j})$ is continuous (all $f \in \mathscr{C}(\mathfrak{g})$); cf. Corollary 8.4.4.2.]

The importance of Theorem 8.4.5.1 cannot be overemphasized; indeed the Plancherel Theorem for G rests ultimately upon it.

We shall begin by establishing the existence of the constant $\mathbf{M}_{\mathfrak{g}}$; after that we shall prove that $\mathbf{M}_{\mathfrak{g}}$ is not zero by actually calculating its value explicitly.

Let us first assume that $\mathfrak{g}$ is semi-simple. To establish the existence of $\mathbf{M}_{\mathfrak{g}}$, it will be enough to show that the distribution $f \mapsto \phi_f(0; \partial(\pi))$ is a scalar multiple of the Haar measure on $\mathfrak{g}$ – in turn this requires some preparation which will now be undertaken.

For any $f \in \mathscr{C}(\mathfrak{g})$, define its Fourier transform by

$$\hat{f}(X) = \frac{1}{(2\pi)^{n/2}} \int_{\mathfrak{g}} f(Y) e^{\sqrt{-1}B(X,Y)} d_{\mathfrak{g}}(Y) \qquad (n = \dim(\mathfrak{g}))$$

where, as always, B denotes the Killing form on $\mathfrak{g}$. Fix an $H_0 \in \mathfrak{j}'$; put

$$T_{H_0}(f) = \phi_f(H_0) = \pi(H_0) \int_{G/J} \hat{f}(\dot{x}H_0) d_{G/J}(\dot{x}) \qquad (f \in \mathscr{C}(\mathfrak{g})).$$

The integral appearing here is absolutely convergent; furthermore the mapping $f \mapsto T_{H_0}(f)$ defines a central, tempered distribution on $\mathfrak{g}$ which satisfies the system of differential equations

$$\partial(p) \cdot T_{H_0} = p(\sqrt{-1}H_0) T_{H_0} \qquad (p \in I(\mathfrak{g}_c)).$$

[All this follows at once from the results sketched out in 8.4.3 (cf., especially, Theorem 8.4.3.1).] But then, thanks to Theorem 8.3.1.1, it must be the case that T_{H_0} coincides with a locally summable function $F_{T_{H_0}} = F_{H_0}$ on $\mathfrak{g}$ – moreover, since there are no real roots present here ($\mathfrak{j}$ is fundamental), the analytic function $F_{H_0, \mathfrak{j}}$ extends to an analytic function on all of $\mathfrak{j}$ (see Proposition 8.3.2.1).

It is not difficult to determine the form of $F_{H_0, \mathfrak{j}}$ – thus:

Lemma 8.4.5.2 There exist unique complex numbers $c_w(H_0)$ $(w \in W$, the Weyl group of the pair $(\mathfrak{g}_c, \mathfrak{j}_c))$ such that

$$F_{H_0, \mathfrak{j}}(H) = \sum_{w \in W} \det(w) c_w(H_0) \exp(\sqrt{-1}B(H, wH_0)) \qquad (H \in \mathfrak{j}).$$

Proof Choose a small open connected neighborhood $\mathcal{O}(H_0)$ of H_0 in $\mathfrak{j}'$ with the property that the map $(\dot{x}, H) \mapsto \dot{x}H$ $(\dot{x} \in G/J,\ H \in \mathcal{O}(H_0))$ is one-to-one. Let $\tilde{T}_{H_0}$ denote the distribution on $\mathcal{O}(H_0)$ which corresponds to $T_{H_0} | G \cdot \mathcal{O}(H_0)$ under the process described in 8.2.1 and put $\sigma_{H_0} = \pi \tilde{T}_{H_0}$. Denote the restriction to $\mathfrak{j}_c$ of an element $p \in I(\mathfrak{g}_c)$ by $\tilde{p}$ – then, using

Theorems 8.2.1.8, 8.2.2.4, we find that

$$\partial(\tilde{p})\cdot\sigma_{H_0} = (\partial(\tilde{p})\circ\pi)\cdot\tilde{T}_{H_0} = (\pi\circ\gamma(\partial(p)))\cdot\tilde{T}_{H_0} = \tilde{p}(\sqrt{-1}H_0)\sigma_{H_0}$$

for all $p \in I(\mathfrak{g}_c)$. Chevalley's Restriction Theorem, in conjunction with Proposition 8.2.2.7, implies that σ_{H_0} is an analytic function on $\mathcal{O}(H_0)$; Corollary 8.2.2.9 then gives unique constants $c_w(H_0)$ $(w \in W)$ such that

$$\sigma_{H_0}(H) = \sum_{w\in W} \det(w)c_w(H_0)\exp(\sqrt{-1}B(H, wH_0)) \qquad (H \in \mathcal{O}(H_0)).$$

Our lemma will thus follow when it is shown that $\sigma_{H_0} = F_{H_0,\mathfrak{j}}$ on $\mathcal{O}(H_0)$. This, however, is clear in view of the Remark following Theorem 8.2.1.8. ☐

The constants $c_w(H_0)$ of Lemma 8.4.5.2 depend, a priori, on H_0; the next result ensures us that the c_w, when viewed as functions on $\mathfrak{j}'$, are actually locally constant.

Lemma 8.4.5.3 Retain the preceding notations; let $\mathscr{C}$ be the Weyl chamber in $\mathfrak{j}$ to which H_0 belongs – then the c_w $(w \in W)$ are constant on $\mathscr{C}$.

Proof It will clearly be enough to prove that the c_w are constant on a small open connected neighborhood $\mathcal{O}(H_0)$ (say) of H_0 in $\mathscr{C}$. To this end, we shall begin by verifying that each c_w is a C^∞ function on $\mathcal{O}(H_0)$; once this is done, our lemma will be a ready consequence of Corollary 8.2.2.11. Because H_0 is regular, there exist polynomials $p_w \in S(\mathfrak{j}_c)$ $(w \in W)$ such that

$$p_w(-\sqrt{-1}\tilde{w}H_0) = \begin{cases} 1 & \text{if} \quad w = \tilde{w} \\ 0 & \text{if} \quad w \neq \tilde{w} \end{cases} \qquad (\tilde{w} \in W).$$

Altering the notation of Lemma 8.4.5.2 a little, fix a small open connected neighborhood $\tilde{\mathcal{O}}(H_0)$ of H_0 in $\mathscr{C}$ with the property that the map $(\dot{x}, \tilde{H}) \mapsto \dot{x}\tilde{H}$ $(\dot{x} \in G/J,\ \tilde{H} \in \tilde{\mathcal{O}}(H_0))$ is one-to-one; for each $w \in W$ select $\beta_w \in C_c^\infty(\tilde{\mathcal{O}}(H_0))$ such that

$$(2\pi)^{-l/2}\int_{\tilde{\mathcal{O}}(H_0)} \beta_w(\tilde{H})\exp(\sqrt{-1}B(\tilde{H}, wH_0))d_{\mathfrak{j}}(\tilde{H}) = \det(w);$$

fix an element $\alpha \in C_c^\infty(G/J)$ such that $\int_{G/J}\alpha = 1$ and define functions $f_w \in C_c^\infty(G\cdot\tilde{\mathcal{O}}(H_0))$ $(w \in W)$ by the requirement

$$f_w(\dot{x}\tilde{H}) = (-1)^{r_I}(2\pi)^{-l/2}\alpha(\dot{x})\beta_w(\tilde{H};\partial(p_w))\pi^{-1}(\tilde{H}),$$

r_I the number of positive imaginary roots of the pair $(\mathfrak{g}_c, \mathfrak{j}_c)$ – then, for $H \in \mathcal{O}(H_0)$, we have that

$$\begin{aligned}\phi_{f_w}(H) &= \int_{\mathfrak{g}} f_w(X)F_H(X)d_{\mathfrak{g}}(X)\\ &= (2\pi)^{-l/2}\int_{\tilde{\mathcal{O}}(H_0)} \beta_w(\tilde{H};\partial(p_w))\pi(\tilde{H})F_H(\tilde{H})d_{\mathfrak{j}}(\tilde{H})\end{aligned}$$

$$= \sum_{\bar{w} \in W} \det(\bar{w}) c_{\bar{w}}(H)(2\pi)^{-l/2} \int_{\tilde{\mathcal{O}}(H_0)} \beta_w(\tilde{H}; \partial(p_w))$$
$$\times \exp(\sqrt{-1}B(\tilde{H}, \bar{w}H)) d_{\mathfrak{j}}(\tilde{H})$$
$$= \sum_{\bar{w} \in W} c_{\bar{w}}(H) a_{w,\bar{w}}(H),$$

where

$$a_{w,\bar{w}}(H) = \det(\bar{w}) p_w(-\sqrt{-1}\bar{w}H)(2\pi)^{-l/2}$$
$$\times \int_{\tilde{\mathcal{O}}(H_0)} \beta_w(\tilde{H}) \exp(\sqrt{-1}B(\tilde{H}, \bar{w}H)) d_{\mathfrak{j}}(\tilde{H}).$$

The $a_{w,\bar{w}}$ are analytic functions on $\mathcal{O}(H_0)$ which, by construction, have the property that

$$a_{w,\bar{w}}(H_0) = \begin{cases} 1 & \text{if } w = \bar{w} \\ 0 & \text{if } w \neq \bar{w}. \end{cases}$$

Consequently, for $\mathcal{O}(H_0)$ sufficiently small, the determinant of the matrix $(a_{w,\bar{w}})$ is nowhere zero (on $\mathcal{O}(H_0)$). If $(b_{w,\bar{w}})$ denotes the inverse matrix, then the $b_{w,\bar{w}}$ are also analytic functions on $\mathcal{O}(H_0)$ and we have that

$$c_{\bar{w}}(H) = \sum_{w \in W} b_{\bar{w},w}(H) \phi_{f_w}(H) \qquad (H \in \mathcal{O}(H_0)).$$

Because the ϕ_{f_w} are C^∞ functions on $\mathcal{O}(H_0)$, it is now clear that the same must be true for the c_w (all $w \in W$). This being so, let us write

$$\sigma_H(\tilde{H}) = \sigma(H : \tilde{H}) = \sum_{w \in W} \det(w) c_w(H) \exp(\sqrt{-1}B(\tilde{H}, wH))$$
$$(H \in \mathcal{O}(H_0),\ \tilde{H} \in \tilde{\mathcal{O}}(H_0)).$$

Elementary considerations, combined with what has been said above, tell us that σ meets all the requirements of Corollary 8.2.2.11 (recall that $\phi_{(pf)^\wedge} = \partial(\bar{p}^\dagger)\phi_f$, $p \in I(\mathfrak{g}_c) \ldots$). Therefore the c_w are constant on $\mathcal{O}(H_0)$, hence on $\mathscr{C}$. □

Let $\tilde{\mathfrak{j}}$ be any Cartan subalgebra of $\mathfrak{g}$; choose an $\tilde{x}$ in Int $(\mathfrak{g}_c)$ such that $\tilde{\mathfrak{j}}_c = \tilde{x}(\mathfrak{j}_c)$ – then there exists an isomorphism $D \mapsto \tilde{x}D$ of $\mathfrak{P}(\mathfrak{j})$ onto $\mathfrak{P}(\tilde{\mathfrak{j}})$ such that $\tilde{x}(p\partial(q)) = \tilde{x}p \cdot \partial(\tilde{x}q)$ $(p, q \in S(\mathfrak{j}_c))$. [Here, of course, $\tilde{x}p$ (respectively $\tilde{x}q$) denotes the polynomial function $\tilde{H} \mapsto p(\tilde{x}^{-1}\tilde{H})$ (respectively $\tilde{H} \mapsto q(\tilde{x}^{-1}\tilde{H})$) on $\tilde{\mathfrak{j}}_c$.] It is obvious that if α is a root of the pair $(\mathfrak{g}_c, \mathfrak{j}_c)$, then its transform under $\tilde{x}$ is a root of the pair $(\mathfrak{g}_c, \tilde{\mathfrak{j}}_c)$ etc.

The two lemmas above admit the following easy generalization.

Lemma 8.4.5.4 Let $\tilde{\mathfrak{j}}$ be a Cartan subalgebra of $\mathfrak{g}$, $\tilde{x}$ an element of Int $(\mathfrak{g}_c)$ such that $\tilde{\mathfrak{j}}_c = \tilde{x}(\mathfrak{j}_c)$; let $\tilde{\mathscr{C}}$ be a Weyl chamber in $\tilde{\mathfrak{j}}$; let H_0 be an element in $\mathfrak{j}'$, $\mathscr{C}$ the Weyl chamber in $\mathfrak{j}$ to which H_0 belongs – then there exist constants $\tilde{c}_w$ $(w \in W)$, depending only on $\mathscr{C}$, such that

$$F_{H_0, \tilde{\mathfrak{j}}}(\tilde{H}) = \sum_{w \in W} \det(w) \tilde{c}_w \exp(\sqrt{-1}B(\tilde{x}^{-1}\tilde{H}, wH_0)) \qquad (\tilde{H} \in \tilde{\mathscr{C}}).$$

Fix a Weyl chamber $\mathscr{C}$ in $\mathfrak{j}$. Given $f \in \mathscr{C}(\mathfrak{g})$, let us agree to write $\lim_{\mathscr{C}} \phi_f(0; \partial(\pi))$ for $\lim_{H_0 \to 0} \phi_f(H_0; \partial(\pi))$ $(H_0 \in \mathscr{C})$.

Let $T_{\mathscr{C}}$ be the tempered distribution on $\mathfrak{g}$ defined by the rule

$$T_{\mathscr{C}}(f) = \lim_{\mathscr{C}} \phi_f(0; \partial(\pi)) \qquad (f \in \mathscr{C}(\mathfrak{g})).$$

[In reality the distribution $T_{\mathscr{C}}$ does *not* depend on our choice of $\mathscr{C}$; cf. Corollary 8.4.4.2.]

Lemma 8.4.5.5 Retain the above notations – then:

(i) The distribution $T_{\mathscr{C}}$ is central;

(ii) There exists an ideal I in $I(\mathfrak{g}_c)$ of finite codimension such that $\partial(p) \cdot T_{\mathscr{C}} = 0$ for all $p \in I$.

Proof That $T_{\mathscr{C}}$ is central is clear. To prove (ii), we shall show that $\partial(p) \cdot T_{\mathscr{C}} = 0$ for every homogeneous polynomial in $I(\mathfrak{g}_c)$ of positive degree. Thus let p be a homogeneous element in $I(\mathfrak{g}_c)$ of degree $m \geqslant 1$. If $f \in C_c^\infty(\mathfrak{g})$, then

$$\partial(p) \cdot T_{\mathscr{C}}(f) = (-1)^m T_{\mathscr{C}}(\partial(p) f) = (-1)^m \lim_{\mathscr{C}} \phi_f(0; \partial(\pi) \circ \bar{p}^\natural),$$

$\bar{p}^\natural$ denoting the restriction of $p^\natural$ to $\mathfrak{j}_c$ (see 8.3.5). But the local expression of $\partial(\pi) \circ \bar{p}^\natural$ at zero is zero (cf. Lemma 1 in the first Appendix to the present number). Hence $\partial(p) \cdot T_{\mathscr{C}} = 0$. $\square$

Lemma 8.4.5.6 Retain the above notations – then, on each connected component of $\mathfrak{g}'$, $T_{\mathscr{C}}$ coincides with a constant.

Proof Let $\tilde{H}_0 \in \mathfrak{g}'$ be arbitrary. To prove the present lemma, it clearly suffices to verify that $T_{\mathscr{C}}$ coincides with a constant on some open neighborhood of $\tilde{H}_0$ in $\mathfrak{g}'$. Since $\tilde{H}_0$ is regular, the centralizer $\tilde{\mathfrak{j}}$ of $\tilde{H}_0$ in $\mathfrak{g}$ is a Cartan subalgebra of $\mathfrak{g}$. Choose a small open connected neighborhood $\mathcal{O}(\tilde{H}_0)$ of $\tilde{H}_0$ in $\tilde{\mathscr{C}}$ (the Weyl chamber in $\tilde{\mathfrak{j}}$ to which $\tilde{H}_0$ belongs) with the property that the map $(\dot{x}, \tilde{H}) \mapsto \dot{x}\tilde{H}$ of $G/\tilde{J} \times \mathcal{O}(\tilde{H}_0)$ onto $G \cdot \mathcal{O}(\tilde{H}_0)$ is one-to-one. Let $d_{\tilde{\mathfrak{j}}}$ denote the canonical Lebesgue measure on $\tilde{\mathfrak{j}}$ induced by $\tilde{\theta}$, $\tilde{\theta}$ a Cartan involution of $\mathfrak{g}$ stabilizing $\tilde{\mathfrak{j}}$. Normalize the invariant measure $d_{G/\tilde{J}}(\dot{x})$ on $G/\tilde{J}$ so that

$$d_{\mathfrak{g}}(X) = |\tilde{\pi}(\tilde{H})|^2 d_{G/\tilde{J}}(\dot{x}) d_{\tilde{\mathfrak{j}}}(\tilde{H}) \qquad (X = \dot{x}\tilde{H}, (\dot{x}, \tilde{H}) \in G/\tilde{J} \times \mathcal{O}(\tilde{H}_0)).$$

Employing the obvious notations, for $H_0 \in \mathscr{C}, f \in C_c^\infty(G \cdot \mathcal{O}(\tilde{H}_0))$, we have

$$\begin{aligned}\phi_f(H_0) &= \int_{\mathfrak{g}} f(X) F_{H_0}(X) d_{\mathfrak{g}}(X) \\ &= (-1)^{\tilde{r}_I} \int_{G/\tilde{J} \times \mathcal{O}(\tilde{H}_0)} \tilde{\pi}(\tilde{H}) F_{H_0}(\tilde{H}) \tilde{\pi}(\tilde{H}) f(\dot{x}\tilde{H}) d_{G/\tilde{J}}(\dot{x}) d_{\tilde{\mathfrak{j}}}(\tilde{H})\end{aligned}$$

$$= (-1)^{\tilde{r}_I} \sum_{w \in W} \tilde{c}_w \int_{G/\mathfrak{J} \times \mathcal{O}(\tilde{H}_0)} \det(w) \exp(\sqrt{-1} B(\tilde{x}^{-1}\tilde{H}, wH_0))$$
$$\times \tilde{\pi}(\tilde{H}) f(\dot{x}\tilde{H}) d_{G/\mathfrak{J}}(\dot{x}) d_{\tilde{\mathfrak{j}}}(\tilde{H}),$$

the last equality following from Lemma 8.4.5.4 above. Let r denote the number of positive roots of the pair $(\mathfrak{g}_c, \mathfrak{j}_c)$ – then

$$T_{\mathscr{C}}(f) = \lim_{\mathscr{C}} \phi_f(0; \partial(\pi))$$
$$= (-1)^{r/2 + \tilde{r}_I} \sum_{w \in W} \tilde{c}_w \int_{G/\mathfrak{J} \times \mathcal{O}(\tilde{H}_0)} \pi(\tilde{x}^{-1}\tilde{H}) \tilde{\pi}(\tilde{H}) f(\dot{x}\tilde{H}) d_{G/\mathfrak{J}}(\dot{x}) d_{\tilde{\mathfrak{j}}}(\tilde{H}).$$

Let II denote the unique element in $I(\mathfrak{g}_c)$ which, on restriction to $\mathfrak{j}_c$ (respectively $\tilde{\mathfrak{j}}_c$), gives π^2 (respectively $\tilde{\pi}^2$) – thus

$$\pi^2(\tilde{x}^{-1}\tilde{H}) = \mathrm{II}(\tilde{x}^{-1}\tilde{H}) = \mathrm{II}(\tilde{H}) = \tilde{\pi}^2(\tilde{H})$$

and so $\pi(\tilde{x}^{-1}\tilde{H})/\tilde{\pi}(\tilde{H})$ is a constant for $\tilde{H} \in \mathcal{O}(\tilde{H}_0)$, call it $\tilde{c}$. Put $\tilde{\mathbf{M}}_{\mathfrak{g}} = (-1)^{r/2}\tilde{c} \sum_{w \in W} \tilde{c}_w$ – then clearly

$$T_{\mathscr{C}}(f) = \tilde{\mathbf{M}}_{\mathfrak{g}} \int_{\mathfrak{g}} f(X) d_{\mathfrak{g}}(X) \qquad (f \in C_c^\infty(G \cdot \mathcal{O}(\tilde{H}_0)))$$

which shows that $T_{\mathscr{C}}$ coincides with $\tilde{\mathbf{M}}_{\mathfrak{g}}$ on $G \cdot \mathcal{O}(\tilde{H}_0)$. □

Proof of Theorem 8.4.5.1 (Existence of $\mathbf{M}_{\mathfrak{g}}$). Let us suppose first that $\mathfrak{g}$ is semi-simple. In view of Lemma 8.4.5.5, Theorem 8.3.1.1 implies that the distribution $T_{\mathscr{C}}$ is actually a locally summable function $F_{\mathscr{C}}$ (say) which is analytic on the regular set; moreover Lemma 8.4.5.6, in conjunction with Proposition 8.3.2.9, tells us that $F_{\mathscr{C}}$ is locally constant on $\mathfrak{g}'$ – hence $T_{\mathscr{C}}$ coincides with $\tilde{\mathbf{M}}_{\mathfrak{g}}$ on $\mathfrak{g}$. Take an arbitrary $f \in \mathscr{C}(\mathfrak{g})$ – then the proof of Lemma 8.4.5.6 shows that

$$\phi_f(0; \partial(\pi)) = \tilde{\mathbf{M}}_{\mathfrak{g}} \int_{\mathfrak{g}} f(X) d_{\mathfrak{g}}(X) = \mathbf{M}_{\mathfrak{g}} \hat{f}(0)$$

where $\mathbf{M}_{\mathfrak{g}} = (2\pi)^{n/2}\tilde{\mathbf{M}}_{\mathfrak{g}}$. Since Fourier transformation is a linear bicontinuous bijection of $\mathscr{C}(\mathfrak{g})$, the existence of the constant $\mathbf{M}_{\mathfrak{g}}$ in the semi-simple case is now clear. [Needless to say the constant $\mathbf{M}_{\mathfrak{g}}$ is independent of the Weyl chamber $\mathscr{C}$ used to define it; cf. Corollary 8.4.4.2.] Turning to the case of a reductive $\mathfrak{g}$, note first that it may be assumed that the derived algebra $\bar{\mathfrak{g}}$ is non-zero (our assertions become trivial if $\mathfrak{g}$ is abelian). This being so, fix a $C \in \mathfrak{c}$ and, for $f \in \mathscr{C}(\mathfrak{g})$, let $\bar{f}_C$ denote the function on $\bar{\mathfrak{g}}$ defined by the rule $\bar{X} \mapsto f(C + \bar{X})$ $(\bar{X} \in \bar{\mathfrak{g}})$ – then, on the basis of the normalizations agreed to in 8.1.2, we have that

$$\phi_f(C; \partial(\pi)) = \phi_{\bar{f}_C}(0; \partial(\pi)).$$

An application of the above discussion to the pair $(\bar{\mathfrak{g}}, \mathfrak{j})$ gives a constant $\mathbf{M}_{\bar{\mathfrak{g}}}$ (independent of C or f) with the property that

$$\phi_{\bar{f}_C}(0; \partial(\pi)) = \mathbf{M}_{\bar{\mathfrak{g}}} \hat{\bar{f}}_C(0) = \mathbf{M}_{\mathfrak{g}} \hat{f}(C).$$

The existence of the constant $\mathbf{M}_{\mathfrak{g}}$ in the reductive case now follows upon taking $C = 0$. □

A priori it is conceivable, even in the fundamental case, that the constant $\mathbf{M}_{\mathfrak{g}}$ is zero; that it is not is the point of the following discussion.

Choose the Cartan subalgebras $\mathfrak{j}^1, \ldots, \mathfrak{j}^r$ in the usual way (this 'r' has nothing to do with the 'r' appearing in the statement of Theorem 8.4.5.1); we may take $\mathfrak{j} = \mathfrak{j}^1$ (as above, $\mathfrak{j}$ is the fixed θ-stable fundamental Cartan subalgebra of $\mathfrak{g}$).

Proof of Theorem 8.4.5.1 (Computation of $\mathbf{M}_{\mathfrak{g}}$: The case rank $(\mathfrak{g}) =$ rank $(\mathfrak{k})$). Let us first assume that $\mathfrak{g}$ is semi-simple; since rank $(\mathfrak{g}) =$ rank $(\mathfrak{k})$, $\mathfrak{j}$ is compact (i.e. $\mathfrak{j}$ is contained in $\mathfrak{k}$). In order to prove that

$$\mathbf{M}_{\mathfrak{g}} = (-1)^{\mathbf{m}_{\mathfrak{g}}}(2\pi)^r \quad (\mathbf{m}_{\mathfrak{g}} = 2^{-1}\dim(G/K)),$$

we shall employ the results of de Rham which were outlined in 8.3.5. As the case when $l = 1$ has already been dealt with (cf. Proposition 8.4.2.4), we may suppose that $l \geqslant 2$. Let ω, $X \mapsto \operatorname{tr}(\operatorname{ad}(X))^2$, denote the Casimir polynomial of $\mathfrak{g}_c$, ω_i the restriction of ω to $\mathfrak{j}_c^i$ $(i = 1, \ldots, r)$. Let $X_1, \ldots, X_n$ be an orthonormal basis for $\mathfrak{g}$ relative to the Euclidean structure $(\,.\,,\,.\,)_\theta$; it may be assumed that $X_i \in \mathfrak{p}$ $(i = 1, \ldots, p = \dim(\mathfrak{p}))$,

$$X_{p+i} \in \mathfrak{j} \quad (i = 1, \ldots, l), \qquad X_{p+i} \in \mathfrak{k} \quad (i = 1, \ldots, k = \dim(\mathfrak{k})).$$

Denoting the corresponding set of coordinates by $x_1, \ldots, x_n$, we have, in the notations of 8.3.5,

$$u = \sum_1^p x_i^2 - \sum_{p+1}^n x_i^2, \ \Box = \partial(\omega), \ \Xi(X) = \xi(\omega(X)) \qquad (X \in \mathfrak{g}).$$

[Thus we are taking $n_1 = p$, $n_2 = k$; of course $n \geqslant 2$, $\mathfrak{g}$ being semi-simple.] Since rank $(\mathfrak{g}) =$ rank $(\mathfrak{k})$ here, p is necessarily even (see the second Appendix to this number), whence $n \equiv k \equiv l \bmod 2$. Write $\tilde{\omega}$ for ω_1 – then it is clear that cases (2) and (3) are applicable simultaneously to both $\partial(\omega)$ and $\partial(\tilde{\omega})$ according to whether l is odd or even, respectively (since, by assumption, $l \geqslant 2$). Let $\tilde{\Xi}$ denote the function corresponding to the pair $(\mathfrak{j}, \partial(\tilde{\omega}))$ – then, for all $f \in \mathscr{C}(\mathfrak{j})$, we have

$$f(0) = \int_{\mathfrak{j}} \tilde{\Xi}(H) f(H; \partial(\tilde{\omega}^{[l/2]})) d_{\mathfrak{j}}(H).$$

On the other hand

$$\int_{\mathfrak{j}} \Xi(H) f(H; \partial(\tilde{\omega}^{[l/2]})) d_{\mathfrak{j}}(H)$$

$$= \begin{cases} (-1)^{p/2} 2^{l-n} \pi^{(l-n)/2} \dfrac{\Gamma((l-1)/2)}{\Gamma((n-1)/2)} f(0) \text{ if } l \text{ is odd,} \\[2ex] (-1)^{p/2} 2^{l-n} \pi^{(l-n)/2} \dfrac{\Gamma(l/2)}{\Gamma(n/2)} f(0) \text{ if } l \text{ is even.} \end{cases}$$

Thus, for example, consider the case when l is odd – then

$$f(0) = \int_{\mathfrak{j}} a_l^{-1} Y(-\tilde{\omega}(H)) \,|\tilde{\omega}(H)|^{-1/2} f(H; \partial(\tilde{\omega}^{[l/2]})) d_{\mathfrak{j}}(H)$$

$$= (-1)^{p/2}a_l^{-1}/a_n^{-1}\int_{\mathfrak{j}}(-1)^{p/2}a_n^{-1}Y(-\tilde{\omega}(H))\,|\tilde{\omega}(H)|^{-1/2}$$
$$\times f(H;\partial(\tilde{\omega}^{[l/2]}))d_{\mathfrak{j}}(H)$$
$$= (-1)^{p/2}a_l^{-1}/a_n^{-1}\int_{\mathfrak{j}}\Xi(H)f(H;\partial(\tilde{\omega}^{[l/2]}))d_{\mathfrak{j}}(H)$$
$$\Rightarrow$$
$$\int_{\mathfrak{j}}\Xi(H)f(H;\partial(\tilde{\omega}^{[l/2]}))d_{\mathfrak{j}}(H)$$
$$= (-1)^{p/2}2^{l-n}\pi^{(l-n)/2}\frac{\Gamma((l-1)/2)}{\Gamma((n-1)/2)}f(0).$$

To complete the proof, we shall use the fact that there exists an $f \in \mathscr{C}(\mathfrak{g})$ with the following properties: (a) $f(0) = 1$; (b) $\phi_f^i = 0$ for $i \neq 1$ ($\phi_f^1 = \phi_f$); (c) The function $H \mapsto \phi_f(H)$ ($H \in \mathfrak{j}'$) admits an extension Φ_f to $\mathfrak{j}$ which lies in $\mathscr{C}(\mathfrak{j})$. [For the details, see Appendix 3 infra.] Using this f, we find that

$$1 = f(0) = \int_{\mathfrak{g}}\Xi(X)f(X;\partial(\omega^{[n/2]}))d_{\mathfrak{g}}(X)$$
$$= (-1)^r\int_{\mathfrak{j}}\pi(H)\Xi(H)\phi_f(H;\partial(\tilde{\omega}^{[n/2]}))d_{\mathfrak{j}}(H)$$
$$= (-1)^r\int_{\mathfrak{j}}\pi(H)\Xi(H)\Phi_f(H;\partial(\tilde{\omega}^{[n/2]}))d_{\mathfrak{j}}(H)$$
$$= (-1)^r\int_{\mathfrak{j}}\Xi(H)\Phi_f(H;\partial(\tilde{\omega}^{[l/2]})\circ\eta)d_{\mathfrak{j}}(H).$$

Here the η in the last expression is as in the first Appendix to the present number; there it is shown that the local expression for η at zero, η_0, is given by

$$\eta_0 = (-1)^r\prod_{i=0}^{r-1}([l/2]+i)2^r\partial(\pi).$$

Plainly $\eta\Phi_f \in \mathscr{C}(\mathfrak{j})$ – hence

$$1 = (-1)^r\int_{\mathfrak{j}}\Xi(H)\Phi_f(H;\partial(\tilde{\omega}^{[l/2]})\circ\eta)d_{\mathfrak{j}}(H)$$
$$= \begin{cases} (-1)^{r+(p/2)}2^{l-n}\pi^{(l-n)/2}\dfrac{\Gamma((l-1)/2)}{\Gamma((n-1)/2)}\Phi_f(0;\eta) \text{ if } l \text{ is odd,} \\[2ex] (-1)^{r+(p/2)}2^{l-n}\pi^{(l-n)/2}\dfrac{\Gamma(l/2)}{\Gamma(n/2)}\Phi_f(0;\eta) \text{ if } l \text{ is even.} \end{cases}$$
$$= \begin{cases} (-1)^{p/2}2^{r+l-n}\pi^{(l-n)/2}\displaystyle\prod_{i=0}^{r-1}([l/2]+i) \\ \quad\times\dfrac{\Gamma((l-1)/2)}{\Gamma((n-1)/2)}\Phi_f(0;\partial(\pi)) \text{ if } l \text{ is odd,} \\[2ex] (-1)^{p/2}2^{r+l-n}\pi^{(l-n)/2}\displaystyle\prod_{i=0}^{r-1}([l/2]+i) \\ \quad\times\dfrac{\Gamma(l/2)}{\Gamma(n/2)}\Phi_f(0;\partial(\pi)) \text{ if } l \text{ is even.} \end{cases}$$

To simplify these formulas, note first that $\Phi_f(0; \partial(\pi)) = \phi_f(0; \partial(\pi))$. Next, if z is a complex number which is neither zero nor a negative integer, then, as is well-known,

$$\prod_{i=0}^{r-1} (z+i) = \frac{\Gamma(z+r)}{\Gamma(z)} \qquad (r \geqslant 1).$$

Using these relations in the last expression for 1 above and bearing in mind that $p = \dim(G/K)$, we deduce without difficulty that

$$\mathbf{M}_{\mathfrak{g}} = (-1)^{\dim(G/K)/2}(2\pi)^r,$$

as desired. The extension to the reductive situation being trivial, the proof of this part of Theorem 8.4.5.1 is now complete. □

Let us drop the assumption that $\mathfrak{j}$ is compact – thus we propose now to discuss the case when rank $(\mathfrak{g}) >$ rank $(\mathfrak{k})$. It will be convenient to divide this discussion into two parts, namely according to whether the set of singular imaginary roots for the pair $(\mathfrak{g}_c, \mathfrak{j}_c)$ is empty or not.

In what follows it may safely be assumed that $\mathfrak{g}$ is semi-simple.

Proof of Theorem 8.4.5.1 (Computation of $\mathbf{M}_{\mathfrak{g}}$: The case rank $(\mathfrak{g}) >$ rank $(\mathfrak{k})$ – I). If the set of singular imaginary roots for the pair $(\mathfrak{g}_c, \mathfrak{j}_c)$ is empty, then $\mathfrak{j}_{\mathfrak{p}}$ is, in addition, a maximal abelian subalgebra of $\mathfrak{p}$ and every Cartan subalgebra of $\mathfrak{g}$ is conjugate to $\mathfrak{j}$. On the other hand, according to the second Appendix of this number, we have $k \equiv \dim(\mathfrak{j}_{\mathfrak{k}}) \bmod 2$, $p \equiv \dim(\mathfrak{j}_{\mathfrak{p}}) \bmod 2$ (as before, $k = \dim(\mathfrak{k})$, $p = \dim(\mathfrak{p})$) and there are four cases:

(1) $\dim(\mathfrak{j}_{\mathfrak{p}})$ odd $\Rightarrow p$ odd, $\dim(\mathfrak{j}_{\mathfrak{k}})$ even $\Rightarrow k$ even;
(2) $\dim(\mathfrak{j}_{\mathfrak{p}})$ even $\Rightarrow p$ even, $\dim(\mathfrak{j}_{\mathfrak{k}})$ odd $\Rightarrow k$ odd;
(3) $\dim(\mathfrak{j}_{\mathfrak{p}})$ even $\Rightarrow p$ even, $\dim(\mathfrak{j}_{\mathfrak{k}})$ even $\Rightarrow k$ even;
(4) $\dim(\mathfrak{j}_{\mathfrak{p}})$ odd $\Rightarrow p$ odd, $\dim(\mathfrak{j}_{\mathfrak{k}})$ odd $\Rightarrow k$ odd.

[Of course this is true whether the pair $(\mathfrak{g}_c, \mathfrak{j}_c)$ admits singular imaginary roots or not.] Take any $f \in C_c^\infty(\mathfrak{g})$ such that $f(0) = 1$ – then the invariant integral ϕ_f of f calculated relative to $\mathfrak{j}$ can be viewed as a rapidly decreasing function on *all* of $\mathfrak{j}$ (and not just $\mathfrak{j}'$) (cf. Corollary 8.4.3.2). This being so, it is then a simple matter to apply the results of de Rham in each of the four cases above and, obtain, thereby the asserted value for $\mathbf{M}_{\mathfrak{g}}$. [Let us consider, for instance Case (1). Proceeding as before, we first derive that

$$\int_{\mathfrak{j}} \Xi(H) f(H; \partial(\tilde{\omega}^{[l/2]})) d_{\mathfrak{j}}(H)$$
$$= (-1)^{2^{-1}(p-\dim(\mathfrak{j}_{\mathfrak{p}}))} 2^{l-n} \pi^{(l-n)/2} \frac{\Gamma((l-1)/2)}{\Gamma((n-1)/2)} f(0) \qquad (\text{all } f \in \mathscr{C}(\mathfrak{j}))$$

where $\mathbf{m}_{\mathfrak{g}} = 2^{-1}(p - \dim(\mathfrak{j}_{\mathfrak{p}}))$ (cf. the second Appendix to this number). Take now any $f \in \mathscr{C}(\mathfrak{g})$ such that $f(0) = 1$ – then

$$1 = f(0) = \int_{\mathfrak{g}} \Xi(X) f(X; \partial(\omega^{[n/2]})) d_{\mathfrak{g}}(X)$$

$$= (-1)^r \int_{\mathfrak{j}} \Xi(H)\phi_f(H; \partial(\tilde{\varpi}^{[l/2]})\circ\eta)d_{\mathfrak{j}}(H)$$

$$= (-1)^{\mathfrak{m}_\mathfrak{g}} 2^{r+l-n}\pi^{(l-n)/2} \prod_{i=0}^{r-1} ([l/2]+i)\frac{\Gamma((l-1)/2)}{\Gamma((n-1)/2)}\phi_f(0; \partial(\pi))$$

$\Rightarrow \mathbf{M}_\mathfrak{g} = (-1)^{\mathfrak{m}_\mathfrak{g}}(2\pi)^r$, as desired.] □

If the pair $(\mathfrak{g}_c, \mathfrak{j}_c)$ admits at least one positive singular imaginary root, then it is necessary to proceed somewhat differently since the results of de Rham are not immediately applicable. [There is no guarantee that we can find an $f \in \mathscr{C}(\mathfrak{g})$, which among other things, has the property that its invariant integral calculated relative to $\mathfrak{j}^i$ $(i = 2, \ldots, r)$ vanishes. . . .] On the other hand it is clear that the derived algebra of the centralizer $\mathfrak{l}$ of $\mathfrak{j}_\mathfrak{p}$ in $\mathfrak{g}$ is a semi-simple Lie algebra which, moreover, possesses a compact Cartan subalgebra; roughly speaking, by a series of transformations, we shall reduce the problem of the calculation of $\mathbf{M}_\mathfrak{g}$ to that of the calculation of the corresponding constant for the derived algebra of $\mathfrak{l}$ – the latter problem was, however, solved above.

So let $\mathfrak{l}$ be the centralizer of $\mathfrak{j}_\mathfrak{p}$ in $\mathfrak{g}$, L the analytic subgroup of G with Lie algebra $\mathfrak{l}$; assign to the symbols $\mathfrak{m}$, M, $\mathfrak{n}^\pm$, $N^\pm$ the usual meanings (so that, in particular, $\mathfrak{l} = \mathfrak{m} + \mathfrak{j}_\mathfrak{p}$ (direct sum) with $[\mathfrak{l}, \mathfrak{l}] = [\mathfrak{m}, \mathfrak{m}]$). We shall assume that the Haar measures on $\mathfrak{l}$, $\mathfrak{m}$, and $\mathfrak{n}^\pm$ are the Euclidean measures derived from the Euclidean structure $(. , .)_\theta$. Since the restriction of the Killing form of $\mathfrak{g}$ to $\mathfrak{l}$ (respectively to $\mathfrak{m}$) is non-degenerate, we shall agree to use it to identify the dual of $\mathfrak{l}$ with $\mathfrak{l}$ itself (respectively the dual of $\mathfrak{m}$ with $\mathfrak{m}$ itself); the elements of $S(\mathfrak{l}_c)$ (respectively $S(\mathfrak{m}_c)$) may then be regarded, in the usual way, as polynomial functions on $\mathfrak{l}_c$ (respectively $\mathfrak{m}_c$).

Put $\mathfrak{m}(\mathfrak{j}_\mathfrak{t}) = \bigcup_{m \in M} m(\mathfrak{j}_\mathfrak{t}'(I))$ ($\mathfrak{j}_\mathfrak{t}'(I)$ the complement in $\mathfrak{j}_\mathfrak{t}$ of the zeros of the imaginary roots of the pair $(\mathfrak{g}_c, \mathfrak{j}_c)$) – then the Haar measure on M will be normalized by the requirement

$$\int_{\mathfrak{m}(\mathfrak{j}_\mathfrak{t})} d_\mathfrak{m}(.) = \int_{M \times \mathfrak{j}_\mathfrak{t}'(I)} |\pi_I(.)|^2 d_M(.) d_{\mathfrak{j}_\mathfrak{t}}(.),$$

π_I denoting the product of the positive imaginary roots of the pair $(\mathfrak{g}_c, \mathfrak{j}_c)$. [Let us bear in mind that J_K is compact and of mass one.] Since the pair $(\mathfrak{m}_c, \mathfrak{j}_{\mathfrak{t}_c})$ admits at least one positive singular imaginary root, the analytic subgroup of M corresponding to $[\mathfrak{m}, \mathfrak{m}]$ is non-compact and, of course, possesses a compact Cartan subalgebra (viz. $\mathfrak{j}_\mathfrak{t} \cap [\mathfrak{m}, \mathfrak{m}]$); the discussion in an earlier paragraph per the computation of $\mathbf{M}_\mathfrak{m}$ is applicable to M in the obvious way (even though $\mathfrak{m}$ may only be reductive); the explicit value of $\mathbf{M}_\mathfrak{m}$ depends, however, on a choice of Euclidean structures for the relevant Lie algebras and upon the corresponding normalizations of the pertinent invariant measures; it is important to note, therefore, that if we agree to equip $\mathfrak{j}$, $\mathfrak{l}$ and $\mathfrak{m}$ with the Euclidean measures derived from the Euclidean structure $(. , .)_\theta$ (which is itself calculated from the

Killing form of $\mathfrak{g}$) and normalize the Haar measure on M per supra, then $\mathbf{M}_{\mathfrak{m}}$ has the hoped for value.

Lemma 8.4.5.7 Retain the above notations and conventions. Given $f \in \mathscr{C}(\mathfrak{l})$, put

$$\phi_f^{\mathfrak{l}}(H) = \pi_I(H) \int_M f(mH) d_M(m) \qquad (H \in \mathfrak{j}'(I)).$$

If r_I (respectively r_{SI}) is the number of positive imaginary roots (respectively positive singular imaginary roots) of the pair $(\mathfrak{g}_c, \mathfrak{j}_c)$, then

$$\phi_f^{\mathfrak{l}}(0; \partial(\pi_I)) = (-1)^{r_{SI}}(2\pi)^{r_I} f(0) \qquad (\text{all } f \in \mathscr{C}(\mathfrak{l})).$$

Proof It is to be emphasized that the Euclidean structures in question are derived from the Killing form B of $\mathfrak{g}$; on the other hand $\partial(\pi_I) = \prod_{\alpha \in \Phi_I^+} H_\alpha$, the H_α being determined through the identification of $\mathfrak{j}_c$ with its dual in the familiar fashion – it is essential to keep these facts in mind. We consider two cases. If dim $(\mathfrak{j}_{\mathfrak{k}}) \geqslant 2$, then, proceeding just as we did above in the case of a compact Cartan subalgebra (which $\mathfrak{j}_{\mathfrak{k}} \subset \mathfrak{m}$ is, after all), one establishes without difficulty that indeed

$$\phi_f^{\mathfrak{l}}(0; \partial(\pi_I)) = (-1)^{r_{SI}}(2\pi)^{r_I} f(0) \qquad (\text{all } f \in \mathscr{C}(\mathfrak{l})).$$

[It is permissible to work directly with $\mathfrak{m}$ (rather than $[\mathfrak{m}, \mathfrak{m}]$), the restriction of B to $\mathfrak{m}$ being non-degenerate.] On the other hand, if dim $(\mathfrak{j}_{\mathfrak{k}}) = 1$, then $\mathfrak{m}$ is a non-compact three dimensional simple Lie algebra; bearing in mind the discussion following Lemma 1.3.4.4, our contention in this case is seen to be a consequence of a computation which was given in 8.4.2 (cf. the discussion following Lemma 8.4.2.5). □

Proof of Theorem 8.4.5.1 (Computation of $\mathbf{M}_{\mathfrak{g}}$: The case rank $(\mathfrak{g}) >$ rank $(\mathfrak{k})$ – II). The Cartan subgroup J of G associated with the fundamental Cartan subalgebra $\mathfrak{j}$ is abelian (cf. Proposition 1.4.1.4); owing to Proposition 8.1.3.4, then, there exists a positive constant c such that for all $H \in \mathfrak{j}'$

$$\int_{G/J} f(\dot{x}H) d_{G/J}(\dot{x}) = c\pi_C(H)^{-1} \iiint_{K \times M \times \mathfrak{n}^+} f(k(mH + Z)) dk\, d_M(m) d_{\mathfrak{n}^+}(Z)$$
$$(f \in C_c(G/J)).$$

[Here π_C denotes the product of the positive complex roots of the pair $(\mathfrak{g}_c, \mathfrak{j}_c)$; evidently π_C assumes only non-negative real values on $\mathfrak{j}$.] Choose functions α, β in $C_c^\infty(G/J)$ and $C_c^\infty(\mathfrak{j}')$, respectively, such that

$$\int_{G/J} \alpha(\dot{x}) d_{G/J}(\dot{x}) = \int_{\mathfrak{j}'} \beta(H) |\pi(H)|^2 d_{\mathfrak{j}}(H) = 1.$$

Introduce, in the usual way, the group $W(G, J) = J^*/J$ (cf. Vol. I, number 1.4.2); we recall that $W(G, J)$ operates on G/J as well as $\mathfrak{j}'$; this being so, put

$$\alpha_0 = [W(G, J)]^{-1} \sum_{w \in W(G,J)} \alpha^w, \quad \beta_0 = [W(G, J)]^{-1} \sum_{w \in W(G,J)} \beta^w.$$

It is clear that there exists a function $F \in C_c^\infty(\mathfrak{g}(\mathfrak{j}))$ such that $F(\dot{x}H) = \alpha_0(\dot{x})\beta_0(H)$ $(\dot{x} \in G/J, H \in \mathfrak{j}')$ – moreover

$$\int_{\mathfrak{g}} F(X)d_{\mathfrak{g}}(X) = \int_{G/J} \alpha_0(\dot{x})d_{G/J}(\dot{x}) \int_{\mathfrak{j}'} \beta_0(H)|\pi(H)|^2 d_{\mathfrak{j}}(H) = 1.$$

Let $\hat{F}$ denote the Fourier transform of F – then, in view of what has been said above, we have

$$\phi_{\hat{F}}(H) = c\pi_I(H) \iiint_{K \times M \times \mathfrak{n}^+} \hat{F}(k(mH + Z))dkd_M(m)d_{\mathfrak{n}^+}(Z) \qquad (H \in \mathfrak{j}').$$

Now put (with $P = MJ_{\mathfrak{p}}N^+ \ldots$)

$$\hat{F}_K(X) = \int_K \hat{F}(kX)dk \qquad (X \in \mathfrak{g})$$

and

$$\hat{F}_{K,P}(X) = \int_{\mathfrak{n}^+} \hat{F}_K(X + Z)d_{\mathfrak{n}^+}(Z) \qquad (X \in \mathfrak{l}).$$

In these notations we can write $\phi_{\hat{F}} = c\phi^{\mathfrak{l}}_{\hat{F}_{K,P}}$; consequently

$$\partial(\pi)\phi_{\hat{F}} = c\partial(\pi_I)\partial(\pi_C)\phi^{\mathfrak{l}}_{\hat{F}_{K,P}} = c\partial(\pi_I)\phi^{\mathfrak{l}}_{\partial(\nabla_{\mathfrak{l}})\hat{F}_{K,P}}$$

on $\mathfrak{j}'$ ($\nabla_{\mathfrak{l}}$ the polynomial function on $\mathfrak{l}_c$ per 8.1.3 – let us keep in mind that $\mathfrak{l}$ has been identified with its dual via the restriction of the Killing form of $\mathfrak{g}$ to $\mathfrak{l}$). We see, therefore, that

$$\phi_{\hat{F}}(0\,;\partial(\pi)) = c\phi^{\mathfrak{l}}_{\partial(\nabla_{\mathfrak{l}})\hat{F}_{K,P}}(0\,;\partial(\pi_I)) = c(-1)^{r_{SI}}(2\pi)^{r_I}\hat{F}_{K,P}(0\,;\partial(\nabla_{\mathfrak{l}})).$$

It remains to compute $\hat{F}_{K,P}(0; \partial(\nabla_{\mathfrak{l}}))$. Because $\mathfrak{g}$ is the orthogonal direct sum of $\mathfrak{n}^+$, $\mathfrak{l}$, and $\mathfrak{n}^-$ $(= \theta(\mathfrak{n}^+))$, the theory of Fourier transforms tells us that

$$\begin{aligned}\hat{F}_{K,P}(X) &= \int_{\mathfrak{n}^+} \hat{F}_K(X + Z)d_{\mathfrak{n}^+}(Z) \\ &= (2\pi)^{-\dim(\mathfrak{l})/2} \int_{\mathfrak{l} \times \mathfrak{n}^+} F_K(\tilde{X} + Z)e^{\sqrt{-1}B(X,\tilde{X})}d_{\mathfrak{l}}(\tilde{X})d_{\mathfrak{n}^+}(Z)\end{aligned}$$

$$(X \in \mathfrak{l})$$

where $F_K(X) = \int_K F(kX)dk$ $(X \in \mathfrak{g})$. It follows easily that

$$\hat{F}_{K,P}(0\,;\partial(\nabla_{\mathfrak{l}})) = (-1)^{r_C/2}(2\pi)^{-\dim(\mathfrak{l})/2} \int_{\mathfrak{l} \times \mathfrak{n}^+} \nabla_{\mathfrak{l}}(X)F_K(X + Z)d_{\mathfrak{l}}(X)d_{\mathfrak{n}^+}(Z),$$

r_C the number of positive complex roots of the pair $(\mathfrak{g}_c, \mathfrak{j}_c)$. Choose $X \in \mathfrak{l}$ and $Z \in \mathfrak{n}^+$ such that $\nabla_{\mathfrak{l}}(X)F_K(X + Z) \neq 0$; owing to the remarks preceding the statement of Proposition 8.1.3.4 (and the definition of F), there exist elements $n \in N^+, x \in G$ and $H \in \mathfrak{j}'$ such that $nX = X + Z =$

xH – thus $X = \tilde{x}H$ if $\tilde{x} = n^{-1}x$. Let $\mathfrak{g}_X$ denote the centralizer of X in $\mathfrak{g}$ – then $\mathfrak{g}_X = \tilde{x}(\mathfrak{j})$ and so, since the ranks of $\mathfrak{g}$ and $\mathfrak{l}$ are the same, it must be the case that $\mathfrak{g}_X = m(\mathfrak{j})$ for some $m \in M$. Because X is a regular element in $\mathfrak{g}_X$, there exists an $H \in \mathfrak{j}'$ such that $X = mH$, whence

$$\nabla_{\mathfrak{l}}(X) = \nabla_{\mathfrak{l}}(H) = \pi_C(H) > 0.$$

Keeping in mind the relationship that exists between the Haar measure on N^+ and the Euclidean measure on $\mathfrak{n}^+$, we then deduce that

$$\hat{F}_{K,P}(0; \partial(\nabla_{\mathfrak{l}})) = (-1)^{rc/2}(2\pi)^{-\dim(\mathfrak{l})/2} \int_{\mathfrak{l} \times N^+} |\nabla_{\mathfrak{l}}(X)|^2 F_K(nX) d_{\mathfrak{l}}(X) d_{N^+}(n).$$

Put $\mathfrak{l}(\mathfrak{j}) = \bigcup_{l \in L} l(\mathfrak{j}')$; in view of the manner in which we have agreed to normalize the Haar measure on M, it is clear that

$$\int_{\mathfrak{l}(\mathfrak{j})} d_{\mathfrak{l}}(.) = \int_{M \times \mathfrak{j}'} |\pi_I(.)|^2 d_M(.) d_{\mathfrak{j}}(.).$$

We have noted above that $\nabla_{\mathfrak{l}}(X) F_K(nX) = 0$ ($n \in N^+$, $X \in \mathfrak{l}$) unless $X \in \mathfrak{l}(\mathfrak{j})$ – therefore

$$\begin{aligned}
\hat{F}_{K,P}(0; \partial(\nabla_{\mathfrak{l}})) &= (-1)^{rc/2}(2\pi)^{-\dim(\mathfrak{l})/2} \int_{\mathfrak{l} \times N^+} |\nabla_{\mathfrak{l}}(X)|^2 F_K(nX) d_{\mathfrak{l}}(X) d_{N^+}(n) \\
&= (-1)^{rc/2}(2\pi)^{-\dim(\mathfrak{l})/2} \\
&\quad \times \iiiint_{K \times N^+ \times M \times \mathfrak{j}'} |\pi(H)|^2 F(knmH) dk d_{N^+}(n) d_M(m) d_{\mathfrak{j}}(H) \\
&= c^{-1}(-1)^{rc/2}(2\pi)^{-\dim(\mathfrak{l})/2} \int_{G/J \times \mathfrak{j}'} |\pi(H)|^2 F(\dot{x}H) d_{G/J}(\dot{x}) d_{\mathfrak{j}}(H) \\
&= c^{-1}(-1)^{rc/2}(2\pi)^{-\dim(\mathfrak{l})/2} \int_{\mathfrak{g}} F(X) d_{\mathfrak{g}}(X) \\
&= c^{-1}(-1)^{rc/2}(2\pi)^{-\dim(\mathfrak{l})/2}.
\end{aligned}$$

Putting everything together, then, we see that

$$\phi_{\hat{F}}(0; \partial(\pi)) = \mathbf{M}_{\mathfrak{g}} \hat{F}(0) = (-1)^{r_{SI} + (r_C/2)}(2\pi)^{r_I}(2\pi)^{-\dim(\mathfrak{l})/2}.$$

In the second Appendix to this number it is shown that

$$r_{SI} + (r_C/2) = 2^{-1}(\dim(G/K) - \operatorname{rank}(G) + \operatorname{rank}(K)) = \mathbf{m}_{\mathfrak{g}};$$

on the other hand $n = \dim(\mathfrak{g}) = l + 2r$, $\dim(\mathfrak{l}) = l + 2r_I$, and $\hat{F}(0) = (2\pi)^{-n/2}$. Therefore $\mathbf{M}_{\mathfrak{g}} = (-1)^{\mathbf{m}_{\mathfrak{g}}}(2\pi)^r$, as desired. □

Appendix 1 We intend to record here two technical results which deal with the local expression at the origin of certain kinds of differential operators on a Cartan subalgebra of a semi-simple Lie algebra.

Let $\mathfrak{g}$ be a semi-simple Lie algebra over $\mathbf{R}$, θ a Cartan involution of $\mathfrak{g}$; let $\mathfrak{j}$ be a θ-stable Cartan subalgebra of $\mathfrak{g}$, π the product of the positive roots of the pair $(\mathfrak{g}_c, \mathfrak{j}_c)$.

Lemma 1 Let $D \in \mathfrak{Z}(\mathfrak{g})$ – then $(\partial(\pi) \circ \tilde{\mathfrak{I}}(D))_0 = (\partial(\pi) \circ \tilde{\mathfrak{I}}(D_0))_0$.

Proof Choose a compact real form $\mathfrak{u}$ for $\mathfrak{g}_c$ such that $\mathfrak{j}_c \cap \mathfrak{u}$ is a Cartan subalgebra of $\mathfrak{u}$; write U for Int $(\mathfrak{u})$ and let du denote the Haar measure on U with $\int_U du = 1$. Given $f \in C^\infty(\mathfrak{u})$, define

$$\phi_f(H) = \pi(H) \int_U f(uH)\, du \qquad (H \in \mathfrak{j}_c \cap \mathfrak{u})$$

as in Vol. I, number 3.2.1. For the purpose of proving the lemma, first suppose that $D_0 = 0$. Owing to Theorem 3.2.1.2, then, it is clear that $\phi_{Df}(0; \partial(\pi)) = 0$ (all $f \in C^\infty(\mathfrak{u})$). But $\phi_{Df} = \tilde{\mathfrak{f}}(D)\phi_f$ (Proposition 3.2.1.1), whence $\partial(\pi)(\tilde{\mathfrak{f}}(D)\phi_f)$ is zero at the origin; in particular take $f \in I(\mathfrak{g}_c)$ – then $\phi_f = \pi\bar{f}$ ($\bar{f}$ the restriction of f to $\mathfrak{j}_c \cap \mathfrak{u}$) and so, in view of Chevalley's Restriction Theorem, we conclude that $(\partial(\pi)\circ\tilde{\mathfrak{f}}(D))(\pi p)$ takes the value zero at the origin for every $p \in I(\mathfrak{j}_c)$ ($I(\mathfrak{j}_c)$ the set of invariants for the Weyl group W of the pair $(\mathfrak{g}_c, \mathfrak{j}_c)$). Choose $q \in S(\mathfrak{j}_c)$ such that $(\partial(\pi)\circ\tilde{\mathfrak{f}}(D))_0 = \partial(q)$. Since $\tilde{\mathfrak{f}}(D)$ is invariant under W, it is clear that $q^w = \det(w)q$ $(w \in W)$. Claim: $q = 0$. For if not, choose a $p_1 \in S(\mathfrak{j}_c)$ such that $\langle q, p_1\rangle \neq 0$ ($\langle\cdot, \cdot\rangle$ as defined in Vol. I, number 2.1.4) – then

$$\langle q, p_1\rangle = \langle q^w, p_1^w\rangle = \det(w)\langle q, p_1^w\rangle \Rightarrow \langle q, p_2\rangle = \langle q, p_1\rangle \neq 0$$

where $p_2 = [W]^{-1} \sum_{w\in W} \det(w)p_1^w$ ($[W]$ the order of W). Thanks to Proposition 2.1.5.11(i), we can write $p_2 = \pi p$ for some $p \in I(\mathfrak{j}_c)$, whence $\langle q, \pi p\rangle \neq 0$. But the definition of q implies that $\langle q, \pi p\rangle$ is precisely the value of

$$(\partial(\pi)\circ\tilde{\mathfrak{f}}(D))(\pi p)$$

at the origin, which, as was noted above, is zero. This contradiction proves that $q = 0$ and therefore that $(\partial(\pi)\circ\tilde{\mathfrak{f}}(D))_0 = 0$. Finally, the general case is reduced to this one by replacing D by $D - D_0$. □

Keeping to the above notations, let ω denote the Casimir polynomial on $\mathfrak{g}_c$, $\tilde{\omega}$ its restriction to $\mathfrak{j}_c$; let R and L, respectively, denote the mappings $D \mapsto D\circ\partial(\tilde{\omega})$, $D \mapsto \partial(\tilde{\omega})\circ D$ $(D \in \mathfrak{P}(\mathfrak{j}))$ of $\mathfrak{P}(\mathfrak{j})$ into itself; set $\nabla = L - R$ (thus $(1/2)\nabla = d_{\partial(\tilde{\omega})}$ in the notations of 8.2.2). Let r denote the number of positive roots of the pair $(\mathfrak{g}_c, \mathfrak{j}_c)$ (which we assume to be non-zero) – then clearly the greatest integer in $n/2$, $[n/2]$, is $[l/2] + r$ (n the dimension of $\mathfrak{g}_c$, l the rank of $\mathfrak{g}_c$). Bearing in mind that π is of degree r and hence that $\nabla^i\pi = 0$ for $i > r$ (cf. Lemma 8.2.2.5), we have

$$\pi\circ\partial(\tilde{\omega}^{[n/2]}) = (L - \nabla)^{[n/2]}\pi = \sum_{i=0}^{[n/2]} \binom{[n/2]}{i}(-1)^i L^{[n/2]-i}\nabla^i\pi$$
$$= \sum_{i=0}^{r} \binom{[l/2]+r}{i}(-1)^i L^{[l/2]+r-i}\nabla^i\pi = \partial(\tilde{\omega}^{[l/2]})\circ\eta$$

where

$$\eta = \sum_{i=0}^{r} \binom{[l/2]+r}{i}(-1)^i L^{r-i}\nabla^i\pi = \sum_{i=0}^{r} \binom{[l/2]+r}{i}(-1)^i (R+\nabla)^{r-i}\nabla^i\pi$$
$$= \sum_{j=0}^{r} \left\{\sum_{i=0}^{j} \binom{[l/2]+r}{i}\binom{r-i}{r-j}(-1)^i\right\} R^{r-j}\nabla^j\pi.$$

Lemma 2 The local expression η_0 for η at zero is given by

$$\eta_0 = (-1)^r[l/2]([l/2]+1)\cdots([l/2]+r-1)2^r\partial(\pi).$$

Proof Thus note that

$$(R^{r-j}\nabla^j\pi)_0 = (\nabla^j\pi)_0\circ\partial(\tilde{\omega}^{r-j}) \qquad (0 \leqslant j \leqslant r)$$

where, moreover $(\nabla^j\pi)_0 = 0$ $(j < r)$. [For, as has been observed in 8.2.2, $(\nabla^j\pi)_0 = \partial(p)$, p being an element of degree $\leqslant j < r$ in $S(\mathfrak{j}_c)$ having the property that $p^w = \det(w)p$ $(w \in W)$, whence $p = 0$ (Proposition 2.1.5.11 (i)).] Therefore

$$\eta_0 = \sum_{i=0}^{r} \binom{[l/2]+r}{i}(-1)^i(\nabla^r\pi)_0.$$

But

$$\sum_{i=0}^{r} \binom{[l/2]+r}{i}(-1)^i = (-1)^r\binom{[l/2]+r-1}{r}$$

as is seen by comparing coefficients of x^r in the identity

$$(1-x)^{[l/2]+r-1} = (1-x)^{[l/2]+r}(1+x+x^2+\cdots).$$

In addition $\nabla^r\pi = 2^r(r!)\partial(\pi)$ (cf. Lemma 8.2.2.5). Finally then,

$$\eta_0 = (-1)^r[l/2]([l/2]+1)\cdots([l/2]+r-1)2^r\partial(\pi),$$

as desired. □

Appendix 2 Let $\mathfrak{g}$ be a semi-simple Lie algebra over $\mathbf{R}$, θ a Cartan involution of $\mathfrak{g}$; let (G, K) be a pair associated with $\mathfrak{g}$ in the usual way. Let $\mathfrak{j}$ be a θ-stable fundamental Cartan subalgebra of $\mathfrak{g}$; let r_{SI} denote the number of positive singular imaginary roots of the pair $(\mathfrak{g}_c, \mathfrak{j}_c)$, r_C the number of positive complex roots of the pair $(\mathfrak{g}_c, \mathfrak{j}_c)$ (we recall that r_C is even).

Lemma Retain the above notations and assumptions – then

$$\mathbf{m}_{\mathfrak{g}} = r_{SI} + (r_C/2) = 2^{-1}(\dim(G/K) - \operatorname{rank}(G) + \operatorname{rank}(K)).$$

Proof We may regard $\mathfrak{g}$, in the usual way, as a real Hilbert space relative to the Euclidean structure $(\,.\,,\,.\,)_\theta$. Let $\mathfrak{m}$ denote the orthogonal complement of $\mathfrak{j}_{\mathfrak{p}}$ in the centralizer $\mathfrak{l}$ of $\mathfrak{j}_{\mathfrak{p}}$ in $\mathfrak{g}$; of course $\mathfrak{m}$ is θ-stable and $\dim(\mathfrak{m}_{\mathfrak{p}}) = 2r_{SI}$. Put $\mathfrak{n}^+ = \mathfrak{g} \cap (\sum_{\alpha\in\Phi_C^+}\mathfrak{g}_c^\alpha)$ (bear in mind that the pair $(\mathfrak{g}_c, \mathfrak{j}_c)$ admits no real roots, $\mathfrak{j}$ being fundamental); plainly $\dim(\mathfrak{n}^+) = r_C$. We have the direct decompositions

$$\mathfrak{g} = \mathfrak{j}_{\mathfrak{p}} + \mathfrak{m} + \mathfrak{n}^+ + \theta(\mathfrak{n}^+), \quad \mathfrak{p} = \mathfrak{j}_{\mathfrak{p}} + \mathfrak{m}_{\mathfrak{p}} + \mathfrak{p} \cap (\mathfrak{n}^+ + \theta(\mathfrak{n}^+)).$$

Since $\operatorname{ad}(X)$ is nilpotent for $X \in \mathfrak{n}^+$, $\mathfrak{k} \cap \mathfrak{n}^+ = \mathfrak{p} \cap \mathfrak{n}^+ = \{0\}$, and the maps $X \mapsto X + \theta(X)$, $X \mapsto X - \theta(X)$ of $\mathfrak{n}^+$ into $\mathfrak{k}$ and $\mathfrak{p}$, respectively, are injections. Hence

$$\dim(\mathfrak{k} \cap (\mathfrak{n}^+ + \theta(\mathfrak{n}^+))) \geqslant \dim(\mathfrak{n}^+), \quad \dim(\mathfrak{p} \cap (\mathfrak{n}^+ + \theta(\mathfrak{n}^+))) \geqslant \dim(\mathfrak{n}^+).$$

But

$$\mathfrak{n}^+ + \theta(\mathfrak{n}^+) = \mathfrak{k} \cap (\mathfrak{n}^+ + \theta(\mathfrak{n}^+)) + \mathfrak{p} \cap (\mathfrak{n}^+ + \theta(\mathfrak{n}^+))$$

and

$$\dim(\mathfrak{n}^+ + \theta(\mathfrak{n}^+)) = 2r_C.$$

Therefore

$$\dim(\mathfrak{k} \cap (\mathfrak{n}^+ + \theta(\mathfrak{n}^+))) = \dim(\mathfrak{p} \cap (\mathfrak{n}^+ + \theta(\mathfrak{n}^+))) = \dim(\mathfrak{n}^+) = r_C.$$

Finally, then, $\dim(\mathfrak{p}) = \dim(\mathfrak{j}_\mathfrak{p}) + 2r_{SI} + r_C$, whence

$$2\mathbf{m}_\mathfrak{g} = 2r_{SI} + r_C = \dim(\mathfrak{p}) - \dim(\mathfrak{j}_\mathfrak{p}) = \dim(\mathfrak{g}/\mathfrak{k}) - \dim(\mathfrak{j}) + \dim(\mathfrak{j}_\mathfrak{k}).$$

Since $\mathfrak{j}$ is fundamental, $\dim(\mathfrak{j}_\mathfrak{k}) = \text{rank}(\mathfrak{k})$. The assertion of the lemma is now clear. □

Appendix 3 Let G be a connected semi-simple Lie group with finite center, K a maximal compact subgroup of G; suppose that rank $(\mathfrak{g})$ = rank $(\mathfrak{k})$ and fix a Cartan subalgebra $\mathfrak{t}$ of $\mathfrak{k}$ – then $\mathfrak{t}$ is also a Cartan subalgebra of $\mathfrak{g}$.

Lemma Retain the above notations and assumptions; fix a K-invariant function f in $C_c^\infty(\mathfrak{g}(\mathfrak{t}))$ and let $\hat{f}$ denote its Fourier transform – then: (1) The invariant integral of $\hat{f}$ calculated relative to every non-compact Cartan subalgebra of $\mathfrak{g}$ vanishes identically; (2) The invariant integral of $\hat{f}$ calculated relative to $\mathfrak{t}$ extends to a rapidly decreasing function on $\mathfrak{t}$.

Proof To begin with observe that we need only establish (1), condition (2) then being an automatic consequence of a theorem which may be found in the second Appendix to 8.4.4 supra. So let $\mathfrak{j}$ be a non-compact θ-stable Cartan subalgebra of $\mathfrak{g}$ (of course the θ-stability assumption on $\mathfrak{j}$ is not restrictive); introduce $\mathfrak{l}$, L, $\mathfrak{m}$, M, $\mathfrak{n}^\pm$, $N^\pm$ etc. in the usual way (relative to $\mathfrak{j}_\mathfrak{p}$) – then, up to a positive scalar factor, $\phi_{\hat{f}}$ and $\phi^{\mathfrak{l}}_{\hat{f}_P}$ agree on $\mathfrak{j}'$ (bear in mind that $f = f_K$). This being the case, the proof of our lemma will be complete when it is shown that $\phi^{\mathfrak{l}}_{\hat{f}_P} = 0$ on $\mathfrak{j}'$; in turn, for this purpose, it will be enough to prove that $\hat{f}_P = 0$, or, thanks to the relation

$$\hat{f}_P(X) = (2\pi)^{-\dim(\mathfrak{l})/2} \int_{\mathfrak{l}\times\mathfrak{n}^+} f(\tilde{X} + Z) e^{\sqrt{-1}B(X,\tilde{X})}\, d_\mathfrak{l}(\tilde{X})\, d_{\mathfrak{n}^+}(Z) \qquad (X \in \mathfrak{l}),$$

that $f_P = 0$. Suppose then, if possible, that there exist $X \in \mathfrak{l}$, $Z \in \mathfrak{n}^+$ such that $f(X + Z) \neq 0$; because $f \in C_c^\infty(\mathfrak{g}(\mathfrak{t}))$, $X + Z$ is, in particular, semi-simple in $\mathfrak{l} + \mathfrak{n}^+$ and thus is conjugate within the identity component of $P(= MJ_\mathfrak{p}N^+)$ to an element Y (say) of $\mathfrak{l}$ (cf. Lemma 1.2.4.13); owing to the regularity of Y in $\mathfrak{g}$, it follows that $\mathfrak{l}$ must necessarily contain a compact Cartan subalgebra, an impossibility .... Hence the lemma. □

Corollary Retain the above notations and assumptions – then there exists an element $F \in \mathscr{C}(\mathfrak{g})$, which does not vanish at the origin, such that: (1) The invariant integral of F calculated relative to every non-compact Cartan subalgebra of $\mathfrak{g}$ vanishes identically; (2) The invariant integral of F calculated relative to $\mathfrak{t}$ extends to a rapidly decreasing function on $\mathfrak{t}$.

Proof Let f be a K-invariant function in $C_c^\infty(\mathfrak{g}(\mathfrak{t}))$ such that $\int_\mathfrak{g} f(X) d_\mathfrak{g}(X) \neq 0$; put $F = \hat{f}$ – then $F(0) \neq 0$ and, in view of the preceding lemma, satisfies the other conditions. □

8.5 The Invariant Integral on a Reductive Lie Group

8.5.1 The Invariant Integral – Definition and Properties

Let (G, K) be a reductive pair satisfying the usual conditions; for simplicity, it will be assumed in all that follows that G is, moreover, *acceptable* (cf. 8.1.1).

Let $\mathfrak{j}$ be a θ-stable Cartan subalgebra of $\mathfrak{g}$, J the Cartan subgroup of G associated with $\mathfrak{j}$, J_0 the center of J.

Definition Let $f \in C_c^\infty(G)$ – then the *invariant integral* of f relative to the Cartan subgroup J is the function Φ_f on J' defined by

$$\Phi_f(j) = \epsilon_R(j)\Delta(j)\int_{G/\mathsf{J}_0} f({}^{\dot{x}}j)d_{G/\mathsf{J}_0}(\dot{x}) \qquad (j \in \mathsf{J}').$$

[Here ϵ_R and Δ have the same meaning as in 8.1.1; of course ${}^{\dot{x}}j = xjx^{-1}$ $(x \in G,\ j \in \mathsf{J}_0)$, $x \mapsto \dot{x}$ denoting, as always, the canonical projection of G onto G/J_0.]

A priori the integral defining Φ_f $(f \in C_c^\infty(G))$ need not even be convergent; that it is may be seen as follows. Fix a point γ in J'; since γ is regular, the Lie algebra of the centralizer G_γ of γ in G is $\mathfrak{j}$ itself (cf. Corollary 1.4.1.8), whence $\mathsf{G}_\gamma/\mathsf{J}_0$ is finite. This being so, put $\omega = \operatorname{spt}(f)$ – then, thanks to the Theorem of Compacity (cf. Theorem 8.1.4.1), there exists an open neighborhood $\mathcal{O}(\gamma)$ of γ in J' and a compact set Ω in G/J_0 such that $xjx^{-1} \in \omega$ $(x \in G, j \in \mathcal{O}(\gamma)) \Rightarrow \dot{x} \in \Omega$. Plainly, then,

$$\int_{G/\mathsf{J}_0} f({}^{\dot{x}}j)d_{G/\mathsf{J}_0}(\dot{x}) = \int_\Omega f({}^{\dot{x}}j)d_{G/\mathsf{J}_0}(\dot{x}) \qquad (j \in \mathcal{O}(\gamma))$$

which shows that Φ_f is in fact defined and even more, namely that $\Phi_f \in C^\infty(\mathsf{J}')$ (the integral defining Φ_f being uniformly convergent in a neighborhood of each point of J'). Finally it should be observed that there exists a compact subset of J off of which Φ_f vanishes (cf. Lemma 8.1.6.1 . . .).

Remarks (1) The invariant measure on G/J_0 is, of course, to be taken in the normalization agreed to in 8.1.2 (see, however, the convention infra). Choose $\mathsf{J}^1, \ldots, \mathsf{J}^r$ in the usual way; assign to Δ^i, ϵ_R^i, Φ_f^i the obvious meanings $(f \in C_c^\infty(G);\ i = 1, \ldots, r)$ – then we have the important formula

$$\int_G f(x)d_G(x) = \sum_{i=1}^r [W(G, \mathsf{J}_0^i)]^{-1}\int_{\mathsf{J}^i} \epsilon_R^i(j)\overline{\Delta^i(j)}\Phi_f^i(j)d_{\mathsf{J}^i}(j)$$
$$(f \in C_c^\infty(G)),$$

the conjugate of Δ^i being computable through Proposition 8.1.1.2.

(2) The definition of Φ_f depends on the choice of an ordering for the roots – this is of little importance however.

(3) The group G operates on itself by inner automorphisms and this action lifts to $C_c^\infty(G)$ (A 2.4); this being so, fix an $f \in C_c^\infty(G)$ – then it is clear that $\Phi_{f^x} = \Phi_f$ (all $x \in G$), whence the term 'invariant integral'.

(4) Fix $f \in C_c^\infty(G)$ – then the invariant integral Φ_f satisfies a simple functional equation. Thus fix a $w \in W(G, \mathsf{J}_0)$ – then, since the operations of $W(G, \mathsf{J}_0)$ preserve the invariant measure on G/J_0, Proposition 8.1.1.3 leads at once to the conclusion that

$$\Phi_f(w \cdot j) = \det(w)\epsilon_R(w)\epsilon_w(j_K)\Phi_f(j) \qquad (j \in \mathsf{J}'),$$

j_K the component of j in K.

Before formulating the basic properties which the invariant integral possesses, it will first be convenient to establish the connection which exists between the invariant integral on the group and in the algebra.

Keeping to the above notations, fix an element γ in J – then γ is, in particular, semi-simple and so its centralizer $\mathfrak{g}_\gamma$ in $\mathfrak{g}$ is reductive and of the same rank as $\mathfrak{g}$. Let G_γ be the centralizer of γ in G, G_γ the identity component of G_γ – then, according to Proposition 1.4.2.4, the space $\mathsf{G}_\gamma/G_\gamma Z$ is finite (Z the center of G). Since $\mathsf{J}_0 \subset \mathsf{G}_\gamma$ and since G_γ is normal in G_γ, $G_\gamma \mathsf{J}_0$ is a group; accordingly, in view of the finiteness of $\mathsf{G}_\gamma/G_\gamma Z$, there exist a finite number of elements $x_\gamma^1 = 1, x_\gamma^2, \ldots, x_\gamma^m$ in G_γ ($m = [\mathsf{G}_\gamma : G_\gamma \mathsf{J}_0]$) such that G_γ is the disjoint union of the cosets $x_\gamma^i G_\gamma \mathsf{J}_0$ ($1 \leqslant i \leqslant m$). Let $\dot{G}_\gamma$ denote the image of G_γ in $\mathsf{G}_\gamma/\mathsf{J}_0$ – then it is clear that $\mathsf{G}_\gamma/\mathsf{J}_0$ is the disjoint union of the open sets $x_\gamma^i \dot{G}_\gamma$ ($1 \leqslant i \leqslant m$) while $\dot{G}_\gamma \sim G_\gamma/G_\gamma \cap \mathsf{J}_0$. [In passing one should note that $G_\gamma \cap \mathsf{J}$ is the Cartan subgroup of G_γ associated with $\mathfrak{j}$; for simplicity, let us write J_γ for $G_\gamma \cap \mathsf{J}$ – then $\mathsf{J}_{\gamma,0} \supset G_\gamma \cap \mathsf{J}_0$, the factor space $\mathsf{J}_{\gamma,0}/G_\gamma \cap \mathsf{J}_0$ being finite.] We shall agree to normalize the relevant invariant measures on G/G_γ and $\mathsf{G}_\gamma/\mathsf{J}_0$ in such a way that $\int_{G/\mathsf{J}_0} = \int_{G/\mathsf{G}_\gamma} \int_{\mathsf{G}_\gamma/\mathsf{J}_0}$; the Haar measure on G_γ is then determined when we require that $\int_G = \int_{G/\mathsf{G}_\gamma} \int_{\mathsf{G}_\gamma}$. [Since $\mathfrak{g}_\gamma$ is reductive and the homogeneous space $\mathsf{G}_\gamma/G_\gamma Z$ is finite, the group G_γ is unimodular, whence G/G_γ admits a G-invariant measure. . . .]

Let Φ denote the set of roots of the pair $(\mathfrak{g}_c, \mathfrak{j}_c)$, Φ_γ the subset of Φ consisting of those α which commute with γ, i.e. for which $\xi_\alpha(\gamma) = 1$ (of course $\mathfrak{g}_{\gamma,c} = \mathfrak{j}_c + \sum_{\alpha \in \Phi_\gamma} \mathfrak{g}_c^\alpha$); assign to the symbols π_γ, Φ_γ^+, $\mathfrak{j}_\gamma'$ ($=\{H \in \mathfrak{j} : \pi_\gamma(H) \neq 0\}$) etc. the obvious meanings. Fix, once and for all, an open, connected, relatively compact neighborhood $\mathcal{O}$ of zero in $\mathfrak{j}$ on which the exponential map is regular and injective and which satisfies the following conditions:

(1) For every α in Φ,

$$|e^{\alpha(H)/2} - \xi_\alpha(\gamma^{-1})e^{-\alpha(H)/2}| \geqslant 2^{-1}|1 - \xi_\alpha(\gamma^{-1})| \qquad (\text{all } H \in \mathcal{O});$$

(2) For every α in Φ, $|\alpha(H)| < 1$ (all $H \in \mathcal{O}$);

(3) If $x \in G$ and $j \in \exp(\mathcal{O})$ vary in such a way that $x(\gamma j)x^{-1}$ stays within a compact subset of G, then the coset $x\mathsf{G}_\gamma$ remains within a compact subset of G/G_γ.

Needless to say, neighborhoods $\mathcal{O}$ with the aforestated properties exist (cf. Theorem 8.1.4.1 . . .). Put $\diamondsuit_\gamma = \prod_{\alpha \in \Phi_\gamma^+}(1 - \xi_\alpha^{-1})$ – then condition (1) implies that $\xi_\alpha(\gamma \exp H) \neq 1$ for all $\alpha \in \Phi^+ - \Phi_\gamma^+$ ($H \in \mathcal{O}$), thus

$$\exp(\mathcal{O}) \cap (\gamma^{-1}\mathsf{J}') = \{j \in \exp(\mathcal{O}) : \diamondsuit_\gamma(j) \neq 0\}$$

and so, in view of (2), $\exp(\mathcal{O}) \cap (\gamma^{-1}\mathsf{J}') = \exp(\mathcal{O}_\gamma')$ where $\mathcal{O}_\gamma' = \mathcal{O} \cap \mathfrak{j}_\gamma'$. It is clear that $\gamma \exp(\mathcal{O}_\gamma') = \gamma \exp(\mathcal{O}) \cap \mathsf{J}'$; this being the case, fix an H in $\mathcal{O}_\gamma'$ – then, since $\mathcal{O}$ is connected, we have

$$\epsilon_R(\gamma \exp H) = \epsilon_\gamma \epsilon_{\gamma,R}(\exp H)$$

where

$$\epsilon_\gamma = \operatorname{sign}\{\prod_{\alpha\in\Phi_R^+-\Phi_{\gamma,R}^+}(1-\xi_\alpha(\gamma^{-1}))\},$$

$$\epsilon_{\gamma,R}(\exp H) = \operatorname{sign}\{\prod_{\alpha\in\Phi_{\gamma,R}^+}(1-e^{-\alpha(H)})\} = \epsilon_{\gamma,R}(H)$$

while

$$\Delta(\gamma\exp H) = \xi_\rho(\gamma)\prod_{\alpha\in\Phi^+-\Phi_\gamma^+}(e^{\alpha(H)/2}-\xi_\alpha(\gamma^{-1})e^{-\alpha(H)/2})\Delta_\gamma(\exp H),$$

where

$$\Delta_\gamma(\exp H) = \prod_{\alpha\in\Phi_\gamma^+}(e^{\alpha(H)/2}-e^{-\alpha(H)/2}).$$

Let ω be a relatively compact open subset of G – then, thanks to condition (3) supra, there exists a compact set Ω in G/G_γ such that $x(\gamma\exp H)x^{-1}\in\omega$ $(x\in G,\ H\in\mathcal{O})\Rightarrow x\mathsf{G}_\gamma\in\Omega$. Fix a function ψ in $C_c^\infty(G)$ with the property that $\int_{\mathsf{G}_\gamma}\psi(xx_\gamma)d_{\mathsf{G}_\gamma}(x_\gamma)=1$ if $x\mathsf{G}_\gamma\in\Omega(x\in G)$; for any $f\in C_c^\infty(\omega)$, put

$$v_f(x_\gamma) = \sum_{i=1}^m\int_G\psi(x)f({}^{xx_{i\gamma}}(\gamma x_\gamma))d_G(x)\qquad(x_\gamma\in G_\gamma).$$

Since ψ has compact support, the assignment $f\mapsto v_f$ is clearly a continuous mapping of $C_c^\infty(\omega)$ into $C_c^\infty(G_\gamma)$. Let us exhibit the connection that exists between the invariant integrals of f and v_f. Thus fix an f in $C_c^\infty(\omega)$ – then, for H in $\mathcal{O}'_\gamma$, we have

$$\begin{aligned}\Phi_f(\gamma\exp H) &= \epsilon_R(\gamma\exp H)\Delta(\gamma\exp H)\int_{G/\mathsf{J}_0}f({}^{\dot{x}}(\gamma\exp H))d_{G/\mathsf{J}_0}(\dot{x})\\ &= \epsilon_R(\gamma\exp H)\Delta(\gamma\exp H)\int_{\dot{G}_\gamma}v_f({}^{\dot{x}_\gamma}(\exp H))d_{\mathsf{G}_\gamma/\mathsf{J}_0}(\dot{x}_\gamma)\\ &= \epsilon_\gamma\xi_\rho(\gamma)[\mathsf{J}_{\gamma,0}:G_\gamma\cap\mathsf{J}_0]\\ &\quad\times\prod_{\alpha\in\Phi^+-\Phi_\gamma^+}(e^{\alpha(H)/2}-\xi_\alpha(\gamma^{-1})e^{-\alpha(H)/2})\Phi_{v_f}(\exp H)\end{aligned}$$

where

$$\Phi_{v_f}(\exp H) = \epsilon_{\gamma,R}(\exp H)\Delta_\gamma(\exp H)\int_{G_\gamma/\mathsf{J}_{\gamma,0}}v_f({}^{\dot{x}_\gamma}(\exp H))d_{G_\gamma/\mathsf{J}_{\gamma,0}}(\dot{x}_\gamma)\qquad(H\in\mathcal{O}'_\gamma).$$

On the other hand, a simple argument shows that there exists a unique G_γ-central analytic function D_γ on $\mathfrak{g}_\gamma$ such that $\Delta(\gamma\exp H)=\pi_\gamma(H)D_\gamma(H)$ for all $H\in\mathfrak{j}$ (cf. Vol. I, number 3.2.1); furthermore, it is possible to produce an open, completely invariant neighborhood Ω_γ of zero in $\mathfrak{g}_\gamma$ on which the exponential mapping (from $\mathfrak{g}_\gamma$ to G_γ) is regular and injective and with the property that there exists a G_γ-central C^∞ function Ψ on $\mathfrak{g}_\gamma$ which is equal to 1 around zero and has its support contained in Ω_γ (cf. Appendix 1 to the present number). This being the case, put

$$\check{v}_f(X_\gamma) = \epsilon_\gamma[\mathsf{J}_\gamma:G_\gamma\cap\mathsf{J}_0]\Psi(X_\gamma)D_\gamma(X_\gamma)v_f(\exp X_\gamma)\qquad(X_\gamma\in\mathfrak{g}_\gamma).$$

Then it is clear that

$$\Phi_f(\gamma \exp H) = \phi_{\breve{v}_f}(H)$$

for all H in $\mathscr{O} \cap \mathscr{O}_\Psi$, $\mathscr{O}_\Psi$ an open neighborhood of zero in $\mathfrak{g}_\gamma$ on which $\Psi = 1$. As we shall see, this result will frequently enable us to reduce questions about the invariant integral on the group to their analogs in the Lie algebra.

As might be expected, it turns out that the invariant integral can actually be defined for all $f \in \mathscr{C}(G)$ (and not just for $f \in C_c^\infty(G)$).

Theorem 8.5.1.1 (Harish-Chandra) Fix $f \in \mathscr{C}(G)$ – then the integral

$$\Phi_f(j) = \epsilon_R(j)\Delta(j) \int_{G/J_0} f({}^{\dot{x}}j) d_{G/J_0}(\dot{x})$$

is absolutely convergent for all $j \in J'$ and the function Φ_f so defined lies in $\mathscr{C}(J')$. Furthermore the mapping $f \mapsto \Phi_f$ of $\mathscr{C}(G)$ into $\mathscr{C}(J')$ is continuous.

Let Φ_{SI}^+ be the set of positive singular imaginary roots of the pair $(\mathfrak{g}_c, \mathfrak{j}_c)$; put

$$J'(SI) = \{j \in J: \prod_{\alpha \in \Phi_{SI}^+} (1 - \xi_\alpha(j^{-1})) \neq 0\}.$$

Then, during the course of the proof of Theorem 8.5.1.1, the following result will be established.

Corollary 8.5.1.2 For every f in $\mathscr{C}(G)$ the function Φ_f admits a unique extension to $J'(SI)$ and the mapping $f \mapsto \Phi_f$ of $\mathscr{C}(G)$ into $\mathscr{C}(J'(SI))$ is continuous.

The proof of Theorem 8.5.1.1 and its corollary require a great deal of preparation; in fact much of the discussion in the remaining numbers of this section is devoted to setting up certain necessary preliminary machinery, the proofs themselves coming in 8.5.8.

Let γ: $\mathfrak{Z} \to I(\mathfrak{j}_c)$ denote the canonical isomorphism of the center $\mathfrak{Z}$ of $\mathfrak{G}$ onto the set $I(\mathfrak{j}_c)$ of Weyl group invariants in $S(\mathfrak{j}_c)$.

Theorem 8.5.1.3 (Harish-Chandra) Fix $f \in \mathscr{C}(G)$ – then

$$\Phi_{Zf} = \gamma(Z)\Phi_f \qquad (\text{all } Z \in \mathfrak{Z}).$$

Proof Since $C_c^\infty(G)$ is dense in $\mathscr{C}(G)$, it can be supposed that $f \in C_c^\infty(G)$. This being so, let ω be a relatively compact open subset of G; fix an arbitrary point j_0 in J' – then we intend to prove that Φ_{Zf} and $\gamma(Z)\Phi_f$ $(Z \in \mathfrak{Z})$ coincide at j_0 for all $f \in C_c^\infty(\omega)$. Of course this will serve to establish the theorem, ω being arbitrary. So fix an open neighborhood

$\mathcal{O}(j_0)$ of j_0 in J' and a compact set Ω in G/J_0 such that $xjx^{-1} \in \omega$ ($x \in G$, $j \in \mathcal{O}(j_0)$) $\Rightarrow \dot{x} \in \Omega$; select a ψ in $C_c^\infty(G)$ such that $\int_{\mathrm{J}_0} \psi(xj)d_{\mathrm{J}_0}(j) = 1$ for all $\dot{x} \in \tilde{\Omega}$, $\tilde{\Omega}$ a compact neighborhood of Ω in G/J_0 – then it is clear that

$$\Phi_{Zf}(j) = \epsilon_R(j)\Delta(j) \int_G \psi(x)f(xjx^{-1}; Z)d_G(x) \qquad (j \in \mathcal{O}(j_0))$$

for all $f \in C_c^\infty(\omega)$. Adopting the notations introduced in 8.2.3, write

$$Z = \mathfrak{f}(Z)_j + \sum_{i=1}^{p} a_i(j)\Gamma_j(S_i \otimes H_i) \qquad (j \in \mathrm{J}',\ a_i \in C^\infty(\mathrm{J}'))$$

where $S_i \in \mathfrak{S}^+$, $H_i \in \mathfrak{Z}$; since there is no loss of generality in taking the S_i and H_i linearly independent, it follows that each S_i is J-invariant – thus

$$\begin{aligned} f(xjx^{-1}; Z) = f(x:j; Z) &= f(x:j; \mathfrak{f}(Z)_j) \\ &+ \sum_{i=1}^{p} a_i(j)f(x; S_i:j; H_i) \qquad (x \in G) \end{aligned}$$

and so

$$\begin{aligned} \Phi_{Zf}(j) &= \epsilon_R(j)\Delta(j) \int_G \psi(x)f(x:j; \mathfrak{f}(Z)_j)d_G(x) \\ &+ \sum_{i=1}^{p} a_i(j)\epsilon_R(j)\Delta(j) \int_G \psi(x; S_i^\natural)f(x:j; H_i)d_G(x) \\ &= \epsilon_R(j)\Delta(j) \int_G \psi(x)f(x:j; \mathfrak{f}(Z)_j)d_G(x) \end{aligned}$$

for $j \in \mathcal{O}(j_0)$ (the integral

$$\int_G \psi(x; S_i^\natural)f(x:j; H_i)d_G(x) \qquad (1 \leqslant i \leqslant p)$$

evidently vanishes . . .). Now, as has been seen in 8.2.3, $\mathfrak{f}(Z)$ and

$$|D_l|^{-1/2}\gamma(Z)\circ|D_l|^{1/2}$$

coincide as differential operators on J'; by choosing $\mathcal{O}(j_0)$ sufficiently small, we are ensured that $\epsilon_R\Delta\mathfrak{f}(Z)$ and $\gamma(Z)\circ(\epsilon_R\Delta)$ agree as differential operators there – but then it is obvious that Φ_{Zf} and $\gamma(Z)\Phi_f$ are equal on $\mathcal{O}(j_0)$, as we wished to prove. □

Theorem 8.5.1.4 (Harish-Chandra) Fix $f \in C_c^\infty(G)$ – then Φ_f can be extended to a C^∞ function on $\mathrm{J}'(SI)$. Let j_0 be a point in J and D an element in $\mathfrak{Z}$ such that $w_\alpha(D) = -D$ for every singular imaginary root α which commutes with j_0 – then $D\Phi_f$ can be extended to a continuous function around j_0.

Proof Invoking the notations which were introduced above, we have $\Phi_f(j_0 \exp H) = \phi_{\tilde{v}_f}(H)$ (all $H \in \mathcal{O} \cap \mathcal{O}_\Psi$); the second assertion of our theorem is therefore an immediate consequence of Theorem 8.4.4.1 (applied to $(\mathfrak{g}_{j_0}, \mathfrak{j})$ and $\check{v}_f$) and, moreover, this obviously implies the first assertion. □

Corollary 8.5.1.5 Set $\Pi = \prod_{\alpha>0} H_\alpha$ – then $\Pi\Phi_f$ can be extended to a continuous function on J (all $f \in \mathscr{C}(G)$).

Theorem 8.5.1.6 (Harish-Chandra) Suppose that the Cartan subalgebra $\mathfrak{j}$ is fundamental – then

$$\Phi_f(1\,;\Pi) = \mathbf{M}_G f(1) \qquad (\text{all } f \in \mathscr{C}(G)).$$

[Here the constant $\mathbf{M}_G (= \mathbf{M}_\mathfrak{g})$ has the same value as in Theorem 8.4.5.1; similarly for $\mathbf{m}_G$ $(= \mathbf{m}_\mathfrak{g})$. . . .]

Proof It will clearly be enough to establish our theorem for elements f in $C_c^\infty(G)$; this being so, fix an f in $C_c^\infty(G)$ – then, on the basis of earlier discussion, there exists a small neighborhood $\mathcal{O}$ of zero in $\mathfrak{j}$ and an element $\check{f}$ (say) in $C_c^\infty(\mathfrak{g})$ with $\check{f}(0) = f(1)$ and such that

$$\Phi_f(\exp H) = \prod_{\alpha>0} F(\alpha(H))\cdot\phi_{\check{f}}(H) \qquad (\text{all } H \in \mathcal{O}')$$

where F denotes the entire function $z \mapsto (e^{z/2} - e^{-z/2})/z$ of the complex variable z. In Appendix 2 of the present number it is shown that the local expression at the origin of the differential operator $\partial(\pi)\circ\mathbf{\Gamma}$ is $\partial(\pi)$ itself, i.e. $(\partial(\pi)\circ\mathbf{\Gamma})_0 = \partial(\pi)$ $(\mathbf{\Gamma} = \prod_{\alpha>0} F\circ\alpha$ on $\mathcal{O})$. Thanks to Corollary 8.5.1.5, $\Pi\Phi_f$ is a continuous function on J – thus the manner in which 1 is approached is immaterial and so our theorem follows upon citing Theorem 8.4.5.1. □

[The preceding result constitutes one of the main steps in the proof of the Plancherel Theorem for G. . . .]

Theorem 8.5.1.7 (Weak Selberg Principle) Let f be a $\mathfrak{Z}$-finite function in $\mathscr{C}(G)$ - then $\Phi_f = 0$ unless $\mathfrak{j}_\mathfrak{p} = \{0\}$.

Proof (Harish-Chandra) Assuming that $\mathfrak{j}_\mathfrak{p} \neq \{0\}$, fix a point $j \in \mathsf{J}'(SI)$; write $j = j_K j_\mathfrak{p}$ $(j_K \in \mathsf{J}_K, j_\mathfrak{p} \in J_\mathfrak{p})$ and choose a small open connected neighborhood $\mathcal{O}_\mathfrak{k}$ of zero in $\mathfrak{j}_\mathfrak{k}$ on which the exponential map is regular and injective with the property that $j_K(\exp \mathcal{O}_\mathfrak{k})J_\mathfrak{p} \subset \mathsf{J}'(SI)$; put $\mathcal{O} = \mathcal{O}_\mathfrak{k} + \mathfrak{j}_\mathfrak{p}$. Let $\mathfrak{I}_f$ be the annihilator of f in $\mathfrak{Z}$; because f is $\mathfrak{Z}$-finite, $\mathfrak{I}_f$ is an ideal in $\mathfrak{Z}$ of finite codimension and so, $S(\mathfrak{j}_c)$ being a finite module over $I(\mathfrak{j}_c) = \gamma(\mathfrak{Z})$, $I_f = S(\mathfrak{j}_c)\gamma(\mathfrak{I}_f)$ has finite codimension m (say) in $S(\mathfrak{j}_c)$. Owing to Theorem 8.5.1.3 it must therefore be the case that Φ_f is annihilated by the elements in I_f, which in turn entails that the function $H \mapsto \Phi_f(j_K \exp H)$ is analytic on $\mathcal{O}$. [Thus let $H_1, \ldots, H_l$ be a basis for $\mathfrak{j}$

over **R**; set $\Delta = \sum_1^l H_i^2 \in S(\mathfrak{j}_c)$ and then choose $c_i \in \mathbf{C}$ such that

$$\square = \Delta^m + \sum_1^m c_i \Delta^{m-i} \in I_f.$$

Then the function in question is annihilated by the elliptic analytic differential operator $\square$.] Therefore, due to the connectedness of $\mathcal{O}$, we conclude that

$$\Phi_f(j_K \exp H) = \sum_{i=1}^r p_i(H) e^{\lambda_i(H)} \qquad (\text{all } H \in \mathcal{O})$$

where the λ_i are linear functions and the p_i polynomial functions on $\mathfrak{j}_c$ (cf. the Appendix to 8.3.1). Now put $F(H) = \Phi_f(j_K \exp H)$ $(H \in \mathfrak{j}_{\mathfrak{p}})$; since $\Phi_f \in \mathscr{C}(\mathsf{J}'(SI))$, we have

$$\sup_{H \in \mathfrak{j}_{\mathfrak{p}}} (1 + \|H\|)^n |F(H)| < \infty$$

for every non-negative integer n and so, as $\mathfrak{j}_{\mathfrak{p}}$ is non-zero, it then follows from elementary considerations that $F = 0$ on $j_K J_{\mathfrak{p}}$ (cf. A3.2). Therefore $\Phi_f(j) = 0 \Rightarrow \Phi_f = 0$, j being an arbitrary element in $\mathsf{J}'(SI)$. $\square$

Corollary 8.5.1.8 Let f be a non-zero $\mathfrak{Z}$-finite function in $\mathscr{C}(G)$ – then, of necessity, rank $(G) =$ rank (K).

Proof Proceeding by contradiction, suppose that rank $(G) >$ rank (K); let $\mathfrak{j}$ be a θ-stable fundamental Cartan subalgebra of $\mathfrak{g}$ – then, in view of our hypotheses, $\mathfrak{j}_{\mathfrak{p}} \neq \{0\}$. Let Φ_f denote the invariant integral of f calculated relative to the Cartan subgroup J of G associated with $\mathfrak{j}$ – then $\Phi_f = 0$ (Weak Selberg Principle), whence $f(1) = 0$ (cf. Theorem 8.5.1.6). Applying the same argument to the right translates of f leads at once to the conclusion that $f = 0$, a contradiction. $\square$

[Thus this result says in particular that there are no eigenfunctions of $\mathfrak{Z}$ in $\mathscr{C}(G)$ unless rank $(G) =$ rank (K), a fact which is basic in the theory of the discrete series for G.]

Theorem 8.5.1.9 (Harish-Chandra) Suppose that G admits a θ-stable compact Cartan subgroup T, say (so that rank $(G) =$ rank (K)); let f be a $\mathfrak{Z}$-finite function in $\mathscr{C}(G)$, Φ_f the invariant integral of f calculated relative to T – then Φ_f extends to a C^∞ function on T.

We shall preface the proof with a lemma.

Let γ_0 be a semi-regular element in G of the non-compact type. [Needless to say this means that γ_0 is a semi-simple element of G whose centralizer $\mathfrak{g}_{\gamma_0}$ in $\mathfrak{g}$ is of dimension $l + 2$ with $\mathfrak{l}_{\gamma_0} = [\mathfrak{g}_{\gamma_0}, \mathfrak{g}_{\gamma_0}]$ a non-compact three dimensional Lie algebra.] Assign to the symbols $\mathfrak{h}^+, \mathfrak{h}^-, \mathsf{H}^+, \mathsf{H}^-, \nu$ their usual meanings in this context; for any $f \in \mathscr{C}(G)$, let $\Phi_f^\pm$ denote the invariant integral of f calculated relative to $\mathsf{H}^{\pm}$ (it makes sense to do this even though $\mathsf{H}^{\pm}$ need not be θ-stable; cf. 8.5.8 infra).

Lemma 8.5.1.10 Retain the above notations – then there exists a locally constant function c on the set of semi-regular elements in $\mathsf{H}^+ \cap \mathsf{H}^-$ which is nowhere zero and has the property that for every $D \in \mathfrak{H}^+$

$$\Phi_f^+(\gamma ; D) = c(\gamma)\{\lim_{t \downarrow 0} \Phi_f^-(\gamma \exp(t(X^* - Y^*)) ; \nu D) - \lim_{t \uparrow 0} \Phi_f^-(\gamma \exp(t(X^* - Y^*)) ; \nu D)\} \qquad (\text{all } f \in \mathscr{C}(G)),$$

γ semi-regular in $\mathsf{H}^+ \cap \mathsf{H}^-$.

[Using an artifice entirely analogous to the one which was introduced in 8.4.4, one is led without difficulty to **SL**(2, **R**) (essentially . . .); we shall leave the details for the reader to work out for himself.]

Proof of Theorem 8.5.1.9 Fix an element $t_0 \in T$. If $t_0 \in T'(SI)$, then Φ_f is C^∞ in a neighborhood of t_0 by virtue of Theorem 8.5.1.4; if t_0 has the property that there exists exactly one positive singular imaginary root of the pair $(\mathfrak{g}_c, \mathfrak{t}_c)$ commuting with it, then, thanks to the Weak Selberg Principle and the preceding lemma, Φ_f can again be extended to a C^∞ function in a neighborhood of t_0. It follows, therefore, that Φ_f extends to a C^∞ function on the set of those points in T which are either regular or semi-regular. Since $D\Phi_f$ is bounded on T' for every $D \in \mathfrak{T}$ (cf. Theorem 8.5.1.1), the proof of the present theorem can now be completed by the usual 'Lipschitz condition' argument (cf. the proofs of Proposition 8.3.3.2 and Theorem 8.4.4.1). □

Convention The invariant measures on G/J_0 and J_0 have thus far been taken in the normalizations which were laid down in 8.1.2. However, in the remaining numbers of the present section, it will be more convenient to utilize the normalizations which were set forth in 8.1.3; since the formula

$$\int_{G(\mathsf{J})} f(x) d_G(x) = [W(G, \mathsf{J}_0)]^{-1} \int_{\mathsf{J}} |\Delta(j)|^2 d_{\mathsf{J}}(j) \int_{G/\mathsf{J}_0} f({}^{\dot{x}}j) d_{G/\mathsf{J}_0}(\dot{x})$$
$$(f \in C_c(G(\mathsf{J})))$$

is still valid, no difficulties will arise from this agreement. [Of course the relation between the invariant integral on the group and in the algebra has to be adjusted by a certain constant factor. . . .]

Appendix 1 Our objective here will be to establish the following result.

Lemma Let $\mathfrak{g}$ be a reductive Lie algebra over **R**, G a connected Lie group with Lie algebra $\mathfrak{g}$ – then there exists an open, completely invariant neighborhood Ω of zero in $\mathfrak{g}$ on which the exponential mapping is regular and injective and with the property that there exists a G-central C^∞ function Ψ on $\mathfrak{g}$ which is equal to 1 around zero and has its support contained in Ω.

Proof For the proof we may obviously assume that $\mathfrak{g}$ is semi-simple. This being so, fix a Cartan involution θ of $\mathfrak{g}$; agreeing to equip $\mathfrak{g}$ with the Hilbert space structure associated with $(\, . \, , . \,)_\theta$, let $\mathscr{S}_r$ $(r > 0)$ denote the subset of $\mathfrak{g}$ comprised of those X such that $\|X\| < r$ and put $\Omega_r = G \cdot \mathscr{S}_r$ – then Ω_r $(r > 0)$ is an open, completely invariant neighborhood of zero in $\mathfrak{g}$. Choose real homogeneous polynomials p_i in $I(\mathfrak{g}_c)$ such that $I(\mathfrak{g}_c) = \mathbf{C}[p_1, \ldots, p_l]$; set $Q(X) = \sum_i p_i(X)^2$ $(X \in \mathfrak{g})$ – then it will be shown below that there exists a number $\delta > 0$ such that $Q(X) < \delta$ $(X \in \mathfrak{g}) \Rightarrow X \in \Omega_r$ $(r > 0$ fixed but arbitrary. . .). Fix now a C^∞ function F on $\mathbf{R}$ such that 1) $F(t) = F(-t)$, 2) $F(t) = 1$ if $|t| \leqslant \delta/3$ and $F(t) = 0$ if $|t| \geqslant \delta/2$ $(t \in \mathbf{R})$ – then the function Ψ defined by the prescription $\Psi(X) = F(Q(X))$ $(X \in \mathfrak{g})$ meets our requirements. . . . □

To verify the unproved contention above, we proceed as follows. Suppose that our assertion is false – then we can choose a sequence $\{X^i\}$ in $\mathfrak{g}$ such that $Q(X^i) \to 0$ while $X^i \notin \Omega_r$. Let $X^i = X_s^i + X_n^i$ be the decomposition of X^i into its semi-simple and nilpotent parts; since $X_s^i \in (G \cdot X^i)^{cl}$, $Q(X^i) = Q(X_s^i)$, whence $Q(X_s^i) \to 0$ and $X_s^i \notin \Omega_r$ (Ω_r being open and invariant). Let $\mathfrak{j}^1, \ldots, \mathfrak{j}^r$ be a maximal set of mutually non-conjugate θ-stable Cartan subalgebras of $\mathfrak{g}$;because X_s^i is semi-simple, we can choose an $x_i \in G$ and an index k_i such that $x_i X_s^i \in \mathfrak{j}^{k_i}$; by passing to a subsequence if necessary, it can be supposed that $H_i = x_i X_s^i \in \mathfrak{j}$ (all i), $\mathfrak{j}$ a fixed Cartan subalgebra of $\mathfrak{g}$ – then $Q(H_i) \to 0$ as $i \to \infty$. Define now homogeneous invariant polynomials q_k by the requirement

$$\det(t - \operatorname{ad}(X)) = t^n + \sum_{k=1}^{n} q_k(X) t^{n-k} \qquad (X \in \mathfrak{g}_c)$$

where t is an indeterminate – then clearly $q_k(H_i) \to 0$ as $i \to \infty$. But

$$\det(t - \operatorname{ad}(H)) = t^l \prod_{\alpha > 0} (t - \alpha(H)^2) \qquad (H \in \mathfrak{j})$$

where $l = \dim(\mathfrak{j})$ and α runs over the positive roots of the pair $(\mathfrak{g}_c, \mathfrak{j}_c)$ – therefore $\alpha(H_i) \to 0$ for every root α and hence $H_i \to 0$. This means that X_s^i lies in Ω_r eventually, a contradiction.

Appendix 2 Let $\mathfrak{g}$ be a reductive Lie algebra over $\mathbf{R}$, $\mathfrak{j}$ a Cartan subalgebra of $\mathfrak{g}$; let F denote the entire function $z \mapsto (e^{z/2} - e^{-z/2})/z$ of the complex variable z – then, on a sufficiently small open neighborhood $\mathcal{O}$ of zero in $\mathfrak{j}$, there exists an analytic function Γ such that

$$\Gamma(H) = \prod_{\alpha > 0} F(\alpha(H)) \qquad (H \in \mathcal{O}).$$

Lemma The local expression at the origin of the differential operator $\partial(\pi) \circ \Gamma$ is $\partial(\pi)$, i.e. $(\partial(\pi) \circ \Gamma)_0 = \partial(\pi)$.

Proof There is obviously no loss of generality in assuming that $\mathfrak{g}$ is semi-simple. If $\mathcal{O}$ is small enough, then $\Gamma = 1 + \sum_{k \geqslant 1} p_k$ where p_k is a homogeneous polynomial in $S(\mathfrak{j}_c)$ of degree k; moreover Γ can be regarded as a holomorphic function on some complex neighborhood $\mathcal{O}_c$ of zero in $\mathfrak{j}_c$, it being assumed that $w(\mathcal{O}_c) = \mathcal{O}_c$ $(w \in W)$ and that the above series converges to Γ on $\mathcal{O}_c$. Therefore, since $\Gamma(wH) = \Gamma(H)$ $(w \in W)$, it follows that $p_k^w = p_k$ $(w \in W, k \geqslant 1)$. Now let q be an arbitrary homogeneous element in $S(\mathfrak{j}_c)$ – then clearly

$$q(0; \partial(\pi) \circ \Gamma) = q(0; \partial(\pi)) + \sum_{1}^{r} q(0; \partial(\pi) \circ p_k),$$

where r is the degree of π. Owing to Lemma 1 in the first Appendix to 8.4.5, we have $q(0; \partial(\pi)\circ p_k) = 0$, whence $q(0; \partial(\pi)\circ\Gamma) = q(0; \partial(\pi))$ – thus

$$\langle p, (\partial(\pi)\circ\Gamma)_0 - \partial(\pi)\rangle = 0 \qquad (\text{all } p \in S(\mathfrak{j}_c))$$

and so $(\partial(\pi)\circ\Gamma)_0 = \partial(\pi)$, as desired. □

8.5.2 The Inequalities of Descent

The proofs of many of the important results in the theory go by induction on the dimension of G. In order to carry out such inductive proofs, one needs a method whereby the problem at hand can be reduced to a similar question on a lower dimensional subgroup; this method will be considered in 8.5.3 infra. However, before taking up the discussion of these 'transformations of descent,' it will first be necessary to deal with certain preliminary questions of a technical nature – this is the objective of the present number.

Let (G, K) be a reductive pair fulfilling the usual hypotheses. Fix a θ-stable Cartan subalgebra $\mathfrak{j}$ of $\mathfrak{g}$ with $\bar{\mathfrak{j}}_{\mathfrak{p}}$ $(= \mathfrak{j}_{\mathfrak{p}} \cap \bar{\mathfrak{g}}) \neq \{0\}$; relative to $\mathfrak{j}$ (or rather $\mathfrak{j}_{\mathfrak{p}} \ldots$), introduce L, L, M, M, $N^{\pm}$, $\mathfrak{l}$, $\mathfrak{m}$, $\mathfrak{n}^{\pm}$ etc. in the customary fashion; write $P = \mathsf{M}J_{\mathfrak{p}}N^+$ and set

$$\mathrm{d}_P(p) = |\det(\mathrm{Ad}(p)|\mathfrak{n}^+)|^{1/2} \qquad (p \in P).$$

Extend $\mathfrak{j}_{\mathfrak{p}}$ to a maximal abelian subspace $\mathfrak{a}_{\mathfrak{p}}$ of $\mathfrak{p}$ and place an order on the dual of $\mathfrak{a}_{\mathfrak{p}}$ which is compatible with the one chosen on the dual of $\mathfrak{j}_{\mathfrak{p}}$; let $\mathscr{C}$ be the positive Weyl chamber in $\mathfrak{a}_{\mathfrak{p}}$ and let d be a non-negative integer such that

$$\sup_{H \in \mathscr{C}cl} e^{\rho(H)} \vdash\!\circ\!\dashv (\exp H)(1 + \sigma(\exp H))^{-d} < \infty,$$

ρ as always (relative to the pair $(\mathfrak{g}, \mathfrak{a}_{\mathfrak{p}})$). Finally let $\vdash\!\circ\!\dashv_L$ denote the function on L corresponding to $\vdash\!\circ\!\dashv$ when the pair (G, K) is replaced by the pair (L, L_K) (here $L_K = L \cap K$).

Here is the main result of the present number.

Theorem 8.5.2.1 (Harish-Chandra) Retain the above notations; fix numbers $r > s \geqslant 0$ – then

$$\mathrm{d}_P(l) \int_{N^+} \vdash\!\circ\!\dashv (ln)(1 + \sigma(ln))^{-(r+2d)} d_{N^+}(n) \lesssim \vdash\!\circ\!\dashv_L (l)(1 + \sigma(l))^{-s}$$

$$(\text{all } l \in L).$$

Furthermore

$$1 + \sigma(l) \lesssim 1 + \sigma(ln) \qquad (\text{all } l \in L, n \in N^+).$$

In passing let us note that upon taking $l = 1$ in Theorem 8.5.2.1 we get the important fact that the function

$$n \mapsto \vdash\!\circ\!\dashv (n)(1 + \sigma(n))^{-(r+2d)} \qquad (n \in N^+)$$

is summable on N^+.

Corollary 8.5.2.2 Retain the above notations – then

$$\mathrm{d}_P(l)^{-1}\int_{N^+}\Xi(nl)(1+\sigma(nl))^{-(r+2d)}d_{N^+}(n) \leq \Xi_L(l)(1+\sigma(l))^{-s}$$

$$(\text{all } l \in L).$$

[One need only observe that

$$\int_{N^+} f(lnl^{-1})d_{N^+}(n) = \mathrm{d}_P(l)^{-2}\int_{N^+} f(n)d_{N^+}(n) \qquad (f \in C_c(N^+)).]$$

The proof of Theorem 8.5.2.1 requires some preparation which will now be undertaken.

For the purposes of the ensuing discussion it will be best to employ the notational conventions which were agreed to in the Introduction (thus we shall be attaching the subscripts 1 and 2 to some of the entities under consideration); in particular, if x is in G and if $\kappa(x)$ and $H(x)$ are the unique elements in K and $\mathfrak{a}_\mathfrak{p}$, respectively, such that $x \in \kappa(x)\exp(H(x))N^+$ ($G = KA_\mathfrak{p}N^+ \ldots$), then $H_i(x)$ will denote the component of $H(x)$ in $\mathfrak{a}_\mathfrak{p}^i$ ($i = 1, 2$) so that $H(x) = H_1(x) + H_2(x)$.

Given H in $\mathfrak{a}_\mathfrak{p}$, set $\lhd(H) = \inf_{\lambda \in \Sigma_2^+} \lambda(H)$.

Lemma 8.5.2.3 Fix an element h_2 in A_2 such that $\lhd(\log h_2) \geqslant 0$ – then

$$\exp\{\rho(H_2(h_2nh_2^{-1}))\} \leqslant 1 + \exp\{-2^{-1}\lhd(\log h_2) + \rho(H_2(n))\}$$

$$(n \in N^-)$$

and

$$\exp\{\rho(H(h_2n_2h_2^{-1}))\} \leqslant 1 + \exp\{-\lhd(\log h_2) + \rho(H(n_2))\}$$

$$(n_2 \in N_2^-).$$

Furthermore we have $\rho(H_2(n)) \geqslant 0$, $\rho(H(n)) \geqslant 0$ (all $n \in N^-$).

[We shall defer the proof of this result to the Appendix of the present number. Let us keep in mind that

$$N^+ = N_1^+N_2^+, \; N^- = \theta(N^+) = N_1^-N_2^-$$

where the present N_2^+ plays the role of the N^+ appearing in the statement of Theorem 8.5.2.1.]

Lemma 8.5.2.4 Let d be the integer of Theorem 8.5.2.1 – then, for any $\epsilon > 0$,

$$\int_{N_2^-} e^{-\rho(H(n_2))}\{1 + \rho(H(n_2))\}^{-(d+\epsilon)}d_{N_2^-}(n_2) < \infty.$$

Proof In view of what has been said in 8.1.3, the Haar measures $d_{N_1^-}$ and d_{N^-} on N_1^- and N^-, respectively, can be normalized in such a way

that

$$\int_{N_1^-} e^{-2\rho(H(n_1))} d_{N_1^-}(n_1) = \int_{N^-} e^{-2\rho(H(n))} d_{N^-}(n) = 1.$$

The Haar measure $d_{N_2^-}$ on N_2^- is thus determined when we require that

$$d_{N^-}(n) = d_{N_2^-}(n_2) d_{N_1^-}(n_1) \qquad (n = n_2 n_1).$$

Given n in N^-, write

$$n = n_2 n_1 \quad (n_2 \in N_2^-, n_1 \in N_1^-);$$

a simple computation then leads without difficulty to the relation

$$H(n) = H(\kappa(n_1)^{-1} n_2 \kappa(n_1)) + H(n_1).$$

Fix an element $H_2 \in \mathfrak{a}_\mathfrak{p}^2$ with the property that

$$\delta = \lhd(H_2) > 0; \text{put } h_2(t) = \exp(tH_2) \qquad (t \in \mathbf{R}).$$

Then Lemma 8.5.2.3 tells us that

$$\exp\{\rho(H(h_2(t) n_2 h_2(-t)))\} \leqslant 1 + \exp\{\rho(H(n_2)) - \delta t\} \qquad (t \geqslant 0)$$

for all $n_2 \in N_2^-$. Since the elements of A_2 commute with those of N_1^-, we thus have the estimate

$$\begin{aligned}
&\int_{N_2^-} e^{-\rho(H(n_2))}[1 + \exp\{\rho(H(n_2)) - \delta t\}]^{-1} d_{N_2^-}(n_2) \\
&\leqslant \int_{N_2^-} \exp\{-\rho(H(h_2(t) n_2 h_2(-t))) - \rho(H(n_2))\} d_{N_2^-}(n_2) \\
&= \int_{N^-} \exp\{-\rho(H(h_2(t) n h_2(-t))) - \rho(H(n))\} d_{N^-}(n) \\
&= e^{t\rho(H_2)} \vdash\!\circ\!\dashv (h_2(t)) \lesssim (1+t)^d \qquad (\text{all } t \geqslant 0).
\end{aligned}$$

For any integer $q \geqslant 0$, let $N_{2,q}^-$ denote the set of all $n_2 \in N_2^-$ such that $\rho(H(n_2)) \leqslant 2^q$; put $t = 2^q \delta^{-1}$ – then $\exp\{\rho(H(n_2)) - \delta t\} \leqslant 1$ for all $n_2 \in N_{2,q}^-$ (with this particular choice of t) and so

$$\int_{N_2^-,q} e^{-\rho(H(n_2))} d_{N_2^-}(n_2) \leqslant D 2^{dq} \qquad (q \geqslant 0)$$

where D is some positive number independent of q. Let $N_2^-(q)$ denote the complement of $N_{2,q-1}^-$ in $N_{2,q}^-$ $(q \geqslant 1)$ – then, for $\epsilon > 0, q \geqslant 1$, we have

$$\begin{aligned}
&\int_{N_2^-(q)} e^{-\rho(H(n_2))}\{1 + \rho(H(n_2))\}^{-(d+\epsilon)} d_{N_2^-}(n_2) \\
&\leqslant D 2^{dq-(q-1)(d+\epsilon)} = D 2^{d-(q-1)\epsilon},
\end{aligned}$$

whence

$$\int_{N_2^- - N_2^-,0} e^{-\rho(H(n_2))}\{1 + \rho(H(n_2))\}^{-(d+\epsilon)} d_{N_2^-}(n_2) \leqslant D 2^d \sum_{q=0}^{\infty} 2^{-q\epsilon} < \infty.$$

In order to complete the proof of the lemma, one need only remark that $N_{2,0}^-$ is compact. . . . □

Corollary 8.5.2.5 Fix $\epsilon > 0$ – then

$$\int_{N_2^-} \exp\{-(1+\epsilon)\rho(H(n_2))\} d_{N_2^-}(n_2) < \infty.$$

Proof Since $\rho(H(n_2)) \geqslant 0$ $(n_2 \in N_2^-)$ and since $e^{-\epsilon t}(1+t)^{d+1}$ remains bounded for $t \geqslant 0$, the assertion of the corollary is clear. □

Note The above results are, of course, applicable in the special case when $\mathfrak{j}_\mathfrak{p} = \mathfrak{a}_\mathfrak{p}$ (the integrals over N_2^- then becoming integrals over N^-).

Lemma 8.5.2.6 There exists a number $M \geqslant 1$ such that

$$1 + \max\{\sigma(h), \rho(H(\theta(n^{-1})))\} \leqslant M(1 + \sigma(hn))$$

and

$$\mapsto\!\!\circ\!\!\dashv(hn) \leqslant M(1+\sigma(hn))^d \exp\{-\rho(\log h) - \rho(H(\theta(n^{-1})))\}$$

for all $h \in A_\mathfrak{p}$, $n \in N^+$.

Proof Let Z be the center of G – then, for the purposes of the proof of this lemma, we can obviously replace G by G/Z. Hence, without any loss of generality, we may agree to subscribe to the assumptions and conventions of Vol. I, number 3.3.2.

(1) It is obviously possible to choose constants $M_1, M_2 > 0$ such that

$$M_1\sigma(h^+) \leqslant \rho(\log h^+) \leqslant M_2\sigma(h^+) \qquad (\text{all } h^+ \in \exp(\mathscr{C}^{cl})).$$

It then follows that there exists a constant $M_3 > 0$ such that

$$1 + \sigma(h) \leqslant M_3(1 + \sigma(hn)) \qquad (\text{all } h \in A_\mathfrak{p},\ n \in N^+).$$

[To see this, fix $h \in A_\mathfrak{p}$ and $n \in N^+$; write $hn = k_1 h^+ k_2$ $(k_1, k_2 \in K;$ $h^+ \in \exp(\mathscr{C}^{cl}))$ – then, since $\theta(n^{-1})h = \theta(hn)^{-1} = k_2^{-1}h^+k_1^{-1}$, the present $\theta(n^{-1})$ can play the role of the n in the Remark following Lemma 3.3.2.6, whence

$$\rho(\log h^+) + \log\sqrt{d_\rho} \geqslant \max_{w \in W} \rho(\log(w \cdot h)) \geqslant |\rho(\log h)|.$$

Therefore

$$M_2\sigma(h^+) + \log\sqrt{d_\rho} \geqslant \max_{w \in W} \rho(\log(w \cdot h)) \geqslant M_1\sigma(h)$$

and so our contention follows from the fact that $\sigma(h^+) = \sigma(hn)$.]

(2) There exists a constant $M_4 > 0$ such that

$$1 + \rho(H(\theta(n^{-1}))) \leqslant M_4(1 + \sigma(hn)) \qquad (\text{all } h \in A_\mathfrak{p},\ n \in N^+).$$

[Fix $h \in A_\mathfrak{p}$, $n \in N^+$; invoking the notations which were introduced in (1), Lemma 3.3.2.6 tells us that $\rho(\log h^+) \geqslant \rho(H(\theta(n^{-1}))) + \rho(\log h)$, hence, by our result above, we deduce that $2\rho(\log h^+) + \log\sqrt{d_\rho} \geqslant \rho(H(\theta(n^{-1})))$. The assertion now follows immediately.]

(3) Select $M_5 > 0$ such that

$$\vdash\!\!\circ\!\!\dashv(h^+) \leqslant M_5(1 + \sigma(h^+))^d e^{-\rho(\log h^+)}$$

for all $h^+ \in \exp(\mathscr{C}^{cl})$ – then it is clear that

$$\vdash\!\!\circ\!\!\dashv(hn) \leqslant M_5(1 + \sigma(hn)))^d \exp\{-\rho(\log h) - \rho(H(\theta(n^{-1})))\}$$
$$(\text{all } h \in A_{\mathfrak{p}},\ n \in N^+).$$

Hence the lemma. □

Given H in $\mathfrak{a}_{\mathfrak{p}}$, set

$$\rho_i(H) = 2^{-1}\operatorname{tr}(\operatorname{ad}(H)\,|\,\mathfrak{n}_i^+) \qquad (i = 1, 2).$$

Lemma 8.5.2.7 Let t_1, t_2 be two non-negative real numbers; put $t = t_1 + t_2$ – then

$$h^{\rho_2} \vdash\!\!\circ\!\!\dashv(hn)(1 + \sigma(hn))^{-(d+t)} \leqslant M^{t+1} h^{-\rho_1}(1 + \sigma(h))^{-t_1} e^{-\rho(H(\theta(n^{-1})))}$$
$$\times (1 + \rho(H(\theta(n^{-1}))))^{-t_2}$$

for all $h \in A_{\mathfrak{p}}$, $n \in N^+$.

[Here M is the constant appearing in the statement of Lemma 8.5.2.6.]

Proof Fix $h \in A_{\mathfrak{p}}$, $n \in N^+$ and, as above, write

$$hn = k_1 h^+ k_2 \quad (k_1, k_2 \in K;\ h^+ \in \exp(\mathscr{C}^{cl})).$$

Then

$$\vdash\!\!\circ\!\!\dashv(hn)(1 + \sigma(hn))^{-(d+t)} = \vdash\!\!\circ\!\!\dashv(h^+)(1 + \sigma(h^+))^{-(d+t)}$$
$$\leqslant M e^{-\rho(\log h^+)}(1 + \sigma(h^+))^{-t}.$$

Since

$$(1 + \sigma(h^+))^{-t} \leqslant M^t(1 + \sigma(h))^{-t_1}(1 + \rho(H(\theta(n^{-1}))))^{-t_2}$$

while $\rho(\log h^+) \geqslant \rho(\log h) + \rho(H(\theta(n^{-1})))$, our assertion is obvious. □

Proof of Theorem 8.5.2.1 Let $\mathscr{C}_1$ be the positive Weyl chamber in $\mathfrak{a}_{\mathfrak{p}}^1$ – then $L = L_K(\exp(\mathscr{C}_1^{cl}))A_2L_K$ and so, for the purposes of the proof of the first inequality, we may assume that $l = h = h_1^+ h_2$ where $h_1^+ \in \exp(\mathscr{C}_1^{cl})$ and $h_2 \in A_2(= J_{\mathfrak{p}} \ldots)$. In the notations of Lemma 8.5.2.7, take $t_1 = s$, $t_2 = r + d - s$, $t = r + d$ – then, bearing in mind Lemma 8.5.2.4, we find that

$$h^{\rho_2} \int_{N_2^+} \vdash\!\!\circ\!\!\dashv(hn_2)(1 + \sigma(hn_2))^{-(r+2d)} d_{N_2^+}(n_2)$$
$$\leqslant M^{t+1} h^{-\rho_1}(1 + \sigma(h))^{-s} \int_{N_2^-} e^{-\rho(H(n_2))}\{1 + \rho(H(n_2))\}^{-t_2} d_{N_2^-}(n_2)$$
$$\leqslant \tilde{M}^{t+1} h^{-\rho_1}(1 + \sigma(h))^{-s},$$

$\tilde{M}$ a certain positive number which is independent of h. According to

Proposition 8.3.7.3 (which is clearly applicable even though $\mathfrak{l}$ isn't semi-simple . . .), we have

$$h^{\rho_1} \mapstochar\!\!\multimap\!\!\mapstochar_L(h) = e^{\rho_1(\log h_1^+)} \mapstochar\!\!\multimap\!\!\mapstochar_L(h_1^+) \geqslant 1.$$

Therefore the inequality in question is obvious. The fact that

$$1 + \sigma(l) \lesssim 1 + \sigma(ln_2) \qquad (\text{all } l \in L,\ n_2 \in N_2^+)$$

is an immediate consequence of Lemma 8.5.2.6 and the relation $L = L_K A_{\mathfrak{p}} L_K$. □

Appendix Here we intend to establish the following result.

Lemma Fix an element h_2 in A_2 such that $\lhd(\log h_2) \geqslant 0$ – then

$$\exp\{\rho(H_2(h_2 n h_2^{-1}))\} \leqslant 1 + \exp\{-2^{-1}\lhd(\log h_2) + \rho(H_2(n))\} \qquad (n \in N^-)$$

and

$$\exp\{\rho(H(h_2 n_2 h_2^{-1}))\} \leqslant 1 + \exp\{-\lhd(\log h_2) + \rho(H(n_2))\} \qquad (n_2 \in N_2^-).$$

Furthermore we have $\rho(H_2(n)) \geqslant 0$, $\rho(H(n)) \geqslant 0$ (all $n \in N^-$).

For the proof of this lemma, there clearly is no loss of generality in assuming that G is semi-simple; in fact we might just as well suppose that G fulfills the assumptions which were set forth in Vol. I, number 3.3.2.

Let Φ denote the set of roots of the pair $(\mathfrak{g}_c, \mathfrak{j}_c)$. Fix a dominant integral linear function Λ on $\mathfrak{j}_c$; let U_Λ denote the corresponding irreducible representation of $\mathfrak{g}_c$ on E_Λ (say) – then it will be assumed in what follows that E_Λ carries the structure of a (complex) Hilbert space relative to which the adjoint of $U_\Lambda(X)$ $(X \in \mathfrak{g}_c)$ is $-U_\Lambda(\tau(X))$ (τ the conjugation of $\mathfrak{g}_c$ with respect to $\mathfrak{u} = \mathfrak{k} + \sqrt{-1}\mathfrak{p}$). This being so, we now claim that $\Lambda(H_2(n)) \geqslant 0$ with

$$\exp\{\Lambda(H_2(h_2 n h_2^{-1}))\} \leqslant 1 + \exp\{-2^{-1}\lhd(\log h_2) + \Lambda(H_2(n))\}$$

for all $n \in N^-$ provided h_2 in A_2 has the property that $\lhd(\log h_2) \geqslant 0$. Granting the claim, we see that two of the assertions of our lemma follow immediately upon taking $\Lambda = 2^{-1}\sum_{\alpha>0}\alpha$. Regarding the claim itself, set $\Lambda_\theta = \Lambda - \theta\Lambda$ – then it is easy to see that Λ_θ is dominant integral. Let $U_{\Lambda\theta}$ be the corresponding irreducible representation of $\mathfrak{g}_c$ on the (complex) Hilbert space $E_{\Lambda\theta}$ (say); let $E_{\Lambda\theta}^0$ be the subspace of $E_{\Lambda\theta}$ consisting of those vectors $a \in E_{\Lambda\theta}$ such that

$$U_{\Lambda\theta}(H)a = \Lambda_\theta(H)a \qquad (\text{all } H \in \mathfrak{a}_2\ (= \mathfrak{j}_{\mathfrak{p}} \ldots)).$$

Then $E_{\Lambda\theta}^0$ is invariant and irreducible under the restriction of $U_{\Lambda\theta}$ to $\mathfrak{l}_c$ and, moreover, in the present situation, $\dim(E_{\Lambda\theta}^0) = 1$ (since $(\Lambda_\theta, \alpha) = 0$ for all $\alpha \in \Phi_I^+ \ldots$). Let $a_{\Lambda\theta}$ be a unit vector in $E_{\Lambda\theta}$ belonging to the highest weight Λ_θ – then, since $G = K\exp(\mathfrak{m}_{\mathfrak{p}})A_2N_2^+$ (cf. Theorem 1.2.4.11) and $\mathfrak{m}_{\mathfrak{p}} \subset [\mathfrak{l}, \mathfrak{l}]$, it must be the case that $U_{\Lambda\theta}(m)a_{\Lambda\theta} = a_{\Lambda\theta}$ for all $m \in \exp(\mathfrak{m}_{\mathfrak{p}})$, whence

$$\| U_{\Lambda\theta}(x)a_{\Lambda\theta} \| = e^{\Lambda_\theta(H_2(x))} = e^{2\Lambda(H_2(x))} \qquad (\text{all } x \in G).$$

Let P_0 denote the orthogonal projection of $E_{\Lambda\theta}$ onto $E_{\Lambda\theta}^0$. In view of the relation

$$\mathsf{P}_0 U_{\Lambda\theta}(n)a_{\Lambda\theta} = a_{\Lambda\theta} \qquad (\text{all } n \in N^- = N_2^- N_1^-),$$

it follows that $\| U_{\Lambda\theta}(n)a_{\Lambda\theta} \| \geqslant \| a_{\Lambda\theta} \| = 1$ (all $n \in N^-$) and so

$$\Lambda(H_2(n)) \geqslant 0 \qquad (\text{all } n \in N^-).$$

As for the second contention of the claim, taking into account the fact that $E^0_{\Lambda\theta} = \mathbf{C}a_{\Lambda\theta}$, we have

$$\begin{aligned}\exp\{4\Lambda(H_2(h_2nh_2^{-1}))\} &= \| U_{\Lambda\theta}(h_2nh_2^{-1})\mathsf{P}_0\|^2_{HS}\\ &= \|\mathsf{P}_0 U_{\Lambda\theta}(h_2nh_2^{-1})\mathsf{P}_0\|^2_{HS}\\ &\quad + \|(1-\mathsf{P}_0)U_{\Lambda\theta}(h_2nh_2^{-1})\mathsf{P}_0\|^2_{HS}\\ &= 1 + \|(1-\mathsf{P}_0)U_{\Lambda\theta}(h_2nh_2^{-1})\mathsf{P}_0\|^2_{HS} \qquad (n \in N^-)\end{aligned}$$

since $\mathsf{P}_0 U_{\Lambda\theta}(h_2nh_2^{-1})\mathsf{P}_0 = \mathsf{P}_0$ $(h_2 \in A_2)$. Because $\mathfrak{G} = \mathfrak{N}_2^-\mathfrak{L}\mathfrak{N}_2^+$ and $E_{\Lambda\theta} = U_{\Lambda\theta}(\mathfrak{N}_2^-\mathfrak{L})a_{\Lambda\theta} = U_{\Lambda\theta}(\mathfrak{N}_2^-)a_{\Lambda\theta}$, the weights of $U_{\Lambda\theta}$, other than Λ_θ itself, are of the form $\Lambda_\theta - \sum_i \alpha_i$ where $\alpha_i \in \Phi_R^+ \cup \Phi_C^+$. Let $\Lambda_\theta = \lambda_0 > \lambda_1 > \cdots > \lambda_r$ be the weights of $U_{\Lambda\theta}$, $E^i_{\Lambda\theta}$ the subspace of $E_{\Lambda\theta}$ consisting of those vectors belonging to the weight $\lambda_i (0 \leqslant i \leqslant r)$; let P_i denote the orthogonal projection of $E_{\Lambda\theta}$ onto $E^i_{\Lambda\theta} (0 \leqslant i \leqslant r)$ – then, since $\mathfrak{j}$ is θ-stable, $E_{\Lambda\theta}$ is the orthogonal direct sum of the $E^i_{\Lambda\theta}$ and $1 - \mathsf{P}_0 = \mathsf{P}_1 + \cdots + \mathsf{P}_r$. Now suppose that $h_2 \in A_2$ has the property that $\lhd(\log h_2) \geqslant 0$ – then

$$\begin{aligned}\exp\{4\Lambda(H_2(h_2nh_2^{-1}))\} &= 1 + \left\| \sum_{i=1}^{r} h_2^{-(\Lambda_\theta-\lambda_i)}\mathsf{P}_i U_{\Lambda\theta}(n)\mathsf{P}_0 \right\|^2_{HS}\\ &\leqslant 1 + e^{-2\lhd(\log h_2)} \sum_{i=1}^{r} \|\mathsf{P}_i U_{\Lambda\theta}(n)\mathsf{P}_0\|^2_{HS}\\ &\leqslant 1 + e^{-2\lhd(\log h_2)} \| U_{\Lambda\theta}(n)a_{\Lambda\theta}\|^2\\ &= 1 + \exp\{-2\lhd(\log h_2) + 4\Lambda(H_2(n))\}\\ &\qquad\qquad (\text{all } n \in N^-).\end{aligned}$$

Hence the claim.

Concerning the other two assertions of our lemma, the fact that $\rho(H(n)) \geqslant 0$ (all $n \in N^-$) has been noted earlier (cf. Vol. I, number 3.3.2). Now let U_Λ be a finite dimensional irreducible representation of $\mathfrak{g}_c$ on E_Λ (say) whose highest weight with respect to $\mathfrak{a}_\mathfrak{p}$ is Λ – then, proceeding as above, one finds without difficulty that

$$\exp\{\Lambda(H(h_2n_2h_2^{-1}))\} \leqslant 1 + \exp\{-\lhd(\log h_2) + \Lambda(H(n_2))\}$$

for all $n_2 \in N_2^-$ provided h_2 in A_2 has the property that $\lhd(\log h_2) \geqslant 0$. Therefore we need only choose our representation in such a way that its highest weight with respect to $\mathfrak{a}_\mathfrak{p}$ is ρ.

8.5.3 The Transformations of Descent

In this number we shall discuss an important technique – the technique of descent – which, as will be seen in due course, enables one to transfer certain problems on G to analogous questions on a proper subgroup.

Let (G, K) be a reductive pair fulfilling the usual hypotheses. Fix a θ-stable Cartan subalgebra $\mathfrak{j}$ of $\mathfrak{g}$ with $\bar{\mathfrak{j}}_\mathfrak{p} (= \mathfrak{j}_\mathfrak{p} \cap \bar{\mathfrak{g}}) \neq \{0\}$; relative to $\mathfrak{j}$ (or rather $\mathfrak{j}_\mathfrak{p} \ldots$), introduce L, L, M, M, $N^\pm$, $\mathfrak{l}$, $\mathfrak{m}$, $\mathfrak{n}^\pm$, etc. in the customary fashion; write $P = \mathsf{M}J_\mathfrak{p}N^+$ and set

$$\mathrm{d}_P(p) = |\det(\mathrm{Ad}(p) | \mathfrak{n}^+)|^{1/2} \qquad (p \in P).$$

We shall begin with an algebraic result. Thus let $\mathfrak{L}$ denote the universal enveloping algebra of $\mathfrak{l}_c$; let $\mathfrak{Q}^+$ denote the left ideal in $\mathfrak{G}$ generated by the complexification in $\mathfrak{g}_c$ of $\mathfrak{n}^+$; let D be an element of $\mathfrak{G}$ which commutes with every element of $\mathfrak{j}_{\mathfrak{p}}$ – then there exists a unique element $D_{\mathfrak{l}} \in \mathfrak{L}$ such that $D - D_{\mathfrak{l}} \in \mathfrak{Q}^+$. [Since $\mathfrak{g}_c = \mathfrak{n}_c^- + \mathfrak{l}_c + \mathfrak{n}_c^+$ (direct sum), this is clear.]

Let $\mathfrak{Z}$ denote the center of $\mathfrak{G}$, $\mathfrak{Z}_{\mathfrak{l}}$ the center of $\mathfrak{L}$; let $\mu_P(= \mu_{\mathfrak{g}/\mathfrak{l}})$ be the canonical isomorphism of $\mathfrak{Z}$ into $\mathfrak{Z}_{\mathfrak{l}}$ (cf. Example 3 following Proposition 2.3.3.6).

Lemma 8.5.3.1 Fix an element Z in $\mathfrak{Z}$ – then $lZ_{\mathfrak{l}} = Z_{\mathfrak{l}}$ for all $l \in L$ (whence $Z_{\mathfrak{l}} \in \mathfrak{Z}_{\mathfrak{l}}$). Furthermore

$$Z_{\mathfrak{l}} = \mathrm{d}_P^{-1}\mu_P(Z)\circ\mathrm{d}_P$$

as differential operators on L.

[This is a simple consequence of the definitions.]

Notation (1) Let $D \mapsto D^\flat$ denote the automorphism of $\mathfrak{L}$ determined by the conditions $H^\flat = H + \rho_P(H)$, $X^\flat = X$ $(H \in \mathfrak{j}_{\mathfrak{p}},\ X \in \mathfrak{m})$ – then $D^\flat = \mathrm{d}_P^{-1}D\circ\mathrm{d}_P$ $(D \in \mathfrak{L})$ as differential operators on L.

(2) Let $D \mapsto {}^\flat D$ denote the automorphism of $\mathfrak{L}$ determined by the conditions ${}^\flat H = H - \rho_P(H)$, ${}^\flat X = X$ $(H \in \mathfrak{j}_{\mathfrak{p}},\ X \in \mathfrak{m})$ – then ${}^\flat D = \mathrm{d}_P D\circ\mathrm{d}_P^{-1}$ $(D \in \mathfrak{L})$ as differential operators on L.

[Here we have set $\rho_P(H) = 2^{-1}\operatorname{tr}(\operatorname{ad}(H)|\mathfrak{n}^+)$ $(H \in \mathfrak{j}_{\mathfrak{p}})$; we remind the reader that $\mathrm{d}_P | M = 1 \ldots$]

Fix an f in $\mathscr{C}(G)$; consider the integral $\int_{N^+}|f(n)|d_{N^+}(n)$ – then this integral is finite. To see this, let r, s, and d be as in Theorem 8.5.2.1 – then

$$\int_{N^+}|f(n)|\,d_{N^+}(n) = \int_{N^+}\frac{|f(n)|}{\Xi(n)}\cdot(1+\sigma(n))^{r+2d}\cdot\frac{\Xi(n)}{(1+\sigma(n))^{r+2d}}d_{N^+}(n) < \infty,$$

the function $n \mapsto \Xi(n)\,(1+\sigma(n))^{-(r+2d)}$ $(n \in N^+)$ being summable on N^+. Now let ω_G be a compact subset of G; consider the integrals $\int_{N^+}|f(xn)|d_{N^+}(n)$, $\int_{N^+}|f(nx)|d_{N^+}(n)$ $(x \in \omega_G)$ – then, for any fixed $x \in \omega_G$, these integrals are, by what we have just seen, finite (here we use the fact that the space $\mathscr{C}(G)$ is stable under both left and right translation by elements of G). However more than this is true, viz. the convergence of the integrals is actually *uniform* in x $(x \in \omega_G)$. Let us verify this for $\int_{N^+}|f(xn)|d_{N^+}(n)$, the proof for the other being similar. Let $\epsilon > 0$ be given. Owing to Corollary 8.1.2.2, there exist numbers $d_1 > 0$, $d_2 > 0$ such that

$$d_1(1+\sigma(xn)) \leqslant (1+\sigma(n)) \leqslant d_2(1+\sigma(xn)) \qquad (x \in \omega_G,\ n \in N^+).$$

Choose $M_1 > 0$ so that $\Xi(xn) \leqslant M_1\,\Xi(n)$ $(x \in \omega_G, n \in N^+)$; let

$$M_2 = \sup_G \frac{(1+\sigma)^{r+2d}}{\Xi}|f|$$

and put $M = M_1M_2$; select a compact set $\omega_{N^+} \subset N^+$ with the property

that

$$\int_{\mathbf{C}\omega_{N^+}} \Xi(n)(1+\sigma(n))^{-(r+2d)} d_{N^+}(n) < \epsilon / M d_2^{r+2d}.$$

Then

$$\begin{aligned}
&\int_{\mathbf{C}\omega_{N^+}} |f(xn)| d_{N^+}(n) \\
&= \int_{\mathbf{C}\omega_{N^+}} \frac{|f(xn)|}{\Xi(xn)} \cdot (1+\sigma(xn))^{r+2d} \cdot \frac{\Xi(xn)}{(1+\sigma(xn))^{r+2d}} d_{N^+}(n) \\
&\leqslant M d_2^{r+2d} \int_{\mathbf{C}\omega_{N^+}} \Xi(n)(1+\sigma(n))^{-(r+2d)} d_{N^+}(n) < \epsilon
\end{aligned}$$

for all $x \in \omega_G$. Of course this gives our contention. . . .

Given an f in $\mathscr{C}(G)$, put

$$f_P(l) = \mathrm{d}_P(l) \int_{N^+} f(ln) d_{N^+}(n) = \mathrm{d}_P(l^{-1}) \int_{N^+} f(nl) d_{N^+}(n) \qquad (l \in \mathsf{L}).$$

In view of what has been said above, $f_P \in C^\infty(\mathsf{L})$; moreover differentiation under the integral sign is legitimate and so we find that

$$\begin{aligned}
f_P(D ; l) &= \mathrm{d}_P(l) \int_{N^+} f(D^\flat ; ln) d_{N^+}(n) \\
f_P(l ; D) &= \mathrm{d}_P(l^{-1}) \int_{N^+} f(nl ; {}^\flat D) d_{N^+}(n) = \mathrm{d}_P(l) \int_{N^+} f(ln ; {}^\flat D) d_{N^+}(n)
\end{aligned}$$

$$(l \in \mathsf{L})$$

for all $D \in \mathfrak{L}$. Hence

$$f_P(D_1 ; l ; D_2) = \mathrm{d}_P(l) \int_{N^+} f(D_1^\flat ; ln ; {}^\flat D_2) d_{N^+}(n) \qquad (l \in \mathsf{L})$$

for all $D_1, D_2 \in \mathfrak{L}$. We claim that f_P actually lies in $\mathscr{C}(\mathsf{L})$, in fact that the assignment $f \mapsto f_P$ is a continuous linear mapping of $\mathscr{C}(G)$ into $\mathscr{C}(\mathsf{L})$ ($f \in \mathscr{C}(G)$). [Since L/L is finite, the space $\mathscr{C}(\mathsf{L})$ is well-defined.] To see this, fix $D_1, D_2 \in \mathfrak{L}$ and $s > 0$; fix an $r > s$ – then, if $f_i \to 0$ in $\mathscr{C}(G)$, we have, for all $l \in L$,

$$\begin{aligned}
&\frac{(1+\sigma(l))^s}{\Xi_L(l)} | f_{i,P}(D_1 ; l ; D_2) | \\
&\leqslant \frac{(1+\sigma(l))^s}{\Xi_L(l)} \left| \mathrm{d}_P(l) \int_{N^+} \frac{f_i(D_1^\flat ; ln ; {}^\flat D_2)}{\Xi(ln)} \cdot (1+\sigma(ln))^{r+2d} \right. \\
&\qquad \left. \cdot \frac{\Xi(ln)}{(1+\sigma(ln))^{r+2d}} d_{N^+}(n) \right| \\
&\leqslant M \frac{(1+\sigma(l))^s}{\Xi_L(l)} \cdot {}_{D_1^\flat}| f_i |_{r+2d, {}^\flat D_2} \cdot \frac{\Xi_L(l)}{(1+\sigma(l))^s}
\end{aligned}$$

(M a positive constant),

whence

$$_{D_1}|f_{i,P}|L|_{s,D_2} \leqslant M \cdot {}_{D_1^\flat}|f_i|_{r+2d,\,{}^\flat D_2} \to 0,$$

as desired (Theorem 8.5.2.1 being used to pass from the first inequality to the second).

Given $f \in \mathscr{C}(G)$, put $f_K(x) = \int_K f(kxk^{-1})dk$ $(x \in G)$ – then it is clear that the mapping $f \mapsto f_K$ is a continuous endomorphism of $\mathscr{C}(G)$ (cf. Lemma 8.3.7.9). Set

$$f_{K,P}(l) = \mathrm{d}_P(l) \int_{N^+} f_K(ln) d_{N^+}(n) \qquad (l \in \mathsf{L}).$$

Then the main result of the present number can be formulated as follows.

Theorem 8.5.3.2 (Harish-Chandra) The assignment $f \mapsto f_{K,P}$ is a continuous linear mapping of $\mathscr{C}(G)$ into $\mathscr{C}(\mathsf{L})$; furthermore

$$(Zf)_{K,P} = \mu_P(Z) f_{K,P} \qquad (Z \in \mathfrak{Z})$$

for all f in $\mathscr{C}(G)$.

Proof In view of the remarks preceding the statement of our theorem, the assignment $f \mapsto f_{K,P}$ is indeed a continuous linear mapping of $\mathscr{C}(G)$ into $\mathscr{C}(\mathsf{L})$. As for the second assertion, it can be supposed that $f \in C_c^\infty(G)$ ($C_c^\infty(G)$ being dense in $\mathscr{C}(G)$); so fix an f in $C_c^\infty(G)$ and write $f(x:n) = f(xn)$ $(x \in G,\ n \in N^+)$ – then, for $Z \in \mathfrak{Z}$, we have

$$\begin{aligned}
(Zf)_{K,P}(l) &= \mathrm{d}_P(l) \int_{N^+} f_K(ln\,; Z) \mathrm{d}_{N^+}(n) = \mathrm{d}_P(l) \int_{N^+} f_K(l\,; Z : n) d_{N^+}(n) \\
&= \mathrm{d}_P(l) \int_{N^+} f_K(l\,; Z_{\mathfrak{l}} + (Z - Z_{\mathfrak{l}}) : n) d_{N^+}(n) \\
&= \mathrm{d}_P(l) \int_{N^+} f_K(l\,; Z_{\mathfrak{l}} : n) d_{N^+}(n) = \mu_P(Z) f_{K,P}(l) \qquad (l \in \mathsf{L}),
\end{aligned}$$

the last equality following from Lemma 8.5.3.1. Hence the theorem. □

[Of course, it is also the case that $(Zf)_P = \mu_P(Z) f_P$ for any $Z \in \mathfrak{Z}$ $(f \in \mathscr{C}(G))$.]

We shall refer to the mappings $f \mapsto f_P$, $f \mapsto f_{K,P}$ $(f \in \mathscr{C}(G))$ as the (*global*) *transformations of descent*.

Note Suppose that $f \in C_c^\infty(G)$ – then, since $G = K \exp(\mathfrak{m}_{\mathfrak{p}}) J_{\mathfrak{p}} N^+$, it follows that $f_{K,P} \in C_c^\infty(\mathsf{L})$ (the assignment $f \mapsto f_{K,P}$ $(f \in C_c^\infty(G))$ evidently being continuous . . .).

8.5.4 The Invariant Integral and the Transformations of Descent

Let (G, K) be a reductive pair fulfilling the usual hypotheses. Fix a θ-stable Cartan subalgebra $\mathfrak{j}$ of $\mathfrak{g}$ with $\mathfrak{j}_{\mathfrak{p}} \neq \{0\}$; let $\mathfrak{l}$ be the centralizer of

$\mathfrak{j}_\mathfrak{p}$ in $\mathfrak{g}$ and introduce L, L, M, M, $N^\pm$, $\mathfrak{m}$, $\mathfrak{n}^\pm$, etc. in the customary fashion; write $P = \mathsf{M}J_\mathfrak{p}N^+$. Let Φ denote the set of roots of the pair $(\mathfrak{g}_c, \mathfrak{j}_c)$, Φ_I the subset of Φ consisting of the imaginary roots – then Φ_I can also be regarded as the set of roots of the pair $(\mathfrak{l}_c, \mathfrak{j}_c)$ (so that $\mathfrak{j}$ is a fundamental Cartan subalgebra of $\mathfrak{l}$). Let J be the Cartan subgroup of G associated with $\mathfrak{j}$, J the identity component of J – then J is the Cartan subgroup of L associated with $\mathfrak{j}$ (cf. Proposition 1.4.1.4). Since G is acceptable, the same must be true of L; therefore, given an $f \in C_c^\infty(L)$, it makes sense to consider the invariant integral Φ_f^L of f calculated relative to the Cartan subgroup J – thus

$$\Phi_f^L(j) = \Delta_I(j)\int_{L/J} f({}^{\dot{l}}j)d_{L/J}(\dot{l}) \qquad (j \in J \cap L')$$

where

$$\Delta_I(j) = \xi_\rho(j_K)\prod_{\alpha\in\Phi_I^+}(1-\xi_\alpha(j^{-1})) \qquad (j \in J \cap L'),$$

j_K the component of j in J_K. [Needless to say, L' denotes the subset of L consisting of those elements which are regular in L.] It will be convenient to assume that the invariant measure on L/J is taken per Proposition 8.1.3.5; in order to simplify the writing, however, we shall agree to multiply $d_{L/J}$ by $\mathbf{c}_G$ and then drop it from the notation (so that we shall be writing $d_{L/J}$ for $\mathbf{c}_G \cdot d_{L/J}$).

Let $\mathsf{Z}(J_\mathfrak{p})$ be the finite subgroup of J given to us by Proposition 1.4.1.3 (hence $\mathsf{J} = \mathsf{Z}(J_\mathfrak{p})J$); in what follows we shall often use without specific mention the obvious fact that the elements of $\mathsf{Z}(J_\mathfrak{p})$ and L commute; note, too, that $\xi_\alpha(a) = 1$ (all $a \in \mathsf{Z}(J_\mathfrak{p})$), α any root of the pair $(\mathfrak{l}_c, \mathfrak{j}_c)$.

As will be seen in the sequel, the following theorem plays a basic role during the discussion of many important questions.

Theorem 8.5.4.1 (Harish-Chandra) Let $f \in C_c^\infty(G)$; for any $a \in \mathsf{Z}(J_\mathfrak{p})$, put $f_{K,P}^a(l) = f_{K,P}(al)$ $(l \in L)$ – then $f_{K,P}^a \in C_c^\infty(L)$ with $(Zf)_{K,P}^a = \mu_P(Z)f_{K,P}^a$ (all $Z \in \mathfrak{Z}$). We have, moreover,

$$\Phi_f(aj) = \xi_\rho(a)\Phi_{f^a_{K,P}}^L(j) \qquad (\text{all } j \in J \cap (a^{-1}G')).$$

[We remind the reader that the invariant measure on G/J_0 is to be taken in the normalization agreed to in 8.1.3 (and not 8.1.2. . .).]

Proof In view of what has been said in 8.5.3, only the last relation need be considered here; but, thanks to Proposition 8.1.3.5, we have

$$\begin{aligned}\Phi_f(aj) &= \epsilon_R(aj)\Delta(aj)\int_{G/\mathsf{J}_0} f({}^{\dot{x}}(aj))d_{G/\mathsf{J}_0}(\dot{x})\\ &= \xi_\rho(a)\Delta_I(j)\xi_\rho(j_\mathfrak{p})\int_{L/J}\left\{\int_{N^+} f_K({}^{\dot{l}}(aj)n)d_{N^+}(n)\right\}d_{L/J}(\dot{l})\\ &= \xi_\rho(a)\Phi_{f^a_{K,P}}^L(j) \qquad (\text{all } j \in J \cap (a^{-1}G')),\end{aligned}$$

as contended. □

8.5.5 Estimation of Φ_f and its Derivatives

Let (G, K) be a reductive pair fulfilling the usual hypotheses. Fix a θ-stable Cartan subalgebra $\mathfrak{j}$ of $\mathfrak{g}$; let Φ denote the set of roots of the pair $(\mathfrak{g}_c, \mathfrak{j}_c)$, Φ_I the subset of Φ consisting of the imaginary roots. Let J be the Cartan subgroup of G associated with $\mathfrak{j}$; write $P = MJ_{\mathfrak{p}}N^+$ and put

$$\Delta_I(j) = \xi_\rho(j_K) \prod_{\alpha \in \Phi_I^+} (1 - \xi_\alpha(j^{-1})) \qquad (j \in J),$$

j_K the component of j in J_K.

Lemma 8.5.5.1 The function $\Delta_I \Phi_f$ is integrable over J (all $f \in C_c^\infty(G)$).

Proof (Mars) Fix $f \in C_c^\infty(G)$ – then, in view of a standard integration formula, it is clear that

$$\int_J |\Delta(j)\Phi_f(j)| \, d_J(j) \leqslant [W(G, J_0)]^{-1} \int_{G(J)} |f(x)| \, d_G(x) < \infty,$$

whence $\Delta\Phi_f$ is summable on J. Therefore, if all the roots of the pair $(\mathfrak{g}_c, \mathfrak{j}_c)$ are imaginary, our assertion follows immediately; on the other hand, if the pair $(\mathfrak{g}_c, \mathfrak{j}_c)$ admits at least one positive root which is not imaginary (so that $\mathfrak{j}_{\mathfrak{p}} \neq \{0\}$), then, thanks to Theorem 8.5.4.1, we have

$$\int_J |\Delta_I(j)\Phi_f(j)| \, d_J(j) \leqslant \sum_{a \in Z(J_{\mathfrak{p}})} \int_J |\Delta_I(aj)\Phi_f(aj)| \, d_J(j)$$

$$= \sum_{a \in Z(J_{\mathfrak{p}})} \int_J |\Delta_I(j)\Phi^L_{f^a_{K,P}}(j)| \, d_J(j) < \infty$$

which serves to establish our contention in this case too. □

The main result of the present number can now be formulated as follows.

Theorem 8.5.5.2 (Harish-Chandra) Let ν be a semi-norm on the complex vector space $C_c^\infty(G)$ such that

$$\int_J |\Delta_I \Phi_f| \, d_J(j) \leqslant \nu(f) \qquad (\text{all } f \in C_c^\infty(G));$$

let $\mathfrak{Z}_0$ be a subalgebra of $\mathfrak{Z}$ (= the center of $\mathfrak{G}$) which contains 1 and over which $\mathfrak{Z}$ is a finite module – then, for any D in $\mathfrak{J}$ ($\sim S(\mathfrak{j}_c)$), there exist a finite number of elements $Z_1, \ldots, Z_m$ in $\mathfrak{Z}_0$ such that

$$\sup_{j \in J'} |\Phi_f(j; D)| \leqslant \sum_{i=1}^m \nu(Z_i f) \qquad (\text{all } f \in C_c^\infty(G)).$$

Note The topology on $C_c^\infty(G)$ is being ignored here, that is the given semi-norm ν is not requested to be continuous.

Our theorem admits the following obvious but important consequence.

Corollary 8.5.5.3 Let $D \in \mathfrak{Z}$ – then $D\Phi_f$ is bounded on J' (all $f \in C_c^\infty(G)$).

The proof of Theorem 8.5.5.2 is rather long and difficult but is highly instructive. The basic idea is to focus attention on the special case when all the roots of the pair $(\mathfrak{g}_c, \mathfrak{j}_c)$ are imaginary; detailed consideration of certain fundamental solutions to a Laplacian then lead to the result. As regards the general situation, observe that all the roots of the pair $(\mathfrak{l}_c, \mathfrak{j}_c)$ are imaginary; this being so, the transformations of descent will enable us to reduce the general case to this one where, by the preceding sentence, the theorem is already known to be valid – such a method of reduction is a very good illustration of Harish-Chandra's *technique of descent*.

Granting for the time being the validity of the theorem in the case when all the roots of the pair $(\mathfrak{g}_c, \mathfrak{j}_c)$ are imaginary, let us proceed to show how the theorem follows in general. So suppose that the pair $(\mathfrak{g}_c, \mathfrak{j}_c)$ admits at least one positive root which is not imaginary (hence $\mathfrak{j}_\mathfrak{p} \neq \{0\}$) – then the notations and conventions which were introduced in 8.5.3 and 8.5.4 are available and will be used without comment in what follows. Define a semi-norm ν_L on the space $C_c^\infty(L)$ by the rule

$$\nu_L(f) = \int_J |\Delta_I(j)\Phi_f^L(j)|\, d_J(j) \qquad (f \in C_c^\infty(L)).$$

Because the center $\mathfrak{Z}_\mathfrak{l}$ of the universal enveloping $\mathfrak{L}$ of $\mathfrak{l}_c$ is a finite module over $\mu_P(\mathfrak{Z})$ (cf. Example 3 following Proposition 2.3.3.6), our theorem holds for $(L, J, \mu_P(\mathfrak{Z}_0), \nu_L)$ in place of $(G, \mathsf{J}, \mathfrak{Z}_0, \nu)$. To exploit this fact, observe that Theorem 8.5.4.1 tells us that

$$\nu_L(f^a_{K,P}) = \int_J |\Delta_I(j)\Phi_f(aj)|\, d_J(j) = \int_J |\Delta_I(aj)\Phi_f(aj)|\, d_J(j) \leqslant \nu(f),$$

whence, given D in $\mathfrak{Z}$, there exist elements $Z_1, \ldots, Z_m$ in $\mathfrak{Z}_0$ such that

$$\sup_{J\cap L'} |D\Phi^L_{f^a_{K,P}}| \leqslant \sum_{i=1}^m \nu_L(\mu_P(Z_i)f^a_{K,P})$$
$$= \sum_{i=1}^m \nu_L((Z_i f)^a_{K,P}) \leqslant \sum_{i=1}^m \nu(Z_i f) \qquad (\text{all } f \in C_c^\infty(G))$$

for any $a \in \mathsf{Z}(J_\mathfrak{p})$ whatsoever. Because

$$aj \in \mathsf{J}' \ (a \in \mathsf{Z}(J_\mathfrak{p}), j \in J) \Rightarrow j \in J \cap L',$$

it therefore follows that

$$\sup_{\mathsf{J}'} |D\Phi_f| \leqslant \sup_{a \in \mathsf{Z}(J_\mathfrak{p})} \sup_{j \in J\cap L'} |\Phi_f(aj; D)|$$
$$\leqslant \sup_{a \in \mathsf{Z}(J_\mathfrak{p})} \sup_{J\cap L'} |D\Phi^L_{f^a_{K,P}}| \leqslant \sum_{i=1}^m \nu(Z_i f) \qquad (\text{all } f \in C_c^\infty(G)),$$

as we wished to prove.

It can now be supposed that all the roots of the pair $(\mathfrak{g}_c, \mathfrak{j}_c)$ are imaginary (thus $\mathsf{J} = J \ldots$); since J/Z is compact (Z the center of G), it will be enough to establish the following assertion: Given an element γ in J, there exists a neighborhood $\mathcal{O}(\gamma)$ of γ in J with the property that for any D in $\mathfrak{J}$ there exist a finite number of elements $Z_1, \ldots, Z_m$ in $\mathfrak{Z}_0$ such that

$$\sup_{j \in \mathcal{O}'(\gamma)} |\Phi_f(j; D)| \leqslant \sum_{i=1}^{m} \nu(Z_i f) \qquad (\text{all } f \in C_c^\infty(G)).$$

So fix an element γ in J; let $\mathfrak{g}_\gamma$ denote the centralizer of γ in $\mathfrak{g}$ – then the notations and conventions which were introduced in 8.5.1 per this situation are available and will be used without comment during the course of the ensuing discussion. Given $\epsilon > 0$ $(0 < \epsilon \leqslant 1)$, let $\mathscr{S}_\epsilon = \{H \in \mathfrak{j}: \|H\| < \epsilon\}$ ($\|.\|$ a norm on $\mathfrak{j}$ with the obvious properties ...); by choosing ϵ sufficiently small, it can be assumed that the exponential mapping is regular and injective on $\mathscr{S}_\epsilon$ with

(1) For every α in Φ,

$$|e^{\alpha(H)/2} - \xi_\alpha(\gamma^{-1})e^{-\alpha(H)/2}| \geqslant 2^{-1}|1 - \xi_\alpha(\gamma^{-1})| \qquad (\text{all } H \in \mathscr{S}_\epsilon);$$

(2) For every α in Φ_γ^+,

$$|(e^{\alpha(H)/2} - e^{-\alpha(H)/2})/\alpha(H)| \geqslant 2^{-1} \qquad (\text{all } H \in \mathscr{S}_\epsilon).$$

The intersection of $\mathfrak{j}_\gamma'$ with $\mathscr{S}_\epsilon$ has but a finite number of connected components, say $\mathscr{C}_1, \ldots, \mathscr{C}_r$; fix a positive number c $(0 < c \leqslant 1)$ and put $\mathcal{O}(\gamma) = \gamma \exp(c\mathscr{S}_\epsilon)$ – then it is clear that $\mathcal{O}'(\gamma) = \bigcup_i \mathcal{O}_i(\gamma)$ where $\mathcal{O}_i(\gamma) = \gamma \exp(c\mathscr{C}_i)$ $(i = 1, \ldots, r)$. To prove our theorem, it will be enough to establish the following assertion: Fix i $(1 \leqslant i \leqslant r)$ – then there exists a number c $(0 < c \leqslant 1)$ with the property that for any D in $\mathfrak{J}$ there exist a finite number of elements $Z_1, \ldots, Z_m$ in $\mathfrak{Z}_0$ such that

$$\sup_{H \in c\mathscr{C}_i} |\Phi_f(H; \partial(D))| \leqslant \sum_{i=1}^{m} \nu(Z_i f) \qquad (\text{all } f \in C_c^\infty(G)).$$

Here we have set $\Phi_f(H) = \Phi_f(\gamma \exp H)$ $(H \in \mathscr{S}_\epsilon \cap \mathfrak{j}_\gamma')$. Throughout the remainder of our discussion, it will be assumed that i $(1 \leqslant i \leqslant r)$ has been fixed once and for all and then dropped from the notation (so that, in particular, we shall be writing $\mathscr{C}$ for $\mathscr{C}_i$).

Notation Let us agree to denote by $\mathfrak{S}$ the set consisting of those (not necessarily continuous) semi-norms σ on $C_c^\infty(G)$ which have the property that there exist a finite number of elements $Z_1, \ldots, Z_m$ in $\mathfrak{Z}_0$ such that

$$\sigma(f) \leqslant \sum_{i=1}^{m} \nu(Z_i f) \qquad (\text{all } f \in C_c^\infty(G)).$$

Let l_γ be the rank of $\bar{\mathfrak{g}}_\gamma$ $(= [\mathfrak{g}_\gamma, \mathfrak{g}_\gamma])$ – then we can choose l_γ roots α_i $(1 \leqslant i \leqslant l_\gamma)$ of the pair $(\mathfrak{g}_{\gamma,c}, \mathfrak{j}_c)$ with the property that if $\alpha \in \Phi_\gamma$ and if $\sqrt{-1}\alpha(H) > 0$ (all $H \in \mathscr{C}$), then $\alpha = \sum_i n_i\alpha_i$ for certain non-negative integers n_i $(1 \leqslant i \leqslant l_\gamma)$. Put $t_i(H) = \sqrt{-1}\alpha_i(H)$ $(1 \leqslant i \leqslant l_\gamma,\ H \in \mathfrak{j})$; choose a basis $\{H_j\}$ for $\bar{\mathfrak{j}}$ such that $t_i(H_j) = \delta_{ij}$; choose an orthonormal basis $\{H_j\}$ for $\mathfrak{c}$ and extend the t_i to a Cartesian coordinate system on all of $\mathfrak{j}$ by the requirement $t_i(H_j) = \delta_{ij}$ $(1 \leqslant i, j \leqslant l;\ l = \dim(\mathfrak{j}))$ – then it is clear that a point H in $\mathscr{S}_\epsilon$ lies in $\mathscr{C}$ iff $t_i(H) > 0$ $(1 \leqslant i \leqslant l_y)$. By a *monomial* T on $\mathfrak{j}$ we shall understand a function of the form $T = \prod_{1 \leqslant i \leqslant l_y} t_i^{m_i}$ where the m_i $(1 \leqslant i \leqslant l_y)$ are non-negative integers; the *degree* deg (T) of T is the integer $\sum_i m_i$. Because $S(\mathfrak{j}_c)$ is a finite module over $I(\mathfrak{j}_c) = \gamma(\mathfrak{Z})$, it is also a finite module over $\gamma(\mathfrak{Z}_0)$; accordingly there exist elements $u_1 = 1, u_2, \ldots, u_r$ (say) such that $S(\mathfrak{j}_c) = \sum_i \gamma(\mathfrak{Z}_0)u_i$. This being so, a monomial T is said to have the property (P) if there exists a number $c = c(T)$ $(0 < c \leqslant 1)$ and an element σ in $\mathfrak{S}$ such that

$$\sup_{H \in c\mathscr{C}} T(H)\,|\Phi_f(H; \partial(u_i))| \leqslant \sigma(f) \qquad (1 \leqslant i \leqslant r)$$

for all $f \in C_c^\infty(G)$. Suppose now that the monomial T has the property (P); put

$$\sigma_T(f) = \max_{1 \leqslant i \leqslant r} \sup_{H \in c\mathscr{C}} T(H)\,|\Phi_f(H; \partial(u_i))| \qquad (f \in C_c^\infty(G)),$$

where $c = c(T)$ – then it is obvious that $\sigma_T \in \mathfrak{S}$. On the other hand, given a $D \in S(\mathfrak{j}_c)$, we can select elements $Z_i \in \mathfrak{Z}_0$ such that $D = \sum_i \gamma(Z_i)u_i$; since $\Phi_{Zf} = \partial(\gamma(Z))\Phi_f$ (cf. Theorem 8.5.1.3), for any $f \in C_c^\infty(G)$, we have

$$\Phi_f(H; \partial(D)) = \sum_i \Phi_{Z_if}(H; \partial(u_i)) \qquad (H \in \mathscr{C})$$

and so it is clear that the semi-norm $\sigma_{T,D}$ defined by the rule

$$\sigma_{T,D}(f) = \sup_{H \in c\mathscr{C}} T(H)\,|\Phi_f(H; \partial(D))| \qquad (f \in C_c^\infty(G))$$

is dominated by $\sum_i \sigma_T(Z_if)$ $(f \in C_c^\infty(G))$. It follows, therefore, that our theorem will be established when it is shown that the monomial 1 has the property (P); in turn, for this purpose, we shall argue essentially as we did in 8.4.3 supra.

Existence of Monomials with the Property (P) Fix a non-zero element D in $S(\mathfrak{j}_c)$ – then we intend to show that there exists an integer $d \geqslant 0$ and a semi-norm $\sigma \in \mathfrak{S}$ such that

$$\sup_{H \in \mathscr{C}/2} \{\prod_{1 \leqslant i \leqslant l_y} t_i(H)\}^d\,|\Phi_f(H; \partial(D))| \leqslant \sigma(f) \qquad (\text{all } f \in C_c^\infty(G)).$$

The proof of the present assertion is based on Lemmas 8.3.5.5 – 8.3.5.10 supra; at this point, then, the reader is asked to familiarize himself with these results (and the notations thereof).

(I) We shall set the stage with a series of technical remarks. Let ω be the Casimir operator in $\mathfrak{Z}$ (cf. the discussion near the end of Example 1 following Proposition 2.3.3.6) – then $\gamma(\omega) + (\rho, \rho)$ is homogeneous of degree 2 and $\Delta = \partial(\gamma(\omega)) + (\rho, \rho)$ is an elliptic differential operator on $\mathfrak{j}$. Because $\mathfrak{Z}$ is a finite module over $\mathfrak{Z}_0$ of degree m_2 (say), for each integer

$m_3 \geqslant 1$ we have a relation of the form

$$\omega_0^{m_2 m_3} + \sum_{k=1}^{m_2} Z_k \omega_0^{m_3(m_2-k)} = 0 \qquad (\omega_0 = \omega + (\rho, \rho)),$$

the Z_k lying in $\mathfrak{Z}_0$; so, if m_1 is the degree of D, then, by taking m_3 sufficiently large, we are ensured of the existence of a function f_0 on $\mathfrak{j}$ of class $C^{2m_3(m_2-1)+m_1}$ such that $\Delta^{m_2 m_3} \cdot f_0 = \delta$ (cf. Lemma 8.5.3.10). Applying γ to the preceding relation yields

$$\Delta^{m_2 m_3} + \sum_{k=1}^{m_2} \partial(\gamma(Z_k))\Delta^{m_3(m_2-k)} = 0,$$

or, upon taking adjoints and bearing in mind that $\Delta = \Delta^\sharp$,

$$\Delta^{m_2 m_3} + \sum_{k=1}^{m_2} \partial(\gamma(Z_k))^\sharp \Delta^{m_3(m_2-k)} = 0.$$

Set

$$f_k = -\Delta^{m_3(m_2-k)}(\partial(D)^\sharp f_0) \qquad (1 \leqslant k \leqslant m_2).$$

Then f_k is of class $C^{2m_3(k-1)}$ (in fact is of class C^∞ everywhere except at zero) with

$$\partial(D)^\sharp \cdot \delta = \sum_{k=1}^{m_2} \partial(\gamma(Z_k))^\sharp \cdot f_k.$$

Suppose that the function h of Lemma 8.3.5.8 has the following additional properties: $h = 1$ on $(-\infty, 0]$, $h = 0$ on $[1, +\infty)$ and $0 \leqslant h \leqslant 1$ everywhere. For any real number r $(0 < r \leqslant 1/2)$ put

$$H_r(H) = h(r^{-1}\|H\| - 2) \qquad (H \in \mathfrak{j}).$$

Then for any element $u \in S(\mathfrak{j}_c)$ of degree $\leqslant m_1$ the entity $r^{m_1}|\partial(u)H_r|$ admits a bound which is uniform in r $(0 < r \leqslant 1/2)$. [In passing let us also note that $H_r(H) = 0$ unless $\|H\| \leqslant 3r$ $(H \in \mathfrak{j})$.] For any r with $0 < r \leqslant 1/3$ put $f_{k,r} = H_r f_k$ – then it is clear that

$$\sum_{k=1}^{m_2} \partial(\gamma(Z_k))^\sharp \cdot f_{k,r} = \partial(D)^\sharp \cdot \delta + g_r$$

where $g_r \in C_c^\infty(\mathfrak{j})$ and $\operatorname{spt}(g_r) \subset \operatorname{spt}(H_r)$. Since $H_r(H) = 1$ whenever $\|H\| \leqslant 2r$, it follows that $g_r(H) = 0$ unless $2r \leqslant \|H\| \leqslant 3r$ and so

$$\sup_H |g_r(H)| = \sup_{2r \leqslant \|H\| \leqslant 3r} |g_r(H)|.$$

Use of the explicit formula for f_0 which was obtained during the proof of Lemma 8.3.5.10 then leads to the conclusion that

$$\sup_{H \in \mathfrak{j}} |g_r(H)| \leqslant M_1 r^{-m+1} |\log r| \leqslant M_2 r^{-m}$$

where M_1, M_2 are positive numbers, m is a non-negative integer and they are all independent of r $(0 < r \leqslant 1/3)$.

(II) Given $H \in \mathfrak{j}$, let us agree to write

$$\tau(H) = \begin{cases} \min\limits_{1 \leqslant i \leqslant l_\gamma} |t_i(H)| & \text{if } l_\gamma > 0, \\ \epsilon & \text{if } l_\gamma = 0. \end{cases}$$

Now fix an $H_0 \in \mathscr{C}/2$; put $r_0 = \tau(H_0)/6, f_{k,H_0} = f_{k,r_0}, g_{H_0} = g_{r_0}$ $(1 \leqslant k \leqslant m_2)$ – then, in view of what was said in (I), we have

$$\sum_{k=1}^{m_2} \partial(\gamma(Z_k))^\sharp \cdot f_{k,H_0} = \partial(D)^\sharp \cdot \delta + g_{H_0}$$

with $\sup |g_{H_0}| \leqslant M_3 \tau(H_0)^{-m}$ where $M_3 = 6^m M_2$. It has been noted above that both spt (f_{k,H_0}) and spt (g_{H_0}) are contained in spt (H_{r_0}); moreover, if H is an element of $\mathfrak{j}$ such that $H - H_0 \in$ spt (H_{r_0}), then clearly $H \in \mathscr{C}$ with $\tau(H) \geqslant \tau(H_0)/2$. Let $\mathscr{S}_\epsilon(H_0)$ denote the subset of $\mathscr{S}_\epsilon$ consisting of those H such that $\tau(H) \geqslant \tau(H_0)/2$ – then $\mathscr{S}_\epsilon(H_0) \subset \mathscr{C}$ and we have

$$\Phi_f(H_0; \partial(D)) = \sum_{k=1}^{m_2} \int_{\mathscr{S}_\epsilon(H_0)} \Phi_f(H; \partial(\gamma(Z_k))) f_{k,H_0}(H - H_0) d_{\mathfrak{j}}(H)$$
$$- \int_{\mathscr{S}_\epsilon(H_0)} \Phi_f(H) g_{H_0}(H - H_0) d_{\mathfrak{j}}(H).$$

By the definition of $\mathscr{S}_\epsilon$ we can find a number $M_4 > 0$ such that

$$|\Delta(\gamma \exp H)| \geqslant M_4 |\pi_\gamma(H)| \qquad (\text{all } H \in \mathscr{S}_\epsilon);$$

since $|\pi_\gamma(H)| \geqslant \tau(H)^{r_\gamma}$ (all $H \in \mathscr{C}$), r_γ the cardinality of Φ_γ^+, it follows that

$$|\Delta(\gamma \exp H)| \geqslant M_4 \tau(H)^{r_\gamma} \qquad (\text{all } H \in \mathscr{C}).$$

Hence

$$\tau(H_0)^{r_\gamma} \left| \int_{\mathscr{S}_\epsilon(H_0)} \Phi_f(H; \partial(\gamma(Z_k))) f_{k,H_0}(H - H_0) d_{\mathfrak{j}}(H) \right|$$
$$\leqslant M_5 \int_{\mathscr{S}_\epsilon(H_0)} |\Phi_{Z_k f}(H) \Delta(\gamma \exp H)| \cdot |f_{k,H_0}(H - H_0)| \, d_{\mathfrak{j}}(H)$$
$$(1 \leqslant k \leqslant m_2)$$

where $M_5 = 2^{r_\gamma} M_4^{-1}$. Because $|H_r| \leqslant 1$, there exists a positive number M_6, independent of H_0 and k, such that

$$\sup_{H \in \mathscr{S}_\epsilon(H_0)} |f_{k,H_0}(H - H_0)| \leqslant \sup_{\|H\| \leqslant 2} |f_k(H)| \leqslant M_6.$$

Therefore we have the estimate

$$\tau(H_0)^{r_\gamma} \left| \int_{\mathscr{S}_\epsilon(H_0)} \Phi_f(H; \partial(\gamma(Z_k))) f_{k,H_0}(H - H_0) d_{\mathfrak{j}}(H) \right| \leqslant M_5 M_6 \nu(Z_k f)$$
$$(1 \leqslant k \leqslant m_2).$$

On the other hand, since $\sup |g_{H_0}| \leqslant M_3 \tau(H_0)^{-m}$, there can be no doubt

that

$$\tau(H_0)^{m+r_\gamma}\left|\int_{\mathscr{S}_\epsilon(H_0)} \Phi_f(H)g_{H_0}(H-H_0)d_{\mathfrak{j}}(H)\right| \leqslant M_3M_5\nu(f).$$

Bearing in mind that $\tau(H_0) \leqslant \epsilon \leqslant 1$, we then arrive at the final estimate, namely

$$\tau(H_0)^{m+r_\gamma}\,|\Phi_f(H_0\,;\partial(D))| \leqslant M_3M_5\nu(f) + M_5M_6\sum_{k=1}^{m_2}\nu(Z_kf)$$
$$(f \in C_c^\infty(G)),$$

valid for all $H_0 \in \mathscr{C}/2$.

Taking into account the relation

$$\prod_{1\leqslant i\leqslant l_\gamma} t_i(H) \leqslant \epsilon^{l_\gamma-1}\tau(H) \qquad (H \in \mathscr{C}),$$

we conclude that the contention supra is met with $d = m + r_\gamma$.

The Monomial 1 has the Property (P) Owing to what we have seen above, monomials T with the property (P) actually do exist; furthermore, the monomial 1 has the property (P) if $l_\gamma = 0$ – therefore it can be assumed in what follows that $l_\gamma \geqslant 1$. This being the case, let T be a monomial with property (P) of the lowest possible degree – then we have to prove that $T = 1$. Assuming to the contrary that $T \neq 1$ (so that $\deg(T) > 0$), we shall produce a monomial with the property (P) whose degree is strictly less than that of T; this however will contradict our original supposition on T. . . . Let i_0 be the least index $(1 \leqslant i_0 \leqslant l_\gamma)$ such that the exponent m_0 of t_{i_0} in T is positive; put $t = t_{i_0}$ – then $T = t^{m_0}T_0$ where T_0 is a monomial in which the exponents of t_i are zero for $1 \leqslant i \leqslant i_0$. Write $\partial_t = \partial/\partial t = \partial(H_{i_0})$; choose elements $Z_{ij} \in \mathfrak{Z}_0$ $(1 \leqslant i, j \leqslant r)$ such that $H_{i_0}u_i = \sum_j \gamma(Z_{ij})u_j$. Now fix an element f in $C_c^\infty(G)$; agreeing to write c for $c(T)$, put $\Phi_i(H) = T_0(H)\Phi_f(H;\partial(u_i))$,

$$\Psi_i(H) = T_0(H)\sum_j \Phi_{Z_{ij}f}(H\,;\partial(u_j)) \qquad (H \in \mathscr{C},\ 1 \leqslant i \leqslant r).$$

Then $\partial_t\Phi_i = \Psi_i$ on $c\mathscr{C}$ (all i). Introduce the semi-norm σ_T as above and put $\sigma = \sum_{i,j}\sigma_T(Z_{ij}\,.)$ – then it is clear that σ lies in $\mathfrak{S}$ and, moreover, we obviously have $|t^{m_0}\partial_t\Phi_i| \leqslant \sigma(f)$. For any $\epsilon > 0$, let $\mathscr{T}_\epsilon$ denote the subset of $\mathfrak{j}$ comprised of those H such that $|t_i(H)| \leqslant \epsilon$ $(1 \leqslant i \leqslant l)$; choose a $(0 < a \leqslant 1)$ small enough to secure the inclusion $\mathscr{T}_a \subset c\mathscr{S}_\epsilon$ and select c^0 $(0 < c^0 \leqslant a)$ such that $c^0\mathscr{S}_\epsilon \subset \mathscr{T}_a$. Suppose that $H \in c^0\mathscr{C}$ – then

$$H^0 = H + (a - t(H))H_{i_0} \in \mathscr{T}_a \subset c\mathscr{S}_\epsilon.$$

Because $t_i(H) > 0$ $(1 \leqslant i \leqslant l_\gamma)$ while $t(H) \leqslant \|H\| < cc^0 \leqslant c^0 \leqslant a$, it is clear that $t_i(H^0) > 0$ $(1 \leqslant i \leqslant l_\gamma)$, whence $H^0 \in c\mathscr{C}$; owing to the convexity of $c\mathscr{C}$, it then follows that the whole line segment joining H to H^0 lies in $c\mathscr{C}$. Accordingly, an integration of the relation $|\partial_t\Phi_i| \leqslant$

$t^{-m_0}\sigma(f)$ along the line segment in question leads at once to the estimate

$$|\Phi_i(H^0) - \Phi_i(H)| \leqslant \sigma(f) \int_{t(H)}^{a} r^{-m_0} dr,$$

$t(H^0)$ being equal to a. Since $|\Phi_i(H^0)| \leqslant a^{-m_0}\sigma_T(f)$, we conclude that

$$|\Phi_i(H)| \leqslant a^{-m_0}\sigma_T(f) + \sigma(f) \int_{t(H)}^{a} r^{-m_0} dr \qquad (\text{all } H \in c^0\mathscr{C}).$$

We shall now distinguish two cases.

(I) The case $m_0 \geqslant 2$: In this situation we have

$$\int_{t(H)}^{a} r^{-m_0} dr = (m_0 - 1)^{-1}(t(H)^{1-m_0} - a^{1-m_0}) \qquad (H \in c^0\mathscr{C}).$$

Put $T^0 = t^{m_0-1}T_0$ – then it is clear that

$$|T^0(H)\Phi_f(H; \partial(u_i))| \leqslant a^{-m_0}\sigma_T(f) + \sigma(f) \qquad (1 \leqslant i \leqslant r)$$

for all $H \in c^0\mathscr{C}$. Since this estimate obtains for every $f \in C_c^\infty(G)$, the monomial T^0 has the property (P); but $\deg(T^0) < \deg(T)$. . . .

(II) The case $m_0 = 1$: In this situation we have

$$\int_{t(H)}^{a} r^{-1} dr = \log(a/t(H)) \qquad (H \in c^0\mathscr{C})$$

and so it is clear that

$$|\Phi_i(H)| \leqslant a^{-1}\sigma_T(f) + \sigma(f)\log(a/t(H)) \qquad (1 \leqslant i \leqslant r)$$

for all $H \in c^0\mathscr{C}$. Put

$$\tilde{\sigma}(f) = \sum_{i,j} \{\sigma(Z_{ij}f) + a^{-1}\sigma_T(Z_{ij}f)\} \qquad (f \in C_c^\infty(G)).$$

Then $\tilde{\sigma} \in \mathfrak{S}$ and we have

$$|\partial_t\Phi_i| \leqslant \tilde{\sigma}(f)\{1 + \log(a/t)\}$$

on $c^0\mathscr{C}$. Choose numbers c_0, b $(0 < c_0 \leqslant b \leqslant c^0)$ such that $\mathscr{T}_b \subset c^0\mathscr{S}_\epsilon$ and $c_0\mathscr{S}_\epsilon \subset \mathscr{T}_b$; given an H in $c_0\mathscr{C}$, set $H_0 = H + (b - t(H))H_{i_0}$ – then $H_0 \in c^0\mathscr{C}$ and so, upon integrating along the line segment joining H and H_0, we deduce that

$$|\Phi_i(H_0) - \Phi_i(H)| \leqslant \tilde{\sigma}(f) \int_{t(H)}^{b} (1 + \log(a/r))dr \leqslant M\tilde{\sigma}(f) \qquad (1 \leqslant i \leqslant r)$$

for all $H \in c_0\mathscr{C}$ (here $M = \int_0^b(1 + \log(a/r))dr < \infty$ . . .). Since $t(H_0)|\Phi_i(H_0)| \leqslant \sigma_T(f)$, it now follows that

$$|\Phi_i(H)| \leqslant b^{-1}\sigma_T(f) + M\tilde{\sigma}(f) \qquad (1 \leqslant i \leqslant r)$$

for all $H \in c_0\mathscr{C}$. Because this estimate holds for every $f \in C_c^\infty(G)$, the monomial T_0 has the property (P), a contradiction.

The proof of Theorem 8.5.5.2 is therefore complete.

Here is an application of the preceding considerations.

Proposition 8.5.5.4 The function $|D_l|^{-1/2}$ is locally summable on G.

Proof Fix a compact subset ω of G; choose an $f \in C_c^\infty(G)$ such that $f \geqslant 0$ everywhere and $f \geqslant 1$ on ω – then it is clear that

$$\int_\omega |D_l(x)|^{-1/2} d_G(x) \leqslant \int_G |D_l(x)|^{-1/2} f(x) d_G(x).$$

Introduce the Cartan subgroups $\mathrm{J}^1, \ldots, \mathrm{J}^r$ in the usual way; since G' is the disjoint union of the $G(\mathrm{J}^i)$ ($G(\mathrm{J}^i)$ as always), our proposition will follow when it is shown that

$$\int_{G(\mathrm{J}^i)} |D_l(x)|^{-1/2} f(x) d_G(x) < \infty \qquad (1 \leqslant i \leqslant r).$$

But for any i,

$$\int_{G(\mathrm{J}^i)} |D_l(x)|^{-1/2} f(x) d_G(x) = [W(G, \mathrm{J}_0^i)]^{-1} \int_{\mathrm{J}^i} |\Phi_f^i(j)| \, d_{\mathrm{J}^i}(j) < \infty,$$

the finiteness of the integral on the right being clear on the basis of Corollary 8.5.5.3 provided we bear in mind that Φ_f^i vanishes off a compact subset of J^i (cf. 8.5.1). □

8.5.6 An Important Inequality

Let (G, K) be a reductive pair fulfilling the usual hypotheses. Fix a θ-stable Cartan subalgebra $\mathfrak{j}$ of $\mathfrak{g}$; let Φ denote the set of roots of the pair $(\mathfrak{g}_c, \mathfrak{j}_c)$, Φ_I the subset of Φ consisting of the imaginary roots. Let J be the Cartan subgroup of G associated with $\mathfrak{j}$; write $P = \mathrm{M}J_\mathfrak{p}N^+$ and put

$$\Delta_I(j) = \xi_\rho(j_K) \prod_{\alpha \in \Phi_I^+} (1 - \xi_\alpha(j^{-1})) \qquad (j \in \mathrm{J}),$$

j_K the component of j in J_K.

The objective of the present number is to establish the following result.

Theorem 8.5.6.1 (Harish-Chandra) There exists a number $M > 0$ such that

$$\int_\mathrm{J} |\Delta_I(j)\Phi_f(j)| \, d_\mathrm{J}(j) \leqslant M \int_G |f(x)| \, \Xi(x) d_G(x)$$

for all $f \in I_c^\infty(G)$.

[Here $I_c^\infty(G)$ denotes the subalgebra of $C_c^\infty(G)$ consisting of those f which are biinvariant under K.]

Just as in 8.5.5, the basic idea lying behind the proof is to reduce everything to the special case when all the roots of the pair $(\mathfrak{g}_c, \mathfrak{j}_c)$ are imaginary; such a reduction will, of course, be carried out via the transformations of descent.

Let us dispose first of the possibility that $\mathfrak{c}_\mathfrak{p} \neq \{0\}$; thus, in this case, $G = \widehat{G} \times C_\mathfrak{p}$ (direct product) with $\dim(\widehat{G}) < \dim(G)$. Granting that the theorem holds for $\widehat{G}$, fix an $f \in I_c^\infty(G)$; given $c \in C_\mathfrak{p}$, let $\hat{f}_c$ denote the function $\hat{x} \mapsto f(\hat{x}c)$ $(\hat{x} \in \widehat{G})$ – then

$$\begin{aligned}\int_{\mathsf{J}} |\Delta_I(j)\Phi_f(j)|\,d_{\mathsf{J}}(j) &= \int_{C_\mathfrak{p}} \left\{\int_{\hat{\mathsf{J}}} |\Delta_I(\hat{j})\hat{\Phi}_{\hat{f}_c}(\hat{j})|\,d_{\hat{\mathsf{J}}}(\hat{j})\right\} d_{C_\mathfrak{p}}(c)\\ &\leqslant M \int_{C_\mathfrak{p}} \left\{\int_{\hat{G}} |\hat{f}_c(\hat{x})|\,\Xi(\hat{x})\,d_{\hat{G}}(\hat{x})\right\} d_{C_\mathfrak{p}}(c)\\ &= M \int_G |f(x)|\,\Xi(x)\,d_G(x),\end{aligned}$$

which shows that the theorem also holds for G.

In what follows, then, it may be supposed that $\mathfrak{c}_\mathfrak{p} = \{0\}$. This being so, let $\mathfrak{l}$ be the centralizer of $\mathfrak{j}_\mathfrak{p}$ in $\mathfrak{g}$; let D_{J} be the analytic function on $G(\mathsf{J})$ defined by the rule

$$D_{\mathsf{J}}(xjx^{-1}) = \det(1 - \mathrm{Ad}(j))_{\mathfrak{g}/\mathfrak{l}} \qquad (j \in \mathsf{J}',\ x \in G).$$

Then

$$|D_{\mathsf{J}}(j)|^{-1/2}\,|\Delta(j)| = |\Delta_I(j)| \qquad (j \in \mathsf{J}'),$$

whence, setting $D_{\mathsf{J}}^{-1} = 0$ on $G - G(\mathsf{J})$, we have

$$\int_G f(x)\,|D_{\mathsf{J}}(x)|^{-1/2} d_G(x) = [W(G, \mathsf{J}_0)]^{-1} \int_{\mathsf{J}} |\Delta_I(j)\Phi_f(j)|\,d_{\mathsf{J}}(j)$$
$$(f \in C_c^\infty(G),\ f \geqslant 0)$$

and so Theorem 8.5.6.1 is equivalent to the assertion that

$$\int_K |D_{\mathsf{J}}(xk)|^{-1/2} dk \leqslant M\,\Xi(x)$$

for almost all x in G (M a positive constant).

Note We admit here the possibility that $\mathfrak{j}_\mathfrak{p} = \{0\}$ (in which case $\mathfrak{l} = \mathfrak{g}$); in this situation, let us agree to write $\mathfrak{t}$ for $\mathfrak{j}$, T for J $(= \mathsf{J}\ldots)$, $\mathscr{E}$ for $G(T)$ (the regular *elliptic* elements in G) – then Theorem 8.5.6.1 is equivalent to the assertion that

$$\int_K \psi_{\mathscr{E}}(xk)\,dk \leqslant M\,\Xi(x)$$

for almost all x in G ($\psi_{\mathscr{E}}$ the characteristic function of $\mathscr{E}$).

We shall now prove the theorem by induction on $\dim(G)$.

To begin with, let us suppose that $\mathfrak{j}_\mathfrak{p} \neq \{0\}$ – then $\dim(\mathfrak{l}) < \dim(\mathfrak{g})$ and our theorem is known to hold for the pair (L, J) through the induction hypothesis. Extend $\mathfrak{j}_\mathfrak{p}$ to a maximal abelian subspace $\mathfrak{a}_\mathfrak{p}$ of $\mathfrak{p}$ – then, for the purposes of the ensuing discussion, it will be best to utilize the notational conventions which were agreed to in the Introduction (cf.

8.5.2 . . .). Now let f be an arbitrary element in $I_c^\infty(G)$ – then it is clear that $f_{K,P}|L \in I_c^\infty(L)$ and, moreover, as can be seen from the proof of Lemma 8.5.5.1, we have

$$\int_{\mathrm{J}} |\Delta_I(j)\Phi_f(j)|\, d_{\mathrm{J}}(j) \leq \int_{L(J)} |f_{K,P}(l)|\, d_L(l) \qquad (\text{all } f \in I_c^\infty(G)),$$

where $L(J) = \bigcup_{l \in L} l(J \cap L')l^{-1}$. It is easy to estimate the integral on the right; in fact, since the theorem holds for (L, J), we have

$$\begin{aligned}
&\int_{L(J)} |f_{K,P}(l)|\, d_L(l) \\
&\leq \iiint_{A_1 \times A_2 \times N_1^+} |f_{K,P}(h_1 n_1 h_2)|\, h_1^{\rho_1} d_{A_1}(h_1) d_{A_2}(h_2) d_{N_1^+}(n_1) \\
&\leq \iiiint_{A_1 \times A_2 \times N_1^+ \times N_2^+} |f(h_1 n_1 h_2 n_2)|\, h_1^{\rho_1} h_2^{\rho_2} d_{A_1}(h_1) d_{A_2}(h_2) d_{N_1^+}(n_1) d_{N_2^+}(n_2) \\
&= \iint_{A_\mathfrak{p} \times N^+} |f(hn)|\, h^\rho d_{A_\mathfrak{p}}(h) d_{N^+}(n) \leq \int_G |f(x)| \mapsto (x) d_G(x) \quad (\text{all } f \in I_c^\infty(G)),
\end{aligned}$$

which serves to establish our contention in this case.

It remains to consider the possibility that $\mathfrak{j}_\mathfrak{p} = \{0\}$. Agreeing to write $\mathfrak{t}$ for $\mathfrak{j}$ and T for J $(= J \ldots)$, let $T = \mathrm{J}^1, \mathrm{J}^2, \ldots, \mathrm{J}^r$ be a maximal collection of mutually non-conjugate Cartan subgroups of G whose Lie algebras are θ-stable. Since $Z \subset T \subset K$ (Z the center of G), there is no loss of generality in assuming that G is the real analytic subgroup of G_c with Lie algebra $\mathfrak{g}$, G_c the simply connected complex analytic group corresponding to $\mathfrak{g}_c$ (therefore Cartan subgroups of G are now abelian . . .). Fix an i $(2 \leqslant i \leqslant r)$ and then drop it from the notation; put

$$\Delta_{R,C}(j) = \xi_\rho(j_\mathfrak{p}) \prod_{\alpha \in \Phi_R^+ \cup \Phi_C^+} (1 - \xi_\alpha(j^{-1})) \qquad (j \in \mathrm{J}),$$

$j_\mathfrak{p}$ the component of j in $J_\mathfrak{p}$ – then clearly $\Delta = \Delta_I \cdot \Delta_{R,C}$. There exists a central $\mathfrak{Z}$-finite distribution Θ on G with the following properties: (1) $\Theta(f) = 0$ (all $f \in I_c^\infty(G)$); (2) The restriction of Θ to $\mathscr{E}$ (qua a locally summable function . . .) is constant; (3) $|\Delta_{R,C} \cdot \Theta|$ remains bounded on J' (any J $(= \mathrm{J}^i, i = 2, \ldots, r)$). Admitting this fact (cf. infra), it is then a simple matter to complete the proof of Theorem 8.5.6.1. Thus, in view of what has been said above, we have

$$\begin{aligned}
&\left| \int_{G(\mathrm{J})} f(x)\Theta(x) d_G(x) \right| \\
&\leq \int_{\mathrm{J}} |\Delta(j)|^2 d_{\mathrm{J}}(j) \int_{G/\mathrm{J}_0} |f({}^*j)\Theta({}^*j)|\, d_{G/\mathrm{J}_0}(\dot{x}) \\
&\leq \int_{\mathrm{J}} |\Delta_{R,C}(j)\Theta(j)| \cdot |\Delta_I(j)| \cdot |\Delta(j)| \left\{ \int_{G/\mathrm{J}_0} |f({}^*j)|\, d_{G/\mathrm{J}_0}(\dot{x}) \right\} d_{\mathrm{J}}(j) \\
&\leq \int_G |f(x)| \mapsto (x) d_G(x) \qquad (\text{all } f \in I_c^\infty(G))
\end{aligned}$$

for any J ($= \mathsf{J}^i$, $i = 2, \ldots, r$). On the other hand

$$\Theta(f) = \sum_{i=1}^{r} \int_{G(\mathsf{J}^i)} f(x)\Theta(x)d_G(x) = 0 \qquad (\text{all } f \in I_c^\infty(G)),$$

whence

$$\left| \int_{G(T)} f(x)d_G(x) \right| \leq \int_G |f(x)| \,\Xi(x)d_G(x) \qquad (\text{all } f \in I_c^\infty(G)),$$

the restriction of Θ to $G(T)$ being constant. Of course this serves to establish the existence of a positive constant M such that

$$\int_K \psi_{\mathscr{E}}(xk)dk \leqslant M\,\Xi(x)$$

for almost all x in G, as desired.

Modulo the Remark infra, the proof of Theorem 8.5.6.1 is now complete.

Remark The existence of the distribution Θ with the aforestated properties lies rather deep. Thus let 2ρ denote the sum of the positive roots of the pair $(\mathfrak{g}_c, \mathfrak{t}_c)$, W the Weyl group of the pair $(\mathfrak{g}_c, \mathfrak{t}_c)$ – then, in the notations of Chapter 10, the distribution

$$\Theta = \sum_{w \in W} \det(w)\Theta_{w(\rho)}$$

turns out to have the required properties. We shall go into this in detail in Volume III.

Proposition 8.5.6.2 Retain the above notations and assumptions – then the map

$$f \mapsto \int_G |f(x)| \cdot |D_{\mathsf{J}}(x)|^{-1/2} d_G(x) \qquad (f \in \mathscr{C}(G))$$

is a continuous semi-norm on $\mathscr{C}(G)$.

Proof Choose $r \geqslant 0$ large enough to ensure that $\Xi^2(1 + \sigma)^{-r}$ lies in $L^1(G)$ – then

$$\int_G |f(x)| \cdot |D_{\mathsf{J}}(x)|^{-1/2} d_G(x) \leqslant |f|_r \cdot \int_G \Xi(x)(1 + \sigma(x))^{-r} |D_{\mathsf{J}}(x)|^{-1/2} d_G(x)$$

and so our assertion follows from the fact that

$$\int_K |D_{\mathsf{J}}(xk)|^{-1/2} dk \leqslant M\,\Xi(x) \qquad (M > 0)$$

for almost all x in G. □

8.5.7 Convergence of Certain Integrals

Let (G, K) be a reductive pair fulfilling the usual hypotheses. Let $I(G)$ be the space of all continuous functions f on G which are biinvariant under K, $I^+(G)$ the subset consisting of the real $f \geqslant 0$. Let $\mathfrak{X}$ be the centralizer of $\mathfrak{k}_c$ in $\mathfrak{G}$ and μ a semi-norm on $I_c^\infty(G)$ satisfying the following conditions:

(a) There exist elements $D_1, \ldots, D_m$ in $\mathfrak{X}$ such that

$$\mu(f) \leqslant \sum_{i=1}^{m} \int_G |D_i f(x)| \Xi(x) d_G(x) \qquad (\text{all } f \in I_c^\infty(G));$$

(b) If f_1, f_2 are two elements in $I_c^\infty(G)$ such that $f_1 \geqslant f_2 \geqslant 0$, then $\mu(f_1) \geqslant \mu(f_2)$.

Note The topology on $I_c^\infty(G)$ is being ignored here, that is we are not requesting that μ be continuous.

Given any element f in $I^+(G)$, let us agree to write $\mu(f) = \sup \mu(f^+)$ where f^+ runs over all functions in $I^+(G) \cap I_c^\infty(G)$ such that $f \geqslant f^+$.

Lemma 8.5.7.1 Retain the above notations and assumptions; fix an $r \geqslant 0$ such that $\Xi^2 (1 + \sigma)^{-r} \in L^1(G)$ – then

$$\mu(\Xi (1 + \sigma)^{-r}) < \infty.$$

Proof (1) Fix an $f^+ \in I^+(G) \cap I_c^\infty(G)$ – then we claim that there exists a positive constant M such that

$$\mu(f^+ * f) \leqslant M \int_G f(x) \Xi(x) d_G(x)$$

for all $f \in I_c^+(G)$. Thus let $f \in I_c^+(G)$ – then $f^+ * f \in I_c^\infty(G)$ and, moreover,

$$D_i(f^+ * f) = D_i(f * f^+) = f * D_i f^+ = D_i f^+ * f \qquad (1 \leqslant i \leqslant m),$$

the convolution being abelian in the case at hand. Therefore

$$\begin{aligned}
\mu(f^+ * f) &\leqslant \sum_{i=1}^{m} \int_G |f^+ * f(x; D_i)| \Xi(x) d_G(x) \\
&\leqslant \sum_{i=1}^{m} \int_G \int_G |f^+(y; D_i)| f(y^{-1}x) \Xi(x) d_G(y) d_G(x) \\
&\leqslant \tilde{M} \int_G f(x) \left\{ \int_{\omega^+} \Xi(yx) d_G(y) \right\} d_G(x)
\end{aligned}$$

$$(\tilde{M} = \sum_i \sup |D_i f^+|),$$

where $\omega^+ = \operatorname{spt}(f^+)$. Since ω^+ is compact, the claim follows upon citing Proposition 8.3.7.2. It is to be noted that, in view of condition (b) above, our inequality is actually valid for all f in $I^+(G)$.

(2) Let us apply the preceding considerations to $f = \Xi (1 + \sigma)^{-r}$ (of course $f \in I^+(G)$). Since Ξ verifies the integral equation

$$\int_K \Xi(xky) dk = \Xi(x) \Xi(y) \qquad (x, y \in G),$$

it is clear that if $f^+ \in I^+(G) \cap I_c^\infty(G)$ is chosen in such a way that

$\int_G f^+ \vdash\!\circ\!\dashv\, d_G(x) = 1$, then $f^+ * \vdash\!\circ\!\dashv\, = \vdash\!\circ\!\dashv$; consequently, for all $x \in G$, we have

$$(f^+ * f)(x) = \int_G f^+(y) \vdash\!\circ\!\dashv (y^{-1}x)(1 + \sigma(y^{-1}x))^{-r} d_G(y)$$
$$\geqslant N^{-1} f(x) \qquad (N = \sup_{x \in \omega^+} (1 + \sigma(x))^r,\ \omega^+ = \operatorname{spt}(f^+)).$$

It therefore follows that

$$\mu(f) \leqslant N\mu(f^+ * f) \leqslant MN \int_G \vdash\!\circ\!\dashv^2 (x)(1 + \sigma(x))^{-r} d_G(x) < \infty.$$

Hence the lemma. □

Let $\mathfrak{j}$ be a θ-stable Cartan subalgebra of $\mathfrak{g}$, J the Cartan subgroup of G associated with $\mathfrak{j}$.

Theorem 8.5.7.2 (Harish-Chandra) Fix $r \geqslant 0$ such that

$$\vdash\!\circ\!\dashv^2 (1 + \sigma)^{-r}$$

belongs to $L^1(G)$ – then

$$\sup_{j \in \mathsf{J}'} |\Delta(j)| \int_{G/\mathsf{J}_0} \vdash\!\circ\!\dashv ({}^*j)(1 + \sigma({}^*j))^{-r} d_{G/\mathsf{J}_0}(\dot{x}) < \infty.$$

Proof Put

$$\mu(f) = \sup_{j \in \mathsf{J}'} |\Phi_f(j)| \qquad (f \in I_c^\infty(G))$$

– then we claim that the semi-norm μ satisfies the conditions (a) and (b) supra. It is obvious that (b) is satisfied; as for (a), put

$$\nu(f) = \int_{\mathsf{J}} |\Delta_I(j)\Phi_f(j)|\, d_{\mathsf{J}}(j) \qquad (f \in I_c^\infty(G)).$$

Then it follows from Theorem 8.5.5.2 that there exist elements $Z_1, \ldots, Z_m$ in $\mathfrak{Z}$ such that $\mu(f) \leqslant \sum_i \nu(Z_i f)$ (all $f \in I_c^\infty(G)$); therefore, on the basis of Theorem 8.5.6.1, it is clear that μ does indeed verify condition (a). On the other hand, in view of the elementary properties of an integral, for any $f \in I^+(G)$, we have

$$\int_{G/\mathsf{J}_0} f({}^*j) d_{G/\mathsf{J}_0}(\dot{x}) = \sup_{f^+} \int_{G/\mathsf{J}_0} f^+({}^*j) d_{G/\mathsf{J}_0}(\dot{x}) \qquad (j \in \mathsf{J}'),$$

f^+ running over all elements in $I^+(G) \cap I_c^\infty(G)$ such that $f^+ \leqslant f$. The present theorem is therefore seen to be an immediate consequence of Lemma 8.5.7.1. □

Proposition 8.5.7.3 There exists a number $r \geqslant 0$ such that

$$\int_G |D_l(x)|^{-1/2} \vdash\!\circ\!\dashv (x)(1 + \sigma(x))^{-r} d_G(x) < \infty.$$

Proof The possibility that the vector part of the center of G is non-zero can be eliminated by the usual argument. This being so, fix a maximal collection $\{\mathsf{J}^i\}$ of mutually non-conjugate θ-stable Cartan subgroups of G – then it will clearly be enough to produce, for each i, a number $r_i \geqslant 0$ such that

$$\int_{\mathsf{J}^i} |\Delta^i(j)| \left\{ \int_{G/\mathsf{J}_0{}^i} \Xi({}^*j)(1 + \sigma({}^*j))^{-r_i} d_{G/\mathsf{J}_0{}^i}(\dot{x}) \right\} d_{\mathsf{J}^i}(j) < \infty.$$

Fix an i and drop it from the notation (thus we shall be writing J for J^i). If J is compact, then we need only quote the preceding theorem; if J is non-compact, then, in view of what has just been said, the result holds for the pair (L, J), i.e. there exists $r_L \geqslant 0$ such that

$$\int_J |\Delta_I(j)| \left\{ \int_{L/J} \Xi_L({}^{l}j)(1 + \sigma({}^{l}j))^{-r_L} d_{L/J}(\dot{l}) \right\} d_J(j) < \infty.$$

Choosing $r \geqslant r_L$ sufficiently large, Theorem 8.5.2.1 then tells us that

$$\begin{aligned}
&\int_{\mathsf{J}} |\Delta(j)| \left\{ \int_{G/\mathsf{J}_0} \Xi({}^*j)(1 + \sigma({}^*j))^{-r} d_{G/\mathsf{J}}(\dot{x}) \right\} d_{\mathsf{J}}(j) \\
&\leq \sum_{a \in \mathsf{Z}(J_{\mathfrak{p}})} \int_J |\Delta(aj)| \left\{ \int_{G/\mathsf{J}_0} \Xi({}^*(aj))(1 + \sigma({}^*(aj)))^{-r} d_{G/\mathsf{J}_0}(\dot{x}) \right\} d_J(j) \\
&\leq \int_J |\Delta_I(j)| \, \mathrm{d}_P(j) \left\{ \int_{L/J} \int_{N^+} \Xi({}^{l}j \cdot n)(1 + \sigma({}^{l}j \cdot n))^{-r} d_{N^+}(n) d_{L/J}(\dot{l}) \right\} d_J(j) \\
&\leq \int_J |\Delta_I(j)| \left\{ \int_{L/J} \Xi_L({}^{l}j)(1 + \sigma({}^{l}j))^{-r_L} d_{L/J}(\dot{l}) \right\} d_J(j) < \infty,
\end{aligned}$$

as desired. □

8.5.8 Continuity of the Map $f \mapsto \Phi_f$

Let (G, K) be a reductive pair fulfilling the usual hypotheses. Fix a θ-stable Cartan subalgebra $\mathfrak{j}$ of $\mathfrak{g}$ and let J be the Cartan subgroup of G associated with $\mathfrak{j}$; write $P = \mathsf{M}J_{\mathfrak{p}}N^+$ and let

$$\mathsf{J}'(SI) = \{j \in \mathsf{J} : \prod_{\alpha \in \Phi_{SI}{}^+} (1 - \xi_\alpha(j^{-1})) \neq 0\}.$$

Theorem 8.5.8.1 (Harish-Chandra) Fix $f \in \mathscr{C}(G)$ – then the integral

$$\Phi_f(j) = \epsilon_R(j)\Delta(j) \int_{G/\mathsf{J}_0} f({}^*j) d_{G/\mathsf{J}_0}(\dot{x})$$

is absolutely convergent for all $j \in \mathsf{J}'$ and the function Φ_f so defined lies in $\mathscr{C}(\mathsf{J}')$. Furthermore the mapping $f \mapsto \Phi_f$ of $\mathscr{C}(G)$ into $\mathscr{C}(\mathsf{J}')$ is continuous.

The following result will emerge during the course of our discussion.

Corollary 8.5.8.2 For every f in $\mathscr{C}(G)$ the function Φ_f admits a unique extension to $\mathsf{J}'(SI)$ and the mapping $f \mapsto \Phi_f$ of $\mathscr{C}(G)$ into $\mathscr{C}(\mathsf{J}'(SI))$ is continuous.

Since the index of J in J is finite, the space $\mathscr{C}(\mathsf{J})$ is well-defined (with $\vdash\!\circ\!\dashv_J = 1 \ldots$); in fact the space $\mathscr{C}(\mathscr{O})$ is defined for any open subset $\mathscr{O}$ of J, in particular for $\mathscr{O} = \mathsf{J}'$ or $\mathsf{J}'(SI)$.

The proof of Theorem 8.5.8.1 (and its corollary) rests on the following lemma.

Lemma 8.5.8.3 The function Φ_f lies in $\mathscr{C}(\mathsf{J}'(SI))$ for every $f \in C_c^\infty(G)$; furthermore, if $|\cdot|_0$ is a given continuous semi-norm on $\mathscr{C}(\mathsf{J}'(SI))$, then there exists a continuous semi-norm $|\cdot|_\alpha$ on $\mathscr{C}(G)$ such that

$$|\Phi_f|_0 \leqslant |f|_\alpha \qquad (\text{all } f \in C_c^\infty(G)).$$

[So as to avoid any confusion, let us recall that Φ_f $(f \in C_c^\infty(G))$ can be regarded as a C^∞ function on $\mathsf{J}'(SI)$ (cf. Theorem 8.5.1.4).]

Proof It is easy to dispense with the case when $\mathfrak{c}_\mathfrak{p} \neq \{0\}$. Without any real loss of generality, therefore, in what follows it may be supposed that $\mathfrak{c}_\mathfrak{p} = \{0\}$. If $\mathfrak{j} \subset \mathfrak{k}$, then

$$\int_{\mathsf{J}} |\Delta_I(j)\Phi_f(j)|\, d_{\mathsf{J}}(j) \leq \int_G |f(x)|\cdot|D_{\mathsf{J}}(x)|^{-1/2} d_G(x) \qquad (f \in C_c^\infty(G))$$

and so the lemma in this case is an immediate consequence of Proposition 8.5.6.2 (in conjunction with Theorem 8.5.5.2). If $\mathfrak{j} \not\subset \mathfrak{k}$, then, in view of what has just been said, the lemma holds for the pair (L, J); upon implementing the transformations of descent, the desired assertion follows readily. The proof of the lemma is therefore complete. □

Because the space $C_c^\infty(G)$ is dense in $\mathscr{C}(G)$, and since $\mathscr{C}(\mathsf{J}'(SI))$ is complete, the mapping $f \mapsto \Phi_f$ can be extended uniquely to a continuous mapping of $\mathscr{C}(G)$ into $\mathscr{C}(\mathsf{J}'(SI))$. Consequently, for every $f \in \mathscr{C}(G)$ we get a function $\Phi_f \in \mathscr{C}(\mathsf{J}'(SI))$; on J', we claim that Φ_f $(f \in \mathscr{C}(G))$ is given by the integral

$$\Phi_f(j) = \epsilon_R(j)\Delta(j) \int_{G/\mathsf{J}_0} f({}^*j)\, d_{G/\mathsf{J}_0}(\dot{x}) \qquad (j \in \mathsf{J}').$$

In fact, thanks to Theorem 8.5.7.2, it is at least clear that the integral on the right exists; on the other hand, if $\{f_n\}$ is a sequence of elements in $C_c^\infty(G)$ such that $f_n \to f$ in $\mathscr{C}(G)$ $(f \in \mathscr{C}(G))$, then Theorem 8.5.7.2 ensures us that

$$\sup_{j \in \mathsf{J}'} |\Delta(j)| \int_{G/\mathsf{J}_0} |(f - f_n)({}^*j)|\, d_{G/\mathsf{J}_0}(\dot{x}) \to 0$$

which suffices, of course, to establish the claim.

The assertions contained in Theorem 8.5.8.1 and its corollary are now obvious.

Note Suppose that $\tilde{\mathrm{J}}$ is another Cartan subgroup of G which is conjugate to J; fix an $\tilde{x}$ in G such that $\tilde{\mathrm{J}} = {}^{\tilde{x}}\mathrm{J}$ – then the isomorphism $j \mapsto \tilde{x}j\tilde{x}^{-1}$ $(j \in \mathrm{J})$ carries $\mathscr{C}(\mathrm{J}'(SI))$ bijectively onto a subspace $\mathscr{C}(\tilde{\mathrm{J}}'(SI))$ of $C^\infty(\tilde{\mathrm{J}}'(SI))$. We shall agree to topologize $\mathscr{C}(\tilde{\mathrm{J}}'(SI))$ by requiring that this bijection be a homeomorphism (the topology thus defined being independent of $\tilde{x}$ . . .). If we then drop the θ-stability assumption on $\mathfrak{j}$, then the invariant integral Φ_f of f relative to J $(f \in \mathscr{C}(G))$ can be defined by the integral formula supra; since $\mathfrak{j}$ is conjugate to some θ-stable Cartan subalgebra of $\mathfrak{g}$, it is obvious that the assignment $f \mapsto \Phi_f$ defines a continuous mapping of $\mathscr{C}(G)$ into $\mathscr{C}(\mathrm{J}'(SI))$.

In conclusion let us mention that the preceding discussion serves to ensure the validity of Theorem 8.5.4.1 for elements f lying in $\mathscr{C}(G)$.

Chapter 9

Spherical Functions on a Semi-Simple Lie Group

Introduction

Let G be a connected semi-simple Lie group with finite center, $G = KA_{\mathfrak{p}}N^+$ an Iwasawa decomposition for G; let $I^2(G)$ be the space of K-biinvariant square integrable functions on G – then, according to well-known generalities, a given element f in $I^2(G)$ can be 'expanded' in terms of zonal spherical functions of positive type. More precisely, let $I_c^\infty(G)$ denote the space of K-biinvariant compactly supported C^∞ functions on G; let $\mathbf{P}$ be the set of positive definite zonal spherical functions on G – then, by the spherical Fourier transform $\hat{f}$ of f ($f \in I_c^\infty(G)$), we shall understand the function on $\mathbf{P}$ defined by the rule

$$\hat{f}(\phi) = \int_G f(x)\phi(x)d_G(x) \qquad (\phi \in \mathbf{P}).$$

This being so, the abstract Plancherel Theorem for $I^2(G)$ asserts that there exists on $\mathbf{P}$ a unique positive measure μ (the Plancherel measure for $I^2(G)$) such that

$$\int_G |f(x)|^2 \, d_G(x) = \int_{\mathbf{P}} |\hat{f}(\phi)|^2 \, d\mu(\phi) \qquad (\text{all } f \in I_c^\infty(G)).$$

The problem is to compute μ explicitly, i.e. to relate μ in a satisfactory manner to the structure of G.

The motivation which lies behind Harish-Chandra's approach and solution to this problem comes from the spectral theory of singular ordinary differential equations of the second order, i.e. from the theory of eigenfunction expansions. Roughly speaking, a certain class of zonal spherical functions of positive type can be parameterized by a space $W\backslash\mathfrak{F}$, $\mathfrak{F}$ a real finite dimensional Euclidean space (W the Weyl group of the pair $(\mathfrak{g}, \mathfrak{a}_{\mathfrak{p}})$). [We have in mind here the zonal spherical functions which arise from the class one principal P-series for G, $P = MA_{\mathfrak{p}}N^+ \ldots$.] Let ν belong to $\mathfrak{F}$ – then the corresponding zonal spherical function ϕ_ν can be expanded in an infinite series, the dominating term of which involves a certain (explicitly computable) coefficient $\mathbf{c}(\nu)$ which, considered as a function of ν, is analytic on $\mathfrak{F}$ modulo certain hyperplanes. On the other hand, the reciprocal $\mathbf{c}^{-1}$ is analytic on $\mathfrak{F}$; moreover, it turns out that the Plancherel measure for $I^2(G)$ is supported by $W\backslash\mathfrak{F}$ (considered as a subset of $\mathbf{P}$) – in fact $d\mu = |\mathbf{c}(\nu)|^{-2} \, d_{\mathfrak{F}}(\nu)$ (provided the Haar measures on G and $\mathfrak{F}$ are suitably normalized). The basis for these results is most easily seen when the split-rank of G is equal to 1; in this case, the ϕ_ν can be identified in a simple way with hypergeometric functions; moreover, the problem of finding the explicit Plancherel measure in this situation is essentially equivalent to a classical eigenvalue problem (limit-point, limit-circle . . .). [In this connection, the reader will find it profitable to consult Titchmarsh [1, pp. 91–96]; in particular one should compare the explicit formula for the $\mathbf{c}$-function in the split-rank 1 case with the measure computed by Titchmarsh on p. 93. The paper of Flensted-Jensen [1] is also pertinent here.]

The first two sections of the present chapter are primarily devoted to the theory of zonal spherical functions on G, in particular to the proof of the Plancherel Theorem for $I^2(G)$. The asymptotic expansion for ϕ_ν arises from a careful study of the equation $\omega\phi_\nu = \gamma_{A_\mathfrak{p}}(\omega : \sqrt{-1}\nu)\phi_\nu$, ω the Casimir operator (cf. 9.1.4 and 9.1.5); the coefficients in the expansion are determined recursively in a manner not unlike that customarily encountered in the theory of ordinary differential equations (method of the 'indicial equation' . . .). [Actually it will be just as easy to carry out this particular aspect of the investigation in a somewhat more general setting; cf. 9.1.] In 9.2 we use the asymptotic formula for ϕ_ν to prove the relation $d\mu = |\mathbf{c}(\nu)|^{-2}d\mathfrak{F}(\nu)$. [To this end, it is necessary to also have at hand information regarding the asymptotic behavior of ϕ_ν qua a function of ν. . . .]

The main problem in 9.3 is to determine the behavior at infinity of a K-finite eigenfunction f of $\mathfrak{Z}$ (the center of $\mathfrak{G}$) which satisfies 'the weak inequality'. The principal result asserts that f lies in $L^2(G)$ iff it lies in $\mathscr{C}(G)$. The proof goes by induction on the dimension of G; a detailed consideration of certain differential equations allows one to reduce the problem from G to a proper subgroup L – this is a good illustration of Harish-Chandra's 'technique of descent via differential equations.'

9.1 Asymptotic Behavior of μ-Spherical Functions on a Semi-Simple Lie Group

9.1.1 The Main Results

Let $\mathfrak{g}$ be a semi-simple Lie algebra over $\mathbf{R}$, $\mathfrak{g} = \mathfrak{k} + \mathfrak{a}_\mathfrak{p} + \mathfrak{n}^+$ an Iwasawa decomposition for $\mathfrak{g}$; let G be a connected Lie group with Lie algebra $\mathfrak{g}$, $G = KA_\mathfrak{p}N^+$ the corresponding Iwasawa decomposition for G – as always we shall assume that G has finite center (so that K is compact). Let Σ denote the set of roots of the pair $(\mathfrak{g}, \mathfrak{a}_\mathfrak{p})$, W_Σ the corresponding Weyl group, $\mathscr{C}$ the positive Weyl chamber in $\mathfrak{a}_\mathfrak{p}$ determined by the given Iwasawa decomposition for $\mathfrak{g}$, $\{\lambda_1, \ldots, \lambda_l\}$ the associated fundamental system of positive roots; let $H_1, \ldots, H_l$ be a basis for $\mathfrak{a}_\mathfrak{p}$ over $\mathbf{R}$ such that $\lambda_i(H_j) = \delta_{ij}$ $(1 \leqslant i, j \leqslant l)$ – then it is clear that an element $H = \sum_i t_i H_i$ $(t_i \in \mathbf{R})$ lies in $\mathscr{C}$ iff $t_i > 0$ $(1 \leqslant i \leqslant l)$. It will be convenient to denote by $\mathscr{C}_c$ the subset in $\mathfrak{a}_{\mathfrak{p}_c}$ comprised of those H with the property that $\mathscr{R}(\lambda_i(H)) > 0$ $(1 \leqslant i \leqslant l)$.

As usual, a linear function λ on $\mathfrak{a}_{\mathfrak{p}_c}$ is said to be *real* provided it takes only real values on $\mathfrak{a}_\mathfrak{p}$; any linear function λ can be written uniquely in the form $\lambda = \mathscr{R}(\lambda) + \sqrt{-1}\mathscr{I}(\lambda)$ where $\mathscr{R}(\lambda)$ and $\mathscr{I}(\lambda)$ are real linear functions on $\mathfrak{a}_{\mathfrak{p}_c}$ which are called the real and imaginary parts of λ, respectively. Given linear functions μ and ν on $\mathfrak{a}_{\mathfrak{p}_c}$, we shall write $\mu > \nu$ (or $\nu < \mu$) if $\mathscr{R}(\mu) > \mathscr{R}(\nu)$ or $\mathscr{R}(\mu) = \mathscr{R}(\nu)$ and $\mathscr{I}(\mu) > \mathscr{I}(\nu)$. Let L be the set of all linear functions λ on $\mathfrak{a}_{\mathfrak{p}_c}$ of the form $\lambda = m_1\lambda_1 + \ldots + m_l\lambda_l$, the $m_i (1 \leqslant i \leqslant l)$ being non-negative integers; put

$$m_\lambda = m_1 + \cdots + m_l \qquad (\lambda \in \mathsf{L}).$$

Given linear functions μ and ν on $\mathfrak{a}_{\mathfrak{p}_c}$, we shall write $\mu \gg \nu$ (or $\nu \ll \mu$) if $\mu \neq \nu$ and $\mu - \nu \in \mathsf{L}$; it is clear that $\mu \gg \nu \Rightarrow \mu > \nu$.

Let $\mathfrak{G}$ denote the universal enveloping algebra of $\mathfrak{g}_c$, $\mathfrak{Z}$ the center of $\mathfrak{G}$. Let $\mu = (\mu_1, \mu_2)$ be a double Banach representation of K on a finite dimensional Banach space E; let κ be a character of $\mathfrak{Z}$ – then by $\mathfrak{S}(\mu, \kappa)$ we shall understand the space of all C^∞ μ-spherical functions Ψ on G which verify the relation $Z\Psi = \kappa(Z)\Psi (Z \in \mathfrak{Z})$. It will be seen in 9.1.3 that the space $\mathfrak{S}(\mu, \kappa)$ is finite dimensional.

A μ-spherical function Ψ on G is completely determined by its restriction to $\exp(\mathscr{C})$; on the other hand, if E_{M} denotes the subspace of E consisting of those elements a with the property that $\mu_1(m)a = a\mu_2(m)$ (all $m \in \mathsf{M}$ (= centralizer of $A_\mathfrak{p}$ in K)), then certainly $\Psi(A_\mathfrak{p}) \subset E_{\mathsf{M}}$. This being so, the following result provides us with a series expansion for each Ψ in $\mathfrak{S}(\mu, \kappa)$, an expansion which is actually valid on *all* of $\exp(\mathscr{C})$.

Theorem 9.1.1.1 (Harish-Chandra) Fix an element Ψ in $\mathfrak{S}(\mu, \kappa)$ – then there exists an at most countable set $\mathscr{E}(\Psi)$ of linear functions on $\mathfrak{a}_{\mathfrak{p}_c}$ and, for each $\lambda \in \mathscr{E}(\Psi)$, a polynomial function $p_\lambda: \mathfrak{a}_{\mathfrak{p}_c} \to E_{\mathsf{M}}$ such that the following conditions are fulfilled:

(i) For every $\lambda \in \mathscr{E}(\Psi)$, $p_\lambda \neq 0$;
(ii) The series

$$\sum_{\lambda \in \mathscr{E}(\Psi)} p_\lambda e^{\langle \cdot, \lambda \rangle}$$

converges decently throughout the region

$$\mathfrak{a}_{\mathfrak{p}_c}(\epsilon, \eta) = \{H \in \mathfrak{a}_{\mathfrak{p}_c} : \mathscr{R}(\lambda_i(H)) > \max(\eta, \epsilon\,|\lambda_1(H)|, \ldots, \epsilon\,|\lambda_l(H)|)\}$$
$$(0 < \epsilon < 1, \eta > 0);$$

(iii) The expansion

$$\Psi(\exp H) = \sum_{\lambda \in \mathscr{E}(\Psi)} p_\lambda(H) e^{\lambda(H)}$$

holds for all $H \in \mathscr{C}$.

Moreover the set $\mathscr{E}(\Psi)$ and the polynomials p_λ $(\lambda \in \mathscr{E}(\Psi))$ are uniquely determined by these three conditions.

[For the notion of 'decent convergence', see Appendix 3.]

Fix a μ-spherical function Ψ in $\mathfrak{S}(\mu, \kappa)$ per Theorem 9.1.1.1 – then the elements of the set $\mathscr{E}(\Psi)$ will be referred to as the *exponents* of Ψ. Let $\mathscr{E}_0(\Psi)$ be the subset of $\mathscr{E}(\Psi)$ comprised of those λ which are maximal with respect to the relation $\gg$; according to Theorem A3.1.3 in Appendix 3, $\mathscr{E}_0(\Psi)$ is a finite set – in addition if λ is an element of $\mathscr{E}(\Psi)$ which is not in $\mathscr{E}_0(\Psi)$, then $\lambda \ll \lambda_0$ for some λ_0 in $\mathscr{E}_0(\Psi)$ (hence $\mathscr{E}_0(\Psi)$ is not empty unless $\Psi = 0$). Elements of $\mathscr{E}_0(\Psi)$ will be called the *principal exponents* of Ψ and the sum $\sum_\lambda p_\lambda e^\lambda$, $\lambda \in \mathscr{E}_0(\Psi)$, the *principal part* of Ψ. Let Λ be the highest principal exponent of Ψ under the relation $>$; Λ will be called the *highest* or *leading exponent* of Ψ while the corresponding term $p_\Lambda e^\Lambda$ will be referred to as the *highest term* of Ψ; if d is the degree of p_Λ and if q_Λ is the homogeneous component of degree d of p_Λ, then $q_\Lambda e^\Lambda$ is said to be the *leading term* of Ψ and will be denoted by Ψ^L. By the degree of Ψ^L we shall understand the degree of q_Λ.

Heuristics The proof of Theorem 9.1.1.1 depends upon the results in Appendix 3; it is based essentially on the following considerations. Fix a double Banach representation $\mu = (\mu_1, \mu_2)$ of K on a finite dimensional Banach space E – then, as will be shown in 9.1.2, for each Z in $\mathfrak{Z}$ there exists a differential operator $\natural_{A_\mathfrak{p}}(Z)$ on $\exp(\mathscr{C})$, whose coefficients are analytic functions on $\exp(\mathscr{C})$ with values in $\mathrm{Hom}_{\mathbf{C}}(E, E)$, such that $(Z\Psi)^\sim = \natural_{A_\mathfrak{p}}(Z)\tilde{\Psi}$ $(\tilde{\Psi} = \Psi|\exp(\mathscr{C})$, Ψ any C^∞ μ-spherical function on G). [We view the correspondence $Z \mapsto \natural_{A_\mathfrak{p}}(Z)$ $(Z \in \mathfrak{Z})$ as assigning to each Z in $\mathfrak{Z}$ its 'μ-radial component'.] Let $l = \mathrm{rank}(G)$ – then $\mathfrak{Z}$ is a polynomial ring in l generators, say

$$\mathfrak{Z} = \mathbf{C}[Z_1, \ldots, Z_l].$$

This being so, fix an element Ψ in $\mathfrak{S}(\mu, \kappa)$ – then the system of differential equations $Z\Psi = \kappa(Z)\Psi$ $(Z \in \mathfrak{Z})$ actually reduces to the finite set

$$Z_i\Psi = \kappa(Z_i)\Psi \qquad (i = 1, \ldots, l),$$

whence

$$(Z_i\Psi)^\sim = \kappa(Z_i)\tilde{\Psi} = \natural_{A_\mathfrak{p}}(Z_i)\tilde{\Psi} \qquad (i = 1, \ldots, l).$$

To illustrate, suppose that $G = \mathbf{SL}(2, \mathbf{R})$, $K = \mathbf{SO}(2)$,

$$A_\mathfrak{p} = \left\{\begin{pmatrix} h & 0 \\ 0 & h^{-1}\end{pmatrix} : h \in \mathbf{R}^+\right\};$$

the map

$$t \mapsto \begin{pmatrix} e^{t/2} & 0 \\ 0 & e^{-t/2}\end{pmatrix}$$

is an isomorphism of the additive group of the line onto $A_\mathfrak{p}$, whence $\exp(\mathscr{C}) \sim \mathbf{R}^+$. Put $H = \left(\begin{smallmatrix} 1 & 0 \\ 0 & -1\end{smallmatrix}\right)$, $X = \left(\begin{smallmatrix} 0 & 1 \\ 0 & 0\end{smallmatrix}\right)$, $Y = \left(\begin{smallmatrix} 0 & 0 \\ 1 & 0\end{smallmatrix}\right)$ – then $\{H, X, Y\}$ is a basis for $\mathfrak{sl}(2, \mathbf{R})$ while $\mathbf{R}\Omega$ $(\Omega = 2^{-1}(X - Y))$ is the Lie algebra of K. In the present case $l = 1$ so that $\mathfrak{Z} = \mathbf{C}[\omega]$ where $\omega = 8^{-1}H^2 + 4^{-1}(XY + YX)$ is the Casimir operator in $\mathfrak{Z}$. Because μ_1 and μ_2 are representations of K, they define canonically representations of $\mathfrak{K}$; this being the case, it then turns out that $\natural_{A_\mathfrak{p}}(\omega)$ is given by

$$\begin{aligned} 2\natural_{A_\mathfrak{p}}(\omega)\tilde{\Psi} &= d^2\tilde{\Psi}/dt^2 + \coth(t)\, d\tilde{\Psi}/dt \\ &\quad - (\sinh(t))^{-2}(\mu_1(\Omega)^2\tilde{\Psi} + \tilde{\Psi}\mu_2(\Omega)^2) \\ &\quad + 2\cosh(t)(\sinh(t))^{-2}\mu_1(\Omega)\tilde{\Psi}\mu_2(\Omega) \qquad (t \in \mathbf{R}^+), \end{aligned}$$

Ψ a C^∞ μ-spherical function on $\mathbf{SL}(2, \mathbf{R})$. The main thing to be noticed here is that this operator has the property that its coefficients can be expanded in powers of $e^{-t}(t > 0)$; in 9.1.2 it will be shown that the same situation obtains in the general case, a fact which will essentially set the stage for an application of the results in Appendix 3 thereby leading to the proof of Theorem 9.1.1.1.

Suppose now that $\mu = (\mu_1, \mu_2)$ is a double *unitary* representation of K on a finite dimensional Hilbert space E; as above, let E_M denote the subspace of E comprised of those elements a which have the property that $\mu_1(m)a = a\mu_2(m)$ (all $m \in M$). Fix $a \in E_M$, $\nu \in \mathfrak{a}_{\mathfrak{p}_c}^\vee$ and consider the function

$$\Psi(a : \nu : x) = \int_K \mu_1(\kappa(xk))a\mu_2(k^{-1}) \exp\{(\sqrt{-1}\nu - \rho)(H(xk))\}\, dk$$
$$(\rho = 2^{-1} \sum_{\lambda > 0} m(\lambda)\lambda)$$

for $x \in G$ (we follow here the usual mode of writing; cf. Vol. I, number 5.5.1) – then, for fixed a and ν, $\Psi(a : \nu : .)$ is a C^∞ μ-spherical function on G which, for suitable $a \in E_{\mathsf{M}}$, is an eigenfunction of $\mathfrak{Z}$ (cf. 9.1.5).

Example To orient ourselves, let us suppose that U is an irreducible unitary representation of G on a Hilbert space E, say; fix a class $\delta \in \hat{K}$ which occurs in $U|K$ and let $\mu_U = (\mu_1, \mu_2)$ be the canonical double representation of K on $\mathrm{Hom}_{\mathbf{C}}(E(\delta), E(\delta))$. [Thus, by definition, we have

$$\mu_1(k_1)T\mu_2(k_2) = U_\delta(k_1)TU_\delta(k_2) \qquad (T \in \mathrm{Hom}_{\mathbf{C}}(E(\delta), E(\delta)); k_1, k_2 \in K));$$

of course we agree to regard $\mathrm{Hom}_{\mathbf{C}}(E(\delta), E(\delta))$ as a (finite dimensional) Hilbert space in the usual way. . . .] Put

$$\Psi_\delta^U(x) = \mathsf{P}(\delta)U(x)\mathsf{P}(\delta) \qquad (x \in G).$$

Then Ψ_δ^U is a μ_U-spherical function on G. Moreover, as has been seen in Theorem 6.2.2.4, there exists a linear function ν on $\mathfrak{a}_{\mathfrak{p}}$ and an endomorphism T of $E(\delta)$, verifying the relation $U_\delta(m)T = TU_\delta(m)$ (all $m \in \mathsf{M}$), such that

$$\Psi_\delta^U(x) = \int_K U_\delta(\kappa(xk))TU_\delta(k^{-1}) \exp\{(\sqrt{-1}\nu - \rho)(H(xk))\}\, dk$$

for all x in G.

The following result will enable us to determine the asymptotic behavior of Ψ on the Weyl chamber $\mathscr{C}$ (for a more precise statement, see 9.1.5).

Theorem 9.1.1.2 (Harish-Chandra) Retain the above notations and assumptions – then there exists an open, connected, dense, W_Σ-stable subset $\mathcal{O}$ of $\mathfrak{a}_{\mathfrak{p}_c}$ and holomorphic functions $\mathbf{c}_w : \mathcal{O} \to \mathrm{Hom}_{\mathbf{C}}(E_{\mathsf{M}}, E_{\mathsf{M}})$ $(w \in W_\Sigma)$ such that for each $\nu \in \mathcal{O}$

$$h^\rho \int_K \mu_1(\kappa(hk))a\mu_2(k^{-1}) \exp\{(\sqrt{-1}\nu - \rho)(H(hk))\}\, dk$$
$$= \sum_{w \in W_\Sigma} \mathbf{\Phi}(w\nu : h)\mathbf{c}_w(\nu)a \qquad (h \in \exp(\mathscr{C}))$$

for all $a \in E_{\mathsf{M}}$.

Here $\mathbf{\Phi}$ is a function on $\mathcal{O} \times \mathscr{C}$ with values in $\mathrm{Hom}_{\mathbf{C}}(E_{\mathsf{M}}, E_{\mathsf{M}})$ whose exact definition may be found in 9.1.4 infra; the point is that the asymptotic behavior of $\mathbf{\Phi}$ on $\mathscr{C}$ (for fixed ν in $\mathcal{O}$) will be obvious from its very construction and so in this way we get an asymptotic formula for Ψ on $\mathscr{C}$ (cf. 9.1.5). It turns out that the $\mathbf{c}_w$ $(w \in W_\Sigma)$ are meromorphic on all of $\mathfrak{a}_{\mathfrak{p}_c}$; in the case when $\mu_1 = \mu_2 =$ trivial one dimensional representation of K (so that Ψ is essentially a zonal spherical function on

G), it is actually possible to explicitly evaluate the $\mathbf{c}_w$ ($w \in W_\Sigma$) in terms of the Gamma function (Theorem of Gindikin and Karpelevič) – cf. 9.1.6.

9.1.2 Analysis in the Universal Enveloping Algebra

The results which are obtained in the present number are essentially algebraic in character; they form, however, an important cornerstone on which the theory of μ-spherical functions on G rests. The readers attention is directed in particular to Theorem 9.1.2.4 which plays a central role in the establishing of certain theorems of finitude (cf. 9.1.3); the other main result in this number is Theorem 9.1.2.9 (whose significance is immediately apparent . . .).

Let $\mathfrak{g}$ be a semi-simple Lie algebra over $\mathbf{R}$, θ a Cartan involution of $\mathfrak{g}$, $\mathfrak{g} = \mathfrak{k} + \mathfrak{p}$ the corresponding Cartan decomposition of $\mathfrak{g}$; fix a maximal abelian subspace $\mathfrak{a}_\mathfrak{p}$ of $\mathfrak{p}$ and let $\mathfrak{l}$ (respectively $\mathfrak{m}$) denote the centralizer of $\mathfrak{a}_\mathfrak{p}$ in $\mathfrak{g}$ (respectively $\mathfrak{k}$); let $\mathfrak{q}$ denote the orthogonal complement of $\mathfrak{m}$ in $\mathfrak{k}$ (relative to the Euclidean structure $(.\,,.)_\theta$). Extend $\mathfrak{a}_\mathfrak{p}$ to a maximal abelian subalgebra $\mathfrak{a}$ of $\mathfrak{g}$; the algebra $\mathfrak{a}_c$ is a Cartan subalgebra of $\mathfrak{g}_c$ and, relative to a choice of compatible orderings in the duals of $\mathfrak{a}_\mathfrak{p}$ and $\mathfrak{a}_\mathfrak{p} + \sqrt{-1}\mathfrak{a}_\mathfrak{k}$, we may introduce Φ, $\Phi^\pm$, $\Phi_\pm$, and $P_\pm$, as usual (cf. Vol. I, number 1.1.3) – in particular we are led to the Iwasawa decomposition $\mathfrak{g} = \mathfrak{k} + \mathfrak{a}_\mathfrak{p} + \mathfrak{n}^+$ for $\mathfrak{g}$.

Let $S(\mathfrak{g}_c)$ denote the symmetric algebra over $\mathfrak{g}_c$, $\lambda : S(\mathfrak{g}_c) \to \mathfrak{G}$ the canonical space bijection; we shall agree to write $\mathfrak{Q}$ for $\lambda(S(\mathfrak{q}_c))$ and will denote the universal enveloping algebras of $\mathfrak{a}_{\mathfrak{p}_c}, \mathfrak{a}_c, \mathfrak{k}_c, \mathfrak{l}_c, \mathfrak{m}_c, \mathfrak{n}_c^+, \mathfrak{n}_c^-$ $(= \theta(\mathfrak{n}_c^+))$, respectively, by $\mathfrak{A}_\mathfrak{p}, \mathfrak{A}, \mathfrak{K}, \mathfrak{L}, \mathfrak{M}, \mathfrak{N}^+, \mathfrak{N}^-$.

Let G be a connected semi-simple Lie group with Lie algebra $\mathfrak{g}$ (so, according to the usual conventions, G has finite center) – then we can write $G = KA_\mathfrak{p}N^+$. Let $\mathfrak{a}_\mathfrak{p}'$ denote the set of all $H \in \mathfrak{a}_\mathfrak{p}$ such that $\alpha(H) \neq 0$ for every $\alpha \in P_+$; put $A_\mathfrak{p}' = \exp(\mathfrak{a}_\mathfrak{p}')$ – then $A_\mathfrak{p}'$ is an open, dense submanifold of $A_\mathfrak{p}$.

Note Unless specific mention is made to the contrary, in what follows the elements of $\mathfrak{G}$ are to be regarded as left invariant differential operators on G.

Convention The group G operates on $\mathfrak{G}$ via the adjoint representation; in order to avoid nests of parentheses, throughout the present number we shall find it convenient to write D^x in place of

$$xD(=\mathrm{Ad}\,(x)D) \qquad (x \in G,\ D \in \mathfrak{G}).$$

Lemma 9.1.2.1 Fix $h \in A_\mathfrak{p}'$ – then $\mathfrak{g}_c = \mathfrak{q}_c^{h^{-1}} + \mathfrak{a}_{\mathfrak{p}_c} + \mathfrak{k}_c$ (direct sum).

Proof Because $\dim(\mathfrak{g}_c) = \dim(\mathfrak{q}_c) + \dim(\mathfrak{a}_{\mathfrak{p}_c}) + \dim(\mathfrak{k}_c)$, it will be sufficient to verify that $\mathfrak{q}_c^{h^{-1}} \cap (\mathfrak{a}_{\mathfrak{p}_c} + \mathfrak{k}_c) = \{0\}$. For this purpose, let

X be an element of $\mathfrak{q}_c$ such that $X^{h^{-1}} \in \mathfrak{a}_{\mathfrak{p}_c} + \mathfrak{k}_c$; let H be an arbitrary element in $\mathfrak{a}_{\mathfrak{p}}$ and put $X_H = \mathrm{ad}\,(H)^2 X$ – then $X_H \in \mathfrak{k}_c$ and $X_H^{h^{-1}} \in \mathfrak{k}_c$, whence

$$X_H^h = \theta(X_H^{h^{-1}}) = X_H^{h^{-1}} \Rightarrow X_H \in \mathfrak{l}_c$$

(since $h \in A'_{\mathfrak{p}}$). Consequently $\mathrm{ad}\,(H)^3 X = \mathrm{ad}\,(H) X_H = 0$; therefore $[H, X] = 0$ ($\mathrm{ad}\,(H)$ is self-adjoint relative to $(\,.\,,\,.\,)_\theta \ldots$). Owing to the arbitrariness of H, it now follows that $X \in \mathfrak{l}_c$; however $\mathfrak{l}_c \cap \mathfrak{q}_c = \{0\}$ and so $X = 0$. Hence the lemma. □

Corollary 9.1.2.2 The mapping

$$D_1 \otimes H \otimes D_2 \mapsto D_1^{h^{-1}} H D_2 \quad (D_1, D_2 \in \mathfrak{K}\,; H \in \mathfrak{A}_{\mathfrak{p}})$$

of $\mathfrak{K} \otimes_{\mathbf{C}} \mathfrak{A}_{\mathfrak{p}} \otimes_{\mathbf{C}} \mathfrak{K}$ into $\mathfrak{G}$, when restricted to $\mathfrak{Q} \otimes_{\mathbf{C}} \mathfrak{A}_{\mathfrak{p}} \otimes_{\mathbf{C}} \mathfrak{K}$, is a linear bijection between $\mathfrak{Q} \otimes_{\mathbf{C}} \mathfrak{A}_{\mathfrak{p}} \otimes_{\mathbf{C}} \mathfrak{K}$ and $\mathfrak{G}$ so long as h lies in $A'_{\mathfrak{p}}$.

Let W_Φ denote the Weyl group of the pair $(\mathfrak{g}_c, \mathfrak{a}_c)$, W_- the subgroup of W_Φ generated by the Weyl reflections $w_\alpha (\alpha \in P_-)$. Let $I(\mathfrak{a}_c)$ (respectively $I_-(\mathfrak{a}_c)$) denote the set of W_Φ-invariants (respectively W_--invariants) in $S(\mathfrak{a}_c)$; let $\mathfrak{Z}$ (respectively $\mathfrak{Z}_-$) denote the center of $\mathfrak{G}$ (respectively $\mathfrak{L}$) – then we have canonical isomorphisms

$$\gamma : \mathfrak{Z} \rightarrow I(\mathfrak{a}_c), \; \gamma_- : \mathfrak{Z}_- \rightarrow I_-(\mathfrak{a}_c)$$

and, as we know, $\mu_- = \gamma_-^{-1} \circ \gamma$ is an algebra isomorphism of $\mathfrak{Z}$ into $\mathfrak{Z}_-$. Indeed $\mathfrak{Z}_-$ is a free abelian module over $\mu_-(\mathfrak{Z})$ of rank $[W_\Phi : W_-]$. [Here we have to bear in mind Example 3 in Vol. I, number 2.3.3; the discussion centering around Lemma 8.5.3.1 is also pertinent for our present purposes.]

Proposition 9.1.2.3 Let $r = [W_\Phi : W_-]$ – then there exist r elements $Z_1 = 1, Z_2, \ldots, Z_r$ in $\mathfrak{Z}_-$ such that the $\gamma_-(Z_i)$ $(1 \leqslant i \leqslant r)$ are homogeneous, and having the property that $\mathfrak{Z}_- = \sum_{i=1}^r \mu_-(\mathfrak{Z}) Z_i$. In fact every element $Z \in \mathfrak{Z}_-$ can be written uniquely in the form

$$Z = \sum_{i=1}^r \mu_-(\tilde{Z}_i) Z_i$$

for certain $\tilde{Z}_i \in \mathfrak{Z}$ with $\deg(Z) = \deg(\tilde{Z}_i) + \deg(Z_i)$ $(1 \leqslant i \leqslant r)$.

[In view of the definition of μ_-, this proposition is an immediate consequence of Theorem 2.1.3.6.]

For $\alpha \in P_+$, let us write

$$f_\alpha^\pm(h) = (h^\alpha \pm 1)^{-1} \qquad (h \in A'_{\mathfrak{p}}).$$

Then it is clear that the $f_\alpha^\pm (\alpha \in P_+)$ are analytic functions on the open submanifold $A'_{\mathfrak{p}}$ of $A_{\mathfrak{p}}$. Let $\mathfrak{F}_0$ denote the algebra generated over $\mathbf{C}$ by the $f_\alpha^\pm (\alpha \in P_+)$ (note that the constant function 1 is not included among the generators of $\mathfrak{F}_0$); let $\mathfrak{F}$ denote the algebra over **C** generated by $\mathfrak{F}_0$ and 1 – then $\mathfrak{F}_0$ is an ideal in $\mathfrak{F}$.

Select elements $Z_1 = 1, Z_2, \ldots, Z_r$ in $\mathfrak{Z}_-$ per Proposition 9.1.2.3 above.

Theorem 9.1.2.4 (Harish-Chandra) Fix $D \in \mathfrak{G}$ – then there exist elements $D_0, D_1, \ldots, D_s$ in $(\sum_{i=1}^r \mathfrak{Z}Z_i^\flat)\mathfrak{K}$, elements $Q_1, \ldots, Q_s$ in $\mathfrak{Q}$, and functions $f_1, \ldots, f_s$ in $\mathfrak{F}_0$ with the following properties:

(i) $D = D_0 + \sum_{j=1}^s f_j(h)Q_j^{h^{-1}}D_j$ (all $h \in A_\mathfrak{p}'$);

(ii) $\deg(D_0) \leqslant \deg(D)$ and $\deg(D_j) + \deg(Q_j) \leqslant \deg(D)$ $(1 \leqslant j \leqslant s)$;

(iii) $D \equiv D_0 \bmod \mathfrak{n}_c^-\mathfrak{G}$.

Corollary 9.1.2.5 Retain the above notations – then

$$\mathfrak{G} = \mathfrak{Q}^{h^{-1}}\Big(\sum_{i=1}^r \mathfrak{Z}Z_i^\flat\Big)\mathfrak{K} \qquad (\text{all } h \in A_\mathfrak{p}').$$

The following lemma will be needed for the proof of our theorem.

Lemma 9.1.2.6 Fix $\alpha \in P_+$; choose $X_\alpha \in \mathfrak{g}_c^\alpha$ as usual and write

$$X_\alpha = Y_\alpha + Z_\alpha \qquad (Y_\alpha \in \mathfrak{k}_c, Z_\alpha \in \mathfrak{p}_c).$$

Then

$$\theta(X_\alpha) = 2(f_\alpha^+(h) + f_\alpha^+(h)f_\alpha^-(h))Y_\alpha^{h^{-1}} - 2f_\alpha^+(h)f_\alpha^-(h)Y_\alpha \qquad (\text{all } h \in A_\mathfrak{p}').$$

Proof Fix $h \in A_\mathfrak{p}'$ and put $H = \log h$; since $Y_\alpha = 2^{-1}(X_\alpha + \theta(X_\alpha))$, we have

$$\begin{aligned} Y_\alpha^{h^{-1}} &= 2^{-1}(e^{-\alpha(H)}X_\alpha + e^{\alpha(H)}\theta(X_\alpha)) \\ &= -\sinh(\alpha(H))Z_\alpha + \cosh(\alpha(H))Y_\alpha \\ &= \sinh(\alpha(H))\theta(X_\alpha) + e^{-\alpha(H)}Y_\alpha \end{aligned}$$

from which we can read off the statement of the lemma. □

Proof of Theorem 9.1.2.4 The proof proceeds by induction on the degree $m = \deg(D)$ of D; the case $D = 0$ being clear, we can suppose that $D \neq 0$.

(1) Because $\mathfrak{g}_c = \mathfrak{n}_c^- + \mathfrak{a}_{\mathfrak{p}_c} + \mathfrak{k}_c$ (direct sum), there exists a unique element D^* in $\mathfrak{A}_\mathfrak{p}\mathfrak{K}$ such that $D - D^* \in \mathfrak{n}_c^-\mathfrak{G}_{m-1}$; it is clear that $\deg(D^*) \leqslant m$. Moreover, since $\mathfrak{A}_\mathfrak{p} \subset \mathfrak{Z}_-$, we can, on the basis of Proposition 9.1.2.3, choose elements $Z_{ij} \in \mathfrak{Z}$, $D_{ij} \in \mathfrak{K}$ having the property that

$$D^* = \sum_{i,j} \mu_-^*(Z_{ij})Z_i^\flat D_{ij}$$

with $\deg(Z_{ij}) + \deg(Z_i^\flat) + \deg(D_{ij}) \leqslant \deg(D^*) \leqslant m$ (we shall agree to write $\mu_-^*(.) = (\mu_-(.))^\flat$). [Here the index i runs between 1 and r while the index j runs over some finite set which need not be specified.] The very definition of the map μ_-^* implies that $D \equiv D^* \equiv D_0 \bmod \mathfrak{n}_c^- \mathfrak{G}_{m-1}$ where $D_0 = \sum_{i,j} Z_{ij} Z_i^\flat D_{ij}$. Since $\mathfrak{n}_c^- = \theta(\mathfrak{n}_c^+)$, we can write $D - D_0 = \sum_{\alpha \in P_+} \theta(X_\alpha) D_\alpha$ where $D_\alpha \in \mathfrak{G}_{m-1}$. On the other hand, thanks to Lemma 9.1.2.6, we have

$$\theta(X_\alpha) = f_\alpha(h) Y_\alpha^{h^{-1}} + g_\alpha(h) Y_\alpha \qquad (h \in A_\mathfrak{p}'),$$

the f_α, g_α ($\alpha \in P_+$) being certain functions in $\mathfrak{F}_0$. This shows, therefore, that

$$D - D_0 = \sum_{\alpha \in P_+} (f_\alpha(h) Y_\alpha^{h^{-1}} D_\alpha + g_\alpha(h) Y_\alpha D_\alpha) \qquad (h \in A_\mathfrak{p}').$$

(2) Since $\deg(D_\alpha) \leqslant m - 1$, the induction hypothesis is applicable to it ($\alpha \in P_+$); this means that we can select a finite set of indices j and elements $D_{\alpha j} \in \sum_{i=1}^r (\mathfrak{Z} Z_i^\flat)\mathfrak{K}$, $Q_{\alpha j} \in \mathfrak{Q}$ and $f_{\alpha j} \in \mathfrak{F}$ such that:

(i) $D_\alpha = \sum_j f_{\alpha j}(h) Q_{\alpha j}^{h^{-1}} D_{\alpha j}$ (all $h \in A_\mathfrak{p}'$);

(ii) $\deg(D_{\alpha j}) + \deg(Q_{\alpha j}) \leqslant \deg(D_\alpha)$.

Consider the products $Y_\alpha Q_{\alpha j}$; because $\mathfrak{K} = \mathfrak{Q}\mathfrak{M}$, we can write

$$Y_\alpha Q_{\alpha j} = \sum_k Q_{\alpha jk} M_{\alpha jk}$$

where k runs through a certain finite set of indices, $M_{\alpha jk} \in \mathfrak{M}$, $Q_{\alpha jk} \in \mathfrak{Q}$ and $\deg(M_{\alpha jk}) + \deg(Q_{\alpha jk}) \leqslant \deg(Q_{\alpha j}) + 1$, whence

$$Y_\alpha^{h^{-1}} D_\alpha = \sum_j \sum_k f_{\alpha j}(h) Q_{\alpha jk}^{h^{-1}} D_{\alpha jk}$$

where $D_{\alpha jk} = M_{\alpha jk} D_{\alpha j}$ and

$$\deg(Q_{\alpha jk}) + \deg(D_{\alpha jk}) \leqslant \deg(Q_{\alpha j}) + \deg(D_{\alpha j}) + 1 \leqslant \deg(D_\alpha) + 1 \leqslant m.$$

Agreeing to write $f_{\alpha j} = f_\alpha f_{\alpha j}$, $g_{\alpha j} = g_\alpha f_{\alpha j}$, we then find that

$$D - D_0 = \sum_{\alpha, j, k} f_{\alpha j}(h) Q_{\alpha jk}^{h^{-1}} D_{\alpha jk} + \sum_{\alpha, j} g_{\alpha j}(h) Q_{\alpha j}^{h^{-1}} D_{\alpha j} Y_\alpha$$
$$+ \sum_\alpha g_\alpha(h)[Y_\alpha, D_\alpha] \qquad (h \in A_\mathfrak{p}').$$

The proof of the theorem can now be completed by applying the induction hypothesis to $[Y_\alpha, D_\alpha]$. □

Application At this juncture we shall indicate a typical application of the preceding theorem. Fix two classes δ_1, δ_2 in $\hat{K}$ and let κ be a homomorphism of $\mathfrak{Z}$ into $\mathbf{C}$; let $\mathfrak{S}(\delta_1, \delta_2, \kappa)$ denote the space of all C^∞ functions f on G which transform to the left under δ_1, to the right under δ_2 and which satisfy the system of differential equations $Zf = \kappa(Z)f$ ($Z \in \mathfrak{Z}$) – then

$$\dim(\mathfrak{S}(\delta_1, \delta_2, \kappa)) \leqslant [W_\Phi : W_-] d(\delta_1)^2 d(\delta_2)^2.$$

The proof goes as follows. Let

$$\mathfrak{K}_1 = \{D \in \mathfrak{K} : (D \otimes 1) f = 0 \text{ (all } f \in \mathfrak{S}(\delta_1, \delta_2, \kappa))\},$$
$$\mathfrak{K}_2 = \{D \in \mathfrak{K} : (1 \otimes D) f = 0 \text{ (all } f \in \mathfrak{S}(\delta_1, \delta_2, \kappa))\}.$$

The group K may be represented by left translation on the space $\mathfrak{S}(\delta_1, \delta_2, \kappa)$; in this way $\mathfrak{S}(\delta_1, \delta_2, \kappa)$ acquires the structure of a semi-simple K-module which has the property that all its irreducible components lie in the class δ_1 – the kernel of the lift of this representation to $\mathfrak{K}$ is the two-sided ideal $\mathfrak{K}_1$. Similar comments apply to the canonical representation of K on $\mathfrak{S}(\delta_1, \delta_2, \kappa)$ by right translation and the two-sided ideal $\mathfrak{K}_2$. In passing observe that $\dim(\mathfrak{K}/\mathfrak{K}_i) = d(\delta_i)^2$ $(i = 1, 2)$ provided that $\mathfrak{S}(\delta_1, \delta_2, \kappa) \neq \{0\}$ (which will be assumed in what follows). Now fix a point $h_0 \in A'_{\mathfrak{p}}$; suppose that f is an element of $\mathfrak{S}(\delta_1, \delta_2, \kappa)$ such that $f(D_1; h_0; Z_i^\dagger D_2) = 0$ for all $D_1, D_2 \in \mathfrak{K}$ $(1 \leqslant i \leqslant r = [W_\Phi : W_-]$; cf. Theorem 9.1.2.4) – then $f = 0$. [Since the space $\mathfrak{S}(\delta_1, \delta_2, \kappa)$ is comprised of analytic functions (cf. the Appendix to number 8.3.9), we need only show that $f(h_0; D) = 0$ (all $D \in \mathfrak{G}$); in view of our hypotheses on f, the desired assertion can be read off from Corollary 9.1.2.5 supra.] This being so, choose a basis D_{1i_1} $(1 \leqslant i_1 \leqslant d(\delta_1)^2)$ for $\mathfrak{K}$ mod $\mathfrak{K}_1$ and a basis D_{2i_2} $(1 \leqslant i_2 \leqslant d(\delta_2)^2)$ for $\mathfrak{K}$ mod $\mathfrak{K}_2$; owing to what was said above, any f in $\mathfrak{S}(\delta_1, \delta_2, \kappa)$ which satisfies the $rd(\delta_1)^2 d(\delta_2)^2$ linear conditions $f(D_{1i_1}; h_0; Z_i^\dagger D_{2i_2}) = 0$ must necessarily vanish identically. Of course this serves to establish our contention.

We shall agree to retain the preceding notations and assumptions.

Let $\mathfrak{D}_F(A'_{\mathfrak{p}})$ denote the algebra of differential operators on $A'_{\mathfrak{p}}$ of finite order; consider the tensor product $\mathfrak{A} = \mathfrak{K} \otimes_{\mathbf{C}} \mathfrak{D}_F(A'_{\mathfrak{p}}) \otimes_{\mathbf{C}} \mathfrak{K}$ – then $\mathfrak{A}$ acquires the structure of an associative algebra upon writing

$$(D_1 \otimes D \otimes D_2)(\tilde{D}_1 \otimes \tilde{D} \otimes \tilde{D}_2) = (D_1\tilde{D}_1 \otimes D\tilde{D} \otimes \tilde{D}_2 D_2)$$
$$(D_1, D_2, \tilde{D}_1, \tilde{D}_2 \in \mathfrak{K}; D, \tilde{D} \in \mathfrak{D}_F(A'_{\mathfrak{p}})).$$

For any integer m, let $\mathfrak{D}_F(A'_{\mathfrak{p}})_m$ denote the subspace of $\mathfrak{D}_F(A'_{\mathfrak{p}})$ comprised of those elements of degree $\leqslant m$; put $\mathfrak{A}_m = \mathfrak{K} \otimes_{\mathbf{C}} \mathfrak{D}_F(A'_{\mathfrak{p}})_m \otimes_{\mathbf{C}} \mathfrak{K}$ – then, if Δ is a non-zero element of $\mathfrak{A}$, $\deg(\Delta)$ is, by definition, the smallest integer m such that $\Delta \in \mathfrak{A}_m$ $(\deg(\Delta) = -\infty$ in case $\Delta = 0)$.

In what follows we shall identify $\mathfrak{D}_F(A'_{\mathfrak{p}})$ with a subalgebra of $\mathfrak{A}$ under the mapping $D \mapsto 1 \otimes D \otimes 1$ $(D \in \mathfrak{D}_F(A'_{\mathfrak{p}}))$ – then $\mathfrak{A}$ becomes a two-sided module over $\mathfrak{D}_F(A'_{\mathfrak{p}})$. [Thus, in particular, since $\mathfrak{D}_F(A'_{\mathfrak{p}})$ contains the C^∞ functions on $A'_{\mathfrak{p}}$, the symbols $f\Delta \circ g$ $(f, g \in C^\infty(A'_{\mathfrak{p}}), \Delta \in \mathfrak{A})$ have a well-defined meaning in $\mathfrak{A}$ (the circle denoting multiplication in $\mathfrak{A}$).]

We may, of course, regard $\mathfrak{A}_{\mathfrak{p}}$ as a subalgebra of $\mathfrak{D}_F(A'_{\mathfrak{p}})$, whence $\mathfrak{K} \otimes_{\mathbf{C}} \mathfrak{A}_{\mathfrak{p}} \otimes_{\mathbf{C}} \mathfrak{K} \subset \mathfrak{A}$. Fix a point $h \in A'_{\mathfrak{p}}$ and for any $D \in \mathfrak{D}_F(A'_{\mathfrak{p}})$ let D_h denote the local expression of D at h – then the assignment $D \mapsto D_h$ is a linear mapping of $\mathfrak{D}_F(A'_{\mathfrak{p}})$ into $\mathfrak{A}_{\mathfrak{p}}$ which admits a unique extension to a linear mapping $\Delta \mapsto \Delta_h$ $(\Delta \in \mathfrak{A})$ of $\mathfrak{A}$ into $\mathfrak{K} \otimes_{\mathbf{C}} \mathfrak{A}_{\mathfrak{p}} \otimes_{\mathbf{C}} \mathfrak{K}$ such that

$$(D_1 \otimes D \otimes D_2)_h = D_1 \otimes D_h \otimes D_2 \quad (D_1, D_2 \in \mathfrak{K}; D \in \mathfrak{D}_F(A'_{\mathfrak{p}})).$$

We shall refer to Δ_h as the *local expression* of Δ at h $(h \in A'_{\mathfrak{p}})$.

An element in $\mathfrak{K} \otimes_{\mathbf{C}} \mathfrak{A}_{\mathfrak{p}} \otimes_{\mathbf{C}} \mathfrak{K}$ is said to be homogeneous of degree m if it lies in $\mathfrak{K} \otimes_{\mathbf{C}} \mathfrak{A}_{\mathfrak{p}}^m \otimes_{\mathbf{C}} \mathfrak{K}$ $(m = 0, 1, \ldots)$; an element in $\mathfrak{K} \otimes_{\mathbf{C}} \mathfrak{A}_{\mathfrak{p}} \otimes_{\mathbf{C}} \mathfrak{K}$

is said to be homogeneous of *total degree m* if it lies in

$$\sum_{m_1+m_2+m_3=m} \mathfrak{K}^{m_1} \otimes_{\mathbf{C}} \mathfrak{A}_{\mathfrak{p}}^{m_2} \otimes_{\mathbf{C}} \mathfrak{K}^{m_3}.$$

The degree of an element $\mathbf{\Delta} \in \mathfrak{K} \otimes_{\mathbf{C}} \mathfrak{A}_{\mathfrak{p}} \otimes_{\mathbf{C}} \mathfrak{K}$ will be denoted by deg $(\mathbf{\Delta})$, its total degree by $\deg_t(\mathbf{\Delta})$ ($\deg(\mathbf{\Delta}) = \deg_t(\mathbf{\Delta}) = -\infty$ in case $\mathbf{\Delta} = 0$).

For any $h \in A'_{\mathfrak{p}}$, let $\mathbf{\Gamma}_h$ denote the linear mapping of $\mathfrak{K} \otimes_{\mathbf{C}} \mathfrak{A}_{\mathfrak{p}} \otimes_{\mathbf{C}} \mathfrak{K}$ into $\mathfrak{G}$ given by

$$\mathbf{\Gamma}_h(D_1 \otimes H \otimes D_2) = D_1^{h^{-1}} H D_2 \quad (D_1, D_2 \in \mathfrak{K}; H \in \mathfrak{A}_{\mathfrak{p}}).$$

As has been seen above, the restriction of $\mathbf{\Gamma}_h$ to $\mathfrak{Q} \otimes_{\mathbf{C}} \mathfrak{A}_{\mathfrak{p}} \otimes_{\mathbf{C}} \mathfrak{K}$ is bijective; let $\mathfrak{P}_{A_{\mathfrak{p}},h}$ denote its inverse (so that $\mathfrak{P}_{A_{\mathfrak{p}},h} : \mathfrak{G} \to \mathfrak{Q} \otimes_{\mathbf{C}} \mathfrak{A}_{\mathfrak{p}} \otimes_{\mathbf{C}} \mathfrak{K}$).

Lemma 9.1.2.7 Fix $D \in \mathfrak{G}_m$ – then there exist elements $f_1, \ldots, f_s$ in $\mathfrak{F}_0$ and $\mathbf{\Delta}_0, \ldots, \mathbf{\Delta}_s$ in $\mathfrak{Q} \otimes_{\mathbf{C}} \mathfrak{A}_{\mathfrak{p}} \otimes_{\mathbf{C}} \mathfrak{K}$ with $\deg_t(\mathbf{\Delta}_i) \leqslant m$ $(0 \leqslant i \leqslant s)$, $\deg(\mathbf{\Delta}_j) \leqslant m - 1$ $(1 \leqslant j \leqslant s)$ and having the property that

$$\mathfrak{P}_{A_{\mathfrak{p}},h}(D) = \mathbf{\Delta}_0 + \sum_{j=1}^{s} f_j(h)\mathbf{\Delta}_j \qquad (\text{all } h \in A'_{\mathfrak{p}}).$$

Corollary 9.1.2.8 Fix $D \in \mathfrak{G}$; let $g_1, \ldots, g_s$ be elements in $\mathfrak{F}_0$ and $\square_0, \ldots, \square_s$ elements in $\mathfrak{Q} \otimes_{\mathbf{C}} \mathfrak{A}_{\mathfrak{p}} \otimes_{\mathbf{C}} \mathfrak{K}$ such that

$$\mathfrak{P}_{A_{\mathfrak{p}},h}(D) = \square_0 + \sum_{j=1}^{s} g_j(h)\square_j \qquad (\text{all } h \in A'_{\mathfrak{p}}).$$

Let D_0 be the unique element in $\mathfrak{A}_{\mathfrak{p}}\mathfrak{K}$ such that $D - D_0 \in \mathfrak{n}_c^- \mathfrak{G}$ – then $\square_0 = \mathfrak{P}_{A_{\mathfrak{p}},h}(D_0)$ and

$$\deg(\mathfrak{P}_{A_{\mathfrak{p}},h}(D - D_0)) \leqslant \deg(D) - 1 \qquad (\text{all } h \in A'_{\mathfrak{p}}).$$

Before giving the proof of Lemma 9.1.2.7 and its corollary, let us explain the significance of the operation $D \mapsto \mathfrak{P}_{A_{\mathfrak{p}}}(D)$ $(D \in \mathfrak{G})$.

Let $\mu = (\mu_1, \mu_2)$ be a double Banach representation of K on a finite dimensional Banach space E.

Let $\mathcal{O}$ be an open set in $A'_{\mathfrak{p}}$ – then $\mathbf{\mathfrak{A}}$ operates on the space $C^\infty(\mathcal{O}; E)$ via the prescription

$$(D_1 \otimes D \otimes D_2)f = \mu_1(D_1)(Df)\mu_2(D_2) \qquad (f \in C^\infty(\mathcal{O}; E))$$

where $D_1, D_2 \in \mathfrak{K}$ and $D \in \mathbf{\mathfrak{D}}_F(A'_{\mathfrak{p}})$. In this fashion, the space $C^\infty(\mathcal{O}; E)$ acquires the structure of a left $\mathbf{\mathfrak{A}}$-module (here we have to bear in mind the definition of multiplication in $\mathbf{\mathfrak{A}}$ . . .). It will frequently be convenient to denote the value of $\mathbf{\Delta}f$ $(\mathbf{\Delta} \in \mathbf{\mathfrak{A}})$ at a point $h \in \mathcal{O}$ by the symbol $f(h; \mathbf{\Delta})$; it is then obvious that $f(h; \mathbf{\Delta}) = f(h; \mathbf{\Delta}_h)$, $\mathbf{\Delta}_h$ the local expression of $\mathbf{\Delta}$ at $h \in \mathcal{O}$ $(f \in C^\infty(\mathcal{O}; E))$.

On the basis of Lemma 9.1.2.7, it is clear that for each $D \in \mathfrak{G}$ there exists an element $\mathfrak{P}_{A_{\mathfrak{p}}}(D) \in \mathbf{\mathfrak{A}}$ such that $\mathfrak{P}_{A_{\mathfrak{p}},h}(D)$ is the local expression of $\mathfrak{P}_{A_{\mathfrak{p}}}(D)$ at any point $h \in A'_{\mathfrak{p}}$.

Theorem 9.1.2.9 (Harish-Chandra) Let Ψ be a C^∞ μ-spherical function on G – then, for all $D \in \mathfrak{G}$, we have

$$(D\Psi)^{\sim} = \mathfrak{P}_{A_\mathfrak{p}}(D)\tilde{\Psi} \qquad (\tilde{\Psi} = \Psi \,|\, A'_\mathfrak{p}).$$

[Roughly speaking, then, the operation $D \mapsto \mathfrak{P}_{A_\mathfrak{p}}(D)$ $(D \in \mathfrak{G})$ may be viewed as assigning to each D in $\mathfrak{G}$ its 'μ-radial component' on $A'_\mathfrak{p}$; cf. with Theorem 8.2.1.7.]

Proof of Theorem 9.1.2.9 Fix an element $D \in \mathfrak{G}$; employing the evident notations, write $\mathfrak{P}_{A_\mathfrak{p}}(D) = \sum_i f_i(Q_i \otimes H_i \otimes D_i)$ – then

$$D = \Gamma_h(\mathfrak{P}_{A_\mathfrak{p},h}(D)) = \sum_i f_i(h)Q_i^{h^{-1}}H_iD_i \qquad (h \in A'_\mathfrak{p}).$$

Consequently

$$\begin{aligned}\Psi(h\,;D) &= \sum_i f_i(h)\Psi(h\,;Q_i^{h^{-1}}H_iD_i)\\ &= \sum_i f_i(h)\mu_1(Q_i)\Psi(h\,;H_i)\mu_2(D_i) = \tilde{\Psi}(h\,;\mathfrak{P}_{A_\mathfrak{p}}(D)) \qquad (h \in A'_\mathfrak{p})\end{aligned}$$

which proves the theorem. $\square$

Corollary 9.1.2.10 Let $\mathfrak{X}$ denote the centralizer of $\mathfrak{k}_c$ in $\mathfrak{G}$; let Ψ be a C^∞ μ-spherical function on G – then, for $D_1, D_2 \in \mathfrak{X}$, we have

$$\mathfrak{P}_{A_\mathfrak{p}}(D_1D_2)\tilde{\Psi} = \mathfrak{P}_{A_\mathfrak{p}}(D_1)(\mathfrak{P}_{A_\mathfrak{p}}(D_2)\tilde{\Psi}) \qquad (\tilde{\Psi} = \Psi \,|\, A'_\mathfrak{p}).$$

[One need only observe that if Ψ is μ-spherical, then so is $D\Psi$ (all $D \in \mathfrak{X}$).]

Proof of Lemma 9.1.2.7 We shall use induction on m; it can be supposed that $D \neq 0$ and $\deg(D) = m \geqslant 1$. Choose $D_0 \in \mathfrak{A}_\mathfrak{p}\mathfrak{K}$ such that $D \equiv D_0 \bmod \mathfrak{n}_c^-\mathfrak{G}$ – then $\deg(D_0) \leqslant m$ and $D - D_0 \in \mathfrak{n}_c^-\mathfrak{G}_{m-1}$. Put $\Delta_0 = \mathfrak{P}_{A_\mathfrak{p},h}(D_0)$ $(h \in A'_\mathfrak{p})$ – then Δ_0 is actually independent of h $(h \in A'_\mathfrak{p})$ and lies in $1 \otimes_C \mathfrak{A}_\mathfrak{p} \otimes_C \mathfrak{K}$. Now choose $D_\alpha \in \mathfrak{G}_{m-1}$ such that $D - D_0 = \sum_{\alpha \in P_+} \theta(X_\alpha)D_\alpha$; in view of Lemma 9.1.2.6, we can find elements f_α, g_α $(\alpha \in P_+)$ in $\mathfrak{F}_0$ having the property that

$$\begin{aligned}D - D_0 &= \sum_{\alpha \in P_+} (f_\alpha(h)Y_\alpha^{h^{-1}} + g_\alpha(h)Y_\alpha)D_\alpha\\ &= \sum_{\alpha \in P_+} f_\alpha(h)Y_\alpha^{h^{-1}}D_\alpha + \sum_{\alpha \in P_+} g_\alpha(h)\{[Y_\alpha, D_\alpha] + D_\alpha Y_\alpha\}\\ &\qquad\qquad\qquad\qquad\qquad\qquad (\text{all } h \in A'_\mathfrak{p}).\end{aligned}$$

Upon applying the induction hypothesis to $[Y_\alpha, D_\alpha]$ and D_α, it follows at once that there exist elements $f_1, \ldots, f_s \in \mathfrak{F}_0$ and elements $\tilde{\Delta}_1, \ldots, \tilde{\Delta}_s$ in $\mathfrak{K} \otimes_C \mathfrak{A}_\mathfrak{p} \otimes_C \mathfrak{K}$ such that $\deg_t(\tilde{\Delta}_j) \leqslant m$, $\deg(\tilde{\Delta}_j) \leqslant m - 1$ $(1 \leqslant j \leqslant s)$ with

$$\sum_{j=1}^{s} f_j(h)\Gamma_h(\tilde{\Delta}_j) = D - D_0 = D - \Gamma_h(\Delta_0) \qquad (\text{all } h \in A'_\mathfrak{p}).$$

Because

$$\Gamma_h(QM \otimes H \otimes D) = \Gamma_h(Q \otimes H \otimes MK)$$
$$(Q \in \mathfrak{Q}, M \in \mathfrak{M}, H \in \mathfrak{A}_\mathfrak{p}, D \in \mathfrak{K})$$

and $\mathfrak{K} = \mathfrak{Q}\mathfrak{M}$, we can select elements $\Delta_j \in \mathfrak{Q} \otimes_C \mathfrak{A}_\mathfrak{p} \otimes_C \mathfrak{K}$ such that $\deg_t(\Delta_j) \leqslant \deg_t(\tilde{\Delta}_j)$, $\deg(\Delta_j) \leqslant \deg(\tilde{\Delta}_j)$ and

$$\Gamma_h(\Delta_j) = \Gamma_h(\tilde{\Delta}_j) \quad (1 \leqslant j \leqslant s)$$

for every $h \in A_\mathfrak{p}'$. Thus

$$D = \Gamma_h(\Delta_0) + \sum_{j=1}^{s} f_j(h)\Gamma_h(\Delta_j)$$

and so

$$\mathfrak{P}_{A_\mathfrak{p},h}(D) = \Delta_0 + \sum_{j=1}^{s} f_j(h)\Delta_j \qquad (\text{all } h \in A_\mathfrak{p}')$$

which serves to complete the proof of the lemma. □

Proof of Corollary 9.1.2.8 The proof of the preceding lemma ensures the existence of elements $f_1, \ldots, f_s$ in $\mathfrak{F}_0$ and $\Delta_0, \ldots, \Delta_s$ in $\mathfrak{Q} \otimes_C \mathfrak{A}_\mathfrak{p} \otimes_C \mathfrak{K}$ such that $\Delta_0 = \mathfrak{P}_{A_\mathfrak{p},h}(D_0)$, $\deg(\Delta_j) \leqslant \deg(D) - 1$ $(1 \leqslant j \leqslant s)$ and having the property that

$$\mathfrak{P}_{A_\mathfrak{p},h}(D) = \Delta_0 + \sum_{j=1}^{s} f_j(h)\Delta_j \qquad (\text{all } h \in A_\mathfrak{p}').$$

It follows, therefore, that $\square_0 - \Delta_0 = \sum_i F_i(h)D_i$ $(h \in A_\mathfrak{p}')$ where $F_i \in \mathfrak{F}_0$ and $D_i \in \mathfrak{Q} \otimes_C \mathfrak{A}_\mathfrak{p} \otimes_C \mathfrak{K}$ (i running through a certain finite set of indices). Now fix an element H in the positive Weyl chamber $\mathscr{C}$; it is clear that $f(\exp(tH)) \to 0$ as $t \to \infty$ for any $f \in \mathfrak{F}_0$. This being so, we see, upon setting $h = h_t = \exp(tH)$ and letting t tend to infinity, that necessarily $\square_0 = \Delta_0 = \mathfrak{P}_{A_\mathfrak{p},h}(D_0)$ (all $h \in A_\mathfrak{p}'$), as desired. The degree assertion is obvious. □

Remark Let M denote the centralizer of $A_\mathfrak{p}$ in K – then M operates on $\mathfrak{K} \otimes_C \mathfrak{A}_\mathfrak{p} \otimes_C \mathfrak{K}$ in the obvious way and it is clear that $\mathfrak{Q} \otimes_C \mathfrak{A}_\mathfrak{p} \otimes_C \mathfrak{K}$ is M-stable. This being so, suppose that our element D in Lemma 9.1.2.7 is left fixed by every element of M – then we can always suppose that the Δ_j $(0 \leqslant j \leqslant s)$ have the same property. In fact for $h \in A_\mathfrak{p}'$ and *any* $D \in \mathfrak{G}$ one has $\mathfrak{P}_{A_\mathfrak{p},h}(D^m) = (\mathfrak{P}_{A_\mathfrak{p},h}(D))^m$ (all $m \in \mathsf{M}$); but $1, f_1, \ldots, f_s$ can be taken linearly independent. . . .

Let ω denote the Casimir element in $\mathfrak{G}$ (hence $\omega \in \mathfrak{Z}$). As we shall see, the asymptotic behavior on $A_\mathfrak{p}'$ of a given μ-spherical function Ψ on G can be determined by studying $\mathfrak{P}_{A_\mathfrak{p}}(\omega)\Psi$; in turn this determination takes for its point of departure the explicit formula for the operator $\mathfrak{P}_{A_\mathfrak{p}}(\omega)$ which we shall now proceed to derive.

Given $\alpha \in P_+$, choose the elements $X_\alpha \in \mathfrak{g}_c^\alpha$ and $X_{-\alpha} \in \mathfrak{g}_c^{-\alpha}$ in the usual way (thus $B(X_\alpha, X_{-\alpha}) = 1$); as above, we shall write $X_{\pm\alpha} = Y_{\pm\alpha} +$

$Z_{\pm\alpha}$ where $Y_{\pm\alpha} \in \mathfrak{k}_c$ and $Z_{\pm\alpha} \in \mathfrak{p}_c$ ($\alpha \in P_+$). Let $X_1, \ldots, X_m$ be a basis for $\mathfrak{m}_c$; put $g_{ij} = B(X_i, X_j)$ $(1 \leqslant i, j \leqslant m)$ and let (g^{ij}) denote the inverse of the matrix (g_{ij}) $(1 \leqslant i, j \leqslant m)$. Choose the basis $H_1, \ldots, H_l$ for $\mathfrak{a}_\mathfrak{p}$ as in 9.1.1; put $h_{ij} = B(H_i, H_j)$ $(1 \leqslant i, j \leqslant l)$ and let (h^{ij}) denote the inverse of the matrix (h_{ij}) $(1 \leqslant i, j \leqslant l)$. Agreeing to write

$$\omega_{\mathfrak{l}} = \omega_{\mathfrak{m}} + \sum_{i,j} h^{ij} H_i H_j \qquad \Big(\omega_{\mathfrak{m}} = \sum_{i,j} g^{ij} X_i X_j\Big),$$

a standard argument tells us that

$$\omega = \omega_{\mathfrak{l}} + \sum_{\alpha \in P_+} (X_\alpha X_{-\alpha} + X_{-\alpha} X_\alpha).$$

Here is the promised formula for $\mathfrak{P}_{A_\mathfrak{p}}(\omega)$.

Proposition 9.1.2.11 Let ω be the Casimir element in $\mathfrak{G}$ – then

$$\begin{aligned}\mathfrak{P}_{A_\mathfrak{p}}(\omega) = \mathfrak{P}_{A_\mathfrak{p}}(\omega_{\mathfrak{m}}) &+ \sum_{i,j} h^{ij} H_i H_j + \sum_{\alpha \in P_+} \coth(\tilde{\alpha}) Q_{\tilde{\alpha}} \\ &- 2 \sum_{\alpha \in P_+} (\sinh(\alpha))^{-2} \{(1 \otimes 1 \otimes Y_\alpha Y_{-\alpha}) + (Y_\alpha Y_{-\alpha} \otimes 1 \otimes 1)\} \\ &+ 4 \sum_{\alpha \in P_+} (\sinh(\alpha))^{-1} \coth(\alpha)(Y_\alpha \otimes 1 \otimes Y_{-\alpha}).\end{aligned}$$

[We assign to the symbols $\tilde{\alpha}$ and $Q_{\tilde{\alpha}}$ ($\alpha \in \Phi_+$) their customary meanings (cf. Vol. I, number 1.1.3); in particular $\tilde{\alpha} = \alpha | \mathfrak{a}_\mathfrak{p}$.]

Proof Fix $h \in A_\mathfrak{p}'$; given $\alpha \in \Phi_+$, let $c = \alpha(\log h)$ – then

$$\begin{aligned}\mathrm{Ad}(h^{-1}) Y_\alpha &= 2^{-1} \mathrm{Ad}(h^{-1})(X_\alpha + \theta(X_\alpha)) = 2^{-1}(e^{-c} X_\alpha + e^c \theta(X_\alpha)) \\ &= 2^{-1}(e^c + e^{-c}) Y_\alpha - 2^{-1}(e^c - e^{-c}) Z_\alpha\end{aligned}$$

and so

$$\begin{aligned}Z_\alpha &= \coth(c) Y_\alpha - (\sinh(c))^{-1} \mathrm{Ad}(h^{-1}) Y_\alpha \Rightarrow \\ X_\alpha &= 2(e^c - e^{-c})^{-1} \{e^c Y_\alpha - Y_\alpha^{h^{-1}}\}.\end{aligned}$$

Consequently

$$-X_\alpha X_{-\alpha} = 4(e^c - e^{-c})^{-2} \{Y_\alpha Y_{-\alpha} + (Y_\alpha Y_{-\alpha})^{h^{-1}} - e^c Y_\alpha Y_{-\alpha}^{h^{-1}} - e^{-c} Y_\alpha^{h^{-1}} Y_{-\alpha}\}.$$

Since

$$\begin{aligned}Y_\alpha Y_{-\alpha}^{h^{-1}} &= Y_{-\alpha}^{h^{-1}} Y_\alpha + [Y_\alpha, Y_{-\alpha}^{h^{-1}}] \\ &= Y_{-\alpha}^{h^{-1}} Y_\alpha + \cosh(c)[Y_\alpha, Y_{-\alpha}] + \sinh(c)[Y_\alpha, Z_{-\alpha}],\end{aligned}$$

we see that

$$\begin{aligned}-X_\alpha X_{-\alpha} = 4(e^c - e^{-c})^{-2} \{&Y_\alpha Y_{-\alpha} + (Y_\alpha Y_{-\alpha})^{h^{-1}} - e^c Y_{-\alpha}^{h^{-1}} Y_\alpha - e^{-c} Y_\alpha^{h^{-1}} Y_{-\alpha} \\ &-2^{-1}(e^{2c} + 1)[Y_\alpha, Y_{-\alpha}] - 2^{-1}(e^{2c} - 1)[Y_\alpha, Z_{-\alpha}]\}.\end{aligned}$$

Applying θ to this equation leads to the relation

$$\begin{aligned}-\theta(X_\alpha X_{-\alpha}) = 4(e^c - e^{-c})^{-2} \{&Y_\alpha Y_{-\alpha} + (Y_\alpha Y_{-\alpha})^h - e^c Y_{-\alpha}^h Y_\alpha - e^{-c} Y_\alpha^h Y_{-\alpha} \\ &-2^{-1}(e^{2c} + 1)[Y_\alpha, Y_{-\alpha}] + 2^{-1}(e^{2c} - 1)[Y_\alpha, Z_{-\alpha}]\}\end{aligned}$$

which in turn implies, upon replacing α and h by $-\alpha$ and h^{-1}, respectively, that

$$\begin{aligned}-\theta(X_{-\alpha}X_{\alpha}) = 4(e^c - e^{-c})^{-2}\{&Y_{-\alpha}Y_{\alpha} + (Y_{-\alpha}Y_{\alpha})^{h^{-1}} \\ &-e^c Y_{\alpha}^{h^{-1}}Y_{-\alpha} - e^{-c}Y_{-\alpha}^{h^{-1}}Y_{\alpha} \\ &+2^{-1}(e^{2c}+1)[Y_{\alpha}, Y_{-\alpha}] + 2^{-1}(e^{2c}-1)[Y_{-\alpha}, Z_{\alpha}]\}.\end{aligned}$$

Therefore

$$\begin{aligned}&-(X_{\alpha}X_{-\alpha} + \theta(X_{-\alpha}X_{\alpha})) \\ &= 4(e^c - e^{-c})^{-2}\{(Y_{\alpha}Y_{-\alpha} + Y_{-\alpha}Y_{\alpha}) + (Y_{\alpha}Y_{-\alpha} + Y_{-\alpha}Y_{\alpha})^{h^{-1}} \\ &\qquad -(e^c + e^{-c})(Y_{\alpha}^{h^{-1}}Y_{-\alpha} + Y_{-\alpha}^{h^{-1}}Y_{\alpha})\} \\ &\quad +2e^c(e^c - e^{-c})^{-1}\{[Y_{-\alpha}, Z_{\alpha}] - [Y_{\alpha}, Z_{-\alpha}]\}.\end{aligned}$$

Because both ω and $\omega_{\mathfrak{l}}$ are θ-stable, the same must be true of the quantity $\sum_{\alpha\in P_+}(X_{\alpha}X_{-\alpha} + X_{-\alpha}X_{\alpha})$ – hence

$$\begin{aligned}&-\sum_{\alpha\in P_+}(X_{\alpha}X_{-\alpha} + X_{-\alpha}X_{\alpha}) \\ &= -2^{-1}\sum_{\alpha\in P_+}\{X_{\alpha}X_{-\alpha} + X_{-\alpha}X_{\alpha} + \theta(X_{\alpha}X_{-\alpha} + X_{-\alpha}X_{\alpha})\} \\ &= \sum_{\alpha\in P_+}4(e^{\alpha(H)} - e^{-\alpha(H)})^{-2}\{(Y_{\alpha}Y_{-\alpha} + Y_{-\alpha}Y_{\alpha}) + (Y_{\alpha}Y_{-\alpha} + Y_{-\alpha}Y_{\alpha})^{h^{-1}} \\ &\qquad -(e^{\alpha(H)} + e^{-\alpha(H)})(Y_{\alpha}^{h^{-1}}Y_{-\alpha} + Y_{-\alpha}^{h^{-1}}Y_{\alpha})\} \\ &\quad + \sum_{\alpha\in P_+}\coth(\alpha(H))\{[Y_{-\alpha}, Z_{\alpha}] - [Y_{\alpha}, Z_{-\alpha}]\} \qquad (H = \log h).\end{aligned}$$

Since $[X_{\alpha}, X_{-\alpha}] = H_{\alpha}$, it must be the case that $[Y_{\alpha}, Z_{-\alpha}] + [Z_{\alpha}, Y_{-\alpha}] = Q_{\alpha}$ – hence

$$\begin{aligned}\sum_{\alpha\in P_+}(X_{\alpha}X_{-\alpha} + X_{-\alpha}X_{\alpha}) = \sum_{\alpha\in P_+}\coth(\alpha(H))Q_{\alpha} - \sum_{\alpha\in P_+}4(e^{\alpha(H)} - e^{-\alpha(H)})^{-2} \\ \times\{(Y_{\alpha}Y_{-\alpha} + Y_{-\alpha}Y_{\alpha}) + (Y_{\alpha}Y_{-\alpha} + Y_{-\alpha}Y_{\alpha})^{h^{-1}} \\ -(e^{\alpha(H)} + e^{-\alpha(H)})(Y_{\alpha}^{h^{-1}}Y_{-\alpha} + Y_{-\alpha}^{h^{-1}}Y_{\alpha})\}.\end{aligned}$$

It is obvious that Y_{α} and $Y_{-\alpha}$ are in $\mathfrak{q}_c$; furthermore

$$Y_{-\alpha\theta}Y_{\alpha\theta} = Y_{-\alpha}Y_{\alpha} \qquad (\text{all } \alpha\in P_+).$$

[We know that the map $\alpha\mapsto -\alpha^{\theta}$ is a permutation of P_+; thus $X_{-\alpha\theta} = c_{-\alpha}(\theta X_{-\alpha})$ and $X_{\alpha\theta} = c_{\alpha}(\theta X_{\alpha})$ $(c_{\alpha}, c_{-\alpha}\in\mathbf{C})$ for $\alpha\in P_+$ – moreover $c_{\alpha}c_{-\alpha} = 1$ $(\alpha\in P_+)$. But this implies that

$$Y_{-\alpha\theta}Y_{\alpha\theta} = c_{\alpha}c_{-\alpha}Y_{-\alpha}Y_{\alpha} = Y_{-\alpha}Y_{\alpha}\ldots.]$$

Since $\alpha = -\alpha^{\theta}$ on $\mathfrak{a}_{\mathfrak{p}_c}$, we then deduce that

$$\begin{aligned}\wp_{A_{\mathfrak{p}}}\Big(\sum_{\alpha\in P_+}(X_{\alpha}X_{-\alpha} + X_{-\alpha}X_{\alpha})\Big) = \sum_{\alpha\in P_+}\coth(\alpha)Q_{\alpha} \\ -2\sum_{\alpha\in P_+}(\sinh(\alpha))^{-2}\{(1\otimes 1\otimes Y_{\alpha}Y_{-\alpha}) + (Y_{\alpha}Y_{-\alpha}\otimes 1\otimes 1)\} \\ +4\sum_{\alpha\in P_+}(\sinh(\alpha))^{-1}\coth(\alpha)(Y_{\alpha}\otimes 1\otimes Y_{-\alpha})\end{aligned}$$

and from this the assertion of our proposition follows at once. □

Corollary 9.1.2.12 Retain the above notations and assumptions – then

$$\begin{aligned}\mathfrak{P}_{A_\mathfrak{p}}(\omega) &= \mathfrak{P}_{A_\mathfrak{p}}(\omega_\mathfrak{m}) + \gamma^*_{A_\mathfrak{p}}(\omega) + 2\sum_{\alpha\in P_+} e^{-2\alpha}(1-e^{-2\alpha})^{-1}Q_{\tilde{\alpha}}\\ &\quad -8\sum_{\alpha\in P_+} e^{-2\alpha}(1-e^{-2\alpha})^{-2}\{(1\otimes 1\otimes Y_\alpha Y_{-\alpha}) + (Y_\alpha Y_{-\alpha}\otimes 1\otimes 1)\}\\ &\quad +8\sum_{\alpha\in P_+} e^{-\alpha}(1+e^{-2\alpha})(1-e^{-2\alpha})^{-2}(Y_\alpha\otimes 1\otimes Y_{-\alpha}).\end{aligned}$$

[Here it is necessary to recall the definition of the map $\gamma^*_{A_\mathfrak{p}}$ (cf. number 6.2.2); in view of the known value for $\gamma^*_{A_\mathfrak{p}}(\omega)$, the corollary follows upon remarking that

$$\coth(\alpha) = 1 + 2e^{-2\alpha}(1-e^{-2\alpha})^{-1},\ (\sinh(\alpha))^{-2} = 4e^{-2\alpha}(1-e^{-2\alpha})^{-2},$$
$$(\sinh(\alpha))^{-1}\coth(\alpha) = 2e^{-\alpha}(1+e^{-2\alpha})(1-e^{-2\alpha})^{-2}.]$$

As in 9.1.1, let Σ denote the set of roots of the pair $(\mathfrak{g}, \mathfrak{a}_\mathfrak{p})$, Σ^+ the set of positive roots singled out by our choice of the Iwasawa decomposition for $\mathfrak{g}$; given $\lambda \in \Sigma$, let $\Sigma(\lambda) = \{\alpha \in \Phi_+ : \tilde{\alpha} = \lambda\}$.

Corollary 9.1.2.13 Retain the above notations and assumptions – then, for any $\lambda \in \Sigma^+$, we have

$$\sum_{\alpha\in\Sigma(\lambda)} (Y^m_\alpha \otimes 1 \otimes Y^m_{-\alpha}) = \sum_{\alpha\in\Sigma(\lambda)} (Y_\alpha \otimes 1 \otimes Y_{-\alpha}) \qquad (\text{all } m \in M).$$

Proof Put

$$f_\lambda = e^{-\lambda}(1+e^{-2\lambda})(1-e^{-2\lambda})^{-2} \quad (\lambda \in \Sigma^+).$$

Then the f_λ are analytic functions on $A'_\mathfrak{p}$ which are, moreover, linearly independent. [To see this, suppose to the contrary that we had a dependence relation $\sum_\lambda c_\lambda f_\lambda = 0$ where not all the c_λ are zero ($c_\lambda \in \mathbf{C}$); let λ_0 be the lowest element in Σ^+ such that $c_{\lambda_0} \neq 0$. On exp ($\mathscr{C}$) we have the expansion

$$f_\lambda = \sum_{n\geqslant 0} (2n+1)e^{-(2n+1)\lambda} \qquad (\lambda \in \Sigma^+).$$

Such a series can be regarded as a power series in $e^{-\lambda_1}, \ldots, e^{-\lambda_l}$ which converges on $\mathscr{C}_{c}$. By equating the coefficient of $e^{-\lambda_0}$ to zero in the corresponding series expansion of $\Sigma_\lambda c_\lambda f_\lambda$, we find that $c_{\lambda_0} = 0.\ldots$] In addition the functions 1 and $f_\lambda(\lambda \in \Sigma^+)$ are also linearly independent (since the f_λ tend to zero when the independent variable tends to infinity on $\mathscr{C}$). On the other hand $\omega^m = \omega$ ($m \in M$); it then follows without difficulty that

$$\mathfrak{P}_{A_\mathfrak{p},h}(\omega) = (\mathfrak{P}_{A_\mathfrak{p},h}(\omega))^m \quad (m \in M)$$

for any $h \in A'_\mathfrak{p}$. Taking into account the linear independence of 1 and the $f_\lambda(\lambda \in \Sigma^+)$, the present corollary is then seen to be a consequence of the preceding one. □

Appendix Let $\mathfrak{M}$ be the subspace of $\mathfrak{A}$ spanned by all elements of the form

$$D_1 M \otimes D \otimes D_2 - D_1 \otimes D \otimes M D_2 \quad (D_1, D_2 \in \mathfrak{K};\, D \in \mathfrak{D}_F(A_{\mathfrak{p}}');\, M \in \mathfrak{M}).$$

Then $\mathfrak{M}$ is a left ideal in $\mathfrak{A}$. Our objective here is to establish the following result.

Proposition Let $D_1, D_2 \in \mathfrak{X}$ – then

$$\mathfrak{P}_{A_{\mathfrak{p}}}(D_1 D_2) - \mathfrak{P}_{A_{\mathfrak{p}}}(D_1)\mathfrak{P}_{A_{\mathfrak{p}}}(D_2) \in \mathfrak{M}.$$

The proof of this proposition depends upon the lemma below.

Lemma Fix a point $h_0 \in A_{\mathfrak{p}}'$ and suppose that Δ_0 is an element in $\mathfrak{Q} \otimes_{\mathbf{C}} \mathfrak{A}_{\mathfrak{p}} \otimes_{\mathbf{C}} \mathfrak{K}$ with the following property: For every (finite dimensional) double representation $\mu = (\mu_1, \mu_2)$ of K and every C^∞ μ-spherical function Ψ on G, $\Psi(h_0; \Delta_0) = 0$. Then it must be the case that $\Delta_0 = 0$.

Granting the lemma, the proof of the proposition goes as follows. Let

$$\square = \mathfrak{P}_{A_{\mathfrak{p}}}(D_1 D_2) - \mathfrak{P}_{A_{\mathfrak{p}}}(D_1)\mathfrak{P}_{A_{\mathfrak{p}}}(D_2);$$

choose elements $Q_i \in \mathfrak{Q}$, $D_j \in \mathfrak{K}$, and

$$\square_{ij} \in \mathfrak{D}_F(A_{\mathfrak{p}}') \qquad (1 \leqslant i \leqslant r, 1 \leqslant j \leqslant s)$$

such that the sets $\{Q_1, \ldots, Q_r\}$, $\{D_1, \ldots, D_s\}$ are separately linearly independent and with

$$\square \equiv \sum_{i,j} Q_i \otimes \square_{ij} \otimes D_j \qquad \text{mod } \mathfrak{M}.$$

Let μ be an arbitrary (finite dimensional) double representation of K, Ψ any C^∞ μ-spherical function on G – then $\square\tilde{\Psi} = 0$ (cf. Corollary 9.1.2.10). On the other hand it is easy to verify that $\tilde{\Psi}$ is annihilated by all the elements of $\mathfrak{M}$. This being so, fix a point $h_0 \in A_{\mathfrak{p}}'$; let H_{ij} be the local expression of $\square_{ij}$ at h_0 – then $\square_0 = \sum_{i,j} Q_i \otimes H_{ij} \otimes D_j$ lies in $\mathfrak{Q} \otimes_{\mathbf{C}} \mathfrak{A}_{\mathfrak{p}} \otimes_{\mathbf{C}} \mathfrak{K}$ and, moreover, $\Psi(h_0; \square_0) = 0$. In view of the arbitrariness of μ and Ψ, we then conclude from the lemma supra that $\square_0 = 0$; but this implies that the local expression of $\square_{ij}$ at h_0 is zero, whence $\square_{ij} = 0$ (h_0 being arbitrary in $A_{\mathfrak{p}}'$). Therefore $\square \in \mathfrak{M}$, as desired.

Let us turn to the proof of the lemma. Suppose to the contrary that $\Delta_0 \neq 0$; put $D_0 = \Gamma_{h_0}(\Delta_0) \in \mathfrak{G}$ – then $D_0 \neq 0$ (since $h_0 \in A_{\mathfrak{p}}'$). Claim: If ψ is any C^∞ function on G which is K-finite to the left and to the right, then $\psi(h_0; D_0) = 0$. [Of course the lemma is a simple consequence of the claim since the space of all doubly K-finite C^∞ functions is dense in $C^\infty(G)$ while on the other hand the map $f \mapsto f(h_0; D_0)$ $(f \in C^\infty(G))$ is a distribution on G with compact support. . . .] In order to establish the claim, we begin by remarking that $C^\infty(K \times K)$ can be regarded as a double K-module in the following manner: Given $f \in C^\infty(K \times K)$, let $\mu_1(k)f$, $f\mu_2(k)$ $(k \in K)$ denote, respectively, the functions

$$(k_1, k_2) \mapsto f(k^{-1}k_1, k_2), \quad (k_1, k_2) \mapsto f(k_1, k_2 k^{-1}) \qquad (k_1, k_2 \in K).$$

Now fix a doubly K-finite C^∞ function ψ on G; for any $x \in G$, let $\Psi(x)$ denote the C^∞ function on $K \times K$ defined by the rule

$$(k_1, k_2) \mapsto \psi(k_1^{-1} x k_2^{-1}) \qquad (k_1, k_2 \in K).$$

Then it is clear that

$$\Psi(k_1 x k_2) = \mu_1(k_1)\Psi(x)\mu_2(k_2) \qquad (x \in G;\, k_1, k_2 \in K).$$

Let E be the subspace of $C^\infty(K \times K)$ spanned by the $\Psi(x)$ ($x \in G$) – then E is μ-stable and dim $(E) < \infty$, ψ being K-finite to the left and to the right. Otherwise said, Ψ is a C^∞ μ-spherical function from G to E. Therefore, in view of our hypotheses, we have $\Psi(h_0; \Delta_0) = 0$; but since $D_0 = \Gamma_{h_0}(\Delta_0)$, it is then clear that $\Psi(h_0; D_0) = 0$, whence $\psi(h_0; D_0) = 0$ which serves to establish the claim and prove the lemma.

9.1.3 The Space $\mathfrak{S}(\mu, \kappa)$

Let G be a connected semi-simple Lie group with finite center, $\mathfrak{g}$ the Lie algebra of G; let $\mathfrak{Z}$ denote the center of $\mathfrak{G}$ (= the universal enveloping algebra of $\mathfrak{g}_c$). Let $\mu = (\mu_1, \mu_2)$ be a double Banach representation of K on a finite dimensional Banach space E; let κ be a character of $\mathfrak{Z}$ – then by $\mathfrak{S}(\mu, \kappa)$ we shall understand the space of all C^∞ μ-spherical functions Ψ on G which verify the relation

$$Z\Psi = \kappa(Z)\Psi \quad (Z \in \mathfrak{Z}).$$

Proposition 9.1.3.1 (Harish-Chandra) Every function in the space $\mathfrak{S}(\mu, \kappa)$ is analytic; furthermore the space $\mathfrak{S}(\mu, \kappa)$ is finite dimensional – in fact dim $(\mathfrak{S}(\mu, \kappa)) \leqslant [W_\Phi : W_-]$ dim (E).

Proof That the elements of $\mathfrak{S}(\mu, \kappa)$ are analytic functions on G can be established by the usual argument (cf. the Appendix to 8.3.9 of Chapter 8); as for the dimension assertion, one may proceed as follows. Fix a point h_0 in $A'_\mathfrak{p}$; suppose that Ψ is an element of $\mathfrak{S}(\mu, \kappa)$ such that

$$\Psi(h_0; Z_i^\natural) = 0$$

($1 \leqslant i \leqslant r = [W_\Phi : W_-]$; cf. Theorem 9.1.2.4) – then $\Psi = 0$. [Since Ψ is analytic, we need only show that $\Psi(h_0; D) = 0$ (all $D \in \mathfrak{G}$); on the basis of Corollary 9.1.2.5, however, this is obvious (Ψ being μ-spherical).] Consequently any Ψ in $\mathfrak{S}(\mu, \kappa)$ which satisfies the r linear conditions $\Psi(h_0; Z_i^\natural) = 0$ must necessarily vanish identically, whence

$$\dim (\mathfrak{S}(\mu, \kappa)) \leqslant [W_\Phi : W_-] \dim (E),$$

as contended. □

Here is the main result of the present number (cf. with Theorem 9.1.1.1).

Theorem 9.1.3.2 (Harish-Chandra) Fix an element Ψ in $\mathfrak{S}(\mu, \kappa)$ – then there exists an at most countable set $\mathscr{E}(\Psi)$ of linear functions on $\mathfrak{a}_{\mathfrak{p}_c}$ and, for each $\lambda \in \mathscr{E}(\Psi)$, a polynomial function $p_\lambda : \mathfrak{a}_{\mathfrak{p}_c} \to E_M$ such that the following conditions are fulfilled:

(i) For every $\lambda \in \mathscr{E}(\Psi)$, $p_\lambda \neq 0$;

(ii) The series

$$\sum_{\lambda \in \mathscr{E}(\Psi)} p_\lambda e^{\langle \cdot, \lambda \rangle}$$

converges decently throughout the region

$$\mathfrak{a}_{\mathfrak{p}_c}(\epsilon, \eta) = \{H \in \mathfrak{a}_{\mathfrak{p}_c} : \mathscr{R}(\lambda_i(H)) > \max(\eta, \epsilon\,|\lambda_1(H)|, \ldots, \epsilon\,|\lambda_l(H)|)\}$$
$$(0 < \epsilon < 1, \eta > 0);$$

(iii) The expansion

$$\Psi(\exp H) = \sum_{\lambda \in \mathscr{E}(\Psi)} p_\lambda(H) e^{\lambda(H)}$$

holds for all $H \in \mathscr{C}$.

Moreover the set $\mathscr{E}(\Psi)$ and the polynomials $p_\lambda(\lambda \in \mathscr{E}(\Psi))$ are uniquely determined by these three conditions.

Proof It is a question here of applying Theorems A3.1.1 and A3.1.2 in Appendix 3. So let H be an element of $\mathfrak{a}_\mathfrak{p}$ and choose, using Theorem 9.1.2.4, elements D_{ij} in

$$\Big(\sum_{i=1}^{r} \mathfrak{Z} Z_i^\natural\Big)\mathfrak{K} \qquad (0 \leqslant j \leqslant s),$$

elements Q_{ij} in $\mathfrak{Q}$ and functions f_{ij} in $\mathfrak{F}_0$ $(1 \leqslant j \leqslant s)$ such that

$$HZ_i^\natural = D_{i0} + \sum_{j=1}^{s} f_{ij}(h) Q_{ij}^{h^{-1}} D_{ij} \qquad (\text{all } h \in A_\mathfrak{p}').$$

Put $\Psi_i = Z_i^\natural \Psi (1 \leqslant i \leqslant r)$ – then

$$\Psi_i(h; H) = \Psi(h; HZ_i^\natural) = \Psi(h; D_{i0}) + \sum_{j=1}^{s} f_{ij}(h)\mu_1(Q_{ij})\Psi(h; D_{ij})$$

for all $h \in A_\mathfrak{p}'$. On the other hand, suppose that $Z \in \mathfrak{Z}$, $D \in \mathfrak{K}$ – then it is clear that

$$\Psi(x; ZZ_i^\natural D) = \kappa(Z)(\Psi(x; Z_i^\natural)\mu_2(D)) \qquad (\text{all } x \in G).$$

We see, therefore, that there exist linear maps ψ_i $(1 \leqslant i \leqslant r)$ of $(\sum_{i=1}^r \mathfrak{Z} Z_i^\natural)\mathfrak{K}$ into $\mathrm{Hom}_{\mathbf{C}}(E, E)$ with the property that

$$\Psi(x; D) = \sum_{i=1}^{r} \psi_i(D)\Psi_i(x) \qquad \Big(\text{all } D \in \Big(\sum_{i=1}^{r} \mathfrak{Z} Z_i^\natural\Big)\mathfrak{K}\Big).$$

Consequently

$$\Psi_i(h; H) = \sum_{k=1}^{r} \psi_k(D_{i0})\Psi_k(h) + \sum_{j=1}^{s}\sum_{k=1}^{r} f_{ij}(h)\mu_1(Q_{ij})(\psi_k(D_{ij})\Psi_k(h))$$
$$(\text{all } h \in A_\mathfrak{p}').$$

Now let E^r denote the direct sum of E with itself r times and define a function $\mathbf{\Psi}$ on G with values in E^r by the rule

$$\mathbf{\Psi}(x) = (\Psi_1(x), \ldots, \Psi_r(x)) \qquad (x \in G).$$

Introduce the basis $\{H_p\}$ for $\mathfrak{a}_\mathfrak{p}$ per 9.1.1 and fix some basis $\{T_q\}$ (say) for $\mathrm{Hom}_{\mathbf{C}}(E^r, E^r)$ – then, in view of what has been said so far, we see that it is possible to choose functions f_{pq} in $\mathfrak{F}$ with the property that

$$\mathbf{\Psi}(h; H_p) = \mathbf{\Gamma}_p(h)\mathbf{\Psi}(h) \qquad (\text{all } h \in A_\mathfrak{p}')$$

where $\mathbf{\Gamma}_p(h) = \sum_q f_{pq}(h) T_q$ $(1 \leqslant p \leqslant l)$. But the very definition of $\mathfrak{F}$ implies that for any given $f \in \mathfrak{F}$, there exist complex numbers $c(\lambda)$ $(\lambda \in \mathsf{L})$ such that:

(1) $\sum_{\lambda \in \mathsf{L}} |c(\lambda)| \, e^{-m_\lambda \eta} < \infty$ (all $\eta > 0$);

(2) $f(\exp H) = \sum_{\lambda \in \mathsf{L}} c(\lambda) e^{-\lambda(H)}$ (all $H \in \mathscr{C}$).

Therefore Theorems A3.1.1 and A3.1.2 in Appendix 3 are applicable to the function $H \mapsto \mathbf{\Psi}(\exp H)$ $(H \in \mathscr{C})$. All the assertions of the present theorem are now clear, $Z_1^{\natural}$ being equal to 1. □

Let $\mathfrak{X}$ denote the centralizer of $\mathfrak{k}_c$ in $\mathfrak{G}$. We shall now define a linear mapping Ω of $\mathfrak{X}$ into the subalgebra $1 \otimes_{\mathbf{C}} \mathfrak{A}_{\mathfrak{p}} \otimes_{\mathbf{C}} \mathfrak{K}$ of $\mathbf{\mathfrak{A}}$ as follows. Given $D \in \mathfrak{X}$, there exists a unique element $D_0 \in \mathfrak{A}_{\mathfrak{p}}\mathfrak{K}$ such that $D - D_0 \in \mathfrak{n}_c^- \mathfrak{G}$: Put $\Omega(D) = \mathfrak{f}_{A\mathfrak{p}}(D_0)$ $(D \in \mathfrak{X})$. If λ is a linear function on $\mathfrak{a}_{\mathfrak{p}_c}$, let us agree to denote by $\mathfrak{e}^{\lambda}$ the anti-homomorphism of $1 \otimes_{\mathbf{C}} \mathfrak{A}_{\mathfrak{p}} \otimes_{\mathbf{C}} \mathfrak{K}$ into $\mathfrak{K}$ such that

$$\langle 1 \otimes H \otimes D, \mathfrak{e}^{\lambda} \rangle = \langle H, \mathbf{e}^{\lambda} \rangle D = \lambda(H) D \quad (H \in \mathfrak{a}_{\mathfrak{p}_c}, D \in \mathfrak{K}).$$

Given $D \in \mathfrak{X}$, we then write $\Omega(D : \lambda)$ for $\langle \Omega(D), \mathfrak{e}^{\lambda} \rangle$; it is not difficult to check that, for fixed λ, the mapping $D \mapsto \Omega(D : \lambda)$ of $\mathfrak{X}$ into $\mathfrak{K}$ is an anti-homomorphism.

Let us now return to the study of the space $\mathfrak{S}(\mu, \kappa)$; we have seen above that it is a finite dimensional space which is obviously $\mathfrak{X}$-stable.

Proposition 9.1.3.3 Let $\mathscr{S}$ be an $\mathfrak{X}$-stable subspace of $\mathfrak{S}(\mu, \kappa)$ which is irreducible as an $\mathfrak{X}$-module; let $\mathbf{\Psi}_1, \mathbf{\Psi}_2$ be two non-zero elements in $\mathscr{S}$ – then the leading exponents and the degrees of the leading terms of $\mathbf{\Psi}_1$ and $\mathbf{\Psi}_2$ are the same.

Corollary 9.1.3.4 Retain the assumptions and notations of the preceding proposition; let Λ denote the leading exponent of any non-zero element in $\mathscr{S}$ and suppose that D is an element of $\mathfrak{X}$ such that

$$\mu_2(\Omega(D : \Lambda)) = 0.$$

Then $D\mathbf{\Psi} = 0$ for every $\mathbf{\Psi} \in \mathscr{S}$.

The proofs depend on the following lemma.

Lemma 9.1.3.5 Fix a non-zero element $\mathbf{\Psi}$ in $\mathfrak{S}(\mu, \kappa)$ and an element D in $\mathfrak{X}$; let Λ be the leading exponent of $\mathbf{\Psi}$, Λ_D the leading exponent of $D\mathbf{\Psi}$; let d and d_D be the degrees of $\mathbf{\Psi}^L$ and $(D\mathbf{\Psi})^L$, respectively – we have:

(i) If $\mathbf{\Psi}^L \mu_2(\Omega(D : \Lambda)) \neq 0$, then $D\mathbf{\Psi} \neq 0$ and $(D\mathbf{\Psi})^L = \mathbf{\Psi}^L \mu_2(\Omega(D : \Lambda))$;

(ii) If $\mathbf{\Psi}^L \mu_2(\Omega(D : \Lambda)) = 0$, but $D\mathbf{\Psi} \neq 0$, then $\Lambda_D \leqslant \Lambda$ and $\Lambda_D = \Lambda \Rightarrow d_D < d$.

Proof Thanks to Theorem 9.1.2.9, we have $(D\Psi)^{\sim} = \gamma_{A_{\mathfrak{p}}}(D)\tilde{\Psi}$; moreover, on the basis of Lemma 9.1.2.7 and its corollary, we can choose elements f_j in $\mathfrak{F}_0$ and

$$\Delta_j \in \mathfrak{Q} \otimes_{\mathbf{C}} \mathfrak{A}_{\mathfrak{p}} \otimes_{\mathbf{C}} \mathfrak{K} \qquad (1 \leqslant j \leqslant s)$$

such that

$$\gamma_{A_{\mathfrak{p}}}(D) = \Omega(D) + \sum_j f_j \Delta_j.$$

Hence

$$(D\Psi)^{\sim} = \Omega(D)\tilde{\Psi} + \sum_{j=1}^{s} f_j \Delta_j \tilde{\Psi}.$$

One may legitimately compare coefficients of the various exponents on both sides of this equation. [Every $f \in \mathfrak{F}_0$ can be written in the form

$$f(\exp H) = \sum_{\lambda \in \mathsf{L}} c(\lambda) e^{-\lambda(H)} \qquad (H \in \mathscr{C})$$

where $c(\lambda) \in \mathbf{C}$, $c(0) = 0$ and the series converges decently on $\mathfrak{a}_{\mathfrak{p}_c}(\epsilon, \eta)$ for all ϵ, η $(0 < \epsilon < 1;\ \eta > 0)$; it therefore follows from Theorem 9.1.3.2 (cf., however, Appendix 3, A3.5) that if Δ is any element in $\mathfrak{K} \otimes_{\mathbf{C}} \mathfrak{A}_{\mathfrak{p}} \otimes_{\mathbf{C}} \mathfrak{K}$, then the series which arises upon multiplying the above series for f with the series obtained by applying Δ term for term to the expansion of $\tilde{\Psi}$ still converges decently on $\mathfrak{a}_{\mathfrak{p}_c}(\epsilon, \eta)$ and its sum is equal to $f\Delta\tilde{\Psi}$.] Consequently, if Ψ^H denotes the highest term of Ψ and $(D\Psi)^H$ the highest term of $D\Psi$ (when $D\Psi \neq 0$), then the term on the right hand side of our equation corresponding to the exponent Λ is $\Omega(D)\Psi^H$ and every other term has a lower exponent (the exponents of f_j being negative ...) – thus

$$\Omega(D)\Psi^H \neq 0 \Rightarrow D\Psi \neq 0$$

and

$$(D\Psi)^H = \Omega(D)\Psi^H.$$

But

$$\Omega(D)\Psi^H = \Psi^L \mu_2(\Omega(D:\Lambda)) + pe^{\Lambda}$$

where p is a polynomial of degree less than d – thus

$$\Psi^L \mu_2(\Omega(D:\Lambda)) \neq 0 \Rightarrow D\Psi \neq 0$$

and

$$(D\Psi)^L = \Psi^L \mu_2(\Omega(D:\Lambda)),$$

proving (i). Now suppose that $\Psi^L \mu_2(\Omega(D:\Lambda)) = 0$ – then $\Omega(D)\Psi^H = pe^{\Lambda}$. If $p \neq 0$, then $(D\Psi)^H = pe^{\Lambda}$, whence $\Lambda_D = \Lambda$ and $d_D < d$; if $p = 0$ (and $D\Psi \neq 0$), then $\Omega(D)\Psi^H = 0$ – thus all the exponents on the right hand side of our equation are lower than Λ and so $\Lambda_D < \Lambda$ in this case. All the assertions of statement (ii) have now been established.... ☐

Proof of Proposition 9.1.3.3 Let Λ_1, Λ_2 be the leading exponents and d_1, d_2 the degrees of the leading terms of Ψ_1, Ψ_2, respectively. Since $\mathscr{S}$ is irreducible as an $\mathfrak{X}$-module, there exists a $D \in \mathfrak{X}$ such that $\Psi_2 = D\Psi_1$;

it then follows from Lemma 9.1.3.5 that $\Lambda_2 \leqslant \Lambda_1$. In a similar way we see that $\Lambda_1 \leqslant \Lambda_2$ – therefore $\Lambda_1 = \Lambda_2$. Quoting Lemma 9.1.3.5 once again, we derive that $d_2 \leqslant d_1$ and so, by symmetry, $d_1 = d_2$. □

Proof of Corollary 9.1.3.4 Suppose to the contrary that $D\Psi \neq 0$ for some $\Psi \in \mathscr{S}$ – then, since Ψ and $D\Psi$ have the same leading exponent, it follows from Lemma 9.1.3.5 that $d_D < d$. This, however, contradicts Proposition 9.1.3.3. □

Given $D \in \mathfrak{X}$, there exists a unique element D_0 in $\mathfrak{A}_\mathfrak{p}\mathfrak{K}$ such that $D - D_0 \in \mathfrak{n}_c^+\mathfrak{G}$: Put

$$\omega(D) = e^{-\rho}\,\gamma_{A_\mathfrak{p}}(D_0)\circ e^{\rho} \qquad (D \in \mathfrak{X}).$$

If λ is a linear function on $\mathfrak{a}_{\mathfrak{p}_c}$, we shall then agree to write $\omega(D:\lambda)$ for $\langle \omega(D), e^{\lambda}\rangle$ $(D \in \mathfrak{X})$; in passing let us note that, for fixed λ, the mapping $D \mapsto \omega(D:\lambda)$ of $\mathfrak{X}$ into $\mathfrak{K}$ is an anti-homomorphism.

Application Let U be a TCI Banach representation of G on a Banach space E; fix a class $\delta \in \hat{K}$ which occurs in $U|K$ – then there exists a linear function λ on $\mathfrak{a}_{\mathfrak{p}_c}$ with the property that

$$\omega(D:\lambda) \in \operatorname{Ker}(\mu_\delta) \Rightarrow D \in \operatorname{Ker}(U_\delta) \qquad (D \in \mathfrak{X}).$$

[Here we agree to employ the customary notations – thus μ_δ denotes some fixed element in the class δ while U_δ denotes the representation of K (and hence of $\mathfrak{K}$) on $E(\delta)$ obtained from $U|K$ in the canonical manner. Let us recall that the

$$U_\delta(D)(=\mathsf{P}(\delta)U_K(D)\mathsf{P}(\delta)) \qquad (D \in \mathfrak{X})$$

comprise the full centralizer of the set $\{U_\delta(k) : k \in K\}$ (cf. number 6.1.2); in particular, then, the algebra $U_\delta(\mathfrak{X})$ is simple.] To see this, let us introduce the μ_U-spherical function Ψ_δ^U (μ_U the usual double representation of K on $\mathrm{Hom}_\mathrm{C}(E(\delta), E(\delta))$) – then clearly $\Psi_\delta^U \in \mathfrak{S}(\mu_U, \kappa_U)$, κ_U the infinitesimal character of U (cf. number 6.1.2). Let $\mathfrak{T}$ be the subspace of $\mathfrak{S}(\mu_U, \kappa_U)$ consisting of all elements of the form $D\Psi_\delta^U$ $(D \in \mathfrak{X})$ – then $\dim(\mathfrak{T}) < \infty$ (cf. Proposition 9.1.3.1). Given $D \in \mathfrak{X}$, let $\tau(D)$ denote the mapping $\Psi \mapsto D\Psi$ of $\mathfrak{T}(\Psi \in \mathfrak{T})$. Evidently the assignment $D \mapsto \tau(D)$ $(D \in \mathfrak{X})$ defines a representation of $\mathfrak{X}$ on $\mathfrak{T}$; on the other hand τ can also be regarded as a representation of the simple algebra $U_\delta(\mathfrak{X})$ (since $U_\delta(D) = 0$ (some $D \in \mathfrak{X}$) $\Rightarrow \tau(D) = 0 \ldots$) – hence τ is semi-simple and all its irreducible components are equivalent. This being so, let $\mathscr{T}$ be a non-zero $\mathfrak{X}$-stable subspace of $\mathfrak{T}$ which is irreducible as an $\mathfrak{X}$-module; reversing the order in the space of real linear functions on $\mathfrak{a}_\mathfrak{p}$, Corollary 9.1.3.4 then provides us with a linear function λ on $\mathfrak{a}_{\mathfrak{p}_c}$ which possesses the following property: If D is any element in $\mathfrak{X}$ such that $TU_\delta(\omega(D:\lambda)) = 0$ (all $T \in \mathrm{Hom}_\mathrm{C}(E(\delta), E(\delta))$), then $D\Psi = 0$ (all $\Psi \in \mathscr{T}$). Because the irreducible components of τ are equivalent, the condition

$$TU_\delta(\omega(D:\lambda)) = 0 \qquad (\text{all } T \in \mathrm{Hom}_\mathrm{C}(E(\delta), E(\delta)))$$

actually implies that $D\Psi = 0$ (all $\Psi \in \mathfrak{T}$) – therefore

$$\begin{aligned} \omega(D:\lambda) \in \operatorname{Ker}(\mu_\delta) \Rightarrow TU_\delta(\omega(D:\lambda)) &= 0 \\ &\qquad (\text{all } T \in \operatorname{Hom}_{\mathbf{C}}(E(\delta), E(\delta))) \\ \Rightarrow D\Psi = 0 &\qquad (\text{all } \Psi \in \mathfrak{T}) \\ \Rightarrow D\Psi_\delta^U = 0 \Rightarrow \Psi_\delta^U(1)U_\delta(D) &= 0 \\ \Rightarrow U_\delta(D) = 0 \Rightarrow D \in \operatorname{Ker}(U_\delta) &\qquad (D \in \mathfrak{X}), \end{aligned}$$

as desired.

9.1.4 The Rational Functions Γ_λ

The primary objective of the present number is to construct, on suitable subsets of $A_\mathfrak{p}$, 'eigenfunctions' for the operators $\wp_{A_\mathfrak{p}}(D)$ $(D \in \mathfrak{X})$. In 9.1.5 we shall express the μ-spherical functions which arise in representation theory as linear combinations of them.

We shall begin by formulating, in a precise fashion, the result toward which we wish to head.

Let G be a connected semi-simple Lie group with finite center, $G = KA_\mathfrak{p}N^+$ an Iwasawa decomposition for G; let Σ denote the set of roots of the pair $(\mathfrak{g}, \mathfrak{a}_\mathfrak{p})$ and assign to the symbols $\lambda_1, \ldots, \lambda_l$, ρ, $\mathscr{C}$, and L their customary meanings. We agree to adhere to the usual conventions as regards the identification of $\mathfrak{a}_{\mathfrak{p}_c}$ and its dual – *thus a function on $\mathfrak{a}_{\mathfrak{p}_c}$ automatically becomes a function on $\mathfrak{a}_{\mathfrak{p}_c}^\vee$ without a change in notation.*

Note Typically H denotes the generic element in $\mathfrak{a}_{\mathfrak{p}_c}$, ν the generic element in $\mathfrak{a}_{\mathfrak{p}_c}^\vee$; in this number (and the next) it will often be convenient to write $\langle \nu, H\rangle$ rather than $\langle H, \nu\rangle$ for the value $\nu(H)$ of ν at H.

Let $\mu = (\mu_1, \mu_2)$ be a double unitary representation of K on a finite dimensional Hilbert space E; let E_M denote the subspace of E comprised of those elements a which have the property that

$$\mu_1(m)a = a\mu_2(m) \qquad (\text{all } m \in \mathsf{M}).$$

Given $T \in \operatorname{Hom}_{\mathbf{C}}(E_\mathsf{M}, E_\mathsf{M})$, put

$$\gamma(T) = [\mu_2(\omega_\mathfrak{m}), T].$$

Then the assignment $T \mapsto \gamma(T)$ determines an endomorphism γ of $\operatorname{Hom}_{\mathbf{C}}(E_\mathsf{M}, E_\mathsf{M})$ which may be extended by linearity over $Q(\mathfrak{a}_{\mathfrak{p}_c})$ to an endomorphism of

$$Q(\mathfrak{a}_{\mathfrak{p}_c}) \otimes_{\mathbf{C}} \operatorname{Hom}_{\mathbf{C}}(E_\mathsf{M}, E_\mathsf{M}) = E_{\mathsf{M},Q}$$

(say). [We agree to denote by $Q(\mathfrak{a}_{\mathfrak{p}_c})$ the quotient field of $S(\mathfrak{a}_{\mathfrak{p}_c})$.]

We come now to the recursive definition of the rational functions Γ_λ $(\lambda \in \mathsf{L})$ (the heuristics lying behind this rather complicated construction appear below). For each $\lambda \in \mathsf{L}$ we shall define an element $\Gamma_\lambda \in E_{\mathsf{M},Q}$ by induction on m_λ as follows: If $\lambda = 0$, set $\Gamma_\lambda = 1$; if $\lambda \neq 0$, then Γ_λ

is to be given by the relation

$$\{2\lambda - (\lambda, \lambda - 2\rho)\}\mathbf{\Gamma}_\lambda - \gamma(\mathbf{\Gamma}_\lambda) = 2\sum_{\alpha\in P_+}\sum_{n\geqslant 1}\{\tilde{\alpha} - (\tilde{\alpha}, \lambda - 2n\tilde{\alpha})\}\mathbf{\Gamma}_{\lambda-2n\tilde{\alpha}}$$
$$+ 8\sum_{\alpha\in P_+}\sum_{n\geqslant 1}(2n-1)\mu_1(Y_\alpha)\mu_2(Y_{-\alpha})\mathbf{\Gamma}_{\lambda-(2n-1)\tilde{\alpha}}$$
$$- 8\sum_{\alpha\in P_+}\sum_{n\geqslant 1}n\{\mu_1(Y_\alpha Y_{-\alpha}) + \mu_2(Y_\alpha Y_{-\alpha})\}\mathbf{\Gamma}_{\lambda-2n\tilde{\alpha}}.$$

Here n runs over all positive integers while $\mathbf{\Gamma}_\lambda$ is to be set equal to zero in the event that $\lambda \notin \mathsf{L}$ (thus all the sums that appear above are actually finite). To verify that this relation really does define (uniquely) an element $\mathbf{\Gamma}_\lambda \in E_{\mathsf{M},Q}$ for each $\lambda \in \mathsf{L}$ one may reason as follows. In view of Corollary 9.1.2.13, E_M is stable under $\sum_{\alpha\in\Sigma(\lambda)}\mu_1(Y_\alpha)\mu_2(Y_{-\alpha})$, $\sum_{\alpha\in\Sigma(\lambda)}\mu_1(Y_\alpha Y_{-\alpha})$, and $\sum_{\alpha\in\Sigma(\lambda)}\mu_2(Y_\alpha Y_{-\alpha})$ for any $\lambda \in \Sigma^+$; therefore the induction hypothesis ensures us that the right hand side of the above equation is in $E_{\mathsf{M},Q}$. On the other hand let $\mathbf{\Lambda}_\lambda$ ($\lambda \in \mathsf{L}$; $\lambda \neq 0$) denote the endomorphism of the vector space $E_{\mathsf{M},Q}$ (over $Q(\mathfrak{a}_{\mathfrak{p}_c})$) defined by the rule

$$\mathbf{\Lambda}_\lambda(f) = \{2\lambda - (\lambda, \lambda - 2\rho)\}f - \gamma(f) \qquad (f \in E_{\mathsf{M},Q}).$$

Then $\mathbf{\Lambda}_\lambda$ is non-singular and so the recursive definition of the $\mathbf{\Gamma}_\lambda$ is indeed meaningful. [To verify that $\mathbf{\Lambda}_\lambda$ ($\lambda \in \mathsf{L}$; $\lambda \neq 0$) is non-singular, suppose that $\mathbf{\Lambda}_\lambda(f) = 0$ ($f \in E_{\mathsf{M},Q}$) – then we are to show that $f = 0$. For this purpose, we may multiply by a non-zero element in $S(\mathfrak{a}_{\mathfrak{p}_c})$ and assume that f is a polynomial function on $\mathfrak{a}_{\mathfrak{p}_c}$ with values in $\mathrm{Hom}_\mathbf{C}(E_\mathsf{M}, E_\mathsf{M})$ – then, for any $H \in \mathfrak{a}_{\mathfrak{p}_c}$,

$$\{2\lambda(H) - (\lambda, \lambda - 2\rho)\}f(H) - \gamma(f(H)) = 0$$

where $f(H)$ is the value of f at H. Let $\gamma_1, \ldots, \gamma_r$ be all the distinct eigenvalues of the endomorphism γ of $\mathrm{Hom}_\mathbf{C}(E_\mathsf{M}, E_\mathsf{M})$ and $m_1, \ldots, m_r$ their respective multiplicities; set

$$p_\lambda = \prod_{i=1}^r \{2\lambda - (\lambda, \lambda - 2\rho) - \gamma_i\}^{m_i}.$$

Because $\lambda \neq 0$, p_λ is a non-zero element in $S(\mathfrak{a}_{\mathfrak{p}_c})$; moreover if H is a point in $\mathfrak{a}_{\mathfrak{p}_c}$ where $p_\lambda(H) \neq 0$, then the linear transformation

$$2\lambda(H) - (\lambda, \lambda - 2\rho) - \gamma$$

of $\mathrm{Hom}_\mathbf{C}(E_\mathsf{M}, E_\mathsf{M})$ is non-singular – therefore

$$p_\lambda(H) \neq 0 \Rightarrow f(H) = 0,$$

whence $f = 0$. . . .]

Let L' denote the set of non-zero elements in L; define $\gamma_1, \ldots, \gamma_r$ as above and for any $\lambda \in \mathsf{L}'$ and i ($1 \leqslant i \leqslant r$), let $\tau_{\lambda,i}$ denote the hyperplane consisting of those points H in $\mathfrak{a}_{\mathfrak{p}_c}$ such that $2\lambda(H) = (\lambda, \lambda - 2\rho) + \gamma_i$ – then it is clear that any compact subset of $\mathfrak{a}_{\mathfrak{p}_c}$ meets $\bigcup_{i=1}^r \tau_{\lambda,i}$ for only a finite number of $\lambda \in \mathsf{L}'$. Let $\mathbf{\Gamma}$ denote the complement of $\bigcup_{\lambda\in\mathsf{L}'}\bigcup_i \tau_{\lambda,i}$ in $\mathfrak{a}_{\mathfrak{p}_c}$ – then $\mathbf{\Gamma}$ is an open, connected, dense subset of $\mathfrak{a}_{\mathfrak{p}_c}$ (or $\mathfrak{a}_{\mathfrak{p}_c}^\vee$ . . .) and the rational functions $\mathbf{\Gamma}_\lambda$ (regarded as functions on $\mathfrak{a}_{\mathfrak{p}_c}$ (or $\mathfrak{a}_{\mathfrak{p}_c}^\vee$ . . .) with values in $\mathrm{Hom}_\mathbf{C}(E_\mathsf{M}, E_\mathsf{M})$) are well-defined on $\mathbf{\Gamma}$ for all $\lambda \in \mathsf{L}$.

Theorem 9.1.4.1 (Harish-Chandra) Fix $\nu \in \mathbf{\Gamma}$; put

$$\mathbf{\Psi}(\nu : h) = e^{\langle \nu, H \rangle} \sum_{\lambda \in \mathsf{L}} \mathbf{\Gamma}_\lambda(\nu) e^{-\lambda(H)} \qquad (h \in \exp(\mathscr{C}))$$

where $H = \log h$ – then the function $h \mapsto \mathbf{\Psi}(\nu : h)$ is analytic on $\exp(\mathscr{C})$ and we have

$$\mathbf{\Psi}(\nu : h ; \mathfrak{f}_{A\mathfrak{p}}(D)) = \mathbf{\Psi}(\nu : h)\mu_2(\Omega(D : \nu)) \qquad (\text{all } D \in \mathfrak{X}).$$

On the other hand, for fixed $h \in \exp(\mathscr{C})$, the function $\nu \mapsto \mathbf{\Psi}(\nu : h)$ is holomorphic on $\mathbf{\Gamma}$.

Note It makes sense to apply the differential operator $\mathfrak{f}_{A\mathfrak{p}}(D)$ $(D \in \mathfrak{X})$ to $\mathbf{\Psi}$. Thus let $\mathfrak{T}$ be the space of all linear mappings of E_{M} into E; given $T \in \mathfrak{T}$, write

$$\|T\| = \sup_{\|a\| \leqslant 1} \|Ta\| \qquad (a \in E_{\mathsf{M}})$$

– then, under this norm, $\mathfrak{T}$ becomes a Banach space of which $\mathrm{Hom}_{\mathbf{C}}(E_{\mathsf{M}}, E_{\mathsf{M}})$ can be regarded as a subspace. In addition the assignments

$$T \mapsto \mu_1(k)T,\ T \mapsto \mu_2(k)T \qquad (k \in K)$$

define on $\mathfrak{T}$ the structure of a double K-module.]

For the applications (9.1.5), it will be best to state the theorem in a slightly different form. Let γ denote the automorphism of $\mathfrak{a}_{\mathfrak{p}_c}$ defined by the rule

$$H \mapsto -\sqrt{-1}(H + Q_\rho) \qquad (H \in \mathfrak{a}_{\mathfrak{p}_c})$$

– then

$$\gamma^{-1}(H) = \sqrt{-1}H - Q_\rho \qquad (H \in \mathfrak{a}_{\mathfrak{p}_c}).$$

Put

$$\sigma_{\lambda,i} = \gamma(\tau_{\lambda,i}) \qquad (\lambda \in \mathsf{L}' ; 1 \leqslant i \leqslant r).$$

Then $\sigma_{\lambda,i}$ is the set of all points H in $\mathfrak{a}_{\mathfrak{p}_c}$ such that

$$2\sqrt{-1}\lambda(H) = (\lambda, \lambda) + \gamma_i.$$

Since the representations μ_1 and μ_2 of K are unitary, every eigenvalue of the transformation $a \mapsto a\mu_2(\omega_{\mathfrak{m}})$ $(a \in E)$ is real, whence the γ_i are real. Consequently, agreeing to write $\mathbf{\Gamma}^\gamma$ for $\gamma(\mathbf{\Gamma})$, we see that

$$\mathbf{\Gamma}^\gamma \cap \mathfrak{a}_{\mathfrak{p}} \qquad (\text{or } \mathbf{\Gamma}^\gamma \cap \mathfrak{a}_{\mathfrak{p}}^\vee \ldots)$$

is the complement in $\mathfrak{a}_{\mathfrak{p}}$ (or $\mathfrak{a}_{\mathfrak{p}}^\vee \ldots$) of a finite number of hyperplanes passing through the origin (there clearly being but a finite number of $\lambda \in \mathsf{L}'$ such that $\prod_i ((\lambda, \lambda) + \gamma_i) = 0$).

Given $\nu \in \mathbf{\Gamma}^\gamma$, let us write

$$\mathbf{\Phi}(\nu : h) = h^\rho \mathbf{\Psi}(\gamma^{-1}(\nu) : h) \qquad (h \in \exp(\mathscr{C})).$$

Theorem 9.1.4.1 can then be stated as follows.

Theorem 9.1.4.1 (bis) (Harish-Chandra) Fix $\nu \in \mathbf{\Gamma}^{\gamma}$; put

$$\mathbf{\Phi}(\nu : h) = e^{\langle\sqrt{-1}\nu, H\rangle} \sum_{\lambda \in \mathsf{L}} \mathbf{\Gamma}_{\lambda}(\sqrt{-1}\nu - \rho)e^{-\lambda(H)} \qquad (h \in \exp(\mathscr{C}))$$

where $H = \log h$ – then the function $h \mapsto \mathbf{\Phi}(\nu : h)$ is analytic on $\exp(\mathscr{C})$ and we have

$$\mathbf{\Phi}(\nu : h; e^{\rho}\text{♀}_{A_{\mathfrak{p}}}(D)\circ e^{-\rho}) = \mathbf{\Phi}(\nu : h)\mu_2(\Omega(D : \sqrt{-1}\nu - \rho)) \qquad (\text{all } D \in \mathfrak{X}).$$

On the other hand, for fixed $h \in \exp(\mathscr{C})$, the function $\nu \mapsto \mathbf{\Phi}(\nu : h)$ is holomorphic on $\mathbf{\Gamma}^{\gamma}$.

Remark Consider the important special case when $\mu_1 = \mu_2 =$ trivial one dimensional representation of K on $\mathbf{C}$ (the case of zonal spherical functions on G) – then the recursive definition of the $\mathbf{\Gamma}_{\lambda}$ ($\lambda \in \mathsf{L}$) simplifies to some extent since the terms involving μ_1 and μ_2 drop out. Moreover in this situation we have $\gamma_1 = \cdots = \gamma_r = 0$ and so here $\mathfrak{a}_{\mathfrak{p}} \subset \mathbf{\Gamma}^{\gamma}$.

Heuristics The recursive definition of the $\mathbf{\Gamma}_{\lambda}$ ($\lambda \in \mathsf{L}$) is suggested by the following considerations. Given an element ν in $\mathfrak{a}_{\mathfrak{p}_c}^{\vee}$, we want to define, for each $\lambda \in \mathsf{L}$, an element $\mathbf{\Gamma}_{\lambda}(\nu) \in \mathrm{Hom}_{\mathbf{C}}(E_{\mathsf{M}}, E_{\mathsf{M}})$ in such a way that the following conditions hold. We require first of all that $\mathbf{\Gamma}_0(\nu) = 1$; secondly we shall want the series

$$H \mapsto e^{\langle\nu, H\rangle} \sum_{\lambda \in \mathsf{L}} \mathbf{\Gamma}_{\lambda}(\nu)e^{-\lambda(H)}$$

to converge in a suitable sense on $\mathscr{C}$ so that its sum $\mathbf{\Psi}(\nu : .)$ is an analytic function from $\mathscr{C}$ to $\mathrm{Hom}_{\mathbf{C}}(E_{\mathsf{M}}, E_{\mathsf{M}})$; finally we shall demand that $\mathbf{\Psi}(\nu : .)$ be an 'eigenfunction' for the Casimir operator ω – more precisely that $\mathbf{\Psi}(\nu : .)$ has the property that

$$\begin{aligned}\mathbf{\Psi}(\nu : H; \text{♀}_{A_{\mathfrak{p}}}(\omega)) &= \mathbf{\Psi}(\nu : H)\mu_2(\Omega(\omega : \nu)) \\ &= \mathbf{\Psi}(\nu : H)\{(\nu, \nu + 2\rho) + \mu_2(\omega_{\mathfrak{m}})\} \qquad (\text{all } H \in \mathscr{C}).\end{aligned}$$

If we differentiate our series term by term and make use of the expansions

$$e^{-2\alpha}(1 - e^{-2\alpha})^{-1} = \sum_{n \geq 1} e^{-2n\alpha}, \quad e^{-2\alpha}(1 - e^{-2\alpha})^{-2} = \sum_{n \geq 1} ne^{-2n\alpha}$$

$$e^{-\alpha}(1 + e^{-2\alpha})(1 - e^{-2\alpha})^{-2} = \sum_{n \geq 0} (2n + 1)e^{-(2n+1)\alpha}$$

which are valid on $\mathscr{C}$ for $\alpha \in P_+$, we obtain from Corollary 9.1.2.12 the following recurrence relation for the $\mathbf{\Gamma}_{\lambda}(\nu)$:

$$\begin{aligned}&\{2(\lambda, \nu) - (\lambda, \lambda - 2\rho)\}\mathbf{\Gamma}_{\lambda}(\nu) - [\mu_2(\omega_{\mathfrak{m}}), \mathbf{\Gamma}_{\lambda}(\nu)] \\ &= 2 \sum_{\alpha \in P_+} \sum_{n \geq 1} \{(\tilde{\alpha}, \nu) - (\tilde{\alpha}, \lambda - 2n\tilde{\alpha})\}\mathbf{\Gamma}_{\lambda - 2n\tilde{\alpha}}(\nu) \\ &\quad + 8 \sum_{\alpha \in P_+} \sum_{n \geq 1} (2n - 1)\mu_1(Y_{\alpha})\mu_2(Y_{-\alpha})\mathbf{\Gamma}_{\lambda - (2n-1)\tilde{\alpha}}(\nu) \\ &\quad - 8 \sum_{\alpha \in P_+} \sum_{n \geq 1} n\{\mu_1(Y_{\alpha}Y_{-\alpha}) + \mu_2(Y_{\alpha}Y_{-\alpha})\}\mathbf{\Gamma}_{\lambda - 2n\tilde{\alpha}}(\nu).\end{aligned}$$

[Here, of course, it is understood that $\mathbf{\Gamma}_{\lambda}(\nu) = 0$ for any λ which does not lie in L so that the sums appearing on the right hand side are all finite.] In order for the above relation to completely determine $\mathbf{\Gamma}_{\lambda}(\nu)$ by recurrence it will be enough to know that the mapping

$$T \mapsto \{2(\lambda, \nu) - (\lambda, \lambda - 2\rho)\}T - [\mu_2(\omega_{\mathrm{m}}), T] \qquad (T \in \mathrm{Hom}_{\mathrm{C}}(E_{\mathrm{M}}, E_{\mathrm{M}}))$$

of $\mathrm{Hom}_{\mathrm{C}}(E_{\mathrm{M}}, E_{\mathrm{M}})$ into itself is non-singular for every $\lambda \in \mathsf{L}'$. . . .]

For any $\eta \geqslant 0$, let $\mathfrak{a}_{\mathfrak{p}_c}(\eta)$ denote the subset of $\mathfrak{a}_{\mathfrak{p}_c}$ comprised of those H such that $\mathscr{R}(\lambda_i(H)) > \eta\ (1 \leqslant i \leqslant l)$; if ϵ is a real number lying between 0 and 1, then $\mathfrak{a}_{\mathfrak{p}_c}(\epsilon, \eta)$ will denote, as before, the set of all H in $\mathfrak{a}_{\mathfrak{p}_c}(\eta)$ such that $\mathscr{R}(\lambda_i(H)) > \epsilon \max_{1 \leqslant i \leqslant l} |\lambda_i(H)|$.

Lemma 9.1.4.2 Let $\mathcal{O}$ be an open, relatively compact subset of $\mathfrak{a}_{\mathfrak{p}_c}^{\vee}$; let E be a finite dimensional Banach space. Suppose that for each $\lambda \in \mathsf{L}$ we are given a holomorphic function f_λ from $\mathcal{O}$ to E which, for some $M > 0$, has the property that

$$\| f_\lambda(\nu) \| \leqslant M^{m_\lambda + 1} \qquad (\text{all } \nu \in \mathcal{O} \text{ and } \lambda \in \mathsf{L}).$$

Then there exists an $\eta \geqslant 0$ such that for every complex number c and every ϵ $(0 < \epsilon < 1)$, the series

$$e^{c\langle \nu, H\rangle} \sum_{\lambda \in \mathsf{L}} f_\lambda(\nu) e^{-\lambda(H)} \qquad (\nu \in \mathcal{O},\ H \in \mathfrak{a}_{\mathfrak{p}_c}(\epsilon, \eta))$$

converges decently on $\mathcal{O} \times \mathfrak{a}_{\mathfrak{p}_c}(\epsilon, \eta)$.

[This statement is a simple consequence of Proposition A3.4.1; the details may be left to the reader.]

We shall now take up the proof of Theorem 9.1.4.1; our first task will be to show that the series which defines $\boldsymbol{\Psi}$ is in fact convergent.

Let $p_0 = 1$ and, as above, write

$$p_\lambda = \prod_{i=1}^{r} \{2\lambda - (\lambda, \lambda - 2\rho) - \gamma_i\}^{m_i} \qquad (\lambda \in \mathsf{L}')$$

where m_i is the multiplicity of the eigenvalue γ_i of $\boldsymbol{\gamma}$ $(1 \leqslant i \leqslant r)$; set

$$P_\lambda = p_\lambda \prod_{\tilde{\lambda} \ll \lambda} p_{\tilde{\lambda}} \quad (\lambda, \tilde{\lambda} \in \mathsf{L})$$

– then, by an easy induction on m_λ, we see that $P_\lambda \boldsymbol{\Gamma}_\lambda$ is a polynomial function for every $\lambda \in \mathsf{L}$.

The following result is actually somewhat more general than that which is needed for our present purposes. . . .

Lemma 9.1.4.3 Let ω be a compact subset of $\mathfrak{a}_{\mathfrak{p}_c}$; let L_ω denote the (finite) set comprised of those $\lambda \in \mathsf{L}'$ for which ω meets $\bigcup_i \tau_{\lambda, i}$ and put $P_\omega = \prod_{\lambda \in \mathsf{L}_\omega} P_\lambda$ ($P_\omega = 1$ if L_ω is empty) – then, for every $\lambda \in \mathsf{L}$, the rational function $\boldsymbol{\Upsilon}_\lambda = P_\omega \boldsymbol{\Gamma}_\lambda$ is defined everywhere on ω. Moreover there exists a number $c_\omega \geqslant 1$ such that

$$\| \boldsymbol{\Upsilon}_\lambda(H) \| \leqslant c_\omega^{m_\lambda + 1} \qquad (H \in \omega)$$

for all $\lambda \in \mathsf{L}$.

[For the present proof it will be convenient to regard the Υ_λ ($\lambda \in \mathsf{L}$) as being defined on $\mathfrak{a}_{\mathfrak{p}_c}$ rather than its dual.]

Proof (1) Fix an element λ_0 in L – then, as has been pointed out above, $P_{\lambda_0}\mathbf{\Gamma}_{\lambda_0}$ is a polynomial function and hence is defined everywhere on $\mathfrak{a}_{\mathfrak{p}_c}$. Let

$$\mathsf{L}_\omega(\lambda_0) = \{\lambda \in \mathsf{L}_\omega : \lambda \ll \lambda_0\};$$

if

$$P'_{\lambda_0} = \prod_{\lambda \in \mathsf{L}_\omega(\lambda_0)} p_\lambda,$$

then it is clear from the definition of P_{λ_0} that

$$P''_{\lambda_0} = P_{\lambda_0}/P'_{\lambda_0}$$

is a polynomial function which does not vanish anywhere on ω and so

$$P'_{\lambda_0}\mathbf{\Gamma}_{\lambda_0} = P_{\lambda_0}\mathbf{\Gamma}_{\lambda_0}/P''_{\lambda_0}$$

is defined everywhere on ω. But P'_{λ_0} divides P_ω in $S(\mathfrak{a}_{\mathfrak{p}_c})$ (since $\mathsf{L}_\omega(\lambda_0) \subset \mathsf{L}_\omega$) – therefore $\Upsilon_{\lambda_0} = P_\omega \mathbf{\Gamma}_{\lambda_0}$ is defined at every point of ω, as contended.

(2) It remains to estimate the $\Upsilon_\lambda(\lambda \in \mathsf{L})$ on ω.

(a) Put $m_0 = \sup_{\lambda \in \mathsf{L}_\omega} m_\lambda$ ($m_0 = 0$ in case L_ω is empty). Because ω is compact, we can choose a number $c_0 \geqslant 1$ such that

$$\|\Upsilon_\lambda(H)\| \leqslant c_0^{m_\lambda + 1} \qquad (\text{all } H \in \omega)$$

whenever $m_\lambda \leqslant m_0$ ($\lambda \in \mathsf{L}$).

(b) For $\lambda \in \mathsf{L}$, $H \in \mathfrak{a}_{\mathfrak{p}_c}$, consider the mapping $\Lambda_{\lambda, H}$ defined by the rule

$$T \mapsto \{2\lambda(H) - (\lambda, \lambda - 2\rho)\}T - \gamma(T) \qquad (T \in \mathrm{Hom}_C(E_M, E_M)).$$

Choose positive numbers r_1, r_2 such that $(\lambda, \lambda) \geqslant r_1 m_\lambda^2$ and

$$2|\lambda(H) + (\lambda, \rho)| \leqslant r_2 m_\lambda \qquad (\text{all } \lambda \in \mathsf{L},\ H \in \omega).$$

Then we can select an integer $m_1 \geqslant m_0$ such that $m_\lambda > m_1$ ($\lambda \in \mathsf{L}$) $\Rightarrow$

$$\begin{aligned} \|\Lambda_{\lambda,H}(T)\| &\geqslant (\lambda, \lambda)\|T\| - \|\gamma(T)\| - 2|\lambda(H) + (\lambda, \rho)|\,\|T\| \\ &\geqslant \{(\lambda, \lambda) - 2|\lambda(H) + (\lambda, \rho)| - \|\gamma\|\}\,\|T\| \\ &\geqslant (r_1 m_\lambda^2 - r_2 m_\lambda - \|\gamma\|)\|T\| \geqslant 2^{-1} r_1 m_\lambda^2 \|T\| \end{aligned}$$
$$(\text{any } T \in \mathrm{Hom}_C(E_M, E_M)),$$

the estimate being independent of $H \in \omega$.

(c) Let $\mathsf{L}(m_0)$ denote the set of all $\lambda \in \mathsf{L}$ for which $m_\lambda > m_0$. Let

$$\Delta_\lambda(H) = \det(\Lambda_{\lambda,H}) \qquad (\lambda \in \mathsf{L});$$

it is clear that Δ_λ is a polynomial function on $\mathfrak{a}_{\mathfrak{p}_c}$ which is nowhere zero on ω if $\lambda \in \mathsf{L}(m_0)$ (since such a λ does not belong to L_ω). Let

$$\delta_\lambda = \inf_{H \in \omega} |\Delta_\lambda(H)|.$$

Because ω is compact, $\delta_\lambda > 0$ for every $\lambda \in \mathsf{L}(m_0)$. Furthermore there exists a polynomial function Ξ_λ from $\mathfrak{a}_{\mathfrak{p}_c}$ to $\mathrm{Hom}_C^2(E_M, E_M)$ (= 'Hom of the Hom') such that

$$\Lambda_{\lambda,H}\Xi_\lambda(H) = \Xi_\lambda(H)\Lambda_{\lambda,H} = \Delta_\lambda(H) \qquad (H \in \mathfrak{a}_{\mathfrak{p}_c}).$$

Consequently, if $\lambda \in \mathsf{L}(m_0)$ and $H \in \omega$, then

$$\|T\| = |\Delta_\lambda(H)|^{-1} \|\Xi_\lambda(H)\Lambda_{\lambda,H}(T)\| \leqslant \delta_\lambda^{-1}\xi_\lambda \|\Lambda_{\lambda,H}(T)\| \qquad (T \in \mathrm{Hom}_C(E_M, E_M))$$

where $\xi_\lambda = \sup_{H\in\omega} \|\Xi_\lambda(H)\| < \infty$. This shows that for each $\lambda \in \mathsf{L}(m_0)$, we can choose a positive number r_λ such that

$$\|\Lambda_{\lambda,H}(T)\| \geqslant r_\lambda m_\lambda^2 \|T\|$$

for all $H \in \omega$ and $T \in \mathrm{Hom}_C(E_M, E_M)$. On the other hand (b) also gives us an estimate on $\|\Lambda_{\lambda,H}(T)\|$ provided $m_\lambda > m_1$; this being so, let c_1 denote the minimum among the numbers $2^{-1}r_1$ and

$$r_\lambda \; (\lambda \in \mathsf{L}(m_0),\ m_\lambda \leqslant m_1).$$

Then c_1 is positive (since there are only a finite number of λ in $\mathsf{L}(m_0)$ with $m_\lambda \leqslant m_1$) and thus we have

$$(*) \quad \|\Lambda_{\lambda,H}(T)\| \geqslant c_1 m_\lambda^2 \|T\| \qquad (T \in \mathrm{Hom}_C(E_M, E_M))$$

for *all* $H \in \omega$ and $\lambda \in \mathsf{L}(m_0)$.

(d) Part (a) provides us with an estimate on $\|\Upsilon_\lambda(H)\|$ $(H \in \omega)$ whenever our λ is such that $m_\lambda \leqslant m_0$; we shall now utilize equation $(*)$ and estimate $\|\Upsilon(H)\|$ $(H \in \omega)$ in the case when $m_\lambda > m_0$ (i.e. when $\lambda \in \mathsf{L}(m_0)$). To this end choose a positive number c_2 such that

$$|\alpha(H) - (\lambda, \tilde{\alpha})| \leqslant c_2(m_\lambda + 1)$$

for $\lambda \in \mathsf{L}$, $\alpha \in P_+$, and $H \in \omega$ – then, if $\lambda \in \mathsf{L}(m_0)$ and $\tilde{\lambda} \ll \lambda$ $(\tilde{\lambda} \in \mathsf{L})$, we have

$$|\alpha(H) - (\tilde{\lambda}, \tilde{\alpha})| \, \|\Lambda_{\lambda,H}^{-1}(T)\| \leqslant c_2(m_\lambda + 1)c_1^{-1}m_\lambda^{-2} \|T\| \leqslant c_2 c_1^{-1} m_\lambda^{-1} \|T\| \qquad (T \in \mathrm{Hom}_C(E_M, E_M))$$

for all $H \in \omega$. Put $c_3 = c_2 c_1^{-1}$ and select c_4 such that

$$\|\mu_1(Y_\alpha)\mu_2(Y_{-\alpha})\| + \|\mu_1(Y_\alpha Y_{-\alpha}) + \mu_2(Y_\alpha Y_{-\alpha})\| \leqslant c_4/8 \qquad (\text{all } \alpha \in P_+)$$

(the norms being taken in $\mathrm{Hom}_C(E, E)$). Now fix a $\lambda \in \mathsf{L}(m_0)$ – then it follows from the recursion formula that

$$\begin{aligned} \|\Upsilon_\lambda(H)\| &\leqslant 2c_3 m_\lambda^{-1} \sum_{\alpha\in P_+}\sum_{n\geqslant 1} \|\Upsilon_{\lambda-2n\alpha}(H)\| \\ &\quad + c_4 c_1^{-1} m_\lambda^{-2} \sum_{\alpha\in P_+}\sum_{n\geqslant 1} (2n-1)\|\Upsilon_{\lambda-(2n-1)\alpha}(H)\| \\ &\quad + c_4 c_1^{-1} m_\lambda^{-2} \sum_{\alpha\in P_+}\sum_{n\geqslant 1} n\,\|\Upsilon_{\lambda-2n\alpha}(H)\| \\ &\leqslant \{2c_3 m_\lambda^{-1} + 2c_4 c_1^{-1} m_\lambda^{-1}\} \sum_{\alpha\in P_+}\sum_{n\geqslant 1} \|\Upsilon_{\lambda-n\alpha}(H)\| \end{aligned}$$

(all $H \in \omega$).

(e) Put $c_\omega = \max\{c_0, 2[P_+](c_3 + c_4 c_1^{-1})\}$; let us prove by induction on m_λ that

$$\|\Upsilon_\lambda(H)\| \leqslant c_\omega^{m_\lambda + 1}$$

for $\lambda \in \mathsf{L}$ and $H \in \omega$. If $m_\lambda \leqslant m_0$, this is obvious since $c_0 \leqslant c_\omega$; so suppose that $m_\lambda > m_0$ – then it follows from the estimate in (d) and the induction hypothesis that

$$\|\Upsilon_\lambda(H)\| \leqslant (2c_3 + 2c_4 c_1^{-1}) m_\lambda^{-1} [P_+] m_\lambda c_\omega^{m_\lambda} \leqslant c_\omega^{m_\lambda + 1}$$

since $c_\omega \geqslant c_0 \geqslant 1$.

The lemma is therefore established. □

As has been noted earlier, $\mathbf{\Gamma}$ is an open, connected, dense subset of $\mathfrak{a}_{\mathfrak{p}_c}$ (or $\mathfrak{a}_{\mathfrak{p}_c}^\vee \ldots$) on which the rational functions $\mathbf{\Gamma}_\lambda$ are well-defined for all $\lambda \in \mathsf{L}$. Let $\mathcal{O}$ be an open, relatively compact subset of $\mathbf{\Gamma}$ – then, on the basis of Lemmas 9.1.4.2 and 9.1.4.3, it is clear that there exists an $\eta \geqslant 0$ such that for every complex number c and every ϵ $(0 < \epsilon < 1)$, the series

$$e^{c\langle \nu, H\rangle} \sum_{\lambda \in \mathsf{L}} \mathbf{\Gamma}_\lambda(\nu) e^{-\lambda(H)} \qquad (\nu \in \mathcal{O},\ H \in \mathfrak{a}_{\mathfrak{p}_c}(\epsilon, \eta))$$

converges decently on $\mathcal{O} \times \mathfrak{a}_{\mathfrak{p}_c}(\epsilon, \eta)$. In particular let us write

$$\psi(\nu : H) = e^{\langle \nu, H\rangle} \sum_{\lambda \in \mathsf{L}} \mathbf{\Gamma}_\lambda(\nu) e^{-\lambda(H)} \qquad (\nu \in \mathcal{O},\ H \in \mathfrak{a}_{\mathfrak{p}_c}(\eta)).$$

Then our discussion thus far serves to ensure that ψ is a holomorphic function from $\mathcal{O} \times \mathfrak{a}_{\mathfrak{p}_c}(\eta)$ to $\mathrm{Hom}_{\mathbf{C}}(E_{\mathsf{M}}, E_{\mathsf{M}})$. We shall now improve this result and show that ψ is actually holomorphic on $\mathbf{\Gamma} \times \mathscr{C}_c$ ($\mathscr{C}_c$ denoting, as in 9.1.1, the subset of $\mathfrak{a}_{\mathfrak{p}_c}$ consisting of those H with the property that

$$\mathscr{R}(\lambda_i(H)) > 0 \qquad (1 \leqslant i \leqslant l)).$$

Of course this fact implies part of Theorem 9.1.4.1.

Lemma 9.1.4.4 (Helgason) Fix $\nu \in \mathbf{\Gamma}$ and $H \in \mathscr{C}$ – then there exists a constant $M(\nu, H)$ such that

$$\|\mathbf{\Gamma}_\lambda(\nu)\| \leqslant M(\nu, H) e^{\lambda(H)} \qquad (\text{all } \lambda \in \mathsf{L}).$$

Proof Let us adopt the notations of part (2) of the proof of the previous lemma; in the present case we may take $m_0 = 0$ (since $\nu \in \mathbf{\Gamma}$) – but then it is clear that there exists a constant $c > 0$ such that for all $\lambda \in \mathsf{L}$,

$$\|\mathbf{\Gamma}_\lambda(\nu)\| \leqslant c m_\lambda^{-1} \sum_{\alpha \in P_+} \sum_{n \geqslant 1} \|\mathbf{\Gamma}_{\lambda - n\bar{\alpha}}(\nu)\|.$$

Let N_0 be an integer such that

$$c \sum_{\alpha \in P_+} (1 - e^{-\bar{\alpha}(H)})^{-1} \leqslant N_0$$

and then select $M(\nu, H)$ such that

$$\|\mathbf{\Gamma}_\lambda(\nu)\| \leqslant M(\nu, H) e^{\lambda(H)}$$

for all $\lambda \in \mathsf{L}$ with $m_\lambda \leqslant N_0$. We now claim that

$$\|\mathbf{\Gamma}_\lambda(\nu)\| \leqslant M(\nu, H)e^{\lambda(H)}$$

for all $\lambda \in \mathsf{L}$. Proceeding by induction on m_λ, let $N > N_0$ be an integer and suppose that our assertion is true for those $\lambda \in \mathsf{L}$ such that $m_\lambda < N$ – then, if λ is an element of L with $m_\lambda = N$, we have

$$\|\mathbf{\Gamma}_\lambda(\nu)\| \leqslant cN^{-1}M(\nu, H) \sum_{\alpha \in P_+} \sum_{n \geqslant 1} e^{(\lambda - n\alpha)(H)} \leqslant M(\nu, H)e^{\lambda(H)}$$

which proves the lemma. □

Fix a $\nu \in \mathbf{\Gamma}$ and consider the series which defines $\psi(\nu : H)$ – then we know that it is convergent provided $H \in \mathfrak{a}_{\mathfrak{p}_c}(\eta)$ (the η depending a priori on ν). Now fix an arbitrary $H \in \mathscr{C}_c$; write

$$H = X + \sqrt{-1}Y \qquad (X \in \mathscr{C}, Y \in \mathfrak{a}_\mathfrak{p})$$

and determine $H^+ \in \mathscr{C}$ such that

$$\lambda_i(H^+) = 4^{-1} \min_{1 \leqslant j \leqslant l} \lambda_j(X) \qquad (1 \leqslant i \leqslant l).$$

Select an open, relatively compact neighborhood $\mathcal{O}$ of X in $\mathscr{C}$ with the property that

$$\lambda_i(A) > 2\lambda_i(H^+) \qquad (A \in \mathcal{O};\ 1 \leqslant i \leqslant l).$$

Let $B \in \mathfrak{a}_\mathfrak{p}$ – then, on the basis of Lemma 9.1.4.4, there exists a constant $M(\nu, H^+)$ such that

$$\sum_{\lambda \in \mathsf{L}} \|\mathbf{\Gamma}_\lambda(\nu)e^{-\lambda(A+\sqrt{-1}B)}\| \leqslant M(\nu, H^+) \sum_{\lambda \in \mathsf{L}} e^{\lambda(H^+ - A)}$$

and consequently the series which defines $\psi(\nu : H)$ converges uniformly on a neighborhood of H in $\mathscr{C}_c$. Otherwise said, for fixed ν in $\mathbf{\Gamma}$, the function $H \mapsto \psi(\nu : H)$ is holomorphic on $\mathscr{C}_c$.

To conclude, therefore, that ψ is holomorphic on the product $\mathbf{\Gamma} \times \mathscr{C}_c$, it will be sufficient, in view of Hartogs' Theorem, to fix an H in $\mathscr{C}_c$ and verify that the function $\nu \mapsto \psi(\nu : H)$ is holomorphic on $\mathbf{\Gamma}$ – but this is straightforward. Thus suppose that ν ranges through an open, relatively compact subset $\mathcal{O}$ of $\mathbf{\Gamma}$; the proof of Lemma 9.1.4.4 shows that the constant $M(.\,, H)$ can be chosen in such a way as to depend *only* on $\mathcal{O}$. This being so, write

$$H = X + \sqrt{-1}Y \quad (X \in \mathscr{C}, Y \in \mathfrak{a}_\mathfrak{p})$$

and let H^+ be as above – then, for any $\nu \in \mathcal{O}$, we have

$$\sum_{\lambda \in \mathsf{L}} \|\mathbf{\Gamma}_\lambda(\nu)e^{-\lambda(H)}\| \leqslant M(\mathcal{O}, H^+) \sum_{\lambda \in \mathsf{L}} e^{\lambda(H^+ - X)}$$

which shows that for fixed $H \in \mathscr{C}_c$ the series defining $\psi(\nu : H)$ is uniformly convergent on $\mathcal{O}$. . . .

As in Theorem 9.1.4.1, let us write

$$\Psi(\nu : h) = e^{\langle \nu, H \rangle} \sum_{\lambda \in \mathsf{L}} \mathbf{\Gamma}_\lambda(\nu)e^{-\lambda(H)} \qquad (\nu \in \mathbf{\Gamma}, h \in \exp(\mathscr{C}))$$

where $H = \log h$ – then, in view of what has been said so far, we know that for fixed ν, Ψ is analytic in h while for fixed h, Ψ is holomorphic in ν. The proof of Theorem 9.1.4.1 will therefore be complete when it is established that

$$\Psi(\nu : h ; \mathfrak{P}_{A\mathfrak{p}}(D)) = \Psi(\nu : h)\mu_2(\Omega(D : \nu)) \qquad (\text{all } D \in \mathfrak{X}).$$

This requires, however, some preparation which will now be undertaken.

Let $\mathfrak{T}$ be the space of all linear mappings of E_M into E – then $\mathfrak{T}$ is a left module over $\mathrm{Hom}_C(E, E)$ and a right module over $\mathrm{Hom}_C(E_M, E_M)$. Given $T \in \mathfrak{T}$, write

$$\|T\| = \sup_{\|a\| \leqslant 1} \|Ta\| \qquad (a \in E_M);$$

under this norm, $\mathfrak{T}$ becomes a Banach space of which $\mathrm{Hom}_C(E_M, E_M)$ may be regarded as a subspace. Let $\mathfrak{t}$ denote the set of all functions t from L to $\mathfrak{T}$ such that

$$\sup_{\lambda \in \mathsf{L}} (m_\lambda + 1)^{-1} \log(\|t(\lambda)\|) < \infty;$$

let $\mathfrak{t}(M)$ be the set of all $t \in \mathfrak{t}$ such that $t(\lambda) \in \mathrm{Hom}_C(E_M, E_M)$ for all $\lambda \in \mathsf{L}$. For any $t \in \mathfrak{t}$ and $T \in \mathrm{Hom}_C(E_M, E_M)$, we denote by tT the function $\lambda \mapsto t(\lambda)T$ ($\lambda \in \mathsf{L}$); it is obvious that tT is also in $\mathfrak{t}$ and thus $\mathfrak{t}$ can be viewed as a right module over $\mathrm{Hom}_C(E_M, E_M)$. Similarly $\mathfrak{t}$ is a left module over $\mathrm{Hom}_C(E, E)$.

Let $t \in \mathfrak{t}$ – then, on the basis of Lemma 9.1.4.2 above, there exists a number $\eta(t) \geqslant 0$ such that:

(i) The series $\sum_{\lambda \in \mathsf{L}} \|t(\lambda)e^{-\lambda(H)}\|$ converges uniformly for $H \in \mathfrak{a}_{\mathfrak{p}_c}(\eta(t))$;

(ii) The series $\sum_{\lambda \in \mathsf{L}} t(\lambda)e^{-\lambda(H)}$ converges decently on $\mathfrak{a}_{\mathfrak{p}_c}(1/2, \eta(t))$.

Fix $t \in \mathfrak{t}$ and $\eta_0 > \eta(t)$; given $\nu \in \mathfrak{a}_{\mathfrak{p}_c}^{\vee}$, write

$$\psi_t(\nu : H) = e^{\langle \nu, H \rangle} \sum_{\lambda \in \mathsf{L}} t(\lambda)e^{-\lambda(H)} \qquad (H \in \mathfrak{a}_{\mathfrak{p}_c}(\eta_0)).$$

Then, for any $p \in S(\mathfrak{a}_{\mathfrak{p}_c})$, it is clear that

$$\psi_t(\nu : H ; \partial(p)) = e^{\langle \nu, H \rangle} \sum_{\lambda \in \mathsf{L}} t(\lambda)p(\nu - \lambda)e^{-\lambda(H)} \qquad (H \in \mathfrak{a}_{\mathfrak{p}_c}(\eta_0))$$

where $p(\nu - \lambda)$ is the value of the polynomial p at the point $\nu - \lambda$. On the other hand, consider an arbitrary element f in $\mathfrak{F}$; as we know, such an f admits an expansion of the form

$$f(\exp H) = \sum_{\lambda \in \mathsf{L}} c(\lambda)e^{-\lambda(H)} \qquad (H \in \mathfrak{a}_{\mathfrak{p}}(\eta_0))$$

where the $c(\lambda)$ ($\lambda \in \mathsf{L}$) are certain complex numbers and the series $\sum_{\lambda \in \mathsf{L}} |c(\lambda)e^{-\lambda(H)}|$ converges uniformly for $H \in \mathfrak{a}_{\mathfrak{p}_c}(\eta_0)$. Now put

$$s(\lambda) = \sum_{\lambda_1 + \lambda_2 = \lambda} c(\lambda_1)p(\nu - \lambda_2)t(\lambda_2) \qquad (\lambda \in \mathsf{L}),$$

the sum being taken over all pairs (λ_1, λ_2) of elements in L such that $\lambda_1 + \lambda_2 = \lambda$. Owing to the estimate

$$\sum_{\lambda \in \mathsf{L}} \|s(\lambda)e^{-\lambda(H)}\| \leqslant \sum_{\lambda \in \mathsf{L}} |c(\lambda)e^{-\lambda(H)}| \sum_{\lambda \in \mathsf{L}} \|t(\lambda)p(\nu - \lambda)e^{-\lambda(H)}\| < \infty$$
$$(\text{all } H \in \mathfrak{a}_{\mathfrak{p}_c}(\eta_0)),$$

it follows easily that $s \in \mathfrak{t}$. [To see this, fix a point $H \in \mathfrak{a}_{\mathfrak{p}_c}(\eta_0)$; choose a positive number M such that (1)

$$|e^{\lambda(H)}| \leqslant M^{m_\lambda} \qquad (\text{all } \lambda \in \mathsf{L})$$

and (2)

$$\sum_{\lambda \in \mathsf{L}} \| s(\lambda) e^{-\lambda(H)} \| \leqslant M.$$

Then

$$\| s(\lambda) \| \leqslant |e^{\lambda(H)}| M \leqslant M^{m_\lambda + 1} \qquad (\text{all } \lambda \in \mathsf{L}) \ldots.]$$

For any $\eta \geqslant 0$, let $A_{\mathfrak{p}}(\eta)$ denote the subset of $A_{\mathfrak{p}}$ consisting of those h such that $\log h \in \mathfrak{a}_{\mathfrak{p}}(\eta)$.

Lemma 9.1.4.5 Fix $t \in \mathfrak{t}$, $D \in \mathfrak{G}$ and let ν be a linear function on $\mathfrak{a}_{\mathfrak{p}_c}$; put

$$\mathbf{\Psi}_t(\nu : h) = e^{\langle \nu, H \rangle} \sum_{\lambda \in \mathsf{L}} t(\lambda) e^{-\lambda(H)} \qquad (h \in A_{\mathfrak{p}}(\eta(t)))$$

where $H = \log h$ – then there exists a unique element $\mathbf{t} \in \mathfrak{t}$ with the property that

$$\mathbf{\Psi}_t(\nu : h ; \mathfrak{P}_{A_{\mathfrak{p}}}(D)) = e^{\langle \nu, H \rangle} \sum_{\lambda \in \mathsf{L}} \mathbf{t}(\lambda) e^{-\lambda(H)}$$

for all $h \in A_{\mathfrak{p}}(\eta)$ ($\eta = \max(\eta(t), \eta(\mathbf{t}))$).

[We observe that $\mathbf{\Psi}_t$ is a $\mathfrak{T}$-valued analytic function on the open set $A_{\mathfrak{p}}(\eta(t))$ of $A_{\mathfrak{p}}'$ – therefore, since $\mathfrak{T}$ is a double K-module, the function $\mathfrak{P}_{A_{\mathfrak{p}}}(D)\mathbf{\Psi}_t$ is well-defined ($D \in \mathfrak{G}$).]

Before giving the proof of Lemma 9.1.4.5, it will be convenient to establish some additional notation. Let $D \in \mathfrak{G}$ – then it follows from Lemma 9.1.2.7 that there exist elements $f_i \in \mathfrak{F}$, $p_i \in S(\mathfrak{a}_{\mathfrak{p}_c})$, and

$$T_i \in \mathrm{Hom}_{\mathbf{C}}(E, E) \qquad (1 \leqslant i \leqslant r)$$

such that

$$f(h ; \mathfrak{P}_{A_{\mathfrak{p}}}(D)) = \sum_i f_i(h) T_i f(h ; \partial(p_i)) \qquad (h \in A_{\mathfrak{p}}')$$

for any $\mathfrak{T}$-valued function f which is of class C^∞ on some open neighborhood in $A_{\mathfrak{p}}'$ of the point h. [The notation $f(h; \partial(p))$ ($p \in S(\mathfrak{a}_{\mathfrak{p}_c})$) is, strictly speaking, not legitimate. . . .] Each f_i has an expansion of the form

$$f_i(h) = \sum_{\lambda \in \mathsf{L}} c_i(\lambda) h^{-\lambda} \qquad (h \in \exp(\mathscr{C}))$$

where the $c_i(\lambda)$ ($\lambda \in \mathsf{L}$) are certain complex numbers and the series

$$\sum_{\lambda \in \mathsf{L}} |c_i(\lambda) e^{-\lambda(H)}|$$

converges uniformly for $H \in \mathfrak{a}_{\mathfrak{p}_c}(\eta)$ (any $\eta > 0$).

Proof of Lemma 9.1.4.5 Relative to the D which appears in the statement of the lemma, let us apply the above remarks. Thus, for $\lambda \in \mathsf{L}$, write

$$\mathbf{t}(\lambda) = \sum_{\lambda_1+\lambda_2=\lambda} \sum_{i=1}^{r} c_i(\lambda_1)p_i(\nu - \lambda_2)T_i t(\lambda_2),$$

the sum being taken over all pairs (λ_1, λ_2) of elements in L such that $\lambda_1 + \lambda_2 = \lambda$ – then, in view of the discussion preceding the statement of our lemma, it is clear that $\mathbf{t} \in \mathsf{t}$ and of course we have

$$\mathbf{\Psi}_t(\nu : h ; \gamma_{A_\mathfrak{p}}(D)) = e^{\langle \nu, H\rangle} \sum_{\lambda \in \mathsf{L}} \mathbf{t}(\lambda)e^{-\lambda(H)} \qquad (H = \log h)$$

for all $h \in A_\mathfrak{p}(\tilde{\eta})$, $\tilde{\eta}$ a sufficiently large positive number. However the left and right sides of this equation are analytic functions on $A_\mathfrak{p}(\eta(t))$ and $A_\mathfrak{p}(\eta(\mathbf{t}))$, respectively – therefore they must coincide on $A_\mathfrak{p}(\eta)$. The uniqueness of $\mathbf{t}$ is an immediate consequence of Lemma A3.2.4. □

The element $\mathbf{t}$ defined above depends on ν, D, and t; we shall therefore agree to write $\mathbf{t}(\nu : D : t)$ instead of $\mathbf{t}$ and $\mathbf{t}(\nu : D : t : \lambda)$ instead of $\mathbf{t}(\lambda)$ $(\lambda \in \mathsf{L})$.

Here are the basic properties which $\mathbf{t}$ possesses.

($\mathbf{t}_1$) Fix D, t, and λ – then $\mathbf{t}(\nu : D : t : \lambda)$ is a polynomial function of ν.

($\mathbf{t}_2$) Let ω be the Casimir operator in $\mathfrak{G}$ – then, for any ν,

$$\begin{aligned} &\mathbf{t}(\nu : \omega : t : \lambda) \\ &= \{(\nu, \nu) + 2(\nu, \rho) - 2(\nu, \lambda) + (\lambda, \lambda - 2\rho)\}t(\lambda) + \mu_2(\omega_\mathfrak{m})t(\lambda) \\ &\quad + 2 \sum_{\alpha \in P_+} \sum_{n \geqslant 1} \{(\tilde{\alpha}, \nu) - (\tilde{\alpha}, \lambda - 2n\tilde{\alpha})\}t(\lambda - 2n\tilde{\alpha}) \\ &\quad + 8 \sum_{\alpha \in P_+} \sum_{n \geqslant 1} (2n - 1)\mu_1(Y_\alpha)\mu_2(Y_{-\alpha})t(\lambda - (2n - 1)\tilde{\alpha}) \\ &\quad - 8 \sum_{\alpha \in P_+} \sum_{n \geqslant 1} n\{\mu_1(Y_\alpha Y_{-\alpha}) + \mu_2(Y_\alpha Y_{-\alpha})\}t(\lambda - 2n\tilde{\alpha}) \qquad (\lambda \in \mathsf{L}) \end{aligned}$$

where $t(\lambda)$ is defined to be zero if $\lambda \notin \mathsf{L}$.

($\mathbf{t}_3$) Let $D \in \mathfrak{X}$ – then

$$\mathbf{t}(\nu : D : t : 0) = \mu_2(\Omega(D : \nu))t(0) \qquad (\text{any } \nu).$$

($\mathbf{t}_4$) Let $D \in \mathfrak{X}$ – then

$$t \in \mathsf{t}(\mathsf{M}) \Rightarrow \mathbf{t}(\nu : D : t) \in \mathsf{t}(\mathsf{M}) \qquad (\text{any } \nu).$$

($\mathbf{t}_5$) Let $D_1, D_2 \in \mathfrak{X}$ and $t \in \mathsf{t}(\mathsf{M})$ – then

$$\mathbf{t}(\nu : D_1 D_2 : t) = \mathbf{t}(\nu : D_1 : t_2) \qquad (\text{any } \nu)$$

where $t_2 = \mathbf{t}(\nu : D_2 : t) \in \mathsf{t}(\mathsf{M})$.

Verification of ($\mathbf{t}_1$) Invoking the notations which were introduced above, observe that

$$(\mathbf{t}) \quad \mathbf{t}(\nu : D : t : \lambda) = \sum_{\lambda_1+\lambda_2=\lambda} \sum_{i=1}^{r} c_i(\lambda_1)p_i(\nu - \lambda_2)T_i t(\lambda_2)$$

as results immediately from the proof of Lemma 9.1.4.5; of course this shows that $\mathbf{t}(. : D : t : \lambda)$ is a polynomial function of ν.

Verification of ($\mathbf{t}_2$) This follows from equation ($\mathbf{t}$) supra and Corollary 9.1.2.12 by direct computation.

Verification of ($\mathbf{t}_3$) We know from Corollary 9.1.2.8 that $\Delta_0 = \Omega(D)$ for all $D \in \mathfrak{X}$ where

$$\gamma_{A_\mathfrak{p}}(D) = \Delta_0 + \sum_{j=1}^{s} f_j \Delta_j$$

in the notations of Lemma 9.1.2.7. Therefore the coefficient of $e^{\langle \nu, H \rangle}$ in $\boldsymbol{\Psi}_t(\nu : h; \Delta_0)$ is $\mu_2(\Omega(D:\nu))t(0)$, that is

$$\mathbf{t}(\nu : D : t : 0) = \mu_2(\Omega(D:\nu))t(0) \qquad (\text{any } \nu).$$

Verification of ($\mathbf{t}_4$) Bearing in mind equation ($\mathbf{t}$) supra, we see that it will be enough to show that $\mathrm{Hom}_{\mathsf{C}}(E_{\mathsf{M}}, E_{\mathsf{M}})$ is stable under multiplication on the left by the $T_i (1 \leqslant i \leqslant r)$. For this purpose let us make the following observation. Suppose that Q_j and D_j $(1 \leqslant j \leqslant s)$ are elements in $\mathfrak{Q}$ and $\mathfrak{K}$, respectively, such that

$$\sum_j mQ_j \otimes mD_j = \sum_j Q_j \otimes D_j$$

in $\mathfrak{Q} \otimes_{\mathsf{C}} \mathfrak{K}$ for all $m \in \mathsf{M}$; let

$$T_0 = \sum_j \mu_1(Q_j)\mu_2(D_j) \in \mathrm{Hom}_{\mathsf{C}}(E, E).$$

Then, if $a \in E_{\mathsf{M}}$ and $T \in \mathrm{Hom}_{\mathsf{C}}(E_{\mathsf{M}}, E_{\mathsf{M}})$, we have

$$T_0 Ta = \sum_j \mu_1(mQ_j)(Ta)\mu_2(mD_j) = \mu_1(m)(T_0 Ta)\mu_2(m^{-1}) \qquad (\text{all } m \in \mathsf{M})$$

which shows that $T_0 Ta \in E_{\mathsf{M}}$ and therefore that

$$T_0 T \in \mathrm{Hom}_{\mathsf{C}}(E_{\mathsf{M}}, E_{\mathsf{M}}).$$

Consequently $\mathrm{Hom}_{\mathsf{C}}(E_{\mathsf{M}}, E_{\mathsf{M}})$ is stable under the mapping $T \mapsto T_0 T$ of $\mathfrak{T}$ $(T \in \mathfrak{T})$. Now suppose that $D \in \mathfrak{X}$ – then, on the basis of the Remark which was made immediately after the proof of Corollary 9.1.2.8, we see that there is indeed no loss of generality in assuming that $\mathrm{Hom}_{\mathsf{C}}(E_{\mathsf{M}}, E_{\mathsf{M}})$ is stable under multiplication on the left by the T_i $(1 \leqslant i \leqslant r)$.

Verification of ($\mathbf{t}_5$) Let $\mathfrak{M}$ be the subspace of $\mathfrak{A}$ spanned by all elements of the form

$$D_1 M \otimes D \otimes D_2 - D_1 \otimes D \otimes MD_2 \qquad (D_1, D_2 \in \mathfrak{K}; D \in \mathfrak{D}_F(A'_\mathfrak{p}); M \in \mathfrak{M}).$$

It has been noted earlier that

$$\gamma_{A_\mathfrak{p}}(D_1 D_2) - \gamma_{A_\mathfrak{p}}(D_1)\gamma_{A_\mathfrak{p}}(D_2) \in \mathfrak{M}$$

for any two elements D_1, D_2 in $\mathfrak{X}$ (see the Appendix to 9.1.2). In addition it is not difficult to see that if f is any C^∞, $\mathrm{Hom}_{\mathsf{C}}(E_{\mathsf{M}}, E_{\mathsf{M}})$-valued function on an open subset of $A'_{\mathfrak{p}}$, then f is necessarily annihilated by the elements of $\mathfrak{M}$. Therefore, for all $h \in A_{\mathfrak{p}}(\eta(t))$, we have

$$\mathbf{\Psi}_t(\nu : h ; \gamma_{A_\mathfrak{p}}(D_1 D_2)) = \mathbf{\Psi}_t(\nu : h ; \gamma_{A_\mathfrak{p}}(D_1)\gamma_{A_\mathfrak{p}}(D_2)).$$

On the other hand, by definition we have

$$\mathbf{\Psi}_t(\nu : h ; \gamma_{A_\mathfrak{p}}(D_2)) = e^{\langle \nu, H\rangle} \sum_{\lambda \in \mathsf{L}} \mathbf{t}(\nu : D_2 : t : \lambda) e^{-\lambda(H)} \qquad (h \in A_\mathfrak{p}(\eta),\ H = \log h)$$

where $\eta = \max(\eta(t)), \eta(t_2))$. Therefore we may conclude from Lemma 9.1.4.5 that

$$\mathbf{t}(\nu : D_1 D_2 : t) = \mathbf{t}(\nu : D_1 : t_2) \qquad (\text{any } \nu),$$

as desired. The fact that t_2 lies in $\mathsf{t}(\mathsf{M})$ is, of course, a consequence of $(\mathbf{t}_4)$.

With this preparation, let us turn back to the problem of showing that $\mathbf{\Psi}(\nu : .)$ $(\nu \in \mathbf{\Gamma})$ is an 'eigenfunction' for the $\gamma_{A_\mathfrak{p}}(D)$ $(D \in \mathfrak{X})$. In precise terms, if

$$\mathbf{\Psi}(\nu : h) = e^{\langle \nu, H\rangle} \sum_{\lambda \in \mathsf{L}} \mathbf{\Gamma}_\lambda(\nu) e^{-\lambda(H)} \qquad (\nu \in \mathbf{\Gamma},\ h \in \exp(\mathscr{C}))$$

where $H = \log h$, then our problem is to prove that

$$\mathbf{\Psi}(\nu : h ; \gamma_{A_\mathfrak{p}}(D)) = \mathbf{\Psi}(\nu : h)\mu_2(\Omega(D : \nu)) \qquad (\text{all } D \in \mathfrak{X}).$$

Let $\mathscr{O}$ be an open, relatively compact subset of $\mathbf{\Gamma}$; for $\nu \in \mathscr{O}$, let $\mathbf{\Gamma}(\nu)$ denote the function $\lambda \mapsto \mathbf{\Gamma}_\lambda(\nu)$ on L – plainly $\mathbf{\Gamma}(\nu) \in \mathsf{t}(\mathsf{M})$ (cf. Lemma 9.1.4.3). This being so, fix a $D \in \mathfrak{X}$ – then we intend to prove that

$$\mathbf{t}(\nu : D : \mathbf{\Gamma}(\nu)) = \mathbf{\Gamma}(\nu)\mu_2(\Omega(D : \nu)) \qquad (\text{all } \nu \in \mathscr{O}).$$

To begin with, observe that $D\omega = \omega D$ – therefore, in view of $(\mathbf{t}_5)$ above, we have

$$\mathbf{t}(\nu : \omega : \mathbf{t}(\nu : D : \mathbf{\Gamma}(\nu))) = \mathbf{t}(\nu : D : \mathbf{t}(\nu : \omega : \mathbf{\Gamma}(\nu))).$$

Using the definition of the $\mathbf{\Gamma}_\lambda$ $(\lambda \in \mathsf{L})$, one sees without difficulty that

$$\mathbf{t}(\nu : \omega : \mathbf{\Gamma}(\nu)) = \{(\nu, \nu) + 2(\rho, \nu)\}\mathbf{\Gamma}(\nu) + \mathbf{\Gamma}(\nu)\mu_2(\omega_{\mathfrak{m}}) \qquad (\text{all } \nu \in \mathscr{O}).$$

But then it follows from $(\mathbf{t}_2)$ that

$$\begin{aligned}
&\{2(\lambda, \nu) - (\lambda, \lambda - 2\rho)\}\mathbf{t}(\nu : D : \mathbf{\Gamma}(\nu) : \lambda) - \mu_2(\omega_{\mathfrak{m}})\mathbf{t}(\nu : D : \mathbf{\Gamma}(\nu) : \lambda) \\
&\quad + \mathbf{t}(\nu : D : \mathbf{\Gamma}(\nu) : \lambda)\mu_2(\omega_{\mathfrak{m}}) \\
&= 2 \sum_{\alpha \in P_+} \sum_{n \geqslant 1} \{(\tilde{\alpha}, \nu) - (\tilde{\alpha}, \lambda - 2n\tilde{\alpha})\}\mathbf{t}(\nu : D : \mathbf{\Gamma}(\nu) : \lambda - 2n\tilde{\alpha}) \\
&\quad + 8 \sum_{\alpha \in P_+} \sum_{n \geqslant 1} (2n - 1)\mu_1(Y_\alpha)\mu_2(Y_{-\alpha})\mathbf{t}(\nu : D : \mathbf{\Gamma}(\nu) : \lambda - (2n - 1)\tilde{\alpha}) \\
&\quad - 8 \sum_{\alpha \in P_+} \sum_{n \geqslant 1} n\{\mu_1(Y_\alpha Y_{-\alpha}) + \mu_2(Y_\alpha Y_{-\alpha})\}\mathbf{t}(\nu : D : \mathbf{\Gamma}(\nu) : \lambda - 2n\tilde{\alpha}) \\
&\qquad\qquad (\lambda \in \mathsf{L}).
\end{aligned}$$

Now

$$\mathbf{t}(\nu : D : \mathbf{\Gamma}(\nu) : 0) = \mu_2(\Omega(D : \nu))\mathbf{\Gamma}_0(\nu)$$

(($\mathbf{t}_3$) above) while $\mathbf{\Gamma}_0(\nu) = 1$ – therefore

$$\mathbf{t}(\nu : D : \mathbf{\Gamma}(\nu) : 0) = \mathbf{\Gamma}_0(\nu)\mu_2(\Omega(D : \nu)).$$

On the other hand the recurrence relation for $\mathbf{t}(\nu : D : \mathbf{\Gamma}(\nu) : \lambda)$ is exactly the same as that satisfied by $\mathbf{\Gamma}_\lambda(\nu)$ ($\lambda \in \mathsf{L}$); furthermore $\mathbf{t}(\nu : D : \mathbf{\Gamma}(\nu))$ belongs to $\mathfrak{t}(\mathsf{M})$ (cf. ($\mathbf{t}_4$)). This being the case, an induction on m_λ then leads readily to the conclusion that

$$\mathbf{t}(\nu : D : \mathbf{\Gamma}(\nu) : \lambda) = \mathbf{\Gamma}_\lambda(\nu)\mu_2(\Omega(D : \nu)) \qquad (\text{all } \lambda \in \mathsf{L}),$$

as desired.

Fix a ν in $\mathbf{\Gamma}$ and a D in $\mathfrak{X}$ – then, in view of what was said above, we have

$$\mathbf{\Psi}(\nu : h ; \mathfrak{l}_{A_\mathfrak{p}}(D)) = \mathbf{\Psi}(\nu : h)\mu_2(\Omega(D : \nu))$$

for all $h \in A_\mathfrak{p}(\eta)$, η a sufficiently large positive number. But since $\mathbf{\Psi}(\nu : .)$ is an analytic function on $\exp(\mathscr{C})$, this relation must actually be valid for all $h \in \exp(\mathscr{C})$.

Theorem 9.1.4.1. is thereby established.

9.1.5 The Expansion of μ-Spherical Functions

Let G be a connected semi-simple Lie group with finite center, $G = KA_\mathfrak{p}N^+$ an Iwasawa decomposition for G; let M be the centralizer of $A_\mathfrak{p}$ in K. Let $\mu = (\mu_1, \mu_2)$ be a double unitary representation of K on a finite dimensional Hilbert space E; let E_M denote the subspace of E comprised of those elements a which have the property that

$$\mu_1(m)a = a\mu_2(m) \qquad (\text{all } m \in \mathsf{M}).$$

Fix $a \in E$, $\nu \in \mathfrak{a}_{\mathfrak{p}_c}^\vee$ and consider the function

$$\Psi(a : \nu : x) = \int_K \mu_1(\kappa(xk))a\mu_2(k^{-1}) \exp\{(\sqrt{-1}\nu - \rho)(H(xk))\}dk$$
$$(\rho = 2^{-1} \sum_{\lambda>0} m(\lambda)\lambda)$$

for $x \in G$ (we follow here the mode of writing introduced in Vol. I, number 5.5.1) – then, for fixed a and ν, $\Psi(a : \nu : .)$ is a C^∞ μ-spherical function on G. [Note too that for fixed a and x, $\Psi(a : . : x)$ is an entire function of the complex variable ν.] Our objective in the present number is to determine the asymptotic behavior of such functions on $\exp(\mathscr{C})$.

We shall keep to the notations which have been employed in the earlier numbers of this section. Let $\mathbf{\Gamma}(\mathbf{c})$ denote the set of all $H \in \mathfrak{a}_{\mathfrak{p}_c}$ with the property that $wH \in \mathbf{\Gamma}^\gamma$ for every $w \in W_\Sigma$ – then $\mathbf{\Gamma}(\mathbf{c})$ is the complement in $\mathfrak{a}_{\mathfrak{p}_c}$ of the hyperplanes $\sigma_{w\lambda,i}$, whence $\mathbf{\Gamma}(\mathbf{c})$ is an open, connected, dense subset of $\mathfrak{a}_{\mathfrak{p}_c}$. Let $\mathbf{\Gamma}'(\mathbf{c})$ denote the subset of $\mathfrak{a}_{\mathfrak{p}_c}'$ comprised of those H such that $wH \in \mathbf{\Gamma}^\gamma$ for every $w \in W_\Sigma$ – then $\mathbf{\Gamma}'(\mathbf{c})$ is contained in $\mathbf{\Gamma}(\mathbf{c})$ and is itself an open, connected, dense subset of $\mathfrak{a}_{\mathfrak{p}_c}$. [Here W_Σ denotes the Weyl group of the pair $(\mathfrak{g}, \mathfrak{a}_\mathfrak{p})$; moreover $\mathfrak{a}_{\mathfrak{p}_c}'$ denotes the complement in $\mathfrak{a}_{\mathfrak{p}_c}$ of the set of zeros of the polynomial function π_Σ (cf. Vol. I, number 2.1.5).]

The main result can be stated as follows.

Theorem 9.1.5.1 (Harish-Chandra) For any $\nu \in \mathbf{\Gamma}'(\mathbf{c})$ there exist uniquely determined elements $\mathbf{c}_w(\nu) \in \mathrm{Hom}_{\mathbf{C}}(E_{\mathrm{M}}, E_{\mathrm{M}})$ $(w \in W_\Sigma)$ such that

$$h^\rho \int_K \mu_1(\kappa(hk))a\mu_2(k^{-1}) \exp\{(\sqrt{-1}\nu - \rho)(H(hk))\}dk$$
$$= \sum_{w \in W_\Sigma} \mathbf{\Phi}(w\nu : h)\mathbf{c}_w(\nu)a \qquad (h \in \exp(\mathscr{C}))$$

for all $a \in E_{\mathrm{M}}$. Furthermore the $\mathbf{c}_w(w \in W_\Sigma)$ are $\mathrm{Hom}_{\mathbf{C}}(E_{\mathrm{M}}, E_{\mathrm{M}})$-valued holomorphic functions on $\mathbf{\Gamma}'(\mathbf{c})$.

[For the definition of $\mathbf{\Phi}$, see Theorem 9.1.4.1 (bis).]

Let $\|\,.\,\|$ be a norm on $\mathfrak{a}_\mathfrak{p}$; we shall say that $h \to \infty$ $(h \in \exp(\mathscr{C}))$ provided $\|\log h\| \to \infty$ under the restriction that

$$\lambda_i(\log h) \geqslant \epsilon \,\|\log(h)\|$$

for some $\epsilon > 0$ $(1 \leqslant i \leqslant l)$.

Suppose now that ν is real, Σ-regular, and in addition that the $w\nu(w \in W_\Sigma)$ do not lie on any member of a certain finite set of hyperplanes passing through the origin (the real points which do not belong to $\mathbf{\Gamma}^\gamma$; cf. 9.1.4) – then, for all $a \in E_{\mathrm{M}}$, we have the asymptotic relation

$$\lim_{h\to\infty} \Big\| h^\rho \int_K \mu_1(\kappa(hk))a\mu_2(k^{-1})$$
$$\times \exp\{(\sqrt{-1}\nu - \rho)(H(hk))\}dk - \sum_{w \in W_\Sigma} \mathbf{c}_w(\nu)h^{\sqrt{-1}w\nu}a \Big\| = 0,$$

as follows at once from the preceding theorem. Thus, for these ν, the asymptotic behavior of $\Psi(a : \nu : .)$ on $\exp(\mathscr{C})$ is determined by the *values* which the $\mathbf{c}_w$ $(w \in W_\Sigma)$ assume at ν (which then suggests the important problem of finding *explicit* formulae for the $\mathbf{c}_w$ $(w \in W_\Sigma)$ . . . ; cf. 9.1.6).

Example By way of illustration, we shall explicate the preceding results in the following important special case. Let us suppose that G is of split-rank 1 (i.e. that $\dim(A_\mathfrak{p}) = 1$) and let us take for μ the trivial double unitary representation of K on a one dimensional space. In order to determine the asymptotic behavior on $\exp(\mathscr{C})$ of the corresponding μ-spherical functions, it will be convenient to have at hand an explicit expression for $\wp_{A_\mathfrak{p}}(\omega)$ relative to a suitable coordinate system on $\exp(\mathscr{C})$. For this purpose, choose an element $\lambda \in \Sigma^+$ such that 2λ is the only other possible element in Σ^+; let $m(\lambda)$ (respectively $m(2\lambda)$) denote the multiplicity of λ (respectively 2λ); let H be the element in $\mathfrak{a}_\mathfrak{p}$ such that $\lambda(H) = 1$ – then

$$(H, H) = 2 \sum_{\alpha \in P_+} \alpha(H)^2 = 2(m(\lambda) + 4m(2\lambda)),$$

hence $Q_\lambda = (2m(\lambda) + 8m(2\lambda))^{-1}H$ while $Q_\rho = 2^{-1}(m(\lambda) + 2m(2\lambda))Q_\lambda$. It therefore follows from Corollary 9.1.2.12 that

$$2(m(\lambda) + 4m(2\lambda))\mathfrak{P}_{A_\mathfrak{p}}(\omega) = H^2 + (m(\lambda) + 2m(2\lambda))H + \{2m(\lambda)(e^{2\lambda} - 1)^{-1} + 4m(2\lambda)(e^{4\lambda} - 1)^{-1}\}H.$$

[Here, of course, we use the conventions laid down in 9.1.2 relative to the trivial double unitary representation μ; this being the case, the above equality is between differential operators on $A'_\mathfrak{p}$.] Put $t(h) = \lambda(\log h)$ $(h \in A_\mathfrak{p})$ – then t can be regarded as a coordinate function on the one dimensional Lie group $A_\mathfrak{p}$ and so it is clear that

$$\begin{aligned}2(m(\lambda) + 4m(2\lambda))\mathfrak{P}_{A_\mathfrak{p}}(\omega) &= d^2/dt^2 + \{m(\lambda)\coth(t) + 2m(2\lambda)\coth(2t)\}\, d/dt\\ &= d^2/dt^2 + \{(m(\lambda) + m(2\lambda))\coth(t)\\ &\quad + m(2\lambda)\tanh(t)\}\, d/dt\end{aligned}$$

since $\coth(2t) = 2^{-1}(\coth(t) + \tanh(t))$. Now let $x = -(\sinh(t))^2$ – then the above relation becomes

$$2^{-1}(m(\lambda) + 4m(2\lambda))\mathfrak{P}_{A_\mathfrak{p}}(\omega) = x(x-1)d^2/dx^2 + 2^{-1}\{(m(\lambda) + 2m(2\lambda) + 2)x - (m(\lambda) + m(2\lambda) + 1)\}\, d/dx.$$

Identifying the space E above with the complex numbers and taking $a = 1$, we see that the μ-spherical functions $\Psi(\nu:.)$ in question are precisely the zonal spherical functions ϕ_ν

$$\left(\phi_\nu(x) = \int_K e^{(\sqrt{-1}\nu-\rho)(H(xk))}\, dk,\ x \in G\right)$$

on G; we intend to show that in the present special case, the ϕ_ν can be identified in a simple way with hypergeometric functions $F(a, b; c; x)$ – once this is done, it will be easy to read off their asymptotic behavior on $\exp(\mathscr{C})$ and at the same time obtain explicit formulae for the $\mathbf{c}_w(\nu)$ $(w \in W_\Sigma)$. To this end, note that here $W_\Sigma = \{1, w\}$ where $wH = -H$ $(H \in \mathfrak{a}_\mathfrak{p})$; in particular, then, the restriction to $A_\mathfrak{p}$ of any zonal spherical function ϕ_ν on G has the property that

$$\phi_\nu(h) = \phi_\nu(w\cdot h) = \phi_\nu(h^{-1}) \qquad (\text{all } h \in A_\mathfrak{p})$$

(this means that ϕ_ν is 'even', i.e. that ϕ_ν can be expanded in powers of t^2 around $h = 1$). Because ϕ_ν is an eigenfunction for ω (with a prescribed eigenvalue . . .), Theorem 9.1.2.9 (in conjunction with the above expression for $\mathfrak{P}_{A_\mathfrak{p}}(\omega)$) implies that

$$\begin{aligned}x(x-1)d^2\phi_\nu/dx^2 &+ 2^{-1}\{(m(\lambda) + 2m(2\lambda) + 2)x\\ &- (m(\lambda) + m(2\lambda) + 1)\}d\phi_\nu/dx\\ &+ 4^{-1}\{\nu(H)^2 + (2^{-1}m(\lambda) + m(2\lambda))^2\}\phi_\nu = 0.\end{aligned}$$

Let

$$\begin{aligned}a &= \{m(\lambda) + 2m(2\lambda) + 2\sqrt{-1}\nu(H)\}/4,\\ b &= \{m(\lambda) + 2m(2\lambda) - 2\sqrt{-1}\nu(H)\}/4,\\ c &= (m(\lambda) + m(2\lambda) + 1)/2.\end{aligned}$$

Then the above equation becomes

$$(*) \quad x(x-1)d^2\phi_\nu/dx^2 + \{(a + b + 1)x - c\}\, d\phi_\nu/dx + ab\phi_\nu = 0.$$

Let $\mathscr{C}$ denote the open positive Weyl chamber in $\mathfrak{a}_\mathfrak{p}$ – thus $\mathscr{C} \sim \{t \in \mathbf{R} : t > 0\}$

and so x may be viewed as a coordinate function on $\exp(\mathscr{C})$. Because $c > 0$ and because we are interested only in those solutions of $(*)$ which are analytic at $x = 0$, it follows from the 'evenness' of ϕ_ν that

$$\phi_\nu(\exp(tH)) = F(a, b; c; -(\sinh(t))^2) \qquad (t \geqslant 0).$$

In order to determine the asymptotic behavior of ϕ_ν as $t \to \infty$, we shall employ the following well-known formula (which is valid provided $a - b = \sqrt{-1}\nu(H)$ is not an integer):

$$\begin{aligned} F(a, b; c; x) = {} & |x|^{-a}\Gamma(c)\Gamma(b-a)\{\Gamma(b)\Gamma(c-a)\}^{-1} \\ & \times F(a, 1+a-c; 1+a-b; x^{-1}) \\ & + |x|^{-b}\Gamma(c)\Gamma(a-b)\{\Gamma(a)\Gamma(c-b)\}^{-1} \\ & \times F(b, 1+b-c; 1+b-a; x^{-1}) \qquad (x < -1), \end{aligned}$$

Γ the classical Gamma function. [See Whittaker and Watson [1, p. 289]; as was pointed out by Bargmann [1, p. 627] the formula given there is not correct – rather the differences $a - b, a - c, b - a, b - c$ which appear as arguments of the Γ-function must be replaced by $b - a, c - a, a - b, c - b$, respectively.] Plainly

$$\begin{cases} |-\sinh(t)|^{-2a} = ((e^t - e^{-t})/2)^{-\rho(H)-\sqrt{-1}\nu(H)} \\ |-\sinh(t)|^{-2b} = ((e^t - e^{-t})/2)^{-\rho(H)+\sqrt{-1}\nu(H)} \end{cases}$$

and so it is clear that for real $\nu \neq 0$,

$$\lim_{t\to\infty} |e^{t\rho(H)}\phi_\nu(\exp(tH)) - \mathbf{c}(\nu)e^{t\sqrt{-1}\nu(H)} - \mathbf{c}(-\nu)e^{-t\sqrt{-1}\nu(H)}| = 0$$

where

$$\begin{aligned} \mathbf{c}(\nu) = {} & 2^{2^{-1}(m(\lambda)+2m(2\lambda))-\sqrt{-1}\nu(H)}\Gamma(2^{-1}(m(\lambda)+m(2\lambda)+1))\Gamma(\sqrt{-1}\nu(H)) \\ & \times \{\Gamma(4^{-1}(m(\lambda)+2m(2\lambda)+2\sqrt{-1}\nu(H))) \\ & \times \Gamma(4^{-1}(m(\lambda)+2+2\sqrt{-1}\nu(H)))\}^{-1}. \end{aligned}$$

[Thus in the present case we have the functional relation $\mathbf{c}_w(\nu) = \mathbf{c}(w\nu)$ ($w \in W_\Sigma$); it will be seen in due course that this phenomenon is true 'in general' (so long as it is a question of zonal spherical functions).] Let us recast the above expression for $\mathbf{c}(\nu)$ in terms of the Beta function B:

$$B(p, q) = \Gamma(p)\Gamma(q)/\Gamma(p+q).$$

This is easily done if one uses Legendre's Duplication Formula for the Gamma function ($\sqrt{\pi}\,\Gamma(2z) = 2^{2z-1}\Gamma(z)\Gamma(z+2^{-1})$) – the result is

$$\mathbf{c}(\nu) = \frac{B\left(\dfrac{m(\lambda)}{2}, \dfrac{(\sqrt{-1}\nu, \lambda)}{(\lambda, \lambda)}\right) B\left(\dfrac{m(2\lambda)}{2}, \dfrac{m(\lambda)}{4} + \dfrac{(\sqrt{-1}\nu, 2\lambda)}{(2\lambda, 2\lambda)}\right)}{B\left(\dfrac{m(\lambda)}{2}, \dfrac{(\rho, \lambda)}{(\lambda, \lambda)}\right) B\left(\dfrac{m(2\lambda)}{2}, \dfrac{m(\lambda)}{4} + \dfrac{(\rho, 2\lambda)}{(2\lambda, 2\lambda)}\right)}$$

(ν real and regular).

[It is more than mere coincidence that this formula for $\mathbf{c}$ is the same as that encountered in the study of intertwining operators for the 'principal P-series' . . . .] It is apparent then that for real non-zero ν (in which case ϕ_ν is positive definite – cf. number 6.2.2), the behavior at infinity of ϕ_ν is 'oscillatory' in character.

The uniqueness assertion in Theorem 9.1.5.1 is straightforward; thus, since ν is Σ-regular, the elements $w\nu$ ($w \in W_\Sigma$) are all distinct – on the other hand we have the following lemma.

Lemma 9.1.5.2 Let $\nu_1, \ldots, \nu_r$ be distinct elements in $\mathbf{\Gamma}^\gamma$; suppose that $a_1, \ldots, a_r$ are elements in E_M such that

$$\sum_{i=1}^{r} \Phi(\nu_i : h) a_i = 0$$

for all $h \in \exp(\mathscr{C})$ – then $a_1 = a_2 = \ldots = a_r = 0$.

Proof Let I be the set of indices i such that $a_i \neq 0$; proceeding by contradiction, let us suppose that I is not empty. Let $\sqrt{-1}\nu_j$ be the highest element among the $\sqrt{-1}\nu_i$ ($i \in I$)–then, in view of what has been said in 9.1.4, we can equate the coefficient of $e^{\sqrt{-1}\langle \nu_j, \cdot \rangle}$ to zero in the equation

$$\sum_{i \in I} \sum_{\lambda \in L} e^{\sqrt{-1}\langle \nu_i, \cdot \rangle - \lambda(\cdot)} \mathbf{\Gamma}_\lambda(\sqrt{-1}\nu_i - \rho) a_i = 0.$$

Since $\mathbf{\Gamma}_0(\sqrt{-1}\nu_j - \rho) = 1$, we deduce that $a_j = 0 \ldots$ □

Consequently the endomorphisms $\mathbf{c}_w$ ($w \in W_\Sigma$) which appear in the statement of Theorem 9.1.5.1 are in any event unique; before taking up the more delicate question of their existence, it will be necessary to digress and develop some additional machinery.

We recall that W_Φ is the Weyl group of the pair $(\mathfrak{g}_c, \mathfrak{a}_c)$. Fix a point $H_0 \in \mathfrak{a}_c$; let W_{H_0} be the subgroup of W_Φ consisting of those w which leave H_0 fixed – then W_{H_0} operates on $S(\mathfrak{a}_c)$ in the obvious way and the corresponding set of invariants $I_{H_0}(\mathfrak{a}_c)$ contains the set $I(\mathfrak{a}_c)$ of all W_Φ-invariants in $S(\mathfrak{a}_c)$. Let $I_{H_0}^+(\mathfrak{a}_c)$ denote the space spanned by the homogeneous elements in $I_{H_0}(\mathfrak{a}_c)$ of degree $\geqslant 1$; let $\mathfrak{E}(H_0)$ be the subspace of $S(\mathfrak{a}_c)$ comprised of those p such that $\langle p, q \rangle = 0$ for every $q \in S(\mathfrak{a}_c) I_{H_0}^+(\mathfrak{a}_c)$ ($\langle \cdot, \cdot \rangle$ as defined in Vol. I, number 2.1.4) – then it is clear that an element $p \in S(\mathfrak{a}_c)$ lies in $\mathfrak{E}(H_0)$ iff $\partial(q)p = 0$ (all $q \in I_{H_0}^+(\mathfrak{a}_c)$).

Lemma 9.1.5.3 Let p_0 be an element in $S(\mathfrak{a}_c)$; let f_0 denote the function $H \mapsto p_0(H)e^{(H_0, H)}$ on $\mathfrak{a}_c$ – then the following conditions are equivalent:

(i) p_0 lies in $\mathfrak{E}(H_0)$;
(ii) f_0 is an eigenfunction of $\partial(p)$ for every p in $I_{H_0}(\mathfrak{a}_c)$;
(iii) f_0 is an eigenfunction of $\partial(p)$ for every p in $I(\mathfrak{a}_c)$.

Furthermore $\dim(\mathfrak{E}(H_0)) = [W_{H_0}]$.

Corollary 9.1.5.4 Retain the above notations and assumptions; suppose that $W_{H_0} = \{1\}$ – then $\mathfrak{E}(H_0) = \mathbf{C}$.

On the basis of the lemma, the corollary is obvious; in turn a proof of the lemma may be found in the Appendix to this number.

Put

$$\mathfrak{E}_{\mathfrak{p}}(H_0) = \mathfrak{E}(H_0) \cap S(\mathfrak{a}_{\mathfrak{p}_c});$$

for any finite dimensional vector space E (over **C**) we shall write $\mathfrak{E}_{\mathfrak{p}}(H_0 : E)$ instead of $\mathfrak{E}_{\mathfrak{p}}(H_0) \otimes_{\mathbf{C}} E$.

Let us return to our double representation $\mu = (\mu_1, \mu_2)$ of K on E. Let $\sigma_1, \ldots, \sigma_r$ be all the distinct classes in $\hat{\mathsf{M}}$ which occur in $\mu_1 | \mathsf{M}$; let $E(\sigma_i)$ denote the subspace of E comprised of those elements in E which transform under $\mu_1 | \mathsf{M}$ according to σ_i, $\mathsf{P}(\sigma_i)$ the canonical projection of E onto $E(\sigma_i)$ $(1 \leqslant i \leqslant r)$. Because $\mu_1(m)$ $(m \in \mathsf{M})$ commutes with $\mathsf{P}(\sigma_i)$ for every i, it is clear that $E_{\mathsf{M}} = \sum_{i=1}^r E_{\mathsf{M}}^i$ where $E_{\mathsf{M}}^i = \mathsf{P}(\sigma_i)(E_{\mathsf{M}})$ $(1 \leqslant i \leqslant r)$.

Fix an i $(1 \leqslant i \leqslant r)$ – then $E(\sigma_i)$ can be written as the direct sum of finitely many irreducible M-modules each of which is in the class σ_i. A representation in the class σ_i gives rise, upon differentiation, to a representation of $\mathfrak{m}_c$; this representation of $\mathfrak{m}_c$ need not, however, be irreducible (for M is not necessarily connected) but fortunately, in the case at hand, it is at least a *multiple* of an irreducible representation. [There are at least two ways to see this; here is one. Fix a class $\sigma \in \hat{\mathsf{M}}$ and a representation μ_σ in the class σ on a space E_σ (say); when restricted to M (the identity component of M), μ_σ breaks up into a finite number of irreducibles – let $E_\sigma = \sum_i E_i$ be the corresponding direct decomposition of E_σ into simple M-modules. Now write $\mathsf{M} = \mathsf{Z}(A_{\mathfrak{p}})M$ (cf. Lemma 1.1.3.8) – then $\mathsf{Z}(A_{\mathfrak{p}})$ is a finite group whose elements commute with the elements of M. This latter property of $\mathsf{Z}(A_{\mathfrak{p}})$ tells us that the $\mu_\sigma(a)$ $(a \in \mathsf{Z}(A_{\mathfrak{p}}))$ are M-module homomorphisms of E_σ. This being so, fix an i – then the $\mu_\sigma(a)E_i$ $(a \in \mathsf{Z}(A_{\mathfrak{p}}))$ are simple M-modules in the same class as E_i and, moreover, they span E_σ (since $\mathsf{M} = \mathsf{Z}(A_{\mathfrak{p}})M$ and μ_σ is irreducible). Our contention is an immediate consequence of these facts.] Therefore $E(\sigma_i)$ is a direct sum of simple $\mathfrak{m}_c$-modules all of which are mutually equivalent; we shall let Λ_i denote the highest weight of any of the irreducible constituents of $E(\sigma_i)$ (when viewed as an $\mathfrak{m}_c$-module). [Let us bear in mind that $\mathfrak{a}_{\mathfrak{k}_c}$ is a Cartan subalgebra of $\mathfrak{m}_c$ ($\mathfrak{m}_c$ itself being reductive); furthermore P_- may be regarded as the set of positive roots of the pair $(\mathfrak{m}_c, \mathfrak{a}_{\mathfrak{k}_c})$.]

In what follows, $\mathfrak{a}_c$ will be identified with its dual in the usual way; similarly for $\mathfrak{a}_{\mathfrak{k}_c}$ and $\mathfrak{a}_{\mathfrak{p}_c}$ – in particular linear functions on $\mathfrak{a}_{\mathfrak{k}_c}$ (or $\mathfrak{a}_{\mathfrak{p}_c}$) may be regarded as linear functions on $\mathfrak{a}_c$ with no change in notation.

Here is the technical keystone for the present investigation.

Proposition 9.1.5.5 Let Λ be a linear function on $\mathfrak{a}_{\mathfrak{p}_c}$; let p be a polynomial mapping of $\mathfrak{a}_{\mathfrak{p}_c}$ into E_{M}^i and suppose that the function ϕ,

$$h \mapsto p(\log h)\, h^{\Lambda - \rho} \qquad (h \in \exp(\mathscr{C})),$$

is an eigenfunction of the operator $\mathfrak{l}_{A\mathfrak{p}}(\mu_-^*(Z))$ for all $Z \in \mathfrak{Z}$ – then

$$p \in \mathfrak{E}_{\mathfrak{p}}(\Lambda + \Lambda_i + \rho_- : E_M^i) \qquad (\rho_- = 2^{-1} \sum_{\alpha \in P_-} \alpha).$$

[In the remainder of the present number we shall often write μ_-^* for $\mu_-(.)^\flat$; cf. 9.1.2.]

Proof Fix a $Z \in \mathfrak{Z}$; write

$$\mu_-^*(Z)(= \mu_-(Z)^\flat) = \sum_{j=1}^{s} H_j Z_j$$

where the $H_j \in \mathfrak{A}_{\mathfrak{p}}$ and the $Z_j \in \mathfrak{Z}_-$ – it can be assumed that the Z_j lie in the center of $\mathfrak{M}$, whence the H_j may be taken linearly independent. Thus

$$\mathfrak{l}_{A\mathfrak{p}}(\mu_-^*(Z)) = \sum_j 1 \otimes H_j \otimes Z_j$$

and so

$$\phi(h; \mathfrak{l}_{A\mathfrak{p}}(\mu_-^*(Z))) = \sum_j \phi(h; H_j)\mu_2(Z_j) = \sum_j \mu_1(Z_j)\phi(h; H_j).$$

As mentioned above, $E(\sigma_i)$ is a direct sum of simple $\mathfrak{M}$-modules; each of the irreducible constituents is mutually equivalent with highest weight Λ_i (say); on the other hand the Z_j lie in the center of $\mathfrak{M}$ and hence are represented by scalars on $E(\sigma_i)$ – in fact, bearing in mind the definition of γ_- (cf. 9.1.2), we have

$$\mu_1(Z_j)\mathsf{P}(\sigma_i) = \langle \gamma_-(Z_j), \mathbf{e}^{\Lambda_i + \rho_-} \rangle \mathsf{P}(\sigma_i)$$

($1 \leqslant j \leqslant s$; cf. Vol. I, number 2.4.3). Taking into account the relations

$$\mu_-(Z) = \sum_j {}^\flat H_j Z_j$$

and

$$\gamma(Z) = \sum_j {}^\flat H_j \gamma_-(Z_j),$$

we then find that

$$\begin{aligned} \phi(h; \mathfrak{l}_{A\mathfrak{p}}(\mu_-^*(Z))) &= \sum_j \langle \gamma_-(Z_j), \mathbf{e}^{\Lambda_i + \rho_-} \rangle \phi(h; H_j) \\ &= \{\langle \gamma(Z), \mathbf{e}^{\Lambda + \Lambda_i + \rho_-} \rangle p(\log h) + q(\log h)\} h^{\Lambda - \rho} \\ &\qquad\qquad (h \in \exp(\mathscr{C})) \end{aligned}$$

where q is a polynomial function from $\mathfrak{a}_{\mathfrak{p}_c}$ to E_M^i with $\deg(q) \leqslant \deg(p) - 1$. But ϕ is by hypothesis an eigenfunction of $\mathfrak{l}_{A\mathfrak{p}}(\mu_-^*(Z))$; this fact implies that $q = 0$, whence

$$\sum_j \langle \gamma_-(Z_j), \mathbf{e}^{\Lambda_i + \rho_-} \rangle H_j \phi = \langle \gamma(Z), \mathbf{e}^{\Lambda + \Lambda_i + \rho_-} \rangle \phi.$$

Now extend p, in the obvious way, to a polynomial function on all of $\mathfrak{a}_c$; put

$$f(H) = p(H)e^{(\Lambda + \Lambda_i + \rho_-)(H)}, \ F(H) = p(H)e^{\Lambda(H)} \qquad (H \in \mathfrak{a}_c),$$

so that $f = Fe^{\Lambda_i+\rho_-}$ – then, in view of what was said above, we have

$$\begin{aligned}\partial(\gamma(Z))f &= \sum_j \langle \gamma_-(Z_j), e^{\Lambda_i+\rho_-}\rangle e^{\Lambda_i+\rho_-}\partial({}^{\flat}H_j)F \\ &= \langle \gamma(Z), e^{\Lambda+\Lambda_i+\rho_-}\rangle Fe^{\Lambda_i+\rho_-} = \langle \gamma(Z), e^{\Lambda+\Lambda_i+\rho_-}\rangle f.\end{aligned}$$

Because γ maps $\mathfrak{Z}$ onto $I(\mathfrak{a}_c)$, it follows that f is an eigenfunction of every element in $I(\mathfrak{a}_c)$. Therefore

$$p \in \mathfrak{E}_{\mathfrak{p}}(\Lambda + \Lambda_i + \rho_- : E^i_M)$$

from Lemma 9.1.5.3. □

Corollary 9.1.5.6 Retain the notations and assumptions of the preceding proposition; suppose that Λ takes only pure imaginary values on $\mathfrak{a}_\mathfrak{p}$ and $w\Lambda \neq \Lambda$ for all $w \neq 1$ in W_Σ – then p is a constant.

Proof Owing to Corollary 9.1.5.4, we need only verify that

$$(\Lambda + \Lambda_i + \rho_-, \alpha) \neq 0$$

for every $\alpha \in \Phi^+$ (Φ^+ the set of positive roots of the pair $(\mathfrak{g}_c, \mathfrak{a}_c)$). So suppose, if possible, that α is a positive root such that $(\Lambda + \Lambda_i + \rho_-, \alpha) = 0$; since (Λ, α) is pure imaginary while $(\Lambda_i + \rho_-, \alpha)$ is real, this implies that $(\Lambda, \alpha) = (\Lambda_i + \rho_-, \alpha) = 0$. From our hypothesis on Λ, it follows that $(\Lambda, \beta) \neq 0$ for any $\beta \in P_+$, whence $\alpha \in P_-$. But Λ_i is the highest weight of an irreducible representation of $\mathfrak{m}_c$ – therefore

$$(\Lambda_i + \rho_-, \alpha) \neq 0$$

for all $\alpha \in P_-$. . . . [The fact that the restriction of the Killing form of $\mathfrak{g}_c$ to $\mathfrak{a}_{\mathfrak{k}_c}$ need not agree with the restriction of the Killing form of $\mathfrak{m}_c$ to $\mathfrak{a}_{\mathfrak{k}_c}$ causes no difficulty here.] □

Let W_σ denote the subgroup of W_Φ consisting of those elements in W_Φ which map $\mathfrak{a}_{\mathfrak{p}_c}$ into itself – then, since the operations of W_Φ preserve the Killing form, it follows that an element $w \in W_\Phi$ lies in W_σ iff

$$w\mathfrak{a}_{\mathfrak{k}_c} = \mathfrak{a}_{\mathfrak{k}_c}.$$

Let $\complement W_\sigma$ denote the complement of W_σ in W_Φ – then

$$\sqrt{-1}(w\mathfrak{a}_\mathfrak{k} + \mathfrak{a}_\mathfrak{k}) \cap \mathfrak{a}_\mathfrak{p} \neq \{0\}$$

for every $w \in \complement W_\sigma$. Given a $w \in \complement W_\sigma$, let us fix once and for all a nonzero element $H(w)$ in $\mathfrak{a}_\mathfrak{p}$ such that $H(w) \equiv wH \bmod \sqrt{-1}\,\mathfrak{a}_\mathfrak{k}$ for some $H \in \sqrt{-1}\mathfrak{a}_\mathfrak{k}$. On the other hand, consider the subgroup W_- of W_Φ which is generated by the Weyl reflections w_α ($\alpha \in P_-$) – then W_- is a normal subgroup of W_σ and W_Σ may be identified with the factor group W_σ/W_- (cf. Proposition 1.1.3.3). Owing to the normality of W_- in W_σ, it is clear that W_σ operates on the space $I_-(\mathfrak{a}_c)$ of W_--invariants in $S(\mathfrak{a}_c)$; moreover if $w_\sigma \in W_\sigma$ and $p \in I_-(\mathfrak{a}_c)$, then p^{w_σ} depends only on the coset $w_\sigma W_-$ –

therefore W_Σ operates on $I_-(\mathfrak{a}_c)$. In view of the isomorphism

$$\gamma_- : \mathfrak{Z}_- \to I_-(\mathfrak{a}_c),$$

W_Σ also operates on $\mathfrak{Z}_-$ (by transport of structure).

Let Λ be a linear function on $\mathfrak{a}_{\mathfrak{p}_c}$ – then, as follows from the proof of Proposition 9.1.5.5, we have

$$\mu_1(\Omega(Z:\Lambda - \rho))\mathsf{P}(\sigma_i) = \langle \gamma(Z), e^{\Lambda+\Lambda_i+\rho_-}\rangle \mathsf{P}(\sigma_i) \qquad (1 \leqslant i \leqslant r;\ Z \in \mathfrak{Z}).$$

Fix i_0 and w_0 $(1 \leqslant i_0 \leqslant r;\ w_0 \in W_\Sigma)$ and let $E_{\mathsf{M}}(i_0, w_0)$ denote the set comprised of those $a \in E_{\mathsf{M}}$ which have the property that

$$\mu_1(\Omega(Z:w_0\Lambda - \rho))a = \langle \gamma(Z), e^{\Lambda+\Lambda_{i_0}+\rho_-}\rangle\, a$$

for all $Z \in \mathfrak{Z}$ and all $\Lambda \in \mathfrak{a}_{\mathfrak{p}_c}^\vee$. Let $I(i_0, w_0)$ denote the set of indices i $(1 \leqslant i \leqslant r)$ such that the homomorphisms

$$Z \mapsto \langle \gamma(Z), e^{w_0\Lambda+\Lambda_i+\rho_-}\rangle,\ Z \mapsto \langle \gamma(Z), e^{\Lambda+\Lambda_{i_0}+\rho_-}\rangle \qquad (Z \in \mathfrak{Z})$$

are equal for every Λ; since

$$\mu_1(\Omega(Z:w_0\Lambda - \rho))\mathsf{P}(\sigma_i) = \langle \gamma(Z), e^{w_0\Lambda+\Lambda_i+\rho_-}\rangle \mathsf{P}(\sigma_i) \qquad (Z \in \mathfrak{Z}),$$

it follows that

$$E_{\mathsf{M}}(i_0, w_0) = \sum_i E_{\mathsf{M}}^i,\ i \in I(i_0, w_0).$$

Lemma 9.1.5.7 Retain the above notations; fix i_0, w_0 $(1 \leqslant i_0 \leqslant r;\ w_0 \in W_\Sigma)$ – then, for any index i $(1 \leqslant i \leqslant r)$, the following statements are equivalent:

(i) The index i belongs to $I(i_0, w_0)$;

(ii) There exists a linear function Λ on $\mathfrak{a}_{\mathfrak{p}_c}$ such that (a) $w\Lambda \neq \Lambda$ for any $w \neq 1$ in W_Σ, (b) Λ takes only pure imaginary values on $\mathfrak{a}_{\mathfrak{p}}$, (c) $\Lambda(H(w)) \neq 0$ (all $w \in \complement W_\sigma$), and (d) the homomorphisms

$$Z \mapsto \langle \gamma(Z), e^{w_0\Lambda+\Lambda_i+\rho_-}\rangle,\ Z \mapsto \langle \gamma(Z), e^{\Lambda+\Lambda_{i_0}+\rho_-}\rangle \qquad (Z \in \mathfrak{Z})$$

are equal;

(iii) For all $Z \in \mathfrak{Z}_-$, $\langle \gamma_-(w_0 Z), e^{\Lambda_i+\rho_-}\rangle = \langle \gamma_-(Z), e^{\Lambda_{i_0}+\rho_-}\rangle$.

Proof (i) ⇒ (ii). This is obvious.

(ii) ⇒ (iii). In view of condition (ii) (d), Theorem 2.4.3.2 implies that there exists a $w \in W_\Phi$ such that $w(\Lambda + \Lambda_{i_0} + \rho_-) = w_0\Lambda + \Lambda_i + \rho_-$; but then $w \in W_\sigma$, whence $w\Lambda = w_0\Lambda$ and $w(\Lambda_{i_0} + \rho_-) = \Lambda_i + \rho_-$. [To see that $w \in W_\sigma$, one may reason as follows. Suppose to the contrary that $w \in \complement W_\sigma$ – then, since W_σ is a group, $w^{-1} \in \complement W_\sigma$ too. Choose elements H and $\tilde{H}$ in $\sqrt{-1}\mathfrak{a}_\mathfrak{k}$ such that $H(w^{-1}) = w^{-1}(H) + \tilde{H}$ – then

$$\begin{aligned}(\Lambda_i + \rho_-)(H) &= (w_0\Lambda + \Lambda_i + \rho_-)(H) = (\Lambda + \Lambda_{i_0} + \rho_-)(w^{-1}(H)) \\ &= \Lambda(H(w^{-1})) - (\Lambda_{i_0} + \rho_-)(\tilde{H}).\end{aligned}$$

Because $\Lambda_i + \rho_-$ and $\Lambda_{i_0} + \rho_-$ take only real values on $\sqrt{-1}\mathfrak{a}_{\mathfrak{k}}$, it follows that $\Lambda(H(w^{-1}))$ is real; condition (ii) (b) on Λ now implies that $\Lambda(H(w^{-1})) = 0$ which, however, contradicts condition (ii) (c). Therefore $w \in W_\sigma$.] Since $w\Lambda = w_0\Lambda$, the image of w in $W_\Sigma = W_\sigma/W_-$ is w_0 itself (condition (ii) (a) on Λ); this being so, let $Z \in \mathfrak{Z}_-$ – then

$$\langle \gamma_-(w_0 Z), \mathrm{e}^{\Lambda_i+\rho_-}\rangle = \langle w(\gamma_-(Z)), \mathrm{e}^{\Lambda_i+\rho_-}\rangle = \langle \gamma_-(Z), \mathrm{e}^{\Lambda_{i_0}+\rho_-}\rangle,$$

as desired.

(iii) $\Rightarrow$ (i). Fix $Z \in \mathfrak{Z}$; write $\mu_-(Z) = \sum_{j=1}^s H_j Z_j$ where the $H_j \in \mathfrak{A}_\mathfrak{p}$ and the $Z_j \in \mathfrak{Z}_- \cap \mathfrak{M}$ – then $\gamma(Z) = \sum_j H_j \gamma_-(Z_j)$ and for any Λ, we have

$$\begin{aligned}\langle \gamma(Z), \mathrm{e}^{w_0\Lambda+\Lambda_i+\rho_-}\rangle &= \sum_j \langle H_j, \mathrm{e}^\Lambda\rangle \langle \gamma_-(w_0 Z_j), \mathrm{e}^{\Lambda_i+\rho_-}\rangle \\ &= \sum_j \langle H_j, \mathrm{e}^\Lambda\rangle \langle \gamma_-(Z_j), \mathrm{e}^{\Lambda_{i_0}+\rho_-}\rangle = \langle \gamma(Z), \mathrm{e}^{\Lambda+\Lambda_{i_0}+\rho_-}\rangle\end{aligned}$$

which shows that (iii) $\Rightarrow$ (i).

Hence the lemma. □

Agreeing to retain the above notations, let us assume that κ is a character of $\mathfrak{Z}$. Put $G^+ = K\exp(\mathscr{C})K$ (whence G^+ is open in G – see below); let $\mathfrak{S}^+(\mu, \kappa)$ denote the set of all E-valued C^∞ functions Ψ on G^+ such that

$$\Psi(k_1 x k_2) = \mu_1(k_1)\Psi(x)\mu_2(k_2) \quad (x \in G^+; k_1, k_2 \in K)$$

and which verify the relation $Z\Psi = \kappa(Z)\Psi$ (all $Z \in \mathfrak{Z}$). Let $\mathfrak{s}^+(\mu, \kappa)$ denote the set of all E_{M}-valued C^∞ functions ψ on $\exp(\mathscr{C})$ such that $\mathfrak{l}_{A_\mathfrak{p}}(Z)\psi = \kappa(Z)\psi$ (all $Z \in \mathfrak{Z}$). For any $\Psi \in \mathfrak{S}^+(\mu, \kappa)$, let ψ denote the restriction of Ψ to $\exp(\mathscr{C})$ – then the map $\Psi \mapsto \psi$ is a one-to-one mapping of $\mathfrak{S}^+(\mu, \kappa)$ onto $\mathfrak{s}^+(\mu, \kappa)$. [It is well-known that the mapping ϕ,

$$(k_1, h, \dot{k}_2) \mapsto k_1 \cdot {}^{\dot{k}_2}h \quad (k_1 \in K, h \in \exp(\mathscr{C}), \dot{k}_2 = k_2\mathsf{M} \in K/\mathsf{M})$$

of $K \times \exp(\mathscr{C}) \times (K/\mathsf{M})$ into G^+ is a surjective analytic diffeomorphism. This being so, let f be any E_{M}-valued C^∞ function on $\exp(\mathscr{C})$; define an E-valued C^∞ function F_ϕ on $K \times \exp(\mathscr{C}) \times (K/\mathsf{M})$ by the rule

$$F_\phi(k_1, h, \dot{k}_2) = \mu_1(k_1)(\mu_1(k_2)f(h)\mu_2(k_2^{-1}))$$

and put $F = F_\phi \circ \phi^{-1}$ – then clearly F is a C^∞ function on G^+ such that

$$F(k_1 h k_2) = \mu_1(k_1)f(h)\mu_2(k_2)\,(h \in \exp(\mathscr{C}); k_1, k_2 \in K).$$

This shows that any E_{M}-valued C^∞ function f on $\exp(\mathscr{C})$ can be extended to an E-valued C^∞ function F on G^+ by writing

$$F(k_1 h k_2) = \mu_1(k_1)f(h)\mu_2(k_2)\,(h \in \exp(\mathscr{C}); k_1, k_2 \in K).]$$

On the other hand, the proof of Theorem 9.1.3.2 is applicable to $\mathfrak{S}^+(\mu, \kappa)$ without any change whatsoever; in particular, for any linear function ν

on $\mathfrak{a}_{\mathfrak{p}_c}$, it makes sense to speak of the principal exponents of

$$e^{\nu}\psi \quad (\psi \in \mathfrak{F}^+(\mu, \kappa)) \ldots .$$

This being so, let us record the following points of information.

($\mathbf{P}_1$) Let Φ be a non-zero E_M-valued C^∞ function on exp ($\mathscr{C}$) which is an eigenfunction of the operator $e^{\rho}\mathfrak{P}_{A_\mathfrak{p}}(Z)\circ e^{-\rho}$ for all $Z \in \mathfrak{Z}$; let Λ be a principal exponent of Φ – then there exists an index i $(1 \leqslant i \leqslant r)$ such that

$$(e^{\rho}\mathfrak{P}_{A_\mathfrak{p}}(Z)\circ e^{-\rho})\Phi = \langle \gamma(Z), \mathbf{e}^{\Lambda+\Lambda_i+\rho_-}\rangle \Phi \qquad (\text{all } Z \in \mathfrak{Z}).$$

($\mathbf{P}_2$) Let Φ be a non-zero E_M-valued C^∞ function on exp ($\mathscr{C}$) which is an eigenfunction of the operator $e^{\rho}\mathfrak{P}_{A_\mathfrak{p}}(Z)\circ e^{-\rho}$ for all $Z \in \mathfrak{Z}$; let Λ and $\tilde{\Lambda}$ be two principal exponents of Φ – then there exist indices i and $\tilde{i}$ $(1 \leqslant i, \tilde{i} \leqslant r)$ and an element $w \in W_\Phi$ such that

$$\tilde{\Lambda} + \Lambda_{\tilde{i}} + \rho_- = w(\Lambda + \Lambda_i + \rho_-).$$

($\mathbf{P}_3$) Fix an index i $(1 \leqslant i \leqslant r)$; let Φ be a non-zero E_M-valued C^∞ function on exp ($\mathscr{C}$) such that

$$(e^{\rho}\mathfrak{P}_{A_\mathfrak{p}}(Z)\circ e^{-\rho})\Phi = \langle \gamma(Z), \mathbf{e}^{\Lambda+\Lambda_i+\rho_-}\rangle \Phi \qquad (\text{all } Z \in \mathfrak{Z})$$

where Λ is a linear function on $\mathfrak{a}_{\mathfrak{p}_c}$. If Λ takes only pure imaginary values on $\mathfrak{a}_\mathfrak{p}$ and has the property that $\Lambda(H(w)) \neq 0$ for all $w \in \complement W_\sigma$, then every principal exponent of Φ is of the form $w\Lambda$ $(w \in W_\Sigma)$.

Verification of ($\mathbf{P}_1$) Let κ be the homomorphism of $\mathfrak{Z}$ into $\mathbf{C}$ such that $(e^{\rho}\mathfrak{P}_{A_\mathfrak{p}}(Z)\circ e^{-\rho})\Phi = \kappa(Z)\Phi$ (all $Z \in \mathfrak{Z}$); let Φ^P be the principal part of Φ – then $(e^{\rho}\mathfrak{P}_{A_\mathfrak{p}}(\mu_-^*(Z))\circ e^{-\rho})\Phi^P = \kappa(Z)\Phi^P$ (all $Z \in \mathfrak{Z}$). [Here we have to bear in mind Corollary 9.1.2.8.] Let $p_\Lambda e^\Lambda$ be the term in Φ^P corresponding to the exponent Λ – then p_Λ is a polynomial function from $\mathfrak{a}_{\mathfrak{p}_c}$ to E_M and $p_\Lambda \neq 0$. This being so, choose an index i such that $\mathsf{P}(\sigma_i)p_\Lambda \neq 0$; put $\phi_i(h) = p_\Lambda^i(\log h)h^\Lambda$ $(h \in \exp(\mathscr{C}))$ where $p_\Lambda^i = \mathsf{P}(\sigma_i)p_\Lambda$ – then clearly

$$(e^{\rho}\mathfrak{P}_{A_\mathfrak{p}}(\mu_-^*(Z))\circ e^{-\rho})\phi_i = \kappa(Z)\phi_i \text{ (all } Z \in \mathfrak{Z}).$$

On the other hand we have

$$(e^{\rho}\mathfrak{P}_{A_\mathfrak{p}}(\mu_-^*(Z))\circ e^{-\rho})\phi_i = \langle \gamma(Z), \mathbf{e}^{\Lambda+\Lambda_i+\rho_-}\rangle \phi_i \text{ (all } Z \in \mathfrak{Z})$$

(cf. the proof of Proposition 9.1.5.5). The contention is now clear, ϕ_i being non-zero.

Verification of ($\mathbf{P}_2$) Let κ be the homomorphism of $\mathfrak{Z}$ into $\mathbf{C}$ such that $(e^{\rho}\mathfrak{P}_{A_\mathfrak{p}}(Z)\circ e^{-\rho})\Phi = \kappa(Z)\Phi$ (all $Z \in \mathfrak{Z}$) – then, in view of what was said in ($\mathbf{P}_1$) above, there exist indices i and $\tilde{i}$ $(1 \leqslant i, \tilde{i} \leqslant r)$ with the property that

$$\langle \gamma(Z), \mathbf{e}^{\Lambda+\Lambda_i+\rho_-}\rangle = \kappa(Z) = \langle \gamma(Z), \mathbf{e}^{\tilde{\Lambda}+\Lambda_{\tilde{i}}+\rho_-}\rangle \qquad (\text{all } Z \in \mathfrak{Z})$$

and so our assertion follows from Theorem 2.4.3.2.

Verification of ($\mathbf{P}_3$) Let $\tilde{\Lambda}$ be any principal exponent of Φ – then there exists an index $\tilde{i}$ and a $w \in W_\Phi$ such that

$$\tilde{\Lambda} + \Lambda_{\tilde{i}} + \rho_- = w(\Lambda + \Lambda_i + \rho_-).$$

The hypotheses on Λ imply, however, that $w \in W_\sigma$ (cf. the proof of Lemma 9.1.5.7). . . .

We now have all the tools which will be needed for the proof of Theorem 9.1.5.1.

As above, for $a \in E$, ν a linear function on $\mathfrak{a}_{\mathfrak{p}_c}$, write

$$\Psi(a:\nu:x) = \int_K \mu_1(\kappa(xk))a\mu_2(k^{-1}) \exp\{(\sqrt{-1}\nu - \rho)(H(xk))\}\, dk \qquad (x \in G).$$

Proposition 9.1.5.8 Let $x, y \in G$ and $a \in E$ – then

$$\int_K \Psi(a:\nu:x^k y) dk = \Psi\Big(\int_K a\mu_2(\kappa({}^k y)) \times \exp\{(\sqrt{-1}\nu - \rho)(H({}^k y))\}\, dk : \nu : x\Big).$$

Proof Since $\kappa(xy) = \kappa(x\kappa(y))$ and $H(xy) = H(x\kappa(y)) + H(y)$, we have

$$\begin{aligned}\int_K \Psi(a:\nu:x^k y)\, dk &= \int_{K\times K} \mu_1(\kappa(x^k y\tilde{k}))a\mu_2(\tilde{k}^{-1}) \\ &\quad \times \exp\{(\sqrt{-1}\nu - \rho)(H(x^k y\tilde{k}))\}\, dk\, d\tilde{k} \\ &= \int_{K\times K} \mu_1(\kappa(x\tilde{k}^k y))a\mu_2)(\tilde{k}^{-1}) \\ &\quad \times \exp\{(\sqrt{-1}\nu - \rho)(H(x\tilde{k}^k y))\}\, dk\, d\tilde{k}.\end{aligned}$$

Taking into account the relations

$$\kappa(x\tilde{k}^k y) = \kappa(x\tilde{k}\kappa({}^k y)),\ H(x\tilde{k}^k y) = H(x\tilde{k}\kappa({}^k y)) + H({}^k y),$$

our assertion then follows without difficulty. □

Corollary 9.1.5.9 Let $D \in \mathfrak{X}$; suppose that

$$D \equiv \textstyle\sum_{j=1}^{s} D_j H_j \bmod \mathfrak{G}\mathfrak{n}_c^+$$

where $D_j \in \mathfrak{K}$ and $H_j \in \mathfrak{A}_\mathfrak{p}$ – then, for any $a \in E$,

$$\Psi(a:\nu:x;D) = \sum_j \langle H_j, \mathbf{e}^{\sqrt{-1}\nu-\rho}\rangle \Psi(a\mu_2(D_j):\nu:x) \qquad (\text{all } x \in G).$$

[This is an immediate consequence of the preceding proposition.]

Fix a vector $a \in E_M^{i_0}$ (say) $(1 \leqslant i_0 \leqslant r)$. Given $Z \in \mathfrak{Z}$, write $\mu_-^*(Z) = \sum_{j=1}^{s} H_j Z_j$ where the $H_j \in \mathfrak{A}_\mathfrak{p}$ and the $Z_j \in \mathfrak{Z}_-$ – it can be assumed that the Z_j lie in the center of $\mathfrak{M}$, whence the H_j may be taken linearly independent. Since $a \in E_{\mathrm{M}}^{i_0}$, we have

$$a\mu_2(Z_j) = \mu_1(Z_j)a = \langle \gamma_-(Z_j), \mathbf{e}^{\Lambda_{i_0}+\rho_-} \rangle \, a$$

(cf. the proof of Proposition 9.1.5.5); therefore, on the basis of Corollary 9.1.5.9, we find that

$$\begin{aligned}\Psi(a:\nu:x;Z) &= \sum_j \langle {}^bH_j, \mathbf{e}^{\sqrt{-1}\nu} \rangle \langle \gamma_-(Z_j), \mathbf{e}^{\Lambda_{i_0}+\rho_-} \rangle \, \Psi(a:\nu:x) \\ &= \langle \gamma(Z), \mathbf{e}^{\sqrt{-1}\nu+\Lambda_{i_0}+\rho_-} \rangle \, \Psi(a:\nu:x) \qquad (\text{all } x \in G).\end{aligned}$$

Now put $\Phi(a:\nu:h) = h^\rho \Psi(a:\nu:h)$ ($h \in \exp(\mathscr{C})$) – then, bearing in mind Theorem 9.1.2.9, we deduce that

$$(e^\rho \mathfrak{I}_{A_\mathfrak{p}}(Z) \circ e^{-\rho}) \Phi(a:\nu:.) = \langle \gamma(Z), \mathbf{e}^{\sqrt{-1}\nu+\Lambda_{i_0}+\rho_-} \rangle \, \Phi(a:\nu:.) \qquad (\text{all } Z \in \mathfrak{Z}).$$

Consequently it makes sense to speak of the principal exponents of $\Phi(a:\nu:.)$ (cf. supra). . . .

Fix a vector $a \in E_{\mathrm{M}}$; for the proof of Theorem 9.1.5.1, it can be assumed that $a \in E_{\mathrm{M}}^{i_0}$ where i_0 is some fixed index lying between 1 and r (hence the immediately preceding notations and observations are in force) – our aim is then to produce (necessarily unique) elements $a_w(\nu) \in E_{\mathrm{M}}$ ($\nu \in \mathbf{\Gamma}'(\mathbf{c})$) such that

$$\Phi(a:\nu:h) = \sum_{w \in W_\Sigma} \Phi(w\nu:h) a_w(\nu) \qquad (\text{all } h \in \exp(\mathscr{C})).$$

Furthermore we shall want to prove that the assignment $\nu \mapsto a_w(\nu)$ ($w \in W_\Sigma$) is holomorphic in ν ($\nu \in \mathbf{\Gamma}'(\mathbf{c})$). To this end, we shall begin by establishing the existence of the $a_w(\nu)$ ($w \in W_\Sigma$) when ν satisfies certain convenient 'regularity' conditions (cf. infra); after that the existence argument will be extended to the general case and at the same time it will be seen that the $a_w(\nu)$ ($w \in W_\Sigma$) do indeed depend on ν ($\nu \in \mathbf{\Gamma}'(\mathbf{c})$) in a holomorphic manner.

Fix an element $\nu \in \mathbf{\Gamma}(\mathbf{c})$; let us suppose that ν is real, Σ-regular, and has the property that $\nu(H(w)) \neq 0$ for all $w \in \complement W_\sigma$ – then, in view of $(\mathbf{P}_3)$ above, every principal exponent of $\Phi(a:\nu:.)$ is of the form $\sqrt{-1}w\nu$ ($w \in W_\Sigma$) (we assume, of course, that $\Phi(a:\nu:.) \neq 0$). [We recall that a is some fixed element of $E_{\mathrm{M}}^{i_0}$.] Let p_w ($w \in W_\Sigma$) be the polynomial mappings of $\mathfrak{a}_{\mathfrak{p}_c}$ into E_{M} such that $\sum_w \Phi_w(a:\nu:.)$ is the principal part of $\Phi(a:\nu:.)$ (here $\Phi_w(a:\nu:h) = p_w(\log h) h^{\sqrt{-1}w\nu}$ ($h \in \exp(\mathscr{C})$)) – then it follows from Corollary 9.1.2.8 that

$$(*) \quad (e^\rho \mathfrak{I}_{A_\mathfrak{p}}(\mu_-^*(Z)) \circ e^{-\rho}) \Phi_w(a:\nu:.) = \langle \gamma(Z), \mathbf{e}^{\sqrt{-1}w\nu+\Lambda_{i_0}+\rho_-} \rangle \, \Phi_w(a:\nu:.) \qquad (\text{all } Z \in \mathfrak{Z})$$

and so, using Proposition 9.1.5.5 and Corollary 9.1.5.6, we deduce that each p_w ($w \in W_\Sigma$) is actually a constant. This shows that we can choose elements $a_w(\nu) \in E_{\mathrm{M}}$ such that $\sum_w e^{\sqrt{-1}w\nu} a_w(\nu)$ is the principal part of

$\mathbf{\Phi}(a:\nu:.)$. Now put

$$\Theta(h)=\Phi(a:\nu:h)-\sum_{w\in W_\Sigma}\mathbf{\Phi}(w\nu:h)a_w(\nu)\qquad(h\in\exp(\mathscr{C}))$$

where $\mathbf{\Phi}$ is defined as in Theorem 9.1.4.1(bis) – then we claim that $\Theta=0$. Thus first note that

$$(e^{\rho}\mathfrak{I}_{A\mathfrak{p}}(Z)\circ e^{-\rho})\Theta=\langle\gamma(Z),\mathbf{e}^{\sqrt{-1}\nu+\Lambda_{i_0}+\rho_-}\rangle\,\Theta\qquad(\text{all } Z\in\mathfrak{Z}).$$

[In view of equation (*) above, we have

$$\begin{aligned}&\mathbf{\Phi}(w\nu:h;e^{\rho}\mathfrak{I}_{A\mathfrak{p}}(Z)\circ e^{-\rho})a_w(\nu)\\&=\mathbf{\Phi}(w\nu:h)a_w(\nu)\mu_2(\Omega(Z:\sqrt{-1}w\nu-\rho))\\&=\langle\gamma(Z),\mathbf{e}^{\sqrt{-1}\nu+\Lambda_{i_0}+\rho_-}\rangle\,\mathbf{\Phi}(w\nu:h)a_w(\nu)\qquad(h\in\exp(\mathscr{C}))\end{aligned}$$

for $Z\in\mathfrak{Z}$.] This being the case, suppose that contrary to our claim $\Theta\neq0$ – then, by ($\mathbf{P}_3$) supra, every principal exponent of Θ is of the form $\sqrt{-1}w\nu$ $(w\in W_\Sigma)$. On the other hand, since $\sum_w e^{\sqrt{-1}w\nu}a_w(\nu)$ is the principal part of both $\Phi(a:\nu:.)$ and $\sum_w\mathbf{\Phi}(w\nu:.)a_w(\nu)$, it is clear that no principal exponent of Θ can be of the form $\sqrt{-1}w\nu$ $(w\in W_\Sigma)$ – thus $\Theta=0$ and so

$$\Phi(a:\nu:h)=\sum_{w\in W_\Sigma}\mathbf{\Phi}(w\nu:h)a_w(\nu)\qquad(h\in\exp(\mathscr{C}))$$

in the present special situation.

It remains to extend the preceding considerations to the general case. For this purpose, select elements $H_1=1,H_2,\ldots,H_s$ in $\mathfrak{A}_\mathfrak{p}$ with the property that $\mathfrak{G}=\mathfrak{K}^{h^{-1}}(\sum_{j=1}^{s}\mathfrak{Z}H_j^{\flat})\mathfrak{K}$ for all $h\in\exp(\mathscr{C})$ (cf. Corollary 9.1.2.5). During the course of the subsequent discussion it will be convenient to have at hand the following criterion.

Criterion Suppose that f is a C^∞ E_M-valued function on $\exp(\mathscr{C})$ which is an eigenfunction of $\mathfrak{I}_{A\mathfrak{p}}(Z)$ for every $Z\in\mathfrak{Z}$; fix a point h_0 in $\exp(\mathscr{C})$ – then the s conditions $f(h_0;H_j^{\flat})=0$ (all j) necessarily imply that $f=0$. Indeed, as has been pointed out above, such an f can be extended to an E-valued 'μ-spherical' C^∞ function F on G^+ which is eigenfunction of $\mathfrak{Z}$; the assumption on f implies that $F(h_0;D)=0$ (all $D\in\mathfrak{G}$). But F is analytic. . . .

Let E^s denote the direct sum of E with itself s times; fix $h_0\in\exp(\mathscr{C})$. Given $a\in E_M$ and $\nu\in\mathbf{\Gamma}(\mathbf{c})$, define vectors $\Phi_0(a:\nu)$ and $\mathbf{\Phi}_0(a:\nu)$ in E^s by the rules $\Phi_0(a:\nu)=(\Phi(a:\nu:h_0;H_1),\ldots,\Phi(a:\nu:h_0;H_s))$, $\mathbf{\Phi}_0(a:\nu)=(\mathbf{\Phi}(\nu:h_0;H_1)a,\ldots,\mathbf{\Phi}(\nu:h_0;H_s)a)$. In addition, for each w in W_Σ, let a_{iw} $(1\leqslant i\leqslant n_w)$ be a basis for the space $E_M(i_0,w)$ (cf. Lemma 9.1.5.7; here i_0 is our fixed index lying between 1 and r).

Lemma 9.1.5.10 Let $n=\sum_{w\in W_\Sigma}n_w$ – then, for any $a\in E_M^{i_0}$ and $\nu\in\mathbf{\Gamma}(\mathbf{c})$, the $n+1$ vectors $\Phi_0(a:\nu)$, $\mathbf{\Phi}_0(a_{iw}:w\nu)$ $(1\leqslant i\leqslant n_w;w\in W_\Sigma)$ are linearly dependent. On the other hand, if $\nu\in\mathbf{\Gamma}'(\mathbf{c})$, then the n vectors $\mathbf{\Phi}_0(a_{iw}:w\nu)$ are linearly independent.

Proof Fix an $a \in E_M^{i_0}$ and set $f_0(\nu) = \mathbf{\Phi}_0(a:\nu)$ $(\nu \in \mathbf{\Gamma}(\mathbf{c}))$; also let us arrange the n functions $\mathbf{\Phi}_0(a_{iw}:w\nu)$ in a sequence $f_1(\nu), \ldots, f_n(\nu)$ $(\nu \in \mathbf{\Gamma}(\mathbf{c}))$ – then the f_q $(0 \leqslant q \leqslant n)$ are E^s-valued *holomorphic* functions on $\mathbf{\Gamma}(\mathbf{c})$ (cf. Theorem 9.1.4.1 (bis)). Choose a basis $\mathbf{z}_p$ $(1 \leqslant p \leqslant m)$ for the space of linear functions on E^s $(m = \dim(E)\cdot s)$; put $D_{pq}(\nu) = \mathbf{z}_p(f_q(\nu))$ $(1 \leqslant p \leqslant m, 0 \leqslant q \leqslant n)$ – then $D = (D_{pq})$ is an m-by-$n+1$ matrix whose coefficients are holomorphic functions on $\mathbf{\Gamma}(\mathbf{c})$.

(1) To establish the first contention of our lemma, it will be enough to show that rank $(D(\nu)) \leqslant n$ (any $\nu \in \mathbf{\Gamma}(\mathbf{c})$); for this latter purpose, it may clearly be assumed that $n < m$. So let Δ be the determinant of an $n+1$-by-$n+1$ submatrix of D – we shall prove that $\Delta = 0$. Choose a non-empty open subset $\mathcal{O}$ of the Σ-regular elements in $\mathfrak{a}_\mathfrak{p}^\vee$ which is contained in $\mathbf{\Gamma}(\mathbf{c})$ and has the property that $\langle \mathcal{O}, H(w)\rangle \neq 0$ for every $w \in \complement W_\sigma$ (such sets clearly exist) – then, in view of what has been said above, for each $\nu \in \mathcal{O}$ we can choose elements $a_w(\nu) \in E_M$ such that $\mathbf{\Phi}(a:\nu:h) = \sum_w \mathbf{\Phi}(w\nu:h)a_w(\nu)$ (all $h \in \exp(\mathscr{C})$). Because $a \in E_M^{i_0}$, we have

$$(e^\rho \mathfrak{l}_{A\mathfrak{p}}(Z)\circ e^{-\rho})\mathbf{\Phi}(a:\nu:.) = \langle \gamma(Z), \mathbf{e}^{\sqrt{-1}\nu+\Lambda_{i_0}+\rho_-}\rangle\, \mathbf{\Phi}(a:\nu:.) \qquad (\text{all } Z \in \mathfrak{Z}).$$

On the other hand

$$\mathbf{\Phi}(w\nu:h;e^\rho \mathfrak{l}_{A\mathfrak{p}}(Z)\circ e^{-\rho})\mathsf{P}(\sigma_i) = \langle \gamma(Z), \mathbf{e}^{\sqrt{-1}w\nu+\Lambda_{i_0}+\rho_-}\rangle\, \mathbf{\Phi}(w\nu:h)\mathsf{P}(\sigma_i) \qquad (h \in \exp(\mathscr{C}))$$

for any i $(1 \leqslant i \leqslant r)$ (since $\mu_1(\Omega(Z:\sqrt{-1}w\nu - \rho))\mathsf{P}(\sigma_i) = \langle \gamma(Z), \mathbf{e}^{\sqrt{-1}w\nu+\Lambda_i+\rho_-}\rangle\, \mathsf{P}(\sigma_i)$). Owing to Lemmas 9.1.5.2 and 9.1.5.7, it must therefore be the case that $\mathsf{P}(\sigma_i)a_w(\nu) = 0$ unless $i \in I(i_0, w)$ and so $a_w(\nu) = \sum_i \mathsf{P}(\sigma_i)a_w(\nu) \in E_M(i_0, w)$. In view of the relation

$$\mathbf{\Phi}(a:\nu:h_0;H_j) = \sum_{w \in W_\Sigma} \mathbf{\Phi}(w\nu:h_0;H_j)a_w(\nu) \qquad (1 \leqslant j \leqslant s),$$

it is now clear that $f_0(\nu)$ is a linear combination of the $f_q(\nu)$ $(1 \leqslant q \leqslant n)$ $(\nu \in \mathcal{O})$, whence Δ vanishes on $\mathcal{O}$. However $\mathbf{\Gamma}(\mathbf{c})$ is connected and Δ is holomorphic – therefore $\Delta = 0$, as desired.

(2) Turning to the second assertion of our lemma, let us suppose to the contrary that there exists a ν in $\mathbf{\Gamma}'(\mathbf{c})$ such that the n vectors $\mathbf{\Phi}_0(a_{iw}:w\nu)$ are linearly dependent. Choose elements $a_w \in E_M(i_0, w)$ $(w \in W_\Sigma)$, not all of which are equal to zero, such that $\sum_w \mathbf{\Phi}_0(a_w:w\nu) = 0$; put $f(h) = h^{-\rho}\sum_w \mathbf{\Phi}(w\nu:h)a_w$ $(h \in \exp(\mathscr{C}))$ – then, for all $Z \in \mathfrak{Z}$, we have

$$\begin{aligned} f(h;\mathfrak{l}_{A\mathfrak{p}}(Z)) &= h^{-\rho}\sum_{w \in W_\Sigma} \mathbf{\Phi}(w\nu:h)\mu_1(\Omega(Z:\sqrt{-1}w\nu - \rho))a_w \\ &= \langle \gamma(Z), \mathbf{e}^{\sqrt{-1}\nu+\Lambda_{i_0}+\rho_-}\rangle f(h) \qquad (h \in \exp(\mathscr{C})) \end{aligned}$$

since $a_w \in E_M(i_0, w)$. On the other hand

$$f(h_0;H_j^\flat) = h_0^{-\rho}\sum_w \mathbf{\Phi}(w\nu:h_0;H_j)a_w = 0$$

(because $\sum_w \mathbf{\Phi}_0(a_w:w\nu) = 0$). Taking into account the Criterion preceding the statement of the present lemma, we then deduce that $f = 0$. But ν is Σ-regular and so the points $w\nu$ ($w \in W_\Sigma$) are all distinct – therefore, in view of Lemma 9.1.5.2, $a_w = 0$ (all $w \in W_\Sigma$), a contradiction. . . .

The proof of the lemma is now complete. $\square$

Corollary 9.1.5.11 Retain the above notations and assumptions; fix a point $a \in E_M^{i_0}$ – then, for any $\nu \in \mathbf{\Gamma}'(\mathbf{c})$, there exist unique elements $a_w(\nu) \in E_M(i_0, w)$ ($w \in W_\Sigma$) such that

$$\Phi_0(a:\nu) = \sum_{w \in W_\Sigma} \Phi_0(a_w(\nu):w\nu).$$

Moreover, for each $w \in W_\Sigma$, the function $\nu \mapsto a_w(\nu)$ is holomorphic on $\mathbf{\Gamma}'(\mathbf{c})$.

Proof We shall keep to the notations which were introduced during the proof of the preceding lemma. Thus, in any event, it is clear that for each $\nu \in \mathbf{\Gamma}'(\mathbf{c})$, $f_0(\nu)$ can be written as a linear combination of the $f_q(\nu)$ ($1 \leqslant q \leqslant n$); this being so, fix a point ν_0 in $\mathbf{\Gamma}'(\mathbf{c})$ – then the matrix $(D_{pq}(\nu_0))$ ($1 \leqslant p \leqslant m, 1 \leqslant q \leqslant n$) is of rank n, hence there exists a submatrix, say (D_{ij}) ($1 \leqslant i, j \leqslant n$), such that $\Delta(\nu_0) \neq 0$ where $\Delta = \det(D_{ij})$. Choosing an open connected neighborhood $\mathcal{O}$ of ν_0 in $\mathbf{\Gamma}'(\mathbf{c})$ on which Δ is nowhere zero, it is then evident that there exist holomorphic functions c_q($1 \leqslant q \leqslant n$) on $\mathcal{O}$ such that

$$f_0(\nu) = \sum_{q=1}^{n} c_q(\nu) f_q(\nu)$$

for all $\nu \in \mathcal{O}$. Hence the corollary. $\square$

Proof of Theorem 9.1.5.1 Fix a vector $a \in E_M^{i_0}$; given $\nu \in \mathbf{\Gamma}'(\mathbf{c})$, put

$$\Theta(h) = \Phi(a:\nu:h) - \sum_{w \in W_\Sigma} \Phi(w\nu:h)a_w(\nu) \qquad (h \in \exp(\mathscr{C}))$$

in the notations of Corollary 9.1.5.11 – then the proof of Theorem 9.1.5.1 will be complete when it is established that $\Theta = 0$. But, on the one hand we have

$$(e^{\rho}\mathfrak{I}_{A_\mathfrak{p}}(Z) \circ e^{-\rho})\Theta = \langle \gamma(Z), \mathbf{e}^{\sqrt{-1}\nu + \Lambda_{i_0} + \rho_-} \rangle \Theta \qquad (\text{all } Z \in \mathfrak{Z}),$$

while on the other $\Theta(h_0; H_j) = 0$ ($1 \leqslant j \leqslant s$) (cf. Corollary 9.1.5.11). Therefore, on the basis of the Criterion supra, we deduce that $\Theta = 0$. $\square$

In conclusion we mention that it is possible to show that the $\mathbf{c}_w$ ($w \in W_\Sigma$) are actually meromorphic on $\mathfrak{a}_{\mathfrak{p}_c}$; as we shall have no need of this fact, we omit the proof (see, however, 9.1.6).

Appendix Let W_Φ denote the Weyl group of the pair $(\mathfrak{g}_c, \mathfrak{a}_c)$. Fix a point $H_0 \in \mathfrak{a}_c$; let W_{H_0} be the subgroup of W_Φ consisting of those w which leave H_0 fixed; introduce $I_{H_0}(\mathfrak{a}_c)$, $I_{H_0}^+(\mathfrak{a}_c)$, and $\mathfrak{F}(H_0)$ per the discussion leading up to the statement of Lemma 9.1.5.3 – then $\dim(\mathfrak{F}(H_0)) = [W_{H_0}]$. [To see this, one may reason as follows. For any $p \in S(\mathfrak{a}_c)$, let $\bar{p}$ denote the polynomial function defined by $\bar{p}(H) = \overline{p(H)}$ $(H \in \mathfrak{a}_\mathfrak{p} + \sqrt{-1}\mathfrak{a}_\mathfrak{k})$ – then $\langle p, \bar{p}\rangle$ is a positive definite Hermitian form on $S(\mathfrak{a}_c)$. Because $\mathfrak{a}_\mathfrak{p} + \sqrt{-1}\mathfrak{a}_\mathfrak{k}$ is W_Φ-stable, it is clear that $I_{H_0}^+(\mathfrak{a}_c)$ is invariant under the mapping $p \mapsto \bar{p}$; on the other hand $\mathfrak{H}(H_0)$ is the orthogonal complement in $S(\mathfrak{a}_c)$ of $S(\mathfrak{a}_c)I_{H_0}^+(\mathfrak{a}_c)$ with respect to the above Hermitian form – hence $\mathfrak{H}(H_0) \cap S(\mathfrak{a}_c)I_{H_0}^+(\mathfrak{a}_c) = \{0\}$. Moreover, if $S^m(\mathfrak{a}_c)$ denotes the space of homogeneous elements in $S(\mathfrak{a}_c)$ of degree m, then $S^m(\mathfrak{a}_c) \cap \mathfrak{H}(H_0)$ is the orthogonal complement of $S^m(\mathfrak{a}_c) \cap S(\mathfrak{a}_c)I_{H_0}^+(\mathfrak{a}_c)$ in $S^m(\mathfrak{a}_c)$ – thus

$$S^m(\mathfrak{a}_c) = S^m(\mathfrak{a}_c) \cap \mathfrak{H}(H_0) + S^m(\mathfrak{a}_c) \cap S(\mathfrak{a}_c)I_{H_0}^+(\mathfrak{a}_c)$$

and so $S(\mathfrak{a}_c) = \mathfrak{H}(H_0) + S(\mathfrak{a}_c)I_{H_0}^+(\mathfrak{a}_c)$. Now W_{H_0} is generated by the Weyl reflections w_α corresponding to those roots α which vanish at H_0 (cf. Theorem 1.1.2.8), whence, on the basis of Proposition 2.1.3.5, we then deduce that

$$\dim(\mathfrak{H}(H_0)) = \dim(S(\mathfrak{a}_c)/S(\mathfrak{a}_c)I_{H_0}^+(\mathfrak{a}_c)) = [W_{H_0}],$$

as contended.]

Let $p \mapsto p^\sigma$ denote the automorphism of $S(\mathfrak{a}_c)$ defined by the rule

$$p^\sigma(H) = p(H + H_0) \qquad (H \in \mathfrak{a}_c).$$

Since H_0 is left fixed by W_{H_0}, $I_{H_0}(\mathfrak{a}_c)$ is σ-stable.

Lemma Let p_0 be an element in $S(\mathfrak{a}_c)$; let f_0 denote the function $H \mapsto p_0(H)e^{(H_0, H)}$ on $\mathfrak{a}_c$ – then the following conditions are equivalent:

(i) p_0 lies in $\mathfrak{H}(H_0)$;

(ii) f_0 is an eigenfunction of $\partial(p)$ for every p in $I_{H_0}(\mathfrak{a}_c)$;

(iii) f_0 is an eigenfunction of $\partial(p)$ for every p in $I(\mathfrak{a}_c)$.

Proof (ii) $\Leftrightarrow$ (i). Let g_0 denote the function $H \mapsto e^{(H_0, H)}$ on $\mathfrak{a}_c$; let $p \in I_{H_0}(\mathfrak{a}_c)$ – then $\partial(p)f_0 = g_0\partial(p^\sigma)p_0$. Now

$$\partial(p^\sigma)p_0 = p^\sigma(0)p_0 + \partial(p^\sigma - p^\sigma(0))p_0,$$

and so, as we can obviously suppose that $p_0 \neq 0$, it is then clear that

$$\deg(\partial(p^\sigma - p^\sigma(0))p_0) < \deg(p_0).$$

Therefore f_0 is an eigenfunction of $\partial(p)$ iff $\partial(p^\sigma - p^\sigma(0))p_0 = 0$. Because $I_{H_0}(\mathfrak{a}_c)$ is σ-stable, this shows that (ii) is equivalent to the condition that $\partial(p)p_0 = 0$ for all $p \in I_{H_0}^+(\mathfrak{a}_c)$ – this latter condition is, however, equivalent to (i).

(ii) $\Rightarrow$ (iii). This is plain (since $I(\mathfrak{a}_c) \subset I_{H_0}(\mathfrak{a}_c)$).

(iii) $\Rightarrow$ (i). Let $r = [W_\Phi : W_{H_0}]$; choose elements w_i $(0 \leqslant i < r,\ w_0 = 1)$ in W_Φ such that $W_\Phi = \bigcup_i w_i W_{H_0}$ and put $H_i = w_i(H_0)$. Let E be the space of all holomorphic functions ϕ on $\mathfrak{a}_c$ such that $\partial(p)\phi = p(H_0)\phi$ for all $p \in I(\mathfrak{a}_c)$ – then $\dim(E) \leqslant [W_\Phi]$ (cf. Proposition 8.2.2.8). Let E_i be the space of all functions on $\mathfrak{a}_c$ of the form $H \mapsto p_i(H)e^{(H_i, H)}$ $(H \in \mathfrak{a}_c)$ where $p_i \in \mathfrak{H}(H_i)$; since $I_{H_i}(\mathfrak{a}_c) \supset I(\mathfrak{a}_c)$, it follows from the equivalences of (i) and (ii) proved above that $E_i \subset E$ – in addition

$$\dim(E_i) = \dim(\mathfrak{H}(H_i)) = [W_{H_i}] = [W_{H_0}].$$

Because the elements H_i are distinct, the functions g_i, $H \mapsto e^{(H_i, H)}$ $(H \in \mathfrak{a}_c)$, are linearly independent over $S(\mathfrak{a}_c)$; consequently the sum $\sum_i E_i$ is direct – but $\sum_i \dim(E_i) = r[W_{H_0}] = [W_\Phi]$, whence $E = \sum_i E_i$. This being so, suppose now that condition (iii) holds for f_0 – then $\partial(p^\sigma - p^\sigma(0))p_0 = 0$ for every $p \in I(\mathfrak{a}_c)$, whence $f_0 \in E$. Owing to the linear independence of the g_i over $S(\mathfrak{a}_c)$, it must therefore be the case that $f_0 \in E_0$. Hence (iii) $\Rightarrow$ (i).

This completes the proof of the lemma. □

Note Let E be the space of all holomorphic functions ϕ on $\mathfrak{a}_c$ such that $\partial(p)\phi = p(H_0)\phi$ for all $p \in I(\mathfrak{a}_c)$ – then it follows from Proposition 8.2.2.8 that $\dim(E) \leqslant [W]$. Because H_0 need not be regular, it is not a priori evident that $\dim(E) = [W]$ (cf. Corollary 8.2.2.9); however, as was seen during the proof of the implication '(iii) $\Rightarrow$ (i)' supra, this is in fact the case (cf. Steinberg [3]).

9.1.6 Investigation of the c-Function

Let G be a connected semi-simple Lie group with finite center, $G = KA_{\mathfrak{p}}N^+$ an Iwasawa decomposition for G; let M be the centralizer of $A_{\mathfrak{p}}$ in K. Let $\mu = (\mu_1, \mu_2)$ be a double unitary representation of K on a finite dimensional Hilbert space E; let E_{M} denote the subspace of E comprised of those elements a which have the property that $\mu_1(m)a = a\mu_2(m)$ (all $m \in \mathsf{M}$) – then, as has been seen in 9.1.5, for any $\nu \in \mathbf{\Gamma}'(\mathbf{c})$, there exist uniquely determined elements $\mathbf{c}_w(\nu) \in \mathrm{Hom}_{\mathbf{C}}(E_{\mathsf{M}}, E_{\mathsf{M}})$ ($w \in W$ (= Weyl group of the pair $(\mathfrak{g}, \mathfrak{a}_{\mathfrak{p}})$)) such that

$$\begin{aligned}\Phi(a:\nu:h) &= h^\rho \int_K \mu_1(\kappa(hk))a\mu_2(k^{-1}) \exp\{(\sqrt{-1}\nu - \rho)(H(hk))\}\,dk \\ &= \sum_{w \in W} \Phi(w\nu:h)\mathbf{c}_w(\nu)a \qquad (h \in \exp(\mathscr{C}))\end{aligned}$$

for all $a \in E_{\mathsf{M}}$. Furthermore the $\mathbf{c}_w$ ($w \in W$) are $\mathrm{Hom}_{\mathbf{C}}(E_{\mathsf{M}}, E_{\mathsf{M}})$-valued holomorphic functions on $\mathbf{\Gamma}'(\mathbf{c})$. This being the case, let us agree to write $\mathbf{c}$ for $\mathbf{c}_1$ – then we have the following result.

Theorem 9.1.6.1 (Harish-Chandra) Retain the above notations and assumptions; suppose that ν is an element in $\mathbf{\Gamma}'(\mathbf{c})$ such that $-\mathscr{I}(\nu) \in \mathscr{C}$ (the positive Weyl chamber in $\mathfrak{a}_{\mathfrak{p}}$) – then

$$\mathbf{c}(\nu)a = \int_{N^-} a\mu_2(\kappa(n)^{-1}) \exp\{-(\sqrt{-1}\nu + \rho)(H(n))\}\,d_{N^-}(n)$$

for all $a \in E_{\mathsf{M}}$.

[Here the Haar measure on N^- is to be taken in the normalization determined by the condition $\int_{N^-} e^{-2\rho(H(n))} d_{N^-}(n) = 1$; cf. the Appendix to number 8.1.3.]

Before giving the proof of our theorem, let us consider the integral which defines Φ; thus fix $a \in E_{\mathsf{M}}$, $\nu \in \mathbf{\Gamma}'(\mathbf{c})$, $h \in \exp(\mathscr{C})$ – then, in view of a lemma which may be found in the Appendix to number 8.1.3, we can write

$$\begin{aligned}\Phi(a:\nu:h) = h^\rho \int_{N^-}\int_{\mathsf{M}} &\mu_1(\kappa(h\kappa(n)m))a\mu_2(m^{-1}\kappa(n)^{-1}) \\ &\times \exp\{(\sqrt{-1}\nu - \rho)(H(h\kappa(n)m))\}\, e^{-2\rho(H(n))}\,dm\,d_{N^-}(n).\end{aligned}$$

Keeping to these notations, it is a simple matter to check that:

$$\text{(1)} \qquad \kappa(h\kappa(n)m) = \kappa(hnm) = \kappa(hn)m\,;$$

$$\text{(2)} \qquad H(h\kappa(n)m) = H(h\kappa(n)) = \log h + H(hnh^{-1}) - H(n).$$

Hence, bearing in mind that $a \in E_M$, we have

$$\Phi(a:\nu:h) = h^{\sqrt{-1}\nu} \int_{N^-} \mu_1(\kappa(hn))a\mu_2(\kappa(n)^{-1}) \times \exp\{(\sqrt{-1}\nu - \rho)(H(hnh^{-1}) - H(n))\}e^{-2\rho(H(n))}d_{N^-}(n).$$

Now put $\nu_+ = \sqrt{-1}\nu + \rho$, $\nu_- = \sqrt{-1}\nu - \rho$ – then it is clear that

$$h^{-\sqrt{-1}\nu}\Phi(a:\nu:h) = \int_{N^-} \mu_1(\kappa(hn))a\mu_2(\kappa(n)^{-1}) \times \exp\{\nu_-(H(hnh^{-1})) - \nu_+(H(n))\}\, d_{N^-}(n).$$

Proof of Theorem 9.1.6.1 Fix $a \in E_M$ – then, on the basis of what has been said above, we have

$$\sum_{w\in W} h^{-\sqrt{-1}\nu}\Phi(w\nu:h)\mathbf{c}_w(\nu)a$$
$$= \int_{N^-} \mu_1(\kappa(hn))a\mu_2(\kappa(n)^{-1}) \exp\{\nu_-(H(hnh^{-1})) - \nu_+(H(n))\}d_{N^-}(n)$$

for all $h \in \exp(\mathscr{C})$. Fix an element $H \in \mathscr{C}$ and put $h_t = \exp(tH)$ $(t \geqslant 0)$ – then, in view of our hypotheses on ν, the real part of $\sqrt{-1}\,\langle w\nu - \nu, H\rangle$ is negative if $w \neq 1$ ($w \in W$; cf. Lemma 3.3.2.1). It therefore follows that

$$\lim_{t\to\infty} h_t^{-\sqrt{-1}\nu}\,\Phi(w\nu:h) = \begin{cases} 1 \text{ if } w = 1 \\ 0 \text{ if } w \neq 1 \end{cases} \qquad (w \in W),$$

whence

$$\mathbf{c}(\nu)a = \lim_{t\to\infty} \int_{N^-} \mu_1(\kappa(h_tn))a\mu_2(\kappa(n)^{-1}) \times \exp\{\nu_-(H(h_tnh_t^{-1})) - \nu_+(H(n))\}d_{N^-}(n).$$

We shall now show that it is permissible to take the limit under the integral sign. Thus choose a positive number $\epsilon < 1$ such that $-\mathscr{I}(\nu) - \epsilon\rho \in \mathscr{C}$; let $\nu_\epsilon = \sqrt{-1}\nu - \epsilon\rho$ – then $\nu_- = \nu_\epsilon - (1 - \epsilon)\rho$, $\nu_+ = \nu_\epsilon + (1 + \epsilon)\rho$ and the real part of ν_ϵ is non-negative on $\mathfrak{a}_\mathfrak{p}^+$ (for notation, see Vol. I, number 3.3.2). On the other hand we have

$$\nu_-(H(h_tnh_t^{-1})) - \nu_+(H(n))$$
$$= \nu_\epsilon(H(h_tnh_t^{-1}) - H(n)) - (1 - \epsilon)\rho(H(h_tnh_t^{-1})) - (1 + \epsilon)\rho(H(n)) \qquad (n \in N^-).$$

Therefore, thanks to Lemma 3.3.2.5, it must be the case that the real part of $\nu_-(H(h_tnh_t^{-1})) - \nu_+(H(n))$ is bounded above by $-(1 + \epsilon)\rho(H(n))$ $(n \in N^-)$ for all $t \geqslant 0$ and so, due to the Dominated Convergence Theorem, we can, by Corollary 8.5.2.5, pass to the limit under the integral sign to conclude that

$$\mathbf{c}(\nu)a = \int_{N^-} \lim_{t\to\infty} \mu_1(\kappa(h_tn))a\mu_2(\kappa(n)^{-1}) \times \exp\{\nu_-(H(h_tnh_t^{-1})) - \nu_+(H\,(n))\}\, d_{N^-}(n).$$

Fix $n \in N^-$; select $X \in \mathfrak{n}^-$ such that $n = \exp X$ – then

$$h_t n h_t^{-1} = \exp(\mathrm{Ad}(h_t)X) \qquad (t \geqslant 0).$$

Write $X = \sum_{\alpha \in P_+} c_\alpha X_{-\alpha} (c_\alpha \in \mathbf{C})$ – then $\mathrm{Ad}(h_t)X = \sum_\alpha c_\alpha e^{-t\alpha(H)} X_{-\alpha}$, hence $h_t n h_t^{-1} \to 1$ as $t \to \infty$. Because $\kappa(h_t n) = \kappa(h_t n h_t^{-1})$ $(t \geqslant 0)$, it is now clear that

$$\mathbf{c}(\nu)a = \int_{N^-} a\mu_2(\kappa(n)^{-1}) \exp\{-(\sqrt{-1}\nu + \rho)(H(n))\} d_{N^-}(n) \qquad (a \in E_M),$$

as we wished to prove. □

Remark It is not difficult to modify the above argument and get an integral formula for $\mathbf{c}_{w^\dagger}$ ($-w^\dagger$ the opposition involution in Σ relative to $\mathscr{C}$); however, except in certain special cases, integral formulae for an arbitrary $\mathbf{c}_w$ ($w \in W$) do not appear to be known.

Fix an irreducible unitary matrix representation μ_δ of K in the class $\delta(\delta \in \hat{K})$ – then the preceding considerations suggest that it might very well be profitable to study the operator

$$\int_{N^-} \mu_\delta(\kappa(n)) \exp\{-(\nu + \rho)(H(n))\} d_{N^-}(n)$$

viewed as an endomorphism of E_δ (the representation space for μ_δ). Here, of course, ν is a (complex valued) linear function on $\mathfrak{a}_\mathfrak{p}$. Needless to say, it will be necessary to place some conditions on ν in order to ensure that our integral is convergent. Actually, with a view toward later applications, it will be convenient to consider a somewhat more general situation. Thus fix an element w in the Weyl group W of the pair $(\mathfrak{g}, \mathfrak{a}_\mathfrak{p})$; let m_w be an element in M^* associated with w and, for any (complex valued) linear function ν on $\mathfrak{a}_\mathfrak{p}$, put

$$T(w, \nu, \delta) = \mu_\delta(m_w) \int_{U_w^-} \mu_\delta(\kappa(u)) \exp\{-(\nu + \rho)(H(u))\} \, d_{U_w^-}(u)$$

where $U_w^- = N^- \cap m_w^{-1} N^+ m_w$ (cf. Vol. I, number 1.1.4). [The definition of T depends upon our choice of a representative for w; this will not cause any difficulty however. Note, too, that $U_{w^\dagger}^- = N^-$ ($w^\dagger$ being, as always, that element in W which takes the positive Weyl chamber to its negative.] Let $C(w)$ be the set comprised of those ν with the property that

$$\int_{U_w^-} \exp\{-(\mathscr{R}(\nu) + \rho)(H(u))\} \, d_{U_w^-}(u) < \infty.$$

Then it is obvious that the integral defining $T(w, \cdot, \delta)$ is absolutely convergent for all ν in $C(w)$.

Note Evidently $C(w^{-1}) = -wC(w)$ $(w \in W)$. . . .

Theorem 9.1.6.2 (Schiffmann) Retain the above notations and assumptions; suppose that w, w', and w'' are three elements in the Weyl group W of the pair $(\mathfrak{g}, \mathfrak{a}_\mathfrak{p})$ such that $w = w'w''$ with

$$L(w) = L(w') + L(w'');$$

choose representatives m_w, $m_{w'}$, and $m_{w''}$ in M^* for w, w', and w'', respectively, such that $m_w = m_{w'}m_{w''}$ – then

$$T(w, \nu, \delta) = T(w', w''(\nu), \delta)T(w'', \nu, \delta)$$

for every ν in $\mathsf{C}(w)$.

Keeping to the supposition that $w = w'w''$ with

$$L(w) = L(w') + L(w''),$$

Proposition 1.1.4.6 tells us that the mapping $(u', u'') \mapsto m_{w''}^{-1}u'm_{w''}u''$ of $U_{w'}^- \times U_{w''}^-$ into U_w^- is an isomorphism of analytic varieties; this being so, in all that follows it will be assumed that the relevant Haar measures have been normalized in such a way that

$$(\dagger)\quad \int_{U_w^-} f(u)d_{U_w^-}(u) = \iint_{U_{w'}^- \times U_{w''}^-} f(m_{w''}^{-1}u'm_{w''}u'')d_{U_{w'}^-}(u')d_{U_{w''}^-}(u'')$$

$$(f \in C_c(U_w^-)).$$

It is to be observed that such a determination does not depend upon the choice of $m_{w''}$.

[To be specific, it will be shown below that if the Haar measure on a given U_w^- is fixed by the requirement

$$\int_{U_w^-} e^{-2\rho(H(u))}d_{U_w^-}(u) = 1,$$

then the integral formula (†) is valid.]

The following lemma is the main step in the proof of Theorem 9.1.6.2.

Lemma 9.1.6.3 Suppose that w, w', w'' are three elements in the Weyl group W of the pair $(\mathfrak{g}, \mathfrak{a}_\mathfrak{p})$ such that $w = w'w''$ with

$$L(w) = L(w') + L(w'').$$

Then

$$\mathsf{C}(w) = \mathsf{C}(w'') \cap w''^{-1}\mathsf{C}(w').$$

Proof Given $x \in G$, let $\kappa(x) \in K$ and $\eta(x) \in A_\mathfrak{p}$ be such that $x \in \kappa(x)\eta(x)N^+$; given a linear function ν on $\mathfrak{a}_\mathfrak{p}$, define f_ν on G by the rule $f_\nu(x) = \eta(x)^{-(\nu+\rho)}$ $(x \in G)$ – then it is clear that $f_\nu(xhn) = h^{-(\nu+\rho)}f_\nu(x)$ (all $x \in G$, $hn \in A_\mathfrak{p}N^+$). Now fix a ν in $\mathsf{C}(w)$; for x in G, put

$$A(w)f_\nu(x) = \int_{U_w^-} f_\nu(xm_wu)d_{U_w^-}(u).$$

Then it is a simple matter to check that

$$\int_{U_w^-} f_\nu(xm_wu)d_{U_w^-}(u) = f_{w(\nu)}(x)\int_{U_w^-} \eta(m_wu)^{-(\nu+\rho)}d_{U_w^-}(u) \qquad (x \in G)$$

from which it follows that the integral defining $A(w)f_\nu$ is absolutely convergent. Note too that $A(w)f_\nu(xhn) = h^{-(w\nu+\rho)}A(w)f_\nu(x)$ (all $x \in G$, $hn \in A_\mathfrak{p}N^+$). With this preparation, let us turn to the proof of our lemma. On the basis of the integral formula supra, we have

$$(*)\quad \int_{U_w^-} f_\nu(xm_w u)d_{U_w^-}(u) = \iint_{U_{w'}^- \times U_{w''}^-} f_\nu(xm_{w'}u'm_{w''}u'')d_{U_{w'}^-}(u')d_{U_{w''}^-}(u'') \qquad (x \in G)$$

for all $\nu \in \mathsf{C}(w)$. Now fix ν in $\mathsf{C}(w)$ – then $\mathscr{R}(\nu) \in \mathsf{C}(w)$ and the integral

$$\int_{U_w^-} f_{\mathscr{R}(\nu)}(xm_w u)d_{U_w^-}(u) \qquad (x \in G)$$

converges. It therefore follows from $(*)$ that the integral

$$\int_{U_{w''}^-} f_{\mathscr{R}(\nu)}(xm_{w'}u'm_{w''}u'')d_{U_{w''}^-}(u'') \qquad (x \in G)$$

converges for almost every u' in $U_{w'}^-$; consequently the integral

$$A(w'')f_{\mathscr{R}(\nu)}(x) = \int_{U_{w''}^-} f_{\mathscr{R}(\nu)}(xm_{w''}u'')d_{U_{w''}^-}(u'') \qquad (x \in G)$$

converges for at least one x in G. But, formally, one has

$$A(w'')f_{\mathscr{R}(\nu)}(khn) = h^{-(w''\nu+\rho)}A(w'')f_{\mathscr{R}(\nu)}(1) \qquad (khn \in KA_\mathfrak{p}N^+)$$

and so the integral in question actually converges for every x in G, in particular for $x = 1$, whence

$$\int_{U_{w''}^-} \exp\{-(\mathscr{R}(\nu) + \rho)(H(u''))\}\, d_{U_{w''}^-}(u'') < \infty$$

which implies that $\nu \in \mathsf{C}(w'')$. Because the (convergent) integral

$$\int_{U_{w'}^-} A(w'')f_{\mathscr{R}(\nu)}(xm_{w'}u')\, d_{U_{w'}^-}(u') \qquad (x \in G)$$

can be written in the form

$$\int_{U_{w'}^-}\left[\int_{U_{w''}^-} \eta(u'')^{-(\mathscr{R}(\nu)+\rho)}d_{U_{w''}^-}(u'')\right] f_{\mathscr{R}(w''\nu)}(xm_{w'}u')\, d_{U_{w'}^-}(u') \qquad (x \in G)$$

and since the inner integral does not vanish, upon taking $x = 1$ we deduce that

$$\int_{U_{w'}^-} \exp\{-(\mathscr{R}(w''\nu) + \rho)(H(u'))\}\, d_{U_{w'}^-}(u') < \infty,$$

whence $\nu \in w''^{-1}\mathsf{C}(w')$. It remains to verify that

$$\nu \in \mathsf{C}(w'') \cap w''^{-1}\mathsf{C}(w') \Rightarrow \nu \in \mathsf{C}(w).$$

Thus fix $\nu \in \mathsf{C}(w'') \cap w''^{-1}\mathsf{C}(w')$ – then, in particular, the integral

$$\int_{U_{w''}^-} f_\nu(xm_{w''}u'')\, d_{U_{w''}^-}(u'') \qquad (x \in G)$$

converges absolutely. But $w''\nu \in \mathsf{C}(w')$ too, hence the integral

$$\int_{U_{w'}^-} A(w'')f_\nu(xm_{w'}u')\,d_{U_{w'}^-}(u') \qquad (x \in G)$$

is also absolutely convergent for every x in G. It then follows at once from (∗) that the integral

$$\int_{U_w^-} f_\nu(xm_w u)\,d_{U_w^-}(u) \qquad (x \in G)$$

converges absolutely for all x in G and so $\nu \in \mathsf{C}(w)$. The proof of the lemma is therefore complete. □

Proof of Theorem 9.1.6.2 The preceding lemma serves to guarantee us that both sides of the functional equation in question make sense whenever $\nu \in \mathsf{C}(w)$; the functional equation itself is an immediate consequence of the integral formula (†) supra. □

Remark It is clear that the map $\nu \mapsto T(w, \nu, \delta)$ is a holomorphic function of ν ($\nu \in \mathsf{C}(w)$); this being so, put

$$\boldsymbol{\Gamma}_w(\nu) = \prod_{\lambda \in \Delta_0(w)} \Gamma\left(2\frac{(\nu, \lambda)}{(\lambda, \lambda)}\right) \qquad (\nu \in \mathsf{C}(w)),$$

Γ the classical Gamma function (the definition of $\Delta_0(w)$ is given below) – then Schiffmann [2] has shown that the map

$$\nu \mapsto T(w, \nu, \delta)/\boldsymbol{\Gamma}_w(\nu)$$

extends to an entire function of ν.

Keeping to the above notations and assumptions, fix a w in W – then one may pose the problem: Describe the region $\mathsf{C}(w)$ explicitly. This will be done by employing a technique due to Gindikin and Karpelevič [1]; the basic idea is to use Lemma 9.1.6.3 and reduce everything to the case when the split-rank of G is equal to 1.

Theorem 9.1.6.4 (Gindikin and Karpelevič) Let G be a connected semi-simple Lie group with finite center, $G = KA_\mathfrak{p}N^+$ an Iwasawa decomposition for G; fix a w in the Weyl group W of the pair $(\mathfrak{g}, \mathfrak{a}_\mathfrak{p})$ and put $U_w^- = N^- \cap m_w^{-1}N^+m_w$ – then the domain of convergence $\mathsf{C}(w)$ for the integral

$$\int_{U_w^-} \exp\{-(\mathscr{R}(\nu) + \rho)(H(u))\}\,d_{U_w^-}(u)$$

consists of those linear functions ν on $\mathfrak{a}_\mathfrak{p}$ with the property that $\mathscr{R}(\nu(Q_\lambda)) > 0$ for every $\lambda \in \Delta_0(w)$.

[Here $\Delta_0(w)$ denotes the positive indivisible roots λ of the pair $(\mathfrak{g}, \mathfrak{a}_\mathfrak{p})$ such that $w(\lambda) < 0$; cf. Vol. I, number 1.1.4.]

Corollary 9.1.6.5 Retain the notations and assumptions of the preceding theorem – then the integral

$$\int_{N^-} \exp\{-(\nu + \rho)(H(n))\}\, d_{N^-}(n)$$

is absolutely convergent iff $\mathscr{R}(\nu) \in \mathscr{C}$ (= the positive Weyl chamber in $\mathfrak{a}_{\mathfrak{p}}^{\vee}$).

Proof of Theorem 9.1.6.4 (Schiffmann) In view of Lemma 9.1.6.3, it is enough to consider the case when $L(w) = 1$, that is to say, we need only deal with the case when w is a Weyl reflection w_i ($= w_{\lambda_i}$) with respect to the simple root λ_i ($1 \leqslant i \leqslant l$); in this situation, $\mathsf{C}(w_i)$ is the domain of convergence for the integral

$$\int_{U_{w_i^-}} \exp\{-(\mathscr{R}(\nu) + \rho)(H(u))\}\, d_{U_{w_i^-}}(u).$$

Let $m(\lambda_i)$ (respectively $m(2\lambda_i)$) denote the multiplicity of λ_i (respectively $2\lambda_i$); set $\rho_i = 2^{-1}(m(\lambda_i) + 2m(2\lambda_i))\lambda_i$ – then, since w_i permutes the positive roots other than λ_i and $2\lambda_i$ amongst themselves, it is clear that

$$(\mathscr{R}(\nu) + \rho)(H(u)) = (\mathscr{R}(\nu) + \rho_i)(H(u)) \qquad (u \in U_{w_i}^-)$$

and so our integral can be written

$$\int_{U_{w_i^-}} \exp\{-(\mathscr{R}(\nu) + \rho_i)(H(u))\}\, d_{U_{w_i^-}}(u).$$

Now, as in the discussion leading up to the statement of Lemma 1.2.3.16, let $\mathfrak{g}(\lambda_i)$ denote the semi-simple subalgebra of $\mathfrak{g}$ generated by $\mathfrak{g}^{\lambda_i} + \mathfrak{g}^{2\lambda_i} + \mathfrak{g}^{-\lambda_i} + \mathfrak{g}^{-2\lambda_i}$; let $G(\lambda_i)$ denote the analytic subgroup of G corresponding to $\mathfrak{g}(\lambda_i)$ – then $G(\lambda_i)$ admits the Iwasawa decomposition

$$G(\lambda_i) = K(\lambda_i)A_{\mathfrak{p}}(\lambda_i)N^+(\lambda_i)$$

where, in particular, $A_{\mathfrak{p}}(\lambda_i)$ is the one parameter subgroup of $G(\lambda_i)$ corresponding to $\mathbf{R}Q_{\lambda_i}$. Therefore the split-rank of $G(\lambda_i)$ is 1; accordingly the results and notations of the Appendix to Vol. I, number 1.1.4 are available. [Of course, the role of 'H' there is played, in the present case, by $2Q_{\lambda_i}/(\lambda_i, \lambda_i)$.] This being so, let us agree to identify $\mathfrak{g}^{-\lambda_i}$ with $\mathbf{R}^{m(\lambda_i)}$ in such a way that

$$\|X_i\|^2 = 2^{-1}Q(X_i) \qquad (X_i \in \mathfrak{g}^{-\lambda_i})$$

and $\mathfrak{g}^{-2\lambda_i}$ with $\mathbf{R}^{m(2\lambda_i)}$ in such a way that

$$\|X_{2i}\|^2 = 2Q(X_{2i}) \qquad (X_{2i} \in \mathfrak{g}^{-2\lambda_i}),$$

the norms being the usual Euclidean norms – then, thanks to the results alluded to above, the integral under investigation can be written in the form

$$\int_{\mathbf{R}^{m(\lambda_i)} \times \mathbf{R}^{m(2\lambda_i)}} [(1 + \|X_i\|^2)^2 + \|X_{2i}\|^2]^{-4^{-1}(\mathscr{R}(\nu_i) + m(\lambda_i) + 2m(2\lambda_i))}\, dX_i\, dX_{2i},$$

where $\nu_i = 2\nu(Q_{\lambda_i})/(\lambda_i, \lambda_i)$. But surely this integral converges iff $\mathscr{R}(\nu_i) > 0$. ☐

Note Keeping to the immediately preceding notations, one finds by direct computation that

$$\int_{U_{w_i}^-} e^{-(\nu+\rho)(H(u))}\, d_{U_{w_i}^-}(u) = \mathbf{c}_{\lambda_i}(\nu) \int_{U_{w_i}^-} e^{-2\rho(H(u))}\, d_{U_{w_i}^-}(u)$$

where

$$\mathbf{c}_{\lambda_i}(\nu) = \frac{\Gamma(m(\lambda_i) + m(2\lambda_i))}{\Gamma(2^{-1}(m(\lambda_i) + m(2\lambda_i)))} \cdot \frac{\Gamma(2^{-1}\nu_i)}{\Gamma(2^{-1}(\nu_i + m(\lambda_i)))} \cdot \frac{\Gamma(4^{-1}(\nu_i + m(\lambda_i)))}{\Gamma(4^{-1}(\nu_i + m(\lambda_i)) + 2^{-1}m(2\lambda_i))},$$

Γ the classical Gamma function. [Here we have to remind ourselves of the familiar rule $\Gamma(p)\Gamma(q)/\Gamma(p+q) = \int_0^\infty t^{p-1}(1+t)^{-p-q}\, dt$ ($\mathscr{R}(p) > 0$, $\mathscr{R}(q) > 0$).]

Let ν be a linear function on $\mathfrak{a}_\mathfrak{p}$; for any positive indivisible root λ of the pair $(\mathfrak{g}, \mathfrak{a}_\mathfrak{p})$, set $\nu_\lambda = 2\nu(Q_\lambda)/(\lambda, \lambda)$ and write

$$\mathbf{c}_\lambda(\nu) = \frac{\Gamma(m(\lambda) + m(2\lambda))}{\Gamma(2^{-1}(m(\lambda) + m(2\lambda)))} \cdot \frac{\Gamma(2^{-1}\nu_\lambda)}{\Gamma(2^{-1}(\nu_\lambda + m(\lambda)))} \cdot \frac{\Gamma(4^{-1}(\nu_\lambda + m(\lambda)))}{\Gamma(4^{-1}(\nu_\lambda + m(\lambda)) + 2^{-1}m(2\lambda))},$$

where, as always, $m(\lambda)$ (respectively $m(2\lambda)$) is the multiplicity of λ (respectively 2λ). For any w in the Weyl group W of the pair $(\mathfrak{g}, \mathfrak{a}_\mathfrak{p})$, put

$$\mathbf{c}_w(\nu) = \prod_{\lambda \in \Delta_0(w)} \mathbf{c}_\lambda(\nu).$$

Proposition 9.1.6.6 Fix a w in the Weyl group W of the pair $(\mathfrak{g}, \mathfrak{a}_\mathfrak{p})$ – then, for any $\nu \in \mathsf{C}(w)$, we have

$$\int_{U_w^-} e^{-(\nu+\rho)(H(u))}\, d_{U_w^-}(u) = \mathbf{c}_w(\nu) \int_{U_w^-} e^{-2\rho(H(u))}\, d_{U_w^-}(u).$$

[This follows at once by induction on the length $L(w)$ of w.]

Remark At this juncture we can clear up a point which was left open earlier. Suppose that we agree to normalize the Haar measure on a given U_w^- by the condition

$$\int_{U_w^-} e^{-2\rho(H(u))}\, d_{U_w^-}(u) = 1.$$

Then we claim that the integral formula (†) obtains with such a determination. Thus suppose that $w = w'w''$ with $L(w) = L(w') + L(w'')$; fix a ν in $\mathsf{C}(w)$ and, as in Lemma 9.1.6.3, put

$$A(w)f_\nu(x) = \int_{U_w^-} f_\nu(xm_w u)\, d_{U_w^-}(u) \qquad (x \in G)$$

where $f_\nu(x) = \eta(x)^{-(\nu+\rho)}$ $(x \in G)$ – then $A(w)f_\nu$ is proportional to $f_{w\nu}$ and, upon calculating its value at the identity, one finds, for the normalization so chosen, that

$$A(w)f_\nu = \mathbf{c}_w(\nu)f_{w\nu}.$$

It therefore follows that

$$\int_{U_{\bar{w}}} A(w'') f_\nu(xm_{w'}u')\, d_{U_{\bar{w}'}}(u') = \mathbf{c}_{w''}(\nu)\mathbf{c}_{w'}(w''\nu) f_{w\nu}(x) \qquad (x \in G).$$

Since $\Delta_0(w) = \Delta_0(w'') \cup w''^{-1}\Delta_0(w')$, the very definition of $\mathbf{c}_w(\nu)$ tells us that $\mathbf{c}_w(\nu) = \mathbf{c}_{w''}(\nu)\mathbf{c}_{w'}(w''\nu)$. Substituting $x = 1$ into the above equation then leads to the relation

$$\int_{U_{\bar{w}}} f_\nu(u)\, d_{U_{\bar{w}}}(u) = \iint_{U_{\bar{w}'} \times U_{\bar{w}''}} f_\nu(m_{w'}^{-1} u' m_{w''} u'')\, d_{U_{\bar{w}'}}(u')\, d_{U_{\bar{w}''}}(u'')$$

from which we can read off the desired assertion (f_ν is never zero).

Given a linear function ν on $\mathfrak{a}_\mathfrak{p}$, set

$$\mathbf{I}(\nu) = \prod_{\lambda \in \Sigma^+} B\left(\frac{m(\lambda)}{2}, \frac{m(\lambda/2)}{4} + \frac{(\nu, \lambda)}{(\lambda, \lambda)}\right)$$

where B is the Beta function – then, as has been seen above, we have

$$\int_{N^-} e^{-(\nu+\rho)(H(n))}\, d_{N^-}(n) = \mathbf{I}(\nu)/\mathbf{I}(\rho)$$

for all ν such that $\mathscr{R}(\nu) \in \mathscr{C}$. This being so, it is then only natural to ask: Is the integral defining $T(w,\nu,\delta)$(say) explicitly computable? Along these lines, one is able to say the following. In the first place it is clearly enough to consider only the case when the split-rank of G is equal to 1; moreover, it will be best to assume in addition that G admits a faithful finite dimensional representation. Making the obvious identifications, let us now distinguish two cases according to whether 2λ is present or not (λ the unique positive indivisible root): (1) If 2λ is a root, then the coefficients of the operator $T(w, \nu, \delta)$ are finite linear combinations of terms of the form

$$\frac{\Gamma(4^{-1}(\nu + r + \rho - 2m))}{\Gamma(4^{-1}(\nu + r + \rho))} \cdot \frac{\Gamma(2^{-1}(\nu + r - 2m - 2n))}{\Gamma(2^{-1}(\nu + r - 2m + m(\lambda)))}$$

where m, n, and r are non-negative integers such that $m + n \leqslant r/2$; (2) If 2λ is not a root, then the coefficients of the operator $T(w, \nu, \delta)$ are finite linear combinations of terms of the form

$$\frac{\Gamma(2^{-1}(\nu + r - 2n))}{\Gamma(2^{-1}(\nu + r + m(\lambda)))}.$$

In particular one establishes that $T(w, \nu, \delta)$ is a rational function of ν iff $m(2\lambda) = 0$ and $m(\lambda)$ is even (this will happen when G has but one conjugacy class of Cartan subgroups -- cf. the second Appendix to Vol. I, number 1.1.3). [The reader will find the details which lie behind the above assertions in Schiffmann [2].]

9.1.7 Applications to Zonal Spherical Functions

Let G be a connected semi-simple Lie group with finite center, $G = KA_\mathfrak{p}N^+$ an Iwasawa decomposition for G; given a (complex valued) linear function ν on $\mathfrak{a}_\mathfrak{p}$, set

$$\phi_\nu(x) = \int_K e^{(\sqrt{-1}\nu - \rho)(H(xk))}\, dk \qquad (x \in G).$$

Then ϕ_ν is a zonal spherical function on G and all such arise in this way for a suitable choice of ν. The theory developed in the earlier numbers of this section is applicable, of course, in the present situation. Thus introduce the semi-lattice L comprised of those linear functions λ on $\mathfrak{a}_\mathfrak{p}$ which are non-negative integral linear combinations of the simple roots of the pair $(\mathfrak{g}, \mathfrak{a}_\mathfrak{p})$; define the rational functions $\mathbf{\Gamma}_\lambda$ $(\lambda \in \mathsf{L})$ on $\mathfrak{a}_{\mathfrak{p}_c}$ (or $\mathfrak{a}_{\mathfrak{p}_c}^\vee \ldots$) recursively by $\mathbf{\Gamma}_0 = 1$ and

$$\{(\lambda, \lambda - 2\rho) - 2\lambda\}\mathbf{\Gamma}_\lambda = 2 \sum_{\alpha \in P_+} \sum_{n \geqslant 1} \{(\tilde{\alpha}, \lambda - 2n\tilde{\alpha}) - \tilde{\alpha}\}\mathbf{\Gamma}_{\lambda - 2n\tilde{\alpha}}$$

where n runs over all integers $\geqslant 1$ such that $\lambda - 2n\tilde{\alpha} \in \mathsf{L}$. Let L' denote the set of non-zero elements in L; for each λ in L', let σ_λ denote the hyperplane in $\mathfrak{a}_{\mathfrak{p}_c}$ (or $\mathfrak{a}_{\mathfrak{p}_c}^\vee \ldots$) consisting of those H such that $2\sqrt{-1}\lambda(H) = (\lambda, \lambda)$; let $\mathbf{\Gamma}(\mathbf{c})$ denote the complement in $\mathfrak{a}_{\mathfrak{p}_c}$ of the $\sigma_{w\lambda}$ $(\lambda \in \mathsf{L}', w \in W$ (= the Weyl group of the pair $(\mathfrak{g}, \mathfrak{a}_\mathfrak{p})))$, $\mathbf{\Gamma}'(\mathbf{c})$ the Σ-regular elements in $\mathbf{\Gamma}(\mathbf{c})$ – then, according to Theorem 9.1.5.1, there exist holomorphic functions $\mathbf{c}_w$ $(w \in W)$ on $\mathbf{\Gamma}'(\mathbf{c})$ such that

$$h^\rho \int_K e^{(\sqrt{-1}\nu - \rho)(H(hk))}\, dk = \sum_{w \in W} \mathbf{\Phi}(w\nu : h)\mathbf{c}_w(\nu) \qquad (\nu \in \mathbf{\Gamma}'(\mathbf{c}))$$

for all $h \in \exp(\mathscr{C})$. [Here $\mathbf{\Phi}(\nu : h) = e^{\langle\sqrt{-1}\nu, H\rangle} \sum_{\lambda \in \mathsf{L}} \mathbf{\Gamma}_\lambda(\sqrt{-1}\nu - \rho)e^{-\lambda(H)}$ $(h \in \exp(\mathscr{C}))$ where $H = \log h$.] Because $\phi_{w\nu} = \phi_\nu$ (all $w \in W$), it must be the case that $\mathbf{c}_w(\nu) = \mathbf{c}_1(w\nu)$ (all $\nu \in \mathbf{\Gamma}'(\mathbf{c})$); agreeing then to write $\mathbf{c}$ for $\mathbf{c}_1$, for fixed ν in $\mathbf{\Gamma}'(\mathbf{c})$ we have the asymptotic relation

$$\lim_{h \to \infty} \left| h^\rho \phi_\nu(h) - \sum_{w \in W} \mathbf{c}(w\nu)h^{\sqrt{-1}w\nu} \right| = 0 \qquad (h \in \exp(\mathscr{C})).$$

The Harish-Chandra $\mathbf{c}$-function is explicitly computable; thus suppose that ν is such that $-\mathscr{I}(\nu) \in \mathscr{C}$ $(\nu \in \mathbf{\Gamma}'(\mathbf{c}))$ – then, according to Theorem 9.1.6.1 and subsequent discussion,

$$\mathbf{c}(\nu) = \int_{N^-} \exp\{-(\sqrt{-1}\nu + \rho)(H(n))\}\, d_{N^-}(n) = \mathbf{I}(\sqrt{-1}\nu)/\mathbf{I}(\rho)$$

where

$$\mathbf{I}(\nu) = \prod_{\lambda \in \Sigma^+} B\left(\frac{m(\lambda)}{2}, \frac{m(\lambda/2)}{4} + \frac{(\nu, \lambda)}{(\lambda, \lambda)}\right).$$

It is easy to check that the singularities of the (meromorphic) function $\nu \mapsto \mathbf{I}(\sqrt{-1}\nu)/\mathbf{I}(\rho)$ lie in the complement of $\mathbf{\Gamma}'(\mathbf{c})$. [Here, of course, one has to use the fact that the Gamma function is never zero and has (simple) poles only at $0, -1, -2, \ldots$.] Consequently, by analytic continuation, it follows that $\mathbf{c}$ is actually a meromorphic function on all of $\mathfrak{a}_{\mathfrak{p}_c}^\vee$ – in fact

$$\mathbf{c}(\nu) = \mathbf{I}(\sqrt{-1}\nu)/\mathbf{I}(\rho) \qquad (\text{all } \nu).$$

Example Let us suppose that G has but one conjugacy class of Cartan subgroups – then every root of the pair $(\mathfrak{g}, \mathfrak{a}_\mathfrak{p})$ is indivisible with even multiplicity (cf. the second Appendix to Vol. I, number 1.1.3). This being so, define the integer $n(\lambda)$ by the rule $m(\lambda) = 2n(\lambda)$ $(\lambda \in \Sigma)$ – then, in view of the general relation

$$\{z\}_n = \prod_{i=1}^{n} (z + i - 1) = \Gamma(z + n)/\Gamma(z) \qquad (z \neq 0, -1, -2, \ldots; n \geqslant 1),$$

we see that here

$$\mathbf{c}(\nu) = \prod_{\lambda \in \Sigma^+} \{(\rho, \lambda)/(\lambda, \lambda)\}_{n(\lambda)}/\{(\sqrt{-1}\nu, \lambda)/(\lambda, \lambda)\}_{n(\lambda)}$$

which shows that $\mathbf{c}^{-1}$ is a polynomial in the present case.

The $\mathbf{c}$-function possesses a 'product formula'; thus, according to Proposition 9.1.6.6 (with $w^\dagger$ substituted for w), we have $\mathbf{c} = \prod_{\lambda \in \Sigma_0^+} \mathbf{c}_\lambda$. Set $\pi(=\pi_\Sigma) = \prod_{\lambda \in \Sigma_0^+} \lambda$ – then it is clear that $\pi\mathbf{c}$ is everywhere meromorphic and, moreover, does not vanish at the origin. On the other hand, $\mathbf{c}^{-1}$ is holomorphic in an open subset of the complex dual of $\mathfrak{a}_\mathfrak{p}$ containing the real dual of $\mathfrak{a}_\mathfrak{p}$.

Here are some additional properties of $\mathbf{c}$.

Proposition 9.1.7.1 Suppose that ν is real and Σ-regular – then
(i) $\mathbf{c}(-\nu) = \overline{\mathbf{c}(\nu)}$; (ii) $|\mathbf{c}(\nu)| = |\mathbf{c}(w\nu)|$
(iii) $|\mathbf{c}(\nu)|^2 = \mathbf{c}(\nu)\mathbf{c}(-\nu) = \mathbf{c}(w\nu)\mathbf{c}(-w\nu)$ $\quad (w \in W)$.
[We remind the reader that $\Gamma'(\mathbf{c})$ contains, as a subset, the real, Σ-regular elements ν.]

Proof Since ν is real, the zonal spherical function ϕ_ν is positive definite (cf. number 6.2.2); therefore $\phi_{-\nu} = \overline{\phi_\nu}$ and so statement (i) follows immediately from the asymptotic expansion for zonal spherical functions (cf. supra). Turning to the proof of (ii), it will clearly be enough to verify that $|\mathbf{c}(\nu)| = |\mathbf{c}(w_i\nu)|$ for a given reflection w_i corresponding to a simple root λ_i $(1 \leqslant i \leqslant l)$; because w_i permutes the elements of Σ^+ different from λ_i or $2\lambda_i$ amongst themselves and sends λ_i, $2\lambda_i$ to $-\lambda_i$, $-2\lambda_i$, respectively, assertion (ii) is then seen to be a consequence of the general rules: $|\Gamma(x + \sqrt{-1}y)| = |\Gamma(x - \sqrt{-1}y)|$ $(x, y \in \mathbf{R}; x > 0)$, $|\Gamma(\sqrt{-1}y)| = |\Gamma(-\sqrt{-1}y)|$ $(y \in \mathbf{R})$ (see Magnus and Oberhettinger [1, p. 3]). It now follows that $\mathbf{c}(\nu)\mathbf{c}(-\nu) = |\mathbf{c}(\nu)|^2 = |\mathbf{c}(w\nu)|^2 = \mathbf{c}(w\nu)\mathbf{c}(-w\nu)$ (all $w \in W$) which proves (iii). □

Proposition 9.1.7.2 There exist positive integers m and M such that

$$\frac{1}{|\mathbf{c}(\nu)|} \leqslant M(1 + \|\nu\|)^m \qquad (\text{all } \nu \in \mathfrak{a}_\mathfrak{p}^\vee).$$

Corollary 9.1.7.3 The function $\mathbf{c}^{-1}$ defines a tempered distribution on the (real) dual of $\mathfrak{a}_\mathfrak{p}$.

Thanks to the 'product formula' for **c**, Proposition 9.1.7.2 is a consequence of the following lemma.

Lemma 9.1.7.4 Let a and b be positive real numbers – then the function f on **R** given by

$$f(x) = \Gamma(a + \sqrt{-1}x)(\Gamma(b + \sqrt{-1}x))^{-1} \qquad (x \in \mathbf{R})$$

has each of its derivatives majorized uniformly by a polynomial.

Proof Using the formula

$$\Gamma'(z)/\Gamma(z) = -\gamma - z^{-1} + z\sum_1^\infty \{n(z+n)\}^{-1}$$

where γ is Euler's constant, we see that

$$\begin{aligned}-\sqrt{-1}\frac{f'(x)}{f(x)} &= -\frac{1}{a+\sqrt{-1}x} + \frac{1}{b+\sqrt{-1}x} \\ &\quad + (a+\sqrt{-1}x)\sum_1^\infty \frac{1}{n(n+a+\sqrt{-1}x)} \\ &\quad - (b+\sqrt{-1}x)\sum_1^\infty \frac{1}{n(n+b+\sqrt{-1}x)}\end{aligned}$$

and so $f' = fg$ where g is a function each of whose derivatives is bounded by a polynomial. On the other hand, for a fixed real number r, it is known that

$$\lim_{|x|\to\infty} |\Gamma(r+\sqrt{-1}x)|\exp(\pi|x|/2)|x|^{2^{-1}-r} = \sqrt{2\pi}$$

(cf. Magnus and Oberhettinger [1, p. 6]). This equation shows that f is uniformly bounded by a polynomial. Hence our lemma follows by induction. □

Example Let us suppose that $\mathfrak{g}$ carries a complex structure – then, in this special case, it is actually possible to obtain explicit formulae for the zonal spherical functions ϕ_ν and the rational functions $\mathbf{\Gamma}_\lambda$. To begin with, observe that here each root λ of the pair $(\mathfrak{g}, \mathfrak{a}_\mathfrak{p})$ has multiplicity 2 (whence $\rho = \sum_{\lambda>0}\lambda$); assigning to $\triangle$ the usual meaning, standard considerations of divisibility imply that $\triangle^{1/2} = \sum_{w\in W}\det(w)e^{w\rho}$, W the Weyl group of the pair $(\mathfrak{g}, \mathfrak{a}_\mathfrak{p})$. This being so, let ω be the Casimir operator – then, by direct computation, we find that

$$\mathfrak{f}_{A_\mathfrak{p}}(\omega) = \triangle^{-1/2}\gamma_{A_\mathfrak{p}}(\omega)\circ\triangle^{1/2}$$

as differential operators on $A'_\mathfrak{p}$ (relative to the trivial double representation of K), hence, as is easily seen,

$$\mathbf{\Phi}(\nu : h) = \triangle(h)^{-1/2}h^{\rho+\sqrt{-1}\nu} \qquad (\nu \in \mathbf{\Gamma}^\gamma, h \in \exp(\mathscr{C})).$$

Suppose now that ν is real and Σ-regular – then, since $\mathbf{c}(\nu) = \pi(\rho)/\pi(\sqrt{-1}\nu)$ (cf. the Example supra), it follows that

$$\pi(\nu)\triangle(h)^{1/2}\phi_\nu(h) = (-1)^{[\Sigma^+]/2}\pi(\rho)\sum_{w\in W}\det(w)h^{\sqrt{-1}w\nu} \qquad (h \in A_\mathfrak{p}).$$

However both sides of this equation are holomorphic in ν and so our formula is actually valid for all ν; consequently

$$\pi(\nu)\phi_\nu(h) = (-1)^{[\Sigma^+]/2}\pi(\rho)\frac{\sum_{w\in W}\det(w)h^{\sqrt{-1}w\nu}}{\sum_{w\in W}\det(w)h^{w\rho}} \qquad (h \in A'_{\mathfrak{p}}),$$

a result which is quite analogous to that of Weyl's for the characters of compact connected semi-simple Lie groups. [Of course if ν is singular, then the preceding formula for ϕ_ν leaves something to be desired. To remedy this, let π'_ν denote the product of the positive roots which vanish at ν, π''_ν the product of the remaining positive roots; applying the differential operator $\partial(\pi'_\nu)$ to the formula

$$\pi'_\nu(\mu)\Big(\sum_{w\in W}\det(w)h^{w\rho}\Big)\phi_\mu(h)$$
$$= (-1)^{[\Sigma^+]/2}\pi(\rho)\sum_{w\in W}\det(w)h^{\sqrt{-1}w\mu}\cdot(\pi''_\nu(\mu))^{-1} \qquad (h \in A_{\mathfrak{p}})$$

and evaluating at $\mu = \nu$ gives the relation

$$\Big(\sum_{w\in W}\det(w)h^{w\rho}\Big)\phi_\nu(h) = C\sum_{w\in W}p_w(\log h)h^{\sqrt{-1}w\nu} \qquad (h \in A_{\mathfrak{p}})$$

where C is a non-zero explicitly computable constant and the p_w $(w \in W)$ are polynomials defined by the prescription

$$p_w(H) = \partial(\pi'_\nu)\{\det(w)(\pi''_\nu(\mu))^{-1}e^{\sqrt{-1}w\mu(H)}\}|_{\mu=\nu}\cdot e^{-\sqrt{-1}w\nu(H)} \qquad (H \in \mathfrak{a}_{\mathfrak{p}_c}).$$

Thus we have a semi-explicit formula for ϕ_ν in the case of singular ν as well.] It is also an easy matter to obtain an explicit formula for the rational functions $\mathbf{\Gamma}_\lambda$ $(\lambda \in \mathsf{L})$. Indeed: $\mathbf{\Gamma}_\lambda$ = the number of partitions of $\lambda/2$ into a sum of positive roots (order immaterial). Thus, for $H \in \mathscr{C}$, $\nu \in \mathbf{\Gamma}'(\mathbf{c})$, we have

$$\phi(\nu:\exp H) = \sum_{w\in W}\mathbf{c}(w\nu)e^{\sqrt{-1}w\nu(H)}\prod_{\lambda\in\Sigma^+}(e^{\lambda(H)} - e^{-\lambda(H)})^{-1}$$

and so

$$e^{\rho(H)}\phi(\nu:\exp H) = \sum_{w\in W}\mathbf{c}(w\nu)e^{\sqrt{-1}w\nu(H)}\prod_{\lambda\in\Sigma^+}(1 + e^{-2\lambda(H)} + e^{-4\lambda(H)} + \cdots)$$
$$= \sum_{w\in W}\mathbf{c}(w\nu)e^{\sqrt{-1}w\nu(H)}\sum_{\lambda\in\mathsf{L}}\mathbf{P}(\lambda/2)e^{-\lambda(H)}$$

where $\mathbf{P}(\lambda/2)$ is the number of partitions of $\lambda/2$ $(\lambda \in \mathsf{L})$ into a sum of positive roots (order immaterial). But, for $H \in \mathscr{C}$, $\nu \in \mathbf{\Gamma}'(\mathbf{c})$, we also have

$$e^{\rho(H)}\phi(\nu:\exp H) = \sum_{w\in W}\mathbf{c}(w\nu)e^{\sqrt{-1}w\nu(H)}\sum_{\lambda\in\mathsf{L}}\mathbf{\Gamma}_\lambda(\sqrt{-1}w\nu - \rho)e^{-\lambda(H)}$$

from which we can read off our contention via the usual argument.

Appendix (*The Gangolli Expansion*) In this Appendix we shall record some technical results which will be needed during the course of the proof of the Paley-Wiener Theorem for zonal spherical functions on G. We shall agree to use without comment the notations and conventions introduced above.

Let ω be the Casimir operator in the center $\mathfrak{Z}$ of $\mathfrak{G}$ – then, in view of Theorem 9.1.2.9 and Corollary 9.1.2.12, it is clear that for any C^∞ K-biinvariant function f on G, we have

$$f(h;\omega) = f(h;\mathfrak{f}_{A_{\mathfrak{p}}}(\omega)) = f\Big(h;\sum_{i=1}^{l}\triangle^{-1}H_i\circ\triangle H_i\Big) \qquad (h \in A'_{\mathfrak{p}}).$$

Here $H_1, \ldots, H_l$ is an orthonormal basis for $\mathfrak{a}_\mathfrak{p}$ while, as always,

$$\triangle(h) = \prod_{\alpha \in P_+} (h^\alpha - h^{-\alpha}) \qquad (h \in A_\mathfrak{p}).$$

For $\nu \in \mathbf{\Gamma}^\gamma$, $h \in \exp(\mathscr{C})$, put

$$\hat{\mathbf{\Psi}}(\nu : h) = \triangle(h)^{1/2}\mathbf{\Psi}(\sqrt{-1}\nu - \rho : h).$$

If we write

$$\triangle(h)^{1/2} = e^{\rho(H)} \sum_{\lambda \in \mathsf{L}} \gamma_\lambda e^{-\lambda(H)} \qquad (h \in \exp(\mathscr{C}))$$

where $H = \log h$, then it follows that

$$\hat{\mathbf{\Psi}}(\nu : h) = e^{\langle\sqrt{-1}\nu, H\rangle} \sum_{\lambda \in \mathsf{L}} \hat{\mathbf{\Gamma}}_\lambda(\nu) e^{-\lambda(H)}$$

where

$$\hat{\mathbf{\Gamma}}_\lambda(\nu) = \sum_{\lambda_1 + \lambda_2 = \lambda} \gamma_{\lambda_1} \mathbf{\Gamma}_{\lambda_2}(\sqrt{-1}\nu - \rho) \qquad (\nu \in \mathbf{\Gamma}^\gamma)$$

which shows that $\hat{\mathbf{\Gamma}}_\lambda$ $(\lambda \in \mathsf{L})$ is a rational function of ν. The function $\hat{\mathbf{\Psi}}$ satisfies a simple differential equation:

Lemma Set $\Delta = \sum_{i=1}^{l} H_i^2$ – then, for all $\nu \in \mathbf{\Gamma}^\gamma$,

$$\hat{\mathbf{\Psi}}(\nu : h; \Delta) = -\{(\nu, \nu) + (\rho, \rho) + \sum_{i=1}^{l} \triangle^{-1/2} H_i \circ \triangle H_i \triangle^{-1/2}\}\hat{\mathbf{\Psi}}(\nu : h)$$
$$(h \in \exp(\mathscr{C})).$$

Proof For $\nu \in \mathbf{\Gamma}^\gamma$, $h \in \exp(\mathscr{C})$, let us agree to write $\hat{\mathbf{\Phi}}(\nu : h)$ in place of $\mathbf{\Psi}(\sqrt{-1}\nu - \rho : h)$ (so that $\hat{\mathbf{\Psi}} = \triangle^{1/2}\hat{\mathbf{\Phi}}$); now fix a ν in $\mathbf{\Gamma}^\gamma$ – then

$$\begin{aligned}
H_i \circ \triangle H_i \hat{\mathbf{\Phi}} &= H_i(\triangle H_i(\triangle^{-1/2}\hat{\mathbf{\Psi}})) \\
&= H_i(\triangle^{1/2}(H_i\hat{\mathbf{\Psi}}) + \triangle(H_i\triangle^{-1/2})\hat{\mathbf{\Psi}}) \\
&= (H_i(\triangle(H_i\triangle^{-1/2})))\hat{\mathbf{\Psi}} + \triangle(H_i\triangle^{-1/2})(H_i\hat{\mathbf{\Psi}}) + (H_i\triangle^{1/2})(H_i\hat{\mathbf{\Psi}}) \\
&\quad + \triangle^{1/2}(H_i^2\hat{\mathbf{\Psi}}) \\
&= (H_i \circ \triangle H_i \triangle^{-1/2})\hat{\mathbf{\Psi}} + \triangle^{1/2}(H_i^2\hat{\mathbf{\Psi}}),
\end{aligned}$$

it being clear that $\triangle(H_i\triangle^{-1/2}) + H_i\triangle^{1/2} = 0$. It therefore follows that

$$\sum_{i=1}^{l} \triangle^{-1} H_i \circ \triangle H_i \hat{\mathbf{\Phi}} = (\sum_{i=1}^{l} \triangle^{-1} H_i \circ \triangle H_i \triangle^{-1/2})\hat{\mathbf{\Psi}} + \triangle^{-1/2}(\Delta\hat{\mathbf{\Psi}}).$$

But, according to Theorem 9.1.4.1,

$$\begin{aligned}
\hat{\mathbf{\Phi}}(\nu : h; \sum_{i=1}^{l} \triangle^{-1} H_i \circ \triangle H_i) &= \mathbf{\Psi}(\sqrt{-1}\nu - \rho : h; \mathfrak{P}_{A_\mathfrak{p}}(\omega)) \\
&= \gamma_{A_\mathfrak{p}}^*(\omega : \sqrt{-1}\nu - \rho)\mathbf{\Psi}(\sqrt{-1}\nu - \rho : h) \\
&= -((\nu, \nu) + (\rho, \rho))\triangle^{-1/2}(h)\hat{\mathbf{\Psi}}(\nu : h) \\
&\qquad (h \in \exp(\mathscr{C}))
\end{aligned}$$

from which the desired assertion follows at once. □

The point to be noticed here is that the preceding differential equation for $\hat{\mathbf{\Psi}}$ has no first order derivatives in it; for this reason we shall be able to estimate $\hat{\mathbf{\Psi}}$ rather closely (cf. infra). [This procedure of 'multiplying up' by $\triangle^{1/2}$ is entirely analogous to a similar device which is employed extensively in the classical theory of second order differential equations with regular singular points.]

In order to make further progress, it will be necessary to compute the function

$$\sum_{i=1}^{l} \triangle^{-1/2} H_i \circ \triangle H_i \triangle^{-1/2}$$

on $\mathscr{C}$. Thus, to begin with, we have

$$\begin{aligned} \triangle^{1/2} H_i \triangle^{-1/2} &= H_i(\log(\triangle^{-1/2})) \\ &= -2^{-1} \sum_{\alpha \in P_+} \coth(\alpha)\alpha(H_i) = \chi \qquad \text{(say)} \end{aligned}$$

and so

$$\begin{aligned} H_i \circ \triangle H_i \triangle^{-1/2} &= H_i \circ \triangle^{1/2} \triangle^{1/2} H_i \triangle^{-1/2} \\ &= H_i(\triangle^{1/2}\chi) = (H_i \triangle^{1/2})\chi + \triangle^{1/2}(H_i\chi), \end{aligned}$$

whence

$$\triangle^{-1/2} H_i \circ \triangle H_i \triangle^{-1/2} = (\triangle^{-1/2} H_i \triangle^{1/2})\chi + H_i\chi \qquad (1 \leqslant i \leqslant l).$$

But

$$H_i\chi = 2^{-1} \sum_{\alpha \in P_+} \operatorname{cosech}^2(\alpha) \cdot \alpha(H_i)^2$$

while

$$\triangle^{-1/2} H_i \triangle^{1/2} = -\chi$$

from which we infer that

$$\begin{aligned} \triangle^{-1/2} H_i \circ \triangle H_i \triangle^{-1/2} &= -\chi^2 + H_i\chi \\ &= -4^{-1} \sum_{\alpha \in P_+} \coth^2(\alpha) \cdot \alpha(H_i)^2 \\ &\quad + 2^{-1} \sum_{\alpha \in P_+} \operatorname{cosech}^2(\alpha) \cdot \alpha(H_i)^2 \\ &\quad - 4^{-1} \sum_{\substack{\alpha, \beta \in P_+ \\ \alpha \neq \beta}} \coth(\alpha)\coth(\beta)\alpha(H_i)\beta(H_i) \\ &\qquad\qquad (1 \leqslant i \leqslant l). \end{aligned}$$

Summing on i and using the relation $\coth^2(x) = 1 + \operatorname{cosech}^2(x)$ $(x \in \mathbf{R})$, we get

$$\begin{aligned} \sum_{i=1}^{l} \triangle^{-1/2} H_i \circ \triangle H_i \triangle^{-1/2} &= -4^{-1} \sum_{\alpha \in P_+} (\tilde{\alpha}, \tilde{\alpha}) + 4^{-1} \sum_{\alpha \in P_+} (\tilde{\alpha}, \tilde{\alpha}) \operatorname{cosech}^2(\alpha) \\ &\quad - 4^{-1} \sum_{\substack{\alpha, \beta \in P_+ \\ \alpha \neq \beta}} (\tilde{\alpha}, \tilde{\beta}) \coth(\alpha)\coth(\beta). \end{aligned}$$

Because $\operatorname{cosech}^2(x) = 4 \sum_{n \geqslant 1} n e^{-2nx}$ $(x > 0)$ while

$$\coth(x) = \sum_{n \geqslant 0} c_n e^{-2nx} \qquad (x > 0;\, c_0 = 1,\, c_n = 2\ (n \geqslant 1)),$$

the preceding expression for $\sum_{i=1}^{l} \triangle^{-1/2} H_i \circ \triangle H_i \triangle^{-1/2}$ can be written

$$\begin{aligned} \sum_{i=1}^{l} \triangle^{-1/2} H_i \circ \triangle H_i \triangle^{-1/2} &= -(\rho, \rho) + \sum_{\alpha \in P_+} (\tilde{\alpha}, \tilde{\alpha}) \sum_{n \geqslant 1} n e^{-2n\alpha} \\ &\quad - 4^{-1} \sum_{\substack{\alpha, \beta \in P_+ \\ \alpha \neq \beta}} (\tilde{\alpha}, \tilde{\beta}) \sum_{m+n \geqslant 1} c_m c_n e^{-2m\alpha - 2n\beta}. \end{aligned}$$

Inserting now this result into the differential equation satisfied by $\hat{\Psi}$ on $\exp(\mathscr{C})$ leads to

$$\Delta\hat{\Psi} = \{-(\nu, \nu) + \sum_{\alpha \in P_+} (\tilde{\alpha}, \tilde{\alpha}) \sum_{n \geq 1} -ne^{-2n\alpha} + 4^{-1} \sum_{\substack{\alpha, \beta \in P_+ \\ \alpha \neq \beta}} (\tilde{\alpha}, \tilde{\beta}) \sum_{m+n \geq 1} c_m c_n e^{-2m\alpha - 2n\beta}\}\hat{\Psi}$$

for fixed ν in Γ^γ, that is, for all $\nu \in \Gamma^\gamma$ and all $h \in \exp(\mathscr{C})$, we have

$$e^{\langle\sqrt{-1}\nu, H\rangle} \sum_{\lambda \in \mathsf{L}} (\sqrt{-1}\nu - \lambda, \sqrt{-1}\nu - \lambda)\hat{\Gamma}_\lambda(\nu)e^{-\lambda(H)}$$

$$= \sum_{\lambda \in \mathsf{L}} \begin{pmatrix} -(\nu, \nu) + \sum_{\alpha \in P_+} (\tilde{\alpha}, \tilde{\alpha}) \sum_n -n\hat{\Gamma}_{\lambda - 2n\tilde{\alpha}}(\nu) \\ + 4^{-1} \sum_{\substack{\alpha, \beta \in P_+ \\ \alpha \neq \beta}} (\tilde{\alpha}, \tilde{\beta}) \sum_{m,n} c_m c_n \hat{\Gamma}_{\lambda - 2m\tilde{\alpha} - 2n\tilde{\beta}}(\nu) \end{pmatrix} e^{(\sqrt{-1}\nu - \lambda)(H)}$$

where $H = \log h$. Here, in the sum $\sum_n$, n runs over the set

$$Q_\nu(\tilde{\alpha}) = \{n : n \geq 1, \lambda - 2n\tilde{\alpha} \in \mathsf{L}\}$$

and, in the sum $\sum_{m,n}$, the indices m, n run through the set

$$Q_\nu(\tilde{\alpha}, \tilde{\beta}) = \{(m, n) : m \geq 0, n \geq 0, m + n \geq 1, \lambda - 2m\tilde{\alpha} - 2n\tilde{\beta} \in \mathsf{L}\}.$$

For the usual reasons it is permissible to equate the coefficients of $e^{\langle\sqrt{-1}\nu - \lambda, \cdot\rangle}$ on both sides of our equation; doing this, we get, for all $\nu \in \Gamma^\gamma$,

$$\begin{aligned}\{(\lambda, \lambda) - 2\sqrt{-1}(\lambda, \nu)\}\hat{\Gamma}_\lambda(\nu) &= \sum_{\alpha \in P_+} (\tilde{\alpha}, \tilde{\alpha}) \sum_n -n\hat{\Gamma}_{\lambda - 2n\tilde{\alpha}}(\nu) \\ &\quad + 4^{-1} \sum_{\substack{\alpha, \beta \in P_+ \\ \alpha \neq \beta}} (\tilde{\alpha}, \tilde{\beta}) \sum_{m,n} c_m c_n \hat{\Gamma}_{\lambda - 2m\tilde{\alpha} - 2n\tilde{\beta}}(\nu) \\ &= S_1 + S_2 \qquad \text{(say).}\end{aligned}$$

The coefficients $\hat{\Gamma}_\lambda$ ($\lambda \in \mathsf{L}$) are determined recursively by this relation (plainly $\hat{\Gamma}_0 = 1$). There are a number of more or less immediate observations to be made here. Agreeing to refer to the integer m_λ as the *level* of λ ($\lambda \in \mathsf{L}$), note first that if λ is of level 1, then the sets $Q_\nu(\tilde{\alpha})$, $Q_\nu(\tilde{\alpha}, \tilde{\beta})$ ($\nu \in \Gamma^\gamma$) are empty; arguing by induction, we then deduce that $\hat{\Gamma}_\lambda = 0$ whenever the level m_λ of λ is odd. So suppose that the level m_λ of λ is even, $= 2M$ (say) – then the only terms that can occur in the recursive definition of $\hat{\Gamma}_\lambda$ are of level $2r$, $r = 0, 1, \ldots, M - 1$. It will be convenient to group together the terms on a given level in the sums S_1, S_2 supra; thus let $S_1(r)$ be the contribution to S_1 from terms of level $2r$ in S_1 and similarly for $S_2(r)$ – clearly

$$S_1 = \sum_{r=0}^{M-1} S_1(r), \qquad S_2 = \sum_{r=0}^{M-1} S_2(r).$$

Moreover if

$$Q_\nu(\tilde{\alpha}, r) = \{n \in Q_\nu(\tilde{\alpha}) : m_{\lambda - 2n\tilde{\alpha}} = 2r\},$$

$$Q_\nu((\tilde{\alpha}, \tilde{\beta}), r) = \{(m, n) \in Q_\nu(\tilde{\alpha}, \tilde{\beta}) : m_{\lambda - 2m\tilde{\alpha} - 2n\tilde{\beta}} = 2r\},$$

then, for a given ν in Γ^γ, we have

$$S_1(r) = \sum_{\alpha \in P_+} (\tilde{\alpha}, \tilde{\alpha}) \sum_{n \in Q_\nu(\tilde{\alpha}, r)} -n\hat{\Gamma}_{\lambda - 2n\tilde{\alpha}}(\nu),$$

$$S_2(r) = 4^{-1} \sum_{\substack{\alpha, \beta \in P_+ \\ \alpha \neq \beta}} (\tilde{\alpha}, \tilde{\beta}) \sum_{(m,n) \in Q_\nu((\tilde{\alpha}, \tilde{\beta}), r)} c_m c_n \hat{\Gamma}_{\lambda - 2m\tilde{\alpha} - 2n\tilde{\beta}}(\nu).$$

We shall now estimate the $\hat{\mathbf{\Gamma}}_\lambda$ ($\lambda \in \mathsf{L}$). Obviously any such function takes on a well-defined value at every point in $\mathbf{\Gamma}^\gamma$. This being so, consider the region

$$\mathscr{R} = \{\xi + \sqrt{-1}\eta : \xi, \eta \in \mathfrak{a}_\mathfrak{p}^\vee, \eta \in \mathscr{C}^{cl}\}.$$

Then $\mathscr{R} \subset \mathbf{\Gamma}^\gamma$. [Thus fix $\lambda \in \mathsf{L}'$; given a ν in $\mathscr{R}$, write $\nu = \xi + \sqrt{-1}\eta$ with $\eta \in \mathscr{C}^{cl}$ – then

$$\begin{aligned}|(\lambda, \lambda) - 2\sqrt{-1}(\lambda, \nu)| &= |(\lambda, \lambda) + 2(\lambda, \eta) - 2\sqrt{-1}(\lambda, \xi)| \\ &\geqslant |(\lambda, \lambda) + 2(\lambda, \eta)| \geqslant (\lambda, \lambda) > 0 \ldots]\end{aligned}$$

Proposition (Gangolli) Fix $\nu \in \mathbf{\Gamma}^\gamma$ – then there exist real numbers $\hat{m}(\nu)$, $\hat{M}(\nu)$ such that

$$|\hat{\mathbf{\Gamma}}_\lambda(\nu)| \leqslant \hat{M}(\nu)(m_\lambda)^{\hat{m}(\nu)} \qquad (\text{all } \lambda \in \mathsf{L}).$$

In addition there exist absolute constants $\hat{d}$, $\hat{D}$ such that

$$|\hat{\mathbf{\Gamma}}_\lambda(\nu)| \leqslant \hat{D} m_\lambda^{\hat{d}} \qquad (\text{all } \nu \in \mathscr{R})$$

uniformly in λ ($\lambda \in \mathsf{L}$).

Proof Fix $\nu \in \mathbf{\Gamma}^\gamma$; we have noted above that $\hat{\mathbf{\Gamma}}_\lambda = 0$ if the level m_λ of λ is odd so we may suppose that m_λ is even, $= 2M$ (say). Put $H_r(\nu) = \sup_{\tilde{\lambda}} |\hat{\mathbf{\Gamma}}_{\tilde{\lambda}}(\nu)|$ where $m_{\tilde{\lambda}} = 2r$ ($\tilde{\lambda} \in \mathsf{L}$); here $r = 0, 1, \ldots, M - 1$. Let us estimate $S_1(r)$. Thus consider the set $Q_\nu(\tilde{\alpha}, r)$; it consists of those positive integers n such that $m_{\lambda - 2n\tilde{\alpha}} = 2r$; since $m_\lambda = 2M$, this means that $2nm_{\tilde{\alpha}} = 2M - 2r$ and so $Q_\nu(\tilde{\alpha}, r)$ is empty unless $m_{\tilde{\alpha}}$ divides $M - r$; assuming then that $m_{\tilde{\alpha}}$ does divide $M - r$, it follows that $Q_\nu(\tilde{\alpha}, r)$ consists of exactly one integer n, viz. $n = (M - r)/m_{\tilde{\alpha}} \leqslant M/m_{\tilde{\alpha}}$. We have, therefore,

$$\begin{aligned}|S_1(r)| &= \Big|\sum_{\alpha \in P_+} (\tilde{\alpha}, \tilde{\alpha}) \sum_{n \in Q_\nu(\tilde{\alpha}, r)} -n\hat{\mathbf{\Gamma}}_{\lambda - 2n\tilde{\alpha}}(\nu)\Big| \\ &\leqslant H_r(\nu) \sum_{\alpha \in P_+} (\tilde{\alpha}, \tilde{\alpha}) \cdot M/m_{\tilde{\alpha}} = c_1 M H_r(\nu)\end{aligned}$$

where $c_1 = \sum_{\alpha \in P_+} (\tilde{\alpha}, \tilde{\alpha})/m_{\tilde{\alpha}}$. Let us estimate $S_2(r)$. Thus consider the set $Q_\nu((\tilde{\alpha}, \tilde{\beta}), r)$; it consists of those pairs (m, n) of non-negative integers such that $m_{\lambda - 2m\tilde{\alpha} - 2n\tilde{\beta}} = 2r$, whence $(m, n) \in Q_\nu((\tilde{\alpha}, \tilde{\beta}), r)$ iff $mm_{\tilde{\alpha}} + nm_{\tilde{\beta}} = M - r$; the number of pairs (m, n) with $m \neq 0$ which solve this equation is at most $(M - r)/m_{\tilde{\alpha}}$ and similarly for pairs (m, n) with $n \neq 0$; consequently the cardinality of $Q_\nu((\tilde{\alpha}, \tilde{\beta}), r)$ cannot exceed

$$(M - r)(1/m_{\tilde{\alpha}} + 1/m_{\tilde{\beta}}) \leqslant M(1/m_{\tilde{\alpha}} + 1/m_{\tilde{\beta}}).$$

Since $|c_m c_n| \leqslant 4$, it therefore follows that

$$\begin{aligned}|S_2(r)| &= 4^{-1}\Big|\sum_{\substack{\alpha, \beta \in P_+ \\ \alpha \neq \beta}} (\tilde{\alpha}, \tilde{\beta}) \sum_{(m,n) \in Q_\nu((\tilde{\alpha}, \tilde{\beta}), r)} c_m c_n \hat{\mathbf{\Gamma}}_{\lambda - 2m\tilde{\alpha} - 2n\tilde{\beta}}(\nu)\Big| \\ &\leqslant M H_r(\nu) \sum_{\substack{\alpha, \beta \in P_+ \\ \alpha \neq \beta}} |(\tilde{\alpha}, \tilde{\beta})|(1/m_{\tilde{\alpha}} + 1/m_{\tilde{\beta}}) = c_2 M H_r(\nu)\end{aligned}$$

where c_2 has the obvious value. Putting these estimates together then leads to the conclusion that

$$|(\lambda, \lambda) - 2\sqrt{-1}(\lambda, \nu)||\hat{\mathbf{\Gamma}}_\lambda(\nu)| \leqslant c_3 M \sum_{r=0}^{M-1} H_r(\nu) \qquad (\lambda \in \mathsf{L}, m_\lambda = 2M)$$

with $c_3 = c_1 + c_2$. There clearly exists a positive constant $c_4(\nu)$ such that

$$|(\lambda, \lambda) - 2\sqrt{-1}(\lambda, \nu)| \geqslant c_4(\nu) m_\lambda^2 \geqslant c_5(\nu) M^2, c_5(\nu) = 4c_4(\nu).$$

So, upon putting $c_6(\nu) = c_3 c_5(\nu)^{-1}$, we get

$$|\hat{\mathbf{\Gamma}}_\lambda(\nu)| \leqslant c_6(\nu) M^{-1} \left(\sum_{r=0}^{M-1} H_r(\nu) \right) \qquad (\lambda \in \mathsf{L}, m_\lambda = 2M),$$

whence, by taking the supremum over all such λ,

$$H_M(\nu) \leqslant c_6(\nu) M^{-1} \left(\sum_{r=0}^{M-1} H_r(\nu) \right).$$

Define now functions H_M via the prescription

$$\mathsf{H}_0(\nu) = 1, \quad \mathsf{H}_M(\nu) = c_6(\nu) M^{-1} \left(\sum_{r=0}^{M-1} \mathsf{H}_r(\nu) \right) \qquad (\nu \in \mathbf{\Gamma}^\gamma).$$

Then it is trivial to check by induction that $H_M(\nu) \leqslant \mathsf{H}_M(\nu)$ for all $M (\nu \in \mathbf{\Gamma}^\gamma)$. Since

$$\begin{aligned} M\mathsf{H}_M(\nu) &= c_6(\nu) \sum_{r=0}^{M-1} \mathsf{H}_r(\nu) \\ &= c_6(\nu) \sum_{r=0}^{M-2} \mathsf{H}_r(\nu) + c_6(\nu) \mathsf{H}_{M-1}(\nu) \\ &= (M-1)\mathsf{H}_{M-1}(\nu) + c_6(\nu)\mathsf{H}_{M-1}(\nu) \\ &= (M-1)\mathsf{H}_{M-1}(\nu)(1 + c_6(\nu)/(M-1)), \end{aligned}$$

we have

$$\begin{aligned} M\mathsf{H}_M(\nu) &= 2\mathsf{H}_2(\nu) \prod_{i=2}^{M-1} (1 + c_6(\nu)/i) \\ &\leqslant 2\mathsf{H}_2(\nu) \exp\left(c_6(\nu) \sum_{i=2}^{M-1} i^{-1} \right) \\ &\leqslant 2\mathsf{H}_2(\nu) \exp(c_6(\nu) \log(M-1)) \\ &\leqslant 2\mathsf{H}_2(\nu) M^{c_6(\nu)} \\ &= c_6(\nu)(1 + c_6(\nu)) M^{c_6(\nu)} \qquad (\nu \in \mathbf{\Gamma}^\gamma). \end{aligned}$$

Set $\hat{m}(\nu) = c_6(\nu) - 1$, $\hat{M}(\nu) = c_6(\nu)(1 + c_6(\nu))2^{-\hat{m}(\nu)}$ $(\nu \in \mathbf{\Gamma}^\gamma)$ – then, for $\nu \in \mathbf{\Gamma}^\gamma$,

$$|\hat{\mathbf{\Gamma}}_\lambda(\nu)| \leqslant H_M(\nu) \leqslant \mathsf{H}_M(\nu) \leqslant \hat{M}(\nu)(m_\lambda)^{\hat{m}(\nu)} \qquad (\lambda \in \mathsf{L}, m_\lambda = 2M)$$

which serves to establish the first contention of our proposition. As for the second assertion, note that the numbers $\hat{m}(\nu)$ and $\hat{M}(\nu)$ depend on ν only because $c_4(\nu)$ does; but if ν lies in $\mathscr{R}$, then, in view of an earlier observation, we have

$$|(\lambda, \lambda) - 2\sqrt{-1}(\lambda, \nu)| \geqslant (\lambda, \lambda) \geqslant c_4 m_\lambda^2$$

where c_4 depends only on the structure of $\mathfrak{g}$. This being the case, we can repeat the above argument word for word to get

$$|\hat{\mathbf{\Gamma}}_\lambda(\nu)| \leqslant \hat{D} m_\lambda^{\hat{d}} \qquad (\text{all } \nu \in \mathscr{R})$$

with $\hat{d} = c_6 - 1$, $\hat{D} = c_6(1 + c_6)2^{-\hat{d}}$ $(c_6 = c_3(4c_4)^{-1})$, the estimate being uniform in λ $(\lambda \in \mathsf{L})$. The proof of the proposition is thereby complete. □

Corollary Retain the above notations – then there exist absolute constants d, D such that

$$|\mathbf{\Gamma}_\lambda(\sqrt{-1}\nu - \rho)| \leqslant D m_\lambda^d \qquad (\text{all } \nu \in \mathscr{R})$$

uniformly in λ $(\lambda \in \mathsf{L})$.

Proof (Helgason) Since $\triangle^{-1/2}$ has an expansion

$$\triangle^{-1/2}(h) = e^{-\rho(H)} \sum_{\lambda \in \mathsf{L}} \hat{\gamma}_\lambda e^{-\lambda(H)} \qquad (h \in \exp(\mathscr{C}))$$

where $H = \log h$, the $\hat{\gamma}_\lambda$ $(\lambda \in \mathsf{L})$ being majorized by Rm_λ^r for certain absolute constants r, R, and since

$$\mathbf{\Gamma}_\lambda(\sqrt{-1}\nu - \rho) = \sum_{\lambda_1+\lambda_2=\lambda} \hat{\gamma}_{\lambda_1} \hat{\mathbf{\Gamma}}_{\lambda_2}(\nu)$$

for all $\nu \in \mathbf{\Gamma}^\nu$, our contention is seen to be an immediate consequence of the preceding proposition. $\square$

9.2 Zonal Spherical Functions on a Semi-Simple Lie Group

9.2.1 Statement of Results—Immediate Applications

Let G be a connected semi-simple Lie group with finite center, $G = KA_\mathfrak{p}N^+$ an Iwasawa decomposition for G; let $\mathbf{\Sigma}$ be the set of roots of the pair $(\mathfrak{g}, \mathfrak{a}_\mathfrak{p})$, W the associated Weyl group (so that $W \sim \mathsf{M}^*/\mathsf{M} \ldots$); assign to the symbols $\mathscr{C}$, ρ etc. their usual meanings.

Note Throughout the present section, we shall write $\mathfrak{F}$ for the real dual of $\mathfrak{a}_\mathfrak{p}$.

Let $I_c^\infty(G)$ be the space of K-biinvariant compactly supported C^∞ functions on G, $I^p(G)$ the closure of $I_c^\infty(G)$ in $L^p(G)$ $(1 \leqslant p < \infty)$.

Consider the space $I^1(G)$ – then, as we know, under the convolution product, $I^1(G)$ is a commutative semi-simple Banach algebra. It is a standard remark that the (non-zero) continuous homomorphisms of the algebra $I^1(G)$ are given by the mappings

$$f \mapsto \int_G f(x)\phi(x)d_G(x) \qquad (f \in I^1(G))$$

where ϕ is a *bounded* zonal spherical function on G.

Lemma 9.2.1.1 Suppose that ϕ is a bounded zonal spherical function on G – then $\|\phi\|_\infty \leqslant 1$.

Proof According to the theory of commutative Banach algebras, we have $|\phi(f)| \leqslant \|f\|_1$ for all $f \in I^1(G)$. This being so, let χ be the character of the trivial one dimensional representation of K – then $\phi(f) = \phi(\chi * f * \chi)$ (all $f \in L^1(G)$), hence $|\phi(f)| = |\phi(\chi * f * \chi)| \leqslant \|\chi * f * \chi\|_1 \leqslant \|f\|_1$ (all $f \in L^1(G)$) from which we can read off our assertion. $\square$

Given a (complex valued) linear function ν on $\mathfrak{a}_\mathfrak{p}$, set

$$\phi_\nu(x)(=\phi(\nu : x)) = \int_K e^{(\sqrt{-1}\nu - \rho)(H(xk))}\, dk \qquad (x \in G).$$

Then ϕ_ν is a zonal spherical function on G and all such arise in this way for a suitable choice of ν; moreover $\phi_\nu = \phi_\mu$ iff $\nu = w\mu$ for some w in W.

The following theorem, whose proof may be found in 9.2.4 infra, gives an explicit parametrization of the set of bounded zonal spherical functions on G, and, in so doing, leads at once to a precise description of the maximal ideal space for $I^1(G)$.

Theorem 9.2.1.2 (Helgason-Johnson) Let $\mathfrak{C}_\rho$ denote the convex hull of the points $w\rho (w \in W)$ – then ϕ_ν if bounded iff $\nu = \xi + \sqrt{-1}\eta$ $(\xi, \eta \in \mathfrak{F})$ belongs to the tube $\mathfrak{T}_\rho = \mathfrak{F} + \sqrt{-1}\mathfrak{C}_\rho$.

Definition Fix an f in $I^1(G)$ – then, by the *spherical Gelfand transform* $\hat{f}$ of f, we shall understand the function defined on the tube $\mathfrak{T}_\rho$ by the rule

$$\hat{f}(\nu) = \int_G f(x)\phi_\nu(x)d_G(x) \qquad (\nu \in \mathfrak{T}_\rho).$$

Let $f \in I^1(G)$ and consider the integral defining its spherical Gelfand transform $\hat{f}$ – thus, on the basis of Lemma 9.2.1.1, this integral converges uniformly in ν whenever ν ranges through a compact subset of the tube $\mathfrak{T}_\rho$. Because $\phi_\nu(.) = \phi(\nu :.)$ is a holomorphic function of ν, it is then clear that $\hat{f}$ is holomorphic throughout the interior of $\mathfrak{T}_\rho$ and continuous on the boundary; moreover $\hat{f}$ is obviously W-invariant. . . .

Lemma 9.2.1.3 (Riemann-Lebesgue) Let $f \in I^1(G)$ – then its spherical Gelfand transform $\hat{f}$ vanishes at infinity in the tube $\mathfrak{T}_\rho$, i.e., given any $\epsilon > 0$, there exists a compact subset ω_ϵ of $\mathfrak{T}_\rho$ such that $\sup_{\nu \in \complement\omega_\epsilon} |\hat{f}(\nu)| < \epsilon$.

Proof We shall agree to use the terminology which is introduced in 9.2.3 infra. So fix an $\epsilon > 0$; choose a function g in $I_c^\infty(G)$ such that $\|f - g\|_1 \leqslant \epsilon/2$ – then, since the Gelfand transform is norm decreasing, we have $|\hat{f}(\nu) - \hat{g}(\nu)| \leqslant \epsilon/2$ (all $\nu \in \mathfrak{T}_\rho$). Let us suppose that g has its support in the ball of radius R in $A_\mathfrak{p}$ – then, according to 9.2.3 below, there exists a constant $C > 0$ such that

$$|\hat{g}(\nu)| \leqslant C(1 + \|\xi + \sqrt{-1}\eta\|)^{-1}e^{R\|\eta\|} \qquad (\nu = \xi + \sqrt{-1}\eta)$$

for all ν in $\mathfrak{T}_\rho$. Because the function $\nu \mapsto e^{R\|\mathscr{I}(\nu)\|}$ is bounded on $\mathfrak{T}_\rho$, it is clear that there exists a compact subset ω_ϵ of $\mathfrak{T}_\rho$ such that $|\hat{g}(\nu)| \leqslant \epsilon/2$ $(\nu \in \complement\omega_\epsilon)$; but then

$$|\hat{f}(\nu)| \leqslant |\hat{f}(\nu) - \hat{g}(\nu)| + |\hat{g}(\nu)| \leqslant \epsilon$$

for all $\nu \in \complement\omega_\epsilon$, as desired. ☐

Let Spec $(I^1(G))$ denote the (regular) maximal ideal space of $I^1(G)$ – then, as is well-known, each $f \in I^1(G)$ defines a function $\hat{f}$ on Spec $(I^1(G))$, namely $\hat{f}(\phi) = \phi(f)$ $(\phi \in \text{Spec}(I^1(G)))$, and, in the weak topology induced by the collection $\{\hat{f}: f \in I^1(G)\}$, Spec $(I^1(G))$ is a locally compact Hausdorff space. [In passing we recall that the set $\{\hat{f}: f \in I^1(G)\}$ is a separating subalgebra of the algebra of continuous functions on Spec $(I^1(G))$ which vanish at infinity.] On the other hand, according to the preceding theorem, there exists a one-to-one correspondence between the points of the orbit space $W\backslash\mathfrak{T}_p$ and the points of Spec $(I^1(G))$; now, in the quotient topology, $W\backslash\mathfrak{T}_p$ is a locally compact Hausdorff space and so, bearing in mind Lemma 9.2.1.3 supra, it follows from a familiar generality that the abstract Gelfand topology on $W\backslash\mathfrak{T}_p$ is the same as the natural Euclidean topology. [Lemma: If $\mathscr{F}$ is a family of complex valued continuous functions vanishing at infinity on a locally compact Hausdorff space X, separating the points of X and not all vanishing at any point of X, then the weak topology induced on X by $\mathscr{F}$ is identical with the given topology of X.]

From what has been said so far, it is clear that the Harmonic Analysis in the algebra $I^1(G)$ is very different from the classical Fourier Analysis in the L^1-algebra of a locally compact abelian group; of course, this is due primarily to the fact that the spherical Gelfand transform of an element in $I^1(G)$ is necessarily holomorphic throughout the interior of $\mathfrak{T}_p$. In particular, it is to be observed that $I^1(G)$ is neither regular nor self-adjoint. [If $I^1(G)$ were regular, then, by definition, for every closed set $T \subset W\backslash\mathfrak{T}_p$ and every point $\nu \notin T$, there would exist an $f \in I^1(G)$ such that $\hat{f} = 0$ on T but $\hat{f}(\nu) \neq 0 \ldots$; if $I^1(G)$ were self-adjoint, then, thanks to the Stone-Weierstrass Theorem, the set $\{\hat{f}: f \in I^1(G)\}$ would be uniformly dense in the algebra of W-invariant continuous functions on $\mathfrak{T}_p$ which vanish at infinity. . . .] Moreover, as can be seen by example, if a given f in $I^1(G)$ lies in no self-adjoint regular maximal ideal of $I^1(G)$, then the closed ideal generated by f need *not* be the whole of $I^1(G)$; otherwise said, Wiener's Tauberian Theorem, confined to the self-adjoint regular maximal ideals of $I^1(G)$, will not be true in general.

Let **P** be the set of positive definite zonal spherical functions on G equipped with the topology of uniform convergence on compacta – then **P** is a locally compact Hausdorff space.

Definition Fix an f in $I^1(G)$ – then, by the *spherical Fourier transform* $\hat{f}$ of f, we shall understand the function defined on **P** by the rule

$$\hat{f}(\phi) = \int_G f(x)\phi(x)d_G(x) \qquad (\phi \in \mathbf{P}).$$

A given $\hat{f}$ $(f \in I^1(G))$ is continuous and vanishes at infinity on **P**; moreover the usual rules of computation are valid:

$$\widehat{f+g} = \hat{f} + \hat{g},\ \widehat{f*g} = \hat{f}\hat{g},\ \widehat{f^*} = \overline{\hat{f}} \qquad (f, g \in I^1(G)).$$

Finally the set $\{\hat{f}: f \in I^1(G)\}$ is uniformly dense in the algebra of continuous functions on **P** which vanish at infinity.

The basic existence theorem in the present circle of ideas may be stated as follows.

Theorem 9.2.1.4 (Godement) Fix a Haar measure on G – then there exists on $\mathbf{P}$ a unique positive measure μ with the following properties:

(1) The functions $\hat{f}$ $(f \in I_c^\infty(G))$ are square integrable for μ and are everywhere dense in $L_\mu^2(\mathbf{P})$;

(2) For all $f, g \in I_c^\infty(G)$ one has

$$\int_G f(x)\overline{g(x)}d_G(x) = \int_{\mathbf{P}} \hat{f}(\phi)\overline{\hat{g}(\phi)}d\mu(\phi).$$

In addition the map $f \mapsto \hat{f}$ $(f \in I_c^\infty(G))$ extends to an isometric isomorphism from $I^2(G)$ onto $L_\mu^2(\mathbf{P})$.

We shall refer to the measure μ as the *Plancherel measure* for $I^2(G)$.

[These results are ready consequences of the theory of commutative unitary algebras (or Theorem 7.2.1.1 . . .); cf. Godement [15] (and Dieudonné [4])].

Suppose that a zonal spherical function ϕ_ν is positive definite – then, since ϕ_ν is bounded in absolute value by 1, ν must lie in the tube $\mathfrak{T}_\rho$. Let $\mathfrak{P}$ be the subset of $\mathfrak{T}_\rho$ consisting of those ν such that ϕ_ν is positive definite – then $\mathfrak{P}$ is W-stable, contains $\mathfrak{F}$ (cf. number 6.2.2), and the points of the orbit space $W\backslash\mathfrak{P}$ are in a one-to-one correspondence with those of $\mathbf{P}$. [Except in a few special cases, no explicit description of $\mathfrak{P}$ is known; this is, of course, tied up with the existence of 'complementary series' . . . .] The Euclidean topology on $W\backslash\mathfrak{F}$ is the same as that which it inherits from $\mathbf{P}$; moreover $W\backslash\mathfrak{F}$ $(\sim \mathscr{C}^{cl} \ldots)$ is closed in $\mathbf{P}$ (cf. Chapter 7).

Notation Suppose that ϕ_ν $(\nu \in \mathfrak{P})$ is positive definite – then, in the sequel, we shall usually write $\hat{f}(\nu)$ for $\hat{f}(\phi_\nu)$ $(f \in I_c^\infty(G))$.

The central problem facing us in the present section is this: Compute the Plancherel measure μ for $I^2(G)$ explicitly. Here, then, is Harish-Chandra's Plancherel Theorem for $I^2(G)$.

Theorem 9.2.1.5 (Harish-Chandra) Let $\mathbf{c}(\nu) = \mathbf{I}(\sqrt{-1}\nu)/\mathbf{I}(\rho)$ $(\nu \in \mathfrak{F})$ where

$$\mathbf{I}(\nu) = \prod_{\lambda \in \Sigma^+} B\left(\frac{m(\lambda)}{2}, \frac{m(\lambda/2)}{4} + \frac{(\nu, \lambda)}{(\lambda, \lambda)}\right) \qquad (\nu \in \mathfrak{F}_c).$$

Then, relative to a suitable normalization of the relevant Haar measures, for all f in $I_c^\infty(G)$, we have

$$(*) \quad \int_G |f(x)|^2\, d_G(x) = [W]^{-1} \int_{\mathfrak{F}} |\hat{f}(\nu)|^2\, |\mathbf{c}(\nu)|^{-2} d_{\mathfrak{F}}(\nu)$$

while

$$(**) \quad f(x) = [W]^{-1} \int_{\mathfrak{F}} \hat{f}(\nu)\overline{\phi_\nu(x)}\, |\mathbf{c}(\nu)|^{-2} d_{\mathfrak{F}}(\nu) \qquad (x \in G).$$

To avoid any confusion, let us recall that $\mathbf{c}^{-1}$ is analytic on $\mathfrak{F}$ (cf. 9.1.7); moreover, its absolute value is W-invariant so that the integrations supra may actually be taken over $\mathscr{C}^{cl}$.

Bearing in mind the uniqueness assertion in Theorem 9.2.1.4 above, we see that the Plancherel measure μ for $I^2(G)$ is absolutely continuous with respect to the Lebesgue measure on $\mathfrak{F}$, its Radon-Nikodym derivative being explicitly computable. Evidently μ is supported by $\mathscr{C}^{cl}$ ($\subset \mathbf{P}\ldots$).

Remark In passing we recall that with each $\nu \in \mathscr{C}^{cl}$ there is associated a class one unitary representation in the principal P-series for G ($P = MA_{\mathfrak{p}}N^+$); according to Kostant's Theorem (Theorem 5.5.2.3), this representation is *always* irreducible.

Example Let us explicate the preceding considerations in the case when $G = \mathbf{SL}(2, \mathbf{R})$. Thus let K, $A_{\mathfrak{p}}$, N^+, respectively, be the one parameter subgroups of G generated by the matrices k_θ, h_t, n_ξ where

$$k_\theta = \begin{pmatrix} \cos(\theta/2) & \sin(\theta/2) \\ -\sin(\theta/2) & \cos(\theta/2) \end{pmatrix}, \quad h_t = \begin{pmatrix} e^{t/2} & 0 \\ 0 & e^{-t/2} \end{pmatrix}, \quad n_\xi = \begin{pmatrix} 1 & \xi \\ 0 & 1 \end{pmatrix}$$

with $0 \leqslant \theta < 4\pi$, t, ξ real – then $d_G(x) = (4\pi)^{-1} e^t d\theta dt d\xi$ if $x = k_\theta h_t n_\xi$ ($x \in G$). For any complex number ν, let χ_ν denote the function on G defined by the rule

$$\chi_\nu(x) = e^{(\sqrt{-1}\nu - 1/2)t} \qquad (x = k_\theta h_t n_\xi).$$

Then the zonal spherical function ϕ_ν associated with ν is given on $A_{\mathfrak{p}}$ by

$$\begin{aligned} \phi_\nu(h_t) &= (4\pi)^{-1} \int_0^{4\pi} \chi_\nu(k_\theta^{-1} h_t k_\theta)\, d\theta \\ &= (2\pi)^{-1} \int_{-\pi}^{\pi} \{\cosh(t) + \sinh(t)\cos(\theta)\}^{\sqrt{-1}\nu - 1/2}\, d\theta \qquad (t \in \mathbf{R}), \end{aligned}$$

which shows that ϕ_ν is expressible in terms of the Legendre function $P_{\sqrt{-1}\nu - 1/2}$. Consider now the Abel transformation F_f of an element f in $I_c^\infty(G)$ (calculated relative to the trivial one dimensional representation of K); thus, by definition, we have

$$F_f(t) = e^{t/2} \int_{-\infty}^{\infty} f(h_t n_\xi)\, d\xi \qquad (t \in \mathbf{R}),$$

it being clear that F_f is an even function of t. We recall from number 6.2.2 (see, too, 9.2.4 infra) that the assignment $f \mapsto F_f$ is one-to-one on $I_c^\infty(G)$; in the present case it is easy to invert the transformation $f \mapsto F_f$ ($f \in I_c^\infty(G)$). For this purpose, it will be convenient to employ the following notation: Given any even function Φ on the line $\mathbf{R}$, let us agree to regard it as a function of $\cosh(t)$:

$$\Phi(t) = \Phi[\cosh(t)] \qquad (t \in \mathbf{R}).$$

Fix $f \in I_c^\infty(G)$; since $G = KA_{\mathfrak{p}}K$, an elementary computation tells us that

$$\begin{aligned} F_f(t) &= e^{t/2} \int_{-\infty}^{\infty} f\left[\cosh(t) + \frac{e^t \xi^2}{2}\right] d\xi \\ &= \int_{-\infty}^{\infty} f\left[\cosh(t) + \frac{\xi^2}{2}\right] d\xi \qquad (t \in \mathbf{R}), \end{aligned}$$

whence

$$-2\pi f[t] = \int_0^\infty \int_0^{2\pi} f'[t + r^2/2] r\,dr\,d\theta$$
$$= \int_{-\infty}^\infty \int_{-\infty}^\infty f'[t + (\xi^2 + \eta^2)/2]\,d\xi\,d\eta = \int_{-\infty}^\infty F_f'[t + \xi^2/2]\,d\xi \qquad (t \geqslant 1),$$

the prime standing for differentiation. This is the inversion formula alluded to above. [As was pointed out by Godement, the problem of inverting the Abel transformation on **SL**(2, **R**) is essentially the same as that of solving the Abel integral equation; for, upon making a simple change of variable, we can write

$$F_f[t] = \sqrt{2} \int_t^\infty f[\xi] \frac{d\xi}{\sqrt{\xi - t}} \qquad (t \geqslant 1),$$

an integral equation of Abel type. . . .] Suppose that ν is real (in which case ϕ_ν is positive definite) – then, for any $f \in I_c^\infty(G)$, we have

$$\hat{f}(\nu) = \int_{-\infty}^\infty F_f(t) e^{\sqrt{-1}\nu t}\,dt = 2\int_0^\infty F_f(t)\cos(\nu t)\,dt$$

and so, by Fourier transformation,

$$F_f(t) = \pi^{-1} \int_0^\infty \hat{f}(\nu)\cos(\nu t)\,d\nu \qquad (t \in \mathbf{R}),$$

whence

$$F_f'[\cosh(t)]\sinh(t) = F_f'(t) = -\pi^{-1}\int_0^\infty \hat{f}(\nu)\nu\sin(\nu t)\,d\nu.$$

On the other hand, according to what has been said above, we can write

$$f(1) = (-2\pi)^{-1} \int_{-\infty}^\infty F_f'[1 + \xi^2/2]\,d\xi$$

or, upon making the change of variable $\xi = 2\sinh(t/2)$,

$$f(1) = (-2\pi)^{-1} \int_{-\infty}^\infty F_f'[\cosh(t)]\cosh(t/2)\,dt.$$

Therefore

$$f(1) = (4\pi^2)^{-1} \int_0^\infty \hat{f}(\nu)\nu \int_{-\infty}^\infty \{\sin(\nu t)/\sinh(t/2)\}\,dt\,d\nu$$
$$= (2\pi)^{-1} \int_0^\infty \hat{f}(\nu)\nu\tanh(\pi\nu)\,d\nu$$

from which it follows that the Plancherel measure for $I^2(G)$ in the present case is given by $(2\pi)^{-1}\nu\tanh(\pi\nu)d\nu$ ($\nu \geqslant 0$). [Needless to say, the L^2-norm of an element f in $I_c^\infty(G)$ can be computed from the quantity $f{*}f^*(1)$; to get the inversion formula for general $x \in G$, consider the function F_x in $I_c^\infty(G)$ defined by the rule $F_x(y) = \int_K f(xky)dk$ ($y \in G$) – then, by a simple calculation, we find that $\hat{F}_x(\nu) = \overline{\phi_\nu(x)}\hat{f}(\nu)$ (ν real). Since $f(x) = F_x(1)$, we then have

$$f(x) = (2\pi)^{-1} \int_0^\infty \hat{f}(\nu)\overline{\phi_\nu(x)}\nu\tanh(\pi\nu)\,d\nu \qquad (x \in G),$$

as desired.] It is to be observed that the positive definite zonal spherical functions associated with the complementary series and the trivial one dimensional representation of G are assigned Plancherel measure zero; in fact, it can be shown that ϕ_ν is positive definite iff: (1) ν is real or (2) ν is pure imaginary and lies between $-\sqrt{-1}/2$ and $\sqrt{-1}/2$. [Here, of course, the real values of the parameter correspond to the representations in the class one principal P-series for

G (P minimal); on the other hand, the endpoints of the interval $[-\sqrt{-1}/2, \sqrt{-1}/2]$ correspond to the trivial one dimensional representation of G while the non-zero points in the interior are associated with the representations in the complementary series.]

The group G can be represented in the obvious fashion on the Hilbert space $L^2(G/K)$; this being so, as an application of Harish-Chandra's Plancherel Theorem we shall now derive the spectral decomposition for $L^2(G/K)$ relative to this action of G, i.e. we shall decompose $L^2(G/K)$ into a direct integral of irreducible unitary representations of G.

Let $C_c^\infty(G/K)$ denote the space of compactly supported C^∞ functions on G which are invariant to the right under K.

Definition Fix an f in $C_c^\infty(G/K)$ – then, by the *Fourier transform* $\hat{f}$ of f, we shall understand the function defined on $\mathfrak{F} \times K/M$ by the rule

$$\hat{f}(\nu, \dot{k}) = \int_G f(x) e^{-(\sqrt{-1}\nu+\rho)(H(x^{-1}k))} d_G(x) \qquad (\nu \in \mathfrak{F}, \dot{k} = kM).$$

[The geometric motivation lying behind this definition will be given in due course.]

Lemma 9.2.1.6 Fix f in $C_c^\infty(G/K)$ – then, for a suitable normalization of the relevant Haar measures, we have

$$f(x) = [W]^{-1} \iint_{\mathfrak{F}\times K/M} \hat{f}(\nu, \dot{k}) e^{-(-\sqrt{-1}\nu+\rho)(H(x^{-1}k))} |\mathbf{c}(\nu)|^{-2} d_{\mathfrak{F}}(\nu) d_{K/M}(\dot{k}) \qquad (x \in G).$$

Proof To start off, let us deal with the case $x = 1$ – thus, since $f_K \in I_c^\infty(G)$, for any $\nu \in \mathfrak{F}$, we have $\hat{f}_K(\nu) = \int_G f_K(x)\phi_\nu(x) d_G(x) = \int_{K/M} \hat{f}(\nu, \dot{k}) d_{K/M}(\dot{k}) (\int_K = \int_{K/M}\int_M \cdots)$, and so, thanks to Theorem 9.2.1.5, we deduce that

$$\begin{aligned} f(1) = f_K(1) &= [W]^{-1} \int_{\mathfrak{F}} \hat{f}_K(\nu) |\mathbf{c}(\nu)|^{-2} d_{\mathfrak{F}}(\nu) \\ &= [W]^{-1} \iint_{\mathfrak{F}\times K/M} \hat{f}(\nu, \dot{k}) |\mathbf{c}(\nu)|^{-2} d_{\mathfrak{F}}(\nu) d_{K/M}(\dot{k}), \end{aligned}$$

as desired. Now fix an $x \neq 1$ in G; let L_x denote left translation by x – then, bearing in mind the elementary change of variable rules which were mentioned at the beginning of Vol. I, number 5.5.1, we find that

$$\begin{aligned} \widehat{f\circ L_x}(\nu, \dot{k}) &= \int_G f(xy) \exp\{-(\sqrt{-1}\nu + \rho)(H(y^{-1}k))\} d_G(y) \\ &= \int_G f(y) \exp\{-(\sqrt{-1}\nu + \rho)(H(y^{-1}xk))\} d_G(y) \\ &= e^{-(\sqrt{-1}\nu+\rho)(H(xk))} \\ &\quad \times \int_G f(y) \exp\{-(\sqrt{-1}\nu + \rho)(H(y^{-1}\kappa(xk)))\} d_G(y) \\ &= e^{-(\sqrt{-1}\nu+\rho)(H(xk))} \hat{f}(\nu, \dot{\widehat{\kappa(xk)}}) \qquad ((\nu, \dot{k}) \in \mathfrak{F} \times K/M), \end{aligned}$$

whence

$$\iint_{\mathfrak{F}\times K/M} \widehat{f\circ L_x}(\nu, \dot{k})\,|\mathbf{c}(\nu)|^{-2}\, d_{\mathfrak{F}}(\nu)\, d_{K/M}(\dot{k})$$

$$= \int_{\mathfrak{F}}\left\{\int_K e^{-(\sqrt{-1}\nu+\rho)(H(xk))}\hat{f}(\nu, \dot{\overbrace{\kappa(xk)}})\, dk\right\}|\mathbf{c}(\nu)|^{-2} d_{\mathfrak{F}}(\nu)$$

$$= \int_{\mathfrak{F}}\left\{\int_{K/M} e^{(\sqrt{-1}\nu-\rho)(H(x^{-1}k))}\hat{f}(\nu, \dot{k})\, d_{K/M}(\dot{k})\right\}|\mathbf{c}(\nu)|^{-2} d_{\mathfrak{F}}(\nu).$$

Therefore

$$f(x) = f\circ L_x(1) = [W]^{-1}\iint_{\mathfrak{F}\times K/M} \widehat{f\circ L_x}(\nu, \dot{k})\,|\mathbf{c}(\nu)|^{-2} d_{\mathfrak{F}}(\nu) d_{K/M}(\dot{k})$$

$$= [W]^{-1}\iint_{\mathfrak{F}\times K/M}\hat{f}(\nu, \dot{k}) e^{-(-\sqrt{-1}\nu+\rho)(H(x^{-1}k))}\,|\mathbf{c}(\nu)|^{-2}\, d_{\mathfrak{F}}(\nu) d_{K/M}(\dot{k})$$

which is the assertion of our lemma. □

Corollary 9.2.1.7 Fix f in $C_c^\infty(G/K)$ – then, for a suitable normalization of the relevant Haar measures, we have

$$\int_G |f(x)|^2\, d_G(x) = [W]^{-1}\iint_{\mathfrak{F}\times K/M} |\hat{f}(\nu, \dot{k})|^2\,|\mathbf{c}(\nu)|^{-2}\, d_{\mathfrak{F}}(\nu) d_{K/M}(\dot{k}).$$

Proof In fact, according to the preceding lemma, we can write

$$\int_G |f(x)|^2\, d_G(x) = \int_G f(x)\overline{f(x)}\, d_G(x)$$

$$= [W]^{-1}\int_G f(x)$$

$$\times\left\{\iint_{\mathfrak{F}\times K/M}\overline{\hat{f}(\nu, \dot{k}) e^{-(\sqrt{-1}\nu+\rho)(H(x^{-1}k))}}\,|\mathbf{c}(\nu)|^{-2}\, d_{\mathfrak{F}}(\nu)\, d_{K/M}(\dot{k})\right\} d_G(x)$$

$$= [W]^{-1}\iint_{\mathfrak{F}\times K/M} |\hat{f}(\nu, \dot{k})|^2\,|\mathbf{c}(\nu)|^{-2}\, d_{\mathfrak{F}}(\nu) d_{K/M}(\dot{k}),$$

as contended. □

Remark It is not difficult to show that the Fourier transform $f \mapsto \hat{f}$ ($f \in C_c^\infty(G/K)$) extends to an isometric isomorphism from $L^2(G/K)$ onto $L^2(\mathfrak{F}\times K/M)$ (with respect to the measure $|\mathbf{c}(\nu)|^{-2} d_{\mathfrak{F}}(\nu) d_{K/M}(\dot{k})\ldots$); for details see Helgason [14].

Given $\nu \in \mathfrak{F}$, let E_ν denote the vector space consisting of those functions a_ν on G which may be written in the form

$$a_\nu(x) = \int_{K/M} e^{-(-\sqrt{-1}\nu+\rho)(H(x^{-1}k))} a(\dot{k})\, d_{K/M}(\dot{k}) \qquad (x \in G)$$

for some a in $L^2(K/M)$; evidently a is uniquely determined by a_ν so that E_ν acquires the structure of a Hilbert space when we write

$$\|a_\nu\| = \left\{\int_{K/M} |a(\dot{k})|^2\, d\dot{k}\right\}^{1/2}.$$

Moreover the natural representation L_ν of G on E_ν by left translation is unitary and irreducible. [To justify these assertions it is necessary to bear in mind that *every* representation in the class one principal P-series for G with P minimal is irreducible (cf. Theorem 5.5.2.3); in the case at hand L_ν is unitarily equivalent to the class one principal P-series representation determined by the character $mhn \mapsto h^{\sqrt{-1}\nu}$. . . .]

Taking into account these remarks, along with the preceding lemma and its corollary, we are then led to the following result.

Theorem 9.2.1.8 (Helgason) Let $L_{G/K}$ denote the left regular representation of G on $L^2(G/K)$ – then we have the direct integral decomposition

$$L^2(G/K) = \int \oplus E_\nu |\mathbf{c}(\nu)|^{-2} \, d\mathfrak{F}(\nu), \; L_{G/K} = \int \oplus L_\nu |\mathbf{c}(\nu)|^{-2} d\mathfrak{F}(\nu),$$

ν running through $\mathfrak{F}$ (mod W).

Problem Characterize the space of analytic vectors for the representation $L_{G/K}$.

Heuristics According to Helgason, a *horocycle* in G/K is, by definition, an orbit in G/K of a subgroup of G conjugate to N^+; the horocycles are all closed submanifolds of G/K and play a role analogous to hyperplanes in $\mathbf{R}^n$. Let Ξ denote the set of all horocycles in G/K; it is easy to see that the natural action of G on Ξ is transitive. If $\mathbf{o}$ denotes the origin in G/K and if $\xi_0 = N^+ \cdot \mathbf{o}$, then the subgroup of G which maps ξ_0 into itself equals $\mathsf{M}N^+$, so there is a natural identification $\Xi \sim G/\mathsf{M}N^+$. Because the map $(k\mathsf{M}, h) \mapsto kh\mathsf{M}N^+$ is a diffeomorphism of $K/\mathsf{M} \times A_\mathfrak{p}$ onto Ξ, a given horocycle ξ can be written in the form $\xi = kh \cdot \xi_0$ where $k\mathsf{M} \in K/\mathsf{M}$ and $h \in A_\mathfrak{p}$ are unique; we shall say that the Weyl chamber $k\mathsf{M}$ is *normal* to the horocycle ξ and that h is the *complex distance* from $\mathbf{o}$ to ξ. [Let us bear in mind that the set of all Weyl chambers in all maximal abelian subspaces of $\mathfrak{p}$ is acted on transitively by K; the subgroup leaving the fixed positive Weyl chamber $\mathscr{C}$ invariant is M.] Now the definition of the Fourier transform for functions on G/K can be motivated as follows. The analog in $\mathbf{R}^n$ of the integral transform figuring in Theorem 9.2.1.5 is an integral transform of radial functions given by the radial eigenfunctions of the Laplacian; when stated in terms of the radial variable this becomes the Hankel type transform $f \mapsto \hat{f}$,

$$\hat{f}(r) = \int_0^\infty f(s) J_\nu(rs)(rs)^{-\nu} s^{2\nu+1} \, ds \qquad (2\nu = n - 2),$$

J_ν the Bessel function of order ν. This being so, let us write the Euclidean Fourier transform in polar coordinate form

$$\hat{f}(r\omega) = \int_{\mathbf{R}^n} f(x) e^{\sqrt{-1} r(x,\omega)} \, dx \qquad (\|\omega\| = 1)$$

so that

$$f(x) = \int_{\mathbf{R}^+} \int_{S^{n-1}} \hat{f}(r\omega) e^{-\sqrt{-1} r(x,\omega)} |r|^{n-1} \, dr d\omega$$

with a suitable normalization of the surface element $d\omega$ of S^{n-1}. Here, of course, (x, ω) gives the distance from the origin to the hyperplane through x with normal vector ω. What is needed now is the symmetric space analog of the quantity (x, ω). But, according to what has been said above, given $\dot{x} = xK \in G/K$ and $\dot{k} = kM \in K/M$, there exists a unique horocycle $\xi(\dot{x}, \dot{k})$ (say) passing through xK with normal kM; let $H(\dot{x}, \dot{k})$ denote the log of the complex distance from $\mathbf{o}$ to $\xi(\dot{x}, \dot{k})$ – then the element $H(\dot{x}, \dot{k})$ $(= -H(x^{-1}k)\ldots)$ plays the role of the inner product (x, ω). Since K/M $(\sim G/MA_{\mathfrak{p}}N^+)$ enters as a (maximal) boundary of G/K, the definition of Fourier transform on G/K is thus seen to be entirely analogous to the classical case of $\mathbf{R}^n$.

9.2.2 The Plancherel Theorem for $I^2(G)$

Let G be a connected semi-simple Lie group with finite center, $G = KA_{\mathfrak{p}}N^+$ an Iwasawa decomposition for G – then the objective of the present number is to derive the Plancherel Theorem for the space $I^2(G)$ of square summable K-biinvariant functions on G. An important role in this study is played by the Abel transformation (taken, of course, relative to the trivial one dimensional representation of K); in fact, as we shall see, the problem of establishing the Plancherel Theorem for $I^2(G)$ is essentially equivalent to that of finding an explicit inversion formula for this transformation.

Convention Let us suppose that the Haar measures on G, $A_{\mathfrak{p}}$, and N^+ are taken per number 8.1.3 – then there exists a positive explicitly computable constant $\mathbf{c}_G$ such that

$$\int_G f(x)d_G(x) = \mathbf{c}_G \iiint_{K\times A_{\mathfrak{p}}\times N^+} f(khn)h^{2\rho}dkd_{A_{\mathfrak{p}}}(h)d_{N^+}(n) \qquad (f \in C_c(G)).$$

In order to avoid carrying along certain constant factors, it will be convenient to make the following agreements. Multiplying the canonical Euclidean measures on $A_{\mathfrak{p}}$, $\mathfrak{a}_{\mathfrak{p}}$, and $\mathfrak{F}(= \mathfrak{a}_{\mathfrak{p}}^{\vee})$ by the factor $(2\pi)^{-l/2}$ $(l = \text{rank}\,(G/K))$, we can arrange that the inversion formula for the Fourier transform

$$\hat{f}(\nu) = \int_{A_{\mathfrak{p}}} f(h)h^{\sqrt{-1}\nu}\,d_{A_{\mathfrak{p}}}(h) \qquad (\nu \in \mathfrak{F})$$

holds without a multiplicative constant

$$f(h) = \int_{\mathfrak{F}} \hat{f}(\nu)h^{-\sqrt{-1}\nu}\,d_{\mathfrak{F}}(\nu) \qquad (f \in \mathscr{C}(A_{\mathfrak{p}})).$$

Define the positive constant $\mathbf{c}_{N^+}$ by the prescription

$$\mathbf{c}_{N^+} = \int_{N^+} \exp\{-2\rho(H(\theta(n^{-1})))\}\,d_{N^+}(n).$$

Then, on the basis of what has been said in 9.1.6, $\mathbf{c}_{N^+}$ is explicitly computable. . . . Now write d_G for $(\mathbf{c}_G\mathbf{c}_{N^+}(2\pi)^{l/2})^{-1}d_G$, $d_{A_{\mathfrak{p}}}$ for $(2\pi)^{-l/2}d_{A_{\mathfrak{p}}}$, and d_{N^+} for $\mathbf{c}_{N^+}^{-1}d_{N^+}$ – then, for any $f \in C_c(G)$, we have

$$\int_G f(x)\,d_G(x) = \iiint_{K\times A_{\mathfrak{p}}\times N^+} f(khn)h^{2\rho}\,dkd_{A_{\mathfrak{p}}}(h)d_{N^+}(n).$$

Let $\mathscr{C}(G)$ be Harish-Chandra's space of rapidly decreasing functions on G; let $\mathscr{I}(G)$ be the (necessarily closed) subspace of $\mathscr{C}(G)$ consisting of those f which are biinvariant under K – then $I_c^\infty(G) = \mathscr{I}(G) \cap C_c^\infty(G)$ is dense in $\mathscr{I}(G)$ (cf. Theorem 8.3.7.13). [In passing, it should also be observed that $\mathscr{I}(G)$ is in fact a topological algebra under convolution; cf. Proposition 8.3.7.14.] Given $f \in \mathscr{I}(G)$, put, by definition,

$$F_f(h) = h^\rho \int_{N^+} f(hn) d_{N^+}(n) \qquad (h \in A_{\mathfrak{p}}).$$

Then, on the basis of what has been said in number 8.5.3, it is clear that the integral defining F_f exists and lies in $\mathscr{C}(A_{\mathfrak{p}})$; furthermore, the assignment (= the Abel transformation) $f \mapsto F_f$ is a continuous linear mapping of $\mathscr{I}(G)$ into $\mathscr{C}(A_{\mathfrak{p}})$.

Note Let $\mathscr{I}(A_{\mathfrak{p}})$ denote the convolution algebra of W-invariants in $\mathscr{C}(A_{\mathfrak{p}})$ (W the Weyl group of the pair $(\mathfrak{g}, \mathfrak{a}_{\mathfrak{p}})$); obviously $\mathscr{I}(A_{\mathfrak{p}})$ is a closed subspace of $\mathscr{C}(A_{\mathfrak{p}})$. Because of the functional relation $w \cdot F_f = F_f$, valid for all $f \in \mathscr{I}(G)$, it follows without difficulty that the assignment $f \mapsto F_f$ is actually a continuous homomorphism of the algebra $\mathscr{I}(G)$ into the algebra $\mathscr{I}(A_{\mathfrak{p}})$.

It has been pointed out in number 6.2.2 (cf., too, 9.2.4 infra) that the Abel transformation is one-to-one on $I_c^\infty(G)$; the extension of this result to $\mathscr{I}(G)$ is of central importance in the theory which is to follow.

Theorem 9.2.2.1 (Harish-Chandra) The Abel transformation $f \mapsto F_f$ is one-to-one on $\mathscr{I}(G)$.

[As will be shown in 9.2.3, this theorem is a ready consequence of the Paley-Wiener Theorem for $I^2(G)$.]

Remark It will be seen in due course that the Abel transformation $f \mapsto F_f$ establishes a *topological isomorphism* between the convolution algebras $\mathscr{I}(G)$ and $\mathscr{I}(A_{\mathfrak{p}})$, a fact which is tantamount to the Plancherel Theorem for $I^2(G)$. . . .

Define the homomorphism $\gamma_{A_{\mathfrak{p}}} : \mathfrak{X} \to I(\mathfrak{a}_{\mathfrak{p}_c})$ in the usual way (cf. number 6.2.2).

Proposition 9.2.2.2 Fix $f \in \mathscr{I}(G)$ – then

$$F_{Df} = \gamma_{A_{\mathfrak{p}}}(D) F_f \qquad (\text{all } D \in \mathfrak{X}).$$

Proof Since $\int_{N^+} f(nh) d_{N^+}(n) = h^{2\rho} \int_{N^+} f(hn) d_{N^+}(n)$ (all $h \in A_{\mathfrak{p}}$), the function F defined by the prescription $F(x) = e^{\rho(H(x^{-1}))} \int_{N^+} f(nx) d_{N^+}(n)$ ($x \in G$) is an extension of F_f to a C^∞ function on G. On the basis of the discussion in number 8.5.3, it is clear that for any $D \in \mathfrak{G}$, the integral $\int_{N^+} Df(nx) d_{N^+}(n)$ ($x \in G$) converges uniformly on compact subsets of

G; hence, in particular,

$$D(e^{-\rho(H(x^{-1}))}F(x)) = \int_{N^+} Df(nx)d_{N^+}(n) \qquad (x \in G).$$

Because the function $x \mapsto e^{-\rho(H(x^{-1}))}F(x)$ is invariant under each substitution $x \mapsto nxk$, for any D in $\mathfrak{X}$, we have

$$e^{\rho}F(h;D) = e^{\rho}F(h; e^{\rho}\circ\gamma_{A_{\mathfrak{p}}}(D)\circ e^{-\rho}) = \int_{N^+} Df(nh)d_{N^+}(n) \qquad (h \in A_{\mathfrak{p}}).$$

Our contention is now evident. □

[It is instructive to compare the statement of Proposition 9.2.2.2 with that of Theorem 8.5.1.3.]

The following result establishes a link between the Harmonic Analysis on $\mathscr{I}(G)$ and the Fourier Analysis on $\mathscr{I}(A_{\mathfrak{p}})$.

Proposition 9.2.2.3 Fix $\nu \in \mathfrak{F}$, $f \in \mathscr{I}(G)$ – then

$$\int_G f(x)\phi_\nu(x)d_G(x) = \int_{A_{\mathfrak{p}}} F_f(h)h^{\sqrt{-1}\nu}d_{A_{\mathfrak{p}}}(h).$$

Proof (1) Claim: $\int_G |f(x)|\, e^{-\rho(H(x))}d_G(x) < \infty$. In fact

$$\begin{aligned}\int_G |f(x)|\, e^{-\rho(H(x))}d_G(x) &= \int_G \left(\int_K |f(xk)|\, e^{-\rho(H(xk))}dk\right) d_G(x)\\ &= \int_G |f(x)|\, \Xi(x)d_G(x)\\ &= \int_G \left\{\frac{(1+\sigma(x))^r}{\Xi(x)}\,|f(x)|\right\}\frac{\Xi^2(x)}{(1+\sigma(x))^r}\,d_G(x)\\ &< \infty\end{aligned}$$

provided r is chosen sufficiently large (cf. Proposition 8.3.7.5).

(2) Due to the Fubini Theorem and (1), it follows immediately that

$$\begin{aligned}\int_G f(x)\phi_\nu(x)d_G(x) &= \int_G f(x)\exp\{\sqrt{-1}\nu(H(x)) - \rho(H(x))\}\,d_G(x)\\ &= \int_{A_{\mathfrak{p}}} h^{\sqrt{-1}\nu+\rho}\left(\int_{N^+} f(hn)d_{N^+}(n)\right)d_{A_{\mathfrak{p}}}(h)\\ &= \int_{A_{\mathfrak{p}}} F_f(h)h^{\sqrt{-1}\nu}\,d_{A_{\mathfrak{p}}}(h).\end{aligned}$$

Hence the proposition. □

Definition Fix f in $\mathscr{I}(G)$ – then, by the *spherical Fourier transform* $\hat{f}$ of f, we shall understand the function defined on $\mathfrak{F}$ by the rule

$$\hat{f}(\nu) = \int_G f(x)\phi_\nu(x)d_G(x) \qquad (\nu \in \mathfrak{F}).$$

According to what we have seen above, the integral defining $\hat{f}$ is absolutely convergent and coincides with the Euclidean Fourier transform $\hat{F}_f$ of F_f ($f \in \mathscr{I}(G)$). [In passing we remark that, relative to a given element of $\mathscr{I}(G)$, it may very well be possible to extend the definition of the spherical Fourier transform to elements ν which need not lie in $\mathfrak{F}$. . . .]

The space $\mathscr{I}(G)$ has the property that it is closed under the formation of 'wave packets'; more precisely one has the following basic theorem.

Theorem 9.2.2.4 (Harish-Chandra) For any rapidly decreasing function a on $\mathfrak{F}$, put

$$\phi_a(x) = \int_{\mathfrak{F}} \pi(\nu)a(\nu)\phi(\nu : x)\, d_{\mathfrak{F}}(\nu) \qquad (x \in G).$$

Then ϕ_a is a K-biinvariant rapidly decreasing function on G, differentiation under the integral sign being permissible. Furthermore the assignment $a \mapsto \phi_a$ defines a continuous mapping of $\mathscr{C}(\mathfrak{F})$ into $\mathscr{I}(G)$.

[Here $\pi(= \pi_\Sigma)$ denotes the product of the positive indivisible roots of the pair $(\mathfrak{g}, \mathfrak{a}_\mathfrak{p})$.]

The key step in the proof of this theorem is the following lemma.

Lemma 9.2.2.5 Given D in $\mathfrak{G}$, there exist non-negative integers m_D, M_D such that

$$|h^\rho \pi(\nu)\phi(\nu : h ; D)| \leqslant M_D(1 + \|\nu\|)^{m_D}$$

for all $\nu \in \mathfrak{F}$, $h \in \exp(\mathscr{C}^{cl})$.

We shall defer the proof of Theorem 9.2.2.4 until Volume III; as will be seen there, this theorem is in fact but a special case of a more general philosophy, namely the theory of the so-called Eisenstein integral (cf. Harish-Chandra [33, Theorem 6]). [To avoid any confusion, it should be stressed that it is perfectly possible to establish Theorem 9.2.2.4 via methods which are entirely internal to the 'theory of zonal spherical functions'; cf. Harish-Chandra [19].]

Note Let $\mathscr{I}_0(G) = \{\phi_a : a \in \mathscr{C}(\mathfrak{F})\}$ – then it will be shown below that it is actually the case that $\mathscr{I}_0(G) = \mathscr{I}(G)$.

Proposition 9.2.2.6 Fix an element $\nu \in \mathfrak{F}$ – then, for any $a \in \mathscr{C}(\mathfrak{F})$, the integral

$$\int_G \phi_a(x)\phi(\nu : x)d_G(x)$$

converges. Moreover the mapping T_ν defined by the rule

$$a \mapsto \int_G \phi_a(x)\phi(\nu : x)d_G(x) \qquad (a \in \mathscr{C}(\mathfrak{F}))$$

is a tempered distribution on $\mathfrak{F}$.

Proof Since $\mathscr{I}_0(G)$ is a subspace of $\mathscr{I}(G)$ (cf. the preceding theorem), the convergence of our integral for any $a \in \mathscr{C}(\mathfrak{F})$ is ensured by the first part of the proof of Proposition 9.2.2.3. This being the case, suppose now that $a_n \to a$ in $\mathscr{C}(\mathfrak{F})$ – then, on the basis of Theorem 9.2.2.4, $\phi_{a_n} \to \phi_a$ in $\mathscr{I}(G)$, whence $T_\nu(a_n) = \hat{\phi}_{a_n}(\nu) \to \hat{\phi}_a(\nu) = T_\nu(a)$ which shows that T_ν is indeed a tempered distribution on $\mathfrak{F}$. □

Our next objective is to obtain a more or less explicit formula for the distribution T_ν ($\nu \in \mathfrak{F}$).

Proposition 9.2.2.7 There exists a unique continuous function $\boldsymbol{\beta}$ on $\mathfrak{F}$ such that

$$\pi(-\nu)T_\nu(a) = \boldsymbol{\beta}(\nu) \sum_{w \in W} \det(w)a(-w\nu) \qquad (\nu \in \mathfrak{F})$$

for all $a \in \mathscr{C}(\mathfrak{F})$; moreover, $\boldsymbol{\beta}$ is W-invariant.

Proof The proof depends upon the following observation: In the sense of distribution theory, for any $p \in I(\mathfrak{a}_{\mathfrak{p}_c})$, we have $p \cdot T_\nu = p(-\nu)T_\nu$ (all $\nu \in \mathfrak{F}$). [Let $\gamma_{A_\mathfrak{p}} : \mathfrak{X} \to I(\mathfrak{a}_{\mathfrak{p}_c})$ be the canonical homomorphism – then, as we know, $D\phi_\nu = \gamma_{A_\mathfrak{p}}(D : \sqrt{-1}\nu)\phi_\nu$ ($D \in \mathfrak{X}$). This being so, fix a D in $\mathfrak{X}$; let p_D be the polynomial function on $\mathfrak{a}_{\mathfrak{p}_c}^\vee$ whose value at ν is $\gamma_{A_\mathfrak{p}}(D : \sqrt{-1}\nu)$ – then, bearing in mind Proposition 9.2.2.2, we find that

$$\begin{aligned} p_D \cdot T_\nu(a) = T_\nu(p_D a) &= \int_G D\phi_a(x)\phi_\nu(x)d_G(x) \\ &= \int_{A_\mathfrak{p}} F_{D\phi_a}(h)h^{\sqrt{-1}\nu}d_{A_\mathfrak{p}}(h) \\ &= \int_{A_\mathfrak{p}} \gamma_{A_\mathfrak{p}}(D)F_{\phi_a}(h)h^{\sqrt{-1}\nu}d_{A_\mathfrak{p}}(h) = p_D(-\nu)T_\nu(a) \qquad (\nu \in \mathfrak{F}), \end{aligned}$$

the 'operational calculus' for Fourier transforms being used to get the last equality. Our contention thus follows from the fact that $\gamma_{A_\mathfrak{p}}(\mathfrak{X}) = I(\mathfrak{a}_{\mathfrak{p}_c})$.] Now *fix* a point ν_0 in $\mathfrak{F}$ – then we claim that the support of the distribution T_{ν_0} is contained in the finite set of points $-w\nu_0$ ($w \in W$). Thus suppose that $\nu \in \mathfrak{F}$ with $\nu \neq -w\nu_0$ for any $w \in W$; choose $p_0 \in I(\mathfrak{a}_{\mathfrak{p}_c})$ such that $p_0(\nu) \neq p_0(-\nu_0)$; let $p = p_0 - p_0(-\nu_0)$ – then it follows from what has been said above that $p \cdot T_{\nu_0} = 0$. On the other hand $p(\nu) \neq 0$; hence ν does not belong to the support of T_{ν_0} which establishes the claim.

(1) Let us verify the uniqueness of $\boldsymbol{\beta}$. Fix a relatively compact open neighborhood $\mathscr{O}(\nu_0)$ of ν_0 in $\mathfrak{F}$ having the property that $w \cdot \mathscr{O}(\nu_0) = \mathscr{O}(\nu_0)$ ($w \in W$). Select a function $a_0 \in C_c^\infty(\mathfrak{F})$ such that $a_0 = 1$ on $-\mathscr{O}(\nu_0)$. Then it is clear that $\sum_{w \in W} \det(w)\pi(-w\nu)a_0(-w\nu) = [W]\pi(-\nu)$ for $\nu \in \mathscr{O}(\nu_0)$. Therefore the uniqueness of $\boldsymbol{\beta}$ follows from its continuity and the equation $\pi(-\nu)T_\nu(\pi a_0) = [W]\boldsymbol{\beta}(\nu)\pi(-\nu)$ ($\nu \in \mathscr{O}(\nu_0)$). Evidently, then, $\boldsymbol{\beta}(w\nu) = \boldsymbol{\beta}(\nu)$ (since $T_{w\nu} = T_\nu$ for all $\nu \in \mathfrak{F}$).

(2) It remains to establish the existence of $\boldsymbol{\beta}$. Keeping to the above notations, it will be shown that for any a in $\mathscr{C}(\mathfrak{F})$

$$\pi(-\nu)T_\nu(a) = [W]^{-1}T_\nu(\pi a_0)\sum_{w\in W}\det(w)a(-w\nu) \qquad (\text{all } \nu \in \mathcal{O}(\nu_0)).$$

In turn the existence of $\boldsymbol{\beta}$ is a consequence of this relation since the map $\nu \mapsto T_\nu(\pi a_0)$ is certainly a continuous function of ν. So fix $a \in \mathscr{C}(\mathfrak{F})$, $\nu \in \mathcal{O}(\nu_0)$; given any integer $m \geqslant 0$, we can choose a polynomial $p \in S(\mathfrak{a}_{\mathfrak{p}_c})$ such that all the derivatives of $a - p$ of order $\leqslant m$ are zero at the points $-w\nu$ ($w \in W$) – hence, by (1) and an elementary result in distribution theory, $T_\nu(a) = T_\nu(pa_0)$ provided m is taken sufficiently large. But $a_0^w - a_0$ vanishes identically on $-\mathcal{O}(\nu_0)$; thus $T_\nu(pa_0) = \det(w)T_\nu(p^w a_0)$ and so $T_\nu(a) = T_\nu(qa_0)$ where $q = [W]^{-1}\sum_{w\in W}\det(w)p^w$. Owing to Proposition 2.1.5.11 (ii), there exists $q_0 \in I(\mathfrak{a}_{\mathfrak{p}_c})$ such that $q = \pi q_0$, whence $T_\nu(a) = T_\nu(\pi q_0 a_0) = q_0(-\nu)T_\nu(\pi a_0)$; on the other hand,

$$\begin{aligned}\pi(-\nu)q_0(-\nu) &= q(-\nu)\\ &= [W]^{-1}\textstyle\sum_{w\in W}\det(w)p^w(-\nu)\\ &= [W]^{-1}\textstyle\sum_{w\in W}\det(w)a(-w\nu).\end{aligned}$$

Therefore, for any $a \in \mathscr{C}(\mathfrak{F})$, we have

$$\pi(-\nu)T_\nu(a) = [W]^{-1}T_\nu(\pi a_0)\sum_{w\in W}\det(w)a(-w\nu) \qquad (\text{all } \nu \in \mathcal{O}(\nu_0)),$$

as desired.

The proof of the proposition is now complete. □

Proposition 9.2.2.8 Let $a \in \mathscr{C}(\mathfrak{F})$ – then

$$F_{\phi_a}(h) = \int_{\mathfrak{F}}\boldsymbol{\beta}(-\nu)\{\pi(\nu)^{-1}\sum_{w\in W}\det(w)a(w\nu)\}h^{\sqrt{-1}\nu}\,d_{\mathfrak{F}}(\nu) \qquad (\text{all } h \in A_{\mathfrak{p}}).$$

Proof Suppose that ν is Σ-regular ($\nu \in \mathfrak{F}$) – then, in view of the preceding proposition, we have

$$\begin{aligned}T_\nu(a) &= \int_{A_{\mathfrak{p}}} F_{\phi_a}(h)h^{\sqrt{-1}\nu}d_{A_{\mathfrak{p}}}(h) \qquad (= \hat{F}_{\phi_a}(\nu))\\ &= \boldsymbol{\beta}(\nu)\{\pi(-\nu)^{-1}\sum_{w\in W}\det(w)a(-w\nu)\}.\end{aligned}$$

But, thanks to the Fourier inversion formula, we also have

$$F_{\phi_a}(h) = \int_{\mathfrak{F}}\hat{F}_{\phi_a}(\nu)h^{-\sqrt{-1}\nu}d_{\mathfrak{F}}(\nu) \qquad (h \in A_{\mathfrak{p}}).$$

It therefore follows that

$$F_{\phi_a}(h) = \int_{\mathfrak{F}}\boldsymbol{\beta}(-\nu)\{\pi(\nu)^{-1}\sum_{w\in W}\det(w)a(w\nu)\}h^{\sqrt{-1}\nu}d_{\mathfrak{F}}(\nu) \qquad (h \in A_{\mathfrak{p}}),$$

as we wished to prove. □

The chief remaining difficulty in the derivation of the Plancherel Theorem for $I^2(G)$ lies in establishing a relationship between $\boldsymbol{\beta}$ and the **c**-function (cf. Harish-Chandra [19]); the result may be stated as follows: For any $a \in \mathscr{C}(\mathfrak{F})$, it can be shown that

$$F_{\phi_a}(h) = \int_{\mathfrak{F}} |\pi(\nu)\mathbf{c}(\nu)|^2 \{\pi(\nu)^{-1} \sum_{w \in W} \det(w) a(w\nu)\} h^{\sqrt{-1}\nu} d_{\mathfrak{F}}(\nu)$$

from which we deduce that

$$\boldsymbol{\beta}(\nu) = |\pi(\nu)\mathbf{c}(\nu)|^2 \qquad (\nu \in \mathfrak{F}).$$

[In passing we remind the reader that

$$|\pi(\nu)\mathbf{c}(\nu)|^2 = |\pi(-\nu)\mathbf{c}(-\nu)|^2 \qquad (\nu \in \mathfrak{F});$$

cf. Proposition 9.1.7.1.]

Heuristics The formula for $\boldsymbol{\beta}$ will be derived in Volume III as a consequence of the general philosophy alluded to earlier; for the time being we shall settle for a plausibility argument. Suppose that the Haar measure on N^- is normalized in such a way that $\int_{N^-} e^{-2\rho(H(n))} d_{N^-}(n) = 1$ (of course this is the normalization relative to which the **c**-function can be computed . . .) – then the formula for $\boldsymbol{\beta}$ is essentially equivalent to the assertion that the relation

$$|\mathbf{c}(\nu)|^{-2} h^{\rho} \int_{N^-} \phi_\nu(nh)\, d_{N^-}(n) = \sum_{w \in W} h^{\sqrt{-1}w\nu} \qquad (h \in A_{\mathfrak{p}})$$

holds in the weak sense in ν, that is, it gives the right result when integrated against any $a \in \mathscr{C}(\mathfrak{F})$. This being so, given $n \in N^-$, $h \in A_{\mathfrak{p}}$, write

$$nh = k_1 h^+ k_2 \qquad (k_1, k_2 \in K;\, h^+ \in \exp(\mathscr{C}^{cl})).$$

Then $\log h^+ \sim \log h + H(n)$ as $h \to \infty$ in $\exp(\mathscr{C})$. Now replace $\phi_\nu(nh)$ $(= \phi_\nu(h^+))$ by

$$h^{-\rho} e^{-\rho(H(n))} \sum_{w \in W} \mathbf{c}(w\nu) h^{\sqrt{-1}w\nu} e^{\sqrt{-1}w\nu(H(n))},$$

the dominating term in the asymptotic expansion of ϕ_ν; integrating this expression over N^- (and ignoring all difficulties) then leads to the relation

$$h^{-\rho} \sum_{w \in W} \mathbf{c}(w\nu)\mathbf{c}(-w\nu) h^{\sqrt{-1}w\nu} = h^{-\rho} |\mathbf{c}(\nu)|^2 \sum_{w \in W} h^{\sqrt{-1}w\nu} \qquad (h \in A_{\mathfrak{p}})$$

which is equivalent to the desired result. . . .

Proposition 9.2.2.9 Let $a \in \mathscr{C}(\mathfrak{F})$ – then

$$\int_G |\phi_a(x)|^2 d_G(x) = [W]^{-1} \int_{\mathfrak{F}} |\pi(\nu)\mathbf{c}(\nu)|^2 \,|\sum_{w \in W} \det(w) a(w\nu)|^2 d_{\mathfrak{F}}(\nu).$$

The proof depends on the following lemma.

Lemma 9.2.2.10 Fix $a \in \mathscr{C}(\mathfrak{F})$ – then

$$\iint_{G \times \mathfrak{F}} |\pi(\nu) a(\nu) \phi(\nu : x) \phi_a(x)|\, d_G(x) d_{\mathfrak{F}}(\nu) < \infty.$$

Proof Choose an integer $m \geqslant 0$ such that

$$M = \int_{\mathfrak{a}_\mathfrak{p}} (1 + \|H\|)^{-m} d_{\mathfrak{a}_\mathfrak{p}}(H) < \infty.$$

Then

$$\begin{aligned}
&|\pi(\nu)| \int_G |\phi(\nu : x)\phi_a(x)| \, d_G(x) \\
&\leq |\pi(\nu)| \int_{\mathscr{C}} |\phi(\nu : H)\phi_a(H)| \bigtriangleup(H) d_{\mathfrak{a}_\mathfrak{p}}(H) \\
&\leq M \sup_{H \in \mathscr{C}} \bigtriangleup(H)(1 + \|H\|)^m |\pi(\nu)\phi(\nu : H)\phi_a(H)|,
\end{aligned}$$

the 'exp' being suppressed from the notation. Now, according to Theorem 8.3.7.4, there exists an integer $d \geqslant 0$ such that $e^\rho \leq \Xi^{-1}(1 + \sigma)^d$ on $\exp(\mathscr{C}^{cl})$; since ϕ_a lies in $\mathscr{I}(G)$ and since $\bigtriangleup$ is dominated by $e^{2\rho}$ on $\mathscr{C}$, in view of Lemma 9.2.2.5, it is then clear that there exists an integer $n \geqslant 0$ such that

$$|\pi(\nu)| \int_G |\phi(\nu : x)\phi_a(x)| \, d_G(x) \leq (1 + \|\nu\|)^n \qquad (\nu \in \mathfrak{F}).$$

Because a is rapidly decreasing on $\mathfrak{F}$, there can now be no doubt as to the integrability of our function over $G \times \mathfrak{F}$. □

Proof of Proposition 9.2.2.9 Owing to the preceding lemma and Fubini's Theorem, the following manipulation is meaningful:

$$\begin{aligned}
&\int_G |\phi_a(x)|^2 d_G(x) \\
&= [W]^{-1} \sum_{w \in W} \int_G |\phi_{a^w}(x)|^2 \, d_G(x) \\
&= [W]^{-1} \sum_{w \in W} \det(w) \int_G \phi_a(x) \left\{ \int_{\mathfrak{F}} \pi(\nu)\overline{a(w\nu)}\phi(-\nu : x) d_{\mathfrak{F}}(\nu) \right\} d_G(x) \\
&= [W]^{-1} \sum_{w \in W} \det(w) \int_{\mathfrak{F}} \pi(\nu)\overline{a(w\nu)} \left\{ \int_G \phi_a(x)\phi(-\nu : x) d_G(x) \right\} d_{\mathfrak{F}}(\nu) \\
&= [W]^{-1} \sum_{w \in W} \det(w) \int_{\mathfrak{F}} |\pi(\nu)\mathbf{c}(\nu)|^2 \overline{a(w\nu)} \sum_{\tilde{w} \in W} \det(\tilde{w}) a(\tilde{w}\nu) d_{\mathfrak{F}}(\nu) \\
&= [W]^{-1} \int_{\mathfrak{F}} |\pi(\nu)\mathbf{c}(\nu)|^2 \left| \sum_{w \in W} \det(w) a(w\nu) \right|^2 d_{\mathfrak{F}}(\nu).
\end{aligned}$$

Of course this serves to establish our proposition. □

Proposition 9.2.2.11 Let $a \in \mathscr{C}(\mathfrak{F})$; put

$$\hat{\phi}_a(\nu) = \int_G \phi_a(x)\phi_\nu(x) d_G(x) \qquad (\nu \in \mathfrak{F}).$$

Then

$$(*) \quad \int_G |\phi_a(x)|^2 d_G(x) = [W]^{-1} \int_{\mathfrak{F}} |\hat{\phi}_a(\nu)|^2 \, |\mathbf{c}(\nu)|^{-2} d_{\mathfrak{F}}(\nu)$$

while

$$(**)\quad \phi_a(x) = [W]^{-1} \int_{\mathfrak{F}} \hat{\phi}_a(\nu)\overline{\phi_\nu(x)}\,|\,\mathbf{c}(\nu)|^{-2}d_{\mathfrak{F}}(\nu) \qquad (x \in G).$$

Proof Consider the first assertion – thus, in view of Proposition 9.2.2.9, we have

$$\begin{aligned}\int_G |\phi_a(x)|^2 d_G(x) &= [W]^{-1} \int_{\mathfrak{F}} \boldsymbol{\beta}(\nu)\,|\sum_{w\in W} \det(w)a(-w\nu)|^2 d_{\mathfrak{F}}(\nu)\\ &= [W]^{-1} \int_{\mathfrak{F}} \boldsymbol{\beta}(\nu)\,|\hat{\phi}_a(\nu)|^2\,|\pi(\nu)|^2 \boldsymbol{\beta}(\nu)^{-2} d_{\mathfrak{F}}(\nu)\\ &= [W]^{-1} \int_{\mathfrak{F}} |\hat{\phi}_a(\nu)|^2\,|\,\mathbf{c}(\nu)|^{-2} d_{\mathfrak{F}}(\nu).\end{aligned}$$

The proof of the second is similar and may be left to the reader. □

In passing, let us recall that $\mathbf{c}^{-1}$ is analytic on $\mathfrak{F}$; cf. 9.1.7.

As above, put $\mathscr{I}_0(G) = \{\phi_a : a \in \mathscr{C}(\mathfrak{F})\}$ – then $\mathscr{I}_0(G) \subset \mathscr{I}(G)$ (cf. Theorem 9.2.2.4). Moreover, the Abel transformation is one-to-one on $\mathscr{I}_0(G)$ (cf. Proposition 9.2.2.11); in fact, it takes $\mathscr{I}_0(G)$ onto *all* of $\mathscr{I}(A_\mathfrak{p})$. This latter point may be seen as follows. Fix $a_0 \in \mathscr{I}(A_\mathfrak{p})$ – then the Fourier transform $\hat{a}_0$ of a_0 is a W-invariant rapidly decreasing function on $\mathfrak{F}$. Consider the function f_0 defined by the integral

$$f_0(x) = [W]^{-1} \int_{\mathfrak{F}} \hat{a}_0(\nu)\overline{\phi(\nu : x)}\,|\,\mathbf{c}(\nu)|^{-2} d_{\mathfrak{F}}(\nu) \qquad (x \in G)$$

whose existence is ensured by Proposition 9.1.7.2 – then we claim that $f_0 \in \mathscr{I}_0(G)$ and $F_{f_0} = a_0$. Thus put

$$a(\nu) = [W]^{-1}\hat{a}_0(-\nu)\pi(\nu)\,|\,\pi(\nu)\mathbf{c}(\nu)|^{-2} \qquad (\nu \in \mathfrak{F}).$$

Then, of course, $a \in \mathscr{C}(\mathfrak{F})$ and $f_0 = \phi_a$, so $f_0 \in \mathscr{I}_0(G)$. In addition we have

$$\hat{\phi}_a(\nu) = \frac{\boldsymbol{\beta}(\nu)}{\pi(-\nu)} \sum_{w\in W} \det(w)a(-w\nu) = \hat{a}_0(\nu) \qquad (\nu \in \mathfrak{F}),$$

whence $\hat{\phi}_a = \hat{F}_{f_0} = \hat{a}_0 \Rightarrow F_{f_0} = a_0 \ldots$ [Incidentally, these considerations provide us with an explicit inversion formula for the Abel transformation.] Owing to Theorem 9.2.2.1, it therefore follows that $\mathscr{I}_0(G) = \mathscr{I}(G)$. Finally we claim that the inverse of the mapping $f \mapsto F_f$ ($f \in \mathscr{I}(G)$) is continuous. Thus, since the Euclidean Fourier transform is a homeomorphism between the spaces of rapidly decreasing functions, it will be sufficient to verify that the mapping $f \mapsto \hat{f}$ of $\mathscr{I}(G)$ onto $\mathscr{I}(\mathfrak{F})$ has a continuous inverse ($\mathscr{I}(\mathfrak{F})$ the W-invariant rapidly decreasing functions in $\mathscr{C}(\mathfrak{F})$). Now $f = \phi_{a_f}$ where, without loss of generality, a_f may be taken to be a skew function in $\mathscr{C}(\mathfrak{F})$ (for we can always replace a by

$$[W]^{-1} \textstyle\sum_{w\in W} \det(w)a^w$$

in the definition of ϕ_a); in addition a_f is certainly unique. . . . Because the mapping $a \mapsto |\pi\mathbf{c}|^2 a$ is a homeomorphism of the space $\mathscr{I}(\mathfrak{F})$, the question obviously boils down to whether or not the mapping $a \mapsto \phi_a$ of $\mathscr{C}(\mathfrak{F})$ into $\mathscr{I}(G)$ is continuous – that it is has been mentioned earlier.

These remarks may be summarized in the following way.

Theorem 9.2.2.12 (Helgason-Harish-Chandra) The Abel transformation $f \mapsto F_f$ sets up a topological isomorphism between the convolution algebra $\mathscr{I}(G)$ and $\mathscr{I}(A_\mathfrak{p})$.

Here, then, is Harish-Chandra's Plancherel Theorem for the space $\mathscr{I}(G)$.

Theorem 9.2.2.13 (Harish-Chandra) The space $\mathscr{I}(G)$ is a commutative topological algebra under convolution. The spherical Fourier transform $\hat{f}$ of a function f in $\mathscr{I}(G)$ is a W-invariant rapidly decreasing function on $\mathfrak{F}$, the map $f \mapsto \hat{f}$ being a topological isomorphism of the algebra $\mathscr{I}(G)$ onto the algebra $\mathscr{I}(\mathfrak{F})$. Moreover, there exists a function $\mathbf{c}$ on $\mathfrak{F}$ such that $\mathbf{c}^{-1}$ is an (analytic) tempered distribution with $|\mathbf{c}(wv)| = |\mathbf{c}(v)|$ $(w \in W)$ and, for all $f \in \mathscr{I}(G)$, we have

$$(*) \quad \int_G |f(x)|^2 d_G(x) = [W]^{-1} \int_{\mathfrak{F}} |\hat{f}(v)|^2 |\mathbf{c}(v)|^{-2} d_{\mathfrak{F}}(v)$$

while

$$(**) \quad f(x) = [W]^{-1} \int_{\mathfrak{F}} \hat{f}(v)\overline{\phi_v(x)}\, |\mathbf{c}(v)|^{-2} d_{\mathfrak{F}}(v) \qquad (x \in G).$$

Because $\mathscr{I}(G)$ is dense in $I^2(G)$, relation $(*)$ can be extended in the usual way to all functions in $I^2(G)$ so that the spherical Fourier transform $f \mapsto \hat{f}$ may be viewed as a unitary equivalence between $I^2(G)$ and the Hilbert space of W-invariant functions on $\mathfrak{F}$ which are square integrable with respect to the measure $|\mathbf{c}(v)|^{-2} d_{\mathfrak{F}}(v)$.

Applications (1) Let $\mathfrak{D}(G/K)$ be the algebra of G-invariant differential operators on the Riemannian symmetric space G/K – then, using Harish-Chandra's Plancherel Theorem, Helgason [5] has shown that every D in $\mathfrak{D}(G/K)$ admits a fundamental solution T_D, say. [Thus T_D is, by definition, a distribution on G/K such that $D \cdot T_D = \delta_K$, δ_K the unit mass at the point in G/K left fixed by K.] Of course, a fundamental solution for D immediately gives a global solution to the differential equation $D\Phi = \phi$, ϕ a C^∞ function on G/K with compact support.

(2) Let Γ be a discrete subgroup of G such that G/Γ is compact – then, using Harish-Chandra's Plancherel Theorem, Gangolli [2] has studied the space $L^2(\Gamma \backslash G/K)$. In particular, when G is complex, he verifies a conjecture of Gelfand's regarding the asymptotic behavior of the spectra in the spectral decomposition of $L^2(\Gamma \backslash G/K)$.

9.2.3 The Paley-Wiener Theorem for $I^2(G)$

The classical Paley-Wiener Theorem characterizes the class of functions which are Fourier transforms of C^∞ functions of compact support on $\mathbf{R}^n$. In the formulation of Hörmander [1], the theorem states that an entire holomorphic function F on $\mathbf{C}^n$ is the Fourier-Laplace transform of a C^∞ function f on $\mathbf{R}^n$ with support in the (closed) ball of radius R iff for each integer $N \geqslant 0$ there exists a constant C_N such that

$$\text{(PW)}\quad |F(X+\sqrt{-1}Y)| \leqslant C_N(1+\|X+\sqrt{-1}Y\|)^{-N}e^{R\|Y\|} \qquad (X, Y \in \mathbf{R}^n).$$

The objective of the present number is to obtain a similar result for the spherical Fourier transform on G/K.

Let G be a connected semi-simple Lie group with finite center, $G = KA_\mathfrak{p}N^+$ an Iwasawa decomposition for G; let $I_c^\infty(G)$ be the algebra (under convolution) of C^∞ functions on G with compact support which are biinvariant under K. We shall agree to say that an element $f \in I_c^\infty(G)$ has support in the ball of radius $R(R > 0)$ if the restriction of f to $A_\mathfrak{p}$ has its support contained in the ball of radius R in $A_\mathfrak{p}$, i.e. in the set $\{h \in A_\mathfrak{p} : \|\log h\| \leqslant R\}$.

Here is the Paley-Wiener Theorem for $I^2(G)$.

Theorem 9.2.3.1 (Gangolli) Suppose that f is an element of $I_c^\infty(G)$ which vanishes outside the ball of radius R in $A_\mathfrak{p}$ – then the spherical Fourier-Laplace transform $\hat{f}$ of f defined by the rule

$$\hat{f}(\nu) = \int_G f(x)\phi_\nu(x)d_G(x) \qquad (\nu \in \mathfrak{F}_c)$$

exists for all $\nu \in \mathfrak{F}_c$ and is a W-invariant entire holomorphic function of $\nu(\nu \in \mathfrak{F}_c)$; moreover, given any integer $N \geqslant 0$, there exists a constant $C_N > 0$ such that

$$|\hat{f}(\xi+\sqrt{-1}\eta)| \leqslant C_N(1+\|\xi+\sqrt{-1}\eta\|)^{-N}e^{R\|\eta\|} \qquad (\xi, \eta \in \mathfrak{F}).$$

Conversely, if F is a W-invariant entire holomorphic function on $\mathfrak{F}_c$ with the property that there exists an $R > 0$ such that for any integer $N \geqslant 0$ there exists a constant $C_N > 0$ for which

$$|F(\xi+\sqrt{-1}\eta)| \leqslant C_N(1+\|\xi+\sqrt{-1}\eta\|)^{-N}e^{R\|\eta\|} \qquad (\xi, \eta \in \mathfrak{F}),$$

then there exists a unique function $f \in I_c^\infty(G)$ such that $\hat{f} = F$; f vanishes outside the ball of radius R in $A_\mathfrak{p}$ and, in fact, is given by the formula

$$f(x) = [W]^{-1}\int_{\mathfrak{F}} F(\nu)\overline{\phi_\nu(x)}\,|\mathbf{c}(\nu)|^{-2}d_{\mathfrak{F}}(\nu) \qquad (x \in G).$$

[The unexplained notations and conventions are as in 9.2.1 and 9.2.2; in particular $\mathfrak{F}_c$ stands for the complexification of $\mathfrak{F}$(= real dual of $\mathfrak{a}_\mathfrak{p}$) while $\|\xi+\sqrt{-1}\eta\| = (\|\xi\|^2+\|\eta\|^2)^{1/2}$ $(\xi, \eta \in \mathfrak{F})$.]

The Paley-Wiener Theorem leads at once to:

Corollary 9.2.3.2 The Abel transformation $f \mapsto F_f$ sets up a topological isomorphism between the convolution algebras $I_c^\infty(G)$ and $I_c^\infty(A_\mathfrak{p})$.

[Since the Abel transformation is one-to-one on the space of C^∞ *compactly supported* K-biinvariant functions on G (cf. number 6.2.2, or better, see 9.2.4 infra), the gist of the result is that the map is onto. . . .]

Application At this juncture, let us clear up a point which was left open in 9.2.2, namely the assertion that the Abel transformation is actually one-to-one on *all* of $\mathscr{I}(G)$ (cf. Theorem 9.2.2.1). The discussion leading up to the Plancherel Theorem for the space

$$\mathscr{I}_0(G)(=\{\phi_a : a \in \mathscr{C}(\mathfrak{F})\}),$$

i.e. Proposition 9.2.2.11, is completely independent of Theorem 9.2.2.1; on the other hand, a priori it is not even clear that $I_c^\infty(G)$ is contained in $\mathscr{I}_0(G)$. However, thanks to the Paley-Wiener Theorem supra (and the remarks prefacing the statement of Theorem 9.2.2.12), it is in fact the case that the inclusion $I_c^\infty(G) \subset \mathscr{I}_0(G)$ obtains; but then, by a simple continuity argument, we deduce that the statements made in Proposition 9.2.2.11 are actually valid for every f in $\mathscr{I}(G)$, whence

$$F_f = 0 \Rightarrow \hat{F}_f = 0 \Rightarrow \hat{f} = 0 \Rightarrow f = 0,$$

as desired.

The proof of the Paley-Wiener Theorem rests on the following lemma.

Lemma 9.2.3.3 Let F be a W-invariant entire holomorphic function on $\mathfrak{F}_c$ with the property that there exists an $R > 0$ such that for any integer $N \geqslant 0$ there exists a constant $C_N > 0$ for which

$$|F(\xi + \sqrt{-1}\eta)| \leqslant C_N(1 + \|\xi + \sqrt{-1}\eta\|)^{-N} e^{R\|\eta\|} \qquad (\xi, \eta \in \mathfrak{F}).$$

Consider the function f defined by the rule

$$f(x) = [W]^{-1} \int_{\mathfrak{F}} F(\nu)\overline{\phi_\nu(x)}\,|\mathbf{c}(\nu)|^{-2} d_{\mathfrak{F}}(\nu) \qquad (x \in G).$$

Then f has support contained in the ball of radius R.

This result is due to Helgason [10] when rank $(G/K) = 1$ and to Gangolli [3] in general; its proof is contained in the following discussion.

Because of our assumptions on F, f must lie in $\mathscr{I}(G)$ (here we have to bear in mind the Cauchy inequalities from classical function theory); furthermore, on the basis of the discussion leading up to the statement of Theorem 9.2.2.12, it is clear that $\hat{f} = F|\mathfrak{F}$. This being the case, fix an $h \in \exp(\mathscr{C}^{cl})$ with $\|\log h\| > R$ – then, since the restriction of f to $A_\mathfrak{p}$ is W-invariant, the proof of our lemma will be complete when it is shown that $f(h) = 0$.

To begin with, let us suppose that $h \in \exp(\mathscr{C})$ – then

$$\begin{aligned} h^\rho f(h) &= [W]^{-1} \int_{\mathfrak{F}} F(-\nu) h^\rho \phi_\nu(h)\,|\mathbf{c}(\nu)|^{-2} d_{\mathfrak{F}}(\nu) \\ &= [W]^{-1} \int_{\mathfrak{F}} F(-\nu) h^\rho \{\sum_{w \in W} \mathbf{\Psi}(\sqrt{-1}w\nu - \rho : h)\mathbf{c}(w\nu)\} \\ &\qquad \times |\mathbf{c}(\nu)|^{-2} d_{\mathfrak{F}}(\nu) \\ &= \int_{\mathfrak{F}} F(-\nu)\mathbf{c}(-\nu)^{-1} h^\rho \mathbf{\Psi}(\sqrt{-1}\nu - \rho : h) d_{\mathfrak{F}}(\nu), \end{aligned}$$

F being W-invariant while $|\mathbf{c}(\nu)|^{-2} = [\mathbf{c}(\nu)\mathbf{c}(-\nu)]^{-1}$ ($\nu \in \mathfrak{F}$; cf. Proposition 9.1.7.1 (iii)). We shall now make use of an estimate which was derived in the Appendix to 9.1.7 supra; thus let $\mathscr{R}$ denote the region

$$\{\xi + \sqrt{-1}\eta : \xi, \eta \in \mathfrak{F}, \eta \in \mathscr{C}^{cl}\}$$

in $\mathfrak{F}_c$ – then there exist absolute constants d, D such that

$$|\mathbf{\Gamma}_\lambda(\sqrt{-1}\nu - \rho)| \leqslant Dm_\lambda^d \qquad (\text{all } \nu \in \mathscr{R})$$

uniformly in $\lambda (\lambda \in \mathsf{L})$. Accordingly

$$\int_{\mathfrak{F}} |F(-\nu)\mathbf{c}(-\nu)^{-1}h^\rho \mathbf{\Psi}(\sqrt{-1}\nu - \rho : h)|\, d_{\mathfrak{F}}(\nu)$$
$$\leqslant \int_{\mathfrak{F}} |F(-\nu)\mathbf{c}(-\nu)^{-1} \sum_{\lambda \in \mathsf{L}} \mathbf{\Gamma}_\lambda(\sqrt{-1}\nu - \rho)h^{\sqrt{-1}\nu - \lambda}|\, d_{\mathfrak{F}}(\nu)$$
$$\leqslant M(h) \int_{\mathfrak{F}} |F(\nu)\mathbf{c}(\nu)^{-1}|\, d_{\mathfrak{F}}(\nu) < \infty$$

where M is the analytic function on $\exp(\mathscr{C})$ defined by the rule

$$M(h) = D \sum_{\lambda \in \mathsf{L}} m_\lambda^d h^{-\lambda} \qquad (h \in \exp(\mathscr{C})).$$

In the last step we used the fact that $F|\mathfrak{F}$ is rapidly decreasing while $\mathbf{c}^{-1}$ is tempered (cf. Proposition 9.1.7.2 and its corollary). It therefore follows that

$$h^\rho f(h) = \sum_{\lambda \in \mathsf{L}} h^{-\lambda} \int_{\mathfrak{F}} F(-\nu)\mathbf{c}(-\nu)^{-1}\mathbf{\Gamma}_\lambda(\sqrt{-1}\nu - \rho)h^{\sqrt{-1}\nu} d_{\mathfrak{F}}(\nu)$$
$$= \sum_{\lambda \in \mathsf{L}} f_\lambda(h)h^{-\lambda}$$

where we have set

$$f_\lambda(h) = \int_{\mathfrak{F}} F(-\nu)\mathbf{c}(-\nu)^{-1}\mathbf{\Gamma}_\lambda(\sqrt{-1}\nu - \rho)h^{\sqrt{-1}\nu} d_{\mathfrak{F}}(\nu).$$

We intend to prove that $f_\lambda(h) = 0$ for every $\lambda \in \mathsf{L}$. As above, let $\mathscr{R}$ be the region $\{\xi + \sqrt{-1}\eta : \xi, \eta \in \mathfrak{F}, \eta \in \mathscr{C}^{cl}\}$ – then the rational function $\nu \mapsto \mathbf{\Gamma}_\lambda(\sqrt{-1}\nu - \rho)$ $(\lambda \in \mathsf{L})$ is well-defined and holomorphic throughout $\mathscr{R}$. On the other hand, due to the formula for the $\mathbf{c}$-function, it is clear that none of the possible singularities of the function $\nu \mapsto \mathbf{c}(-\nu)^{-1}$ lie in $\mathscr{R}$, whence $\nu \mapsto \mathbf{c}(-\nu)^{-1}$ is holomorphic there. Because F is, by hypothesis, everywhere holomorphic, we see that the integrand defining f_λ is a holomorphic function of ν in $\mathscr{R}$. Now let η be an element of $\mathfrak{F}$ such that $\eta \in \mathscr{C}^{cl}$; the set $\{\xi + \sqrt{-1}\eta : \xi \in \mathfrak{F}\}$ is a hyperplane in $\mathfrak{F}_c$, lying in $\mathscr{R}$, and, moreover, is homotopic to the hyperplane $\{\xi + \sqrt{-1}0 : \xi \in \mathfrak{F}\}$ via a homotopy within $\mathscr{R}$. Subject, then, to suitable behavior of the integrand at infinity, we can apply Cauchy's Theorem to the integral defining f_λ and write

$$f_\lambda(h) = h^{-\eta} \int_{\mathfrak{F}} F(-\xi - \sqrt{-1}\eta)\mathbf{c}(-\xi - \sqrt{-1}\eta)^{-1}$$
$$\times \mathbf{\Gamma}_\lambda(\sqrt{-1}\xi - \eta - \rho)h^{\sqrt{-1}\xi} d_{\mathfrak{F}}(\xi).$$

In order to show that our integrand does behave suitably at infinity, recall first that $\mathbf{\Gamma}_\lambda$ ($\lambda \in \mathsf{L}$) admits the estimate $|\mathbf{\Gamma}_\lambda(\sqrt{-1}\nu - \rho)| \leqslant Dm_\lambda^d$ (all $\nu \in \mathscr{R}$). Furthermore, there exists a non-negative integer M and a positive constant C_M such that

$$|\mathbf{c}(-\xi - \sqrt{-1}\eta)|^{-1} \leqslant C_M(1 + \|\xi + \sqrt{-1}\eta\|)^M \qquad (\text{all } \xi + \sqrt{-1}\eta \in \mathscr{R}).$$

To see this, we shall use the formula (see Magnus and Oberhettinger [1, p. 6])

$$\lim_{|z| \to \infty} \frac{\Gamma(z + a)}{\Gamma(z)} e^{-a \log z} = 1, \ |\arg(z)| \leqslant \pi - \delta \qquad (\delta > 0)$$

where a is any constant, log being the principal value of the logarithm. Write

$$\begin{aligned} \mathbf{c}(\nu)^{-1} &= \mathbf{I}(\rho) / \prod_{\lambda \in \Sigma^+} B\left(\frac{m(\lambda)}{2}, \frac{m(\lambda/2)}{4} + \sqrt{-1}\frac{(\nu, \lambda)}{(\lambda, \lambda)}\right) \\ &= \mathbf{I}(\rho) / \prod_{\lambda \in \Sigma^+} \mathbf{c}_\lambda(\nu) \qquad (\text{say}). \end{aligned}$$

Fix $\lambda \in \Sigma^+$; take $a = m(\lambda)/2$, $z = 4^{-1}m(\lambda/2) + \sqrt{-1}\,(\nu, \lambda)/(\lambda, \lambda)$ – then

$$\left|\frac{\Gamma(m(\lambda)/2 + z)}{\Gamma(z)}\right| \leqslant K_1 + K_2 |z|^{m(\lambda)/2} \qquad (K_1, K_2 \text{ constants})$$

whenever the real part of z is $\geqslant 0$ which is the case if $\nu = \xi + \sqrt{-1}\eta$ lies in the set $\{\xi + \sqrt{-1}\eta : \xi, \eta \in \mathfrak{F}, -\eta \in \mathscr{C}^{cl}\}$. It therefore follows that

$$|\mathbf{c}(\xi + \sqrt{-1}\eta)|^{-1} \leqslant C_1 + C_2 \prod_{\lambda \in \Sigma^+} \|\xi + \sqrt{-1}\eta\|^{m(\lambda)/2} \qquad (C_1, C_2 \text{ constants})$$

provided $-\eta \in \mathscr{C}^{cl}$. Of course this suffices to establish the desired estimate for $|\mathbf{c}(-\xi - \sqrt{-1}\eta)|^{-1}$ ($\xi + \sqrt{-1}\eta \in \mathscr{R}$). Finally, on $\mathfrak{F}$ itself, we have the growth condition per the hypothesis of our lemma. Clearly, then, the above application of Cauchy's Theorem is valid. So, for a sufficiently large integer $N \geqslant 0$, we have

$$\begin{aligned} |f_\lambda(h)| &\leqslant h^{-\eta} \int_{\mathfrak{F}} |F(-\xi - \sqrt{-1}\eta)|\,|\mathbf{c}(-\xi - \sqrt{-1}\eta)|^{-1} \\ &\qquad \times |\mathbf{\Gamma}_\lambda(\sqrt{-1}\xi - \eta - \rho)|\, d_{\mathfrak{F}}(\xi) \\ &\leqslant C_M C_N D m_\lambda^d e^{R\|\eta\|} h^{-\eta} \int_{\mathfrak{F}} (1 + \|\xi + \sqrt{-1}\eta\|)^{M-N} d_{\mathfrak{F}}(\xi) \\ &\leqslant C_\lambda e^{R\|\eta\|} h^{-\eta}, \end{aligned}$$

C_λ a positive constant depending only on $\lambda (\lambda \in \mathsf{L})$. Of course this estimate holds for all $\eta \in \mathscr{C}$; this being the case, put $H = \log h$ – then

$$|f_\lambda(h)| \leqslant C_\lambda e^{t\|H\|(R - \|H\|)} \qquad (t > 0)$$

so, upon letting $t \to \infty$ and remembering that $\|H\| > R$, we deduce that $f_\lambda(h) = 0$. Because $h^\rho \neq 0$ (all $h \in \exp(\mathscr{C})$), it now follows that $f(h) = 0$

whenever $h \in \exp(\mathscr{C})$ is such that $\|\log h\| > R$. But since f is continuous and W-invariant, this implies that f vanishes identically outside the ball of radius R in $A_{\mathfrak{p}}$, thereby completing the proof of the lemma.

Proof of Theorem 9.2.3.1 (1) Suppose that f is an element of $I_c^\infty(G)$ which vanishes outside the ball of radius R in $A_{\mathfrak{p}}$ – then there can be no question but that the spherical Fourier-Laplace transform $\hat{f}$ of f defined by the rule

$$\hat{f}(\nu) = \int_G f(x)\phi_\nu(x)d_G(x) \qquad (\nu \in \mathfrak{F}_c)$$

exists for all $\nu \in \mathfrak{F}_c$ and is a W-invariant holomorphic function of $\nu(\nu \in \mathfrak{F}_c)$. It is not difficult to see that the Abel transform F_f of f has support in the ball of radius R in $A_{\mathfrak{p}}$; this being so, the easy half of the classical Paley-Wiener Theorem tells us that the Euclidean Fourier transform $\hat{F}_f$ of F_f is holomorphic throughout all of $\mathfrak{F}_c$ and has the property that for any given integer $N \geqslant 0$ there exists a constant $C_N > 0$ such that

$$|\hat{F}_f(\xi + \sqrt{-1}\eta)| \leqslant C_N(1 + \|\xi + \sqrt{-1}\eta\|)^{-N}e^{R\|\eta\|} \qquad (\xi, \eta \in \mathfrak{F}).$$

The first contention of our theorem thus follows from the fact that $\hat{f} = \hat{F}_f$ on $\mathfrak{F}$. . . .

(2) Suppose now that F is a W-invariant entire holomorphic function on $\mathfrak{F}_c$ with the property that there exists an $R > 0$ such that for any integer $N \geqslant 0$ there exists a constant $C_N > 0$ for which

$$|F(\xi + \sqrt{-1}\eta)| \leqslant C_N(1 + \|\xi + \sqrt{-1}\eta\|)^{-N}e^{R\|\eta\|} \qquad (\xi, \eta \in \mathfrak{F}).$$

Define a function f on G by the rule

$$f(x) = [W]^{-1}\int_{\mathfrak{F}} F(\nu)\overline{\phi_\nu(x)}|\mathbf{c}(\nu)|^{-2}d_{\mathfrak{F}}(\nu) \qquad (x \in G).$$

Then, according to Lemma 9.2.3.3, f belongs to $I_c^\infty(G)$ and, in fact, vanishes outside the ball of radius R in $A_{\mathfrak{p}}$; moreover, by analytic continuation, $\hat{f} = F$ on $\mathfrak{F}_c$. It remains only to show that f is unique; so suppose that g is another element of $I_c^\infty(G)$ with the asserted properties – then

$$\hat{f} = \hat{g} \Rightarrow \hat{F}_f = \hat{F}_g \Rightarrow F_f = F_g \Rightarrow f = g,$$

the Abel transformation being one-to-one on $I_c^\infty(G)$ (cf. number 6.2.2 or 9.2.4 infra).

The proof of the Paley-Wiener Theorem for $I^2(G)$ is therefore complete. □

Definition A holomorphic function F on $\mathfrak{F}_c$ is said to be a *rapidly decreasing holomorphic function of exponential type* if it satisfies the condition: There exists a constant $R > 0$ such that for each holomorphic differential operator D on $\mathfrak{F}_c$ with polynomial coefficients

$$\sup_{\nu \in \mathfrak{F}_c} e^{-R\|\eta\|}|F(\nu; D)| < \infty \qquad (\nu = \xi + \sqrt{-1}\eta).$$

The space of all rapidly decreasing holomorphic functions on $\mathfrak{F}_c$ of exponential type is, in the natural way, a locally convex, Hausdorff, topological vector space which is, moreover, complete. This being so, it is then clear that we have proved the following theorem.

Theorem 9.2.3.1 (bis) The spherical Fourier-Laplace transform $f \mapsto \hat{f}$ sets up a topological isomorphism between the space $I_c^\infty(G)$ and the space of W-invariant rapidly decreasing holomorphic functions on $\mathfrak{F}_c$ of exponential type.

9.2.4 Harmonic Analysis in $I^1(G)$

Let G be a connected semi-simple Lie group with finite center, $G = KA_{\mathfrak{p}}N^+$ an Iwasawa decomposition for G; let $I^1(G)$ be the space of K-biinvariant integrable functions on G – then, under the convolution product, $I^1(G)$ is a commutative semi-simple Banach algebra. Our first objective in the present number will be to obtain an explicit description of the set of $\nu(\nu \in \mathfrak{F}_c)$ for which the associated zonal spherical function ϕ_ν is bounded; as has been pointed out in 9.2.1, this result will serve to determine the maximal ideal space for $I^1(G)$.

Theorem 9.2.4.1 (Helgason-Johnson) Let $\mathfrak{C}_\rho$ denote the convex hull of the points $w\rho(w \in W)$ – then ϕ_ν is bounded iff $\nu = \xi + \sqrt{-1}\eta$ $(\xi, \eta \in \mathfrak{F})$ belongs to the tube $\mathfrak{T}_\rho = \mathfrak{F} + \sqrt{-1}\mathfrak{C}_\rho$.

[All unexplained notations and conventions are per 9.2.1 and 9.2.2.]

As was remarked by E. M. Stein, the boundedness of ϕ_ν for $\nu \in \mathfrak{T}_\rho$ is an immediate consequence of the Maximum Principle applied to the entire function $\nu \mapsto \phi(\nu : x)$ $(x \in G)$ and the fact that

$$|\phi(\xi + \sqrt{-1}w\rho : x)| \leqslant 1 \qquad (x \in G)$$

for all $w \in W$. This being so, let us turn now to the proof of the necessity.

Lemma 9.2.4.2 Suppose that ϕ_ν is bounded and that $-Q_\eta \in \mathscr{C}^{cl}$ – then $\eta + \rho$ is non-negative on $\mathscr{C}$.

Proof Assume that the Haar measure on N^- is normalized in such a way that

$$\int_{N^-} e^{-2\rho(H(n))} d_{N^-}(n) = 1.$$

Then, as we know, the zonal spherical function ϕ_ν admits the integral representation

$$\phi_\nu(h) = h^{\sqrt{-1}\nu - \rho} \int_{N^-} \exp\{(\sqrt{-1}\nu - \rho)(H(hnh^{-1}))\}$$
$$\times \exp\{-(\sqrt{-1}\nu + \rho)(H(n))\}\, d_{N^-}(n) \qquad (h \in A_{\mathfrak{p}}).$$

(1) To begin with, let us suppose that $-Q_\eta \in \mathscr{C}$. Fix an element $H \in \mathscr{C}$ and put $h_t = \exp(tH)$ $(t \geqslant 0)$ – then, as can be seen from the

proof of Theorem 9.1.6.1, we have

$$\begin{aligned}\lim_{t\to\infty}\int_{N^-} &\exp\{(\sqrt{-1}\nu-\rho)(H(h_t n h_t^{-1}))\}\\ &\times\exp\{-(\sqrt{-1}\nu+\rho)(H(n))\}\,d_{N^-}(n)\\ =\int_{N^-}&\lim_{t\to\infty}\exp\{(\sqrt{-1}\nu-\rho)(H(h_t n h_t^{-1}))\}\\ &\times\exp\{-(\sqrt{-1}\nu+\rho)(H(n))\}\,d_{N^-}(n),\end{aligned}$$

the interchange being justified on the basis of the relations

$$H(n)\in\mathfrak{a}_{\mathfrak{p}}^{+},\ H(n)-H(h_t n h_t^{-1})\in\mathfrak{a}_{\mathfrak{p}}^{+}\qquad(n\in N^-).$$

For later purposes, it will be convenient to give a different justification of the interchange. Thus choose ϵ, $0<\epsilon<1$, such that $Q_\rho+\epsilon Q_\eta\in\mathscr{C}$ – then the integrand on the left above is majorized by

$$\exp\{-(\eta+\rho)(H(h_t n h_t^{-1}))\}\exp\{(\eta-\rho)(H(n))\}$$

and this is

$$\leqslant\exp\{(\epsilon\eta-\rho)(H(n))\},$$

which is integrable (cf. Corollary 9.1.6.5). It therefore follows that

$$\begin{aligned}\lim_{t\to\infty}\int_{N^-}&\exp\{(\sqrt{-1}\nu-\rho)(H(h_t n h_t^{-1}))\}\\ &\times\exp\{-(\sqrt{-1}\nu+\rho)(H(n))\}\,d_{N^-}(n)\\ =\int_{N^-}&\exp\{-(\sqrt{-1}\nu+\rho)(H(n))\}\,d_{N^-}(n)=\mathbf{c}(\nu)\neq 0,\end{aligned}$$

whence

$$\lim_{t\to\infty}\phi_\nu(h_t)h_t^{\rho-\sqrt{-1}\nu}$$

exists and is non-zero. Taking absolute values, we deduce that

$$(\eta+\rho)(H)\geqslant 0$$

which is equivalent to our contention, H being arbitrary in $\mathscr{C}$.

(2) It remains to consider the possibility that $-Q_\eta$ lies on the boundary of $\mathscr{C}$. Following the usual practice, let us divide the positive roots Σ^+ of the pair $(\mathfrak{g},\mathfrak{a}_{\mathfrak{p}})$ into two disjoint subsets Σ_1^+ and $\Sigma_2^+(=\Sigma^+-\Sigma_1^+)$, a root λ being placed in Σ_1^+ if it annihilates $-Q_\eta$; assign to the symbols ρ_1,ρ_2,N_1^-, and N_2^- the obvious meanings (so that, in particular, $N^-=N_1^-N_2^-$).

(a) By a simple computation, we find that

$$\begin{aligned}(\dagger)\quad\phi_\nu(h)=Mh^{\sqrt{-1}\nu-\rho}\int_{N_2^-}&\exp\{(\sqrt{-1}\nu-\rho)(H(hn_2h^{-1}))\}\\ &\times\exp\{-(\sqrt{-1}\nu+\rho)(H(n_2))\}\,d_{N_2^-}(n_2)\end{aligned}$$

for all h in $A_{\mathfrak{p}}$ with the property that $h^\lambda=1(\lambda\in\Sigma_1^+)$ (here $M=\int_{N_1^-}\exp\{-2\rho_1(H(n_1))\}\,d_{N_1^-}(n_1)$). This being the case, let H be an element in $\mathscr{C}^{cl}$ such that $\lambda(H)=0$ for $\lambda\in\Sigma_1^+$ but such that $\lambda(H)>0$ for $\lambda\in\Sigma_2^+$;

insert $h_t = \exp(tH)$ $(t \geqslant 0)$ into the integral formula (†) and let $t \to \infty$ – then, in view of the inequalities supra and the integrability of the function $n_2 \mapsto \exp\{(\epsilon\eta - \rho)(H(n_2))\}$ $(n_2 \in N_2^-)$, we can take the limit under the integral sign and infer as before that $(\eta + \rho)(H) \geqslant 0$.

(b) Now let H be an arbitrary element in $\mathscr{C}$; its projection H_1 onto the plane $\sigma_1 = \{H \in \mathfrak{a}_{\mathfrak{p}} : \lambda(H) = 0$ (all $\lambda \in \dot{\Sigma}_1^+)\}$ is given by $H_1 = [W_1]^{-1} \sum_{w_1 \in W_1} w_1 H$ where W_1 is the subgroup of W leaving σ_1 pointwise fixed. Owing then to (a) and the relation $\rho(H - wH) \geqslant 0$ $(w \in W$; cf. Lemma 3.3.2.1), we have

$$\begin{aligned}(\eta + \rho)(H) &= (\eta + \rho)(H_1) + (\eta + \rho)(H - H_1)\\ &= (\eta + \rho)(H_1) + \rho(H - H_1) \geqslant 0,\end{aligned}$$

as desired.

The proof of the lemma is therefore complete. □

Let $\mathscr{C}^\vee$ denote the dual cone of $\mathscr{C}$, i.e. the subset of $\mathfrak{a}_{\mathfrak{p}}$ comprised of those H such that $(H, \mathscr{C}) > 0$ (thus the closure of $\mathscr{C}^\vee$ is $\mathfrak{a}_{\mathfrak{p}}^+$); in the Appendix to the present number it will be shown that the closure of the set

$$\bigcup_{w \in W} w\{\mathscr{C} \cap (-\mathscr{C}^\vee + Q_\rho)\}$$

is the convex hull of the set of points $Q_{w\rho}$ $(w \in W)$. Because ϕ_ν is bounded iff $\phi_{-\nu}$ is bounded, on the basis of Lemma 9.2.4.2 it is now clear that if ϕ_ν is bounded, then, of necessity, η lies in $\mathfrak{C}_\rho$ $(\nu = \xi + \sqrt{-1}\eta)$ which thereby completes the proof of Theorem 9.2.4.1.

Remark It is a standard observation that the commutative Banach algebra $I^1(G)$ is semi-simple, i.e. that the spherical Gelfand transform $f \mapsto \hat{f}$ is injective (cf. Helgason [2, p. 453] or Tamagawa [1]); using this fact, it is then a simple matter to show that the Abel transform $f \mapsto F_f$ is one-to-one on the space $I_c^\infty(G)$. [Thus suppose that $F_f = 0$ – then, for *any* ν in the tube $\mathfrak{T}_\rho$, it must be the case that $\hat{f}(\nu)$ $(= \int_{A_{\mathfrak{p}}} F_f(h) h^{\sqrt{-1}\nu} d_{A_{\mathfrak{p}}}(h)) = 0$, whence $f = 0$. . . .]

Given x in G, put

$$\phi_\nu(x) = \phi(\nu : x) = \int_K e^{(\sqrt{-1}\nu - \rho)(H(xk))}\, dk,$$

ν a linear function on $\mathfrak{a}_{\mathfrak{p}}$. The following estimate will play an important role in the discussion which is to follow.

Lemma 9.2.4.3 Fix $D \in \mathfrak{G}$ – then there exists $R_D > 0$ and a non-negative integer r_D such that

$$|\phi(\nu : x; D)| \leqslant R_D \phi(\sqrt{-1}\mathscr{I}(\nu) : x)(1 + \|\nu\|)^{r_D}$$

for all $\nu \in \mathfrak{F}_c$ and $x \in G$.

[Given $\nu \in \mathfrak{F}_c$, we write $\|\nu\| = (\|\mathscr{R}(\nu)\|^2 + \|\mathscr{I}(\nu)\|^2)^{1/2}$.]

We shall preface the proof with a preliminary observation. Extend $\mathfrak{a}_{\mathfrak{p}}$ to a Cartan subalgebra $\mathfrak{a}$ of $\mathfrak{g}$; let Λ_i be the highest weight of the ith

basic irreducible module of the pair $(\mathfrak{g}_c, \mathfrak{a}_c)$, U_i the corresponding representation of $\mathfrak{g}_c$ on E_i, say $(1 \leqslant i \leqslant l)$. [Here, of course, $l = \operatorname{rank}(\mathfrak{g}_c)$.] It will be tacitly assumed that E_i has been supplied with the structure of a Hilbert space relative to which $U_i | \mathfrak{u} (= \mathfrak{k} + \sqrt{-1}\mathfrak{p})$ is skew-adjoint; this being so, fix once and for all a unit vector a_i in E_i belonging to Λ_i. Now suppose that we are given a linear function ν on $\mathfrak{a}_\mathfrak{p}$; write $\nu = \sum_i n_i \Lambda_i (n_i \in \mathbf{C})$ – then it is easy to see that

$$\phi(\nu : x) = \int_K \prod_{i=1}^{l} \| U_i(x^{-1}k)a_i \|^{-\sqrt{-1}n_i - r_i} dk \qquad (\text{all } x \in G)$$

where the r_i are real numbers such that $\rho = \sum_i r_i \Lambda_i$.

Proof of Lemma 9.2.4.3 Let $n = \dim(\mathfrak{g})$; given $t = (t_1, \ldots, t_n) \in \mathbf{R}^n$, put $|t| = \max |t_i| (1 \leqslant i \leqslant n)$ and, for any ordered set $M = (m_1, \ldots, m_n)$ of non-negative integers, write $t^M = t_1^{m_1} \ldots t_n^{m_n}$. Fix a basis $X_1, \ldots, X_n$ for $\mathfrak{g}$ over $\mathbf{R}$; let $X(M)$ denote the coefficient in $\mathfrak{G}$ of t^M in the expression $(|M|!)^{-1}(t_1X_1 + \cdots + t_nX_n)^{|M|}$ (here $|M| = m_1 + \cdots + m_n$) – then, as we know, the $X(M)$ form a basis for $\mathfrak{G}$. Let τ denote the conjugation of $\mathfrak{g}_c$ with respect to the compact real form $\mathfrak{u} = \mathfrak{k} + \sqrt{-1}\mathfrak{p}$; let $D \mapsto D^\tau$ $(D \in \mathfrak{G})$ be the anti-automorphism of $\mathfrak{G}$ over $\mathbf{R}$ which coincides with $-\tau$ on $\mathfrak{g}_c$. Bearing in mind the nature of the Hilbert space structure with which each of the E_i above is equipped, it is clear that the Hilbert space adjoint of $U_i(D)$ is $U_i(D^\tau)$ (all $D \in \mathfrak{G}$). The space $\operatorname{Hom}_{\mathbf{C}}(E_i, E_i)$ is a Banach space under the usual operator norm. We have

$$U_i(\exp\{-(t_1X_1 + \cdots + t_nX_n)\}) = \sum_M t^M(-1)^{|M|}U_i(X(M)),$$

the series converging absolutely and uniformly in $\operatorname{Hom}_{\mathbf{C}}(E_i, E_i)$ provided $|t|$ remains bounded. For any M put

$$D_M = \sum_{M_1 + M_2 = M} (-1)^{|M|}(X(M_1))^\tau X(M_2).$$

Given x in G, let us agree to write x_t for $x \exp(t_1X_1 + \ldots + t_nX_n)$ $((t_1, \ldots, t_n) \in \mathbf{R}^n)$ – then it is obvious that

$$\| U_i(x_t^{-1}k)a_i \|^2 = \| U_i(x^{-1}k)a_i \|^2 + \sum_{|M| \geqslant 1} t^M (U_i(x^{-1}k)a_i, U_i(D_M)U_i(x^{-1}k)a_i)$$

for all x, k and t $(1 \leqslant i \leqslant l)$. Set

$$\Phi_{M,i}(x) = \| U_i(x)a_i \|^{-2}(U_i(x)a_i, U_i(D_M)U_i(x)a_i) \qquad (x \in G).$$

Evidently $\Phi_{M,i}$ is a bounded analytic function on G and

$$\| U_i(x_t^{-1}k)a_i \|^2 = \| U_i(x^{-1}k)a_i \|^2 \{1 + \sum_{|M| \geqslant 1} t^M \Phi_{M,i}(x^{-1}k)\},$$

the series on the right converging uniformly with respect to x, k and t so long as x stays within a compact subset of G and $|t|$ remains bounded.

Thus, for $|t|$ sufficiently small,

$$\begin{aligned}
&\exp\{-\sqrt{-1}\nu(H(x_t^{-1}k)) - \rho(H(x_t^{-1}k))\} \\
&= \prod_1^l \| U_i(x_t^{-1}k)a_i \|^{-\sqrt{-1}n_i - r_i} \\
&= \exp\{-\sqrt{-1}\nu(H(x^{-1}k)) - \rho(H(x^{-1}k))\} \sum_M t^M \Phi_M(x^{-1}k : \nu).
\end{aligned}$$

Here $\Phi_M(x:\nu)$ is a function on $G \times \mathfrak{F}_c$ which can be written as a polynomial in $(\sqrt{-1}n_i + r_i)$ and $\Phi_{M',i}(1 \leqslant i \leqslant l, |M'| \leqslant |M|)$ with constant coefficients. This means that it is possible to pick a positive number R_M and an integer $r_M \geqslant 0$ such that

$$|\Phi_M(x:\nu)| \leqslant R_M(1 + \|\nu\|)^{r_M} \qquad (x \in G, \nu \in \mathfrak{F}_c).$$

In addition it is clear that for fixed x the above series converges uniformly with respect to k and t provided $|t|$ is sufficiently small. Consequently, if we integrate over K, then we get an expansion of $\phi(-\nu : x_t^{-1}) = \phi(\nu : x_t)$ in powers of $t_1, \ldots, t_n$. Therefore, if $M = (m_1, \ldots, m_n)$, then

$$\begin{aligned}
&\phi(\nu : x ; X(M)) \\
&= (m_1! \ldots m_n!)^{-1}(\partial^{m_1 + \cdots + m_n}/\partial t_1^{m_1} \cdots \partial t_n^{m_n})\phi(\nu : x_t)|_{t=0} \\
&= \int_K \Phi_M(x^{-1}k : \nu) \exp\{-\sqrt{-1}\nu(H(x^{-1}k)) - \rho(H(x^{-1}k))\}\, dk
\end{aligned}$$

and so

$$|\phi(\nu : x ; X(M))| \leqslant R_M \phi(\sqrt{-1}\mathscr{I}(\nu) : x)(1 + \|\nu\|)^{r_M} \qquad (\nu \in \mathfrak{F}_c, x \in G).$$

The assertion of the lemma now follows from the fact that the $X(M)$ constitute a basis for $\mathfrak{G}$. □

Let $\mathfrak{I}(G)$ denote the set of all C^∞ functions f on G which are biinvariant under K and satisfy the condition:

$$|f|_{r,D} = \sup_{x \in G} \frac{(1 + \sigma(x))^r}{\Xi^2(x)} |f(x ; D)| < \infty$$

for all $r \geqslant 0$, $D \in \mathfrak{G}$. Evidently $\mathfrak{I}(G)$ is a subset of $\mathscr{I}(G)$; on the other hand, the elements of $\mathfrak{I}(G)$ also lie in $L^1(G)$ (cf. Proposition 8.3.7.5). Finally, when equipped with the evident topology, $\mathfrak{I}(G)$ acquires the structure of a Fréchet space. Our objective now will be to study the spherical Gelfand transform $\hat{f}$ of a given element f in $\mathfrak{I}(G)$.

Definition Let $\mathfrak{T}_\rho^*$ denote the interior of the tube $\mathfrak{T}_\rho = \mathfrak{F} + \sqrt{-1}\mathfrak{C}_\rho$ – then a holomorphic function F on $\mathfrak{T}_\rho^*$ is said to be a *rapidly decreasing holomorphic function in the tube* $\mathfrak{T}_\rho^*$ if it satisfies the condition: For each holomorphic differential operator D on $\mathfrak{F}_c$ with polynomial coefficients

$$\sup_{\nu \in \mathfrak{T}_\rho^*} |F(\nu ; D)| < \infty.$$

Let $\mathfrak{H}(\mathfrak{T}_\rho^*)$ denote the space of W-invariant rapidly decreasing holomorphic functions in the tube $\mathfrak{T}_\rho^*$; when equipped with the evident topology, it is clear that $\mathfrak{H}(\mathfrak{T}_\rho^*)$ acquires the structure of a Fréchet algebra, i.e. $\mathfrak{H}(\mathfrak{T}_\rho^*)$ is a Fréchet space with (pointwise) multiplication being jointly continuous.

Proposition 9.2.4.4 Suppose that f belongs to $\mathsf{I}(G)$ – then the restriction of its spherical Gelfand transform to $\mathfrak{T}_\rho^*$ lies in $\mathfrak{H}(\mathfrak{T}_\rho^*)$.

Proof Let $\lambda_1, \ldots, \lambda_l$ be the set of simple roots for the pair $(\mathfrak{g}, \mathfrak{a}_\mathfrak{p})$ singled out by $\mathscr{C}$; determine elements $\Lambda_1, \ldots, \Lambda_l$ in $\mathfrak{F}$ by the condition $(\lambda_i, \Lambda_j) = \delta_{ij}$ $(1 \leqslant i, j \leqslant l)$. Given $\nu \in \mathfrak{F}_c$, $\nu = \xi + \sqrt{-1}\eta$ $(\xi, \eta \in \mathfrak{F})$, write $\nu = \sum_i \zeta_i \Lambda_i$ $(\zeta_i \in \mathbf{C})$ and $\zeta_i = \xi_i + \sqrt{-1}\eta_i$ $(\xi_i, \eta_i \in \mathbf{R})$. We shall agree to express the holomorphic differential operators on $\mathfrak{F}_c$ by means of the coordinates $\zeta_1, \ldots, \zeta_l$.

(1) Let p be a polynomial in l variables – then we claim that

$$p(\partial/\partial\zeta_1, \ldots, \partial/\partial\zeta_l)\hat{f}$$

is bounded on $\mathfrak{T}_\rho^*$. By definition

$$\hat{f}(\nu) = \int_G f(x)\phi_\nu(x)d_G(x) \qquad (\nu \in \mathfrak{T}_\rho)$$

and

$$\frac{\partial}{\partial\zeta_i}\int_K e^{(\sqrt{-1}\nu-\rho)(H(xk))}\,dk = \int_K e^{(\sqrt{-1}\nu-\rho)(H(xk))}\sqrt{-1}\,\langle H(xk), \Lambda_i\rangle\,dk \qquad (x \in G).$$

Suppose that $h \in \exp(\mathscr{C})$ – then, according to Lemmas 3.3.2.2 and 3.3.2.3, we have

$$(\log h - H(hk), \mathscr{C}) \geqslant 0, \; (-w^\dagger(\log h) + H(hk), \mathscr{C}) \geqslant 0 \qquad (k \in K).$$

Thus

$$\Lambda_i(w^\dagger(\log h)) \leqslant \Lambda_i(H(hk)) \leqslant \Lambda_i\,(\log h)$$

and so

$$|\Lambda_i(H(hk))| \leqslant ||\Lambda_i||\cdot||\log h|| \qquad (k \in K;\, 1 \leqslant i \leqslant l).$$

It therefore follows that

$$\left|\frac{\partial}{\partial\zeta_i}\phi_\nu(h)\right| \leqslant \phi_{\sqrt{-1}\eta}(h)\,||\Lambda_i||\cdot||\log h|| \qquad (h \in \exp(\mathscr{C})).$$

Because $f \in \mathsf{I}(G)$, the integral

$$\int_{\mathscr{C}} |f(H)|\cdot||H||^n \bigtriangleup(H)\,d_{\mathfrak{a}_\mathfrak{p}}(H)$$

is finite for any $n \in \mathbf{N}$ (the 'exp' being dropped from the notation ...); this being the case, fix an i lying between 1 and l – then

$$\int_G \left| f(x)\frac{\partial}{\partial \zeta_i}\phi_\nu(x) \right| d_G(x) \leq \|\Lambda_i\| \int_G |f(H)|\cdot\|H\| \frown (H) d_{\mathfrak{a}_\mathfrak{p}}(H) < \infty$$
$$(\text{all } \nu \in \mathfrak{T}_p^*)$$

which shows that the integral $\int_G f(x)(\partial/\partial\zeta_i)\phi_\nu(x)d_G(x)$ converges uniformly for $\nu \in \mathfrak{T}_p^*$. By iteration, we see that for each polynomial p in l variables the integral

$$\int_G f(x)\{p(\partial/\partial\zeta_1, \ldots, \partial/\partial\zeta_l)\phi_\nu(x)\}\, d_G(x)$$

converges uniformly for $\nu \in \mathfrak{T}_p^*$. Consequently the derivatives of $\hat{f}\,|\,\mathfrak{T}_p^*$ can be computed by differentiating under the integral sign, whence

$$p(\partial/\partial\zeta_1, \ldots, \partial/\partial\zeta_l)\hat{f}$$

is bounded on $\mathfrak{T}_p^*$, as claimed.

(2) Let p be a polynomial in l variables – then we claim that

$$\sup_{\nu\in\mathfrak{T}_p^*} |\, p(\partial/\partial\zeta_1, \ldots, \partial/\partial\zeta_l)\{((\nu, \nu) + (\rho, \rho))^n\hat{f}(\nu)\}\,| < \infty$$
$$(n = 0, 1, 2 \ldots).$$

To prove this, we shall need a preliminary remark. Suppose that $D \in \mathfrak{G}$ – then, since $f \in \mathsf{I}(G)$, Df is integrable on G. On the other hand, according to Lemma 9.2.4.3, there exists $R_D > 0$ and a non-negative integer r_D such that

$$|\,\phi(\nu : x; D)\,| \leqslant R_D\phi(\sqrt{-1}\eta : x)\,(1 + \|\nu\|)^{r_D}$$

for all $\nu \in \mathfrak{F}_c$ and $x \in G$. Consequently, if $\nu \in \mathfrak{T}_p^*$, then $D(f\phi_\nu)$ is also integrable. Let ω be the Casimir operator in the center $\mathfrak{Z}$ of $\mathfrak{G}$ – then, on the basis of what has just been said, the following manipulations are permissible:

$$\widehat{\omega^n f}(\nu) = \int_G f(x; \omega^n)\phi_\nu(x)d_G(x)$$
$$= \int_G f(x)\phi_\nu(x; \omega^n)d_G(x) = (-(\nu, \nu) - (\rho, \rho))^n\hat{f}(\nu)$$
$$(\nu \in \mathfrak{T}_p^*).$$

Consequently, for each polynomial p in l variables, we have

$$\sup_{\nu\in\mathfrak{T}_p^*} |\, p(\partial/\partial\zeta_1, \ldots, \partial/\partial\zeta_l)\{((\nu, \nu) + (\rho, \rho))^n\hat{f}(\nu)\}\,| < \infty$$
$$(n = 0, 1, 2, \ldots),$$

as desired.

The fact that $\hat{f}$ lies in $\mathfrak{H}(\mathfrak{T}_p^*)$ is now clear. □

Note The considerations supra serve to show that the map which assigns to each f in $\mathsf{I}(G)$ the restriction of its spherical Gelfand transform to $\mathfrak{T}_p^*$ is actually a *continuous* mapping of $\mathsf{I}(G)$ into $\mathfrak{H}(\mathfrak{T}_p^*)$. . . .

Here is the central result as regards $\mathsf{I}(G)$.

Theorem 9.2.4.5 (Trombi-Varadarajan) The space $\mathsf{I}(G)$ is a convolution algebra, the mapping $f \mapsto \hat{f} \mid \mathfrak{T}_\rho^*$ setting up a topological isomorphism between $\mathsf{I}(G)$ and $\mathfrak{H}(\mathfrak{T}_\rho^*)$.

The proof of this theorem for an arbitrary G, due to Trombi and Varadarajan [1], is difficult (see Helgason [14] for the case G complex or rank $(G/K) = 1$); the key step is the following lemma (cf. Lemma 9.2.3.3).

Lemma 9.2.4.6 Given $F \in \mathfrak{H}(\mathfrak{T}_\rho^*)$, let ϕ_F be the function on G defined by the rule

$$\phi_F(x) = [W]^{-1} \int_{\mathfrak{F}} F(\nu)\overline{\phi_\nu(x)}\,|\,\mathbf{c}(\nu)\,|^{-2} d_{\mathfrak{F}}(\nu) \qquad (x \in G).$$

Then ϕ_F lies in $\mathsf{I}(G)$ and the mapping $F \mapsto \phi_F$ of $\mathfrak{H}(\mathfrak{T}_\rho^*)$ into $\mathsf{I}(G)$ is continuous.

Granting the lemma (which takes 180 typed pages to establish . . .) and bearing in mind Proposition 9.2.4.4, the theorem then follows in the usual way.

Application As has been pointed out to the author by R. Gangolli, one may use Theorem 9.2.4.5 to prove 'Tauberian Theorems' for $I^1(G)$, i.e. to give conditions on $\hat{f}$ which serve to ensure that the closed ideal generated by f is all of $I^1(G)$. [Let g_t $(t > 0)$ be the Gauss kernel on G/K (cf. Gangolli [2]) – then, for a given f in $I^1(G)$, it is a question of finding f_t $(t > 0)$ in $\mathsf{I}(G)$ which verify the relation $f * f_t = g_t$. . . .]

Appendix Here we intend to establish the following result.

Lemma (Helgason) Let $\mathfrak{g}$ be a semi-simple Lie algebra over $\mathbf{R}$, $\mathfrak{g} = \mathfrak{k} + \mathfrak{a}_\mathfrak{p} + \mathfrak{n}^+$ an Iwasawa decomposition for $\mathfrak{g}$; let $\mathscr{C}$ be the positive Weyl chamber in $\mathfrak{a}_\mathfrak{p}$, $\mathscr{C}^\vee$ the dual cone of $\mathscr{C}$ – then the closure of the set

$$\bigcup_{w \in W} w\{\mathscr{C} \cap (-\mathscr{C}^\vee + Q_\rho)\} \qquad (\rho = 2^{-1} \sum_{\lambda > 0} m(\lambda)\lambda)$$

is the convex hull of the set of points $Q_{w\rho}$ $(w \in W)$.

Proof Let $\lambda_1, \ldots, \lambda_l$ denote the fundamental system of roots for Σ determined by $\mathscr{C}$; let $H_1, \ldots, H_l$ be a basis for $\mathfrak{a}_\mathfrak{p}$ which is dual to the basis $\lambda_1, \ldots, \lambda_l$ of $\mathfrak{F}$ – then the H_i $(1 \leqslant i \leqslant l)$ form the edges of the closed Weyl chamber $\mathscr{C}^{cl}$. Let $\mathscr{H}_i$ denote the open half-space containing the origin and bounded by the hyperplane σ_i through Q_ρ perpendicular to H_i $(1 \leqslant i \leqslant l)$; let

$$\mathfrak{H}_w = w\mathscr{H}_1 \cap \cdots \cap w\mathscr{H}_l \cap w\mathscr{C} \qquad (w \in W)$$

and put

$$\mathfrak{H} = \bigcap_{w \in W} (w\mathscr{H}_1 \cap \cdots \cap w\mathscr{H}_l).$$

Then $\mathfrak{H}_w = w\mathsf{H}$ $(\mathsf{H} = \mathfrak{H}_1)$ and since

$$-\mathscr{C}^\vee = \left\{\sum_{i=1}^{l} t_i Q_{\lambda_i} : t_1 < 0, \ldots, t_l < 0\right\}$$

while

$$\mathsf{H} = \{H \in \mathscr{C} : (Q_p - H, H_i) > 0, i = 1, \ldots, l\},$$

we have $\mathsf{H} = \mathscr{C} \cap (-\mathscr{C}^\vee + Q_p)$. We claim next that $\mathsf{H} = \mathfrak{H} \cap \mathscr{C}$; thus, the inclusion $\mathfrak{H} \cap \mathscr{C} \subset \mathsf{H}$ being clear, let us suppose that $H \in \mathsf{H}$ – then

$$(H_i, H) < (H_i, Q_p) \qquad (1 \leqslant i \leqslant l)$$

so, in view of Lemma 3.3.2.1, we have $(wH_i, H) < (wH_i, Q_{wp})$ $(1 \leqslant i \leqslant l)$, whence the claim. Because $\mathfrak{a}_\mathfrak{p} = \bigcup_{w \in W} w\mathscr{C}^{cl}$, it therefore follows that $\mathfrak{H}^{cl} = \bigcup_{w \in W} \mathfrak{H}_w^{cl}$. Consequently, due to the relation $\mathsf{H} = \mathscr{C} \cap (-\mathscr{C}^\vee + Q_p)$, the proof of our lemma will be complete when it is shown that the closure of $\mathfrak{H}$ is equal to the convex hull of the set $\{Q_{wp} : w \in W\}$. Since $\mathfrak{H}^{cl}$ clearly contains the convex hull in question, we need only establish the reverse inclusion; for this purpose it will be enough, on the basis of what has been said above, to show that the part of the boundary of $\mathfrak{H}$ which is inside $\mathscr{C}$ belongs to the convex hull of the points $Q_p, Q_{w_1 p}, \ldots, Q_{w_l p}$ (w_i the Weyl reflection associated with λ_i $(1 \leqslant i \leqslant l)$). But the hyperplane σ_i passes through the points

$$Q_p, Q_{w_1 p}, \ldots, Q_{w_{i-1} p}, Q_{w_{i+1} p}, \ldots, Q_{w_l p}$$

and the planes $\sigma_1, \ldots, \sigma_l$ meet at Q_p, the planes $\sigma_1, \ldots, \sigma_{l-1}$ meet in the line joining Q_p and $Q_{w_l p}$ etc. Hence the lemma. □

9.3 Spherical Functions and Differential Equations

9.3.1 The Weak Inequality and Some of its Implications

Let (G, K) be a reductive pair fulfilling the usual hypotheses (cf. Vol. I, number 1.1.5); assign to the symbols σ, $\vdash\!\!\circ\!\!\dashv$ etc. their usual meanings (cf. number 8.3.7).

Definition Let E be a Banach space.

(i) A function $f : G \to E$ is said to satisfy the *weak inequality* if there exist $r \geqslant 0$, $R > 0$ such that

$$\|f(x)\| \leqslant R \vdash\!\!\circ\!\!\dashv(x)(1 + \sigma(x))^r \qquad (x \in G).$$

(ii) A function $f : G \to E$ is said to satisfy the *strong inequality* if there exist $r > 0$, $R > 0$ such that

$$\|f(x)\| \leqslant R \vdash\!\!\circ\!\!\dashv(x)(1 + \sigma(x))^{-r} \qquad (x \in G).$$

Notation Let $r \in \mathbf{R}$ – then, for brevity, we shall often write $\vdash\!\!\circ\!\!\dashv_r$ for $\vdash\!\!\circ\!\!\dashv(1 + \sigma)^r$.

Example Let T be a central, $\mathfrak{Z}$-finite, tempered distribution on G – then, according to Theorem 8.3.8.6, each Fourier component of T satisfies the weak inequality. On the other hand, the Fourier components of the characters of the discrete series for G will satisfy the strong inequality (cf. Chapter 10).

Let E be a finite dimensional (complex) Hilbert space, $\mu = (\mu_1, \mu_2)$ a double unitary representation of K on E.

Theorem 9.3.1.1 (Harish-Chandra) Let Ψ be a (non-zero) C^∞, $\mathfrak{Z}$-finite, μ-spherical function from G to E such that

$$\|\Psi(x)\| \leqslant R \vdash\!\circ\!\dashv_r(x) \qquad (x \in G)$$

for some $r \geqslant 0$ – then there exists a unique integer $\imath$ ($\geqslant \dim(C_{\mathfrak{p}})$) such that

$$0 < \liminf_{t\to\infty} t^{-\imath/2} \|\Psi\|_t \leqslant \limsup_{t\to\infty} t^{-\imath/2} \|\Psi\|_t < \infty$$

where

$$\|\Psi\|_t = \left\{\int_{\sigma(x)\leqslant t} \|\Psi(x)\|^2 d_G(x)\right\}^{1/2} \qquad (t \geqslant 0).$$

We shall refer to (the obviously unique) integer $\imath$ as the *index* of Ψ.

Fix a maximal abelian subspace $\mathfrak{a}_{\mathfrak{p}}$ of $\mathfrak{p}$; let $\mathfrak{g} = \mathfrak{k} + \mathfrak{a}_{\mathfrak{p}} + \mathfrak{n}^+$ be an Iwasawa decomposition for $\mathfrak{g}$, $G = KA_{\mathfrak{p}}N^+$ the corresponding Iwasawa decomposition for G; assign to the symbols $\rho, \mathscr{C}, \Sigma^+$ etc. their usual meanings per the present context; let $\mathscr{S}^+$ denote the subset of $\mathscr{C}^{cl}$ comprised of those elements H such that $\|H\| = 1$.

As above, let $\mu = (\mu_1, \mu_2)$ be a double unitary representation of K on the finite dimensional Hilbert space E.

Theorem 9.3.1.2 (Harish-Chandra) Let Ψ be a (non-zero) C^∞, $\mathfrak{Z}$-finite, μ-spherical function from G to E satisfying the weak inequality; let $\imath$ be the index of Ψ – then the following three conditions on Ψ are mutually equivalent:

(i) The index of Ψ is equal to zero;

(ii) For all $h \in \exp(\mathscr{C}^{cl})$ and $H \in \mathscr{S}^+$,

$$\lim_{t\to\infty} e^{t\rho(H)}\Psi(h \exp(tH)) = 0;$$

(iii) Ψ lies in $\mathscr{C}(G; E)$.

[In other words, under the given conditions, $\|\Psi\|$ is square integrable iff Ψ is rapidly decreasing. . . .]

Note Needless to say, the Ψ which figures in Theorems 9.3.1.1 and 9.3.1.2 is analytic (cf. the Appendix to number 8.3.9).

The proof of these theorems goes by induction on the dimension of G; roughly speaking, we shall analyze certain differential equations and, in so doing, reduce the problem to a lower dimensional subgroup of G (cf. 9.3.2 and 9.3.3 infra). In what follows it can be assumed that $\dim(\mathfrak{p}) \geqslant 1$; this being so, our work then splits into two cases, $\mathfrak{c}_{\mathfrak{p}} \neq \{0\}$ ($\mathfrak{c}_{\mathfrak{p}} = \{0\}$) handled in 9.3.2 (9.3.3) below.

Here is a simple but important fact which will play an important role in the subsequent discussion.

Lemma 9.3.1.3 Keeping to the above notations, suppose that Ψ is a C^∞, $\mathfrak{Z}$-finite, μ-spherical function from G to E such that

$$\|\Psi(x)\| \leqslant R \vdash\!\circ\!\dashv_r(x) \qquad (x \in G)$$

for some $r \geqslant 0$ – then, given $D_1, D_2 \in \mathfrak{G}$, there exists a number $R(D_1, D_2) > 0$ such that

$$\|\Psi(D_1; x; D_2)\| \leqslant R(D_1, D_2) \vdash\!\circ\!\dashv_r(x)$$

for all x in G.

Proof In view of a lemma which may be found in the Appendix to number 8.3.7 there is no loss of generality in assuming that $D_1 = 1$, $D_2 = D$ (say); this being the case, choose a $\psi \in C_c^\infty(G)$ such that $\Psi * \psi = \Psi$ (cf. Corollary 8.3.9.2) – then $D\Psi = \Psi*(D\psi)$ and so our assertion follows immediately from Propositions 8.1.2.1 and 8.3.7.2. □

For the purposes of the ensuing discussion, we shall assume in addition that G is acceptable.

Let f be a measurable function on G – then we shall agree to say that f is *K-finite* if the left and right translates of f under K span a finite dimensional space.

Lemma 9.3.1.4 Let f be a function in $C^\infty(G) \cap L^2(G)$ which is K-finite as well as $\mathfrak{Z}$-finite – then f satisfies the weak inequality.

Proof Since convergence in $\mathscr{C}(G)$ implies convergence in $L^2(G)$ (cf. Proposition 8.3.7.5), the Schwarz inequality implies that f defines a tempered distribution on G. Hence our lemma follows immediately from Lemma 8.3.8.7. □

As an application of the preceding results on spherical functions, we shall now derive the following fundamental theorem.

Theorem 9.3.1.5 (Harish-Chandra) Let f be a (non-zero) function in $L^2(G)$ which is both K-finite and $\mathfrak{Z}$-finite – then $f \in \mathscr{C}(G)$.

Proof Given F in $L^2(K \times K)$, let $\mu_1(k)F$, $F\mu_2(k)$ $(k \in K)$ denote, respectively, the functions

$$(k_1, k_2) \mapsto F(k^{-1}k_1, k_2),\ (k_1, k_2) \mapsto F(k_1, k_2k^{-1}) \qquad (k_1, k_2 \in K).$$

Then $\mu = (\mu_1, \mu_2)$ obviously defines a double unitary representation of K on $L^2(K \times K)$. This being so, define now a function Ψ_f from G to $L^2(K \times K)$ as follows: For a given x in G, $\Psi_f(x)$ is to be that function in $L^2(K \times K)$ determined by the prescription

$$(k_1, k_2) \mapsto f(k_1^{-1}xk_2^{-1}) \qquad (k_1, k_2 \in K).$$

It is clear that

$$\Psi_f(k_1xk_2) = \mu_1(k_1)\Psi_f(x)\mu_2(k_2) \qquad (x \in G; k_1, k_2 \in K);$$

moreover the space E_f (say) spanned by the $\Psi_f(x)$ $(x \in G)$ is finite dimensional, f being K-finite. Because E_f is μ-stable and f is analytic, it therefore follows that Ψ_f is an E_f-valued C^∞ μ-spherical function on G which in fact satisfies the weak inequality (cf. Lemma 9.3.1.4); since f lies in $L^2(G)$, the index of Ψ_f must be zero, whence $\Psi_f \in \mathscr{C}(G; E_f)$ – but surely this means that f belongs to $\mathscr{C}(G)$. . . . □

Corollary 9.3.1.6 Retain the notations and assumptions of Theorem 9.3.1.5 – then, of necessity, rank (G) = rank (K).

[By supposition f is non-zero; our contention thus follows from Corollary 8.5.1.8.]

Definition An element γ of G is said to be *elliptic* if it is contained in some compact Cartan subgroup.

Theorem 9.3.1.7 (The Selberg Principle) Let γ be a semi-simple element of G, G_γ the centralizer of γ in G; let f be a (non-zero) K-finite, $\mathfrak{Z}$-finite function in $L^2(G)$ – then the integral

$$\int_{G/\mathsf{G}_\gamma} f({}^x\gamma)\, d_{G/\mathsf{G}_\gamma}(\dot{x})$$

exists and, if γ is not elliptic, its value is zero.

It will be convenient to preface the proof with a couple of lemmas. So fix a semi-simple element γ in G; let $\mathfrak{g}_\gamma$ be the centralizer of γ in $\mathfrak{g}$, G_γ the centralizer of γ in G; let $\mathfrak{j}$ be a Cartan subalgebra of $\mathfrak{g}_\gamma$ which is *fundamental* in $\mathfrak{g}_\gamma$; let J be the Cartan subgroup of G associated with $\mathfrak{j}$ – then $\gamma \in \mathsf{J}$. [Needless to say, $\mathfrak{j}$, when viewed as a Cartan subalgebra of $\mathfrak{g}$, need not be fundamental.] Upon replacing γ by ${}^x\gamma$ for some $x \in G$, we can assume in addition that $\mathfrak{j}$ is θ-stable; accordingly we shall now feel free to use the notations introduced in number 8.5.1 relative to this situation.

Lemma 9.3.1.8 (Langlands) Let γ be a semi-simple element in G per the preceding agreements – then there exists a non-zero number $\mathbf{M}_\gamma$ such that

$$\Phi_f(\gamma; \Pi_\gamma) = \mathbf{M}_\gamma \int_{G/\mathsf{G}_\gamma} f({}^x\gamma)\, d_{G/\mathsf{G}_\gamma}(\dot{x}) \qquad \Bigl(\Pi_\gamma = \prod_{\alpha \in \Phi_\gamma^+} H_\alpha\Bigr)$$

for all $f \in C_c^\infty(G)$.

[Here, of course, the invariant integral is calculated relative to J. One should observe that, in view of Theorem 8.5.1.4, the left hand side of the asserted relation is meaningful; on the other hand, the Theorem of Compacity (Theorem 8.1.4.1) serves to guarantee us that the integral on the right is finite.]

Proof of Lemma 9.3.1.8 According to what has been said in number 8.5.1, for any H in $\mathcal{O}'_\gamma$, we have

$$\begin{aligned}
&\Phi_f(\gamma \exp H)\\
&= \epsilon_\gamma \xi_\rho(\gamma)[\mathrm{J}_{\gamma,0} : G_\gamma \cap \mathrm{J}_0] \prod_{\alpha \in \Phi^+ - \Phi_{\gamma^+}} (e^{\alpha(H)/2} - \xi_\alpha(\gamma^{-1})e^{-\alpha(H)/2})\Phi_{v_f}(\exp H)
\end{aligned}$$

where

$$\begin{aligned}
&\Phi_{v_f}(\exp H)\\
&= \epsilon_{\gamma,R}(\exp H)\Delta_\gamma(\exp H) \int_{G_\gamma/\mathrm{J}_{\gamma,0}} v_f({}^{\dot{x}_\gamma}(\exp H))\, d_{G_\gamma/\mathrm{J}_{\gamma,0}}(\dot{x}_\gamma)\\
&\qquad\qquad\qquad\qquad\qquad\qquad\qquad\qquad (H \in \mathcal{O}'_\gamma).
\end{aligned}$$

Let D denote the differential operator on $\mathfrak{j}$ given by

$$\partial(\Pi_\gamma) \circ \prod_{\alpha \in \Phi^+ - \Phi_{\gamma^+}} (e^{\alpha/2} - \xi_\alpha(\gamma^{-1})e^{-\alpha/2}).$$

Then it is a simple matter to check that the local expression of D at the origin is

$$\prod_{\alpha \in \Phi^+ - \Phi_{\gamma^+}} (1 - \xi_\alpha(\gamma^{-1}))\cdot\partial(\Pi_\gamma).$$

Because $\mathfrak{j}$ is, by assumption, a fundamental Cartan subalgebra of $\mathfrak{g}_\gamma$, the contention of the lemma is now seen to be a consequence of Theorem 8.5.1.6 (and the proof thereof . . .). □

Lemma 9.3.1.9 Let γ be a semi-simple element in G per the preceding agreements; fix $r \geqslant 0$ such that $\Xi^2(1 + \sigma)^{-r} \in L^1(G)$ – then

$$\int_{G/G_\gamma} \Xi({}^{\dot{x}}\gamma)(1 + \sigma({}^{\dot{x}}\gamma))^{-r}\, d_{G/G_\gamma}(\dot{x}) < \infty.$$

Proof Keeping to the above notations and assumptions, let

$$\mu(f) = \left| \int_{G/G_\gamma} f({}^{\dot{x}}\gamma)\, d_{G/G_\gamma}(\dot{x}) \right| = |\mathbf{M}_\gamma^{-1}\Phi_f(\gamma; \Pi_\gamma)| \qquad (f \in C_c^\infty(G))$$

and

$$\nu(f) = \int_{\mathrm{J}} |\Delta_I(j)\Phi_f(j)|\, d_{\mathrm{J}}(j) \qquad (f \in I_c^\infty(G)).$$

Then it follows from Theorem 8.5.5.2 that there exist elements $Z_1, \ldots, Z_m$ in $\mathfrak{Z}$ such that $\mu(f) \leqslant \sum_i \nu(Z_i f)$ (all $f \in I_c^\infty(G)$); therefore, on the basis of Theorem 8.5.6.1, it is clear that the semi-norm μ meets the conditions (a) and (b) which were set down at the beginning of number 8.5.7. On the other hand, in view of the elementary properties of an integral, for any $f \in I^+(G)$, we have

$$\int_{G/G_\gamma} f({}^{\dot{x}}\gamma)\, d_{G/G_\gamma}(\dot{x}) = \sup_{f^+} \int_{G/G_\gamma} f^+({}^{\dot{x}}\gamma)\, d_{G/G_\gamma}(\dot{x}),$$

f^+ running over all elements in $I^+(G) \cap I_c^\infty(G)$ such that $f^+ \leqslant f$. This being so, the present lemma then follows at once from Lemma 8.5.7.1. □

Let γ be a semi-simple element in G per the preceding agreements – then the significance of the last lemma lies in the fact that it serves to ensure the existence of the integral

$$\int_{G/G_\gamma} f({}^*\gamma)\, d_{G/G_\gamma}(\dot{x})$$

for *all* f in $\mathscr{C}(G)$. This being the case, a simple continuity argument then leads to the conclusion that

$$(*)\quad \Phi_f(\gamma; \Pi_\gamma) = \mathbf{M}_\gamma \int_{G/G_\gamma} f({}^*\gamma)\, d_{G/G_\gamma}(\dot{x}) \qquad (\text{all } f \in \mathscr{C}(G)).$$

Consequently the assignment

$$f \mapsto \int_{G/G_\gamma} f({}^*\gamma)\, d_{G/G_\gamma}(\dot{x}) \qquad (f \in \mathscr{C}(G))$$

defines a central, tempered distribution on G. [Question: Is this true if γ is not semi-simple? . . .]

Proof of Theorem 9.3.1.7 (Harish-Chandra) By conjugating γ if necessary, we can assume that the agreements supra are in force. According to Theorem 9.3.1.5, f must necessarily lie in $\mathscr{C}(G)$, whence, in view of what has been said above, the integral

$$\int_{G/G_\gamma} f({}^*\gamma)\, d_{G/G_\gamma}(\dot{x})$$

exists. Suppose now that γ is not elliptic; assign to the symbols $\mathfrak{j}$, J etc. the preceding connotations – then $\gamma \in \mathsf{J}$ and, of necessity, J is not compact, hence, thanks to the Weak Selberg Principle (Theorem 8.5.1.7), the invariant integral Φ_f of f calculated relative to J vanishes. But then, in view of relation $(*)$, the integral of f over G/G_γ must vanish too. The proof of the Selberg Principle is therefore complete. □

Remark The Selberg Principle was verified by Selberg himself in a few special cases (cf. Selberg [1, p. 70]) and proved in complete generality by Harish-Chandra [30, p. 78]; the paper of Langlands [1] is also pertinent in this connection. The integrals

$$\int_{G/G_\gamma} f({}^*\gamma)\, d_{G/G_\gamma}(\dot{x}) \qquad (\gamma \text{ semi-simple})$$

arise in virtually any application of the Selberg Trace Formula. Thus suppose that Γ is a discrete subgroup of G such that G/Γ is compact – then the natural representation $L_{G/\Gamma}$ of G on $L^2(G/\Gamma)$ is discretely decomposable, a given irreducible unitary representation of G occurring but finitely many times. Let U be an irreducible *integrable* unitary representation of G (such exist only if rank (G) = rank (K) . . .) – then, using the Selberg Principle, Langlands [1, 4] obtained an explicit formula for the multiplicity of U in $L_{G/\Gamma}$. [For a different approach, see Hirzebruch [2,4] and Schmid [2].]

9.3.2 Existence and Uniqueness of the Indices I

Let (G, K) be a reductive pair fulfilling the usual hypotheses; in the present number it will be assumed that dim $(C_\mathfrak{p}) \geqslant 1$. Let E be a finite dimensional (complex) Hilbert space, $\mu = (\mu_1, \mu_2)$ a double unitary representation of K on E – then the primary objective of the present number is to establish, in the case at hand, the following result.

Theorem 9.3.2.1 (Harish-Chandra) Let Ψ be a (non-zero) C^∞, $\mathfrak{Z}$-finite, μ-spherical function from G to E such that

$$\|\Psi(x)\| \leqslant R \mathrel{\vdash\!\!\circ\!\!\dashv}_r(x) \qquad (x \in G)$$

for some $r \geqslant 0$ – then there exists a unique integer $\imath$ ($\geqslant \dim(C_{\mathfrak{p}})$) such that

$$0 < \liminf_{t \to \infty} t^{-\imath/2} \|\Psi\|_t \leqslant \limsup_{t \to \infty} t^{-\imath/2} \|\Psi\|_t < \infty$$

where

$$\|\Psi\|_t = \left\{\int_{\sigma(x) \leqslant t} \|\Psi(x)\|^2 \, d_G(x)\right\}^{1/2} \qquad (t \geqslant 0).$$

[As in 9.3.1, we shall call (the obviously unique) integer $\imath$ the *index* of Ψ.]

We shall now proceed to establish the existence of $\imath$.

Fix a point $C_0 \in \mathfrak{c}_{\mathfrak{p}}$ with $\|C_0\| = 1$ and let $\mathfrak{p}^*$ be the orthogonal complement of the line $\mathbf{R}C_0$ in $\mathfrak{p}$ so that $\mathfrak{g}^* = \mathfrak{k} + \mathfrak{p}^*$ is an ideal in $\mathfrak{g}$; let G^* be the analytic subgroup of G corresponding to $\mathfrak{g}^*$ – then the mapping $(t, x^*) \mapsto \exp(tC_0) \cdot x^*$ ($t \in \mathbf{R}, x^* \in G^*$) sets up an analytic diffeomorphism between $\mathbf{R} \times G^*$ and G. Put $\Psi(t : x^*) = \Psi(\exp(tC_0) \cdot x^*)$ ($t \in \mathbf{R}, x^* \in G^*$) – then, in view of the $\mathfrak{Z}$-finiteness of Ψ, there exist complex numbers c_i ($0 \leqslant i \leqslant m$) such that

$$\sum_{i=0}^{m} c_i C_0^{m-i} \Psi = 0 \qquad (c_0 = 1).$$

Because Ψ is analytic, on the basis of what has been said in the Appendix to number 8.3.1, it is therefore clear that there exist distinct complex numbers $\lambda_1, \ldots, \lambda_p$ such that

$$\begin{aligned} \Psi(t : x^*) &= \sum_{i=1}^{p} e^{\sqrt{-1}\lambda_i t} \Psi_i(t : x^*) \\ &= \sum_{i=1}^{p} e^{\sqrt{-1}\lambda_i t} \sum_{j=0}^{q} t^j \Psi_{ij}(x^*) \\ &= \sum_{i=1}^{p} \sum_{j=0}^{q} (e^{\sqrt{-1}\lambda_i t} t^j) \Psi_{ij}(x^*) \end{aligned}$$

where

$$\Psi_i(t : x^*) = \sum_{j=0}^{q} t^j \Psi_{ij}(x^*) \qquad (t \in \mathbf{R}, x^* \in G^*).$$

Since the functions $t \mapsto e^{\sqrt{-1}\lambda_i t} t^j$ are linearly independent (for distinct λ_i), the Ψ_{ij} are C^∞ functions on G^*. It can be assumed that $\Psi_i \neq 0$; due to the weak inequality supposition on Ψ, the λ_i are real (cf. A 3.2). Furthermore, as $K \subset G^*$ with $\dim(G^*) < \dim(G)$, those Ψ_{ij} which are non-zero fall within the domain of applicability of the ambient induction hypothesis. This being so, let ι_{ij} denote the index of Ψ_{ij} on G^* (set $\iota_{ij} = -\infty$ if $\Psi_{ij} = 0$) and put

$$\iota = 1 + \max_{i,j}(2j + \iota_{ij}).$$

Then we intend to prove that ι is the index of Ψ.

It is easy to verify that

$$\limsup_{T \to \infty} T^{-\iota/2} \| \Psi \|_T < \infty.$$

In fact, since the λ_i are real, for any $T \geqslant 0$, we have

$$\| \Psi \|_T \leqslant \sum_{i,j} \left\{ \int_0^T t^{2j} \, dt \int_{\sigma(x^*) \leqslant T} \| \Psi_{ij}(x^*) \|^2 \, d_{G^*}(x^*) \right\}^{1/2}$$
$$\leqslant \sum_{i,j} \{ (A_{ij}/(2j+1)) T^{2j+1}(1 + T)^{\iota_{ij}} \}^{1/2}$$
$$\leqslant A T^\iota \qquad (A > 0),$$

the A_{ij} being positive real numbers such that

$$\int_{\sigma(x^*) \leqslant T} \| \Psi_{ij}(x^*) \|^2 \, d_{G^*}(x^*) \leqslant A_{ij}(1 + T)^{\iota_{ij}} \qquad (T \geqslant 0).$$

Because $\iota \geqslant 1 + 2j + \iota_{ij}$, it is then clear that

$$\limsup_{T \to \infty} T^{-\iota/2} \| \Psi \|_T < \infty,$$

as desired.

On the other hand, in order to show that

$$\liminf_{T \to \infty} T^{-\iota/2} \| \Psi \|_T > 0,$$

it will clearly be enough to verify that

$$\liminf_{T \to \infty} T^{-\iota} \left\{ \int_{\substack{t^2 + \sigma(x^*)^2 \leqslant T \\ \sigma(x^*) \leqslant \delta t}} \| \Psi(t : x^*) \|^2 \, dt d_{G^*}(x^*) \right\} > 0$$

where $0 < \delta \leqslant 1$; in turn, for this purpose, it suffices to establish that

$$\liminf_{T \to \infty} T^{-\iota} I(T) > 0$$

where

$$I(T) = \int_{T/2}^{T/\sqrt{2}} dt \int_{G^*(\delta T/2)} \| \Psi(t : x^*) \|^2 \, d_{G^*}(x^*) \qquad (T \geqslant 0).$$

Here $G^*(T) = \{ x^* \in G^* : \sigma(x^*) \leqslant T \}$ (any $T \geqslant 0$). Let $(.\,,.)$ denote the inner product on our Hilbert space E. For $1 \leqslant i, j \leqslant p$, put

$$I_{ij}(T) = \int_{T/2}^{T/\sqrt{2}} dt \int_{G^*(\delta T/2)} e^{\sqrt{-1}(\lambda_i - \lambda_j)t}(\Psi_i(t : x^*), \Psi_j(t : x^*)) \, d_{G^*}(x^*)$$
$$(T \geqslant 0)$$

so that

$$I(T) = \sum_{i=1}^p I_{ii}(T) + 2\mathscr{R}\left(\sum_{1 \leqslant i < j \leqslant p} I_{ij}(T) \right) \qquad (T \geqslant 0).$$

Fix m, n $(0 \leqslant m, n \leqslant q)$ and set

$$I_{im,jn}(T) = \int_{T/2}^{T/\sqrt{2}} dt \int_{G^*(\delta T/2)} e^{\sqrt{-1}(\lambda_i - \lambda_j)t} t^{m+n} (\Psi_{im}(x^*), \Psi_{jn}(x^*))\, d_{G^*}(x^*) \qquad (T \geqslant 0)$$

for $1 \leqslant i, j \leqslant p$. An integration by parts together with induction on $m+n$ shows that if $i \neq j$, then there exists a constant $A(im, jn) > 0$ such that

$$\left| \int_{T/2}^{T/\sqrt{2}} t^{m+n} e^{\sqrt{-1}(\lambda_i - \lambda_j)t}\, dt \right| \leqslant A(im, jn)(1+T)^{m+n} \qquad (T \geqslant 0)$$

and so it is clear that

$$|I_{im,jn}(T)| \leqslant A(im, jn)(1+T)^{m+n} \|\Psi_{im}\|_{\delta T/2} \|\Psi_{jn}\|_{\delta T/2} \qquad (T \geqslant 0).$$

Therefore, if $\Psi_{im} \neq 0$, $\Psi_{in} \neq 0$, then there exists a positive number $\tilde{A}(im, jn)$, independent of T, such that

$$|I_{im,jn}(T)| \leqslant \tilde{A}(im, jn)(1+T)^{m+n+(\iota_{im}+\iota_{jn})/2} \leqslant \tilde{A}(im, jn)(1+T)^{\iota-1} \qquad (T \geqslant 0).$$

Accordingly, in view of the above formula for $I(T)$ $(T \geqslant 0)$, we shall now have to verify that

$$\liminf_{T\to\infty} T^{-\iota} \sum_{i=1}^{p} I_i(T) > 0 \qquad (I_i(T) = I_{ii}(T),\ T \geqslant 0).$$

To this end, *fix* an index i such that $\iota = 1 + 2j + \iota_{ij}$ for some j $(0 \leqslant j \leqslant q)$ – then we need only show that

$$\liminf_{T\to\infty} T^{-\iota} I_i(T) > 0$$

for this particular i. Let Q_i be the set of all j $(0 \leqslant j \leqslant q)$ such that $\iota = 1 + 2j + \iota_{ij}$. If $j \notin Q_i$, then $1 + 2j + \iota_{ij} < \iota$, whence

$$T^{-\iota} \int_{T/2}^{T/\sqrt{2}} t^{2j}\, dt \int_{G^*(\delta T/2)} \|\Psi_{ij}(x^*)\|^2\, d_{G^*}(x^*) \lesssim T^{1+2j+\iota_{ij}-\iota} \to 0$$

as $T \to \infty$. Thanks to the Schwarz inequality, this observation serves to eliminate from our considerations various terms which arise when we expand the integral defining $I_i(T)$ $(T \geqslant 0)$ in the obvious way. Put

$$\Psi_i^0(t : x^*) = \sum_{j \in Q_i} t^j \Psi_{ij}(x^*) \qquad (t \in \mathbf{R},\ x^* \in G^*)$$

and

$$I_i^0(T) = \int_{T/2}^{T/\sqrt{2}} dt \int_{G^*(\delta T/2)} \|\Psi_i^0(t : x^*)\|^2\, d_{G^*}(x^*) \qquad (T \geqslant 0).$$

Then we are reduced to proving that

$$\liminf_{T\to\infty} T^{-\iota} I_i^0(T) > 0.$$

Fix an ϵ $(0 < \epsilon \leqslant \delta/2)$ and set

$$I(\epsilon : T) = \int_{T/2}^{T/\sqrt{2}} dt \int_{G^*(\epsilon T)} \|\Psi_i^0(t : x^*)\|^2\, d_{G^*}(x^*) \qquad (T \geqslant 0).$$

Because $I(\epsilon : T) \leqslant I_i^0(T)$ $(T \geqslant 0)$, our discussion will be complete when it is shown that

$$\liminf_{T \to \infty} T^{-\imath} I(\epsilon : T) > 0$$

for some ϵ $(0 < \epsilon \leqslant \delta/2)$. This will now be done. Given $m, n \in Q_i$, write

$$I_{mn}(\epsilon : T) = \int_{T/2}^{T/\sqrt{2}} dt \int_{G^*(\epsilon T)} t^{m+n}(\Psi_{im}(x^*), \Psi_{in}(x^*))\, d_{G^*}(x^*) \qquad (T \geqslant 0)$$

so that

$$I(\epsilon : T) = \sum_{j \in Q_i} I_{jj}(\epsilon : T) + 2\mathscr{R}\Big(\sum_{\substack{m,n \in Q_i \\ m<n}} I_{mn}(\epsilon : T)\Big) \qquad (T \geqslant 0).$$

Since $\Psi_{ij} \neq 0$ for $j \in Q_i$, we can choose, by the induction hypothesis, positive numbers A and B such that

$$\|\Psi_{ij}\|_T \leqslant A(1+T)^{\imath_{ij}/2}, \; \|\Psi_{ij}\|_T \geqslant BT^{\imath_{ij}/2} \qquad (T \geqslant 0)$$

for all sufficiently large T; thus, for large T,

$$|I_{mn}(\epsilon : T)| \leqslant A^2 T^{m+n+1}(1+\epsilon T)^{(\imath_{im}+\imath_{in})/2} \qquad (m, n \in Q_i)$$

and so

$$I(\epsilon : T) \geqslant \sum_{j \in Q_i} M_j T^{2j+1}(\epsilon T)^{\imath_{ij}} - 2 \sum_{\substack{m,n \in Q_i \\ m<n}} A^2 T^{m+n+1}(1+\epsilon T)^{(\imath_{im}+\imath_{in})/2}$$

where

$$M_j = B^2(2^{-(j+(1/2))} - 2^{-(2j+1)})/(2j+1) > 0 \qquad (j \in Q_i).$$

Let $k = \max\{j : j \in Q_i\}$. Since $\imath_{ij} - \imath_{ik} = 2(k-j)$, it then follows that

$$I(\epsilon : T) \geqslant T^{\imath} \epsilon^{\imath_{ik}} \Big\{\sum_{j \in Q_i} M_j \epsilon^{2(k-j)} - 2 \sum_{\substack{m,n \in Q_i \\ m<n}} A^2 \epsilon^{2k-m-n}(1+(\epsilon T)^{-1})^{(\imath_{im}+\imath_{in})/2}\Big\}$$

provided T is large enough – therefore

$$\liminf_{T \to \infty} T^{-\imath} I(\epsilon : T) \geqslant \epsilon^{\imath_{ik}}\Big\{M_k + \sum_{j \in Q_i'} M_j \epsilon^{2(k-j)} - 2 \sum_{\substack{m,n \in Q_i \\ m<n}} A^2 \epsilon^{2k-m-n}\Big\}$$

where Q_i' is the set of all j in Q_i which are not equal to k. Because $M_k > 0$ while $2k - m - n \geqslant 1$ $(m < n)$, it now follows that for ϵ sufficiently small

$$\liminf_{T \to \infty} T^{-\imath} I(\epsilon : T) > 0,$$

as desired.

In summary, then, we have accomplished what we set out to do, namely we have shown that

$$0 < \liminf_{T \to \infty} T^{-\imath/2} \|\Psi\|_T \leqslant \limsup_{T \to \infty} T^{-\imath/2} \|\Psi\|_T < \infty.$$

Note too that, in view of the induction hypothesis, we have $\imath_{ij} \geqslant \dim(C_{\mathfrak{p}}) - 1$ whenever $\Psi_{ij} \neq 0$, whence $\imath = 1 + \max_{i,j}(2j + \imath_{ij}) \geqslant \dim(C_{\mathfrak{p}})$. . . .

The proof of Theorem 9.3.2.1 is therefore complete.

Fix a maximal abelian subspace $\mathfrak{a}_{\mathfrak{p}}$ of $\mathfrak{p}$; let $\mathfrak{g} = \mathfrak{k} + \mathfrak{a}_{\mathfrak{p}} + \mathfrak{n}^+$ be an Iwasawa decomposition for $\mathfrak{g}$, $G = KA_{\mathfrak{p}}N^+$ the corresponding Iwasawa decomposition for G; assign to the symbols $\rho, \mathscr{C}, \Sigma^+$ etc. their usual meanings in the present context; let $\mathscr{S}^+$ denote the subset of $\mathscr{C}^{cl}$ comprised of those points H such that $\|H\| = 1$.

As above, let $\mu = (\mu_1, \mu_2)$ be a double unitary representation of K on the finite dimensional Hilbert space E.

Theorem 9.3.2.2 (Harish-Chandra) Let Ψ be a (non-zero) C^∞, $\mathfrak{Z}$-finite, μ-spherical function from G to E satisfying the weak inequality; let $\imath$ be the index of Ψ – then the following three conditions on Ψ are mutually equivalent:

(i) The index of Ψ is equal to zero;

(ii) For all $h \in \exp(\mathscr{C}^{cl})$ and $H \in \mathscr{S}^+$,

$$\lim_{t \to \infty} e^{t\rho(H)} \Psi(h \exp(tH)) = 0;$$

(iii) Ψ lies in $\mathscr{C}(G; E)$.

Proof Since $\imath \geqslant \dim(C_{\mathfrak{p}}) \geqslant 1$, condition (i) cannot be fulfilled in the present case; because (iii) $\Rightarrow$ (ii) trivially, it remains only to verify that (ii) is never satisfied here; but if (ii) held, then, since $\rho | \mathfrak{c}_{\mathfrak{p}} = 0$, it would necessarily follow that $\Psi = 0$, contrary to assumption. Hence the theorem. □

9.3.3 Existence and Uniqueness of the Indices II

Let (G, K) be a reductive pair fulfilling the usual hypotheses, $G = KA_{\mathfrak{p}}N^+$ an Iwasawa decomposition for G; in the present number it will be assumed that $\dim(C_{\mathfrak{p}}) = 0$. Let E be a finite dimensional (complex) Hilbert space, $\mu = (\mu_1, \mu_2)$ a double unitary representation of K on E – then, as in 9.3.2, the primary objective of the present number is to establish, in the case at hand, the following result.

Theorem 9.3.3.1 (Harish-Chandra) Let Ψ be a (non-zero) C^∞, $\mathfrak{Z}$-finite, μ-spherical function from G to E such that

$$\|\Psi(x)\| \leqslant R \,\Xi\!\!-\!\!\sigma_r(x) \qquad (x \in G)$$

for some $r \geqslant 0$ – then there exists a unique non-negative integer $\imath$ such that

$$0 < \liminf_{t \to \infty} t^{-\imath/2} \|\Psi\|_t \leqslant \limsup_{t \to \infty} t^{-\imath/2} \|\Psi\|_t < \infty$$

where

$$\|\Psi\|_t = \left\{\int_{\sigma(x) \leqslant t} \|\Psi(x)\|^2 \, d_G(x)\right\}^{1/2} \qquad (t \geqslant 0).$$

[Just as in 9.3.1, we shall call (the obviously unique) integer $\imath$ the *index* of Ψ.]

We shall first state a lemma and then see how our theorem follows from it.

Let $\mathscr{S}^+$ denote the subset of $\mathscr{C}^{cl}$ consisting of those H such that $\|H\| = 1$ ($\mathscr{C}$ the positive Weyl chamber in $\mathfrak{a}_\mathfrak{p}$); given any open subset $\mathscr{O}$ of $\mathscr{S}^+$, put

$$F_{\mathscr{O}}(T) = \int_{\mathscr{O}_T} \|\Psi(\exp H)\|^2 \bigtriangleup(\exp H)\, d_{\mathfrak{a}_\mathfrak{p}}(H) \qquad (T \geqslant 0)$$

where $\mathscr{O}_T$ is the set of all elements in $\mathscr{C}^{cl}$ of the form tH ($0 \leqslant t \leqslant T$; $H \in \mathscr{O}$).

Lemma 9.3.3.2 Fix a point H_0 in $\mathscr{S}^+$ – then we can select an open neighborhood $\mathscr{O}$ $(= \mathscr{O}(H_0))$ of H_0 in $\mathscr{S}^+$ and a unique non-negative integer $\imath$ $(= \imath(H_0))$ with the following property: There exist numbers A and B $(0 < B \leqslant A < \infty)$ such that

$$BT^\imath \leqslant F_{\mathscr{O}}(T) \leqslant AT^\imath$$

for all $T \geqslant 1$.

Proof of Theorem 9.3.3.1 Since $\mathscr{S}^+$ is compact, we can find open sets $\mathscr{O}_i$ in $\mathscr{S}^+$ and integers $\imath_i \geqslant 0$ $(1 \leqslant i \leqslant n$, say) such that $\mathscr{S}^+ = \bigcup_i \mathscr{O}_i$ and with the assertion of Lemma 9.3.3.2 holding for $(\mathscr{O}_i, \imath_i)$ in place of $(\mathscr{O}, \imath)$; put $F_i = F_{\mathscr{O}_i}$, $\imath = \max_i \imath_i$ – then, in view of a standard integration formula (and a suitable normalization of the relevant invariant measures), it is clear that

$$\max_i F_i(T) \leqslant \|\Psi\|_T^2 \leqslant \sum_{i=1}^n F_i(T) \qquad (T \geqslant 0),$$

whence

$$0 < \liminf_{T \to \infty} T^{-\imath/2} \|\Psi\|_T \leqslant \limsup_{T \to \infty} T^{-\imath/2} \|\Psi\|_T < \infty,$$

as desired. □

Thus our main problem is to prove the 'local' lemma supra; in turn, for this purpose, we shall use Harish-Chandra's 'technique of descent via differential equations.'

Let P $(= P_\Theta)$ be a (proper) standard parabolic subgroup of G, $P = \mathsf{M}AN^+$ a Langlands decomposition for P 'compatible' with the given Iwasawa decomposition of G (thus the present N^+ will differ, in general, from the N^+ appearing in the relation $G = KA_\mathfrak{p}N^+ \ldots$); assign to the symbols L $(= \mathsf{M}A)$, L, M, M, $N^\pm$, $\mathfrak{l}$, $\mathfrak{m}$, $\mathfrak{n}^\pm$ etc. their customary meanings per the context at hand; set

$$\mathrm{d}_P(p) = |\det(\mathrm{Ad}(p)|\mathfrak{n}^+)|^{1/2} \qquad (p \in P).$$

Let $\mathfrak{Z}$ denote the center of $\mathfrak{G}$, $\mathfrak{Z}_\mathfrak{l}$ the center of $\mathfrak{L}$; let μ_P $(= \mu_{\mathfrak{g}/\mathfrak{l}})$ be the canonical isomorphism of $\mathfrak{Z}$ into $\mathfrak{Z}_{\mathfrak{l}}$ (cf. Example 3 following Proposition 2.3.3.6).

Let Ψ be the μ-spherical function on G appearing in the statement of Theorem 9.3.3.1; let $\mathfrak{J}$ be the annihilator of Ψ in $\mathfrak{Z}$ and put $\mathfrak{J}_{\mathfrak{l}} = \mathfrak{Z}_{\mathfrak{l}}\mu_P(\mathfrak{J})$ – then, since Ψ is $\mathfrak{Z}$-finite, it must be the case that both $\mathfrak{Z}/\mathfrak{J}$ and $\mathfrak{Z}_{\mathfrak{l}}/\mathfrak{J}_{\mathfrak{l}}$ are finite dimensional. Let us agree to write $E_{\mathfrak{l}}$ for the factor space $\mathfrak{Z}_{\mathfrak{l}}/\mathfrak{J}_{\mathfrak{l}}$ – then $E_{\mathfrak{l}}$ and its dual $E_{\mathfrak{l}}^{\vee}$ can be regarded as left $\mathfrak{Z}_{\mathfrak{l}}$-modules in the natural way. Fix elements $Z_1 = 1, Z_2, \ldots, Z_p$ ($p = \dim(E_{\mathfrak{l}})$) in $\mathfrak{Z}_{\mathfrak{l}}$ with the property that their images in $E_{\mathfrak{l}}$, under the canonical projection, form a basis for $E_{\mathfrak{l}}$; let $\mathbf{z}_1, \ldots, \mathbf{z}_p$ be the corresponding dual basis for $E_{\mathfrak{l}}^{\vee}$ – then, in what follows, we shall consider $E_{\mathfrak{l}}^{\vee}$ as a Hilbert space with $\mathbf{z}_1, \ldots, \mathbf{z}_p$ as an orthonormal basis. This being so, put $\mathbf{E} = E \otimes_{\mathbf{C}} E_{\mathfrak{l}}^{\vee}$ and let $\mathbf{\Gamma}$ denote the representation of $\mathfrak{Z}_{\mathfrak{l}}$ on $\mathbf{E}$ defined by the rule $\mathbf{\Gamma}(Z)(a \otimes \mathbf{z}) = a \otimes Z\mathbf{z}$ ($Z \in \mathfrak{Z}_{\mathfrak{l}}$). By making K act trivially on the dual of $E_{\mathfrak{l}}$, $\mathbf{E}$ can also be viewed as a (unitary) double K-module; with this agreement, it is then clear that the $\mathbf{\Gamma}(Z)$ ($Z \in \mathfrak{Z}_{\mathfrak{l}}$) commute with the operations of K on $\mathbf{E}$.

Let $D \mapsto D^{\flat}$ denote the automorphism of $\mathfrak{L}$ determined by the conditions $H^{\flat} = H + \rho_P(H)$, $X^{\flat} = X$ ($H \in \mathfrak{a}, X \in \mathfrak{m}$) – then $D^{\flat} = \mathrm{d}_P^{-1} D \circ \mathrm{d}_P$ ($D \in \mathfrak{L}$) as differential operators on L (cf. number 8.5.3). [Here we have set $\rho_P(H) = 2^{-1} \operatorname{tr}(\operatorname{ad}(H) | \mathfrak{n}^+)$ ($H \in \mathfrak{a}$); we remind the reader that $\mathrm{d}_P | \mathsf{M} = 1 \ldots$.]

We shall now begin the descent from G to L (note that $\dim(L) < \dim(G)$, the parabolic subgroup P being, by assumption, proper). The first step along this road is the definition of a certain $\boldsymbol{\mu}$-spherical function $\mathbf{\Psi}$ on L with values in the (finite dimensional) Hilbert space $\mathbf{E}$. As we shall see, it turns out that for any $Z \in \mathfrak{Z}_{\mathfrak{l}}$, one has $Z\mathbf{\Psi} = \mathbf{\Gamma}(Z)\mathbf{\Psi} + \mathbf{\Psi}_Z$ so that $\mathbf{\Psi}$ is not necessarily a $\mathbf{\Gamma}$-eigenfunction; however the 'error term' $\mathbf{\Psi}_Z$ ($Z \in \mathfrak{Z}_{\mathfrak{l}}$) can be handled without difficulty. The next step in the construction consists in replacing $\mathbf{\Psi}$ by a $\boldsymbol{\mu}$-spherical $\mathbf{\Gamma}$-eigenfunction $\mathbf{\Psi}_P$ (on L) and then determining the growth of the difference $\mathbf{\Psi} - \mathbf{\Psi}_P$ at infinity. Finally it will be shown that the 'first component' Ψ_P of $\mathbf{\Psi}_P$ can be used to approximate Ψ on a suitable subset of $\mathscr{C}^{cl}$; however, relative to the pair (L, L_K), Ψ_P satisfies the same general conditions as Ψ and so its asymptotic behavior is known through the induction hypothesis (cf. 9.3.2). . . .

Thus define a function $\mathbf{\Psi}$ on L by the prescription

$$\mathbf{\Psi}(l) = \sum_{i=1}^{p} \mathbf{\Psi}_i(l) \otimes \mathbf{z}_i \qquad (l \in L)$$

where

$$\mathbf{\Psi}_i(l) = \mathrm{d}_P(l)\Psi(l; Z_i^{\flat}) \qquad (l \in L; 1 \leqslant i \leqslant p).$$

Then, relative to the obvious (unitary) double representation $\boldsymbol{\mu} = (\boldsymbol{\mu}_1, \boldsymbol{\mu}_2)$ (say) of $L_K (= L \cap K)$ on $\mathbf{E}$, it is clear that $\mathbf{\Psi}$ is $\boldsymbol{\mu}$-spherical. Now, for any Z in $\mathfrak{Z}_{\mathfrak{l}}$, there exist unique complex numbers c_{ij} such that

$$I_i(Z) = ZZ_i - \sum_j c_{ij} Z_j \in \mathfrak{J}_{\mathfrak{l}} \qquad (1 \leqslant i, j \leqslant p).$$

this being so, put

$$\boldsymbol{\Psi}_Z(l) = \sum_{i=1}^{p} \mathrm{d}_P(l)\Psi(l; I_i(Z)^\flat) \otimes \mathbf{z}_i \qquad (l \in L).$$

Then it is clear that

$$\boldsymbol{\Psi}(l; Z) = \boldsymbol{\Gamma}(Z)\boldsymbol{\Psi}(l) + \boldsymbol{\Psi}_Z(l) \qquad (l \in L)$$

for all $Z \in \mathfrak{Z}_{\mathfrak{l}}$. Because $\mathfrak{a}$ (the Lie algebra of the split component A of P) lies in the center of $\mathfrak{L}$, it must therefore be the case that

$$d(e^{-t\boldsymbol{\Gamma}(H)}\boldsymbol{\Psi}(l \exp(tH)))/dt = e^{-t\boldsymbol{\Gamma}(H)}\boldsymbol{\Psi}_H(l \exp(tH)) \qquad (l \in L)$$

for all $H \in \mathfrak{a}$, whence

$$(\boldsymbol{\Psi}) \quad \boldsymbol{\Psi}(l \exp(TH)) = e^{T\boldsymbol{\Gamma}(H)}\boldsymbol{\Psi}(l) + \int_0^T e^{(T-t)\boldsymbol{\Gamma}(H)}\boldsymbol{\Psi}_H(l \exp(tH))\, dt \qquad (l \in L)$$

for all $H \in \mathfrak{a}$ $(T \in \mathbf{R})$.

Note Equation ($\boldsymbol{\Psi}$) supra will play an important role in the discussion which is to follow.

Our next objective will be to obtain some estimates for $\boldsymbol{\Psi}$ and $\boldsymbol{\Psi}_Z$ $(Z \in \mathfrak{Z}_{\mathfrak{l}})$.

Let $\vdash\!\!\circ\!\!\dashv_L$ denote the function on L corresponding to $\vdash\!\!\circ\!\!\dashv$ when the pair (G, K) is replaced by the pair (L, L_K) – then there exists an integer $d \geqslant 0$ such that

$$\mathrm{d}_P(\exp H) \vdash\!\!\circ\!\!\dashv (\exp H) \preceq \vdash\!\!\circ\!\!\dashv_L (\exp H)(1 + \|H\|)^d$$

for all $H \in \mathscr{C}^{cl}$. Set $L^+ = L_K \exp(\mathscr{C}^{cl})L_K$ – then, since L_K lies in the kernel of d_P, it must be the case that

$$\mathrm{d}_P(l) \vdash\!\!\circ\!\!\dashv (l) \preceq \vdash\!\!\circ\!\!\dashv_L (l)(1 + \sigma(l))^d \qquad (l \in L^+).$$

According to our assumption on Ψ, there exists an $r \geqslant 0$ such that $\|\Psi(x)\| \leqslant R \vdash\!\!\circ\!\!\dashv_r(x)$ $(x \in G)$; moreover, as was pointed out in 9.3.1, the various derivatives of Ψ are also bounded in norm by a multiple of $\vdash\!\!\circ\!\!\dashv_r$. Let $s = r + d$; put $\vdash\!\!\circ\!\!\dashv_{L,s}(l) = \vdash\!\!\circ\!\!\dashv_L(l)(1 + \sigma(l))^s$ $(l \in L)$ – then, on the basis of what has been said above, it is clear that for each $D \in \mathfrak{L}$ there exists a positive number S_D such that

$$\|\boldsymbol{\Psi}(l; D)\| \leqslant S_D \vdash\!\!\circ\!\!\dashv_{L,s}(l) \qquad (\text{all } l \in L^+).$$

Otherwise said, $\boldsymbol{\Psi}$ (and its derivatives) satisfy the weak inequality on L^+. On the other hand, as regards $\boldsymbol{\Psi}_Z$ $(Z \in \mathfrak{Z}_{\mathfrak{l}})$, one can say the following.

Lemma 9.3.3.3 Fix $D \in \mathfrak{L}$ and $Z \in \mathfrak{Z}_{\mathfrak{l}}$ – then there exists a positive number S $(= S_{D,Z})$ such that

$$\|\boldsymbol{\Psi}_Z(l \exp H; D)\| \leqslant S e^{-\triangleleft(H)} \vdash\!\!\circ\!\!\dashv_{L,s}(l)(1 + \|H\|)^s \qquad (l \in L^+)$$

for all $H \in \mathfrak{a} \cap \mathscr{C}^{cl}$.

[Let $\Sigma^+(\mathfrak{g}, \mathfrak{a})$ denote the positive roots of the pair $(\mathfrak{g}, \mathfrak{a})$ (cf. Vol. I, number 1.2.4) – then, for any H in $\mathfrak{a}$, we shall agree to write $\lhd(H)$ for $\inf_\sigma \sigma(H)$, $\sigma \in \Sigma^+(\mathfrak{g}, \mathfrak{a})$.]

Proof of Lemma 9.3.3.3 Fix $D \in \mathfrak{L}$ and $Z \in \mathfrak{Z}_{\mathfrak{l}}$ per the statement of our lemma – then

$$\mathbf{\Psi}_Z(l; D) = \sum_{i=1}^{p} \mathrm{d}_P(l)\Psi(l; D^\flat \cdot I_i(Z)^\flat) \otimes \mathbf{z}_i \qquad (l \in L).$$

Because each $I_i(Z)$ lies in $\mathfrak{J}_{\mathfrak{l}}$ which is itself contained in $\mathfrak{Z}_{\mathfrak{l}}$, we have

$$DI_i(Z) = I_i(Z)D \in \mu_P(\mathfrak{J})\mathfrak{L} \qquad (1 \leqslant i \leqslant p);$$

consequently there exist elements $D_i \in \theta(\mathfrak{n}^+)\mathfrak{G}$ ($\mathfrak{n}^+$ the Lie algebra of the unipotent radical of P) such that

$$\Psi(l; D^\flat \cdot I_i(Z)^\flat) = \Psi(l; D_i) \qquad (l \in L; 1 \leqslant i \leqslant p).$$

But, for any $D \in \mathfrak{G}$, $X \in \mathfrak{g}^\sigma$ $(\sigma \in \Sigma^+(\mathfrak{g}, \mathfrak{a}))$, we clearly have

$$\|\mathrm{d}_P(h)\Psi(h; \theta(X)D)\| \lesssim h^{-\sigma} \mapsto_{L,s}(h) \qquad (\text{all } h \in \exp(\mathscr{C}^{cl})).$$

Since $L^+ = L_K \exp(\mathscr{C}^{cl})L_K$, our lemma now follows without difficulty. □

Fix a point H_0 on $\mathscr{S}^+$ whose centralizer in $\mathfrak{g}$ is precisely $\mathfrak{l}$ – then H_0 necessarily belongs to $\mathfrak{a}$. Let $\mathscr{E}$ be the set of all eigenvalues of the endomorphism $\mathbf{\Gamma}(H_0)$ – then, by linear algebra, $\mathbf{E} = \sum_{\gamma \in \mathscr{E}} \mathbf{E}_\gamma$ (direct sum) where $\mathbf{E}_\gamma$ is the subspace of $\mathbf{E}$ consisting of all vectors $\mathbf{a}$ such that

$$(\mathbf{\Gamma}(H_0) - \gamma)^m \mathbf{a} = 0$$

for some integer $m \geqslant 0$. Let P_γ denote the projection of $\mathbf{E}$ onto $\mathbf{E}_\gamma$ $(\gamma \in \mathscr{E})$ corresponding to this direct decomposition. Divide $\mathscr{E}$ into three disjoint sets $\mathscr{E}^+$, $\mathscr{E}^-$, and $\mathscr{E}^0$, a given element γ in $\mathscr{E}$ lying in $\mathscr{E}^+$, $\mathscr{E}^-$, or $\mathscr{E}^0$ according to whether $\mathscr{R}(\gamma) > 0$, $\mathscr{R}(\gamma) < 0$, or $\mathscr{R}(\gamma) = 0$, respectively; put $\mathsf{P}^\pm = \sum_{\gamma \in \mathscr{E}^\pm} \mathsf{P}_\gamma$, $\mathsf{P}^0 = \sum_{\gamma \in \mathscr{E}^0} \mathsf{P}_\gamma$ and $\mathbf{E}^\pm = \mathsf{P}^\pm\mathbf{E}$, $\mathbf{E}^0 = \mathsf{P}^0\mathbf{E}$. Let

$$4\epsilon_0 = \inf_\gamma |\mathscr{R}(\gamma)|, \quad \gamma \in \mathscr{E}^+ \cup \mathscr{E}^-$$

(take $\epsilon_0 = 1$ if $\mathscr{E}^+ \cup \mathscr{E}^-$ is empty) – then ϵ_0 is positive. Fix a number ϵ $(0 < \epsilon < \epsilon_0)$ and an open, relatively compact, convex neighborhood Ω of H_0 in $\mathfrak{a} \cap \mathscr{C}^{cl}$; by selecting them sufficiently small, it can be assumed that $\lhd(H_0) \geqslant 5\epsilon$ while $\lhd(H) \geqslant 4\epsilon$ with $\|\mathbf{\Gamma}(H) - \mathbf{\Gamma}(H_0)\| < \epsilon/2$ (all $H \in \Omega$). Since, for any $\gamma \in \mathscr{E}$, we can write

$$\mathbf{\Gamma}(H)\mathsf{P}_\gamma = \{\gamma + (\mathbf{\Gamma}(H) - \mathbf{\Gamma}(H_0)) + (\mathbf{\Gamma}(H_0) - \gamma)\}\mathsf{P}_\gamma \qquad (H \in \Omega)$$

and since $(\mathbf{\Gamma}(H_0) - \gamma)\mathsf{P}_\gamma$ is nilpotent, it is clear that there exists a positive number M with the property that

$$(\mathsf{P}^\pm) \quad \|e^{-t\mathbf{\Gamma}(H)}\mathsf{P}^+\| + \|e^{t\mathbf{\Gamma}(H)}\mathsf{P}^-\| \leqslant Me^{-2\epsilon_0 t} \qquad (t \geqslant 0),$$

$$(\mathsf{P}^0) \quad \|e^{t\mathbf{\Gamma}(H)}\mathsf{P}^0\| \leqslant Me^{\epsilon|t|} \qquad (t \in \mathbf{R}),$$

for all H in Ω.

(**ER**) Given $D \in \mathfrak{L}$, there exists a positive number $R(D, \epsilon)$ such that

$$\|\mathbf{\Psi}_H(l \exp(TH); D)\| \leqslant R(D, \epsilon)e^{-3\epsilon T} \vdash\!\circ\!\dashv_{L,s}(l)$$

for all $l \in L^+$, $H \in \Omega$, and $T \geqslant 0$.

Verification of (**ER**) Because $\mathbf{\Psi}_H$ $(H \in \mathfrak{a})$ depends linearly on H and Ω is relatively compact, this estimate is seen to be a simple consequence of Lemma 9.3.3.3.

Now put, by definition, $\mathbf{\Psi}^{\pm} = \mathsf{P}^{\pm}\mathbf{\Psi}$, $\mathbf{\Psi}^0 = \mathsf{P}^0\mathbf{\Psi}$ – then, of course, $\mathbf{\Psi} = \mathbf{\Psi}^+ + \mathbf{\Psi}^- + \mathbf{\Psi}^0$ and, moreover, the following estimates obtain.

(**ES**)$^+$ Given $D \in \mathfrak{L}$, there exists a positive number $M^+(D, \epsilon)$ such that

$$\|\mathbf{\Psi}^+(l \exp(TH); D)\| \leqslant M^+(D, \epsilon)e^{-\epsilon T} \vdash\!\circ\!\dashv_{L,s}(l)$$

for all $l \in L^+$, $H \in \Omega$, and $T \geqslant 0$.

(**ES**)$^-$ Given $D \in \mathfrak{L}$, there exists a positive number $M^-(D, \epsilon)$ such that

$$\|\mathbf{\Psi}^-(l \exp(TH); D)\| \leqslant M^-(D, \epsilon)e^{-\epsilon T} \vdash\!\circ\!\dashv_{L,s}(l)$$

for all $l \in L^+$, $H \in \Omega$, and $T \geqslant 0$.

Verification of (**ES**)$^+$ Owing to equation ($\mathbf{\Psi}$) supra, we have

$$\mathbf{\Psi}^+(l; D) = e^{-T\Gamma(H)}\mathbf{\Psi}^+(l \exp(TH); D) - \int_0^T e^{-t\Gamma(H)}\mathsf{P}^+\mathbf{\Psi}_H(l \exp(tH); D)\, dt$$

for all $T \geqslant 0$ $(H \in \Omega)$. Now fix $l \in L^+$, $H \in \Omega$, and let T approach ∞ – then, since $D\mathbf{\Psi}$ satisfies the weak inequality on L^+, it follows from ($\mathsf{P}^{\pm}$) supra that

$$\mathbf{\Psi}^+(l; D) = -\int_0^\infty e^{-t\Gamma(H)}\mathsf{P}^+\mathbf{\Psi}_H(l \exp(tH); D)\, dt \qquad (l \in L^+)$$

and so

$$\mathbf{\Psi}^+(l \exp(TH); D) = -\int_T^\infty e^{-(t-T)\Gamma(H)}\mathsf{P}^+\mathbf{\Psi}_H(l \exp(tH); D)\, dt \qquad (l \in L^+).$$

Our contention thus follows from (**ER**) above (modulo (P^+) . . .).

Verification of (**ES**)$^-$ The discussion here is entirely analogous to that for (**ES**)$^+$; the details may be left to the reader.

It remains to consider $\mathbf{\Psi}^0$; this requires, however, some additional preparation.

Lemma 9.3.3.4 Let ω be a compact subset of L – then we can choose $T_0 \geqslant 0$ such that $l \exp(TH) \in L^+$ for $l \in \omega$, $H \in \Omega$, and $T \geqslant T_0$.

Proof We may obviously assume that $L_K \omega L_K = \omega$. This being so, let $\tilde{\omega}$ be the set of all h in $\omega \cap A_\mathfrak{p}$ such that $\lambda(\log h) \geqslant 0$ for every positive root λ of the pair $(\mathfrak{l}, \mathfrak{a}_\mathfrak{p})$ – then $\tilde{\omega}$ is also compact and, moreover, $\omega = L_K \tilde{\omega} L_K$. Now choose $T_0 \geqslant 0$ with the property that $\lambda(\log h) + 4\epsilon T_0 \geqslant 0$ for all $h \in \tilde{\omega}$ and for all $\lambda \in \Sigma^+$ which are not roots of the pair $(\mathfrak{l}, \mathfrak{a}_\mathfrak{p})$ – then certainly $\tilde{\omega} \exp(TH) \subset \mathscr{C}^{cl}$ $(H \in \Omega, T \geqslant T_0)$, whence $\omega \exp(TH) \subset L^+$ $(H \in \Omega, T \geqslant T_0)$, as desired. ☐

Now fix $D \in \mathfrak{L}$ and $H \in \Omega$; consider the integral

$$\int_0^\infty \| e^{-t\Gamma(H)} \mathrm{P}^0 \boldsymbol{\Psi}_H(l \exp(tH); D) \| \, dt \qquad (l \in L).$$

Then, on the basis of Lemma 9.3.3.4 and the earlier estimates, it is clear that this integral converges uniformly on compact subsets of L and so the function $\boldsymbol{\Theta}_H$ defined by the prescription

$$\boldsymbol{\Theta}_H(l) = \boldsymbol{\Psi}^0(l) + \int_0^\infty e^{-t\Gamma(H)} \mathrm{P}^0 \boldsymbol{\Psi}_H(l \exp(tH)) \, dt \qquad (l \in L)$$

is a C^∞ function on L with

$$\boldsymbol{\Theta}_H(l; D) = \boldsymbol{\Psi}^0(l; D) + \int_0^\infty e^{-t\Gamma(H)} \mathrm{P}^0 \boldsymbol{\Psi}_H(l \exp(tH); D) \, dt \qquad (l \in L).$$

Owing to equation $(\boldsymbol{\Psi})$ supra,

$$\boldsymbol{\Theta}_H(l; D) = \lim_{T \to \infty} e^{-T\Gamma(H)} \boldsymbol{\Psi}^0(l \exp(TH); D) \qquad (l \in L).$$

Hence, in particular, we have

$$\boldsymbol{\Theta}_H(l \exp(tH)) = e^{t\Gamma(H)} \boldsymbol{\Theta}_H(l) \qquad (l \in L, t \in \mathbf{R}).$$

The function $\boldsymbol{\Theta}_H$ is actually independent of H $(H \in \Omega)$; to see this, fix $H_1, H_2 \in \Omega$ – then it will be shown below that

$$\boldsymbol{\Theta}_{H_2}(l) = \lim_{T_1, T_2 \to \infty} e^{-\Gamma(T_1 H_1 + T_2 H_2)} \boldsymbol{\Psi}^0(l \exp(T_1 H_1 + T_2 H_2)) \qquad (l \in L).$$

Since the right hand side is symmetrical in H_1 and H_2, this will serve to establish our contention. So fix $l \in L$; choose $T_0 \geqslant 0$ such that $l \exp(tH) \in L^+$ for $t \geqslant T_0$ and $H \in \Omega$ (cf. Lemma 9.3.3.4); put

$$l_2 = l \exp(T_2 H_2) \qquad (T_2 \geqslant T_0).$$

Then, thanks to equation $(\boldsymbol{\Psi})$, we have

$$e^{-\Gamma(T_1 H_1)} \boldsymbol{\Psi}^0(l_2 \exp(T_1 H_1)) = \boldsymbol{\Psi}^0(l_2) + \int_0^{T_1} e^{-t\Gamma(H_1)} \mathrm{P}^0 \boldsymbol{\Psi}_{H_1}(l_2 \exp(tH_1)) \, dt$$

and therefore

$$e^{-\Gamma(T_1 H_1 + T_2 H_2)} \boldsymbol{\Psi}^0(l \exp(T_1 H_1 + T_2 H_2)) - e^{-\Gamma(T_2 H_2)} \boldsymbol{\Psi}^0(l \exp(T_2 H_2))$$
$$= \int_0^{T_1} e^{-\Gamma(t_1 H_1 + T_2 H_2)} \mathrm{P}^0 \boldsymbol{\Psi}_{H_1}(l \exp(t_1 H_1 + T_2 H_2)) \, dt_1 \qquad (T_1 \geqslant 0).$$

It follows from estimate (P^0) above and Lemma 9.3.3.3 that

$$\| e^{-\Gamma(t_1H_1+T_2H_2)}\mathsf{P}^0\boldsymbol{\Psi}_{H_1}(l\exp(t_1H_1+T_2H_2))\| \leq e^{-\epsilon(t_1+T_2-T_0)}$$

for $t_1 \geqslant 0$ and $T_2 \geqslant T_0$. Hence, upon letting T_1 and T_2 approach ∞, we get the limiting relation stated above.

Note In what follows, we shall write $\boldsymbol{\Psi}_P$ for $\boldsymbol{\Theta}_H$ $(H \in \Omega)$.

(ES)^0 Given $D \in \mathfrak{L}$, there exists a positive number $M^0(D, \epsilon)$ such that

$$\| \boldsymbol{\Psi}_P(l; D) - e^{-T\Gamma(H)}\boldsymbol{\Psi}^0(l\exp(TH); D)\| \leqslant M^0(D, \epsilon)e^{-2\epsilon T} \vdash\!\circ\!\dashv_{L,s}(l)$$

for all $l \in L^+$, $H \in \Omega$, and $T \geqslant 0$.

Verification of **(ES)^0** Since

$$\boldsymbol{\Psi}_P(l; D) - e^{-T\Gamma(H)}\boldsymbol{\Psi}^0(l\exp(TH); D)$$
$$= \int_T^\infty e^{-t\Gamma(H)}\mathsf{P}^0\boldsymbol{\Psi}_H(l\exp(tH); D)\,dt \qquad (l \in L^+),$$

the desired estimate follows immediately from **(ER)** supra (modulo (P^0) . . .).

Putting everything together then leads to:

(ES) Given $D \in \mathfrak{L}$, there exists a positive number $M(D, \epsilon)$ such that

$$\| \boldsymbol{\Psi}(l\exp(TH); D) - \boldsymbol{\Psi}_P(l\exp(TH); D)\| \leqslant M(D, \epsilon)e^{-\epsilon T} \vdash\!\circ\!\dashv_{L,s}(l)$$

for all $l \in L^+$, $H \in \Omega$, and $T \geqslant 0$.

Verification of **(ES)** Fix $l \in L^+$, $H \in \Omega$, and $T \geqslant 0$ – then we have

$$\| \boldsymbol{\Psi}(l\exp(TH); D) - \boldsymbol{\Psi}_P(l\exp(TH); D)\|$$
$$\leqslant \| \boldsymbol{\Psi}^+(l\exp(TH); D)\| + \| \boldsymbol{\Psi}^-(l\exp(TH); D)\|$$
$$+ \| e^{T\Gamma(H)}\mathsf{P}^0\| \cdot \| e^{-T\Gamma(H)}\boldsymbol{\Psi}^0(l\exp(TH); D) - \boldsymbol{\Psi}_P(l; D)\|$$

and so our assertion is seen to follow from the earlier estimates.

Here are the basic properties which $\boldsymbol{\Psi}_P$ possesses qua an **E**-valued function on L.

Lemma 9.3.3.5 Retain the above notations and assumptions – then:
(i) $\boldsymbol{\Psi}_P(l\exp H) = e^{\Gamma(H)}\boldsymbol{\Psi}_P(l)$ $(l \in L, H \in \mathfrak{a})$;
(ii) $\boldsymbol{\Psi}_P(k_1lk_2) = \boldsymbol{\mu}_1(k_1)\boldsymbol{\Psi}_P(l)\boldsymbol{\mu}_2(k_2)$ $(k_1, k_2 \in L_K; l \in L)$;
(iii) $\boldsymbol{\Psi}_P(l; Z) = \boldsymbol{\Gamma}(Z)\boldsymbol{\Psi}_P(l)$ $(l \in L, Z \in \mathfrak{Z}_{\mathfrak{l}})$.
Furthermore $\boldsymbol{\Psi}_P$ satisfies the weak inequality on L.

Proof Relation (i) is trivial. As for (ii), fix $H \in \Omega$ – then, since $\boldsymbol{\Gamma}(H)$ commutes with the operations of K on **E**, the second assertion follows from the fact that

$$\mathbf{\Psi}_P(l) = \lim_{T\to\infty} e^{-T\Gamma(H)}\mathbf{\Psi}^0(l\exp(TH)) \qquad (l \in L).$$

To prove (iii), fix once again an $H \in \Omega$ – then, as has been seen above,

$$\mathbf{\Psi}(l\exp(TH); Z) = \mathbf{\Gamma}(Z)\mathbf{\Psi}(l\exp(TH)) + \mathbf{\Psi}_Z(l\exp(TH)) \qquad (l \in L)$$

for all $Z \in \mathfrak{Z}_{\mathfrak{l}}$. Now P^0, $\mathbf{\Gamma}(H)$, and $\mathbf{\Gamma}(Z)$ ($Z \in \mathfrak{Z}_{\mathfrak{l}}$) all commute with one another; thus, for any $Z \in \mathfrak{Z}_{\mathfrak{l}}$, we have

$$\begin{aligned}\mathbf{\Psi}_P(l; Z) &= \lim_{T\to\infty} e^{-T\Gamma(H)}\mathbf{\Psi}^0(l\exp(TH); Z)\\ &= \mathbf{\Gamma}(Z)\mathbf{\Psi}_P(l) + \lim_{T\to\infty} e^{-T\Gamma(H)}\mathsf{P}^0\mathbf{\Psi}_Z(l\exp(TH)) \qquad (l \in L)\end{aligned}$$

and so (iii) is clear, the limit on the right being zero (here, of course, we have to keep in mind our basic estimates . . .). It remains to show that $\mathbf{\Psi}_P$ verifies the weak inequality on L; in any event, it is clear that this is so on L^+ (take $D = 1$ and $T = 0$ in $(\mathbf{ES})^0$ and recall that $\mathbf{\Psi}$ itself verifies the weak inequality on L^+). To establish the validity of a similar estimate for all $l \in L$, we proceed as follows. Choose a number $M > 0$ such that

$$|\lambda(\log h)| \leqslant M\sigma(h) \qquad (h \in A_{\mathfrak{p}})$$

for every $\lambda \in \Sigma^+$ which is not a root of the pair $(\mathfrak{l}, \mathfrak{a}_{\mathfrak{p}})$; put

$$N = \max(1, M/4\epsilon).$$

Then, as can be seen from the proof of Lemma 9.3.3.4, the entity $l\exp(TH)$ ($l \in L$, $H \in \Omega$) lies in L^+ provided $T \geqslant N\sigma(l)$. This being the case, fix an $l \in L$ and put $T_0 = N\sigma(l)$, $l_0 = l\exp(T_0H_0)$ – then $l_0 \in L^+$ and $\mathbf{\Psi}_P(l) = e^{-T_0\Gamma(H_0)}\mathbf{\Psi}_P(l_0)$ (cf. (i)), whence

$$\|\mathbf{\Psi}_P(l)\| \leqslant C\|e^{-T_0\Gamma(H_0)}\mathsf{P}^0\| \vdash\!\circ\!\dashv_{L,s}(l_0)$$

where C is a positive constant which does not depend on l. Because $\vdash\!\circ\!\dashv_L(l_0) = \vdash\!\circ\!\dashv_L(l)$, $\sigma(l_0) \leqslant \sigma(l) + T_0 = (N + 1)\sigma(l)$ and $\mathbf{\Gamma}(H_0)\mathsf{P}^0$ has only pure imaginary eigenvalues, the last statement of our lemma is now clear. $\square$

Write

$$\mathbf{\Psi}_P(l) = \sum_{i=1}^{p} \Psi_P^i(l) \otimes \mathbf{z}_i \qquad (l \in L)$$

and put $\Psi_P = \Psi_P^1$; suppose that $\Psi_P \neq 0$ – then, in view of Lemma 9.3.3.5, it is clear that Ψ_P satisfies the same general conditions as Ψ (relative, of course, to the pair (L, L_K)) and in addition, thanks to the ambient induction hypothesis, the index of Ψ_P (on L) is positive (cf. 9.3.2).

With this preparation, we can now give the proof of Lemma 9.3.3.2. So fix a point H_0 in $\mathscr{S}^+$; let P be the standard parabolic subgroup of G

canonically associated with H_0 (cf. the Example following Proposition 1.2.4.10) – then the preceding discussion is applicable in the obvious way. It will be best, however, to alter our notation a little and attach the subscripts 1 and 2 to some of the entities under consideration; thus the following conventions should be self-explanatory (cf. the Introduction to Chapter 8): $\Sigma^+ = \Sigma_1^+ \cup \Sigma_2^+$, $A_\mathfrak{p} = A_1 A_2$, $N^\pm = N_1^\pm N_2^\pm$ $(G = KA_\mathfrak{p}N^\pm \ldots)$, $L = MA_2$, $M = M_K A_1 N_1^+$, $P = \mathsf{M}A_2 N_2^+$.

Let $\mathscr{C}_1$ be the positive Weyl chamber in $\mathfrak{a}_\mathfrak{p}^1$; fix a number $N \geqslant 1$ such that if $h_1 \in \exp(\mathscr{C}_1^{cl})$ and $T \geqslant N\sigma(h_1)$, then $h_1 \exp(TH) \in \exp(\mathscr{C}^{cl})$ for all $H \in \Omega$ (cf. the proof of Lemma 9.3.3.5). Because $Z_1 = 1$, it follows without difficulty from the basic estimate (**ES**) supra that there exists a positive number M_ϵ with the property that

$$(*)\quad |d_P(h_1 \exp(TH))\Psi(h_1 \exp(TH)) - \Psi_P(h_1 \exp(TH))| \leqslant M_\epsilon e^{-\epsilon(T - N\sigma(h_1))} \,⊢\!\circ\!⊣_{L,s}(h_1)$$

for all $h_1 \in \exp(\mathscr{C}_1^{cl})$, $H \in \mathbf{\Omega}$, and $T \geqslant N\sigma(h_1)$.

Let $\mathfrak{h}$ be the orthogonal complement in $\mathfrak{a}_\mathfrak{p}$ of the line determined by H_0; let $\mathfrak{h}^+$ be the subset of $\mathfrak{h}$ comprised of those H such that $\lambda(H) \geqslant 0$ for every $\lambda \in \Sigma_1^+$; let P and P_i denote the orthogonal projections of $\mathfrak{a}_\mathfrak{p}$ onto $\mathfrak{h}$ and $\mathfrak{a}_i$ $(i = 1, 2)$, respectively. Fix a number $\delta > 0$ and let $\mathscr{O}$ $(= \mathscr{O}(H_0))$ be the set of all $H \in \mathscr{S}^+$ such that $\|\mathsf{P}H\| \leqslant \delta(H_0, H)$; we shall assume that δ is so small that $\delta N \leqslant 1/2$, $\mathsf{P}_2\mathscr{O} \subset \Omega$, and $\lambda(\mathscr{O}) \geqslant 3\epsilon$ (all $\lambda \in \Sigma_2^+$). It will be shown in a moment that $\mathscr{O}$ meets the requirements of Lemma 9.3.3.2.

Lemma 9.3.3.6 Retain the above notations and assumptions – then there exists a positive number $M_{\epsilon,\delta}$ with the property that

$$|e^{\rho(tH_0 + \mathsf{P}_2 H)}\Psi(\exp(tH_0 + H)) - \Psi_P(\exp(tH_0 + H))| \leqslant M_{\epsilon,\delta} e^{-\epsilon t/3} e^{-\rho(\mathsf{P}_1 H)}. \qquad (H \in \mathfrak{h}^+)$$

provided $\|H\| \leqslant \delta t$ $(t \geqslant 0)$.

[This is a straightforward consequence of estimate (*) supra.]

Proof of Lemma 9.3.3.2 Since the uniqueness of $\imath$ is obvious from its definition, it is the existence which requires proof. Let $\mathscr{O}_0(T)$ denote the subset of $\mathfrak{a}_\mathfrak{p}$ consisting of those elements of the form $tH_0 + H$ where $0 \leqslant t \leqslant T$, $H \in \mathfrak{h}^+$, and $\|H\| \leqslant \delta t$ – then it is clear that

$$\mathscr{O}_T \subset \mathscr{O}_0(T) \subset \mathscr{O}_{\sqrt{1+\delta^2}T} \qquad (T \geqslant 0).$$

Agreeing to supress 'exp' from the notation, set

$$F_0(T) = \int_{\mathscr{O}_0(T)} \|\Psi(H)\|^2 \triangle(H)\, d_{\mathfrak{a}_\mathfrak{p}}(H) \qquad (T \geqslant 0).$$

Then $F_0(T/\sqrt{1+\delta^2}) \leqslant F_\mathscr{O}(T) \leqslant F_0(T)$ $(T \geqslant 0)$ and so it will be enough to establish the existence of an integer $\imath \geqslant 0$ such that

$$0 < \liminf_{T \to \infty} T^{-\imath} F_0(T) \leqslant \limsup_{T \to \infty} T^{-\imath} F_0(T) < \infty.$$

For any $d \geqslant 0$, let $\mathfrak{h}^+(d)$ denote the set of all $H \in \mathfrak{h}^+$ with $\|H\| \leqslant d$ – then we have

$$F_0(T) = \int_0^T dt \int_{\mathfrak{h}^+(\delta t)} \| e^{\rho(tH_0 + \mathrm{P}_2 H)} \Psi(tH_0 + H) \|^2 \triangle_L(H) \times \prod_{\lambda \in \Sigma_2^+} (1 - e^{-2\lambda(tH_0 + H)})^{m(\lambda)} \, d_{\mathfrak{h}}(H) \qquad (T \geqslant 0)$$

where $\triangle_L$ is the function on $\mathfrak{a}_\mathfrak{p}$ corresponding to $\triangle$ when the pair $(\mathfrak{g}, \mathfrak{a}_\mathfrak{p})$ is replaced by the pair $(\mathfrak{l}, \mathfrak{a}_\mathfrak{p})$. Choose T_0 large enough to guarantee that $2\lambda(tH_0 + H) \geqslant \log 2$ for $\lambda \in \Sigma_2^+$, $H \in \mathfrak{h}^+(\delta t)$, and $t \geqslant T_0$; put

$$I(T) = \int_{T_0}^T dt \int_{\mathfrak{h}^+(\delta t)} \| e^{\rho(tH_0 + \mathrm{P}_2 H)} \Psi(tH_0 + H) \|^2 \triangle_L(H) \, d_{\mathfrak{h}}(H) \qquad (T \geqslant T_0),$$

$$I_P(T) = \int_{T_0}^T dt \int_{\mathfrak{h}^+(\delta t)} \| \Psi_P(tH_0 + H) \|^2 \triangle_L(H) \, d_{\mathfrak{h}}(H) \qquad (T \geqslant T_0).$$

Then

$$| I(T)^{1/2} - I_P(T)^{1/2} |^2 \leqslant \int_{T_0}^T dt \int_{\mathfrak{h}^+(\delta t)} \| e^{\rho(tH_0 + \mathrm{P}_2 H)} \Psi(tH_0 + H) - \Psi_P(tH_0 + H) \|^2 \triangle_L(H) \, d_{\mathfrak{h}}(H)$$

provided $T \geqslant T_0$. On the basis of Lemma 9.3.3.6, it is therefore clear that

$$| I(T)^{1/2} - I_P(T)^{1/2} | \leqslant M \qquad (T \geqslant T_0)$$

where M is a fixed positive number. Now we distinguish two cases. Thus first suppose that $\Psi_P = 0$ – then $I_P = 0$ and so it is clear that we can take $\imath = 0$. On the other hand, if $\Psi_P \neq 0$, then, as has been mentioned above, the index $\imath_P$ of Ψ_P (on L) is positive; moreover (cf. 9.3.2)

$$0 < \liminf_{T \to \infty} T^{-\imath_P} I_P(T) \leqslant \limsup_{T \to \infty} T^{-\imath_P} I_P(T) < \infty$$

with similar inequalities holding if I_P is replaced by I – thus we can take $\imath = \imath_P$ in this case. Hence the lemma. □

Note Keeping to the above notations, we thus see that $\imath = 0$ iff $\Psi_P = 0$.

Theorem 9.3.3.7 (Harish-Chandra) Let Ψ be a (non-zero) C^∞, $\mathfrak{Z}$-finite, μ-spherical function from G to E satisfying the weak inequality; let $\imath$ be the index of Ψ – then the following three conditions on Ψ are mutually equivalent:

(i) The index of Ψ is equal to zero;

(ii) For all $h \in \exp(\mathscr{C}^{cl})$ and $H \in \mathscr{S}^+$,

$$\lim_{t \to \infty} e^{t\rho(H)} \Psi(h \exp(tH)) = 0;$$

(iii) Ψ lies in $\mathscr{C}(G; E)$.

Proof (i) $\Rightarrow$ (iii). Suppose that $\imath = 0$ – then, in the notations which were employed during the proof of Theorem 9.3.3.1, each $\imath_i = 0$ too ($1 \leqslant i \leqslant p$). This being so, fix a point H_0 on $\mathscr{S}^+$; introduce $\mathscr{O}$ ($= \mathscr{O}(H_0)$) as above relative to H_0 and let $A_{\mathfrak{p}}(\mathscr{O})$ denote the set of all $h \in A_{\mathfrak{p}}$ of the form $h = \exp(tH)$ ($t \geqslant 0$, $H \in \mathscr{O}$). Since there is no loss of generality in assuming that $\mathscr{O}$ is a subset of one of the $\mathscr{O}_i$ ($1 \leqslant i \leqslant p$), it follows from what has been said above that $\Psi_P = 0$; therefore, thanks to Lemma 9.3.3.6, we have

$$|e^{\rho(tH)}\Psi(\exp(tH))| \leqslant M_{\epsilon,\delta}e^{-\tilde{\epsilon}t} \qquad (H \in \mathscr{O}, t \geqslant 0)$$

where $\tilde{\epsilon} = \epsilon/3\cdot(1 + \delta^2)^{-1/2}$; thus (cf. the proof of Proposition 8.3.7.3)

$$\sup_{h \in A_{\mathfrak{p}}(\mathscr{O})} \|\Psi(h)\| \, \Xi(h)^{-1}(1 + \sigma(h))^r < \infty$$

for every $r \geqslant 0$ and so, $\mathscr{S}^+$ being compact,

$$\sup_{x \in G} \|\Psi(x)\| \, \Xi(x)^{-1}(1 + \sigma(x))^r < \infty$$

for every $r \geqslant 0$. But then it follows without difficulty that $\Psi \in \mathscr{C}(G; E)$ (cf. the proof of Lemma 9.3.1.3).

(iii) $\Rightarrow$ (i), (ii) – these implications are clear. . . .

(ii) $\Rightarrow$ (i). Fix a point H_0 on $\mathscr{S}^+$ – then, as can be seen from Lemma 9.3.3.6,

$$\lim_{t \to \infty} \Psi_P(\exp(tH_0 + H)) = 0 \qquad (\text{all } H \in \mathfrak{h}^+).$$

Now fix a point H in $\mathfrak{h}^+$; set $f(t) = \Psi_P(\exp(tH_0 + H))$ ($t \in \mathbf{R}$) – then, since

$$\Psi_P(\exp(tH_0 + H)) = e^{t\Gamma(H_0)}\Psi_P(\exp H)$$

and since the eigenvalues of $\Gamma(H_0)\mathsf{P}^0$ are pure imaginary, it is clear that

$$f(t) = \sum_{i=1}^{r} p_i(t)e^{\sqrt{-1}\lambda_i t} \qquad (t \in \mathbf{R})$$

where $\lambda_1, \ldots, \lambda_r$ are distinct real numbers and the p_i are E-valued polynomial functions. Because $f(t) \to 0$ as $t \to \infty$, f must vanish identically (cf. A3.2). But $\mathfrak{h}^+ + \mathbf{R}H_0$ is open in $\mathfrak{a}_{\mathfrak{p}}$ while Ψ_P is analytic; thus $\Psi_P = 0$ and so the integer $\imath$ ($= \imath(H_0)$) appearing in the statement of Lemma 9.3.3.2 must be zero; this being true for every H_0 in $\mathscr{S}^+$, we conclude that the index of Ψ is zero, as desired.

The proof of the theorem is therefore complete. □

Chapter 10

The Discrete Series for a Semi-Simple Lie Group – Existence and Exhaustion

Introduction

Let G be an acceptable connected semi-simple Lie group with finite center, $\hat{G}$ its unitary dual. Let $\hat{G}_d$ be the discrete series for G. Given $\hat{U} \in \hat{G}_d$, let $T_{\hat{U}}$ denote its character, $d_{\hat{U}}$ its formal dimension – then, as is known, the distribution

$$T_d = \sum_{\hat{U} \in \hat{G}_d} d_{\hat{U}} T_{\hat{U}}$$

represents the contribution of the discrete series to the Plancherel formula for G. This being so, the primary objectives of the present chaper are as follows:

(1) Determine when the set $\hat{G}_d$ is non-empty;

(2) Describe explicitly the entities $d_{\hat{U}}$, $T_{\hat{U}}$ $(\hat{U} \in \hat{G}_d)$ and T_d.

Concerning the first problem, the answer is simple enough: Fix a maximal compact subgroup K of G – then G has a non-empty discrete series iff rank (G) = rank (K). Let us assume, therefore, that rank (G) = rank (K). Fix a maximal torus T in K; let $\hat{T}$ be the unitary character group of T – then $\hat{T}$ can be identified with a lattice L_T in the dual space of $\sqrt{-1}\mathfrak{t}$. Fix a regular element τ in L_T and let ξ_τ denote the corresponding character of T – then it can be shown a priori that there exists one and only one central eigendistribution Θ_τ on G with the following properties:

(i) $Z \cdot \Theta_\tau = \gamma_\tau(Z)\Theta_\tau$ (all $Z \in \mathfrak{Z}$);

(ii) $\sup_{x \in G'} |D_l(x)|^{1/2} |\Theta_\tau(x)| < \infty$;

(iii) $\Theta_\tau = \Delta_T^{-1} \sum_{w \in W(G,T)} \det(w)\xi_{w\tau}$ on T' $(= T \cap G')$.

Here γ_τ is the character of $\mathfrak{Z}$ which is attached to τ in a standard way (cf. 10.1.1). One can then say the following: To each $\tau \in \mathsf{L}_T'$ (= regular elements in L_T) there corresponds a unique element $\hat{U}_\tau \in \hat{G}_d$ whose character is given by

$$(-1)^{\mathrm{m}_G}\epsilon(\tau)\Theta_\tau \qquad (\epsilon(\tau) = \operatorname{sign}\{\prod_{\alpha>0} (\tau, \alpha)\});$$

the mapping $\tau \mapsto \hat{U}_\tau$ of L_T' into $\hat{G}_d$ is surjective with $\hat{U}_{\tau_1} = \hat{U}_{\tau_2}$ $(\tau_1, \tau_2 \in \mathsf{L}_T')$ iff τ_1 and τ_2 are conjugate under $W(G, T)$; the formal dimension $d_{\hat{U}_\tau}$ of the class $\hat{U}_\tau$ is given by

$$\frac{[W(G,T)]}{(2\pi)^r} \Big| \prod_{\alpha>0} (\tau, \alpha) \Big| \qquad (r = 2^{-1} \dim (G/T));$$

finally

$$T_d = \mathbf{M}_G^{-1} \sum_{\tau \in \mathsf{L}_T'} \prod_{\alpha>0} (\tau, \alpha)\Theta_\tau.$$

Granted the existence of the Θ_τ, the proof of the results just listed parallels to a certain extent Weyl's original method of determining the characters in the case of a compact G. Thus fix a class $\hat{U} \in \hat{G}_d$ – then the character $T_{\hat{U}}$ of $\hat{U}$ is tempered and, moreover, there exists a $\tau \in \mathsf{L}_T'$ such that the infinitesimal character of $\hat{U}$ is equal to γ_τ. This being the case, let $w_1 = 1, w_2, \ldots, w_n$ be a complete set of representatives for $W(G, T)\backslash W(\tau)$ in $W(\tau)$ ($W(\tau)$ being the subset of the Weyl group W of the pair $(\mathfrak{g}_c, \mathfrak{t}_c)$ consisting of those w such that $w\tau \in \mathsf{L}_T$) – then, using the differential equations, one can prove that $T_{\hat{U}} = \sum_{i=1}^n c_i \Theta_{w_i\tau}$ for certain complex numbers c_i. The Schur Orthogonality Relations can be used to establish that $\sum_{i=1}^n |c_i|^2 = 1$; on the other hand, by an independent argument, one verifies that the c_i are actually integers – therefore $T_{\hat{U}}$ is, up to a sign (which can be made precise), equal to $\Theta_{w_i\tau}$ for some i. . . .

Note Our discussion of the discrete series is continued in Volume III; in particular the following points are dealt with there: (1) The a priori construction of the Θ_τ; (2) The extension of the theory to disconnected groups (which is necessary for the theory of the 'continuous series'); (3) The Langlands conjecture (i.e. the infinite dimensional analogue of the Borel-Weil Theorem . . .).

10.1 The Role of the Distributions Θ_τ in the Harmonic Analysis on G

10.1.1 Existence and Uniqueness of the Θ_τ

Let (G, K) be a reductive pair satisfying the usual conditions; in all that follows, it will be assumed that G is, in addition, acceptable.

Suppose now that rank (G) = rank (K); fix once and for all a maximal torus T in K – then, of course, T is a compact Cartan subgroup of G. The unitary character group $\hat{T}$ of T may be identified with a lattice L_T (say) in the dual space of $\sqrt{-1}\mathfrak{t}$ ($\mathfrak{t}$ the Lie algebra of T): Thus a given τ in L_T gives rise to a unitary character ξ_τ of T by the prescription

$$\xi_\tau(\exp H) = e^{\tau(H)} \qquad (H \in \mathfrak{t})$$

and all such arise in this fashion. A unitary character ξ_τ ($\tau \in \mathsf{L}_T$) is said to be *regular* or *singular* according to whether τ is regular or singular, i.e. according to whether $\tau^w \neq \tau$ for every non-trivial element $w \in W$ or not (W the Weyl group of the pair $(\mathfrak{g}_c, \mathfrak{t}_c)$). The primary objective of the present number is to assign to every τ in L_T' (= the regular elements in L_T) a central, tempered, eigendistribution Θ_τ having certain properties. The significance of this result for Harmonic Analysis on G is simply this: The irreducible characters of G which correspond to the elements of the discrete series for G turn out to be constant multiples of these distributions.

Note For the most part, the results in the present number will be stated without proof; complete details will be provided, however, in Volume III.

We shall agree to retain the above notations and assumptions. Let $\mathfrak{Z}$ denote the center of $\mathfrak{G}$, $I(\mathfrak{t}_c)$ the Weyl group invariants in $S(\mathfrak{t}_c)$, $\gamma : \mathfrak{Z} \to I(\mathfrak{t}_c)$ the canonical isomorphism; for any linear function λ on $\mathfrak{t}_c$, let γ_λ denote the homomorphism $p \mapsto p(\lambda)$ ($p \in I(\mathfrak{t}_c)$) – then the assignment γ_λ

(say) given by the prescription $Z \mapsto \gamma(Z : \lambda)$ (= the value of $\gamma(Z)$ ($Z \in \mathfrak{Z}$) at λ) determines a character of $\mathfrak{Z}$ and, moreover, as we know, all such arise in this way for a suitable choice of λ. [In passing, let us also recall that linear functions λ_1 and λ_2 determine the same character of $\mathfrak{Z}$ (per the above agreement) iff there exists a w in W such that $w\lambda_1 = \lambda_2$.]

Theorem 10.1.1.1 (Harish-Chandra) Fix an element $\tau \in \mathsf{L}'_T$ – then there exists one and only one central eigendistribution Θ_τ on G with the following properties:

(i) $Z \cdot \Theta_\tau = \gamma_\tau(Z)\Theta_\tau$ (all $Z \in \mathfrak{Z}$);

(ii) $\sup_{x \in G'} |D_l(x)|^{1/2} |\Theta_\tau(x)| < \infty$;

(iii) $\Theta_\tau = \Delta_T^{-1} \sum_{w \in W(G,T)} \det(w)\xi_{w\tau}$ on T' ($= T \cap G'$).

[Here, of course, Δ_T is to be taken relative to the pair (G, T). In passing we remind the reader that $W(G, T)$ is the quotient by T of the normalizer of T in G (i.e. in K . . .); evidently $W(G, T)$ may be regarded as the subgroup of the Weyl group W of the pair $(\mathfrak{g}_c, \mathfrak{t}_c)$ generated by the compact roots. . . .]

Let us explicitly point out that it makes sense to speak of the value of Θ_τ ($\tau \in \mathsf{L}'_T$) at a point x in G'; cf. Theorem 8.3.3.1.

Corollary 10.1.1.2 The distribution Θ_τ is tempered (all $\tau \in \mathsf{L}'_T$).

[In view of property (ii) supra, this is an immediate consequence of Theorem 8.3.8.2.]

The Θ_τ ($\tau \in \mathsf{L}'_T$) were constructed a priori by Harish-Chandra [28], the method used being completely independent of (infinite dimensional) representation theory; in fact the point of departure for their construction is the following infinitesimal result.

Let $\mathfrak{J}$ denote the set of all linear functions τ on $\mathfrak{t}_c$ which take only pure imaginary values on $\mathfrak{t}$; let $\mathfrak{J}'$ denote the subset of $\mathfrak{J}$ comprised of those τ which are regular.

Lemma 10.1.1.3 Fix an element $\tau \in \mathfrak{J}'$ – then there exists a unique, central, tempered distribution θ_τ on $\mathfrak{g}$ such that:

(i) $\partial(p) \cdot \theta_\tau = \tilde{p}(\tau)\theta_\tau$ ($p \in I(\mathfrak{g}_c)$; $\tilde{p} = p | \mathfrak{t}_c$);

(ii) $\theta_\tau(H) = \pi(H)^{-1} \sum_{w \in W(G,T)} \det(w) e^{w\tau(H)}$ (all $H \in \mathfrak{t}'$).

Proof (Sketch) Admitting the uniqueness of the θ_τ ($\tau \in \mathfrak{J}$) (which really isn't terribly difficult to see), let us indicate how one may establish their existence; for this purpose there is no essential loss of generality in assuming the $\mathfrak{g}$ is semi-simple. Given any $f \in \mathscr{C}(\mathfrak{g})$, define its Fourier transform $\hat{f}$ by

$$\hat{f}(X) = \frac{1}{(2\pi)^{n/2}} \int_{\mathfrak{g}} f(Y) e^{\sqrt{-1}B(X,Y)} \, d_{\mathfrak{g}}(Y) \qquad (n = \dim(\mathfrak{g}))$$

where, as always, B denotes the Killing form of $\mathfrak{g}$. Fix an $H_0 \in \mathfrak{t}'$; put

$$T_{H_0}(f) = \phi_f(H_0) = \pi(H_0)\int_{G/T} \hat{f}(\dot{x}H_0)d_{G/T}(\dot{x}) \qquad (f \in \mathscr{C}(\mathfrak{g})).$$

The integral appearing here is absolutely convergent; furthermore, the mapping $f \mapsto T_{H_0}(f)$ defines a central, tempered distribution on $\mathfrak{g}$ which satisfies the system of differential equations

$$\partial(p)\cdot T_{H_0} = p(\sqrt{-1}H_0)T_{H_0} \qquad (p \in I(\mathfrak{g}_c)).$$

[All of this is by now familiar; cf. number 8.4.5.] Let $\mathbf{T}_{H_0}$ denote the space of tempered, central distributions T on $\mathfrak{g}$ such that

$$\partial(p)\cdot T = p(\sqrt{-1}H_0)T \qquad (\text{all } p \in I(\mathfrak{g}_c));$$

let

$$r = [W(\mathfrak{g}_c, \mathfrak{t}_c) : W(G, T)].$$

Then $\dim(\mathbf{T}_{H_0}) \leqslant r$. [For any $T \in \mathbf{T}_{H_0}$, introduce the analytic function $F_{T,\mathfrak{t}}$ on $\mathfrak{t}$ which is given on $\mathfrak{t}'$ by

$$F_{T,\mathfrak{t}}(H) = \pi(H)F_T(H) \qquad (H \in \mathfrak{t}')$$

(cf. Proposition 8.3.2.1) – then, in view of Lemma 8.4.5.2, we have

$$F_{T,\mathfrak{t}}(H) = \sum_w \det(w)c_w(T)\exp(\sqrt{-1}B(H, wH_0)) \qquad (H \in \mathfrak{t})$$

where the $c_w(T)$ ($w \in W(\mathfrak{g}_c, \mathfrak{t}_c)$) are uniquely determined complex numbers ($\sum_w$ denoting a sum over the elements w in $W(\mathfrak{g}_c, \mathfrak{t}_c)$). It is clear that

$$t\cdot F_{T,\mathfrak{t}} = \det(t)F_{T,\mathfrak{t}} \qquad (\text{all } t \in W(G, T))$$

and so

$$c_{tw}(T) = c_w(T)$$

for $t \in W(G, T)$, $w \in W(\mathfrak{g}_c, \mathfrak{t}_c)$. Since the linear map $T \mapsto F_{T,\mathfrak{t}}$ can be shown to be injective, it must therefore be the case that $\dim(\mathbf{T}_{H_0}) \leqslant r$.] On the other hand the T_{wH_0} belong to $\mathbf{T}_{H_0}$ for all $w \in W(\mathfrak{g}_c, \mathfrak{t}_c)$; this being so, choose elements $w_1, \ldots, w_r$ such that

$$W(\mathfrak{g}_c, \mathfrak{t}_c) = \bigcup_i W(G, T)w_i$$

and put $T_i = T_{w_iH_0}$ – then the T_i ($1 \leqslant i \leqslant r$) are linearly independent over $\mathbf{C}$, whence $\dim(\mathbf{T}_{H_0}) = r$. [Define tempered distributions S_i ($1 \leqslant i \leqslant r$) on $\mathfrak{g}$ by the prescription

$$S_i(f) = \phi_f(w_iH_0) \qquad (f \in \mathscr{C}(\mathfrak{g})).$$

Then the linear independence of the T_i will follow when it is shown that the S_i themselves are linearly independent. Because w_iH_0 is semi-simple, the orbit $G\cdot(w_iH_0)$ is closed in $\mathfrak{g}$ (cf. Proposition 1.3.5.5) – therefore

$$\operatorname{spt}(S_i) = G\cdot(w_iH_0) \qquad (i = 1, \ldots, r).$$

In view of the regularity of H_0 and the definition of the w_i, it is obvious that

$$G\cdot(w_iH_0) \cap G\cdot(w_jH_0) = \varnothing$$

if $i \neq j$ – thus the sets spt (S_i) are disjoint and non-empty, hence the distributions S_i are indeed linearly independent.] Let c_w $(w \in W(\mathfrak{g}_c, \mathfrak{t}_c))$ be given complex numbers such that $c_{tw} = c_w$ for all $t \in W(G, T)$ – then, since dim $(\mathbf{T}_{H_0}) = r$, it follows that there exists a unique element $T \in \mathbf{T}_{H_0}$ such that $c_w = c_w(T)$. Therefore, in particular, there exists a distribution T in $\mathbf{T}_{H_0}$ such that

$$F_{T,\mathfrak{t}}(H) = \sum_{w \in W(G,T)} \det(w) \exp(\sqrt{-1}B(H, wH_0)) \qquad (H \in \mathfrak{t})$$

which serves to ensure the existence of θ_τ $(\tau \in \mathfrak{F}')$. □

Roughly speaking, the construction of the Θ_τ $(\tau \in \mathsf{L}_T')$ is a purely local affair and goes as follows. One first defines Θ_τ in a neighborhood of the identity by making use of θ_τ and the exponential mapping; next, for an arbitrary semi-simple element $\gamma \neq 1$, the question of constructing Θ_τ locally around γ is reduced to a similar question in a neighborhood of the identity in the centralizer G_γ of γ; finally (and this is the most difficult step), one shows that these local data fit together and have the required properties.

We shall now record, via a series of lemmas, a number of important facts concerning the Θ_τ.

Lemma 10.1.1.4 Fix $\tau \in \mathsf{L}_T'$ – then

$$\Theta_\tau(x^{-1}) = \overline{\Theta_\tau(x)} = (-1)^r\Theta_{-\tau}(x) \qquad (x \in G')$$

where r is the number of positive roots of the pair $(\mathfrak{g}_c, \mathfrak{t}_c)$.

[This is a direct consequence of the definitions.]

Lemma 10.1.1.5 Fix $\tau \in \mathsf{L}_T'$ – then

$$\Theta_\tau(xz) = (\xi_\tau(z)/\xi_\rho(z))\Theta_\tau(x) \qquad (x \in G')$$

for all z in the center of G.

[Needless to say, the character ξ_ρ $(\rho = 2^{-1} \sum_{\alpha>0} \alpha)$ is calculated relative to the pair (G, T).]

Let σ be an automorphism of G which leaves T stable – then σ operates on $\hat{T}$ in the obvious way. Define a number $\epsilon(\sigma) = \pm 1$ by the rule

$$\sigma\cdot\Pi = \epsilon(\sigma)\Pi \qquad (\Pi = \prod_{\alpha>0} H_\alpha\ldots).$$

Lemma 10.1.1.6 Retain the immediately preceding assumptions and notations – then

$$\sigma(\Theta_\tau) = \epsilon(\sigma)\Theta_{\sigma\tau} \qquad (\text{all } \tau \in \mathsf{L}_T').$$

The following result represents an important strengthening of Theorem 10.1.1.1 (ii).

Lemma 10.1.1.7 There exists an absolute constant M with the property that

$$\sup_{x \in G'} |D_l(x)|^{1/2} |\Theta_\tau(x)| \leqslant M$$

for all τ in L_T'.

Remark It is an important, albeit difficult, problem to determine the behavior of the Θ_τ ($\tau \in \mathsf{L}_T'$) on the *non-compact* Cartan subgroups of G; as we shall see, this study can be carried out via Harish-Chandra's so-called Principle of Propagation (cf. Volume III). [In passing we point out that the results so obtained will be used to complete the discussion in number 8.5.6. . . .]

Let us agree to the convention that $\prod_{\alpha>0} (\tau, \alpha)\Theta_\tau = 0$ if τ is singular ($\tau \in \mathsf{L}_T$) (see, however, the Appendix to the present number).

Theorem 10.1.1.8 (Harish-Chandra) The assignment

$$f \mapsto \sum_{\tau \in \mathsf{L}_T} \prod_{\alpha > 0} (\tau, \alpha)\Theta_\tau(f) \qquad (f \in \mathscr{C}(G))$$

defines a central, tempered distribution on G, the series being absolutely convergent.

Note It will be shown in 10.2.4 that the distribution figuring in Theorem 10.1.1.8 is, apart from an explicitly computable constant factor, the contribution of the discrete series to the Plancherel formula for G (cf. with Vol. I, number 3.2.2).

The proof of our theorem depends on the following lemma.

Lemma 10.1.1.9 Let p be a polynomial function on $\mathfrak{J}$ ($\sim$ the real dual of $\sqrt{-1}\mathfrak{t}$) – then there exists an element Z_p in $\mathfrak{Z}$ with the property that

$$|p(\tau)\Theta_\tau(f)| \leqslant \sum_{i=1}^{r} [W(G, \mathsf{J}_0^i)]^{-1} \int_{\mathsf{J}^i} |\Phi^i_{Z_p f}(j)| \, d_{\mathsf{J}^i}(j) \qquad (f \in C_c^\infty(G))$$

for all $\tau \in \mathsf{L}_T'$.

[Here $\mathsf{J}^1 = T$, $\mathsf{J}^2, \ldots, \mathsf{J}^r$ is a maximal family of mutually non-conjugate θ-stable Cartan subgroups of G. . . .]

Proof of Lemma 10.1.1.9 Adopting the notations which were introduced in Example 1 following Proposition 2.3.3.6, put $\omega = \bar{\omega} - \sum_i C_i^2$ – then, according to what has been said there, we have

$$\gamma_\tau(\omega) = \|\tau\|^2 - R \qquad (\tau \in \mathsf{L}_T)$$

where R is a certain real number which does not depend on τ ($\|.\|$ a

Euclidean norm on $\mathfrak{J}$). This being so, set $\Omega = 1 + R + \omega$ – then

$$\gamma_\tau(\Omega) = 1 + \|\tau\|^2 \qquad (\tau \in \mathsf{L}_T)$$

and Ω is a self-adjoint differential operator lying in the center $\mathfrak{Z}$ of $\mathfrak{G}$. Now define the constant M per Lemma 10.1.1.7 – then, for all $\tau \in \mathsf{L}'_T$, we have

$$|\Theta_\tau(f)| \leqslant \sum_{i=1}^{r} [W(G, \mathsf{J}_0^i)]^{-1} \int_{\mathsf{J}^i} |\Delta^i(j)\Theta_\tau(j)\Phi_f^i(j)| \, d_{\mathsf{J}^i}(j)$$

$$\leqslant M \sum_{i=1}^{r} [W(G, \mathsf{J}_0^i)]^{-1} \int_{\mathsf{J}^i} |\Phi_f^i(j)| \, d_{\mathsf{J}^i}(j) \qquad (f \in C_c^\infty(G)).$$

In this inequality, replace f by $\Omega^n f$ ($n \in \mathbf{N}$) – then, thanks to the relations,

$$\Theta_\tau(\Omega^n f) = \gamma_\tau(\Omega^n)\Theta_\tau(f) = (1 + \|\tau\|^2)^n \Theta_\tau(f) \qquad (\tau \in \mathsf{L}'_T),$$

for all $\tau \in \mathsf{L}'_T$, we have

$$(1 + \|\tau\|^2)^n |\Theta_\tau(f)| \leqslant \sum_{i=1}^{r} [W(G, \mathsf{J}_0^i)]^{-1} \int_{\mathsf{J}^i} |\Phi_{Z_n f}^i(j)| \, d_{\mathsf{J}^i}(j)$$
$$(f \in C_c^\infty(G))$$

where $Z_n = M\Omega^n$. The assertion of the lemma is now obvious. □

Proof of Theorem 10.1.1.8 Since L_T is a lattice, there exists an integer $n \geqslant 0$ such that

$$N = \sum_{\tau \in \mathsf{L}_T} (1 + \|\tau\|)^{-n} < \infty.$$

Using the preceding lemma, choose an element Z in $\mathfrak{Z}$ with the property that

$$(1 + \|\tau\|)^n \Big| \prod_{\alpha > 0} (\tau, \alpha)\Theta_\tau(f) \Big| \leqslant \sum_{i=1}^{r} [W(G, \mathsf{J}_0^i)]^{-1} \int_{\mathsf{J}^i} |\Phi_{Zf}^i(j)| \, d_{\mathsf{J}^i}(j)$$
$$(f \in C_c^\infty(G))$$

for all $\tau \in \mathsf{L}'_T$ – then, for a given f in $C_c^\infty(G)$, our series is majorized by

$$N \sum_{i=1}^{r} [W(G, \mathsf{J}_0^i)]^{-1} \int_{\mathsf{J}^i} |\Phi_{Zf}^i(j)| \, d_{\mathsf{J}^i}(j)$$

and so the theorem is seen to be a consequence of Lemma 8.5.8.3. □

Appendix For certain purposes (other than the theory of the discrete series), it is necessary to have at hand an alternative description of the Θ_τ ($\tau \in \mathsf{L}'_T$); to this end, one needs to define the regularization of an integral (possibly divergent) over G. The process is familiar in $\mathbf{R}^n$; on the group G it goes as follows. Let f be a measurable function on G' which is integrable over every $G(\epsilon)$ where, for any $\epsilon > 0$, $G(\epsilon)$ denotes the set of those $x \in G$ such that $|D_l(x)| > \epsilon^2$ – then we shall agree to write

$$\text{P.V.} \int_G f(x) d_G(x) = \lim_{\epsilon \to 0} \int_{G(\epsilon)} f(x) d_G(x)$$

provided the limit on the right exists and is finite (P.V. = Principal Value).

Theorem (Harish-Chandra) Fix $\tau \in \mathsf{L}'_T$ – then there exists a central analytic function Ψ_τ on G' with the property that

$$\prod_{\alpha>0} (\tau, \alpha)\Theta_\tau(f) = \text{P.V.} \int_G D_I^{-1}(x)\Psi_\tau(x)f(x; \nabla_G)d_G(x) \qquad (\text{all } f \in C_c^\infty(G)).$$

[Here the differential operator ∇_G is to be taken per Lemma 8.3.4.6.]

Lemma Fix $\tau \in \mathsf{L}'_T$ – then Ψ_τ can be extended to a continuous function on all of G. Moreover one has

$$\Psi_\tau(xz) = \xi_\tau(z)\Psi_\tau(x) \qquad (x \in G)$$

for every z in the center of G.

Example To illustrate, let us place ourselves in the setting of Vol. I, number 3.2.2; thus G will now stand for a compact, simply connected, semi-simple Lie group (so that T is then a maximal torus in G). Let τ be a regular element in $\hat{T}$ – then, as we know, there exists a unique pair $(\Lambda, w) \in \mathfrak{W}^+ \times W$ such that $\tau = w(\Lambda + \rho)$. Fix an f in $C^\infty(G)$ – then we have

$$\begin{aligned}
\prod_{\alpha>0} (\tau, \alpha)\Theta_\tau(f) &= \prod_{\alpha>0} (\Lambda + \rho, \alpha) \int_G \operatorname{tr}(U_\Lambda(x))d_G(x) \\
&= ((-1)^r/[W]) \sum_{w\in W} \int_T \prod_{\alpha>0} (w(\Lambda + \rho), \alpha)\xi_{w(\Lambda+\rho)}(t)\Phi_f(t)d_T(t) \\
&= ((-1)^r/[W]) \sum_{w\in W} \int_T \xi_{w(\Lambda+\rho)}(t; \Pi)\Phi_f(t)d_T(t) \\
&= (1/[W]) \int_T \{\sum_{w\in W} \xi_{w(\Lambda+\rho)}(t)\}\Phi_f(t; \Pi)d_T(t) \\
&= \text{P.V.} \int_G D_I^{-1}(x)\Psi_\tau(x)f(x; \nabla_G)d_G(x)
\end{aligned}$$

where Ψ_τ is the central extension of

$$\sum_{w\in W} \xi_{w(\Lambda+\rho)} = \sum_{w\in W} \xi_{w\tau}$$

to G.

In conclusion it should be pointed out that it is actually possible to define Θ_τ and Ψ_τ in the case of singular τ as well. Thus fix a Weyl chamber $\mathscr{C}$ in the dual of $\sqrt{-1}\mathfrak{t}$ – then, during the course of the investigation, it turns out that one may define functions $\Theta_{\tau,\mathscr{C}}$ and $\Psi_{\tau,\mathscr{C}}$ for any τ in $\mathscr{C}^{cl} \cap \mathsf{L}_T$. Moreover the results described earlier remain valid as they stand in this somewhat more general setting. Now for a singular τ, the entities $\Theta_{\tau,\mathscr{C}}$, $\Psi_{\tau,\mathscr{C}}$ will in general vary with $\mathscr{C}$; to remedy this, one proceeds as follows. Given $\tau \in \mathsf{L}_T$, let $\mathfrak{C}(\tau)$ denote the set comprised of those Weyl chambers $\mathscr{C}$ to which τ belongs – then the functions Θ_τ and Ψ_τ ($\tau \in \mathsf{L}_T$) are to be defined by the prescription

$$\Theta_\tau = [\mathfrak{C}(\tau)]^{-1} \sum_{\mathscr{C}\in\mathfrak{C}(\tau)} \Theta_{\tau,\mathscr{C}}, \quad \Psi_\tau = [\mathfrak{C}(\tau)]^{-1} \sum_{\mathscr{C}\in\mathfrak{C}(\tau)} \Psi_{\tau,\mathscr{C}}.$$

Remark Suppose that $G = \mathbf{SL}(2, \mathbf{R})$ – then

$$\mathsf{L}_T \sim \{0, \pm 1, \pm 2, \ldots\} \qquad (T(= K) = \mathbf{SO}(2))$$

and so the only singular element in L_T is $\tau = 0$. Thus, in this case, we get two 'singular' distributions Θ_0^+ and Θ_0^-; it is well-known that Θ_0^+ and Θ_0^- are intimately related with the two irreducible characters which arise upon breaking the exceptional principal P-series representation $U^{-,0}$ (cf. Vol. I, number 5.5.1) into its irreducible constituents. . . . [For a general discussion of this phenomenon, we refer the reader to the paper of Knapp and Okamoto [1].]

10.1.2 Expansion of $\mathfrak{Z}$-Finite Functions in $\mathscr{C}(G)$

Let (G, K) be a reductive pair satisfying the usual conditions (with G acceptable); throughout the present number it will be assumed in addition that rank (G) = rank (K). Fix once and for all a maximal torus T in K – then the notations and conventions introduced in 10.1.1 are now in force.

Given $\delta \in \hat{K}$, let $\Theta_{\tau,\delta}$ denote the δth Fourier component of Θ_τ $(\tau \in \mathsf{L}'_T)$ – then $\Theta_{\tau,\delta}$ is an analytic function on G which satisfies the weak inequality. [Let us recall that Θ_τ is tempered (cf. Corollary 10.1.1.2); the fact that its Fourier components satisfy the weak inequality then follows from Theorem 8.3.8.6.]

Theorem 10.1.2.1 (Harish-Chandra) Fix an element τ in L'_T – then the Fourier components of Θ_τ lie in $\mathscr{C}(G)$ (and hence satisfy the strong inequality).

The present theorem is an immediate consequence of the following lemma.

Lemma 10.1.2.2 Let $\mu = (\mu_1, \mu_2)$ be a double unitary representation of K on a finite dimensional Hilbert space E; suppose that Ψ is a μ-spherical function from G to E satisfying the weak inequality and such that

$$Z\Psi = \gamma_\tau(Z)\Psi \qquad (Z \in \mathfrak{Z})$$

for some regular element τ in $\mathfrak{J}$ – then Ψ belongs to $\mathscr{C}(G; E)$.

[This lemma follows without essential difficulty from Theorem 9.3.1.2; it will be best, however, to defer its proof until Volume III (there we shall give a proof based on the 'philosophy of cusp forms').]

Proposition 10.1.2.3 Fix $\delta \in \hat{K}$ – then there are but a finite number of $\tau \in \mathsf{L}'_T$ such that $\Theta_{\tau,\delta} \neq 0$.

Proof Let ω be the Casimir operator in $\mathfrak{Z}$; write, in the usual way, $\omega = -\omega_- + \omega_+$ (cf. Example 2 following Proposition 2.3.3.6). Let $\gamma : \mathfrak{Z} \to I(\mathfrak{t}_c)$ be the canonical isomorphism of $\mathfrak{Z}$ onto the Weyl group invariants $I(\mathfrak{t}_c)$ in $S(\mathfrak{t}_c)$ – then the assignment

$$\tau \mapsto \|\tau\|^2 = \gamma(\omega : \tau) + 2^{-1} \sum_{\alpha>0} \rho(H_\alpha) \qquad (\tau \in \mathfrak{J})$$

determines a positive definite quadratic form on $\mathfrak{J}$. Suppose now that τ is an element in L'_T such that $\Theta_{\tau,\delta} \neq 0$ – then we claim that necessarily

$$\|\tau\|^2 \leqslant r_\delta + \|\rho\|^2 \qquad (r_\delta \geqslant 0)$$

from which we can read off the assertion of the proposition (L_T being a lattice in $\mathfrak{I}$). [Here r_δ is the non-negative real number defined by the relation $\omega_-\chi_\delta = (-r_\delta)\chi_\delta$, χ_δ as always. . . .] Put $f = \Theta_{\tau,\delta}$ – then, according to the preceding theorem, f lies in $\mathscr{C}(G)$, hence is square integrable on G. Let R denote the right regular representation of G on $L^2(G)$, E the smallest closed R-stable subspace of $L^2(G)$ containing f, and U the restriction of R to E; since convergence in $\mathscr{C}(G)$ implies convergence in $L^2(G)$, f is differentiable under U with

$$U_\infty(\omega)f = \gamma_\tau(\omega)f, \qquad U_\infty(\omega_-)f = (-r_\delta)f$$

(bear in mind that $f = \chi_\delta * f = f * \chi_\delta \ldots$) – thus

$$\{\|\tau\|^2 - \|\rho\|^2\}\|f\|^2 = (U_\infty(\omega)f, f) = r_\delta\|f\|^2 + (U_\infty(\omega_+)f, f)$$
$$\leqslant r_\delta\|f\|^2$$

and so the claim is seen to be a consequence of the fact that $\|f\| > 0$. □

Remark It is possible to prove a priori that the Fourier component of Θ_τ corresponding to the class of the trivial one dimensional representation of K is necessarily zero (all $\tau \in \mathsf{L}_T'$) (cf. Harish-Chandra [29]); on the other hand, since the Θ_τ ($\tau \in \mathsf{L}_T'$) are, to within a constant factor, the characters of the discrete series and since the Plancherel measure for the space $I^2(G)$ has no 'atoms', it is easy to read off this result directly. . . .

Theorem 10.1.2.4 (Harish-Chandra) Let f be a $\mathfrak{Z}$-finite function in $\mathscr{C}(G)$ – then

$$f(1) = \mathbf{M}_G^{-1} \sum_{\tau \in \mathsf{L}_T} \prod_{\alpha > 0} (\tau, \alpha)\Theta_\tau(f).$$

[Here the constant $\mathbf{M}_G$ has the same value as in Theorem 8.5.1.6.]

Proof Let r denote the number of positive roots of the pair $(\mathfrak{g}_c, \mathfrak{t}_c)$. Fix a τ in L_T' – then we claim that

$$\Theta_\tau(f) = (-1)^r \hat{\Phi}_f(\tau),$$

Φ_f the invariant integral of f calculated relative to the compact Cartan subgroup T. [Here $\hat{\Phi}_f$ is the Fourier transform of Φ_f; let us bear in mind that Φ_f is, in the case at hand, a C^∞ function on T (cf. Theorem 8.5.1.9).] Thus, thanks to the Weak Selberg Principle (and a familiar integration formula . . .), we have

$$\begin{aligned}
\Theta_\tau(f) &= \int_G f(x)\Theta_\tau(x)d_G(x) \\
&= (-1)^r[W(G, T)]^{-1} \int_T \Delta_T(t)\Theta_\tau(t)\Phi_f(t)d_T(t) \\
&= (-1)^r[W(G, T)]^{-1} \sum_{w \in W(G,T)} \det(w) \int_T \xi_{w\tau}(t)\Phi_f(t)d_T(t) \\
&= (-1)^r \hat{\Phi}_f(\tau),
\end{aligned}$$

as claimed. Owing to Theorem 8.5.1.6, we have

$$f(1) = \mathbf{M}_G^{-1}\Phi_f(1; \Pi).$$

But, according to the theory of Fourier series (and the claim supra),

$$\Phi_f(1;\Pi) = (-1)^r \int_{\hat{T}} \prod_{\alpha>0} (\tau, \alpha)\hat{\Phi}_f(\tau)d_{\hat{T}}(\tau)$$
$$= \sum_{\tau\in \mathsf{L}_T} \prod_{\alpha>0} (\tau, \alpha)\Theta_\tau(f).$$

It therefore follows that

$$f(1) = \mathbf{M}_G^{-1} \sum_{\tau\in \mathsf{L}_T} \prod_{\alpha>0} (\tau, \alpha)\Theta_\tau(f),$$

as we wished to prove. □

Corollary 10.1.2.5 Let f be a $\mathfrak{Z}$-finite function in $\mathscr{C}(G)$ – then

$$f(x) = \mathbf{M}_G^{-1} \sum_{\tau\in \mathsf{L}_T} \prod_{\alpha>0} (\tau, \alpha)\Theta_\tau(R(x)f) \qquad (x \in G).$$

[Here R denotes the right regular representation of G on $L^2(G)$.]

Theorem 10.1.2.6 (Harish-Chandra) Let f be a function in $L^2(G)$ which is both K-finite and $\mathfrak{Z}$-finite – then

$$f(1) = \mathbf{M}_G^{-1} \sum_{\tau\in \mathsf{L}_T} \prod_{\alpha>0} (\tau, \alpha)\Theta_\tau(f),$$

the sum on the right being finite.

Proof In view of Theorem 9.3.1.5, f lies in $\mathscr{C}(G)$ (and is actually analytic); accordingly, due to Theorem 10.1.2.4, we have only to verify that the sum on the right is finite. Using the K-finiteness of f, select a finite subset F of $\hat{K}$ such that

$$f = \chi_F * f * \chi_F \qquad (\chi_F = \sum_{\delta\in F} \chi_\delta \cdots);$$

upon replacing F by $F \cup \check{F}$, we can assume that $F = \check{F}$. [We recall that $\check{\delta}$ is the class in $\hat{K}$ contragredient to δ ($\delta \in \hat{K}$).] Given $\tau \in \mathsf{L}_T'$, put

$$f_\tau(x) = \Theta_\tau(R(x)f) \qquad (x \in G).$$

Then it is clear that

$$f_\tau(x) = \Theta_{\tau,F}(R(x)f) \qquad (x \in G)$$

where $\Theta_{\tau,F} = \sum_{\delta\in F} \Theta_{\tau,\delta}$. It now follows from Proposition 10.1.2.3 that $f_\tau = 0$ for all but finitely many τ in L_T'. Hence the theorem. □

Corollary 10.1.2.7 Let f be a function in $L^2(G)$ which is both K-finite and $\mathfrak{Z}$-finite – then

$$f(x) = \mathbf{M}_G^{-1} \sum_{\tau\in \mathsf{L}_T} \prod_{\alpha>0} (\tau, \alpha)\Theta_\tau(R(x)f) \qquad (x \in G),$$

the sum on the right being finite.

Given τ in L_T, let $\mathscr{C}_\tau(G)$ denote the subspace of $\mathscr{C}(G)$ comprised of those functions f such that $Zf = \gamma_\tau(Z)f$ (all $Z \in \mathfrak{Z}$).

Proposition 10.1.2.8 Let f be a (non-zero) eigenfunction of $\mathfrak{Z}$ in $\mathscr{C}(G)$ – then there exists a τ in L_T' such that f belongs to $\mathscr{C}_\tau(G)$.

Proof Let κ be the homomorphism of $\mathfrak{Z}$ into $\mathbf{C}$ such that $Zf = \kappa(Z)f$ $(Z \in \mathfrak{Z})$ – then we have to prove that $\kappa = \gamma_\tau$ for some τ in L_T'. Suppose that this is false. Let $\tau \in \mathsf{L}_T'$ – then

$$\kappa(Z)\Theta_\tau(f) = \Theta_\tau(Zf) = \gamma_\tau(Z^\sharp)\Theta_\tau(f) = \gamma_{-\tau}(Z)\Theta_\tau(f) \qquad (\text{all } Z \in \mathfrak{Z}),$$

whence

$$\Theta_\tau(f) = 0 \qquad (\text{all } \tau \in \mathsf{L}_T').$$

Owing to Theorem 10.1.2.4, this means that $f(1) = 0$. Applying the same argument to the right translates of f, we deduce that $f = 0$, in contradiction with our initial supposition. □

10.2 Theory of the Discrete Series

10.2.1 Existence of the Discrete Series

Let (G, K) be a reductive pair satisfying the usual conditions (with G acceptable).

Let $\hat{G}$ be the unitary dual of G, $\hat{G}_d$ the discrete series for G.

Theorem 10.2.1.1 (Harish-Chandra) Let U be an (irreducible) square integrable unitary representation of G on a Hilbert space E, T_U its character – then T_U is tempered.

Proof Choose an orthonormal basis $\{a_i : i \in I\}$ for E by selecting an orthonormal basis $\{a_i : i \in I(\delta)\}$ for each $E(\delta)$ $(\delta \in \hat{K})$; put

$$\phi_i(x) = (U(x)a_i, a_i) \qquad (x \in G).$$

Then, as we know,

$$T_U(f) = \sum_i \int_G f(x)\phi_i(x)d_G(x) \qquad (f \in C_c^\infty(G)),$$

the series being absolutely convergent. Invoking now the notations centering around Theorem 4.4.2.1 and supporting lemmas, choose a non-negative integer m such that $\sum_{\delta\in\hat{K}} d(\delta)^2c(\delta)^{-m} < \infty$ – then, bearing in mind that ϕ_i is analytic, for any $f \in C_c^\infty(G)$, we have

$$\begin{aligned}\int_G f(x)\phi_i(x)d_G(x) &= c(\delta)^{-m}\int_G f(x)\phi_i(x;\Omega^m)d_G(x)\\ &= c(\delta)^{-m}\int_G f(x;\Omega^m)\phi_i(x)d_G(x) \qquad (i \in I(\delta)),\end{aligned}$$

whence, in view of the Schwarz inequality,

$$\left|\int_G f(x)\phi_i(x)d_G(x)\right| \leqslant c(\delta)^{-m}(d_U)^{-1/2}\,\|\Omega^m f\|_2 \qquad (i \in I(\delta)).$$

Here, of course, d_U is the formal dimension of U. Select a non-negative integer m_U with the property that $\dim(E(\delta)) \leqslant m_U d(\delta)^2$ (all $\delta \in \hat{K}$) – then

$$|T_U(f)| \leqslant (d_U)^{-1/2} \|\Omega^m f\|_2 \sum_{\delta \in \hat{K}} \dim(E(\delta))c(\delta)^{-m}$$
$$\leqslant \{(d_U)^{-1/2} m_U \sum_{\delta \in \hat{K}} d(\delta)^2 c(\delta)^{-m}\} \|\Omega^m f\|_2 \quad (\text{all } f \in C_c^\infty(G)).$$

The contention of the theorem is thus seen to be a consequence of the fact that the rule $f \mapsto \|\Omega^m f\|_2$ ($f \in C_c^\infty(G)$) defines a continuous semi-norm on $\mathscr{C}(G)$ (cf. Proposition 8.3.7.5). □

Theorem 10.2.1.2 (Harish-Chandra) The reductive Lie group G has a (non-empty) discrete series iff rank (G) = rank (K).

Proof Suppose that $\hat{G}_d \neq \varnothing$ – then the (non-zero) spherical trace functions associated with any class $\hat{U}$ in $\hat{G}_d$ are square integrable, as well as being both K-finite and $\mathfrak{Z}$-finite. The fact that rank (G) = rank (K) then follows from Theorem 9.3.1.5 (and its corollary). Conversely, suppose that rank (G) = rank (K); fix a maximal torus T in K together with an element τ in L'_T (cf. 10.1.1); choose a class $\delta \in \hat{K}$ such that the Fourier component $\Theta_{\tau,\delta}$ of the distribution Θ_τ does not vanish identically – then $\Theta_{\tau,\delta}$ lies in $L^2(G)$ (cf. Theorem 10.1.2.1) and, according to a principle enunciated in number 8.3.9, the smallest closed L-stable subspace of $L^2(G)$ containing $\Theta_{\tau,\delta}$ breaks up into a finite sum of non-zero, mutually orthogonal, closed, L-stable subspaces on each of which G acts irreducibly, whence $\hat{G}_d$ is not empty. . . . □

10.2.2 The Characters of the Discrete Series I-Implication of the Orthogonality Relations

Let (G, K) be a reductive pair satisfying the usual conditions (with G acceptable); throughout the present number it will be assumed in addition that rank (G) = rank (K). Fix once and for all a maximal torus T in K – then the notations and conventions introduced in 10.1 are now in force.

Owing to our assumptions, the discrete series $\hat{G}_d$ for G is not empty; cf. Theorem 10.2.1.2. This being so, let U be an (irreducible) square integrable unitary representation of G on a Hilbert space E; let $\mathscr{E}_U$ be the closed subspace of $L^2(G)$ spanned by the conjugates of the coefficients of U – then, as we know, $\mathscr{E}_U$ is stable with respect to both left and right translation by elements of G and, moreover, depends only on the unitary equivalence class $\hat{U}$ of U. The restriction to $\mathscr{E}_U$ of the left regular representation L of G on $L^2(G)$ is unitarily equivalent to $\dim(U)U$ while the restriction to $\mathscr{E}_U$ of the right regular representation R of G on $L^2(G)$ is unitarily equivalent to $\dim(\bar{U})\bar{U}$ ($\bar{U}$ the conjugate of U). Finally let us recall that

$$L^2(G)_d = \sum_{\hat{U} \in \hat{G}_d} \mathscr{E}_{\hat{U}},$$

the (orthogonal) sum being taken in the sense of Hilbert space theory. [For a complete discussion of these results, see Vol. I, number 4.5.9.]

Proposition 10.2.2.1 Let U be an (irreducible) square integrable unitary representation of G on a Hilbert space E; put

$$\mathscr{C}_U(G) = \mathscr{C}(G) \cap \mathscr{E}_U.$$

Then $\mathscr{C}_U(G)$ is a closed subspace of $\mathscr{C}(G)$ and, moreover, is dense in $\mathscr{E}_U$.

Proof On the basis of Proposition 8.3.7.5, it is clear that $\mathscr{C}_U(G)$ is a closed subspace of $\mathscr{C}(G)$. Now choose an orthonormal basis $\{a_i : i \in I\}$ for E by selecting an orthonormal basis $\{a_i : i \in I(\delta)\}$ for each $E(\delta)$ $(\delta \in \hat{K})$; put $\phi_{ij}(x) = (U(x)a_i, a_j)$ $(x \in G)$ – then, thanks to Theorem 9.3.1.5, each $\overline{\phi_{ij}}$ lies in $\mathscr{C}_U(G)$, whence $\mathscr{C}_U(G)$ is dense in $\mathscr{E}_U$. □

Let U be an (irreducible) square integrable unitary representation of G on a Hilbert space E, F_U its character (qua a locally summable function on G); let $F_{U,T}$ be the analytic function on T which agrees on T' with $\Delta_T F_U$ (cf. Proposition 8.3.4.3). Let $\bar{U}$ be the conjugate (= contragredient) of U, $F_{\bar{U}}$ its character, $d_{\bar{U}}$ its formal dimension – then, as we know, $F_{\bar{U}} = \overline{F_U}$ while $d_U = d_{\bar{U}}$.

Proposition 10.2.2.2 Let U be an (irreducible) square integrable unitary representation of G on a Hilbert space E, d_U its formal dimension; let F_U be the character of U – then

$$\Phi_f = (d_{\bar{U}})^{-1} f(1) F_{\bar{U},T}$$

for all f in $\mathscr{C}_U(G)$.

[Here, of course, Φ_f $(f \in \mathscr{C}_U(G))$ is the invariant integral of f calculated relative to the compact Cartan subgroup T; because the elements of $\mathscr{C}_U(G)$ are eigenfunctions of $\mathfrak{Z}$, Φ_f $(f \in \mathscr{C}_U(G))$ is a C^∞ function on T (cf. Theorem 8.5.1.9).]

We shall preface the proof with a lemma.

Lemma 10.2.2.3 Let U be an (irreducible) square integrable unitary representation of G on a Hilbert space E, d_U its formal dimension; let F_U be the character of U; fix two K-finite vectors a and b in E and put $f(x) = (U(x)a, b)$ $(x \in G)$ – then

$$\Phi_f = (d_U)^{-1}(a, b) F_{U,T}.$$

Proof To begin with, observe that f lies in $\mathscr{C}(G)$ (cf. Theorem 9.3.1.5) and so Φ_f is well-defined; moreover, in view of the $\mathfrak{Z}$-finiteness of f, Φ_f is actually a C^∞ function on T (cf. Theorem 8.5.1.9). In order to establish the asserted equality, then, it will be enough to show that the entities in question are the same when regarded as distributions on T'

(the regular elements in T); because both sides are skew relative to $W(G, T)$, we need only verify that

$$\int_T \Phi_\alpha(t)\Phi_f(t)d_T(t) = (d_U)^{-1}(a, b)\int_T \Phi_\alpha(t)F_{U,T}(t)dt$$

for all $\alpha \in C_c^\infty(G(T))$ ($G(T)$ = regular elliptic elements in G). This being the case, fix an $\alpha \in C_c^\infty(G(T))$ – then, in view of Theorem 4.5.9.7 (bis), we have

$$\begin{aligned}
&(d_U)^{-1}(a, b)\int_G \alpha(x)F_U(x)d_G(x)\\
&= \int_G (U(x)U(\alpha)U(x^{-1})a, b)d_G(x)\\
&= (-1)^r[W(G, T)]^{-1}\int_G d_G(x)\\
&\quad\times\left\{\int_T \Delta_T(t)^2\, d_T(t)\int_{G/T} \alpha({}^{\dot{y}}t)(U(\dot{\overline{{}^{xy}t}})a, b)d_{G/T}(\dot{y})\right\},
\end{aligned}$$

r as always. Let us grant for the moment that it is permissible to interchange the order of integrations on the right – then

$$\int_G (U({}^{xy}t)a, b)d_G(x) = \int_G (U({}^{x}t)a, b)d_G(x) = \Delta_T(t)^{-1}\Phi_f(t) \qquad (t \in T')$$

and so the right hand side can be written as

$$(-1)^r[W(G, T)]^{-1}\int_T \Delta_T(t)\Phi_f(t)\left\{\int_{G/T} \alpha({}^{\dot{y}}t)d_{G/T}(\dot{y})\right\} d_T(t),$$

whence

$$(d_U)^{-1}(a, b)\int_G \alpha(x)F_U(x)d_G(x) = (-1)^r[W(G, T)]^{-1}\int_T \Phi_\alpha(t)\Phi_f(t)d_T(t).$$

On the other hand, it is obvious that

$$\int_G \alpha(x)F_U(x)d_G(x) = (-1)^r[W(G, T)]^{-1}\int_T \Phi_\alpha(t)F_{U,T}(t)d_T(t).$$

Therefore the proof of the lemma will be complete when the application supra of Fubini's Theorem is justified. For this purpose, choose $\alpha_0 \in C_c^\infty(G(T))$ such that $\alpha_0 \geqslant |\alpha|$ – then, by obvious majorizations, we are reduced to showing that the quantity

$$\int_T |\Phi_{\alpha_0}(t)|\left\{|\Delta_T(t)|\int_G |f({}^{x}t)|\, d_G(x)\right\} d_T(t)$$

is finite. But $|\Phi_{\alpha_0}|$ is bounded on T' (cf. Lemma 8.5.8.3) and, thanks to Theorem 8.5.7.2, the same is true of the function

$$t \mapsto |\Delta_T(t)|\int_G |f({}^{x}t)|\, d_G(x) \qquad (t \in T').$$

Hence the lemma. □

Proof of Proposition 10.2.2.2 Let $\mathscr{S}$ be the set of all $f \in \mathscr{C}_U(G)$ such that

$$\Phi_f = (d_{\bar{U}})^{-1} f(1) F_{\bar{U},T}.$$

Then, bearing in mind Lemma 8.5.8.3, it is clear that $\mathscr{S}$ is a closed subspace of $\mathscr{C}_U(G)$ (the topology on $\mathscr{C}_U(G)$ being the one inherited from $\mathscr{C}(G)$) and so it will be enough to prove that $\mathscr{S}$ is dense in $\mathscr{C}_U(G)$. For this purpose, let us agree to use the notations which were introduced during the proof of Proposition 10.2.2.1 – then certainly the $\overline{\phi_{ij}}$ span a dense subspace of $\mathscr{C}_U(G)$. On the other hand, owing to the preceding lemma (with U replaced by $\bar{U}$), the asserted relation is known to hold for the $\overline{\phi_{ij}}$. . . . □

Proposition 10.2.2.4 Let U and V be (irreducible) square integrable unitary representations of G – then

$$\int_G f(x) F_U(x) d_G(x) = \begin{cases} (d_U)^{-1} f(1) & \text{if } \hat{U} = \hat{V}, \\ 0 & \text{otherwise}. \end{cases} \qquad (f \in \mathscr{C}_V(G))$$

Proof On account of the fact that the character of an (irreducible) square integrable unitary representation of G is tempered (cf. Theorem 10.2.1.1), it can be assumed that f is K-finite (the K-finite functions being dense in $\mathscr{C}_V(G)$); but in this case, the assertion of our proposition follows without difficulty from the Schur Orthogonality Relations (cf. Theorem 4.5.9.3). □

Proposition 10.2.2.5 Let U and V be (irreducible) square integrable unitary representations of G – then

$$\int_T F_{U,T}(t) \overline{F_{V,T}(t)}\, d_T(t) = \begin{cases} [W(G, T)] & \text{if } \hat{U} = \hat{V}, \\ 0 & \text{otherwise}. \end{cases}$$

Proof Choose a function f in $\mathscr{C}_V(G)$ such that $f(1) \neq 0$ – then, since f is $\mathfrak{Z}$-finite, we have

$$\int_G f(x) F_U(x) d_G(x) = (-1)^r [W(G, T)]^{-1} \int_T \Phi_f(t) F_{U,T}(t) d_T(t).$$

On the other hand, thanks to Proposition 10.2.2.2, Φ_f is given by the relation

$$\Phi_f = (d_{\bar{V}})^{-1} f(1) F_{\bar{V},T} = (d_V)^{-1} (-1)^r f(1) \overline{F_{V,T}}$$

and so the contention of the present proposition is thus seen to be a consequence of Proposition 10.2.2.4. □

10.2.3 The Characters of the Discrete Series II - Application of the Differential Equations

Let (G, K) be a reductive pair meeting the usual requirements (with G acceptable); throughout the present number it will be assumed in addition that rank (G) = rank (K) (thus the discrete series $\hat{G}_d$ for G is not empty). Fix once and for all a maximal torus T in K. We recall that the unitary character group $\hat{T}$ of T has been identified with a lattice L_T in the (real) dual of $\sqrt{-1}\mathfrak{t}$. Evidently L_T is stable under $W(G, T)$; however, L_T will not, in general, be invariant under the action of the full Weyl group W of the pair $(\mathfrak{g}_c, \mathfrak{t}_c)$; accordingly, given $\tau \in \mathsf{L}_T$, let us agree to denote by $W(\tau)$ the subset of W consisting of those $w \in W$ such that $w\tau \in \mathsf{L}_T$.

Relative to a given (irreducible) square integrable unitary representation U of G, assign to the symbols F_U, $F_{U,T}$, d_U etc. their customary meanings (cf. 10.2.2).

Proposition 10.2.3.1 Let U be an (irreducible) square integrable unitary representation of G – then the infinitesimal character of U is of the form γ_τ for some τ in L_T'.

[This follows at once from Proposition 10.1.2.8.]

Note For any $\tau \in \mathsf{L}_T'$, let $\hat{G}_d(\tau)$ denote the set of all classes $\hat{U}$ in $\hat{G}_d$ with infinitesimal character γ_τ – then, according to Proposition 10.2.3.1, we have

$$\hat{G}_d = \bigcup_{\tau \in \mathsf{L}_T'} \hat{G}_d(\tau).$$

In passing we recall that the cardinality of a given $\hat{G}_d(\tau)$ $(\tau \in \mathsf{L}_T')$ is finite (cf. Theorem 4.5.8.11); it will be shown below that

$$[\hat{G}_d(\tau)] = [W(\tau)]/[W(G, T)] \qquad (\tau \in \mathsf{L}_T').$$

Introduce the distributions Θ_τ $(\tau \in \mathsf{L}_T')$ per 10.1.1; let $\Theta_{\tau,T}$ denote the analytic extension to T (from T') of $\Delta_T\Theta_\tau$ $(\tau \in \mathsf{L}_T')$.

Proposition 10.2.3.2 Let U be an (irreducible) square integrable unitary representation of G with infinitesimal character γ_τ $(\tau \in \mathsf{L}_T')$ – then there exist unique complex numbers $c_w(U)$ $(w \in W(\tau))$ such that

$$c_{tw}(U) = c_w(U) \qquad (t \in W(G, T))$$

and with the property that

$$F_U = [W(G, T)]^{-1} \sum_{w \in W(\tau)} \det(w) c_w(U) \Theta_{w\tau}.$$

Proof Let $\gamma : \mathfrak{Z} \to I(\mathfrak{t}_c)$ denote the canonical isomorphism of $\mathfrak{Z}$ onto the Weyl group invariants $I(\mathfrak{t}_c)$ in $S(\mathfrak{t}_c)$ – then, in view of Theorem 8.2.3.9, we have

$$\gamma(Z)F_{U,T} = \gamma_\tau(Z)F_{U,T}, \qquad \gamma(Z)\Theta_{\tau,T} = \gamma_\tau(Z)\Theta_{\tau,T} \qquad (\tau \in \mathsf{L}_T')$$

for all $Z \in \mathfrak{Z}$. Now both $F_{U,T}$ and $\Theta_{\tau,T}$ are skew relative to the action of $W(G, T)$; so, keeping in mind Theorem 10.1.1.1 (iii), it follows from Corollary 8.2.2.9 that there exist unique complex numbers $c_w(U)$ $(w \in W(\tau))$ such that $c_{tw}(U) = c_w(U)$ $(t \in W(G, T))$ and with the property that the (central) distribution

$$\Theta = F_U - [W(G, T)]^{-1} \sum_{w \in W(\tau)} \det(w) c_w(U) \Theta_{w\tau}$$

vanishes pointwise on T'. Evidently Θ is a tempered, eigendistribution of $\mathfrak{Z}$; in order to show that Θ vanishes identically, we need only verify that each of its Fourier components Θ_δ ($\delta \in \hat{K}$) is null. So fix a $\delta \in \hat{K}$ and consider Θ_δ; it is a square integrable function on G which is both K-finite and $\mathfrak{Z}$-finite, hence lies in $\mathscr{C}(G)$ (cf. Theorem 9.3.1.5). Let f be a $\mathfrak{Z}$-finite element in $\mathscr{C}(G)$ – then

$$\Theta_\delta(f) = \Theta(f * \bar{\chi}_\delta) = (-1)^r \int_T \Delta_T(t) \Theta(t) \Phi_{f_{\breve{\delta}}}(t) d_T(t) = 0,$$

r as always. In this relation, take $f = \bar{\Theta}_\delta$ – then, due to Proposition 8.3.8.8, we have $\| \Theta_\delta \|_2^2 = 0$, whence $\Theta_\delta = 0$, as desired. □

Let $\mathbf{T}(\tau)$ ($\tau \in \mathsf{L}_T'$) be the space of all central, tempered distributions Θ on G such that $Z \cdot \Theta = \gamma_\tau(Z) \Theta$ ($Z \in \mathfrak{Z}$).

Proposition 10.2.3.3 Fix $\tau \in \mathsf{L}_T'$ – then the $F_{\hat{U}}$ ($\hat{U} \in \hat{G}_d(\tau)$) form a basis for $\mathbf{T}(\tau)$ over $\mathbf{C}$.

Proof Owing to Theorem 10.2.1.1, the $F_{\hat{U}}$ ($\hat{U} \in \hat{G}_d(\tau)$) lie in $\mathbf{T}(\tau)$; moreover they are linearly independent (cf. Proposition 10.2.2.4). Now fix a $\Theta \in \mathbf{T}(\tau)$ – then we have to prove that Θ is a linear combination of the $F_{\hat{U}}$ ($\hat{U} \in \hat{G}_d(\tau)$). Put

$$\mathbf{\Theta}_T = \Delta_T \Theta - \sum_{\hat{U}} c(\hat{U}) F_{\hat{U},T} \qquad (\hat{U} \in \hat{G}_d(\tau))$$

where

$$c(\hat{U}) = [W(G, T)]^{-1} \int_T \Delta_T \Theta \cdot \overline{F_{\hat{U},T}} \, d_T(t).$$

Then, on the basis of Proposition 10.2.2.5, it is clear that $\mathbf{\Theta}_T$ is orthogonal to $F_{\hat{U},T}$ ($\hat{U} \in \hat{G}_d(\tau)$) in $L^2(T)$. This being so, let

$$\mathbf{\Theta} = \Theta - \sum_{\hat{U}} c(\hat{U}) F_{\hat{U}} \qquad (\hat{U} \in \hat{G}_d(\tau)).$$

Then the proof of the present proposition will be complete when it is shown that $\mathbf{\Theta} = 0$. In turn, for this purpose, it will be enough to verify that each of the Fourier components $\mathbf{\Theta}_\delta$ ($\delta \in \hat{K}$) of $\mathbf{\Theta}$ vanishes identically. This being the case, fix a $\delta \in \hat{K}$ – then, due to Lemma 10.1.2.2, $\mathbf{\Theta}_\delta \in \mathscr{C}_\tau(G)$. Put $f = \bar{\mathbf{\Theta}}_\delta$ – then $f \in \mathscr{C}_{-\tau}(G)$ and verifies the relation $f = f * \bar{\chi}_\delta$. Owing to a principle which was spelled out in number 8.3.9, the smallest closed L-stable subspace E (say) of $L^2(G)$ containing f can be written as a finite sum of non-zero, mutually orthogonal, closed, L-stable subspaces E_i ($1 \leqslant i \leqslant p$) on each of which G acts irreducibly. Let P_i be the orthogonal projection of E onto E_i ($1 \leqslant i \leqslant p$); set $f_i = \mathsf{P}_i f$ – then $f = f_1 + \cdots + f_p$ and every f_i belongs to $\mathscr{C}_{-\tau}(G)$. Let U_i

denote the restriction of L to E_i – then, of course, U_i is an (irreducible) square integrable unitary representation of G on E_i with $\hat{U}_i \in \hat{G}_d(\tau)$. Because $f_i \in \mathscr{C}_{U_i}(G)$ $(1 \leqslant i \leqslant p)$, it follows that

$$\begin{aligned} \|\mathbf{\Theta}_\delta\|_2^2 = \mathbf{\Theta}(f) &= (-1)^r[W(G, T)]^{-1} \int_T \mathbf{\Theta}_T(t)\Phi_f(t)d_T(t) \\ &= (-1)^r[W(G, T)]^{-1} \sum_{i=1}^{p} \int_T \mathbf{\Theta}_T(t)\Phi_{f_i}(t)d_T(t) \\ &= \sum_{i=1}^{p} c_i \int_T \mathbf{\Theta}_T(t)\overline{F_{U_i,T}}(t)d_T(t) = 0 \qquad (c_i = \text{constant}), \end{aligned}$$

where, in passing from the second to the third line, we invoked Proposition 10.2.2.2. This shows that $\mathbf{\Theta}_\delta = 0$ (all $\delta \in \hat{K}$); thus $\mathbf{\Theta}$ itself must vanish identically and so

$$\Theta = \sum_{\hat{U}} c(\hat{U})F_{\hat{U}} \qquad (\hat{U} \in \hat{G}_d(\tau)),$$

as desired. □

Corollary 10.2.3.4 Fix $\tau \in \mathsf{L}_T'$ – then

$$[\hat{G}_d(\tau)] = [W(\tau)]/[W(G, T)].$$

Proof According to the preceding proposition we have $\dim(\mathbf{T}(\tau)) = [\hat{G}_d(\tau)]$; on the other hand, let $w_1 = 1, w_2, \ldots, w_n$ be a complete set of representatives for $W(G, T)\backslash W(\tau)$ in $W(\tau)$ – then, in view of Proposition 10.2.3.2 (or rather the argument thereof), the distributions $\Theta_{w_i\tau}$ $(1 \leqslant i \leqslant n)$ also constitute a basis for $\mathbf{T}(\tau)$, whence $\dim(\mathbf{T}(\tau)) = [W(\tau)]/[W(G, T]. \ldots$ □

10.2.4 The Theorem of Harish-Chandra

Let (G, K) be a reductive pair fulfilling the usual hypotheses (with G acceptable); in what follows it will be assumed in addition that rank (G) = rank (K); this being so, fix a maximal torus T in K; agreeing to employ the customary notations and conventions, we have the following fundamental theorem.

Theorem 10.2.4.1 (Harish-Chandra) The discrete series $\hat{G}_d$ for G is not empty. To each $\tau \in \mathsf{L}_T'$, there corresponds a unique element $\hat{U}_\tau \in \hat{G}_d$ whose character is given by $(-1)^{\mathbf{m}_G}\epsilon(\tau)\Theta_\tau$ $(\epsilon(\tau) = \text{sign}\{\prod_{\alpha>0}(\tau, \alpha)\})$; the mapping $\tau \mapsto \hat{U}_\tau$ of L_T' into $\hat{G}_d$ is surjective and the formal dimension $d_{\hat{U}_\tau}$ of the class $\hat{U}_\tau$ is given by

$$d_{\hat{U}_\tau} = \frac{[W(G, T)]}{(2\pi)^r} \Big|\prod_{\alpha>0}(\tau, \alpha)\Big| \qquad (r = 2^{-1}\dim(G/T)).$$

Finally $\hat{U}_{\tau_1} = \hat{U}_{\tau_2}$ $(\tau_1, \tau_2 \in \mathsf{L}_T')$ iff τ_1 and τ_2 are conjugate under $W(G, T)$ – symbolically $\hat{G}_d \sim W(G, T)\backslash\hat{T}'$.

[Here, of course, $\mathbf{m}_G$ $(= \mathbf{m}_{\mathfrak{g}}) = 2^{-1}\dim(G/K)$; cf. Theorem 8.5.1.6.]

Note Evidently this result also provides us with explicit formulas for the central and infinitesimal characters of a given element in $\hat{G}_d$ (cf. Theorem 10.1.1.1 (i) and Lemma 10.1.1.5).

Corollary 10.2.4.2 The (tempered) distribution

$$\mathbf{M}_G^{-1} \sum_{\tau \in \mathsf{L}_T} \prod_{\alpha > 0} (\tau, \alpha)\Theta_\tau$$

represents the contribution of the discrete series to the Plancherel formula for G.

[One need only observe that

$$\mathbf{M}_G^{-1} \sum_{\tau \in \mathsf{L}_T} \prod_{\alpha > 0} (\tau, \alpha)\Theta_\tau = \sum_{\hat{U} \in \hat{G}_d} d_{\hat{U}} T_{\hat{U}}$$

as distributions on G ($T_{\hat{U}}$ the character of $\hat{U}$ ($\hat{U} \in \hat{G}_d$)).]

Remark (Lipsman) Consider the unitary dual $\hat{G}$ of G qua a topological space (cf. number 7.1.1) – then the hull-kernel topology, relativized to $\hat{G}_d$, is in fact discrete. [Thus let $\mathscr{S}$ be an arbitrary subset of $\hat{G}_d$ – then it will be enough to show that $\mathscr{S}$ is closed in $\hat{G}$. This being so, let $\hat{U}$ be a point in the hull-kernel closure of $\mathscr{S}$; choose a net $\{\hat{U}_n\}$ in $\mathscr{S}$ such that $\hat{U}_n \to \hat{U}$ hull-kernelwise – then, in view of Lemma 10.1.1.9 (= Riemann-Lebesgue Lemma for $\hat{G}_d$) and Lemma 7.1.2.3, no subnet of $\{\hat{U}_n\}$ can converge to the point at infinity in $\hat{G}_d$ per the discrete topology (cf. part (2) of the proof of Theorem 7.1.2.1). Consequently the natural cluster points of the net $\{\hat{U}_n\}$ lie in the finite part of $\hat{G}_d$. Let $\hat{V} \in \hat{G}_d$ be one of them – then there exists a subnet $\{\hat{V}_n\}$ of $\{\hat{U}_n\}$ which converges to $\hat{V}$ in the discrete topology, thus $\hat{V}_n = \hat{V}$ eventually and so $\hat{V} \in \mathscr{S}$. But $\hat{V}_n \to \hat{U}$ in the hull-kernel topology; therefore $\hat{U}$ is in the hull-kernel closure of $\{\hat{V}\}$, whence $\hat{U} = \hat{V} \in \mathscr{S}$, $\hat{G}$ being a T_1-space (here we have to use the fact that G is liminaire . . .).] This means that the correspondence $\hat{G}_d \leftrightarrow W(G, T)\backslash\hat{T}'$ is topological. . . .

We shall agree to retain the above notations and assumptions.

Lemma 10.2.4.3 Let U be an irreducible unitary representation of G on a Hilbert space E; let $f \in C^\infty(K)$ – then the operator

$$U_K(f) = \int_K f(k)U(k)dk$$

is of the trace class.

[This follows by applying Theorem 4.5.7.6 to the representation U_K $(= U|K)$.]

Lemma 10.2.4.4 Let U be an irreducible unitary representation of G on a Hilbert space E, F_U its character (qua a locally summable function on G); let m_δ be the multiplicity of δ ($\delta \in \hat{K}$) in $U|K$ – then

$$\int_T g(t)F_{U,T}(t)d_T(t) = \sum_{\delta \in \hat{K}} m_\delta/d(\delta) \cdot \int_T g(t)\chi_\delta(t)\Delta_T(t)d_T(t)$$

for all $g \in C_c^\infty(T')$, the series being absolutely convergent.

Proof Fix a function $\alpha \in I_c^\infty(G)$ such that $\int_G \alpha(x)d_G(x) = 1$. Given $g \in C_c^\infty(T')$, define a function $f_g \in C_c^\infty(G(T))$ by the prescription

$$f_g({}^x t) = \alpha(x)\Delta_T(t)^{-1} \sum_{w \in W(G,T)} \det(w) g^w(t) \qquad (t \in T', x \in G).$$

Then it is clear that

$$\int_G f_g(x)F_U(x)d_G(x) = (-1)^r \int_T g(t)F_{U,T}(t)d_T(t) \qquad (g \in C_c^\infty(T')),$$

r the number of positive roots of the pair $(\mathfrak{g}_c, \mathfrak{t}_c)$. We shall now compute the entity $\int_G f_g F_U \, d_G(x)$ $(g \in C_c^\infty(T'))$ in a different way. Consider the function defined by the prescription

$$\Delta_K^2(t) = \xi_{2\rho_K}(t) \prod_\alpha (1 - \xi_\alpha(t^{-1}))^2 \qquad (t \in T),$$

the product being taken over the positive compact roots of the pair $(\mathfrak{g}_c, \mathfrak{t}_c)$, i.e. the positive roots of the pair $(\mathfrak{k}_c, \mathfrak{t}_c)$ (here $2\rho_K$ denotes the sum of the roots in question); certainly Δ_K^2 is a well-defined function on T which is, moreover, invariant under $W(G, T)$. This being so, given $g \in C_c^\infty(T')$, define a function $h_g \in C^\infty(K)$ by the prescription

$$h_g({}^k t) = \Delta_T(t) \cdot \Delta_K(t)^{-2} \sum_{w \in W(G,T)} \det(w) g^w(t) \qquad (t \in T', k \in K).$$

Then, for any $g \in C_c^\infty(T')$, we have

$$\begin{aligned}
U(f_g) &= \int_G f_g(x)U(x)d_G(x) \\
&= (-1)^r[W(G, T)]^{-1} \int_T \Delta_T(t) \sum_{w \in W(G,T)} \det(w) g^w(t) d_T(t) \\
&\quad \times \int_G \alpha(x)U({}^x t)d_G(x) \\
&= (-1)^r[W(G, T)]^{-1} \int_G \alpha(x)U(x) \cdot \left\{\int_T \int_K h_g({}^k t)\Delta_K^2(t)U({}^k t) \, dk d_T(t)\right\} \\
&\quad \times U(x^{-1})d_G(x) \\
&= (-1)^r[W(G, T)]^{-1} \int_G \alpha(x)U(x) \cdot \{(-1)^{r_K}[W(G, T)]U_K(h_g)\} \\
&\quad \times U(x^{-1})d_G(x) \\
&= (-1)^{r+r_K} \int_G \alpha(x)U(x)U_K(h_g)U(x^{-1})d_G(x),
\end{aligned}$$

r_K the number of positive roots of the pair $(\mathfrak{k}_c, \mathfrak{t}_c)$. It therefore follows that

$$\begin{aligned}
\int_G f_g(x)F_U(x)d_G(x) &= \operatorname{tr}(U(f_g)) \\
&= (-1)^{r+r_K} \int_G \alpha(x) \operatorname{tr}(U_K(h_g))d_G(x) \qquad \text{(Lemma 10.2.4.3)} \\
&= (-1)^{r+r_K} \operatorname{tr}(U_K(h_g)) \\
&= (-1)^{r+r_K} \sum_{\delta \in \hat{K}} m_\delta/d(\delta) \cdot \int_K h_g(k)\chi_\delta(k)dk \\
&= (-1)^r \sum_{\delta \in \hat{K}} m_\delta/d(\delta) \cdot \int_T g(t)\chi_\delta(t)\Delta_T(t)d_T(t) \qquad (g \in C_c^\infty(T')).
\end{aligned}$$

Hence the lemma. □

Note Suppose given a sequence $\{m_\delta : \delta \in \hat{K}\}$ of non-negative integers m_δ $(\delta \in \hat{K})$ which has the property that there exists a positive integer M such that $m_\delta \leqslant Md(\delta)$ (all $\delta \in \hat{K}$) – then the series

$$\sum_{\delta \in \hat{K}} m_\delta / d(\delta) \cdot \chi_\delta \Delta_T$$

represents a distribution on T. To see this, let $\tilde{K}$ be a finite acceptable covering of K, $\pi : \tilde{K} \to K$ the covering map, $\tilde{T} = \pi^{-1}(T)$ – then the function $\Delta_{\tilde{K}}$ given by the rule

$$\Delta_{\tilde{K}}(\tilde{t}) = \xi_{\rho_{\tilde{K}}}(\tilde{t}) \prod_\alpha (1 - \xi_\alpha(\tilde{t}^{-1})) \qquad (\tilde{t} \in \tilde{T})$$

is a well-defined function on $\tilde{T}$. [Here $2\rho_{\tilde{K}}$ stands for the sum of the positive roots of the pair $(\mathfrak{k}_c, \mathfrak{t}_c)$, the product being taken over the roots in question.] Evidently it will be enough to show that the series

$$\sum_{\delta \in \hat{K}} m_\delta / d(\delta) \cdot \chi_\delta \circ \pi (\Delta_T \circ \pi)$$

represents a distribution on $\tilde{T}$; in turn, for this purpose, we need only verify that the series

$$\sum_{\delta \in \hat{K}} m_\delta / d(\delta) \cdot \chi_\delta \circ \pi \cdot \Delta_{\tilde{K}}$$

represents a distribution on $\tilde{T}$. Let us now agree to use the notations centering around Theorem 4.4.2.1 and supporting lemmas; thus choose a non-negative integer m such that

$$\sum_{\delta \in \hat{K}} d(\delta)^2 c(\delta)^{-m} < \infty$$

and let D denote the image of Ω^m in $S(\mathfrak{t}_c)$ under the canonical isomorphism from $\mathfrak{Z}_\mathfrak{k}$ to the $W(\mathfrak{k}_c, \mathfrak{t}_c)$-invariants in $S(\mathfrak{t}_c)$ – then

$$\int_{\tilde{T}} g(\tilde{t}) \chi_\delta \circ \pi(\tilde{t}) \Delta_{\tilde{K}}(\tilde{t}) d_{\tilde{T}}(\tilde{t}) = c(\delta)^{-m} \int_{\tilde{T}} g(\tilde{t}; D^t) \chi_\delta \circ \pi(\tilde{t}) \Delta_{\tilde{K}}(\tilde{t}) d_{\tilde{T}}(\tilde{t}) \qquad (g \in C^\infty(\tilde{T}))$$

and so it follows that

$$\sum_{\delta \in \hat{K}} m_\delta / d(\delta) \left| \int_{\tilde{T}} g(\tilde{t}) \chi_\delta \circ \pi(\tilde{t}) \Delta_{\tilde{K}}(\tilde{t}) d_{\tilde{T}}(\tilde{t}) \right| \leqslant \{M[W(\tilde{K}, \tilde{T})] \sum_{\delta \in \hat{K}} d(\delta)^2 c(\delta)^{-m}\} \| D^t g \|_\infty$$

for all $g \in C^\infty(\tilde{T})$. Of course this serves to establish our contention.

We shall now give the proof of Theorem 10.2.4.1; it will be convenient to divide the discussion into several parts.

Fix an (irreducible) square integrable unitary representation U of G with character F_U – then we claim that there exists a $\tau \in \mathsf{L}'_T$ such that $F_U = \pm \Theta_\tau$. Thus, according to Proposition 10.2.3.2, $F_{U,T}$ can be written as a finite linear combination of the characters of T; let τ be the highest element in L_T such that

$$c_0 = \int_T F_{U,T}(t)\overline{\xi_\tau(t)}\, d_T(t) \neq 0.$$

Then certainly τ belongs to L'_T. Let m_δ be the multiplicity of δ ($\delta \in \hat{K}$) in $U|K$ – then, on the basis of what has been said above, the series

$$\sum_{\delta \in \hat{K}} m_\delta / d(\delta) \cdot \chi_\delta \Delta_T$$

represents a distribution ∇ (say) on T, the difference $F_{U,T} - \nabla$ being a distribution (on T) with support in $T - T'$. Because T is compact, there exists a non-negative integer m such that

$$\Delta_T^m \cdot (F_{U,T} - \nabla) = 0$$

(cf. Lemma A2.1.3), i.e. such that

$$\int_T g(t)\Delta_T^m(t)F_{U,T}(t)d_T(t) = \sum_{\delta \in \hat{K}} m_\delta/d(\delta) \cdot \int_T g(t)\chi_\delta(t)\Delta_T^{m+1}(t)d_T(t)$$

for all $g \in C_c^\infty(T)$. Into this expression, let us substitute $g = \bar{\xi}_{\tau+m\rho}$ – then it is clear that the left hand side is equal to c_0 whereas the right hand side is an integer. On the other hand, let $w_1 = 1, w_2, \ldots, w_n$ be a complete set of representatives for $W(G, T)\backslash W(\tau)$ in $W(\tau)$. Owing to Proposition 10.2.3.2, there exist unique complex numbers c_i $(1 \leqslant i \leqslant n)$ such that

$$F_U = \sum_{i=1}^n \det(w_i)c_i\Theta_{w_i\tau}.$$

A simple calculation (using Proposition 10.2.2.5) tells us that $\sum_i |c_i|^2 = 1$; because $c_0 = c_1 \neq 0$, we then deduce that $c_0 = c_1 = \pm 1$ while $c_i = 0$ for $i \geqslant 2$. Therefore $F_U = \pm\Theta_\tau$, as we wished to prove.

Fix a τ in L'_T – then we claim that there exists an (irreducible) square integrable unitary representation U (say) of G such that $\Theta_\tau = \pm F_U$. To establish this assertion, it will be enough to show that if $\Theta_\tau \neq \pm F_U$ for every such U, then of necessity $\Theta_\tau = 0$; to this latter end, we shall verify that the Fourier components $\Theta_{\tau,\delta}$ ($\delta \in \hat{K}$) of Θ_τ all vanish identically. This being the case, fix a $\delta \in \hat{K}$ – then, as we know, $f = \bar{\Theta}_{\tau,\delta}$ is a K-finite, $\mathfrak{Z}$-finite, rapidly decreasing function on G. According to a by now familiar principle, the smallest closed L-stable subspace E (say) of $L^2(G)$ containing f can be written as a finite sum of non-zero, mutually orthogonal, closed, L-stable subspaces E_i $(1 \leqslant i \leqslant p)$ on each of which G acts irreducibly. Let P_i be the orthogonal projection of E onto E_i $(1 \leqslant i \leqslant p)$; set $f_i = \mathsf{P}_i f$ – then $f = \sum_i f_i$ and each f_i is a K-finite, $\mathfrak{Z}$-finite function lying in $\mathscr{C}(G)$. Let U_i denote the restriction of L to E_i – then, of course, U_i is an (irreducible) square integrable unitary representation of G on E_i with character F_{U_i}. Owing to what was said above, there exist $\tau_i \in \mathsf{L}'_T$ such that $F_{U_i} = \pm\Theta_{\tau_i}$ $(1 \leqslant i \leqslant n)$. Plainly $\Theta_{\tau,T}$ is orthogonal to $\Theta_{\tau_i,T}$ in $L^2(T)$ $(1 \leqslant i \leqslant n)$. [Here, of course, $\Theta_{\tau,T}$ is the analytic extension (from T') to T of $\Delta_T\Theta_\tau$; similarly for $\Theta_{\tau_i,T}$ $(1 \leqslant i \leqslant n)$.] Using Proposition 10.2.2.2, determine constants c_i such that $\Phi_f = \sum_i c_i \bar{\Theta}_{\tau_i, T}$ – then we have

$$\|\Theta_{\tau,\delta}\|_2^2 = \Theta_\tau(f) = (-1)^r [W(G, T)]^{-1} \int_T \Theta_{\tau,T}(t) \Phi_f(t) d_T(t)$$

$$= (-1)^r [W(G, T)]^{-1} \sum_i \int_T \Theta_{\tau,T}(t) \bar{\Theta}_{\tau_i,T}(t) d_T(t) = 0,$$

whence $\Theta_{\tau,\delta} = 0$. . . .

Summary Given $\tau \in \mathsf{L}'_T$, there exists one and only one constant $c(\tau) = \pm 1$ such that $c(\tau)\Theta_\tau$ is a character of the discrete series. Because

$$\Theta_{w\tau} = \det(w)\Theta_\tau \qquad (\text{all } w \in W(G, T)),$$

one has

$$c(w(\tau)) = \det(w)c(\tau) \qquad (\text{all } w \in W(G, T)).$$

Let U be an (irreducible) square integrable unitary representation of G, F_U its character; write

$$F_U = c(\tau)\Theta_\tau \qquad (\tau \in \mathsf{L}'_T).$$

Fix a function f in $\mathscr{C}_U(G)$ such that $f(1) \neq 0$ – then, taking into account Theorem 10.1.2.4 and Proposition 10.2.2.4, we have

$$f(1) = \mathbf{M}_G^{-1}[W(G, T)] \{\prod_{\alpha>0} (\tau, \alpha)\} c(\tau)^{-1}(d_U)^{-1} f(1).$$

On the one hand this shows that

$$c(\tau) = \operatorname{sign}\{\mathbf{M}_G\} \operatorname{sign}\{\prod_{\alpha>0} (\tau, \alpha)\} = (-1)^{\mathbf{m}_G}\epsilon(\tau)$$
$$(\mathbf{m}_G = 2^{-1} \dim (G/K))$$

while on the other that

$$d_U = \frac{[W(G, T)]}{(2\pi)^r}\epsilon(\tau) \prod_{\alpha>0} (\tau, \alpha) \qquad (r = 2^{-1} \dim (G/T)).$$

The assertions of Theorem 10.2.4.1 are now obvious or immediate. [Let us bear in mind that an element in $\hat{G}$ is completely determined by its character. . . .]

Problem The following important question had not been resolved at the time of this writing: For what values of the parameter τ is the associated element $\hat{U}_\tau \in \hat{G}_d$ not only square integrable but actually integrable? [Partial results in this direction can be found in Harish-Chandra [10], [11]; see also Trombi and Varadarajan [2].] Here is a related question: Suppose that $\hat{U}_\tau \in \hat{G}_d$ is actually integrable – then is every element of $\hat{G}_d$ with the same infinitesimal character as $\hat{U}_\tau$ also integrable? [The answer is 'no'; cf. Trombi and Varadarajan [2].]

Remark In connection with the problem posed above, it should be pointed out that one cannot expect that (irreducible) square integrable unitary representations are always integrable. [Thus, for instance, consider $G = \mathbf{SL}(2, \mathbf{R})$ with $T(= K) = \mathbf{SO}(2)$ – then

$$\mathsf{L}_T \sim \{0, \pm 1, \pm 2, \ldots\}$$

while

$$\mathsf{L}_T' \sim \{\pm 1, \pm 2, \ldots\}.$$

The representations indexed by $\pm 2, \pm 3, \ldots$ are actually integrable but the two corresponding to ± 1 are not; incidentally, these considerations serve to show that the elements of $\mathscr{C}(G)$ do not, in general, belong to $L^1(G)$. . . .] There exist interesting relations between the topology on the dual and the question of the integrability of a given irreducible unitary representation of G (cf. Dixmier [12]).

Epilogue

Let G be a connected semi-simple Lie group with finite center, $G = KA_{\mathfrak{p}}N^+$ an Iwasawa decomposition for G; in all that follows it will be assumed that rank $(G/K) = 1$ with rank $(G) =$ rank (K) (so that G admits a compact Cartan subgroup T, say); moreover, for the sake of simplicity, it will be convenient to assume in addition that G is the real analytic subgroup of G_c corresponding to $\mathfrak{g}$, G_c the simply connected complex Lie group with Lie algebra $\mathfrak{g}_c$ (therefore G is, in particular, acceptable). Here, then, we intend to consider the following points:

(1) The Plancherel Theorem for G;

(2) The topology on the reduced dual $\hat{G}_r$ of G;

(3) The Paley-Wiener Theorem for $\mathscr{C}(G)$.

Before embarking on the analytical aspects of our discussion, we shall first dispense with the necessary structural preliminaries. Fix a singular imaginary root $\alpha_{\mathfrak{t}}$ of the pair $(\mathfrak{g}_c, \mathfrak{t}_c)$; let Γ be an element in $\mathfrak{t}$ such that $\pm\alpha_{\mathfrak{t}}$ are the only roots which vanish at Γ – then Γ is a semi-regular element of the non-compact type. Let $\mathfrak{g}_\Gamma$ be the centralizer of Γ in $\mathfrak{g}$; write, in the usual way, $\mathfrak{g}_\Gamma = \mathfrak{c}_\Gamma + \mathfrak{l}_\Gamma$; choose a basis H^*, X^*, Y^* for $\mathfrak{l}_\Gamma$ over $\mathbf{R}$ such that

$$[H^*, X^*] = 2X^*, \quad [H^*, Y^*] = -2Y^*, \quad [X^*, Y^*] = H^*.$$

Then

$$\mathfrak{a}\,(= \mathfrak{h}_\Gamma^+ \cdots) = \mathbf{R}H^* + \mathfrak{c}_\Gamma$$

and

$$\mathfrak{t}\,(= \mathfrak{h}_\Gamma^- \cdots) = \mathbf{R}(X^* - Y^*) + \mathfrak{c}_\Gamma$$

form a complete set of non-conjugate Cartan subalgebras of $\mathfrak{g}$ (cf. Lemma 1.3.4.3). Put

$$\mu = \exp\{\sqrt{-1}(\pi/4)(X^* + Y^*)\} \in G_c.$$

Then $\mu(\mathfrak{t}_c) = \mathfrak{a}_c$ and, moreover, the μ-transform $\alpha_{\mathfrak{a}}$ of $\alpha_{\mathfrak{t}}$ takes on the value 2 at H^* but vanishes identically on $\mathfrak{c}_\Gamma$. Introduce an order into the space of real linear functions on $\mathbf{R}H^* + \sqrt{-1}\mathfrak{c}_\Gamma$ by stipulating that such a function is to be declared positive whenever it assumes a positive value at H^*; under this order, $\alpha_{\mathfrak{a}}$ is certainly a positive root of the pair $(\mathfrak{g}_c, \mathfrak{a}_c)$. A set of positive roots for the pair $(\mathfrak{g}_c, \mathfrak{t}_c)$ is then singled out upon demanding that the μ-transform of such a root must be positive when considered as a root of the pair $(\mathfrak{g}_c, \mathfrak{a}_c)$. Let A be the Cartan subgroup of G associated with $\mathfrak{a}$ – then $\mathsf{A} = \mathsf{Z}(A_{\mathfrak{p}})A$ where

$$\mathsf{Z}(A_{\mathfrak{p}}) = K \cap \exp(\sqrt{-1}\mathbf{R}H^*).$$

It is clear that every element of $\mathsf{Z}(A_{\mathfrak{p}})$ is of order 2; in fact, in the present special situation, $\mathsf{Z}(A_{\mathfrak{p}}) = \{1, \gamma\}$ where

$$\gamma = \exp(\pi(X^* - Y^*)) = \exp(\sqrt{-1}\pi H^*) \neq 1.$$

Put

$$\mathfrak{t}_1 = \mathfrak{c}_\Gamma, \quad \mathfrak{t}_2 = \mathbf{R}(X^* - Y^*)$$

and let T_1, T_2 be the analytic subgroups of T corresponding to $\mathfrak{t}_1$, $\mathfrak{t}_2$, respectively – then both T_1 and T_2 are compact with $T_1 \cap T_2 \subset \mathsf{Z}(A_\mathfrak{p})$. Because

$$\mathsf{A}_K = T_1 \cup \gamma T_1,$$

it follows that A_K has one or two connected components according to whether γ lies in $T_1 \cap T_2$ or not. Let M be the centralizer of $A_\mathfrak{p}$ in K – then, as we know, $\mathsf{M} = M \cup \gamma M$ (M the identity component of M), A_K ($= T_1$) being a maximal torus in M. Set $P = \mathsf{M}A_\mathfrak{p}N^+$ – then P is a minimal (maximal . . .) parabolic subgroup of G.

The group G admits two series of unitary representations which will enter into the Plancherel formula (cf. infra); let us briefly recall their basic properties.

The Discrete Series The unitary character group $\hat{T}$ of T can be identified with a lattice L_T (say) in the dual space of $\sqrt{-1}\mathfrak{t}$. According to Theorem 10.2.4.1, the elements of $\hat{G}_d$ are in a natural one-to-one correspondence with the points of the orbit space $W(G, T)\backslash\hat{T}'$ ($\hat{T}' \sim \mathsf{L}_T'$); moreover

$$\mathbf{M}_G^{-1} \sum_{\tau \in \mathsf{L}_T} \prod_{\alpha > 0} (\tau, \alpha)\Theta_\tau = \sum_{\hat{U} \in \hat{G}_d} d_{\hat{U}} T_{\hat{U}}$$

as (tempered) distributions on G.

The Principal *P*-series The representations $U^{\sigma,\nu}$ in this series are parameterized by the points $(\sigma, \nu) \in \hat{\mathsf{M}} \times \hat{A}_\mathfrak{p}$, two such being unitarily equivalent iff the corresponding parameters lie on the same W-orbit (W the Weyl group of the pair $(\mathfrak{g}, \mathfrak{a}_\mathfrak{p})$). Let $W(M, A_K)$ be the Weyl group of the pair (M, A_K); let $\hat{A}_K$ be the unitary character group of A_K ($=$ the identity component of A_K) – then $\hat{A}_K$ can be identified with a lattice L_{A_K} in the dual space of $\sqrt{-1}\mathfrak{a}_\mathfrak{k}$ on which the group $W(M, A_K)$ operates in the obvious way and, of course,

$$\hat{M} \sim W(M, A_K)\backslash\hat{A}_K' \qquad (\hat{A}_K' \sim \mathsf{L}_{A_K}').$$

The group $W(M, A_K)$ also operates on the unitary dual $\hat{\mathsf{A}}_K$ of A_K; assigning to the symbol $\hat{\mathsf{A}}_K'$ the only possible interpretation, we have $\hat{\mathsf{M}} \sim W(M, A_K)\backslash\hat{\mathsf{A}}_K'$. [The point here is that M may not be connected; but M is at least the central product of M and $\mathsf{Z}(A_\mathfrak{p})$ with respect to $\mathsf{Z}(A_\mathfrak{p}) \cap M$ (cf. part (2) of the proof of Theorem 5.5.4.1).] To explicate this correspondence, fix $\sigma \in \hat{\mathsf{M}}$ – then we may associate with σ (in $[W(M, A_K)]$ ways) a regular element χ_σ (say) in $\hat{\mathsf{A}}_K$ and a regular element $\log \chi_\sigma$ (say) in L_{A_K} such that for all $a \in \mathsf{Z}(A_\mathfrak{p})$

$$\xi_\sigma(ah_K) = \operatorname{sign}\{\prod_{\alpha \in \Phi_I^+} (\log \chi_\sigma, \alpha)\}\xi_\rho(a) \frac{\sum_{w \in W(M, A_K)} \det(w)\langle ah_K, w\chi_\sigma\rangle}{\Delta_I(h_K)} \quad (h_K \in A_K'),$$

where, as always, ξ_σ denotes the character of $\sigma(\Phi_I^+ = P_- \ldots)$. In such a situation it will be convenient to write $|\sigma|$ for $(B(\log \chi_\sigma, \log \chi_\sigma))^{1/2}$; obviously $|\sigma|$ is well defined. [We shall employ this convention of writing in other cases as well; e.g. $|\delta|$ (any $\delta \in \hat{K}$). . . .] Suppose now that $\nu \in \hat{A}_\mathfrak{p}$ – then the character $T^{\sigma,\nu}$ of $U^{\sigma,\nu}$ is given by (cf. Vol. I, number 5.5.4)

$$T^{\sigma,\nu}(f) = (-1)^{r_I} \operatorname{sign}\{\prod_{\alpha \in \Phi_I^+} (\log \chi_\sigma, \alpha)\} \iint_{\mathsf{A}_K \times A_\mathfrak{p}} \chi_\sigma(h_K)\, h_\mathfrak{p}^{\sqrt{-1}\nu}$$
$$\times\, \Phi_f^{\mathsf{A}}(h_K h_\mathfrak{p})\, d_{\mathsf{A}_K}(h_K)\, d_{A_\mathfrak{p}}(h_\mathfrak{p}),$$

Φ_f^{A} the invariant integral of f ($f \in C_c^\infty(G)$) calculated relative to the Cartan subgroup A.

With this preparation, the Plancherel Theorem for G reads as follows (cf. Harish-Chandra [29]).

Theorem E 1 (Harish-Chandra) Fix $f \in L^1(G) \cap L^2(G)$ – then

$$\int_G |f(x)|^2 \, d_G(x) = \sum_{\hat{U} \in \hat{G}_d} d_{\hat{U}} \operatorname{tr}(T_{\hat{U}}(f) T_{\hat{U}}(f)^*)$$
$$+ 2^{-1} \sum_{\sigma \in \hat{\mathsf{M}}} \int_{-\infty}^{\infty} \mathbf{Q}(\sigma, \nu) \operatorname{tr}(U^{\sigma,\nu}(f) U^{\sigma,\nu}(f)^*) \, d\nu.$$

The non-negative function $\mathbf{Q}$ has the following properties:

(i) For any w in W (the Weyl group of the pair $(\mathfrak{g}, \mathfrak{a}_{\mathfrak{p}})$),

$$\mathbf{Q}(\sigma, \nu) = \mathbf{Q}(\sigma, -\nu) = \mathbf{Q}(w\sigma, \nu) \qquad ((\sigma, \nu) \in \hat{\mathsf{M}} \times \hat{A}_{\mathfrak{p}});$$

(ii) For any σ in $\hat{\mathsf{M}}$, $\mathbf{Q}(\sigma, .)$ (qua a function of ν) is the restriction to $\mathbf{R}(\sim \hat{A}_{\mathfrak{p}})$ of a meromorphic function on $\mathbf{C}$ with no real poles;

(iii) For any σ in $\hat{\mathsf{M}}$ and any non-zero ν, $\mathbf{Q}(\sigma, \nu) \neq 0$;

(iv) For every non-negative integer n, there are polynomials u and v such that

$$\left| \left(\frac{d}{d\nu}\right)^n \mathbf{Q}(\sigma, \nu) \right| \leqslant u(|\sigma|) v(|\nu|) \qquad ((\sigma, \nu) \in \hat{\mathsf{M}} \times \hat{A}_{\mathfrak{p}}).$$

Note Property (i) serves to ensure us that $\mathbf{Q}$ is W-invariant (as it should be).

[In passing observe that the integration over $\hat{\mathsf{M}} \times (-\infty, \infty)$ can be replaced, if desired, by one over $\hat{\mathsf{M}} \times [0, \infty)$.]

It will be convenient to preface the proof with some preliminary remarks.

Fix a τ in L_T – then, according to a theorem which was formulated in the Appendix to number 10.1.1 (and valid even if τ is singular ...), there exists a central analytic function Ψ_τ on G' (which can actually be extended continuously to all of G) such that

$$\prod_{\alpha > 0} (\tau, \alpha) \Theta_\tau(f) = \text{P.V.} \int_G D_I^{-1}(x) \Psi_\tau(x) f(x; \nabla_G) \, d_G(x) \qquad (\text{all } f \in C_c^\infty(G)).$$

In the case at hand, one can compute Ψ_τ $(\tau \in \mathsf{L}_T)$ explicitly (cf. Harish-Chandra [29, p. 122]) – in fact:

$$(T) \quad \Psi_\tau(t) = \sum_{w \in W(G,T)} \xi_{w\tau}(t) \qquad (t \in T),$$
$$(\mathsf{A}) \quad \Psi_\tau(h_K h_{\mathfrak{p}}) = \sum_{w \in W(G,T)} \xi_{w\tau}(h_K) \exp(-|(w\tau)^\mu (\log h_{\mathfrak{p}})|)$$
$$(h_K \in \mathsf{A}_K, h_{\mathfrak{p}} \in A_{\mathfrak{p}}).$$

Note For a 'self-contained' derivation of the above formulae, see Okamoto [1].

Let d_{A_K} denote the normalized Haar measure on A_K; put $h_t^* = \exp(tH^*)$ $(t \in \mathbf{R})$ and then normalize the Haar measure d_{A} on A in such a way that

$$d_{\mathsf{A}}(h) = d_{\mathsf{A}_K}(h_K) \, dt$$

if $h = h_K h_t^*$. [Needless to say, the current normalization of the Haar measure on A differs by a positive explicitly computable factor $\mathbf{c}_{\mathsf{A}}$ (say) from the normalization agreed to in Vol. I, number 5.5.3 (one can calculate $\mathbf{c}_{\mathsf{A}}$ by the procedure outlined in the discussion following Lemma 8.4.2.5).] Let d_T denote the normalized

Haar measure on T – then, for any $\tau \in \mathsf{L}_T$, we have

$$\prod_{\alpha>0} (\tau, \alpha)\Theta_\tau(f) = [W(G, T)]^{-1} \int_T \Psi_\tau \cdot \Pi^T \Phi_f^T \, d_T(t)$$
$$- \mathbf{c}_\mathsf{A}[W(G, \mathsf{A})]^{-1} \int_\mathsf{A} \epsilon_R^\mathsf{A} \Psi_\tau \cdot \Pi^\mathsf{A} \Phi_f^\mathsf{A} \, d_\mathsf{A}(h) \qquad (f \in C_c^\infty(G)).$$

We propose now to consider the series

$$\sum_{\tau \in \mathsf{L}_T} \int_\mathsf{A} \epsilon_R^\mathsf{A} \Psi_\tau \cdot \Pi^\mathsf{A} \Phi_f^\mathsf{A} \, d_\mathsf{A}(h) \qquad (f \in C_c^\infty(G)).$$

To begin with, observe that the series in question is actually absolutely convergent. [On the basis of Theorem 10.1.1.8 and the relation

$$\sum_{\tau \in \mathsf{L}_T} \int_T \Psi_\tau \cdot \Pi^T \Phi_f^T \, d_T(t) = [W(G, T)] \Phi_f^T(1; \Pi^T) \qquad (f \in C_c^\infty(G)),$$

this is clear.] Keeping in mind the formula for $\Psi_\tau | \mathsf{A}$ ($\tau \in \mathsf{L}_T$), it then follows that

$$\sum_{\tau \in \mathsf{L}_T} \int_\mathsf{A} \epsilon_R^\mathsf{A} \Psi_\tau \cdot \Pi^\mathsf{A} \Phi_f^\mathsf{A} \, d_\mathsf{A}(h)$$
$$= 2[W(G, T)] \sum_{\tau \in \mathsf{L}_T} \int_0^\infty e^{-t|\tau(\sqrt{-1}(X^* - Y^*))|} \, dt \int_{\mathsf{A}_K} \xi_\tau(h_K) \Phi_f^\mathsf{A}(h_K h_t^*; \Pi^\mathsf{A}) \, d_{\mathsf{A}_K}(h_K)$$
$$(f \in C_c^\infty(G)).$$

Let $\tilde{\mathsf{L}}_T$ be the sublattice of L_T consisting of those τ in L_T which vanish at $\sqrt{-1}(X^* - Y^*)$; let L_0 be the lattice generated by $\tilde{\mathsf{L}}_T$ and α_t – then it is not difficult to see that there exists an element $\tau_0 \in \mathsf{L}_T$ having the property that

$$\tau_0(\sqrt{-1}(X^* - Y^*)) = 1.$$

In addition $\mathsf{L}_T/\mathsf{L}_0$ is an additive group of order 2 and $\tau_0 \notin \mathsf{L}_0$. [Fix $\tau \in \mathsf{L}_T$ – then, since $\alpha_\mathrm{t}(\sqrt{-1}(X^* - Y^*)) = 2$, there exists an integer n_τ such that

$$(\tau - n_\tau \alpha_\mathrm{t})(\sqrt{-1}(X^* - Y^*)) = 0 \text{ or } 1,$$

whence either $\tau \in \mathsf{L}_0$ or $\tau - \tau_0 \in \mathsf{L}_0$. The fact that $\tau_0 \notin \mathsf{L}_0$ is equally obvious.] We claim that for all $f \in C_c^\infty(G)$

$$\sum_{\tau \in \mathsf{L}_T} \int_\mathsf{A} \epsilon_R^\mathsf{A} \Psi_\tau \cdot \Pi^\mathsf{A} \Phi_f^\mathsf{A} \, d_\mathsf{A}(h)$$
$$= [W(G, T)] \sum_{a \in Z(A_\mathfrak{p})} \int_0^\infty \Phi_f^\mathsf{A}(a h_t^*; \Pi^\mathsf{A}) \{\coth(t) + \xi_{\tau_0}(a)(\sinh(t))^{-1}\} \, dt.$$

Because $\alpha_\mathrm{t}(\sqrt{-1}(X^* - Y^*)) = 2$ while

$$\coth(t) = 1 + 2 \sum_{n=1}^\infty e^{-2nt}, \quad (\sinh(t))^{-1} = 2 \sum_{n=0}^\infty e^{-(2n+1)t} \qquad (t > 0),$$

the claim is (evidently . . .) a consequence of the following statement: Fix $\tau \in \mathsf{L}_T$ – then, for any $f \in C_c^\infty(G)$, we have

$$\sum_{\tilde{\tau} \in \tilde{\mathsf{L}}_T} \int_{\mathsf{A}_K} \xi_{\tau+\tilde{\tau}}(h_K) \Phi_f^\mathsf{A}(h_K h_t^*; \Pi^\mathsf{A}) \, d_{\mathsf{A}_K}(h_K) = 2^{-1} \sum_{a \in Z(A_\mathfrak{p})} \xi_\tau(a) \Phi_f^\mathsf{A}(a h_t^*; \Pi^\mathsf{A})$$
$$(t \in \mathbf{R}).$$

To verify this assertion, it is best to distinguish two cases: (1) $\gamma \in T_1$ (so that $\mathsf{A}_K = T_1$); (2) $\gamma \notin T_1$ (so that $\mathsf{A}_K = T_1 \cup \gamma T_1$ (disjoint union)). Consider, for instance, the first case. Fix $f \in C_c^\infty(G)$; given $h_K \in \mathsf{A}_K$, put

$$F(h_K) = \sum_{a \in Z(A_\mathfrak{p})} \xi_\tau(ah_K)\Phi_f^{\mathsf{A}}(ah_K h_t^*; \Pi^{\mathsf{A}}) \qquad (h_t^* \text{ fixed}).$$

Then F is a continuous function on A_K (cf. Corollary 8.5.1.5). Moreover it is not difficult to see that the Fourier transform $\hat{F}$ of F lies in $L^1(\hat{\mathsf{A}}_K)$ so that in particular,

$$F(1) = \int_{\hat{\mathsf{A}}_K} \hat{F}(\chi)\, d_{\hat{\mathsf{A}}_K}(\chi).$$

The very definition of F tells us that $\hat{F}(\chi) = 0$ unless $\chi \,|\, T_1 \cap T_2 = 1$; since $T/T_2 \sim \mathsf{A}_K/T_1 \cap T_2$, it therefore follows that

$$\begin{aligned}
&\sum_{a \in Z(A_\mathfrak{p})} \xi_\tau(a)\Phi_f^{\mathsf{A}}(ah_t^*; \Pi^{\mathsf{A}}) \\
&= F(1) = \sum_{\tilde{\tau} \in \tilde{L}_T} \hat{F}(\xi_{\tilde{\tau}}) \\
&= 2 \sum_{\tilde{\tau} \in \tilde{L}_T} \int_{\mathsf{A}_K} \xi_{\tau+\tilde{\tau}}(h_K)\Phi_f^{\mathsf{A}}(h_K h_t^*; \Pi^{\mathsf{A}})\, d_{\mathsf{A}_K}(h_K),
\end{aligned}$$

as desired. The second possibility can be treated in a similar way.

Proof of Theorem E.1 Fix $f \in C_c^\infty(G)$; set

$$\varphi_f^{\pm}(t) = 2^{-1}\{\Phi_f^{\mathsf{A}}(h_t^*; \Pi^{\mathsf{A}}/H_{\alpha_\mathfrak{a}}) \pm \Phi_f^{\mathsf{A}}(\gamma h_t^*; \Pi^{\mathsf{A}}/H_{\alpha_\mathfrak{a}}\} \qquad (t \in \mathbf{R}).$$

Then both φ_f^+ and φ_f^- are even functions of t. Because $\xi_{\tau_0}(\gamma) = -1$ while

$$H_{\alpha_\mathfrak{a}} = (|\alpha_\mathfrak{a}|^2/2)H^* \qquad (|\alpha_\mathfrak{a}|^2 = B(H_{\alpha_\mathfrak{a}}, H_{\alpha_\mathfrak{a}})),$$

it follows from what has been said above that

$$\begin{aligned}
&\sum_{\tau \in L_T} \int_{\mathsf{A}} \epsilon_R^{\mathsf{A}} \Psi_\tau \cdot \Pi^{\mathsf{A}} \Phi_f^{\mathsf{A}}\, d_{\mathsf{A}}(h) \\
&= [W(G,T)]\,|\alpha_\mathfrak{a}|^2 \int_0^\infty \{\coth(t) \cdot d\varphi_f^+/dt + (\sinh(t))^{-1} \cdot d\varphi_f^-/dt\}\, dt.
\end{aligned}$$

Put

$$\hat{\varphi}_f^{\pm}(\nu) = \int_{-\infty}^{\infty} \varphi_f^{\pm}(t) e^{\sqrt{-1}\nu t}\, dt \qquad (\nu \in \mathbf{R}).$$

Then it follows from the theory of Fourier transforms that

$$\begin{aligned}
2\int_0^\infty d\varphi_f^+/dt \cdot \coth(t)\, dt &= \int_{-\infty}^{\infty} d\varphi_f^+/dt \cdot \coth(t)\, dt \\
&= -\int_{-\infty}^{\infty} \frac{\nu}{2} \coth\left(\frac{\pi\nu}{2}\right) \cdot \hat{\varphi}_f^+(\nu)\, d\nu \\
&= -\int_0^\infty \nu \coth\left(\frac{\pi\nu}{2}\right) \cdot \hat{\varphi}_f^+(\nu)\, d\nu,
\end{aligned}$$

φ_f^+ being an even function; in a similar way we find that

$$2\int_0^\infty d\varphi_f^-/dt \cdot (\sinh(t))^{-1}\, dt = -\int_0^\infty \nu \tanh\left(\frac{\pi\nu}{2}\right) \cdot \hat{\varphi}_f^-(\nu)\, d\nu.$$

Let

$$\hat{\mathsf{A}}_K^{\pm} = \{\chi \in \hat{\mathsf{A}}_K : \chi(\gamma) = \pm 1\}$$

(since $\gamma^2 = 1$, every element in $\hat{\mathsf{A}}_K$ lies in either $\hat{\mathsf{A}}_K^+$ or $\hat{\mathsf{A}}_K^-$) – then it is clear that

$$\sum_{\chi \in \hat{\mathsf{A}}_K \pm} \int_{\mathsf{A}_K} \int_{-\infty}^{\infty} \Phi_f^{\mathsf{A}}(h_K h_t^*; \Pi^{\mathsf{A}}/H_{\alpha_{\mathfrak{a}}}) \chi(h_K) e^{\sqrt{-1}\nu t} \, d_{\mathsf{A}_K}(h_K) \, dt = \hat{\varphi}_f^{\pm}(\nu)$$

for all real ν. Given $\chi \in \hat{\mathsf{A}}_K$, let us agree to denote by $\log \chi$ the linear function on $\mathfrak{a}_c$ determined by the requirement

$$\chi(\exp H) = e^{\log \chi(H)} \qquad (H \in \mathfrak{a}_{\mathfrak{k}})$$

(it being understood, of course, that $\log \chi \,|\, \mathfrak{a}_{\mathfrak{p}_c} = 0$). Putting everything together then leads to the conclusion that

$$\begin{aligned}
f(1) &= \mathbf{M}_G^{-1} \Phi_f^T(1; \Pi^T) \\
&= \mathbf{M}_G^{-1} [W(G,T)]^{-1} \sum_{\tau \in \mathsf{L}_T} \int_T \Psi_\tau \cdot \Pi^T \Phi_f^T \, d_T(t) \\
&= \mathbf{M}_G^{-1} \sum_{\tau \in \mathsf{L}_T} \prod_{\alpha > 0} (\tau, \alpha) \Theta_\tau(f) + \mathbf{M}_G^{-1} \mathbf{c}_{\mathsf{A}} [W(G, \mathsf{A})]^{-1} \\
&\quad \times \sum_{\tau \in \mathsf{L}_T} \int_{\mathsf{A}} \epsilon_R^{\mathsf{A}} \Psi_\tau \cdot \Pi^{\mathsf{A}} \Phi_f^{\mathsf{A}} \, d_{\mathsf{A}}(h) \\
&= \sum_{\hat{U} \in \hat{G}_d} d_{\hat{U}} T_{\hat{U}}(f) + \sqrt{-1}(-1)^{r+1} \mathbf{M}_G^{-1} \mathbf{c}_{\mathsf{A}} ([W(G,T)]/[W(G,\mathsf{A})]) \\
&\quad \times \Bigg\{ \sum_{\chi \in \hat{\mathsf{A}}_K^+} \int_0^\infty \coth\left(\frac{\pi\nu}{2}\right) \prod_{\alpha > 0} \left(\log \chi + \frac{\sqrt{-1}\nu}{2} \alpha_{\mathfrak{a}}, \alpha\right) \\
&\quad \times \left[\iint_{\mathsf{A}_K \times \mathbf{R}} \Phi_f^{\mathsf{A}}(h_K h_t^*) \chi(h_K) e^{\sqrt{-1}\nu t} \, d_{\mathsf{A}_K}(h_K) \, dt \right] d\nu \\
&\quad + \sum_{\chi \in \hat{\mathsf{A}}_K^-} \int_0^\infty \tanh\left(\frac{\pi\nu}{2}\right) \prod_{\alpha > 0} \left(\log \chi + \frac{\sqrt{-1}\nu}{2} \alpha_{\mathfrak{a}}, \alpha\right) \\
&\quad \times \left[\iint_{\mathsf{A}_K \times \mathbf{R}} \Phi_f^{\mathsf{A}}(h_K h_t^*) \chi(h_K) e^{\sqrt{-1}\nu t} \, d_{\mathsf{A}_K}(h_K) \, dt \right] d\nu \Bigg\}.
\end{aligned}$$

Here r denotes the number of positive roots of the pair $(\mathfrak{g}_c, \mathfrak{a}_c)$, i.e.

$$r = 2^{-1} (\dim(\mathfrak{g}) - \operatorname{rank}(\mathfrak{g})).$$

We have seen above that one can attach to each χ in $\hat{\mathsf{A}}_K'$ an element σ_χ (say) in $\hat{\mathsf{M}}$; so, in view of the fact that

$$(-1)^r = (-1)^{r_I}(-1)^{r_R}(-1)^{r_C} = (-1)^{r_I}(-1)^1,$$

we can write

$$\begin{aligned}
f(1) = {} & \sum_{\hat{U} \in \hat{G}_d} d_{\hat{U}} T_{\hat{U}}(f) + \sqrt{-1} \mathbf{M}_G^{-1} ([W(G,T)]/[W(G,\mathsf{A})]) \\
& \times \Bigg\{ \sum_{\chi \in \hat{\mathsf{A}}_K^+} \int_0^\infty \coth\left(\frac{\pi\nu}{2}\right) \prod_{\alpha > 0} \left(\log \chi + \frac{\sqrt{-1}\nu}{2} \alpha_{\mathfrak{a}}, \alpha\right) \\
& \times \operatorname{sign}\{ \prod_{\alpha \in \Phi_I^+} (\log \chi, \alpha) \} T^{\sigma_\chi, \nu}(f) \, d\nu \\
& + \sum_{\chi \in \hat{\mathsf{A}}_K^-} \int_0^\infty \tanh\left(\frac{\pi\nu}{2}\right) \prod_{\alpha > 0} \left(\log \chi + \frac{\sqrt{-1}\nu}{2} \alpha_{\mathfrak{a}}, \alpha\right) \\
& \times \operatorname{sign}\{ \prod_{\alpha \in \Phi_I^+} (\log \chi, \alpha) \} T^{\sigma_\chi, \nu}(f) \, d\nu \Bigg\}.
\end{aligned}$$

Define now

$$\mathbf{q}(\chi, \nu) = \sqrt{-1}\mathbf{M}_G^{-1}([W(G, T)]/[W(G, \mathsf{A})])$$
$$\times \begin{cases} \coth\left(\frac{\pi\nu}{2}\right) \prod_{\alpha>0}\left(\log\chi + \frac{\sqrt{-1}\nu}{2}\alpha_{\mathfrak{a}}, \alpha\right) \cdot \operatorname{sign}\{\prod_{\alpha\in\Phi_I^+}(\log\chi, \alpha)\} \\ \tanh\left(\frac{\pi\nu}{2}\right) \prod_{\alpha>0}\left(\log\chi + \frac{\sqrt{-1}\nu}{2}\alpha_{\mathfrak{a}}, \alpha\right) \cdot \operatorname{sign}\{\prod_{\alpha\in\Phi_I^+}(\log\chi, \alpha)\} \end{cases}$$
$$(\nu \in \mathbf{R})$$

where coth or tanh is used depending on whether χ is in $\hat{\mathsf{A}}_K^+$ or $\hat{\mathsf{A}}_K^-$. If χ is regular, then certainly

$$\frac{\prod_{\alpha>0}\left(\log\chi + \frac{\sqrt{-1}\nu}{2}\alpha_{\mathfrak{a}}, \alpha\right)}{\left|\prod_{\alpha>0}\left(\log\chi + \frac{\sqrt{-1}\nu}{2}\alpha_{\mathfrak{a}}, \alpha\right)\right|}$$
$$= \sqrt{-1}(-1)^{\mathbf{m}_G+1}\operatorname{sign}\{\nu\}\cdot\operatorname{sign}\{\prod_{\alpha\in\Phi_I^+}(\log\chi, \alpha)\} \qquad (\nu \neq 0).$$

Because

$$\mathbf{M}_G = (2\pi)^r(-1)^{\mathbf{m}_G}$$

(cf. Theorem 8.4.5.1), it is then clear that $\mathbf{q}$ is non-negative; we have, moreover,

$$f(1) = \sum_{\hat{U}\in\hat{G}_d} d_{\hat{U}} T_{\hat{U}}(f) + \sum_{\chi\in\hat{\mathsf{A}}_K} \int_0^\infty \mathbf{q}(\chi, \nu) T^{\sigma_\chi, \nu}(f)\, d\nu.$$

Fix ν – then $\mathbf{q}$, qua a function of χ, is $W(M, A_K)$-invariant. This being the case, let $\mathbf{Q}$ be the function on $\hat{\mathsf{M}} \times \hat{A}_{\mathfrak{p}}$ defined by the rule

$$\mathbf{Q}(\sigma, \nu) = [W(M, A_K)]\mathbf{q}(\chi_\sigma, \nu).$$

Then Q is well-defined and

$$f(1) = \sum_{\hat{U}\in\hat{G}_d} d_{\hat{U}} T_{\hat{U}}(f) + \sum_{\sigma\in\hat{\mathsf{M}}} \int_0^\infty \mathbf{Q}(\sigma, \nu) T^{\sigma, \nu}(f)\, d\nu$$
$$= \sum_{\hat{U}\in\hat{G}_d} d_{\hat{U}} T_{\hat{U}}(f) + 2^{-1} \sum_{\sigma\in\hat{\mathsf{M}}} \int_{-\infty}^\infty \mathbf{Q}(\sigma, \nu) T^{\sigma, \nu}(f)\, d\nu.$$

Since f is an arbitrary element in $C_c^\infty(G)$ and since $\mathbf{Q}$ obviously possesses the properties (i)–(iv) supra, the proof of our theorem is complete (cf. number 7.2.2). □

We shall leave it to the reader to explicate the above result when $G = \mathbf{SL}(2, \mathbf{R})$ (take care to observe that under the identification $\hat{\mathsf{M}} \sim W(M, A_K)\backslash\hat{\mathsf{A}}'_K$, the class of the trivial one dimensional representation of M corresponds to the character $1 \mapsto 1, \gamma \mapsto -1 \ldots$).

Remark Recently P. Sally and the author have given a different proof of the theorem supra; in particular this proof avoids the use of Harish-Chandra's 'principal value integral'.

Given $\hat{U} \in \hat{G}_d$, fix a representation U in the class $\hat{U}$ which operates on the Hilbert space E_U (say); let $\mathfrak{L}^2(E_U; E_U)$ be the space of Hilbert-Schmidt operators on E_U with the Hilbert-Schmidt norm $\| \,.\, \|_{HS}$. Let $L^2(\hat{G}_d)$ be the set of all functions

$$f_d : \hat{G}_d \to \sum_{\hat{U} \in \hat{G}_d} \mathfrak{L}^2(E_U; E_U) \qquad \text{(Hilbert space direct sum)}$$

such that:

(i) $f_d(\hat{U}) \in \mathfrak{L}^2(E_U; E_U)$ for each $\hat{U}$ in $\hat{G}_d$;

(ii) $\|f_d\|^2 = \sum_{\hat{U} \in \hat{G}_d} \|f_d(\hat{U})\|_{HS}^2 \, d_{\hat{U}} < \infty$.

Then $L^2(\hat{G}_d)$ is a Hilbert space.

Let $E^{\sigma,\nu}$ be the representation space for the principal P-series representation $U^{\sigma,\nu}$; let $I^{\sigma,\nu}$ be a fixed unitary intertwining operator between $U^{\sigma,\nu}$ and $U^{w(\sigma,\nu)}$ (w the unique non-trivial element in the Weyl group W of the pair $(\mathfrak{g}, \mathfrak{a}_{\mathfrak{p}})$) – thus, by definition,

$$I^{\sigma,\nu}U^{\sigma,\nu}(x) = U^{w(\sigma,\nu)}(x)I^{\sigma,\nu} \qquad (\text{all } x \in G).$$

Let E^σ be the representation space for the representation of K unitarily induced by σ – then, as we know, restricting functions in $E^{\sigma,\nu}$ to K allows us to realize $U^{\sigma,\nu}$ on E^σ (cf. Vol. I, number 5.5.1). Let $\mathfrak{L}^2(E^\sigma; E^\sigma)$ be the space of Hilbert-Schmidt operators on E^σ with the Hilbert-Schmidt norm $\| \,.\, \|_{HS}$. Let $L^2(\hat{G}_P)$ be the set of all functions

$$f_P : \hat{M} \times \hat{A}_{\mathfrak{p}} \to \sum_{\sigma \in \hat{M}} \mathfrak{L}^2(E^\sigma; E^\sigma) \qquad \text{(Hilbert space direct sum)}$$

such that:

(i) $f_P(\sigma, \nu) \in \mathfrak{L}^2(E^\sigma; E^\sigma)$ for $(\sigma, \nu) \in \hat{M} \times \hat{A}_{\mathfrak{p}}$;

(ii) $f_P(w(\sigma, \nu)) = I^{\sigma,\nu} f_P(\sigma, \nu)(I^{\sigma,\nu})^{-1}$ for $(\sigma, \nu) \in \hat{M} \times \hat{A}_{\mathfrak{p}}$;

(iii) $f_P(\sigma, .)$ is a Borel function of ν for each $\sigma \in \hat{M}$;

(iv) $\|f_P\|^2 = 2^{-1} \sum_{\sigma \in \hat{M}} \int_{-\infty}^{\infty} \|f_P(\sigma, \nu)\|_{HS}^2 \mathbf{Q}(\sigma, \nu) \, d\nu < \infty$.

Then $L^2(\hat{G}_P)$ is a Hilbert space. [Here, of course, it is understood that we make the usual agreements as regards functions which are defined (or equal) almost everywhere. . . .]

Now let $L^2(\hat{G})$ be the Hilbert space direct sum of $L^2(\hat{G}_d)$ and $L^2(\hat{G}_P)$; given $f \in C_c^\infty(G)$, define its 'operator valued' Fourier transform $\mathfrak{F}f$ by the prescription $\mathfrak{F}f = (\mathfrak{F}_d f, \mathfrak{F}_P f)$ where

$$(\mathfrak{F}_d f)(\hat{U}) = \int_G f(x)U(x) \, d_G(x) \qquad (\hat{U} \in \hat{G}_d),$$

$$(\mathfrak{F}_P f)(\sigma, \nu) = \int_G f(x)U^{\sigma,\nu}(x) \, d_G(x) \qquad ((\sigma, \nu) \in \hat{M} \times \hat{A}_{\mathfrak{p}}).$$

[Needless to say, we can regard $(\mathfrak{F}_P f)(\sigma, \nu)$ $((\sigma, \nu) \in \hat{M} \times \hat{A}_{\mathfrak{p}})$ as an operator on E^σ.]

Theorem E 2 (Arthur) The map $f \mapsto \mathfrak{F}f \; (f \in C_c^\infty(G))$ extends to an isometric isomorphism from $L^2(G)$ onto $L^2(\hat{G})$.

Proof Owing to Theorem E1, the map in question is an isometry; our main problem, therefore, is to show that it is surjective. Since the Schur Orthogonality Relations serve to ensure the surjectivity on the 'discrete spectrum', we need only show that our map goes onto all of $L^2(\hat{G}_P)$. Let $\mathfrak{U}_P$ be the unitary representation of $G \times G$ on $L^2(\hat{G}_P)$ defined by the rule

$$\mathfrak{U}_P(x, y)f_P(\sigma, \nu) = U^{\sigma,\nu}(x)f_P(\sigma, \nu)U^{\sigma,\nu}(y^{-1}) \qquad (x, y \in G).$$

Since $G \times G$ is type I, the representation $\mathfrak{U}_P$ is type I. Let $\mathfrak{C}$ be the measure class on $\hat{M} \times \mathbf{R}^+$ defined by the discrete measure on $\hat{M}$ and the Lebesgue measure on $\mathbf{R}^+$; the density $\mathbf{Q}$ does not vanish on $\hat{M} \times \mathbf{R}^+$ and the representations

$$U^{\sigma,\nu} \times U^{\sigma,\nu} \qquad ((\sigma, \nu) \in \hat{M} \times \mathbf{R}^+)$$

of $G \times G$ are all irreducible and inequivalent (cf. Theorem 5.5.2.1). Because $\mathfrak{U}_P$ is the direct integral of these representations of $G \times G$ with respect to the measure class $\mathfrak{C}$, it follows from standard generalities (cf. Mackey [6, p. 160]) that $\mathfrak{U}_P$ is multiplicity free, hence that the algebra of $\mathfrak{U}_P$-intertwining operators is commutative. The map

$$f \mapsto \mathfrak{F}_P f \qquad (f \in L^2(G))$$

interwines the two-sided regular representation of $G \times G$ on $L^2(G)$ with $\mathfrak{U}_P$; so, if $\mathscr{S}_P$ is the closure of the set $\{\mathfrak{F}_P f : f \in L^2(G)\}$ and if P is the orthogonal projection of $L^2(\hat{G}_P)$ onto $\mathscr{S}_P$, then P intertwines $\mathfrak{U}_P$; as is well-known, this means that P is of the form $\mathsf{P}_{\mathfrak{B}}$ where $\mathfrak{B}$ is a Borel subset of $\hat{M} \times \mathbf{R}^+$ (the range of $\mathsf{P}_{\mathfrak{B}}$ consisting of those f_P in $L^2(\hat{G}_P)$ which vanish off of $\mathfrak{B}$). The proof of our theorem will therefore be complete when it is shown that the complement of $\mathfrak{B}$ in $\hat{M} \times \mathbf{R}^+$ is a null set with respect to $\mathfrak{C}$. If the contrary were true, then there would necessarily exist a σ in $\hat{M}$ and a subset $\mathscr{S}^+$ of $\mathbf{R}^+$ of positive Lebesgue measure such that

$$(\mathfrak{F}_P f)(\sigma, \nu) = 0 \qquad (\text{any } f \in C_c^\infty(G))$$

for almost all ν in $\mathscr{S}^+$; a little reflection leads without difficulty to the conclusion that this is manifestly an impossibility. . . . □

Our next objective will be to compute the topology on the reduced dual $\hat{G}_r$ of G (cf. Lipsman [6]). Agreeing to assign to the symbols $\hat{G}_P$, $\check{G}_P$ the same connotations as in number 7.1.2, note that

$$\hat{G}_d \cap (\hat{G}_P \cup \check{G}_P) = \varnothing.$$

[Since the character of an element in $\hat{G}_P$ cannot live on the elliptic set, it is enough to prove that $\hat{G}_d \cap \check{G}_P = \varnothing$; in turn, for this purpose, one need only remark that if the intersection $\hat{G}_d \cap \check{G}_P$ were not empty, then some discrete series representation of G would necessarily have a *singular* infinitesimal character. . . .]

Theorem E 3 (Lipsman) Let $\hat{G}_r$ be the reduced dual of G – then

$$\hat{G}_r = \hat{G}_d \cup \hat{G}_P \cup \check{G}_P \qquad (\text{disjoint union}),$$

both $\hat{G}_d$ and $\hat{G}_P \cup \check{G}_P$ being open in $\hat{G}_r$. The hull-kernel topology on $\hat{G}_d$ is discrete; the hull-kernel topology on $\hat{G}_P$ is the same as that provided by Theorem 7.1.2.1; the hull-kernel topology on $\check{G}_P$ is discrete.

Proof The thrust of the present theorem lies in the assertion that the hull-kernel topology on $\check{G}_P$ is discrete. In fact $\hat{G}_d$ is always closed in $\hat{G}$ (cf. the Remark following Corollary 10.2.4.2) as is $\hat{G}_P \cup \check{G}_P$ (being the hull-kernel closure of $\hat{G}_P$); on the other hand, thanks to Theorem E1,

$$\hat{G}_r = \hat{G}_d \cup \hat{G}_P \cup \check{G}_P. \ldots$$

Let us show, then, that the hull-kernel topology on $\check{G}_P$ is discrete. Evidently

$$\mathscr{Q}_P - \hat{G}_P \subset \{W \cdot (\sigma, 0) : \sigma \in \hat{M}\}$$

is a discrete, countable, and therefore closed subset of the orbit space

$$\mathscr{Q}_P = W\backslash(\hat{M} \times \hat{A}_{\mathfrak{p}})$$

in the natural topology (W the Weyl group of the pair $(\mathfrak{g}, \mathfrak{a}_{\mathfrak{p}})$). If $\mathscr{Q}_P - \hat{G}_P$ is empty, then $\check{G}_P = \varnothing$ and there is nothing to prove; otherwise, let $\mathscr{S}$ be an arbitrary subset of $\check{G}_P$ – then we intend to prove that $\mathscr{S}$ is closed in $\hat{G}$. Let $\hat{U}$ be a point in the hull-kernel closure of $\mathscr{S}$; choose a net $\{\hat{U}_n\}$ in $\mathscr{S}$ such that $\hat{U}_n \to \hat{U}$ in the hull-kernel topology – then no subnet of $\{\hat{U}_n\}$ can converge in the discrete topology to the point at infinity of $\check{G}_P$. [Without altering the notation, suppose, if possible, that $\hat{U}_n \to \infty$ in the discrete topology; select an element $q_n \in \mathscr{Q}_P - \hat{G}_P$ such that $\hat{U}_n$ is a direct summand of q_n (q_n is *unique;* cf. the discussion following Corollary 5.5.2.2) – then, by a simple argument, we deduce that $q_n \to \infty$ in the natural topology of $\mathscr{Q}_P$. This being the case, Theorem 5.5.4.1 now implies that

$$\lim_n \int_G f(x) F_{q_n}(x)\, d_G(x) = 0$$

for every $f \in C_c^\infty(G)$ (F_{q_n} the character of q_n). Because $\hat{U}_n$ is a direct summand of q_n, it follows that no subnet of $\{\hat{U}_n\}$ can converge hull-kernelwise to any limit whatsoever (cf. Fell [1, p. 382 and p. 390]), a contradiction. . . .] Accordingly, let $\hat{V}$ be a discrete cluster point of $\{\hat{U}_n\}$ in $\check{G}_P$; let $\{\hat{V}_n\}$ be a subnet of $\{\hat{U}_n\}$ which converges to $\hat{V}$ in the discrete topology – then $\hat{V} \in \mathscr{S}$ and $\hat{U}$ is in the hull-kernel closure of $\{\hat{V}\}$. Since $\hat{G}$ is a T_1-space (G is liminaire . . .), it therefore follows that $\hat{U} = \hat{V} \in \mathscr{S}$, as we wished to prove. □

Remarks (1) It can very well happen that the cardinality of $\check{G}_P$ is infinite. [This phenomenon is present in the universal covering group of $\mathbf{SO}_e(4, 1)$; cf. Takahashi [2].]

(2) The set $\hat{G}_d$ is not, in general, open in $\hat{G}$. [For example, in $G = \mathbf{SL}(2, \mathbf{R})$, a certain member in the discrete series is a limit point of representations in the complementary series.]

Note In the case at hand, one can say the following as regards the irreducibility of the $U^{\sigma,0}$ when $w\sigma = \sigma$ ($\sigma \in \hat{M}$) (these, of course, are the members of the principal P-series to which the Bruhat Criterion is not applicable): Suppose that $w\sigma = \sigma$ ($w \neq 1$) – then $U^{\sigma,0}$ is reducible iff $\mathbf{Q}(\sigma, 0) \neq 0$ (cf. Knapp and Stein [4]).

Theorem E2 supra serves to establish an isometric isomorphism between $L^2(G)$ and $L^2(\hat{G})$; it is thus only natural to pose the following problem: Characterize the image in $L^2(\hat{G})$ of the space $\mathscr{C}(G)$ under the Fourier transform $f \mapsto \mathfrak{F}f$. For this purpose we shall need to introduce a space of 'rapidly decreasing' functions $\mathscr{C}(\hat{G})$ (say) on $\hat{G}$. . . .

Given $\hat{U} \in \hat{G}_d$, fix, as above, a representation U in the class $\hat{U}$ which operates on the Hilbert space E_U (say); choose an orthonormal basis $\{a_i^U : i \in I\}$ for E_U by selecting an orthonormal basis $\{a_{\delta,i}^U : i \in I(\delta)\}$ for each $E_U(\delta)$ ($\delta \in \hat{K}$). It will also be convenient to attach to $\hat{U}$ the positive real number $|\hat{U}|$ defined by the prescription

$$|\hat{U}| = (B(\tau, \tau))^{1/2}$$

(τ any element in L_T' associated with $\hat{U}$ under the correspondence

$$\hat{G}_d \leftrightarrow W(G, T)\backslash \mathsf{L}_T').$$

We come now to the definition of the space $\mathscr{C}(\hat{G}_d)$. For each triple (p, q_1, q_2) of polynomials, let ${}_p|\cdot|_{q_1,q_2}$ be the semi-norm on $L^2(\hat{G}_d)$ defined by the rule

$$ {}_p|f_d|_{q_1,q_2} = \sup |(f_d(\hat{U})a^U_{\delta_1,i_1}, a^U_{\delta_2,i_2})| \cdot |p(|\hat{U}|)q_1(|\delta_1|)q_2(|\delta_2|)| \qquad (f_d \in L^2(\hat{G}_d)), $$

the sup being taken over all $\hat{U}$, (δ_1, i_1), (δ_2, i_2). Then, by definition, $\mathscr{C}(\hat{G}_d)$ consists of those f_d in $L^2(\hat{G}_d)$ having the property that

$$ {}_p|f_d|_{q_1,q_2} < \infty $$

for every triple (p, q_1, q_2).

To define the space $\mathscr{C}(\hat{G}_P)$, it will be necessary to choose a basis in each E^σ $(\sigma \in \hat{M})$. So fix $\sigma \in \hat{M}$ – then, by the Frobenius Reciprocity Theorem, a given δ in $\hat{K}$ occurs exactly $[\delta : \sigma]$ times in the restriction of $U^{\sigma,\nu}$ $(\nu \in \hat{A}_\mathfrak{p})$ to K. [Here, of course, $[\delta : \sigma]$ is the multiplicity of σ in $\delta|M$.] Let μ_σ be an irreducible unitary representation in the class σ which operates on the Hilbert space E_σ (say); agreeing to employ a similar notation for a class $\delta \in \hat{K}$, let $\mathfrak{I}(\delta; \sigma)$ be the set of intertwining operators from E_δ to E_σ per $\mu_\delta|M$ and μ_σ – then the Hilbert-Schmidt norm makes $\mathfrak{I}(\delta; \sigma)$ into a Hilbert space of dimension $[\delta : \sigma]$. Let $\{a_i : 1 \leqslant i \leqslant d(\delta)\}$ be an orthonormal basis for E_δ; let

$$ \{T_j : 1 \leqslant j \leqslant [\delta : \sigma]\} $$

be an orthogonal basis for $\mathfrak{I}(\delta; \sigma)$, each T_j having norm equal to $\sqrt{d(\delta)}$; put

$$ a^\sigma_{\delta, i+(j-1)d(\delta)}(k) = T_j(\mu_\delta(k^{-1})a_i) \qquad (k \in K) $$

– then the elements of the set

$$ \{a^\sigma_{\delta,i} : \delta \in \hat{K}, 1 \leqslant i \leqslant d(\delta)[\delta : \sigma]\} $$

comprise an orthonormal basis for E^σ. We can now define the space $\mathscr{C}(\hat{G}_P)$. Thus for each set of polynomials (u, v, q_1, q_2) and for each non-negative integer n, let ${}_{u,v}|\cdot|^n_{q_1,q_2}$ be the semi-norm on $L^2(\hat{G}_P)$ defined as follows: Put

$$ {}_{u,v}|f_P|^n_{q_1,q_2} = \infty $$

if, for some $\sigma \in \hat{M}$ and some $a^\sigma_{\delta_1,i_1}$ and some $a^\sigma_{\delta_2,i_2}$, the function

$$ \nu \mapsto (f_P(\sigma, \nu)a^\sigma_{\delta_1,i_1}, a^\sigma_{\delta_2,i_2}) \qquad (\nu \in \hat{A}_\mathfrak{p}) $$

is not n times continuously differentiable in ν; otherwise put

$$ {}_{u,v}|f_P|^n_{q_1,q_2} = \sup \left| \left(\frac{d}{d\nu}\right)^n (f_P(\sigma, \nu)a^\sigma_{\delta_1,i_1}, a^\sigma_{\delta_2,i_2}) \right| \times |u(|\sigma|)v(|\nu|)q_1(|\delta_1|)q_2(|\delta_2|)| \qquad (f_P \in L^2(\hat{G}_P)), $$

the sup being taken over all (σ, ν), (δ_1, i_1), (δ_2, i_2). By definition, then, the space $\mathscr{C}(\hat{G}_P)$ consists of all those f_P in $L^2(\hat{G}_P)$ having the property that

$$ {}_{u,v}|f_P|^n_{q_1,q_2} < \infty $$

for every $n \geqslant 0$ and every set (u, v, q_1, q_2).

Let $\mathscr{C}(\hat{G}) = \mathscr{C}(\hat{G}_d) + \mathscr{C}(\hat{G}_P)$ (direct sum) – then $\mathscr{C}(\hat{G})$ is a Fréchet space and we have:

Theorem E 4 (Arthur) The map $f \mapsto \mathfrak{F}f$ sets up a topological isomorphism between $\mathscr{C}(G)$ and $\mathscr{C}(\hat{G})$.

Historically this result was first proved by Langlands in the case $G = \mathbf{SL}(2, \mathbf{R})$; the extension to general split-rank 1 groups was made by Arthur [1] in his Yale thesis.

The proof of Theorem E4 is long and difficult; we shall not stop to reproduce the details here. [At the core of the matter are a number of estimates on certain spherical functions; these estimates are, however, but special cases of Harish-Chandra's 'philosophy of the Eisenstein integral.']

Appendix

3 Some Results on Differential Equations

3.1 The Main Theorems

Fix an integer $l \geqslant 1$; by $\mathbf{C}^l$ we shall understand, as usual, the Cartesian product of $\mathbf{C}$ with itself l times – similarly for $\mathbf{R}^l$ and $\mathbf{N}^l$. Given $\mathbf{z} = (z_1, \ldots, z_l)$ and $\tilde{\mathbf{z}} = (\tilde{z}_1, \ldots, \tilde{z}_l)$ in $\mathbf{C}^l$, put

$$\langle \mathbf{z}, \tilde{\mathbf{z}} \rangle = z_1\tilde{z}_1 + \cdots + z_l\tilde{z}_l;$$

we shall also write

$$\mathscr{R}(\mathbf{z}) = \mathscr{R}(z_1 + \cdots + z_l), \quad \|\mathbf{z}\| = \sup_i |z_i|,$$

and

$$\mathbf{z}^{\mathbf{n}} = z_1^{n_1} \cdots z_l^{n_l} \qquad (\mathbf{n} = (n_1, \ldots, n_l) \in \mathbf{N}^l).$$

As regards the elements $\mathbf{n}$ of $\mathbf{N}^l$, it will be convenient to employ the familiar multi-index notation. Thus, if $\mathbf{n} = (n_1, \ldots, n_l) \in \mathbf{N}^l$, then we shall write $|\mathbf{n}|$ for the sum $n_1 + \cdots + n_l$ and $\mathbf{n}!$ for the product $n_1! \cdots n_l!$; $D^{\mathbf{n}}$ will stand for the symbol of partial differentiation

$$D^{\mathbf{n}} = \partial^{n_1 + \cdots + n_l}/\partial z_1^{n_1} \cdots \partial z_l^{n_l}.$$

For any $\eta \geqslant 0$, $\mathbf{C}^l(\eta)$ will denote the set comprised of those $\mathbf{z}$ in $\mathbf{C}^l$ such that $\mathscr{R}(z_i) > \eta$ $(1 \leqslant i \leqslant l)$; for any ϵ, $0 < \epsilon < 1$, $\mathbf{C}^l(\epsilon, \eta)$ will denote the set consisting of those $\mathbf{z}$ in $\mathbf{C}^l$ with the property that $\mathscr{R}(z_i) > \max(\epsilon\|\mathbf{z}\|, \eta)$ $(1 \leqslant i \leqslant l)$ – obviously $\mathbf{C}^l(\epsilon, \eta)$ is a non-empty open subset of $\mathbf{C}^l$.

Let E be a finite dimensional (complex) Banach space. By an *E-polynomial* we shall mean a polynomial function on $\mathbf{C}^l$ with values in E. If

$$p(\mathbf{z}) = \sum_{\mathbf{n}} a(\mathbf{n})\mathbf{z}^{\mathbf{n}} \qquad (a(\mathbf{n}) \in E; \mathbf{n} \in \mathbf{N}^l)$$

is an E-polynomial on $\mathbf{C}^l$, then we define

$$\|p\| = \sum_{\mathbf{n}} \|a(\mathbf{n})\|.$$

Theorem A 3.1.1 Fix $\eta \geqslant 0$; let $\mathbf{\Gamma}_i(1 \leqslant i \leqslant l)$ be given holomorphic functions on $\mathbf{C}^l(\eta)$ with values in $\mathrm{Hom}_{\mathbf{C}}(E, E)$. Let $\mathscr{O}$ be a non-empty open connected set in $\mathbf{R}^l(\eta)$ $(= \mathbf{R}^l \cap \mathbf{C}^l(\eta))$ and let ψ be an E-valued function on $\mathscr{O}$ such that $\partial\psi/\partial x_i = \mathbf{\Gamma}_i\psi$ $(1 \leqslant i \leqslant l)$ on $\mathscr{O}$ – then there exists a unique holomorphic function Ψ on $\mathbf{C}^l(\eta)$ with values in E such that:

(i) $\Psi = \psi$ on $\mathscr{O}$;
(ii) $\partial\Psi/\partial z_i = \mathbf{\Gamma}_i\Psi$ $(1 \leqslant i \leqslant l)$ on $\mathbf{C}^l(\eta)$.

Let $c_i(\mathbf{n})$ $(1 \leqslant i \leqslant l; \mathbf{n} \in \mathbf{N}^l)$ be elements in $\mathrm{Hom}_{\mathbf{C}}(E, E)$ such that

$$\sum_i \sum_{\mathbf{n}} \|c_i(\mathbf{n})\| e^{-|\mathbf{n}|\eta} < \infty$$

for every $\eta > 0$; put

$$(\boldsymbol{\Gamma}) \quad \boldsymbol{\Gamma}_i(\mathbf{z}) = \sum_{\mathbf{n}} c_i(\mathbf{n}) e^{-\langle \mathbf{z}, \mathbf{n} \rangle} \qquad (\mathbf{z} \in \mathbf{C}^{l,+} (= \mathbf{C}^l(0)).$$

Then it is clear that each $\boldsymbol{\Gamma}_i$ $(1 \leqslant i \leqslant l)$ is a holomorphic function on $\mathbf{C}^{l,+}$ with values in $\text{Hom}_{\mathbf{C}}(E, E)$.

Theorem A 3.1.2 (Harish-Chandra) Define holomorphic functions $\boldsymbol{\Gamma}_i$ on $\mathbf{C}^{l,+}$ by condition $(\boldsymbol{\Gamma})$ supra; suppose that Ψ is an E-valued holomorphic function on $\mathbf{C}^{l,+}$ such that $\partial\Psi/\partial z_i = \boldsymbol{\Gamma}_i \Psi$ $(1 \leqslant i \leqslant l)$ there – then there exists an at most countable set $\mathscr{E}(\Psi)$ of points in $\mathbf{C}^l$ and, for each $\lambda \in \mathscr{E}(\Psi)$, an E-polynomial p_λ such that the following conditions are fulfilled:

(i) For every $\lambda \in \mathscr{E}(\Psi)$, $p_\lambda \neq 0$;

(ii) The series

$$\sum_{\lambda \in \mathscr{E}(\Psi)} p_\lambda e^{\langle \cdot, \lambda \rangle}$$

converges decently (cf. A3.4) on the set $\mathbf{C}^l(\epsilon, \eta)$ for any ϵ, η with $0 < \epsilon < 1$, $\eta > 0$;

(iii) The expansion

$$\Psi(\mathbf{z}) = \sum_{\lambda \in \mathscr{E}(\Psi)} p_\lambda(\mathbf{z}) e^{\langle \mathbf{z}, \lambda \rangle}$$

holds for all $\mathbf{z} \in \mathbf{C}^{l,+}$.

The set $\mathscr{E}(\Psi)$ and the polynomials $p_\lambda (\lambda \in \mathscr{E}(\Psi))$ are uniquely determined by these conditions; moreover there exists an integer $d \geqslant 0$ and two positive numbers M and N such that $\deg(p_\lambda) \leqslant d$ and $\| p_\lambda \| \leqslant MN^{-\mathscr{R}(\lambda)}$ (all $\lambda \in \mathscr{E}(\Psi)$).

Let us suppose that Ψ verifies the assumptions of Theorem A 3.1.2 – then we shall refer to the elements λ in the set $\mathscr{E}(\Psi)$ as the *exponents* of Ψ. An element $\lambda \in \mathscr{E}(\Psi)$ is said to be *maximal* if $\lambda + \mathbf{n}$ does not lie in $\mathscr{E}(\Psi)$ for any $\mathbf{n} \neq \mathbf{0}$ in $\mathbf{N}^l$.

Theorem A 3.1.3 (Harish-Chandra) Define holomorphic functions $\boldsymbol{\Gamma}_i$ on $\mathbf{C}^{l,+}$ by condition $(\boldsymbol{\Gamma})$ supra; suppose that Ψ is an E-valued holomorphic function on $\mathbf{C}^{l,+}$ such that $\partial\Psi/\partial z_i = \boldsymbol{\Gamma}_i \Psi$ $(1 \leqslant i \leqslant l)$ there – then the set $\mathscr{E}_0(\Psi)$ of maximal elements in $\mathscr{E}(\Psi)$ is finite. Furthermore every exponent of Ψ is of the form $\lambda_0 - \mathbf{n}$ where $\lambda_0 \in \mathscr{E}_0(\Psi)$ and $\mathbf{n} \in \mathbf{N}^l$.

The elements of the set $\mathscr{E}_0(\Psi)$ will be called the *principal exponents* of Ψ and the sum $\sum_\lambda p_\lambda e^\lambda$, $\lambda \in \mathscr{E}_0(\Psi)$, the *principal part* of Ψ.

The remaining numbers of this Appendix will be devoted to giving the proofs of these results. Thus in A3.2 we assemble some simple analytic tools which will be used off and on in the subsequent discussion. In A3.3 we give the proof of Theorem A 3.1.1; although this result is well-known, for the sake of completeness it seems best to go through the details of the argument. [Of course the point of Theorem A 3.1.1 is that it ensures that solutions of certain differential equations which may initially be defined only on a 'small' open connected set can actually always be extended by analytic continuation to a much larger set; its proof depends on a rather standard lemma (Lemma A 3.3.2) which reduces a differential equation to an integral equation.] The notion of 'decent convergence' is introduced in A3.4; Proposition A 3.4.1 provides a useful criterion. The proofs of Theorems A 3.1.2 and A 3.1.3 are presented in A 3.5 via a sequence of propositions; the arguments proceed by induction on l, the main tool being Proposition A 3.5.4.

3.2 Lemmas from Analysis

In this number we shall record a few elementary lemmas which, however, will be used rather frequently in what follows.

Lemma A 3.2.1 Let t be a real variable and $r_1, \ldots, r_m$ distinct real numbers – then

$$\limsup_{t\to\infty}\left|\sum_{i=1}^{m} c_i \exp\{\sqrt{-1}r_i t\}\right| \geqslant \left\{\sum_{i=1}^{m} |c_i|^2\right\}^{1/2}$$

for any complex numbers $c_1, \ldots, c_m$.

Proof Set $f(t) = \sum_{i=1}^{m} c_i \exp\{\sqrt{-1}r_i t\}$ – then, by direct computation, we find that $\lim_{T\to\infty} T^{-1}\int_0^T |f(t)|^2\,dt = \sum_i |c_i|^2$. On the other hand, if

$$M = \limsup_{t\to\infty} |f(t)|,$$

then it is clear that $\lim_{T\to\infty} T^{-1}\int_0^T |f(t)|^2\,dt \leqslant M^2$ and so $\sum_i |c_i|^2 \leqslant M^2$. □

Lemma A 3.2.2 Let $c_1, \ldots, c_m$ be non-zero complex numbers and $p_0, p_1, \ldots, p_m$ polynomials in t with complex coefficients. Suppose that

$$\limsup_{t\to\infty} |p_0(t) + p_1(t)e^{c_1 t} + \cdots + p_m(t)e^{c_m t}| \leqslant M$$

for some real number M – then p_0 is a constant and $|p_0| \leqslant M$.

Proof Let r_i denote the real part of c_i. If $r_i < 0$, then it is obvious that

$$\lim_{t\to\infty} |p_i(t)e^{c_i t}| = 0.$$

Hence, if we assume that $r_i \geqslant 0$ for $1 \leqslant i \leqslant n$ (say) while $r_i < 0$ for $i > n$, then it follows that

$$\limsup_{t\to\infty}\left|p_0(t) + \sum_{i=1}^{n} p_i(t)e^{c_i t}\right| \leqslant M.$$

Therefore, without loss of generality, we may assume that $n = m$, that $c_1, \ldots, c_m$ are distinct, and that $p_i \neq 0$ $(0 \leqslant i \leqslant m)$.

(1) Claim: $r_1 = r_2 = \cdots = r_m = 0$. For otherwise suppose that

$$r = \max(r_1, \ldots, r_m) > 0.$$

We may assume that $r_1 = r_2 = \cdots = r_{i'} = r$ while $r_i < r$ for $i > i'$. Let d' be the highest among the degrees of $p_1, \ldots, p_{i'}$ and put $P'_i = \lim_{t\to\infty} p_i(t)/t^{d'}$ – then $P'_1, \ldots, P'_{i'}$ cannot all be zero and so it is clear that

$$\begin{aligned} 0 &= \limsup_{t\to\infty} t^{-d'}e^{-rt}|p_0(t) + p_1(t)e^{c_1 t} + \cdots + p_m(t)e^{c_m t}| \\ &= \limsup_{t\to\infty} |P'_1 e^{(c_1 - r)t} + \cdots + P'_{i'}e^{(c_{i'} - r)t}|. \end{aligned}$$

Since the $c_i - r$ $(1 \leqslant i \leqslant i')$ are distinct pure imaginary numbers, this contradicts Lemma A 3.2.1, whence $r = 0$, as claimed.

(2) Let d be the maximum of the degrees of $p_0, p_1, \ldots, p_m$. Claim: $d = 0$. For if not, then, since $r_i = 0$ $(1 \leqslant i \leqslant m)$, it follows that

$$\lim_{t\to\infty} t^{-d}\left|p_0(t) + \sum_{i=1}^{m} p_i(t)e^{c_i t}\right| = 0$$

$$\Rightarrow$$

$$\lim_{t\to\infty}\left|P_0 + \sum_{i=1}^{m} P_i e^{c_i t}\right| = 0$$

where $P_i = \lim_{t\to\infty} p_i(t)/t^d$ $(0 \leqslant i \leqslant m)$. In view of the definition of d, not all the P_i are zero and so once again we get a contradiction with Lemma A 3.2.1, whence $d = 0$, as claimed.

Because of (2), $p_0, \ldots, p_m$ are constants. Moreover, by Lemma A 3.2.1, $|p_0| \leqslant M$. □

Let E be a finite dimensional vector space over $\mathbf{R}$. We shall say that a subset $\mathscr{S}$ of E is *full* provided $tX \in \mathscr{S}$ whenever $X \in \mathscr{S}$ and $t \geqslant 1$ $(t \in \mathbf{R})$.

Lemma A 3.2.3 Let $\lambda_i \neq 0$ $(1 \leqslant i \leqslant m)$ be a finite set of linear functions and $p_0, \ldots, p_m$ polynomial functions on E. Suppose that $\mathcal{O}$ is a non-empty, open and full subset of E and M a real number such that

$$|p_0(X) + p_1(X)e^{\lambda_1(X)} + \cdots + p_m(X)e^{\lambda_m(X)}| \leqslant M$$

for all $X \in \mathcal{O}$ – then p_0 is a constant and $|p_0| \leqslant M$.

Proof Let d be the degree of p_0. Since $\mathcal{O}$ is open, we can choose $X_0 \in \mathcal{O}$ in such a way that $\lambda_i(X_0) \neq 0$ $(1 \leqslant i \leqslant m)$ and $q_0(t) = p_0(tX_0)$ is a polynomial of degree d in the real variable t. Put $q_i(t) = p_i(tX_0)$ and $c_i = \lambda_i(X_0)$ $(1 \leqslant i \leqslant m)$. Because $\mathcal{O}$ is full, it follows that

$$|q_0(t) + q_1(t)e^{c_1 t} + \cdots + q_m(t)e^{c_m t}| \leqslant M \qquad (t \geqslant 1).$$

We may then conclude from Lemma A 3.2.2 that q_0 is a constant and $|q_0| \leqslant M$. Therefore $d = 0$, whence $|p_0| = |q_0| \leqslant M$. □

Lemma A 3.2.4 Let λ_i $(i \geqslant 1)$ be a sequence of distinct linear functions, p_i $(i \geqslant 1)$ a sequence of polynomial functions on E and $\mathcal{O}$ a non-empty, open and full subset of E. Suppose that the following two conditions hold:

(1) For each linear function λ on E the series

$$\sum_{i=1}^{\infty} |p_i(X)\exp(\lambda(X) + \lambda_i(X))|$$

converges uniformly for $X \in \mathcal{O}$;

(2) $\sum_{i=1}^{\infty} p_i(X)e^{\lambda_i(X)} = 0$ for $X \in \mathcal{O}$.

Then $p_i = 0$ for every i.

Proof Proceeding by contradiction, select an index j such that $p_j \neq 0$. Given $\epsilon > 0$, choose an integer $N \geqslant j$ such that

$$\sum_{i>N} |p_i(X)\exp(\lambda_i(X) - \lambda_j(X))| \leqslant \epsilon$$

for all $X \in \mathcal{O}$ (this is possible by condition (1)). In view of condition (2) it is then clear that

$$\left|\sum_{i=1}^{N} p_i(X)\exp(\lambda_i(X) - \lambda_j(X))\right| \leqslant \epsilon \qquad (X \in \mathcal{O}).$$

Since $\lambda_i - \lambda_j \neq 0$ for $i \neq j$, Lemma A 3.2.3 implies that p_j is a constant and $|p_j| \leqslant \epsilon$. But ϵ is arbitrary, whence $p_j = 0$. As this contradicts the choice of j, the lemma follows. □

Let $z_1, \ldots, z_l$ denote the Cartesian coordinates of a point $\mathbf{z}$ in $\mathbf{C}^l$; put $\|\mathbf{z}\| = \sup_i |z_i|$. The distance between two sets $\mathcal{O}$ and $\tilde{\mathcal{O}}$ in $\mathbf{C}^l$ is defined to be $\inf \|\mathbf{z} - \tilde{\mathbf{z}}\|$ $(\mathbf{z} \in \mathcal{O}, \tilde{\mathbf{z}} \in \tilde{\mathcal{O}})$.

Lemma A 3.2.5 Let $\mathcal{O}$ be an open set in $\mathbf{C}^l$ and $\tilde{\mathcal{O}}$ a subset of $\mathcal{O}$; let ϵ denote the distance between $\tilde{\mathcal{O}}$ and the complement of $\mathcal{O}$ in $\mathbf{C}^l$ – then, if f is any holomorphic function on $\mathcal{O}$,

$$\sup_{\tilde{\mathbf{z}} \in \tilde{\mathcal{O}}} |D^{\mathbf{n}} f(\tilde{\mathbf{z}})| \leqslant \mathbf{n}!(2\pi/\epsilon)^{|\mathbf{n}|} \sup_{\mathbf{z} \in \mathcal{O}} |f(\mathbf{z})| \qquad (\mathbf{n} \in \mathbf{N}^l).$$

Proof We may obviously assume that $\epsilon > 0$. Let $\tilde{\epsilon}$ be any positive number less than ϵ. If $\tilde{\mathbf{z}} \in \tilde{\mathcal{O}}$ and if $\|\mathbf{z} - \tilde{\mathbf{z}}\| \leqslant \tilde{\epsilon}$, then $\mathbf{z} \in \mathcal{O}$ – hence

$$D^{\mathbf{n}} f(\tilde{\mathbf{z}}) = \mathbf{n}!\{2\pi\sqrt{-1}\}^{-1} \int_1 \cdots \int_l f(\mathbf{z}) \left\{\prod_{i=1}^{l} (z_i - \tilde{z}_i)^{n_i+1}\right\}^{-1} dz_1 \cdots dz_l$$

where $\int_i$ denotes complex integration with respect to z_i on the circle $|z_i - \tilde{z}_i| = \tilde{\epsilon}$. It is therefore clear that

$$|D^{\mathbf{n}} f(\tilde{\mathbf{z}})| \leqslant \mathbf{n}!\tilde{\epsilon}^{-|\mathbf{n}|}(2\pi)^{|\mathbf{n}|-1} \sup_{\mathbf{z} \in \mathcal{O}} |f(\mathbf{z})|.$$

This being true for every $\tilde{\epsilon} < \epsilon$, our assertion now follows at once. □

Lemma A 3.2.6 Let $\mathcal{S}$ be a subset of $\mathbf{C}^l$; let $\{f_i : i \in I\}$, $\{g_j : j \in J\}$ be two indexed families of complex valued functions on $\mathcal{S}$ such that:

(1) The series $\sum_i |f_i e^{\langle \cdot, \lambda \rangle}|$, $\sum_j |g_j e^{\langle \cdot, \lambda \rangle}|$ converge uniformly on $\mathcal{S}$ for every λ in $\mathbf{C}^l$;

(2) There exist elements μ, ν in $\mathbf{C}^l$ and positive constants r, s such that

$$\sum_i |f_i(\mathbf{z})| \leqslant r|e^{\langle \mathbf{z}, \mu \rangle}|, \quad \sum_j |g_j(\mathbf{z})| \leqslant s|e^{\langle \mathbf{z}, \nu \rangle}| \qquad (\text{all } \mathbf{z} \in \mathcal{S}).$$

Then, for every $\lambda \in \mathbf{C}^l$, the series $\sum_{i,j} |f_i g_j e^{\langle \cdot, \lambda \rangle}|$ converges uniformly on $\mathcal{S}$.

[We shall leave it to the reader to supply the proof of this elementary fact.]

3.3 Analytic Continuation of Solutions

Let E be a finite dimensional (complex) Banach space.

Theorem A 3.3.1 Fix $\eta \geqslant 0$; let Γ_i $(1 \leqslant i \leqslant l)$ be given holomorphic functions on $\mathbf{C}^l(\eta)$ with values in $\mathrm{Hom}_{\mathbf{C}}(E, E)$. Let $\mathcal{O}$ be a non-empty open connected set in $\mathbf{R}^l(\eta)$ $(= \mathbf{R}^l \cap \mathbf{C}^l(\eta))$ and let ψ be an E-valued function on $\mathcal{O}$ such that $\partial\psi/\partial x_i = \Gamma_i \psi$ $(1 \leqslant i \leqslant l)$ on $\mathcal{O}$ – then there exists a unique holomorphic function Ψ on $\mathbf{C}^l(\eta)$ with values in E such that:

(i) $\Psi = \psi$ on $\mathcal{O}$;

(ii) $\partial\Psi/\partial z_i = \Gamma_i \Psi$ $(1 \leqslant i \leqslant l)$ on $\mathbf{C}^l(\eta)$.

The proof depends in an essential way on the following well-known fact.

Lemma A 3.3.2 Let a and b be real numbers with $a \leqslant b$; let

$$I = \{t \in \mathbf{R} : a \leqslant t \leqslant b\}$$

and let Γ be a continuous function from I to $\mathrm{Hom}_{\mathbf{C}}(E, E)$. Suppose that ψ is an E-valued function on I such that $d\psi/dt = \Gamma\psi$ on I; put

$$H_0 = 1 \in \mathrm{Hom}_{\mathbf{C}}(E, E),$$
$$H_n = \int_{b \geqslant t_1 \geqslant \cdots \geqslant t_n \geqslant a} \Gamma(t_1) \cdots \Gamma(t_n)\, dt_1 \cdots dt_n \qquad (n \in \mathbf{N}^+)$$

and $H = \sum_{n=0}^{\infty} H_n$ – then $\sum_{n=0}^{\infty} \|H_n\| < \infty$ and $\psi(b) = H\psi(a)$.

Proof Given $t \in I$, let us agree to write

$$H_n(t) = \int_{t \geqslant t_1 \geqslant \cdots \geqslant t_n \geqslant a} \Gamma(t_1) \cdots \Gamma(t_n)\, dt_1 \cdots dt_n \qquad (n \in \mathbf{N}^+).$$

In view of our hypotheses on ψ, it is clear that ψ is continuous; moreover

$$\psi(t) = \psi(a) + \int_a^t \Gamma(t_1)\psi(t_1)\, dt_1 \qquad (\text{all } t \in I).$$

We now claim that

$$\psi(t) = \psi(a) + \sum_{i=1}^{n-1} H_i(t)\psi(a) + \int_{t \geqslant t_1 \geqslant \cdots \geqslant t_n \geqslant a} \Gamma(t_1) \cdots \Gamma(t_n)\psi(t_n)\, dt_1 \cdots dt_n$$

for every $n \geqslant 1$ (all $t \in I$). Indeed, assuming that the relation is true for n, one may prove it for $n + 1$ by replacing $\psi(t_n)$ with the expression

$$\psi(a) + \int_a^{t_n} \Gamma(t_{n+1})\psi(t_{n+1})\, dt_{n+1}.$$

Let $M_\Gamma = \sup_{a \leqslant t \leqslant b} \|\Gamma(t)\|$, $M_\psi = \sup_{a \leqslant t \leqslant b} \|\psi(t)\|$ – then

$$\|H_n\| \leqslant M_\Gamma^n (b - a)^n/n!$$

and

$$\left\| \int_{b \geqslant t_1 \geqslant \cdots \geqslant t_n \geqslant a} \Gamma(t_1) \cdots \Gamma(t_n)\psi(t_n)\, dt_1 \cdots dt_n \right\| \leqslant M_\psi M_\Gamma^n (b - a)^n/n!.$$

The first of these inequalities implies that $\sum_{n=0}^{\infty} \|H_n\| \leqslant e^{M_\Gamma(b-a)}$ while the second shows that

$$\left\| \psi(b) - \sum_{i=0}^{n-1} H_i\psi(a) \right\| \to 0$$

as $n \to \infty$. Hence the lemma. □

Remark Proceeding as above, one may show that

$$\psi(a) = \psi(b) + \sum_{n=1}^{\infty} (-1)^n \int_{a \leqslant t_1 \leqslant \cdots \leqslant t_n \leqslant b} \Gamma(t_1) \cdots \Gamma(t_n)\psi(b)\, dt_1 \cdots dt_n.$$

By a domain in $\mathbf{C}^l$ we understand, as usual, a non-empty open connected subset of $\mathbf{C}^l$.

Lemma A 3.3.3 Let $\mathcal{O}$ be a simply connected domain in $\mathbf{C}$; let Γ be a holomorphic function on $\mathcal{O}$ with values in $\mathrm{Hom}_{\mathbf{C}}(E, E)$. Fix a point z_0 in $\mathcal{O}$ – then there exists exactly one holomorphic function H on $\mathcal{O}$ with values in $\mathrm{Hom}_{\mathbf{C}}(E, E)$ such that $H(z_0) = 1$ and $dH(z)/dz = \Gamma(z)H(z)$ $(z \in \mathcal{O})$.

[For obvious reasons, we need not stop to prove this lemma here.]

Remark Suppose that Λ is an open subset of $\mathbf{C}^l$ ($l \geqslant 1$). Keeping to the notations of the preceding lemma, let us assume that $\Gamma(z, \lambda)$ ($z \in \mathcal{O}, \lambda \in \Lambda$) is a holomorphic function on $\mathcal{O} \times \Lambda$ with values in $\mathrm{Hom}_{\mathbf{C}}(E, E)$; for a fixed $\lambda \in \Lambda$, let $H(z, \lambda)$ denote the function H of Lemma A 3.3.3 corresponding to $\Gamma(z, \lambda)$ – then $H(z, \lambda)$ is a holomorphic function of (z, λ) on $\mathcal{O} \times \Lambda$ (holomorphic dependence on the parameters . . .).

Proof of Theorem A 3.3.1 Since a holomorphic function on $\mathbf{C}^l(\eta)$ which vanishes on $\mathcal{O}$ must vanish identically, the uniqueness of Ψ is clear; it is therefore the existence of Ψ which requires proof. Fix a point $\mathbf{x}^0 = (x_1^0, \ldots, x_l^0)$ in $\mathcal{O}$; owing to Lemma A 3.3.3, there exists, for each i lying between 1 and l, a holomorphic function H_i from $\mathbf{C}^l(\eta)$ to $\mathrm{Hom}_{\mathbf{C}}(E, E)$ such that

$$\partial H_i(\mathbf{z})/\partial z_i = \Gamma_i(\mathbf{z})H_i(\mathbf{z})$$

with $H_i(\mathbf{z}) = 1$ if $z_i = x_i^0$. Given $\mathbf{z} \in \mathbf{C}^l(\eta)$, let us agree to denote by $\mathbf{z}^r$ the point $(z_1, \ldots, z_r, x_{r+1}^0, \ldots, x_l^0)$ ($1 \leqslant r \leqslant l$); put

$$\Psi(\mathbf{z}) = H_l(\mathbf{z}^l)H_{l-1}(\mathbf{z}^{l-1}) \cdots H_1(\mathbf{z}^1)\psi(\mathbf{x}^0) \qquad (\mathbf{z} \in \mathbf{C}^l(\eta)).$$

Then it is obvious that Ψ is holomorphic on $\mathbf{C}^l(\eta)$ with $\Psi(\mathbf{x}^0) = \psi(\mathbf{x}^0)$. Fix a number $\epsilon > 0$ and let $\mathscr{S}(\epsilon)$ be the set of all points $\mathbf{x}$ in $\mathbf{R}^l$ such that $\|\mathbf{x} - \mathbf{x}^0\| < \epsilon$; by taking ϵ sufficiently small, it can be assumed that $\mathscr{S}(\epsilon) \subset \mathcal{O}$. [Note too that $\mathbf{x} \in \mathscr{S}(\epsilon) \Rightarrow \mathbf{x}^r \in \mathscr{S}(\epsilon)$.] Now fix r and choose real numbers x_j ($1 \leqslant j < r$) such that $|x_j - x_j^0| < \epsilon$; put

$$\psi_r(x) = \psi(x_1, \ldots, x_{r-1}, x, x_{r+1}^0, \ldots, x_l^0) \qquad (|x - x_r^0| < \epsilon; x \in \mathbf{R}).$$

Then it is clear that

$$d\psi_r(x)/dx = \Gamma_r(x_1, \ldots, x_{r-1}, x, x_{r+1}^0, \ldots, x_l^0)\psi_r(x) \qquad (|x - x_r^0| < \epsilon; x \in \mathbf{R}).$$

On the other hand let

$$\phi_r(x) = H_r(x_1, \ldots, x_{r-1}, x, x_{r+1}^0, \ldots, x_l^0)\psi(x_1, \ldots, x_{r-1}, x_r^0, x_{r+1}^0, \ldots, x_l^0)$$

where x is real and $|x - x_r^0| < \epsilon$ – then it follows from the definition of H_r that

$$d\phi_r(x)/dx = \Gamma_r(x_1, \ldots, x_{r-1}, x, x_{r+1}^0, \ldots, x_l^0)\phi_r(x) \qquad (|x - x_r^0| < \epsilon; x \in \mathbf{R}).$$

Because $\psi_r(x_r^0) = \phi_r(x_r^0)$, we conclude from Lemma A 3.3.2 that actually $\psi_r = \phi_r$ – thus $\psi(\mathbf{x}^r) = H_r(\mathbf{x}^r)\psi(\mathbf{x}^{r-1})$ ($1 \leqslant r \leqslant l$) for $\mathbf{x} \in \mathscr{S}(\epsilon)$ and so $\psi = \Psi$ on $\mathscr{S}(\epsilon)$. In particular, then, Ψ satisfies the requisite differential equations on $\mathscr{S}(\epsilon)$, hence satisfies them throughout $\mathbf{C}^l(\eta)$ (Ψ being holomorphic). It remains to show that ψ and Ψ agree on all of $\mathcal{O}$. So fix another point $\mathbf{y}^0$ in $\mathcal{O}$; choose a differentiable path $\mathbf{p}(t)$ ($a \leqslant t \leqslant b$) in $\mathcal{O}$ such that $\mathbf{p}(a) = \mathbf{x}^0$ and $\mathbf{p}(b) = \mathbf{y}^0$; put $f(t) = \psi(\mathbf{p}(t))$ and $F(t) = \Psi(\mathbf{p}(t))$ ($a \leqslant t \leqslant b$) – then we have $df/dt = \Gamma f$ and $dF/dt = \Gamma F$ where $\Gamma(t) = \sum_{i=1}^{l} \Gamma_i(\mathbf{p}(t))\, dp_i(t)/dt$ ($a \leqslant t \leqslant b$). Since $f(a) = \psi(\mathbf{x}^0) = \Psi(\mathbf{x}^0) = F(a)$, it follows from Lemma A 3.3.2 that $f(b) = \psi(\mathbf{y}^0) = \Psi(\mathbf{y}^0) = F(b)$. Therefore $\psi = \Psi$ on all of $\mathcal{O}$, as desired. □

3.4 Decent Convergence

Let $\mathcal{O}$ be an open set in $\mathbf{C}^l$; let $\{f_i : i \in I\}$ be an indexed family of holomorphic functions on $\mathcal{O}$ with values in a finite dimensional (complex) Banach space E – then the series $\sum_i f_i$ is said to converge *decently* on $\mathcal{O}$ if, for every $\mathbf{n} \in \mathbf{N}^l$ and every $\lambda \in \mathbf{C}^l$, the series $\sum_i \|(D^{\mathbf{n}} f_i)e^{\langle \cdot, \lambda \rangle}\|$ converges uniformly on $\mathcal{O}$.

Let $\mathscr{S}$ be a subset of $\mathbf{C}^l$; let $\{f_i : i \in I\}$ be an indexed family of functions on

$\mathscr{S}$ with values in a finite dimensional (complex) Banach space E – then the series $\sum_i f_i$ is said to converge decently on $\mathscr{S}$ provided there exists an open neighborhood $\mathcal{O}$ of $\mathscr{S}$ in $\mathbf{C}^l$ and, for each $i \in I$, a holomorphic function F_i on $\mathcal{O}$ such that (1) $F_i = f_i$ on $\mathscr{S}$ and (2) the series $\sum_i F_i$ converges decently on $\mathcal{O}$.

Proposition A 3.4.1 Let $\mathcal{O}$ and $\tilde{\mathcal{O}}$ be open sets in $\mathbf{C}^l$ such that $\tilde{\mathcal{O}} \subset \mathcal{O}$; let d denote the distance between $\tilde{\mathcal{O}}$ and the complement of $\mathcal{O}$ in $\mathbf{C}^l$; let $\{f_i : i \in I\}$ be an indexed family of holomorphic functions on $\mathcal{O}$ with values in a finite dimensional (complex) Banach space E; suppose that for every λ in $\mathbf{C}^l$, the series $\sum_i \|f_i e^{\langle \cdot, \lambda \rangle}\|$ converges uniformly on $\mathcal{O}$ – then, if $d > 0$, the series $\sum_i f_i$ converges decently on $\tilde{\mathcal{O}}$.

Proof This is a simple consequence of Lemma A 3.2.5. □

3.5 Normal Sequences of *E*-Polynomials

Let E be a finite dimensional (complex) Banach space.

Definition A sequence $\{p_{\mathbf{n}} : \mathbf{n} \in \mathbf{N}^l\}$ of E-polynomials is said to be *normal* if:

(i) The degrees of the $p_{\mathbf{n}}$ ($\mathbf{n} \in \mathbf{N}^l$) admit a finite upper bound;
(ii) There exists a positive number M such that

$$\|p_{\mathbf{n}}\| \leqslant M^{|\mathbf{n}|+1} \qquad (\text{all } \mathbf{n} \in \mathbf{N}^l).$$

[Let p be an E-polynomial. If d is any integer exceeding the degree of p, then clearly $\|p(\mathbf{z})\| \leqslant \|p\|(1 + \|\mathbf{z}\|)^d$ (all $\mathbf{z} \in \mathbf{C}^l$).]

Lemma A 3.5.1 Let $\{p_{\mathbf{n}} : \mathbf{n} \in \mathbf{N}^l\}$ be a normal sequence of E-polynomials on $\mathbf{C}^l$ – then, for any ϵ with $0 < \epsilon < 1$, there exists an $\eta \geqslant 0$ such that the series

$$\sum_{\mathbf{n}} p_{\mathbf{n}} e^{-\langle \cdot, \mathbf{n} \rangle}$$

converges decently on $\mathbf{C}^l(\epsilon, \eta)$.

Proof Choose $\eta > 1 + 2\epsilon^{-1}$ – then $\mathbf{C}^l(\epsilon/2, \eta - 1) \supset \mathbf{C}^l(\epsilon, \eta)$. Therefore, thanks to Proposition A 3.4.1, it will be enough to verify that the series $\sum_{\mathbf{n}} \|p_{\mathbf{n}} e^{\langle \cdot, \lambda - \mathbf{n} \rangle}\|$ converges uniformly on $\mathbf{C}^l(\epsilon/2, \eta - 1)$ for η sufficiently large (all $\lambda \in \mathbf{C}^l$). Now $|\langle \mathbf{z}, \lambda \rangle| \leqslant l\|\mathbf{z}\|\|\lambda\|$ (all $\mathbf{z}, \lambda$); on the other hand, if $\mathbf{z} \in \mathbf{C}^l(\epsilon/2, \eta - 1)$, then $\mathscr{R}(\langle \mathbf{z}, \mathbf{n} \rangle) \geqslant \epsilon 2^{-1}\|\mathbf{z}\||\mathbf{n}|$, whence $\mathscr{R}(\langle \mathbf{z}, \lambda - \mathbf{n} \rangle) \leqslant \{l\|\lambda\| - \epsilon 2^{-1}|\mathbf{n}|\}\|\mathbf{z}\|$ (any $\mathbf{z} \in \mathbf{C}^l(\epsilon/2, \eta - 1)$). Therefore, since $l\|\lambda\| < 4^{-1}\epsilon|\mathbf{n}|$ for all but finitely many $\mathbf{n}$, it will be sufficient to choose η in such a way as to ensure that the series

$$\sum_{\mathbf{n}} \|p_{\mathbf{n}}(\mathbf{z})\| e^{-\epsilon\|\mathbf{z}\||\mathbf{n}|/4} \qquad (\mathbf{z} \in \mathbf{C}^l)$$

converges uniformly on $\mathbf{C}^l(\epsilon/2, \eta - 1)$. Owing to the normality of the $p_{\mathbf{n}}$ ($\mathbf{n} \in \mathbf{N}^l$), there exists an integer $d \geqslant 0$ and a number $M \geqslant 1$ such that $\deg(p_{\mathbf{n}}) \leqslant d$ and $\|p_{\mathbf{n}}\| \leqslant M^{|\mathbf{n}|+1}$ for all $\mathbf{n} \in \mathbf{N}^l$; obviously

$$\|p_{\mathbf{n}}(\mathbf{z})\| \leqslant \|p_{\mathbf{n}}\|(1 + \|\mathbf{z}\|)^d \leqslant M^{2|\mathbf{n}|}(1 + \|\mathbf{z}\|)^{d|\mathbf{n}|}$$

provided $|\mathbf{n}| \geqslant 1$. This being so, choose $\eta > 1 + 2\epsilon^{-1}$ large enough to secure the inequality $M^2(1 + t)^d e^{-\epsilon t/4} \leqslant 2^{-1}$ (all $t \geqslant \eta - 1$) – then

$$\|p_{\mathbf{n}}(\mathbf{z})\| e^{-\epsilon\|\mathbf{z}\||\mathbf{n}|/4} \leqslant 2^{-|\mathbf{n}|}$$

if $|\mathbf{n}| \geqslant 1$ and $\|\mathbf{z}\| \geqslant \eta - 1$. Since $\sum_{\mathbf{n}} 2^{-|\mathbf{n}|} = 2^l$, it now follows that the series

$$\sum_{\mathbf{n}} \|p_{\mathbf{n}}(\mathbf{z})\| e^{-\epsilon\|\mathbf{z}\||\mathbf{n}|/4} \qquad (\mathbf{z} \in \mathbf{C}^l)$$

converges uniformly on $\mathbf{C}^l(\epsilon/2, \eta - 1)$, as desired. □

Corollary A 3.5.2 Let $\{p_{\mathbf{n}} : \mathbf{n} \in \mathbf{N}^l\}$ be a normal sequence of E-polynomials on $\mathbf{C}^l$; fix ϵ $(0 < \epsilon < 1)$ – then, for any $\eta > 0$, there exists a number $P > 0$ such that

$$\sum_{\mathbf{n}} \|p_{\mathbf{n}}(\mathbf{z})e^{-\langle \mathbf{z}, \mathbf{n}\rangle}\| \leqslant Pe^{\mathscr{R}(\mathbf{z})}$$

for all $\mathbf{z} \in \mathbf{C}^l(\epsilon, \eta)$.

Proof Relative to our fixed ϵ, let us choose $\eta_0 > 1 + 2\epsilon^{-1}$ in such a way that the series

$$\sum_{\mathbf{n}} \|p_{\mathbf{n}}(\mathbf{z})\| e^{-\epsilon\|\mathbf{z}\||\mathbf{n}|/4} \qquad (\mathbf{z} \in \mathbf{C}^l)$$

converges uniformly on $\mathbf{C}^l(\epsilon/2, \eta_0 - 1)$ (per the proof of the preceding lemma) – then, with $\lambda = 0$ in the above proof, we find that

$$\sum_{\mathbf{n}} \|p_{\mathbf{n}}(\mathbf{z})e^{-\langle \mathbf{z}, \mathbf{n}\rangle}\| \leqslant \|p_0\|(1 + \|\mathbf{z}\|)^d + \sum_{|\mathbf{n}| \geqslant 1} 2^{-|\mathbf{n}|} \leqslant e^{\epsilon\|\mathbf{z}\|/4} + (2^l - 1)$$
$$\leqslant 2^l e^{\epsilon\|\mathbf{z}\|/4} \leqslant 2^l e^{\mathscr{R}(\mathbf{z})} \qquad (\text{all } \mathbf{z} \in \mathbf{C}^l(\epsilon/2, \eta_0 - 1)).$$

Our assertion now follows upon remarking that there exists a compact subset Ω of $\mathbf{C}^{l,+}$ such that $\mathbf{C}^l(\epsilon, \eta) \subset \Omega \cup \mathbf{C}^l(\epsilon, \eta_0)$. □

Lemma A 3.5.3 Let r be an integer lying between 1 and l; suppose that $\{P_{\mathbf{n}} : \mathbf{n} \in \mathbf{N}^l\}$ is a normal sequence of $\mathrm{Hom}_{\mathbf{C}}(E, E)$-polynomials on $\mathbf{C}^l$ and that $\{q_{\mathbf{n}} : \mathbf{n} \in \mathbf{N}^r\}$ is a normal sequence of E-polynomials on $\mathbf{C}^r$; put

$$p_{\mathbf{n}} = \sum P_{\mathbf{n}_1} q_{\mathbf{n}_2}$$

where the sum is over all pairs $(\mathbf{n}_1, \mathbf{n}_2) \in \mathbf{N}^l \times \mathbf{N}^r$ such that $\mathbf{n}_1 + \mathbf{n}_2 = \mathbf{n}$ – then $\{p_{\mathbf{n}} : \mathbf{n} \in \mathbf{N}^l\}$ is a normal sequence of E-polynomials on $\mathbf{C}^l$.

[We agree to regard $q_{\mathbf{n}}$ $(\mathbf{n} \in \mathbf{N}^r)$ as an E-polynomial on $\mathbf{C}^l$ in the obvious way; likewise it makes sense to add elements of $\mathbf{N}^l$ and $\mathbf{N}^r$.]

Proof Choose an integer $d \geqslant 0$ and a number $M \geqslant 1$ such that $\deg(P_{\mathbf{n}}) \leqslant d$, $\|P_{\mathbf{n}}\| \leqslant M^{|\mathbf{n}|+1}$ (all $\mathbf{n} \in \mathbf{N}^l$) and $\deg(q_{\mathbf{n}}) \leqslant d$, $\|q_{\mathbf{n}}\| \leqslant M^{|\mathbf{n}|+1}$ (all $\mathbf{n} \in \mathbf{N}^r$) – then it is obvious that $\deg(p_{\mathbf{n}}) \leqslant 2d$ and

$$\|p_{\mathbf{n}}\| \leqslant M^2 \sum M^{|\mathbf{n}_1|+|\mathbf{n}_2|} \leqslant M^2(2M)^{|\mathbf{n}|} \sum 2^{-|\mathbf{n}_1|-|\mathbf{n}_2|} \leqslant M^2(2M)^{|\mathbf{n}|}2^{l+r},$$

the sum being taken over all pairs $(\mathbf{n}_1, \mathbf{n}_2)$ with $\mathbf{n}_1 + \mathbf{n}_2 = \mathbf{n}$. The normality of the $p_{\mathbf{n}}$ is therefore clear. □

Before taking up the main results on normal sequences of E-polynomials, it will be best to make a technical digression.

As above, let E be a finite dimensional (complex) Banach space. Let $P(\mathbf{C}^l; E)$ denote the space of E-polynomials on $\mathbf{C}^l$; for any complex number γ, we denote by $(\partial/\partial z_i + \gamma)^{-1}$ $(1 \leqslant i \leqslant l)$ the linear mapping of $P(\mathbf{C}^l; E)$ into itself defined as follows: If $\gamma = 0$,

$$(\partial/\partial z_i)^{-1}(a\mathbf{z}^{\mathbf{n}}) = a\mathbf{z}^{\mathbf{n}}(z_i/(n_i + 1))$$

and if $\gamma \neq 0$,

$$(\partial/\partial z_i + \gamma)^{-1}(a\mathbf{z}^{\mathbf{n}}) = \sum_{r\geqslant 0} (-1)^r\gamma^{-r-1}(\partial^r/\partial z_i^r)(a\mathbf{z}^{\mathbf{n}}) \qquad (a \in E, \mathbf{n} \in \mathbf{N}^l).$$

It is easy to see that $(\partial/\partial z_i + \gamma)(\partial/\partial z_i + \gamma)^{-1}p = p$ $(p \in P(\mathbf{C}^l; E))$; however, in general, $(\partial/\partial z_i + \gamma)^{-1}(\partial/\partial z_i + \gamma)p \neq p$ $(p \in P(\mathbf{C}^l; E))$.

Let T be a linear transformation on E; we wish to define a linear mapping $(\partial/\partial z_i + T)^{-1}$ of $P(\mathbf{C}^l; E)$ into itself such that

$$(\partial/\partial z_i + T)(\partial/\partial z_i + T)^{-1}p = p \qquad (\text{all } p \in P(\mathbf{C}^l; E)).$$

First suppose that T has only one eigenvalue (γ, say) – then $T - \gamma$ is nilpotent and so $e^{(T-\gamma)z_i}$ is a $\mathrm{Hom}_{\mathbf{C}}(E, E)$-polynomial on $\mathbf{C}^l$. Put

$$(\partial/\partial z_i + T)^{-1}p = e^{-(T-\gamma)z_i}(\partial/\partial z_i + \gamma)^{-1}(e^{(T-\gamma)z_i}p) \qquad (p \in P(\mathbf{C}^l; E)).$$

Since $(\partial/\partial z_i)(e^{-(T-\gamma)z_i}) = -(T-\gamma)e^{-(T-\gamma)z_i}$, it is clear that

$$\begin{aligned}(\partial/\partial z_i + \gamma)(\partial/\partial z_i + T)^{-1}p &= -(T-\gamma)(\partial/\partial z_i + T)^{-1}p \\ &\quad + e^{-(T-\gamma)z_i}(\partial/\partial z_i + \gamma)\{(\partial/\partial z_i + \gamma)^{-1}(e^{(T-\gamma)z_i}p)\} \\ &= -(T-\gamma)(\partial/\partial z_i + T)^{-1}p + p\end{aligned}$$

and so, in this case, $(\partial/\partial z_i + T)(\partial/\partial z_i + T)^{-1}p = p$ for all $p \in P(\mathbf{C}^l; E)$. Turning to the general situation, let $\gamma_1, \ldots, \gamma_r$ be the distinct eigenvalues of T; let E_j be the set of all $a \in E$ such that $(T - \gamma_j)^k a = 0$ for some integer $k \geqslant 1$ $(1 \leqslant j \leqslant r)$ – then E is the direct sum of the E_j. Let P_j denote the projection of E onto E_j corresponding to this direct sum decomposition of E – then, as is well-known, T commutes with each of the P_j. Put

$$(\partial/\partial z_i + T)^{-1}p = \sum_{j=1}^{r} e^{-(T-\gamma_j)z_i}(\partial/\partial z_i + \gamma_j)^{-1}(e^{(T-\gamma_j)z_i}\mathsf{P}_j p)$$
$$(p \in P(\mathbf{C}^l; E)).$$

Since T admits only γ_j as its eigenvalue on E_j, $e^{(T-\gamma_j)z_i}\mathsf{P}_j$ is a $\mathrm{Hom}_{\mathbf{C}}(E, E)$-polynomial on $\mathbf{C}^l$; it then follows from what was said above that

$$(\partial/\partial z_i + T)(\partial/\partial z_i + T)^{-1}p = p \qquad (\text{all } p \in P(\mathbf{C}^l; E)).$$

The following inequalities will prove to be of value in the subsequent discussion.

($\mathbf{I}_1$) Let d be a non-negative integer; let p be an element of $P(\mathbf{C}^l; E)$ such that $\deg(p) \leqslant d$ – then
(i) $\|(\partial/\partial z_i)^{-1}p\| \leqslant \|p\|$;
(ii) $\|(\partial/\partial z_i + \gamma)^{-1}p\| \leqslant (d+1)!\max(|\gamma|^{-1}, |\gamma|^{-(d+1)})\|p\|$ $(\gamma \in \mathbf{C}; \gamma \neq 0)$.

($\mathbf{I}_2$) Let T be a linear transformation on E, $t \mapsto m_T(t)$ its minimal polynomial (t an indeterminate); let t^m be the highest power of t which divides m_T – then

$$\deg((\partial/\partial z_i + T)^{-1}p) \leqslant \deg(p) + 2m \qquad (\text{all } p \in P(\mathbf{C}^l; E)).$$

($\mathbf{I}_3$) Let T be a linear transformation on E; let d be a non-negative integer – then there exist numbers $M > 0$ and $N \geqslant 1$ such that

$$\|(\partial/\partial z_i + T)^{-1}p\| \leqslant d!\,MN^d\|p\|$$

for all $p \in P(\mathbf{C}^l; E)$ with $\deg(p) \leqslant d$.

Verification of $(\mathbf{I}_1)$ Suppose that $p = a\mathbf{z}^{\mathbf{n}}$ $(a \in E)$ where $\mathbf{n} \in \mathbf{N}^l$ is such that $|\mathbf{n}| \leqslant d$ – then $\|(\partial/\partial z_i)^{-1} p\| = (n_i + 1)^{-1} \|p\| \leqslant \|p\|$ while

$$\|(\partial/\partial z_i + \gamma)^{-1} p\| = \sum_{r=0}^{n_i} n_i(n_i - 1) \cdots (n_i - r + 1) |\gamma|^{-(r+1)} \|p\|$$

$$\leqslant d! \sum_{r=0}^{d} |\gamma|^{-(r+1)} \|p\| \leqslant (d+1)! \sup_{0 \leqslant r \leqslant d} |\gamma|^{-(r+1)} \|p\|$$

and so the required inequalities are now obvious.

Verification of $(\mathbf{I}_2)$ Relative to our given endomorphism T, let us decompose the space E into the E_j $(1 \leqslant j \leqslant r)$ as above; put $q_j = (\partial/\partial z_i + T)^{-1} \mathsf{P}_j p$ – then q_j is an E_j-polynomial such that $(\partial/\partial z_i + T) q_j = \mathsf{P}_j p$. If $\gamma_j \neq 0$, then T is non-singular on E_j and so $\deg(q_j) \leqslant \deg(\mathsf{P}_j p)$ (since in this case

$$q_j = \sum_{k \geqslant 0} (-1)^k (T\mathsf{P}_j)^{-k-1} (\partial/\partial z_i)^k (\mathsf{P}_j p));$$

on the other hand if $\gamma_j = 0$, then

$$q_j = e^{-Tz_i} (\partial/\partial z_i)^{-1} (e^{Tz_i} \mathsf{P}_j p),$$

whence

$$\deg(q_j) \leqslant 2(m-1) + 1 + \deg(\mathsf{P}_j p) \leqslant 2m + \deg(\mathsf{P}_j p).$$

Therefore

$$\deg((\partial/\partial z_i + T)^{-1} p) \leqslant \sup_j \deg((\partial/\partial z_i + T)^{-1} \mathsf{P}_j p) \leqslant \deg(p) + 2m \qquad (\text{all } p \in P(\mathbf{C}^l; E)),$$

as claimed.

Verification of $(\mathbf{I}_3)$ Let us keep to the earlier notations – then, for any $p \in P(\mathbf{C}^l; E)$, we have the estimate

$$\|(\partial/\partial z_i + T)^{-1} p\| \leqslant \sum_{j=1}^{r} \|e^{-(T-\gamma_j)z_i} \mathsf{P}_j\| \, \|(\partial/\partial z_i + \gamma_j)^{-1} (e^{(T-\gamma_j)z_i} \mathsf{P}_j p)\|.$$

It is clear that $\|e^{-(T-\gamma_j)z_i} \mathsf{P}_j\| \leqslant \|\mathsf{P}_j\| e^{\|T-\gamma_j\|}$; moreover, on the basis of $(\mathbf{I}_1)$ supra, there exists an $N \geqslant 1$ with the property that

$$\|(\partial/\partial z_i + \gamma_j)^{-1} p\| \leqslant (n+1)! N^{n+1} \|p\| \qquad (1 \leqslant j \leqslant r)$$

for all $p \in P(\mathbf{C}^l; E)$ such that $\deg(p) \leqslant n$ (say); this being so, let m_j be the least positive integer for which $(T - \gamma_j)^{m_j} \mathsf{P}_j = 0$ – then, if $\deg(p) \leqslant d$, we obviously have $\deg(e^{(T-\gamma_j)z_i} \mathsf{P}_j p) \leqslant d + m_j - 1$, thus

$$\|(\partial/\partial z_i + \gamma_j)^{-1} (e^{(T-\gamma_j)z_i} \mathsf{P}_j p)\| \leqslant (d + m_j)! N^{d+m_j} \|e^{(T-\gamma_j)z_i} \mathsf{P}_j p\|$$

$$\leqslant (d + m_j)! N^{d+m_j} e^{\|T-\gamma_j\|} \|\mathsf{P}_j\| \|p\|$$

$$\leqslant d! m_j! (2N)^{d+m_j} e^{\|T-\gamma_j\|} \|\mathsf{P}_j\| \|p\|$$

and so

$$\|(\partial/\partial z_i + T)^{-1} p\| \leqslant d! \sum_j m_j! (2N)^{d+m_j} e^{2\|T-\gamma_j\|} \|\mathsf{P}_j\|^2 \|p\| \leqslant d! M N^d \|p\|$$

if $M = 2^d r \sup_j m_j! (2N)^{m_j} e^{2\|T-\gamma_j\|} \|\mathsf{P}_j\|^2$.

We come now to the main results.

Let E be a finite dimensional (complex) Banach space. Fix a number $\eta_0 > 0$.

Let us suppose that we are given elements $c(\mathbf{n}) \in \mathrm{Hom}_{\mathbf{C}}(E, E)$ ($\mathbf{n} \in \mathbf{N}^l$) such that (i) $\sum_{\mathbf{n}} \|c(\mathbf{n})\| e^{-|\mathbf{n}|\eta_0} < \infty$ and (ii) $c(\mathbf{n}) = 0$ if $n_l = 0$ and $\mathbf{n} \neq \mathbf{0}$. Put

$$\Gamma(\mathbf{z}) = \sum_{\mathbf{n}} c(\mathbf{n}) e^{-\langle \mathbf{z}, \mathbf{n} \rangle} \qquad (\mathbf{z} \in \mathbf{C}^l(\eta_0)).$$

Then it is clear that Γ is a holomorphic function on $\mathbf{C}^l(\eta_0)$. Fix a number $a > \eta_0$; let H_a denote the holomorphic function from $\mathbf{C}^l(\eta_0)$ to $\mathrm{Hom}_{\mathbf{C}}(E, E)$ such that $H_a(\mathbf{z}) = 1$ if $z_l = a$ and

$$\partial H_a(\mathbf{z})/\partial z_l = \Gamma(\mathbf{z}) H_a(\mathbf{z}) \qquad (\mathbf{z} \in \mathbf{C}^l(\eta_0))$$

(in view of Lemma A 3.3.3 (and subsequent Remark) we know that such a function exists and is unique).

The following result is at the basis of the present investigation.

Proposition A 3.5.4 Retain the above notations and assumptions – then there exists a finite number of points $\lambda_1, \ldots, \lambda_r$ ($r = \dim(E)$) in $\mathbf{C}^l$ and for each j ($1 \leqslant j \leqslant r$) a normal sequence $\{P_{j,\mathbf{n}} : \mathbf{n} \in \mathbf{N}^l\}$ of $\mathrm{Hom}_{\mathbf{C}}(E, E)$-polynomials on $\mathbf{C}^l$ such that the following condition holds. Fix ϵ ($0 < \epsilon < 1$) and choose $\eta \geqslant \eta_0$ large enough to ensure that the series

$$\sum_{j=1}^{r} \sum_{\mathbf{n}} P_{j,\mathbf{n}} e^{\langle \cdot, \lambda_j - \mathbf{n} \rangle}$$

converges decently on $\mathbf{C}^l(\epsilon, \eta)$ – then its sum coincides with H_a there.

The proof requires some preparation which we shall now undertake.

Let us agree to write $\mathbf{n} \succ \mathbf{n}'$ (or $\mathbf{n}' \prec \mathbf{n}$) if $\mathbf{n} - \mathbf{n}' \in \mathbf{N}^l$ and $n_l > n_l'$. Put $\Xi = c(\mathbf{0})$ and fix an eigenvalue ξ of Ξ; let a be a vector in E such that

$$(\Xi - \xi)^k a = 0$$

for some integer $k \geqslant 1$; set

$$p_0(\mathbf{z}) = a + \sum_{k=1}^{m-1} z_l^k \frac{(\Xi - \xi)^k}{k!} a \qquad (\mathbf{z} \in \mathbf{C}^l)$$

where m is the least positive integer such that $(\Xi - \xi)^m a = 0$ – then $p_0 \in P(\mathbf{C}^l; E)$ and one verifies by direct computation that

$$(\partial/\partial z_l + \xi - \Xi) p_0 = 0.$$

Proceeding by induction on n_l, for each $\mathbf{n} \in \mathbf{N}^l$ we shall define an element $p_{\mathbf{n}} \in P(\mathbf{C}^l; E)$ as follows. If $n_l = 0$ and $\mathbf{n} \neq \mathbf{0}$, set $p_{\mathbf{n}} = 0$; if $n_l > 0$, set

$$p_{\mathbf{n}} = (\partial/\partial z_l + \xi - n_l - \Xi)^{-1} q_{\mathbf{n}}$$

where $q_{\mathbf{n}} = \sum_{\mathbf{n}' \prec \mathbf{n}} c(\mathbf{n} - \mathbf{n}') p_{\mathbf{n}'}$.

We claim that the degrees of the $p_{\mathbf{n}}$ ($\mathbf{n} \in \mathbf{N}^l$) admit a finite upper bound. Thus, in the first place we have,

$$\deg(p_{\mathbf{n}}) \leqslant \deg(q_{\mathbf{n}}) + 2\dim(E) \leqslant \sup_{\mathbf{n}' \prec \mathbf{n}} \deg(p_{\mathbf{n}'}) + 2\dim(E)$$

(cf. (I_2) supra); since $\deg(p_0) \leqslant \dim(E)$, an easy induction on n_l then shows that $\deg(p_{\mathbf{n}}) \leqslant (2n_l + 1)\dim(E)$. If $k \geqslant 1$, then

$$\xi - k - \Xi = -k\left(1 + \frac{\Xi - \xi}{k}\right);$$

this being so, choose an integer $k_0 \geqslant 2$ such that $k^{-1}\|\Xi - \xi\| \leqslant 2^{-1}$ for $k \geqslant k_0$ – then $1 + (\Xi - \xi)/k$ is *non-singular* and

$$\left\| \left(1 + \frac{\Xi - \xi}{k}\right)^{-1} \right\| \leqslant 1 + 2^{-1} + 2^{-2} + \cdots = 2 \qquad (k \geqslant k_0),$$

whence

$$\|(\xi - k - \Xi)^{-1}\| \leqslant 2/k \leqslant 1$$

for $k \geqslant k_0$. Consequently, if $n_l \geqslant k_0$, then actually

$$\deg(p_{\mathbf{n}}) \leqslant \deg(q_{\mathbf{n}}) \leqslant \sup_{\mathbf{n}' < \mathbf{n}} \deg(p_{\mathbf{n}'}).$$

Therefore it is obvious that $\deg(p_{\mathbf{n}}) \leqslant (2k_0 + 1)\dim(E)$ for all $\mathbf{n}$ $(\mathbf{n} \in \mathbf{N}^l)$ which serves to establish the claim. Now let $d = \sup_{\mathbf{n}} \deg(p_{\mathbf{n}})$; in view of $(\mathbf{I}_3)$ above, there exists $M > 0$ such that $\|(\partial/\partial z_l + \xi - k - \Xi)^{-1}p\| \leqslant M\|p\|$ if $1 \leqslant k < k_0$ and $\deg(p) \leqslant d$ $(p \in P(\mathbf{C}^l; E))$; since it can be assumed that $M \geqslant (d+1)!$, we also have

$$\|(\partial/\partial z_l + \xi - k - \Xi)^{-1}p\| \leqslant \sum_{r \geqslant 0} \|(\partial/\partial z_l)^r p\| \leqslant (d+1)!\,\|p\| \leqslant M\|p\|$$

for $k \geqslant k_0$ under the same supposition on p. Hence

$$\|p_{\mathbf{n}}\| \leqslant M\|q_{\mathbf{n}}\| \qquad (\mathbf{n} \in \mathbf{N}^l;\, n_l \geqslant 1).$$

Let $N = \sum_{\mathbf{n}} \|c(\mathbf{n})\| e^{-|\mathbf{n}|\eta_0}$ $(< \infty)$ – then

$$\|q_{\mathbf{n}}\| e^{-|\mathbf{n}|\eta_0} \leqslant e^{-|\mathbf{n}|\eta_0} \sum_{\mathbf{n}' < \mathbf{n}} \|c(\mathbf{n} - \mathbf{n}')\|\,\|p_{\mathbf{n}'}\| \leqslant N \sup_{\mathbf{n}' < \mathbf{n}} \|p_{\mathbf{n}'}\| e^{-|\mathbf{n}'|\eta_0}$$

$$(n_l \geqslant 1).$$

Choose $s > 0$ such that $e^s \geqslant MN$ – then

$$\|p_{\mathbf{n}}\| e^{-|\mathbf{n}|\eta_0} \leqslant e^s \sup_{\mathbf{n}' < \mathbf{n}} \|p_{\mathbf{n}'}\| e^{-|\mathbf{n}'|\eta_0}$$

$$\Rightarrow$$

$$\|p_{\mathbf{n}}\| e^{-|\mathbf{n}|\eta_0 - n_l s} \leqslant \sup_{\mathbf{n}' < \mathbf{n}} \|p_{\mathbf{n}'}\| e^{-|\mathbf{n}'|\eta_0 - n_l' s} \qquad (n_l \geqslant 1)$$

which proves that $\|p_{\mathbf{n}}\| \leqslant \exp(|\mathbf{n}|\eta_0 + n_l s)\|p_0\|$ (all $\mathbf{n} \in \mathbf{N}^l$).

Summary Fix an eigenvalue ξ of $\Xi = c(\mathbf{0})$; let a be a vector in E such that $(\Xi - \xi)^k a = 0$ for some integer $k \geqslant 1$ – then there exists a sequence $\{p_{\mathbf{n}} : \mathbf{n} \in \mathbf{N}^l\}$ of E-polynomials on $\mathbf{C}^l$ which has the following properties:
(1) $p_{\mathbf{n}} = 0$ if $n_l = 0$ and $\mathbf{n} \neq \mathbf{0}$;
(2) $d = \sup_{\mathbf{n}} \deg(p_{\mathbf{n}}) < \infty$;
(3) $(\partial/\partial z_l + \xi - n_l - \Xi)p_{\mathbf{n}} = \sum_{\mathbf{n}' < \mathbf{n}} c(\mathbf{n} - \mathbf{n}')p_{\mathbf{n}'}$ $(n_l \geqslant 1)$;
(4) $\|p_{\mathbf{n}}\| \leqslant \exp(|\mathbf{n}|\eta_0 + n_l s)\|p_0\|$ (all $\mathbf{n} \in \mathbf{N}^l$) ($s$ a certain positive number);
(5) $p_{\mathbf{n}}(\mathbf{z})$ depends only on z_l $(\mathbf{z} \in \mathbf{C}^l)$.

Let us keep to the above notations. Fix $S \geqslant \eta_0 + s$; let $\mathbf{C}^l(\eta_0 : S)$ denote the subset of $\mathbf{C}^l$ comprised of those $\mathbf{z}$ such that $\mathscr{R}(z_i) > \eta_0$ $(1 \leqslant i < l)$ and $\mathscr{R}(z_l) > S$. Since $\|p_{\mathbf{n}}(\mathbf{z})\| \leqslant \|p_{\mathbf{n}}\|(1 + |z_l|)^d$ $(\mathbf{z} \in \mathbf{C}^l)$, it is clear that the series $\sum_{\mathbf{n}} \|p_{\mathbf{n}} e^{-\langle \cdot, \mathbf{n} \rangle}\|$ converges uniformly on any compact subset of $\mathbf{C}^l(\eta_0 : S)$. Put

$$\Psi(\mathbf{z}) = \sum_{\mathbf{n}} p_{\mathbf{n}}(\mathbf{z}) e^{\xi z_l - \langle \mathbf{z}, \mathbf{n} \rangle} \qquad (\mathbf{z} \in \mathbf{C}^l(\eta_0 : S)).$$

Then Ψ is an E-valued holomorphic function on $\mathbf{C}^l(\eta_0 : S)$. Moreover, in view of Proposition A 3.4.1, the series defining Ψ converges decently on every compact subset of $\mathbf{C}^l(\eta_0 : S)$. Because the same is true of the series $\sum_{\mathbf{n}} c(\mathbf{n})e^{-\langle \cdot, \mathbf{n} \rangle}$, it follows from the relation

$$(\partial/\partial z_l + \xi - n_l - \Xi)p_{\mathbf{n}} = \sum_{\mathbf{n}'<\mathbf{n}} c(\mathbf{n}-\mathbf{n}')p_{\mathbf{n}'} \qquad (n_l \geqslant 1)$$

that $\partial\Psi/\partial z_l = \mathbf{\Gamma}\Psi$ on $\mathbf{C}^l(\eta_0 : S)$.

For any number $b > \eta_0$, let us denote by $H(\mathbf{z} : b)$ the holomorphic function from $\mathbf{C}^l(\eta_0)$ to $\operatorname{Hom}_{\mathbf{C}}(E, E)$ such that $H(\mathbf{z} : b) = 1$ if $z_l = b$ and

$$\partial H(\mathbf{z}:b)/\partial z_l = \mathbf{\Gamma}(\mathbf{z})H(\mathbf{z}:b) \qquad (\mathbf{z} \in \mathbf{C}^l(\eta_0)).$$

Now fix $b > S$ and put

$$\Phi(\mathbf{z}) = H(\mathbf{z}:b)\Psi(z_1, \ldots, z_{l-1}, b) \qquad (\mathbf{z} \in \mathbf{C}^l(\eta_0)).$$

Then Φ is an E-valued holomorphic function on $\mathbf{C}^l(\eta_0)$ which satisfies the differential equation $\partial\Phi/\partial z_l = \mathbf{\Gamma}\Phi$ there and has the property that

$$\Phi(z_1, \ldots, z_{l-1}, b) = \Psi(z_1, \ldots, z_{l-1}, b)$$

if $\mathscr{R}(z_i) > \eta_0$ $(1 \leqslant i < l)$; it therefore follows from Lemma A 3.3.2 that $\Phi = \Psi$ on $\mathbf{C}^l(\eta_0 : S)$. We shall then agree to denote this holomorphic extension of Ψ to $\mathbf{C}^l(\eta_0)$ again by Ψ.

Summary Fix ξ, a and define the polynomials $p_{\mathbf{n}}$ $(\mathbf{n} \in \mathbf{N}^l)$ as above – then there exists an E-valued holomorphic function Ψ on $\mathbf{C}^l(\eta_0)$ such that:
(1) $\partial\Psi/\partial z_l = \mathbf{\Gamma}\Psi$ on $\mathbf{C}^l(\eta_0)$;
(2) $\Psi(\mathbf{z}) = \sum_{\mathbf{n}} p_{\mathbf{n}}(\mathbf{z})e^{\xi z_l - \langle \mathbf{z}, \mathbf{n}\rangle}$ (all $\mathbf{z} \in \mathbf{C}^l(\eta_0 : S)$).

Let $r = \dim(E)$ – then we may clearly find a basis $a_1, \ldots, a_r$ for E which has the property that there exists an eigenvalue ξ_j of Ξ such that

$$(\Xi - \xi_j)^r a_j = 0 \qquad (1 \leqslant j \leqslant r).$$

This being the case, let $\{p_{j,\mathbf{n}} : \mathbf{n} \in \mathbf{N}^l\}$ denote the sequence of polynomials corresponding to the pair (ξ_j, a_j) per the construction supra and let Ψ_j denote the associated holomorphic function on $\mathbf{C}^l(\eta_0)$ $(1 \leqslant j \leqslant r)$; it can be assumed that S has been chosen so large that

$$\|p_{j,\mathbf{n}}\| \leqslant \exp(|\mathbf{n}|\eta_0 + n_l(S - \eta_0))\|p_{j,\mathbf{0}}\| \qquad (1 \leqslant j \leqslant r)$$

for all $\mathbf{n} \in \mathbf{N}^l$ – then

$$\Psi_j(\mathbf{z}) = \sum_{\mathbf{n}} p_{j,\mathbf{n}}(\mathbf{z})e^{\xi_j z_l - \langle \mathbf{z}, \mathbf{n}\rangle} \qquad (\text{all } \mathbf{z} \in \mathbf{C}^l(\eta_0 : S)).$$

Suppose now that $l > 1$. We can then write

$$\Psi_j(\mathbf{z}, z) = \sum_n \sum_{\mathbf{n}} p_{j,\mathbf{n},n}(z)e^{(\xi_j - n)z - \langle \mathbf{z}, \mathbf{n}\rangle} \qquad (1 \leqslant j \leqslant r)$$

for $\mathbf{z} \in \mathbf{C}^{l-1}(\eta_0)$ and $z \in \mathbf{C}^1(S)$. [Here $\mathbf{n}$ runs through $\mathbf{N}^{l-1}$ while n ranges over $\mathbf{N}$; the pair $(\mathbf{n}, n)$ is to be considered as an element of $\mathbf{N}^l$.] Choose an integer $d \geqslant 0$ and a number $A \geqslant 0$ such that $\deg(p_{j,\mathbf{n},n}) \leqslant d$ and

$$\|p_{j,\mathbf{n},n}\| \leqslant A\exp(|\mathbf{n}|\eta_0 + nS)$$

for $1 \leqslant j \leqslant r$ (all $\mathbf{n} \in \mathbf{N}^{l-1}$, $n \in \mathbf{N}$); put

$$v_j(\mathbf{n} : z) = \sum_n p_{j,\mathbf{n},n}(z)e^{(\xi_j - n)z} \qquad (\mathbf{n} \in \mathbf{N}^{l-1}; z \in \mathbf{C}^1(S)).$$

Then $v_j(\mathbf{n} : .)$ is an E-valued holomorphic function on $\mathbf{C}^1(S)$. It is clear that

$$\partial\Psi_j(\mathbf{z}, z)/\partial z = \sum_{\mathbf{n}} \{\partial v_j(\mathbf{n} : z)/\partial z\}e^{-\langle \mathbf{z}, \mathbf{n}\rangle} \qquad (\mathbf{z} \in \mathbf{C}^{l-1}(\eta_0); z \in \mathbf{C}^1(S)).$$

On the other hand we also have the relation $\partial\Psi_j(\mathbf{z}, z)/\partial z = \Gamma(\mathbf{z}, z)\Psi_j(\mathbf{z}, z)$. Put $w_i = e^{-z_i}$ $(1 \leqslant i < l)$ – then $\mathbf{w} = (w_1, \ldots, w_{l-1})$ is a point in $\mathbf{C}^{l-1}$. Moreover, for fixed z in $\mathbf{C}^1(S)$, $\Psi_j(\mathbf{z}, z)$ and $\Gamma(\mathbf{z}, z)$ can be regarded as power series in $\mathbf{w}$ which converge absolutely for $\|\mathbf{w}\| < e^{-\eta_0}$. It is therefore permissible to equate the coefficients of $\mathbf{w}^{\mathbf{n}}$ in the two sides of the equation

$$\partial\Psi_j(\mathbf{z}, z)/\partial z = \Gamma(\mathbf{z}, z)\Psi_j(\mathbf{z}, z).$$

In particular, upon considering the case $\mathbf{n} = \mathbf{0}$, we get

$$\partial v_j(\mathbf{0} : z)/\partial z = c(\mathbf{0} : z)v_j(\mathbf{0} : z) \qquad (\text{all } z \in \mathbf{C}^1(S)).$$

[Given $\mathbf{n} \in \mathbf{N}^{l-1}$ and $z \in \mathbf{C}^1(S)$, we shall agree to write

$$c(\mathbf{n} : z) = \sum_n c(\mathbf{n}, n)e^{-nz};$$

plainly $\Gamma(\mathbf{z}, z) = \sum_{\mathbf{n}} c(\mathbf{n} : z)e^{-\langle \mathbf{z}, \mathbf{n}\rangle}$ $(\mathbf{z} \in \mathbf{C}^{l-1}(\eta_0))$.] Suppose that $b > S$ – then the vectors $v_j(\mathbf{0} : b)$ $(1 \leqslant j \leqslant r)$ are linearly independent. [If the contrary were true, then there would exist complex numbers c_j, not all of which are zero, such that $\sum_j c_j v_j(\mathbf{0} : b) = 0$; put $v(z) = \sum_j c_j v_j(\mathbf{0} : z)$ $(z \in \mathbf{C}^1(S))$ – then

$$\partial v(z)/\partial z = c(\mathbf{0} : z)v(z) \qquad (z \in \mathbf{C}^1(S))$$

and $v(b) = 0$, whence $v(x) = 0$ for all real $x \geqslant b$ (cf. Lemma A 3.3.2). Because $\deg(p_{j,0,n}) \leqslant d$ and $\|p_{j,0,n}\| \leqslant A \exp(nS)$, it follows from Lemma A3.5.1 that the series

$$\sum_n p_{j,0,n}(z)e^{(\xi_j - n)z} \qquad (z \in \mathbf{C}^1)$$

converges decently on $\mathbf{C}^1(1/2, T)$, T a sufficiently large positive number $\geqslant b$ $(1 \leqslant j \leqslant r)$. Without loss of generality we may assume that

$$\mathscr{R}(\xi_1) \geqslant \mathscr{R}(\xi_2) \geqslant \cdots \geqslant \mathscr{R}(\xi_r).$$

Let j_0 be the least index $(1 \leqslant j_0 \leqslant r)$ such that $c_{j_0} \neq 0$ – then since

$$\sum_j c_j \sum_n p_{j,0,n}(x)e^{(\xi_j - n)x} = v(x) = 0$$

for all $x \geqslant b$, we can, by Lemma A 3.2.4, equate the coefficient of $\exp(\xi_{j_0}x)$ in the left hand side of this equation to zero. Let J_0 be the set of all indices j such that $\xi_j = \xi_{j_0}$ – then $\sum_{j \in J_0} c_j p_{j,0,0} = 0$ which implies that $\sum_{j \in J_0} c_j a_j = 0$ (the constant term of $p_{j,0,0}$ being a_j), a contradiction. . . .] Put

$$v_j(\mathbf{n}) = v_j(\mathbf{n} : b) \qquad (1 \leqslant j \leqslant r; \mathbf{n} \in \mathbf{N}^{l-1}).$$

Then

$$\begin{aligned}\|v_j(\mathbf{n})\| &\leqslant (1 + b)^d \sum_n \|p_{j,\mathbf{n},n}\| \, |e^{(\xi_j - n)b}| \leqslant A(1 + b)^d |e^{\xi_j b}| e^{|\mathbf{n}|\eta_0} \sum_n e^{-n(b-S)} \\ &= A_j e^{|\mathbf{n}|\eta_0}\end{aligned}$$

where

$$A_j = A(1 + b)^d |e^{\xi_j b}| \sum_n e^{-n(b-S)} < \infty.$$

Let α_i be the linear function on E such that $\alpha_i(a_j) = \delta_{ij}$ $(1 \leqslant i, j \leqslant r)$ – then

$$\alpha_i(\Psi_j(\mathbf{z}, b)) = \sum_{\mathbf{n}} \alpha_i(v_j(\mathbf{n}))\mathbf{w}^{\mathbf{n}} \qquad (\mathbf{z} \in \mathbf{C}^{l-1}(\eta_0)),$$

the series being absolutely convergent for $\|\mathbf{w}\| \leqslant e^{-\eta_0}$ ($\mathbf{w}$ as above). Put

$$\Delta(\mathbf{z}) = \det(\alpha_i(\Psi_j(\mathbf{z}, b))) \qquad (\mathbf{z} \in \mathbf{C}^{l-1}(\eta_0)).$$

Then it is clear that we have an expansion of the form

$$\Delta(\mathbf{z}) = \sum_{\mathbf{n}} \delta(\mathbf{n}) e^{-\langle \mathbf{z}, \mathbf{n} \rangle} \qquad (\mathbf{z} \in \mathbf{C}^{l-1}(\eta_0))$$

where the $\delta(\mathbf{n})$ ($\mathbf{n} \in \mathbf{N}^{l-1}$) are certain complex numbers and the series converges absolutely for $\mathbf{z} \in \mathbf{C}^{l-1}(\eta_0)$. In view of the linear independence of the $v_j(\mathbf{0} : b)$ (cf. supra), $\Delta(\mathbf{z})^{-1}$ can also be expanded as a power series in $\mathbf{w}$ for sufficiently small values of $\|\mathbf{w}\|$; this means that we can choose $T \geqslant \eta_0$ and $\omega_{ij}(\mathbf{n}) \in \mathbf{C}$ ($\mathbf{n} \in \mathbf{N}^{l-1}$) such that the series

$$\Omega_{ij}(\mathbf{z}) = \sum_{\mathbf{n}} \omega_{ij}(\mathbf{n}) e^{-\langle \mathbf{z}, \mathbf{n} \rangle} \qquad (\mathbf{z} \in \mathbf{C}^{l-1})$$

converges absolutely for $\mathbf{z} \in \mathbf{C}^{l-1}(T)$ and $(\Omega_{ij}(\mathbf{z}))$ is the inverse of the matrix $(\alpha_i(\Psi_j(\mathbf{z}, b)))$ $(1 \leqslant i, j \leqslant r)$ – hence

$$a_i = \sum_{j=1}^{r} \Omega_{ij}(\mathbf{z}) \Psi_j(\mathbf{z}, b) \qquad (\mathbf{z} \in \mathbf{C}^{l-1}(T)).$$

Put $\Phi_j(\mathbf{z}) = \Psi_j(\mathbf{z}) - H(\mathbf{z}: b)\Psi_j(z_1, \ldots, z_{l-1}, b)$ $(\mathbf{z} \in \mathbf{C}^l(\eta_0))$. Because Φ_j is a holomorphic function on $\mathbf{C}^l(\eta_0)$, verifies the relation $\partial\Phi_j/\partial z_l = \Gamma\Phi_j$ there and has the property that $\Phi_j(\mathbf{z}) = 0$ if $z_l = b$, it follows from Lemma A 3.3.2 that $\Phi_j = 0$; consequently

$$H(\mathbf{z} : b) a_i = \sum_{j=1}^{r} \Omega_{ij}(z_1, \ldots, z_{l-1}) \Psi_j(\mathbf{z}) \qquad (\mathbf{z} \in \mathbf{C}^l(T)).$$

Put

$$q_{ij,\mathbf{n},n}(\mathbf{z}) = \sum_{\mathbf{n}_1 + \mathbf{n}_2 = \mathbf{n}} \omega_{ij}(\mathbf{n}_1) p_{j,\mathbf{n}_2,n}(z_l) \qquad (\mathbf{z} \in \mathbf{C}^l).$$

Since the series defining Ω_{ij} converges absolutely for $\mathbf{z} = (T+1)\mathbf{1}$, it is clear that $B = \sup_{i,j} \sum_{\mathbf{n}} |\omega_{ij}(\mathbf{n})| \exp(-|\mathbf{n}|(T+1)) < \infty$, whence

$$\|q_{ij,\mathbf{n},n}\| \leqslant AB \exp(|\mathbf{n}|(T+1) + nS);$$

as the $q_{ij,\mathbf{n},n}$ are evidently bounded in degree by d, we conclude that $\{q_{ij,\mathbf{n}} : \mathbf{n} \in \mathbf{N}^l\}$ is a normal sequence of E-polynomials on $\mathbf{C}^l$. Define now $\mathrm{Hom}_{\mathbf{C}}(E, E)$-polynomials $Q_{j,\mathbf{n}}$ on $\mathbf{C}^l$ by the prescription $Q_{j,\mathbf{n}}(\mathbf{z})a_i = q_{ij,\mathbf{n}}(\mathbf{z})$ $(\mathbf{z} \in \mathbf{C}^l)$. Obviously $\deg(Q_{j,\mathbf{n}}) \leqslant d$; moreover

$$\|Q_{j,\mathbf{n}}\| \leqslant C \sum_i \|q_{ij,\mathbf{n}}\| \leqslant rABCe^{|\mathbf{n}|R}$$

where $R = \max(S+1, T+1)$ (C a certain absolute constant with the property that $\|T\| \leqslant C \sum_i \|Ta_i\|$ for all $T \in \mathrm{Hom}_{\mathbf{C}}(E, E)$) – therefore $\{Q_{j,\mathbf{n}} : \mathbf{n} \in \mathbf{N}^l\}$ is a normal sequence of $\mathrm{Hom}_{\mathbf{C}}(E, E)$-polynomials on $\mathbf{C}^l$. Plainly

$$H(\mathbf{z} : b) = \sum_{j=1}^{r} \sum_{\mathbf{n}} Q_{j,\mathbf{n}}(\mathbf{z}) e^{\xi_j z_l - \langle \mathbf{z}, \mathbf{n} \rangle} \qquad (\text{all } \mathbf{z} \in \mathbf{C}^l(R)).$$

Summary Suppose that $l > 1$. Let $a_1, \ldots, a_r$ be a basis for E with the property that there exists an eigenvalue ξ_j of Ξ $(= c(\mathbf{0}))$ such that

$$(\Xi - \xi_j)^r a_j = 0 \qquad (1 \leqslant j \leqslant r);$$

let $H(\mathbf{z} : b)$ $(b > \eta_0)$ denote the holomorphic function from $\mathbf{C}^l(\eta_0)$ to $\mathrm{Hom}_{\mathbf{C}}(E, E)$ such that $H(\mathbf{z} : b) = 1$ if $z_l = b$ and $\partial H(\mathbf{z} : b)/\partial z_l = \Gamma(\mathbf{z}) H(\mathbf{z} : b)$ $(\mathbf{z} \in \mathbf{C}^l(\eta_0))$; let $S \geqslant \eta_0 + s$ be sufficiently large, $R = \max(S+1, T+1)$ (T as above) – then there exists a normal sequence $\{Q_{j,\mathbf{n}} : \mathbf{n} \in \mathbf{N}^l\}$ of $\mathrm{Hom}_{\mathbf{C}}(E, E)$-polynomials on $\mathbf{C}^l$ such that for any $b > S$

$$H(\mathbf{z} : b) = \sum_{j=1}^{r} \sum_{\mathbf{n}} Q_{j,\mathbf{n}}(\mathbf{z}) e^{\xi_j z_l - \langle \mathbf{z}, \mathbf{n} \rangle} \qquad (\text{all } \mathbf{z} \in \mathbf{C}^l(R)).$$

Proof of Proposition A 3.5.4 (1) *The Case $l > 1$.* As in the statement of the proposition, fix a number $a > \eta_0$; except for the requirement $b > S$, the number b above was arbitrary – therefore we may assume that $b > a$. Let

$$G(\mathbf{z}) = H(\mathbf{z}:a) - H(\mathbf{z}:b)H(z_1, \ldots, z_{l-1}, b:a) \qquad (\mathbf{z} \in \mathbf{C}^l(\eta_0)).$$

Because G is a holomorphic function on $\mathbf{C}^l(\eta_0)$, satisfies the differential equation $\partial G/\partial z_l = \boldsymbol{\Gamma} G$ there and has the property that $G(\mathbf{z}) = 0$ if $z_l = b$, it follows from Lemma A 3.3.2 that $G = 0$ – hence

$$H(\mathbf{z}, z:a) = H(\mathbf{z}, z:b)H(\mathbf{z}, b:a) \qquad (\mathbf{z} \in \mathbf{C}^{l-1}(\eta_0), z \in \mathbf{C}^1(\eta_0)).$$

Given $\mathbf{z} \in \mathbf{C}^{l-1}(\eta_0)$, set $H_0(\mathbf{z}:b, a) = 1 \in \mathrm{Hom}_{\mathbf{C}}(E, E)$ and put

$$H_n(\mathbf{z}:b, a) = \int_{b \geqslant t_1 \geqslant \cdots \geqslant t_n \geqslant a} \boldsymbol{\Gamma}(\mathbf{z}, t_1) \cdots \boldsymbol{\Gamma}(\mathbf{z}, t_n)\, dt_1 \cdots dt_n \qquad (n \in \mathbf{N}^+).$$

Then Lemma A 3.3.2 implies that

$$H(\mathbf{z}, b:a) = \sum_n H_n(\mathbf{z}:b, a) \qquad (\mathbf{z} \in \mathbf{C}^{l-1}(\eta_0)).$$

Now

$$\boldsymbol{\Gamma}(\mathbf{z}, z) = \sum_{\mathbf{n}} c(\mathbf{n}:z)e^{-\langle \mathbf{z}, \mathbf{n}\rangle} \qquad (\mathbf{z} \in \mathbf{C}^{l-1}(\eta_0); z \in \mathbf{C}^1(\eta_0))$$

and for $x > \eta_0$ we have

$$\sum_{\mathbf{n}} \| c(\mathbf{n}:x)\| e^{-|\mathbf{n}|\eta_0} \leqslant \sum_n \sum_{\mathbf{n}} \| c(\mathbf{n}, n)\| e^{-|\mathbf{n}|\eta_0 - n\eta_0} = D\,(\text{say}) < \infty.$$

Set $h_0(\mathbf{0}:b, a) = 1 \in \mathrm{Hom}_{\mathbf{C}}(E, E)$ and $h_0(\mathbf{n}:b, a) = 0$ if $\mathbf{n} \neq \mathbf{0}$ $(\mathbf{n} \in \mathbf{N}^{l-1})$; put

$$h_n(\mathbf{n}:b, a) = \sum_{\mathbf{n}_1 + \cdots + \mathbf{n}_n = \mathbf{n}} \int_{b \geqslant t_1 \geqslant \cdots \geqslant t_n \geqslant a} c(\mathbf{n}_1:t_1) \cdots c(\mathbf{n}_n:t_n)\, dt_1 \cdots dt_n \qquad (\mathbf{n} \in \mathbf{N}^{l-1})$$

for $n > 0$. Since $\| c(\mathbf{n}:x)\| \leqslant D \exp(|\mathbf{n}|\eta_0)$ $(x > \eta_0)$, we have

$$\| h_n(\mathbf{n}:b, a)\| \leqslant D^n e^{|\mathbf{n}|\eta_0}(b - a)^n/n!$$

and so $\sum_{\mathbf{n}} \sum_n \| h_n(\mathbf{n}:b, a)\| e^{-|\mathbf{n}|\eta_0} \leqslant e^{D(b-a)} < \infty$. Write

$$h(\mathbf{n}:b, a) = \sum_n h_n(\mathbf{n}:b, a) \qquad (\mathbf{n} \in \mathbf{N}^{l-1}).$$

Then it is clear that

$$H_n(\mathbf{z}:b, a) = \sum_{\mathbf{n}} h_n(\mathbf{n}:b, a)e^{-\langle \mathbf{z}, \mathbf{n}\rangle} \qquad (\mathbf{z} \in \mathbf{C}^{l-1}(\eta_0))$$

$$\Rightarrow$$

$$H(\mathbf{z}, b:a) = \sum_{\mathbf{n}} h(\mathbf{n}:b, a)e^{-\langle \mathbf{z}, \mathbf{n}\rangle} \qquad (\mathbf{z} \in \mathbf{C}^{l-1}(\eta_0)).$$

We come at last to the definition of the $P_{j,\mathbf{n}}$ $(1 \leqslant j \leqslant r; \mathbf{n} \in \mathbf{N}^l)$. Thus, for $\mathbf{z} \in \mathbf{C}^l$, let

$$P_{j,\mathbf{n}}(\mathbf{z}) = \sum_{\mathbf{n}_1 + \mathbf{n}_2 = \mathbf{n}} Q_{j,\mathbf{n}_1}(\mathbf{z})h(\mathbf{n}_2:b, a)$$

where the sum is over all pairs $(\mathbf{n}_1, \mathbf{n}_2)$ $(\mathbf{n}_1 \in \mathbf{N}^l, \mathbf{n}_2 \in \mathbf{N}^{l-1})$ such that $\mathbf{n}_1 + \mathbf{n}_2 = \mathbf{n}$. The sequence $\{P_{j,\mathbf{n}}:\mathbf{n} \in \mathbf{N}^l\}$ of $\mathrm{Hom}_{\mathbf{C}}(E, E)$-polynomials is normal; in fact the $P_{j,\mathbf{n}}$ are obviously bounded in degree by d and, moreover,

$$\begin{aligned} \| P_{j,\mathbf{n}}\| &\leqslant \sum_{\mathbf{n}_1 + \mathbf{n}_2 = \mathbf{n}} \| Q_{j,\mathbf{n}_1}\| \, \| h(\mathbf{n}_2:b, a)\| \\ &\leqslant rABCe^{D(b-a)} \sum_{\mathbf{n}_1 + \mathbf{n}_2 = \mathbf{n}} e^{(|\mathbf{n}_1| + |\mathbf{n}_2|)R} \leqslant rABCe^{D(b-a)}2^{2l-1}(2e^R)^{|\mathbf{n}|} \end{aligned}$$

which serves to establish the normality of the $P_{j,\mathbf{n}}$. We have seen above that $H(\mathbf{z}:a) = H(\mathbf{z}:b)H(z_1, \ldots, z_{l-1}, b:a)$ $(\mathbf{z} \in \mathbf{C}^l(\eta_0))$; it is therefore clear that

$$H(\mathbf{z}:a) = \sum_{j=1}^{r} \sum_{\mathbf{n}} P_{j,\mathbf{n}}(\mathbf{z})e^{\xi_j z_l - \langle \mathbf{z}, \mathbf{n} \rangle} \qquad (\text{all } \mathbf{z} \in \mathbf{C}^l(R)).$$

The contention of the proposition in the present case now follows upon citing Lemma A 3.5.1.

(2) *The Case $l = 1$.* Except for some minor changes, this case is treated in the same way as the case $l > 1$. Thus one first shows that the vectors $\Psi_1(b)$, $\ldots, \Psi_r(b)$ are linearly independent; this fact then implies that

$$a_i = \sum_j \Omega_{ij}\Psi_j(b) \qquad (1 \leqslant i \leqslant r)$$

where the Ω_{ij} are certain complex numbers; consequently

$$H(z:b)a_i = \sum_j \Omega_{ij}\Psi_j(z) \qquad (z \in \mathbf{C}^1(\eta_0)).$$

Since

$$H(z:a) = H(z:b)H(b:a) \qquad (z \in \mathbf{C}^1(\eta_0)),$$

the sought for $\mathrm{Hom}_{\mathbf{C}}(E, E)$-polynomials are to be defined by the prescription $P_{j,n}(z) = Q_{j,n}(z)H(b:a)$ where $Q_{j,n}(z)a_i = p_{j,n}(z)a_i$ $(z \in \mathbf{C}^1)$; it is easy to verify that the sequence $\{P_{j,n} : n \in \mathbf{N}\}$ is normal and has the property that

$$H(z:a) = \sum_{j=1}^{r} \sum_{n} P_{j,n}(z)e^{(\xi_j - n)z}$$

for all $z \in \mathbf{C}^1(\eta)$, η some number $\geqslant \eta_0$. The statement of the proposition in this case now follows readily from Lemma A 3.5.1.

The proof of Proposition A 3.5.4 is therefore complete. □

Let $c_i(\mathbf{n})$ $(1 \leqslant i \leqslant l;\ \mathbf{n} \in \mathbf{N}^l)$ be elements in $\mathrm{Hom}_{\mathbf{C}}(E, E)$ such that

$$\sum_i \sum_{\mathbf{n}} \| c_i(\mathbf{n}) \| e^{-|\mathbf{n}|\eta} < \infty$$

for every $\eta > 0$; put

$$(\boldsymbol{\Gamma}) \quad \boldsymbol{\Gamma}_i(\mathbf{z}) = \sum_{\mathbf{n}} c_i(\mathbf{n})e^{-\langle \mathbf{z}, \mathbf{n} \rangle} \qquad (\mathbf{z} \in \mathbf{C}^{l,+}).$$

Then it is clear that each $\boldsymbol{\Gamma}_i$ $(1 \leqslant i \leqslant l)$ is a holomorphic function on $\mathbf{C}^{l,+}$ with values in $\mathrm{Hom}_{\mathbf{C}}(E, E)$.

Proposition A 3.5.5 Define holomorphic functions $\boldsymbol{\Gamma}_i$ on $\mathbf{C}^{l,+}$ by condition $(\boldsymbol{\Gamma})$ supra; suppose that Ψ is an E-valued holomorphic function on $\mathbf{C}^{l,+}$ such that $\partial\Psi/\partial z_i = \boldsymbol{\Gamma}_i\Psi$ $(1 \leqslant i \leqslant l)$ there – then there exists a finite number of points $\lambda_1, \ldots, \lambda_r$ in $\mathbf{C}^l$ and for each j $(1 \leqslant j \leqslant r)$ a normal sequence $\{p_{j,\mathbf{n}} : \mathbf{n} \in \mathbf{N}^l\}$ of E-polynomials on $\mathbf{C}^l$ such that the following condition holds. Fix ϵ $(0 < \epsilon < 1)$ and choose a number $\eta \geqslant 0$ such that the series $\sum_{\mathbf{n}} p_{j,\mathbf{n}}e^{-\langle \cdot, \mathbf{n} \rangle}$ $(1 \leqslant j \leqslant r)$ converges decently on $\mathbf{C}^l(\epsilon, \eta)$ – then

$$\Psi(\mathbf{z}) = \sum_{j=1}^{r} \sum_{\mathbf{n}} p_{j,\mathbf{n}}(\mathbf{z})e^{\langle \mathbf{z}, \lambda_j - \mathbf{n} \rangle} \qquad (\text{all } \mathbf{z} \in \mathbf{C}^l(\epsilon, \eta)).$$

[The number 'r' appearing here has no connection with $r = \dim(E)$.]

Proof (1) Let us first suppose that the c_i satisfy an auxilliary requirement, namely that $c_i(\mathbf{n}) = 0$ if $n_i = 0$ and $\mathbf{n} \neq \mathbf{0}$ $(1 \leqslant i \leqslant l)$. Fix a number $a > 0$; let

$H_{a,l}$ be the holomorphic function from $\mathbf{C}^{l,+}$ to $\mathrm{Hom}_{\mathbf{C}}(E, E)$ such that

$$\partial H_{a,l}/\partial z_l = \mathbf{\Gamma}_l H_{a,l} \qquad (\mathbf{z} \in \mathbf{C}^{l,+})$$

with $H_{a,l}(\mathbf{z}) = 1$ if $z_l = a$. We distinguish two cases. If $l = 1$, then

$$\Psi(z) = H_{a,1}(z)\Psi(a) \qquad (z \in \mathbf{C}^{1,+})$$

and so in this situation the present proposition is seen to be an immediate consequence of Proposition A 3.5.4. To handle the general case, we shall use induction. Thus put $\Phi(\mathbf{z}) = \Psi(\mathbf{z}, a)$ $(\mathbf{z} \in \mathbf{C}^{l-1,+})$ $(l > 1)$; on the basis of Lemma A 3.3.2, it is clear that $\Psi(\mathbf{z}) = H_{a,l}(\mathbf{z})\Phi(z_1, \ldots, z_{l-1})$ (all $\mathbf{z} \in \mathbf{C}^{l,+}$); applying now the induction hypothesis to Φ, our proposition is seen to follow once again from Proposition A 3.5.4 (recall Lemma A 3.5.3 . . .).

(2) It remains to remove the restriction which was imposed on the c_i in (1). For this we proceed as follows. Let τ denote the endomorphism of $\mathbf{C}^l$ which is defined by the rule $\tau(\mathbf{z}) = \mathbf{z} + \langle \mathbf{z}, \mathbf{1}\rangle \mathbf{1}$ $(\mathbf{z} \in \mathbf{C}^l)$; τ is obviously non-singular, maps $\mathbf{C}^l(\eta)$ into itself (any $\eta \geqslant 0$) and leaves $\mathbf{N}^l$ stable. Put

$$\Psi'(\mathbf{z}) = \Psi(\tau(\mathbf{z})), \quad \mathbf{\Gamma}_i'(\mathbf{z}) = \mathbf{\Gamma}_i(\tau(\mathbf{z})) + \sum_{j=1}^{l} \mathbf{\Gamma}_j(\tau(\mathbf{z}))$$
$$(\mathbf{z} \in \mathbf{C}^{l,+};\ 1 \leqslant i \leqslant l).$$

Then Ψ' and $\mathbf{\Gamma}_i'$ are holomorphic functions on $\mathbf{C}^{l,+}$ and Ψ' satisfies the differential equations

$$\partial\Psi'/\partial z_i = \mathbf{\Gamma}_i'\Psi' \qquad (1 \leqslant i \leqslant l)$$

there. Fix $\mathbf{n} \in \mathbf{N}^l$. If $\mathbf{n} \notin \tau(\mathbf{N}^l)$, put $c_i'(\mathbf{n}) = 0$; if $\mathbf{n}' = \tau^{-1}(\mathbf{n}) \in \mathbf{N}^l$, put

$$c_i'(\mathbf{n}) = c_i(\mathbf{n}') + \sum_{j=1}^{l} c_j(\mathbf{n}').$$

Then it is clear that the elements c_i' $(1 \leqslant i \leqslant l)$ meet the auxiliary condition which was laid down in (1) above and, of course,

$$\mathbf{\Gamma}_i'(\mathbf{z}) = \sum_{\mathbf{n}} c_i'(\mathbf{n})e^{-\langle \mathbf{z}, \mathbf{n}\rangle} \qquad (\text{all } \mathbf{z} \in \mathbf{C}^{l,+}).$$

This means that our proposition is, in any event, valid for Ψ'; thus there exists a finite number of points $\lambda_1', \ldots, \lambda_{r'}'$ in $\mathbf{C}^l$ and for each j $(1 \leqslant j \leqslant r')$ a normal sequence $\{p_{j,\mathbf{n}}' : \mathbf{n} \in \mathbf{N}^l\}$ of E-polynomials on $\mathbf{C}^l$ such that the following condition holds: Fix ϵ $(0 < \epsilon < 1)$ and choose $\eta' \geqslant 0$ large enough to ensure that the series $\sum_{\mathbf{n}} p_{j,\mathbf{n}}' e^{-\langle \cdot, \mathbf{n}\rangle}$ $(1 \leqslant j \leqslant r')$ converges decently on $\mathbf{C}^l(\epsilon, \eta')$ – then

$$\Psi'(\mathbf{z}) = \sum_{j=1}^{r'} \sum_{\mathbf{n}} p_{j,\mathbf{n}}'(\mathbf{z})e^{\langle \mathbf{z}, \lambda_j' - \mathbf{n}\rangle} \qquad (\text{all } \mathbf{z} \in \mathbf{C}^l(\epsilon, \eta')).$$

Let Λ'' be the set of all points in $\mathbf{C}^l$ of the form $\lambda_j' - \mathbf{n}$ $(1 \leqslant j \leqslant r';\ \mathbf{n} \in \mathbf{N}^l)$. Given $\lambda'' \in \Lambda''$, set $p_{\lambda''}'' = \sum p_{j,\mathbf{n}}'$ where the sum is over all pairs $(j, \mathbf{n})$ such that $\lambda_j' - \mathbf{n} = \lambda''$; denoting by Λ' the subset of Λ'' consisting of those λ' for which $p_{\lambda'}' = \sum p_{j,\mathbf{n}}' \neq 0$, we evidently have

$$\Psi'(\mathbf{z}) = \sum_{\lambda' \in \Lambda'} p_{\lambda'}'(\mathbf{z})e^{\langle \mathbf{z}, \lambda'\rangle} \qquad (\mathbf{z} \in \mathbf{C}^l(\epsilon, \eta')),$$

the series converging decently on $\mathbf{C}^l(\epsilon, \eta')$. Let Λ_0' be the subset of Λ' comprised of those $\lambda' \in \Lambda'$ such that $\lambda' + \tau(\mathbf{n}) \notin \Lambda'$ for every $\mathbf{n} \neq \mathbf{0}$ in $\mathbf{N}^l$. It is easy to see that every element in Λ' is of the form $\lambda_0' - \tau(\mathbf{n})$ where $\lambda_0' \in \Lambda_0'$ and $\mathbf{n} \in \mathbf{N}^l$. Furthermore, it will be shown below that Λ_0' is a finite set; this being so, let us arrange the distinct elements of Λ_0' in a sequence $\lambda_1', \ldots, \lambda_r'$ and put $\lambda_j = \tau^{-1}(\lambda_j')$ $(1 \leqslant j \leqslant r)$. Fix $\mathbf{n} \in \mathbf{N}^l$: If $\lambda_j' - \tau(\mathbf{n}) \in \Lambda'$ and $\lambda_j' - \tau(\mathbf{n}) \neq \lambda_{j'}' -$

$\tau(\mathbf{n}')$ for any $j' < j$ and $\mathbf{n}' \in \mathbf{N}^l$, put $q'_{j,\mathbf{n}} = p'_{\lambda'_j - \tau(\mathbf{n})}$; otherwise put $q'_{j,\mathbf{n}} = 0$ $(1 \leqslant j \leqslant r)$ – then we have

$$\Psi'(\mathbf{z}) = \sum_{\lambda' \in \Lambda'} p'_{\lambda'}(\mathbf{z}) e^{\langle \mathbf{z}, \lambda' \rangle} = \sum_{j=1}^{r} \sum_{\mathbf{n}} q'_{j,\mathbf{n}}(\mathbf{z}) e^{\langle \mathbf{z}, \lambda_j' - \tau(\mathbf{n}) \rangle}$$

for all $\mathbf{z} \in \mathbf{C}^l(\epsilon, \eta')$. Define now E-polynomials $p_{j,\mathbf{n}}$ on $\mathbf{C}^l$ by the rule

$$p_{j,\mathbf{n}}(\mathbf{z}) = q'_{j,\mathbf{n}}(\tau^{-1}(\mathbf{z})) \qquad (\mathbf{z} \in \mathbf{C}^l).$$

It is easy to check that the sequence $\{p_{j,\mathbf{n}} : \mathbf{n} \in \mathbf{N}^l\}$ is normal $(1 \leqslant j \leqslant r)$. [In view of the normality of the $p'_{j,\mathbf{n}}$, there exists an integer $d' \geqslant 0$ and positive numbers M' and N' such that $\deg(p'_{\lambda'}) \leqslant d'$ and $\| p'_{\lambda'} \| \leqslant M' N'^{-\mathscr{R}(\lambda')}$ for all $\lambda' \in \Lambda' \ldots$.] This being the case, choose $\eta \geqslant 0$ in such a way as to ensure that the series $\sum_{\mathbf{n}} p_{j,\mathbf{n}} e^{-\langle \cdot, \mathbf{n} \rangle}$ $(1 \leqslant j \leqslant r)$ converges decently on $\mathbf{C}^l(\epsilon, \eta)$ (cf. Lemma A 3.5.1) – then we claim that

$$\Psi(\mathbf{z}) = \sum_{j=1}^{r} \sum_{\mathbf{n}} p_{j,\mathbf{n}}(\mathbf{z}) e^{\langle \mathbf{z}, \lambda_j - \mathbf{n} \rangle} \qquad (\text{all } \mathbf{z} \in \mathbf{C}^l(\epsilon, \eta)).$$

In view of the fact that both sides are holomorphic functions on $\mathbf{C}^l(\epsilon, \eta)$, the claim will follow when it is shown that they are equal on some non-empty open subset of $\mathbf{R}^l \cap \mathbf{C}^l(\epsilon, \eta)$. For this latter purpose, choose $\eta' \geqslant \eta$ large enough to secure the equality

$$\Psi(\tau(\mathbf{z})) = \Psi'(\mathbf{z}) = \sum_{j=1}^{r} \sum_{\mathbf{n}} p_{j,\mathbf{n}}(\tau(\mathbf{z})) e^{\langle \tau(\mathbf{z}), \lambda_j - \mathbf{n} \rangle}$$

for all $\mathbf{z} \in \mathbf{C}^l(\epsilon, \eta')$. Since $\tau(\mathbf{R}^l \cap \mathbf{C}^l(\epsilon, \eta'))$ is obviously a non-empty open subset of $\mathbf{R}^l \cap \mathbf{C}^l(\epsilon, \eta)$, the contention of the claim is now evident.

This completes the proof of Proposition A 3.5.5. □

Remark The following observations are immediate consequences of the definition of the λ_j and $p_{j,\mathbf{n}}$:

(1) If $i \neq j$, then $\lambda_i \neq \lambda_j - \mathbf{n}$ for any $\mathbf{n} \in \mathbf{N}^l$;

(2) Fix j $(1 \leqslant j \leqslant r)$ – then $p_{j,\mathbf{0}} \neq 0$.

Retain the notations which were introduced in part (2) of the proof of the preceding proposition. Let us verify that Λ'_0 (which we shall assume is non-empty) is a finite set. Put

$$N' = \sum_{i=1}^{l} \sum_{\mathbf{n}} \| c'_i(\mathbf{n}) \| e^{-|\mathbf{n}|}.$$

Then N' is finite and $\| c'_i(\mathbf{n}) \| \leqslant N' e^{|\mathbf{n}|}$ (all $\mathbf{n} \in \mathbf{N}^l$); thus, for each i, the sequence $\{c'_i(\mathbf{n}) : \mathbf{n} \in \mathbf{N}^l\}$ of $\mathrm{Hom}_{\mathbf{C}}(E, E)$-polynomials is normal and so by choosing η' sufficiently large, it can be assumed that the series

$$\sum_{\mathbf{n}} c'_i(\mathbf{n}) e^{-\langle \mathbf{z}, \mathbf{n} \rangle} \qquad (1 \leqslant i \leqslant l)$$

converges decently on $\mathbf{C}^l(\epsilon, \eta')$. Set

$$\gamma'_i(\mathbf{n}) = c_i(\mathbf{n}) + \sum_{j=1}^{l} c_j(\mathbf{n}) \qquad (\mathbf{n} \in \mathbf{N}^l).$$

Then clearly

$$\Gamma'_i(\mathbf{z}) = \sum_{\mathbf{n}} c'_i(\mathbf{n}) e^{-\langle \mathbf{z}, \mathbf{n} \rangle} = \sum_{\mathbf{n}} \gamma'_i(\mathbf{n}) e^{-\langle \mathbf{z}, \tau(\mathbf{n}) \rangle} \qquad (\mathbf{z} \in \mathbf{C}^l(\epsilon, \eta'); 1 \leqslant i \leqslant l),$$

both series converging decently on $\mathbf{C}^l(\epsilon, \eta')$. On the other hand

$$\partial \Psi' / \partial z_i = \Gamma'_i \Psi';$$

since the series defining $\boldsymbol{\Psi}'$ converges decently on $\mathbf{C}^l(\epsilon, \eta')$, it may be differentiated term by term – we obtain

$$\sum_{\lambda' \in \Lambda'} (\lambda'_i p'_{\lambda'}(\mathbf{z}) + \partial p'_{\lambda'}(\mathbf{z})/\partial z_i) e^{\langle \mathbf{z}, \lambda' \rangle} = \sum_{\mathbf{n}} \sum_{\lambda' \in \Lambda'} \gamma'_i(\mathbf{n}) p'_{\lambda'}(\mathbf{z}) e^{\langle \mathbf{z}, \lambda' - \tau(\mathbf{n}) \rangle}$$

for $\mathbf{z} \in \mathbf{C}^l(\epsilon, \eta')$. [Here λ'_i denotes the ith coordinate of the point λ'.] Because

$$\sum_{\mathbf{n}} \| \gamma'_i(\mathbf{n}) e^{-\langle \mathbf{z}, \tau(\mathbf{n}) \rangle} \| < \infty \qquad (\text{any } \mathbf{z} \in \mathbf{C}^l(\epsilon, \eta')),$$

it follows from Corollary A.3.5.2 and Lemma A 3.2.6 that the series

$$\sum_{\mathbf{n}} \sum_{\lambda' \in \Lambda'} \| \gamma'_i(\mathbf{n}) p'_{\lambda'}(\mathbf{z}) e^{\langle \mathbf{z}, \lambda' + \lambda - \tau(\mathbf{n}) \rangle} \|$$

converges uniformly on $\mathbf{C}^l(\epsilon, \eta')$ for every $\lambda \in \mathbf{C}^l$. This being the case, fix an element λ'_0 in Λ'_0 – then it follows from Lemma A 3.2.4 that the coefficients of $e^{\langle \cdot, \lambda_0' \rangle}$ on the two sides of the equation above must be equal. Since $\lambda'_0 \in \Lambda'_0$, $\lambda'_0 \neq \lambda' - \tau(\mathbf{n})$ for any $\lambda' \in \Lambda'$ and $\mathbf{n} \in \mathbf{N}^l$ unless $\lambda'_0 = \lambda'$ and $\mathbf{n} = \mathbf{0}$; thus $\lambda'_{0,i} p'_{\lambda'_0} + \partial p'_{\lambda'_0}/\partial z_i = \gamma'_i(\mathbf{0}) p'_{\lambda'_0}$ and so, by repeated differentiation, we find that $\partial^n p'_{\lambda'_0}/\partial z_i^n = (\gamma'_i(\mathbf{0}) - \lambda'_{0,i})^n p'_{\lambda'_0}$ $(n \geqslant 1)$; the left hand side of this last relation vanishes for large n, whence $\lambda_{i,0}$ is an eigenvalue of $\gamma'_i(\mathbf{0})$. As this is true for every i, it follows that the cardinality of Λ'_0 cannot exceed $(\dim(E))^l$. . . .

Proposition A 3.5.6 Define holomorphic functions $\boldsymbol{\Gamma}_i$ on $\mathbf{C}^{l,+}$ by condition $(\boldsymbol{\Gamma})$ supra; suppose that $\boldsymbol{\Psi}$ is an E-valued holomorphic function on $\mathbf{C}^{l,+}$ such that $\partial \boldsymbol{\Psi}/\partial z_i = \boldsymbol{\Gamma}_i \boldsymbol{\Psi}$ $(1 \leqslant i \leqslant l)$ there; select points λ_j in $\mathbf{C}^l$ and E-polynomials $p_{j,\mathbf{n}}$ on $\mathbf{C}^l$ $(1 \leqslant j \leqslant r; \mathbf{n} \in \mathbf{N}^l)$ per Proposition A 3.5.5 above; let ϵ, η be any real numbers such that $0 < \epsilon < 1$, $\eta > 0$ – then the series

$$\sum_{j=1}^{r} \sum_{\mathbf{n}} \| p_{j,\mathbf{n}} e^{\langle \cdot, \lambda_j - \mathbf{n} \rangle} \|$$

converges decently on $\mathbf{C}^l(\epsilon, \eta)$ and, moreover,

$$\boldsymbol{\Psi}(\mathbf{z}) = \sum_{j=1}^{r} \sum_{\mathbf{n}} p_{j,\mathbf{n}}(\mathbf{z}) e^{\langle \mathbf{z}, \lambda_j - \mathbf{n} \rangle} \qquad (\text{all } \mathbf{z} \in \mathbf{C}^{l,+}).$$

Proof Owing to Proposition A 3.5.5, there exists an $\eta_0 \geqslant \eta$ with the property that the series $\sum_{\mathbf{n}} p_{j,\mathbf{n}} e^{-\langle \cdot, \mathbf{n} \rangle}$ $(1 \leqslant j \leqslant r)$ converges decently on $\mathbf{C}^l(\epsilon, \eta_0)$; on the other hand it is obvious that there exists a compact subset Ω of $\mathbf{C}^{l,+}$ such that $\mathbf{C}^l(\epsilon, \eta) \subset \Omega \cup \mathbf{C}^l(\epsilon, \eta_0)$ – therefore, as far as the first assertion of our proposition is concerned, it will be enough to show that the series

$$\sum_{j} \sum_{\mathbf{n}} \| p_{j,\mathbf{n}} e^{\langle \cdot, \lambda_j - \mathbf{n} \rangle} \|$$

converges uniformly on compact subsets of $\mathbf{C}^{l,+}$. Because a given compact subset of $\mathbf{C}^{l,+}$ is contained in $\mathbf{C}^l(\epsilon, \eta)$ for suitable ϵ, η, we actually need only verify that the series at hand converges uniformly on compact subsets of $\mathbf{C}^l(\epsilon, \eta)$ (ϵ and η subject to the restrictions supra but otherwise arbitrary). As regards the second assertion of our proposition, it is, of course, an immediate consequence of the first; in fact both sides of the equation in question are then *holomorphic* on $\mathbf{C}^{l,+}$ and agree on a non-empty open subset of $\mathbf{C}^{l,+}$ (cf. Proposition A 3.5.5). . . .

(1) So let us fix ϵ, η and choose η_0 as above; plainly

$$\mathbf{C}^l(\epsilon, \eta) + \eta_0 \mathbf{1} \subset \mathbf{C}^l(\epsilon, \eta_0).$$

Put

$$c(\mathbf{n}) = \sum_{i=1}^{l} c_i(\mathbf{n}) \qquad (\mathbf{n} \in \mathbf{N}^l)$$

and

$$\mathbf{\Gamma}(\mathbf{z}) = \sum_{i=1}^{l} \mathbf{\Gamma}_i(\mathbf{z}) \qquad (\mathbf{z} \in \mathbf{C}^{l,+});$$

obviously

$$d\Psi(\mathbf{z} + t\mathbf{1})/dt = \mathbf{\Gamma}(\mathbf{z} + t\mathbf{1})\Psi(\mathbf{z} + t\mathbf{1}) \qquad (\mathbf{z} \in \mathbf{C}^{l,+}; t \geqslant 0).$$

Given $\mathbf{z} \in \mathbf{C}^{l,+}$, put $H_0(\mathbf{z}) = 1 \in \mathrm{Hom}_{\mathbf{C}}(E, E)$ and

$$H_n(\mathbf{z}) = (-1)^n \int_{0 \leqslant t_1 \leqslant \cdots \leqslant t_n \leqslant \eta_0} \mathbf{\Gamma}(\mathbf{z} + t_1\mathbf{1}) \cdots \mathbf{\Gamma}(\mathbf{z} + t_n\mathbf{1})\, dt_1 \cdots dt_n \qquad (n \in \mathbf{N}^+).$$

Then it is clear that the series $\sum_n \|H_n(.)\|$ converges uniformly on compact subsets of $\mathbf{C}^{l,+}$. Let

$$c(\mathbf{n} : t) = c(\mathbf{n}) \exp(-|\mathbf{n}|t) \qquad (t \geqslant 0).$$

Then we have

$$\mathbf{\Gamma}(\mathbf{z} + t\mathbf{1}) = \sum_{\mathbf{n}} c(\mathbf{n} : t) e^{-\langle \mathbf{z}, \mathbf{n} \rangle} \qquad (t \geqslant 0)$$

for all $\mathbf{z} \in \mathbf{C}^{l,+}$. Set $h_0(\mathbf{0}) = 1 \in \mathrm{Hom}_{\mathbf{C}}(E, E)$ and $h_0(\mathbf{n}) = 0$ if $\mathbf{n} \neq \mathbf{0}$ $(\mathbf{n} \in \mathbf{N}^l)$; for $n > 0$, put

$$h_n(\mathbf{n}) = (-1)^n \sum_{\mathbf{n}_1 + \cdots + \mathbf{n}_n = \mathbf{n}} \int_{0 \leqslant t_1 \leqslant \cdots \leqslant t_n \leqslant \eta_0} c(\mathbf{n}_1 : t_1) \cdots c(\mathbf{n}_n : t_n)\, dt_1 \cdots dt_n \qquad (\mathbf{n} \in \mathbf{N}^l).$$

Then it follows at once that $H_n(\mathbf{z}) = \sum_{\mathbf{n}} h_n(\mathbf{n}) e^{-\langle \mathbf{z}, \mathbf{n} \rangle}$ (all $\mathbf{z} \in \mathbf{C}^l(\eta)$). Now let $H(\mathbf{z}) = \sum_n H_n(\mathbf{z})$ $(\mathbf{z} \in \mathbf{C}^{l,+})$, $h(\mathbf{n}) = \sum_n h_n(\mathbf{n})$ $(\mathbf{n} \in \mathbf{N}^l)$ – then we find without difficulty that

$$H(\mathbf{z}) = \sum_n H_n(\mathbf{z}) = \sum_n \sum_{\mathbf{n}} h_n(\mathbf{n}) e^{-\langle \mathbf{z}, \mathbf{n} \rangle} = \sum_{\mathbf{n}} h(\mathbf{n}) e^{-\langle \mathbf{z}, \mathbf{n} \rangle} \qquad (\text{all } \mathbf{z} \in \mathbf{C}^l(\eta)).$$

(2) On the basis of what has been said in A3.3, it is clear that

$$\Psi(\mathbf{z}) = H(\mathbf{z})\Psi(\mathbf{z} + \eta_0\mathbf{1}) \qquad (\text{all } \mathbf{z} \in \mathbf{C}^{l,+});$$

on the other hand, since $\mathbf{C}^l(\epsilon, \eta) + \eta_0\mathbf{1} \subset \mathbf{C}^l(\epsilon, \eta_0)$, we have

$$\Psi(\mathbf{z} + \eta_0\mathbf{1}) = \sum_{j=1}^{r} \sum_{\mathbf{n}} p_{j,\mathbf{n}}(\mathbf{z} + \eta_0\mathbf{1}) e^{\langle \mathbf{z} + \eta_0\mathbf{1}, \lambda_j - \mathbf{n} \rangle} \qquad (\text{all } \mathbf{z} \in \mathbf{C}^l(\epsilon, \eta)).$$

This being so, put

$$v_{j,\mathbf{n}}(\mathbf{z}) = \sum_{\mathbf{n}_1 + \mathbf{n}_2 = \mathbf{n}} h(\mathbf{n}_1) p_{j,\mathbf{n}_2}(\mathbf{z} + \eta_0\mathbf{1}) e^{\eta_0 \langle \mathbf{1}, \lambda_j - \mathbf{n}_2 \rangle} \qquad (\mathbf{z} \in \mathbf{C}^l).$$

Then the $v_{j,\mathbf{n}}$ are E-polynomials on $\mathbf{C}^l$. Let Ω be a compact subset of $\mathbf{C}^l(\epsilon, \eta)$. In view of the discussion in (1), it is clear that the series $\sum_{\mathbf{n}} \|h(\mathbf{n}) e^{-\langle \cdot, \mathbf{n} \rangle}\|$ converges uniformly on Ω (in fact on $\mathbf{C}^l(\eta)$); since $\Omega + \eta_0\mathbf{1} \subset \mathbf{C}^l(\epsilon, \eta_0)$, the series

$$\sum_{j=1}^{r} \sum_{\mathbf{n}} \|p_{j,\mathbf{n}}(\mathbf{z} + \eta_0\mathbf{1}) e^{\langle \mathbf{z} + \eta_0\mathbf{1}, \lambda_j - \mathbf{n} \rangle}\| \qquad (\mathbf{z} \in \mathbf{C}^l)$$

also converges uniformly on Ω – thus the series

$$\sum_{j=1}^{r} \sum_{\mathbf{n}} \|v_{j,\mathbf{n}}(\mathbf{z}) e^{\langle \mathbf{z}, \lambda_j - \mathbf{n} \rangle}\| \qquad (\mathbf{z} \in \mathbf{C}^l)$$

is uniformly convergent on Ω and so

$$\Psi(\mathbf{z}) = H(\mathbf{z})\Psi(\mathbf{z} + \eta_0\mathbf{1}) = \sum_{j=1}^{r} \sum_{\mathbf{n}} v_{j,\mathbf{n}}(\mathbf{z}) e^{\langle \mathbf{z}, \lambda_j - \mathbf{n} \rangle} \qquad (\text{all } \mathbf{z} \in \Omega).$$

It is not hard to see that the sequence $\{v_{j,\mathbf{n}} : \mathbf{n} \in \mathbf{N}^l\}$ $(1 \leqslant j \leqslant r)$ is normal. Define now E-polynomials $q_{j,\mathbf{n}}$ $(1 \leqslant j \leqslant r;\ \mathbf{n} \in \mathbf{N}^l)$ as follows: Write $q_{j,\mathbf{n}} = 0$ if there exists an index $j' < j$ such that $\lambda_j - \mathbf{n} = \lambda_{j'} - \mathbf{n}'$ for some $\mathbf{n} \in \mathbf{N}^l$; otherwise let $q_{j,\mathbf{n}} = \sum v_{j',\mathbf{n}'}$, the sum being taken over all pairs $(j', \mathbf{n}')$ such that $\lambda_j - \mathbf{n} = \lambda_{j'} - \mathbf{n}'$. Fix a point λ in $\mathbf{C}^l$ and let $\mathscr{S}$ be a subset of $\mathbf{C}^{l,+}$ on which the series $\sum_j \sum_{\mathbf{n}} \|v_{j,\mathbf{n}} e^{\langle \cdot, \lambda + \lambda_j - \mathbf{n} \rangle}\|$ is uniformly convergent – then it is obvious that the series $\sum_j \sum_{\mathbf{n}} \|q_{j,\mathbf{n}} e^{\langle \cdot, \lambda + \lambda_j - \mathbf{n} \rangle}\|$ also converges uniformly on $\mathscr{S}$ and there $\sum_j \sum_{\mathbf{n}} v_{j,\mathbf{n}} e^{\langle \cdot, \lambda_j - \mathbf{n} \rangle} = \sum_j \sum_{\mathbf{n}} q_{j,\mathbf{n}} e^{\langle \cdot, \lambda_j - \mathbf{n} \rangle}$. All this holds in particular for $\mathscr{S} = \Omega$, Ω a compact subset of $\mathbf{C}^l(\epsilon, \eta)$.

(3) Owing to the remarks which were made at the beginning, our proposition will follow when it is established that the series $\sum_j \sum_{\mathbf{n}} \|p_{j,\mathbf{n}} e^{\langle \cdot, \lambda_j - \mathbf{n} \rangle}\|$ converges uniformly on compact subsets of $\mathbf{C}^l(\epsilon, \eta)$; for this purpose we need only show, in view of (2), that $p_{j,\mathbf{n}} = q_{j,\mathbf{n}}$ $(1 \leqslant j \leqslant r; \mathbf{n} \in \mathbf{N}^l)$. Because the sequence $\{v_{j,\mathbf{n}} : \mathbf{n} \in \mathbf{N}^l\}$ is normal for every j, there exists $\eta^0 \geqslant \eta_0$ with the property that the series $\sum_{\mathbf{n}} v_{j,\mathbf{n}} e^{-\langle \cdot, \mathbf{n} \rangle}$ $(1 \leqslant j \leqslant r)$ converges decently on $\mathbf{C}^l(\epsilon, \eta^0)$; as has been pointed out in (2), this implies that the series $\sum_j \sum_{\mathbf{n}} \|q_{j,\mathbf{n}} e^{\langle \cdot, \lambda + \lambda_j - \mathbf{n} \rangle}\|$ converges uniformly on $\mathbf{C}^l(\epsilon, \eta^0)$ for every $\lambda \in \mathbf{C}^l$. Now, on the one hand we have

$$\Psi(\mathbf{z}) = \sum_{j=1}^{r} \sum_{\mathbf{n}} v_{j,\mathbf{n}}(\mathbf{z}) e^{\langle \mathbf{z}, \lambda_j - \mathbf{n} \rangle} = \sum_{j=1}^{r} \sum_{\mathbf{n}} q_{j,\mathbf{n}}(\mathbf{z}) e^{\langle \mathbf{z}, \lambda_j - \mathbf{n} \rangle} \qquad (\mathbf{z} \in \mathbf{C}^l(\epsilon, \eta^0)),$$

while on the other

$$\Psi(\mathbf{z}) = \sum_{j=1}^{r} \sum_{\mathbf{n}} p_{j,\mathbf{n}}(\mathbf{z}) e^{\langle \mathbf{z}, \lambda_j - \mathbf{n} \rangle} \qquad (\mathbf{z} \in \mathbf{C}^l(\epsilon, \eta_0)).$$

Therefore

$$\sum_j \sum_{\mathbf{n}} p_{j,\mathbf{n}} e^{\langle \cdot, \lambda_j - \mathbf{n} \rangle} = \sum_j \sum_{\mathbf{n}} q_{j,\mathbf{n}} e^{\langle \cdot, \lambda_j - \mathbf{n} \rangle}$$

on $\mathbf{C}^l(\epsilon, \eta^0)$, whence $p_{j,\mathbf{n}} = q_{j,\mathbf{n}}$ (all j, $\mathbf{n}$) (cf. Lemma A 3.2.4).

This completes the proof of the proposition. $\square$

Let us retain the above assumptions on c_i, Γ_i, and Ψ. Let $\mathscr{E}(\Psi)$ be the set of all points λ in $\mathbf{C}^l$ for which we can select a j and an $\mathbf{n}$ $(1 \leqslant j \leqslant r; \mathbf{n} \in \mathbf{N}^l)$ such that $\lambda = \lambda_j - \mathbf{n}$ and $p_{j,\mathbf{n}} \neq 0$; it follows from the definition of the $p_{j,\mathbf{n}}$ that j and $\mathbf{n}$ are uniquely determined by the conditions $\lambda = \lambda_j - \mathbf{n}$ and $p_{j,\mathbf{n}} \neq 0$. Given $\lambda \in \mathscr{E}(\Psi)$, set $p_\lambda = p_{j,\mathbf{n}}$ – then it is clear that the series $\sum_{\lambda \in \mathscr{E}(\Psi)} p_\lambda e^{\langle \cdot, \lambda \rangle}$ is obtained from the series $\sum_j \sum_{\mathbf{n}} p_{j,\mathbf{n}} e^{\langle \cdot, \lambda_j - \mathbf{n} \rangle}$ by simply omitting those terms which are zero.

We have now proved the following theorems.

Theorem A 3.5.7 (Harish-Chandra) Define holomorphic functions Γ_i on $\mathbf{C}^{l,+}$ by condition (Γ) supra; suppose that Ψ is an E-valued holomorphic function on $\mathbf{C}^{l,+}$ such that $\partial\Psi/\partial z_i = \Gamma_i \Psi$ $(1 \leqslant i \leqslant l)$ there – then there exists an at most countable set $\mathscr{E}(\Psi)$ of points in $\mathbf{C}^l$ and, for each $\lambda \in \mathscr{E}(\Psi)$, an E-polynomial p_λ such that the following conditions are fulfilled:

(i) For every $\lambda \in \mathscr{E}(\Psi)$, $p_\lambda \neq 0$;

(ii) The series

$$\sum_{\lambda \in \mathscr{E}(\Psi)} p_\lambda e^{\langle \cdot, \lambda \rangle}$$

converges decently on the set $\mathbf{C}^l(\epsilon, \eta)$ for any ϵ, η with $0 < \epsilon < 1$, $\eta > 0$;

(iii) The expansion

$$\Psi(\mathbf{z}) = \sum_{\lambda \in \mathscr{E}(\Psi)} p_\lambda(\mathbf{z}) e^{\langle \mathbf{z}, \lambda \rangle}$$

holds for all $\mathbf{z} \in \mathbf{C}^{l,+}$.

The set $\mathscr{E}(\Psi)$ and the polynomials p_λ ($\lambda \in \mathscr{E}(\Psi)$) are uniquely determined by these conditions; moreover there exists an integer $d \geqslant 0$ and two positive numbers M and N such that $\deg(p_\lambda) \leqslant d$ and $\|p_\lambda\| \leqslant MN^{-\mathscr{R}(\lambda)}$ (all $\lambda \in \mathscr{E}(\Psi)$).

[The uniqueness of $\mathscr{E}(\Psi)$ and the E-polynomials p_λ ($\lambda \in \mathscr{E}(\Psi)$) is an immediate consequence of Lemma A 3.2.4. The assertions concerning the degrees and norms of the p_λ are plain provided we bear in mind that the sequence $\{p_{j,\mathbf{n}} : \mathbf{n} \in \mathbf{N}^l\}$ $(1 \leqslant j \leqslant r)$ is normal.]

An element $\lambda \in \mathscr{E}(\Psi)$ is said to be *maximal* if $\lambda + \mathbf{n} \notin \mathscr{E}(\Psi)$ for any $\mathbf{n} \neq \mathbf{0}$ in $\mathbf{N}^l$. Since $p_{\lambda_j} = p_{j,\mathbf{0}} \neq 0$ $(1 \leqslant j \leqslant r)$, it follows from the Remark given after the proof of Proposition A 3.5.5 that the λ_j $(1 \leqslant j \leqslant r)$ are distinct maximal elements in $\mathscr{E}(\Psi)$; moreover every element in $\mathscr{E}(\Psi)$ is of the form $\lambda_j - \mathbf{n}$ for some j and $\mathbf{n}$. Therefore $\mathscr{E}(\Psi)$ admits no other maximal elements besides the λ_j.

Theorem A 3.5.8 (Harish-Chandra) Define holomorphic functions $\mathbf{\Gamma}_i$ on $\mathbf{C}^{l,+}$ by condition ($\mathbf{\Gamma}$) supra; suppose that Ψ is an E-valued holomorphic function on $\mathbf{C}^{l,+}$ such that $\partial\Psi/\partial z_i = \mathbf{\Gamma}_i\Psi$ $(1 \leqslant i \leqslant l)$ there – then the set $\mathscr{E}_0(\Psi)$ of maximal elements in $\mathscr{E}(\Psi)$ is finite. Furthermore every exponent of Ψ (i.e. every element of $\mathscr{E}(\Psi)$) is of the form $\lambda_0 - \mathbf{n}$ where $\lambda_0 \in \mathscr{E}_0(\Psi)$ and $\mathbf{n} \in \mathbf{N}^l$.

Remark Retain the above notations and assumptions. Fix ϵ $(0 < \epsilon < 1)$ and $\eta > 0$ – then there exist positive numbers P and Q such that

$$\sum_{\lambda \in \mathscr{E}(\Psi)} \|p_\lambda(\mathbf{z})e^{\langle \mathbf{z}, \lambda\rangle}\| \leqslant P|e^{Q\langle \mathbf{z}, \mathbf{1}\rangle}|$$

for all $\mathbf{z} \in \mathbf{C}^l(\epsilon, \eta)$. [In fact the sequence $\{p_{j,\mathbf{n}} : \mathbf{n} \in \mathbf{N}^l\}$ is normal for each j; therefore we may quote Corollary A 3.5.2.]

General Notational Conventions

(1) By **Z**, **N**, $\mathbf{N}^+$, **Q**, **R**, $\mathbf{R}^+$, $\mathbf{R}'$, **C**, $\mathbf{C}^*$, **T**, **H**, $\mathbf{Z}_n$ ($n \geqslant 1$) we shall understand, respectively, the integers, the non-negative integers, the strictly positive integers, the rational numbers, the real numbers, the strictly positive real numbers, the non-zero real numbers, the complex numbers, the non-zero complex numbers, the complex numbers of absolute value 1, Hamilton's quaternions, the cyclic group of order n.

(2) If z is a complex number, then $\mathscr{R}(z)$ (respectively $\mathscr{I}(z)$) denotes the real (respectively imaginary) part of z.

(3) By $\varnothing$ we shall understand the empty set. If S and T are sets, then we shall sometimes write $S \subset T$ when S is a proper subset of T. If F is a finite set, then $[F]$ denotes the cardinality of F. Set theoretic differences are indicated by a minus sign, set theoretic complements by $\complement$.

(4) If E is a locally convex topological vector space, then, unless stated to the contrary, the base field is **C**; this convention is applicable, in particular, to all Banach spaces and Hilbert spaces. The symbol $\mathfrak{L}(E; E)$ stands for the space of continuous linear operators on E; we write $\mathfrak{L}_s(E; E)$ (respectively $\mathfrak{L}_b(E; E)$) when we wish to emphasize that $\mathfrak{L}(E; E)$ is carrying the topology of simple convergence (respectively the topology of bounded convergence). [If F is another space satisfying the same general conditions as E, then the symbol $\mathfrak{L}(E; F)$ is to be assigned the obvious connotation; similarly in other cases.] By E^* we shall understand the topological dual of E (or simply dual); the symbols E_s^*, E_c^*, or E_b^* stand, respectively, for E^* equipped with the topology of simple, compact, or bounded convergence.

(5) If $\{E_i: i \in I\}$ is a collection of Hilbert spaces, then the orthogonal direct sum of the E_i is denoted by $\sum_{i \in I} \oplus E_i$.

(6) Let X be a topological space – then X is said to be *locally compact* if every point in X has a neighborhood basis comprised of compact subspaces. [Note: It is not assumed that the topology on X is Hausdorff.] According to this convention, a compact space X may not be locally compact (since X need not be Hausdorff). A locally compact Hausdorff space X is said to be *countable at infinity* if X can be written as the countable union of compact subspaces.

(7) Let G be a group. If $x \in G$, then L_x (respectively R_x) denotes the left (respectively right) translation by x; the effect of applying the inner automorphism of G per x to the element y ($y \in G$) is denoted either by ${}^x y$ or xyx^{-1}.

(8) For us the term 'locally compact topological group G' means: G is a topological group whose underlying topology is Hausdorff and which, as a locally compact topology, is countable at infinity. [In making this latter assumption, we thus avoid certain measure theoretic pathologies; in certain instances, it will be dropped.] Let, then, G be a locally compact topological group. By $C(G)$, $C_c(G)$, $C^+(G)$, and $C_c^+(G)$ we shall understand, respectively, the space of continuous complex valued functions on G, the space of continuous compactly supported complex valued functions on G, the space of continuous functions on G whose range lies in the set of non-negative real numbers, and the space of continuous compactly supported functions on G whose range lies in the set of non-negative real numbers. More generally, if E is a space per (4) supra, then

$C(G; E)$ (respectively $C_c(G; E)$) denotes the space of continuous E-valued functions on G (respectively the space of continuous compactly supported E-valued functions on G); if $f \in C_c(G; E)$, then $\mathrm{spt}(f)$ is the support of f. The space of complex Radon measures on G is denoted by $M(G)$ while the space of compactly supported complex Radon measures on G is denoted by $M_c(G)$; if $x \in G$, then δ_x is the Dirac measure concentrated at x.

(9) Let G be a Lie group – then, according to the convention laid down in (8) above, G is countable at infinity. Typically Lie groups are denoted by upper case Roman letters $G, H, \ldots$, their (real) Lie algebras by lower case German letters $\mathfrak{g}, \mathfrak{h}, \ldots$, the complexifications of the latter by $\mathfrak{g}_c, \mathfrak{h}_c, \ldots$, and their universal enveloping algebras by $\mathfrak{G}, \mathfrak{H}, \ldots$ (so we drop the subscript 'c'). The adjoint group of a Lie group G is denoted by $\mathrm{Ad}(G)$ or $\mathrm{Int}(\mathfrak{g})$ (if G is connected); G operates on its Lie algebra $\mathfrak{g}$ via the adjoint action which we shall agree to write as: $\mathrm{Ad}(x)X = xX$ ($x \in G$, $X \in \mathfrak{g}$). The word 'analytic', when used without qualification, means 'real analytic'.

(10) If G is a connected semi-simple Lie group with finite center, $G = KA_{\mathfrak{p}}N^+$ an Iwasawa decomposition for G, then, per a given x in G, we shall denote by $\kappa(x)$ and $H(x)$ the unique elements in K and $\mathfrak{a}_{\mathfrak{p}}$ (the Lie algebra of $A_{\mathfrak{p}}$), respectively, such that $x \in \kappa(x) \exp (H(x))N^+$. If λ is a (complex valued) linear function on $\mathfrak{a}_{\mathfrak{p}}$, then we shall often write h^λ for $\exp (\lambda(\log h))$ ($h \in A_{\mathfrak{p}}$, $\log h$ the unique element in $\mathfrak{a}_{\mathfrak{p}}$ such that $h = \exp (\log h)$). [A similar mode of writing will be used in other cases as well.]

(11) Unless specifically stated to the contrary, it is understood that all Lie algebras are finite dimensional over the base field; thus, whenever it is a question of a semi-simple Lie algebra, or a solvable Lie algebra, or a nilpotent Lie albegra, etc., then the dimension of this Lie algebra is tacitly finite.

(12) A *character* of a topological group G is any continuous homomorphism of G into $\mathbf{C}^*$; if the range of this character is a subset of $\mathbf{T}$ (the complex numbers of absolute value 1), then the character is said to be *unitary*. Unless stated to the contrary, all representations of G are on complex vector spaces; all finite dimensional representations of G are understood to be continuous (and hence analytic if G is a Lie group). If G is a locally compact abelian group, then $\hat{G}$ is its dual group (in the sense of Pontryagin).

(13) Let f, g be two functions defined on the topological space X with values in the set of non-negative real numbers.

(i) We write $f \asymp g$ if there exist $r_1, r_2 > 0$ such that $r_1 f(x) \leqslant g(x) \leqslant r_2 f(x)$ for all $x \in X$.

(ii) We write $f \preceq g$ (or $g \succeq f$) if there exists $r > 0$ such that $f(x) \leqslant rg(x)$ for all $x \in X$.

(14) In so far as the classical groups are concerned, we shall agree to use the customary notations as set forth, e.g. in Helgason [2]. The following notations will be employed in particular.

I_n: Identity matrix of degree n.
$D(a_1, \ldots, a_n)$: Diagonal matrix with diagonal elements $a_1, \ldots, a_n$.
$\mathfrak{gl}(n,\mathbf{C})(\mathfrak{gl}(n,\mathbf{R}))$: The set of all complex (real) n by n matrices.
$\mathfrak{sl}(n,\mathbf{C})(\mathfrak{sl}(n,\mathbf{R}))$: The set of all complex (real)n by n matrices of zero trace.
$\mathbf{S}(n,\mathbf{C})(\mathbf{S}(n,\mathbf{R}))$: The set of all complex (real) n by n symmetric matrices.
$\mathfrak{o}(n,\mathbf{C})(\mathfrak{o}(n,\mathbf{R}))$: The set of all complex (real) n by n skew symmetric matrices.
$\mathbf{H}(n)$: The set of all n by n Hermitian matrices.
$\mathfrak{u}(n)$: The set of all n by n skew Hermitian matrices.
$\mathfrak{su}(n)$: The set of all n by n skew Hermitian matrices of zero trace.

Note Other commonly used notations are introduced in Appendices 1 and 2.

List of Notations

Chapter 1

1.1.1 $\mathfrak{g}, \mathfrak{k}, \mathfrak{p}, \mathfrak{a}_{\mathfrak{k}}, \mathfrak{a}_{\mathfrak{p}}, \mathfrak{a}, \mathfrak{m}, \mathfrak{n}^+; \mathfrak{g}_c, \mathfrak{g}_c^{\alpha}, \boldsymbol{\Phi}, B, \mathfrak{j}_c, \mathfrak{j}_c^{\vee}, \mathfrak{j}^*; H_\alpha, X_\alpha, N_{\alpha,\beta}, \{\alpha_1, \ldots \alpha_l\}, X_i, Y_i, H_i, A_{ij}(1 \leqslant i, j \leqslant l)$; $\mathrm{Int}(\mathfrak{g})$; $\sigma, \tau, \theta, P_+, P_-$

1.1.2 $\Phi, W(=W_\Phi), l(=l_\Phi), \mathfrak{W}(=\mathfrak{W}_\Phi), \mathfrak{w}(=\mathfrak{w}_\Phi); \alpha, \beta, \alpha^\vee, \beta^\vee, A_{\alpha,\beta}; \boldsymbol{\Phi}^+, \Psi, \Theta, \langle\Theta\rangle, \Xi, \Xi_s, \Xi_u, W_s, W_u, \boldsymbol{\Phi}_w^+ (w \in W); \mathscr{C}, w^{\dagger}, w_\alpha, \Delta(w), |\alpha|$

1.1.3 $(\boldsymbol{\Phi}, \sigma), \boldsymbol{\Phi}_+, \boldsymbol{\Phi}_-, E_+, E_-, l_+, l_-; \Sigma, \Sigma^+, \tilde{\alpha} = \lambda, \Sigma(\lambda), m(\lambda), Q_\lambda, W_\Sigma (= W_\sigma/W_-), l_\Sigma, \Upsilon = \{\lambda_1, \ldots, \lambda_l\} (l = l_\Sigma); \mathfrak{g}^\lambda, G, K, A_{\mathfrak{p}}, N^+, N^-, G = KA_{\mathfrak{p}}N^+, M, \mathsf{M}, \mathsf{M}^*, \mathsf{Z}(A_{\mathfrak{p}}), W = \mathsf{M}^*/\mathsf{M}$

1.1.4 $\mathfrak{a}_{\mathfrak{p}}', m_w, \mathfrak{n}_w^+, \mathfrak{u}_w^+, \mathfrak{u}_w^-, N_w^+, U_w^+, U_w^-; \Sigma_0, \Delta_0(w)$

1.1.5 $(\mathfrak{g},\mathfrak{k}), \bar{\mathfrak{g}}, \hat{\mathfrak{g}}, \mathfrak{c}, \mathfrak{c}_{\mathfrak{k}}, \mathfrak{c}_{\mathfrak{p}}; (G, K), \bar{G}, \widehat{G}, C, C_{\mathfrak{k}}, C_{\mathfrak{p}}, \bar{Z}, \widehat{Z}, Z = \widehat{Z} \times C_{\mathfrak{p}}; i_c, G_c$

1.2.1 $(G, P, M), P_\Theta, W, W_\Theta, L(w)$

1.2.2 $\mathfrak{g}_c, \mathfrak{j}_c, \mathfrak{b}_c, \mathfrak{p}_c, \mathfrak{l}_c, \mathfrak{n}_c^\pm; G_c, J_c, B_c, P_c, L_c, N_c^\pm; \mathfrak{w}, \mathfrak{W}, \hat{\mathfrak{w}}, \widehat{\mathfrak{W}}, \hat{\mathsf{L}}(G_c)$

1.2.3 $P = \mathsf{M}A_{\mathfrak{p}}N^+$, $\mathfrak{p} = \mathfrak{m} + \mathfrak{a}_{\mathfrak{p}} + \mathfrak{n}^+$, $G = \bigcup_{w\in W} PwP = \bigcup_{w \in W} U_w^+ wP$, $N^-P, w^{\dagger}, (G, P, \mathsf{M}^*), P_\Theta; \mathfrak{g}(\Theta), \mathfrak{k}(\Theta), \mathfrak{p}(\Theta), \mathfrak{a}_{\mathfrak{p}}(\Theta), \mathfrak{n}^+(\Theta), \mathfrak{n}^-(\Theta), \mathfrak{a}_\Theta$

1.2.4 $\mathfrak{p}_\Theta, \mathfrak{l}_\Theta, \mathfrak{m}_\Theta, \mathfrak{n}_\Theta^+, \mathfrak{n}_\Theta^-, \Sigma_\Theta; P_\Theta, \mathsf{L}_\Theta, L_\Theta, \mathsf{M}_\Theta, M_\Theta, \mathsf{M}_\Theta(K), M_\Theta(K), A_{\mathfrak{p}}(\Theta), A_\Theta, N_\Theta^+, N_\Theta^-; P_\Theta = \mathsf{M}_\Theta(K)A_{\mathfrak{p}}N^+ = \mathsf{M}_\Theta A_\Theta N_\Theta^+; P$, $\mathrm{prk}\,(P)$, $R_u(P), A, \mathsf{L}, \mathsf{M}, P = \mathsf{M}AR_u(P), W(A), \mathfrak{a}, \Sigma^+(\mathfrak{g}, \mathfrak{a}); (P_1, A_1) \succ (P_2, A_2), W(A_1, A_2); \mathfrak{P}, \mathfrak{C}_{\mathfrak{P}}, \mathfrak{P}_A$

1.3.1 $\mathfrak{j}, \mathfrak{j}_I, \mathfrak{j}_R, \mathfrak{j}_{\mathfrak{k}}, \mathfrak{j}_{\mathfrak{p}}; \mathfrak{j}^1, \ldots, \mathfrak{j}^r$

1.3.2 $\boldsymbol{\Phi}, \boldsymbol{\Phi}^+, \boldsymbol{\Phi}_R^+, \boldsymbol{\Phi}_C^+, \boldsymbol{\Phi}_{CP}^+, \boldsymbol{\Phi}_{SI}^+, \boldsymbol{\Phi}_I^+; E_\alpha, E_{-\alpha}(\alpha \in \boldsymbol{\Phi}^+); \mathfrak{l}_c^\alpha, \mathfrak{l}^\alpha(\alpha \in \boldsymbol{\Phi}^+)$

1.3.3 $l_-(\mathfrak{j}), l_-; \omega, X \mapsto \mathrm{tr}(\mathrm{ad}(X))^2 (X \in \mathfrak{g})$

1.3.4 $l, d_l, \mathfrak{g}', \mathfrak{j}' = \mathfrak{j} \cap \mathfrak{g}', \mathfrak{g}(\mathfrak{j}) = \bigcup_{x\in G} x(\mathfrak{j}')(G = \mathrm{Int}(\mathfrak{g})), \pi = \prod_{\alpha>0} \alpha; \boldsymbol{\Gamma}, \mathfrak{g}_\Gamma = \mathfrak{l}_\Gamma + \mathfrak{c}_\Gamma, \mathfrak{h}_\Gamma^+, \mathfrak{h}_\Gamma^-, \{H^*, X^*, Y^*\}, \nu = \exp\{-\sqrt{-1}\,(\pi/4)\,\mathrm{ad}(X^* + Y^*)\}, \nu(\mathfrak{h}_\Gamma^+) = \mathfrak{h}_\Gamma^-, \alpha_+, \alpha_- = \alpha_+^\nu, \mathfrak{g}^\sharp$

1.3.5 $\mathscr{S}, \mathscr{N}, X = X_s + X_n(X_s \in \mathscr{S}, X_n \in \mathscr{N}; [X_s, X_n] = 0), \boldsymbol{\Gamma} \in \mathscr{S}, \mathfrak{g}_\Gamma$

1.4.1 $\mathsf{J}, \mathsf{J}_0, J, \mathsf{J}_K, J_K, J_{\mathfrak{p}}, \mathsf{Z}(J_{\mathfrak{p}}); D_l, G', \mathsf{J}' = \mathsf{J} \cap G', G(\mathsf{J}) = \bigcup_{x\in G} x\mathsf{J}'x^{-1}; \mathsf{J}^1, \ldots, \mathsf{J}^r$

1.4.2 $\mathsf{J}^*, \mathsf{J}_K^*, W(G, \mathsf{J}), W(G, \mathsf{J}_0)$

1.4.3 $\mathscr{N}_G, x = x_s x_u, \gamma, \mathfrak{g}_\gamma, G_\gamma, \mathsf{G}_\gamma$

Chapter 2

2.1.1 $E, X_E(X \in \mathfrak{g}), U, \delta, E(\delta), \hat{\mathfrak{g}}, S = S(E), T = T(E), \boldsymbol{\Lambda} = \boldsymbol{\Lambda}(E); F(\mathfrak{g}), R(\mathfrak{g}), K(\mathfrak{g}) = F(\mathfrak{g})/R(\mathfrak{g}), \mathfrak{R}(\mathfrak{g}), \mathrm{ch}_{\mathfrak{g}}(E), \lambda_t$, Aug

2.1.2 $S(E), P(E), E_I, S, S_I, S^+, S_I^+$

2.1.3 $P, Q(P), P_I, P_I^+, Q(P_I), [\mathbf{k}_1 : \mathbf{k}_2]$

2.1.4 $\mathfrak{D}(E), \mathfrak{P}(E), \mathbf{e}^b, S^*(F)$

2.1.5 $I(\mathfrak{g}_c), I(\mathfrak{p}_c), I(\mathfrak{a}_c), I(\mathfrak{a}_{\mathfrak{p}_c}), I(\mathfrak{g}_c; \mathfrak{p}_c), I(\mathfrak{g}_c; \mathfrak{p}_c; \mathfrak{a}_{\mathfrak{p}_c}), I(\mathfrak{g}_c; \mathfrak{a}_c), I(\mathfrak{g}_c; \mathfrak{a}_c; \mathfrak{a}_{\mathfrak{p}_c}) (= I(\mathfrak{a}_c; \mathfrak{a}_{\mathfrak{p}_c})); W(\mathfrak{g}_c, \mathfrak{j}_c), W(\mathfrak{h}_c, \mathfrak{j}_c), I(\mathfrak{g}_c; \mathfrak{h}_c); H(\mathfrak{g}_c); \boldsymbol{\pi}_\Phi, \boldsymbol{\pi}_\Sigma$

2.2.1 $\hat{\mathfrak{k}}_c, \mathfrak{G}, \mathfrak{K}, \mathfrak{Z}$

2.2.2 $m_U, \mathfrak{P} = \lambda(S(\mathfrak{p}_c)), \mathfrak{P}^m = \lambda(S^m(\mathfrak{p}_c)), H(\mathfrak{p}_c)$

2.3.1 $\mathfrak{g}, \mathfrak{G}, \mathfrak{G}^+, \nabla, \lambda : S = S(\mathfrak{g}) \to \mathfrak{G}, \mathfrak{G}^m, \mathfrak{G}_m, \mathfrak{X}$

2.3.2 $\mathfrak{sl}(E)$

2.3.3 $\epsilon^-, \epsilon^+, \epsilon = \epsilon^- \otimes 1 \otimes \epsilon^+, \gamma^*, \gamma_1, \gamma_2, \gamma_3, \gamma = \gamma_1^{-1} \circ \gamma^*; \mathfrak{n}^-, \mathfrak{n}^+, \mathfrak{N}^-, \mathfrak{N}^+, \mathfrak{Q}^+ = \mathfrak{G}\mathfrak{n}^+; \omega = \sum_{i,j} g^{ij} X_i X_j = -\omega_- + \omega_+; \mu_{\mathfrak{g}/\mathfrak{h}} : \mathfrak{Z} \to \mathfrak{Z}_{\mathfrak{h}}$

2.4.1 $E^\lambda, a_\Lambda, E_\Lambda, \{\boldsymbol{\Lambda}_1, \ldots, \boldsymbol{\Lambda}_l\}, \{E_1, \ldots, E_l\}, \rho = \frac{1}{2} \sum_{\alpha>0} \alpha$

2.4.2 $\mathbf{Z}[\mathfrak{W}], \mathbf{Z}[\mathfrak{j}^\vee] \sim \mathfrak{R}(\mathfrak{b}), \mathfrak{W}^+(\rho), \mathfrak{W}_s^+(\rho), \rho_s, \rho_u, \rho = \rho_s + \rho_u, \boldsymbol{\Delta} = \prod_{\alpha>0} (e^{\alpha/2} - e^{-\alpha/2})$

2.4.3 κ, κ_Λ, $m(\lambda)$, $\mathbf{P}$, $\mathbf{e}^\lambda$, f_λ
2.5.1 $C^q(\mathfrak{g}, E)$, $C(\mathfrak{g}, E)$, $H^q(\mathfrak{g}, E)$, $H(\mathfrak{g}, E)$, $\theta = \theta_{\mathfrak{g}}$, $i = i_E \sim i_{\mathbf{k}} \otimes 1$, $d = d_{\mathfrak{g}}$, Ω_E, $\mathbf{X}_{\mathfrak{s}}(\mathrm{t}; E)$, $\chi_{\mathfrak{s}}(\mathrm{t}; \mathrm{ch}_{\mathfrak{g}}(E))$
2.5.2 $W(q)$, $W_u(q)$, $\square$, $>F<$

Chapter 3
3.1.1 $\mathfrak{g}_c^+$, $\mathfrak{g}_c^-$, $\mathfrak{u}_c^+$, $\mathfrak{u}_c^-$, $N(G)$, $\mathsf{L}(G_c)$, $\hat{J}_c$
3.1.2 $\mathsf{W}_s^+ = \mathfrak{W}/W_s$
3.2.1 ϕ_f, $\langle \pi, \pi \rangle = [W] \prod_{\alpha>0} (\rho, \alpha)$
3.2.2 χ_Λ, d_Λ, $d\mu(\Lambda)$, Δ_T, Φ_f, $\check{f}$, Θ_τ, ξ_τ
3.3.1 L_0, $\mathfrak{L}$
3.3.2 $\mathfrak{a}_{\mathfrak{p}}^+$, $\rho = 2^{-1} \sum_{\lambda>0} m(\lambda)\lambda$

Chapter 4
4.1.1 E, $\{|\cdot|_\alpha : \alpha \in \mathfrak{A}\}$, $M_c(G)$, $|\mu|$, $\|\mu\|$, $\{g_\mathscr{O}\}$, d_G, U, $U(x)$, $U(f)$, $U(\mu)$
4.1.2 E^*, E_s^*, E_c^*, E_b^*, $E^\vee$, $\bar{E}$, $E_1 \hat{\otimes} E_2$, $E_1 \times E_2$; U^*, $U^\vee$, $\bar{U}$, $U_1 \hat{\otimes} U_2$, $U^1 \times U^2$
4.1.3 $\mathbf{B}$, $I(U, V)$, $\mathfrak{J}(U, V)$, $i(U, V)$, $\mathrm{i}(U, V)$
4.2.1 TCI, FDS; A, $\tilde{A}$, $I(U)$, $\mathrm{Ker}(U)$, E_0, E_0^*, U_0, U_0^*, $\mathscr{E}$
4.2.2 ζ_U, ρ, $C_\rho(G)$, $\|\mu\|_\rho$, $\|f\|_\rho$
4.3.1 $\check{f} = \delta_G^{-1/2} f'$, $f^* = \delta_G^{-1} \bar{f}'$; $\phi_{a,b}$, μ^*
4.3.2 $\mu_\delta(\delta \in \hat{K})$, sp
4.4.1 E_∞, E_α, U_∞, $a \mapsto \tilde{a}$ $(a \in E_\infty)$, $D_c(G)$, κ_U
4.4.2 $\delta \in \hat{K}$, $d(\delta)$, ξ_δ, χ_δ, $\mathsf{P}(\delta)$, $[\delta:\delta_0]$, $D_\infty \in \mathfrak{R}$
4.4.3 $\mathsf{P}_L(\delta)f = \bar{\chi}_\delta * f$, $\mathsf{P}_R(\delta)f = f * \chi_\delta$, $T_\delta = T * \chi_\delta$
4.4.4 $D^*(D \in \mathfrak{G})$, $\Delta = X_1^2 + \ldots + X_n^2$, $dU(X)$, $\sum_{(m)}$, $\mathrm{Dom}(T)$
4.4.5 $C^\omega(M; E)$, E_ω, U_ω, $\mathfrak{G}a$, $p_t = e^{t\Delta} f$, $d(x, y)$, D^R, D^L, ∇f, $E_\omega(\delta)(\delta \in \hat{K})$, E_K, T_K, D_K
4.4.6 $A = \overline{U_\infty(1 - \Delta)}$, $B = A^{1/2}$, $|a|_m$
4.4.7 $E_{PW}(T)$, $E_n(T) = \mathsf{P}([-n, n])(E)$ $(n \in \mathbf{N})$
4.5.1 ${}^\delta C_c(G)$, $C_c^\delta(G)$, $C_{c,\delta}(G)$, $I_{c,\delta}(G)$, $I(\delta)$, ${}_\delta f$, f_δ, f_K, U_δ, $m(\delta)$; $\mathfrak{J}_c(G)$, $I_c(G)$; $\bar{G}$, $\bar{G}(\delta)$, $\bar{C}_{c,\delta}(G)$, $\bar{I}_{c,\delta}(G)$; $C_{\rho,\delta}(G)$, $I_{\rho,\delta}(G)$; $\mu_K(f) = \mu(f_K)$, $\phi(\mu) = \mu(\phi)$, $\phi(f) = f(\phi)$
4.5.2 $\hat{\xi}$, ${}^\xi C(G)$, U^ξ
4.5.3 m_U, n_U
4.5.4 $C_{c,F}(G)$ $(F \subset \hat{K}$, F finite$)$, $C_*(G)$, U_F, E_F, $\bar{\chi}_F$, U_K, $E_K = \sum_{\delta \in \hat{K}} E(\delta)$
4.5.5 $U_K = U_\omega | E_K$, $I_\infty = \mathrm{Ker}(U_\infty)$, Prim $(\mathfrak{G})$
4.5.6 $m(\mathscr{E}; \bar{V})$, $m(U; \bar{V})$, $m(U; V)$
4.5.7 $C^*(G)$, $\|\ \|_{HS}$
4.5.8 $C_*^\infty(G)$, T_U, $T_{\bar{U}}$, $T_{\hat{U}}$, F_U, $\mathfrak{S}(\kappa, \zeta)$
4.5.9 d_U, $\mathscr{E}_U$, $d_{\hat{U}}$, $\mathscr{E}_{\hat{U}}$, $L^2(G)_d$

Chapter 5
5.1.1 $\mathscr{E}^L$, E^L, ${}_GE^L$, U^L, ${}_GU^L$, $\mu_{f,g}$; ${}^LC^\infty(G; E)$, ${}^{L\infty}C^\infty(G; E_\infty)$, ${}^LC(G; E)$, π^L, f^L, ${}_Lh$; $\mathfrak{B} = (B, \pi)$, $\mathfrak{B}_x(x \in X)$, $L^2(\mathfrak{B}; \mu)$; $\mathbf{E} = \mathbf{E}(G, H, L)$, $\mathfrak{E}(\rho_H) = (\mathbf{E}, \pi_{\mathbf{E}})$, ρ_H, μ_H, s_H, $\dot{\phi}$, r_ω
5.1.2 α, β, I_α, I_β, E_α, E_β, U^A, B, μ_ρ, ϕ_x^f
5.1.3 None
5.2.1 $A: G \times M \to \mathfrak{L}(E; E)$, $T^x = {}^tA(x^{-1}, .)T$
5.2.2 $A(x, \dot{y})$, $\mathscr{P}$
5.2.3 Q, $\mathbf{T}(Q)$, $\mathbf{T}^n(Q)$, $D^I\delta_p(p \in M)$, $\Lambda_n(x, p)$, $i(E; Q, n)$
5.2.4 $\{S_\alpha\}$, Q, Ω_Q; $\mathbf{T}$, $\mathbf{T}_\alpha$, $\mathbf{T}[Q]$, T_Q, $\mathbf{Q}_\alpha$
5.3.1 U^{L_∞}, U^L
5.3.2 $Q = H_2xH_1(x \in G)$, $\mathbf{B}$, $\mathbf{B}^\sharp$, $T^\sharp$, T; $\mathbf{T}$, $\mathbf{T}(Q)$, $\mathbf{T}(Q_x)$; H_1, H_2, δ_1, δ_2, d_1, d_2, ρ_1, ρ_2, $H_x = H_1 \cap (x^{-1}H_2x)(x \in G)$, σ_x, ν_x, δ_x; A_n, $i(L^1, L^2; Q, n)$, Φ, S, $\mathscr{O} = Q \cup (G - Q^{cl})$, $\mathbf{B}_z$, T_z; $i(L_\infty, \bar{L}_\infty; Q, n)$
5.3.3 $i(U^L, M)$, $i(U^{L_\infty}, M_\infty)$

5.3.4 $\tilde{V}_x$, $\tilde{V}_D$, V_{x_1,x_2}, V_D, M_{x_1,x_2}, $\mathfrak{D}$
5.4.1 P, P^L, $\mathfrak{B}$, $\hat{\xi} \in \hat{H}$, H_ξ, $\mathscr{L}(\hat{\xi})$, P_Q, $\hat{G}(Q)$, $\hat{\xi} \times L$
5.4.2 '$\alpha t + \beta$', U^+, U^-; U^r; $\mathbf{E}(n)$, $\mathbf{L}(4)$
5.5.1 κ_x, H_x, $H(x, k)$, $\rho = 2^{-1}\sum_{\lambda>0} m(\lambda)\lambda$, δ_P, ρ_P, $L^\mu(\mu\colon \mathfrak{a}_\mathfrak{p} \to \mathbf{C})$, ${}^{L^\mu}C(G)$, U^μ, $L^2(K;\delta)(\delta \in \hat{K})$, $\hat{\mathsf{M}} = \{\sigma\}$, μ_σ, χ_σ, ξ_σ, $d(\sigma)$, $\mathsf{Q}_R(\sigma)$, $\mathsf{Q}_R(\sigma_{ij})$, $L^{\sigma,\mu}$, $U^{\sigma,\mu}$, $E^{\sigma,\mu}$, ${}^{\sigma,\mu}C(G; \mathbf{C}^{d(\sigma)})$, $\mathscr{S}^{\sigma,\mu}$, $[\delta\colon\sigma]$, π_P, π_{N^-}, N_x^-, $\mathfrak{X}$
5.5.2 wL, $P_w = P \cap m_w P m_w^{-1}$, $i(L^1, \bar{L}^2; w, n)$, W_L
5.5.3 $U^{\sigma,\mu}$, $T^{\sigma,\mu}$, $F^{\sigma,\mu}$, $W(M, A_K)$, Φ_f
5.5.4 d_P, $f_{K,P}$, $U^{\sigma,\nu}$, $T^{\sigma,\nu}(\mu + \rho = \sqrt{-1}\nu)$, $\hat{f}_P$, Λ_σ, ρ_I, $\tilde{\mathsf{Z}}(A_\mathfrak{p})$

Chapter 6

6.1.1 $\mu = (\mu_1, \mu_2)$, $\sigma(T)$, $I_{c,\delta}(G) \sim \mathfrak{A}_{c,\delta}(G)$ $(f \mapsto \Psi_f^\delta)$, μ_U, ψ_δ^U, Ψ_δ^U, $\Psi_{\delta,K}^U$, Φ_δ^U; $\bar{G}_\rho$, $\bar{G}_1$, $\bar{U}_\delta$, $\bar{\mu}_\delta$, E_δ $(E(\delta) = E_\delta \otimes_{\mathbf{C}} \mathsf{E}_\delta)$
6.1.2 $\mathfrak{X}$, D_K $(= \int_K Ad(k)\cdot D\, dk,\ D \in \mathfrak{G})$, μ_Φ
6.2.1 $f \mapsto F_f^\delta$ $(f \in I_{c,\delta}(G))$, μ_ξ, U^ξ, $U^{\sigma,\xi}$
6.2.2 ϕ_ν, $\gamma_{A\mathfrak{p}}^*$, $\gamma_{A\mathfrak{p}}$, $f \mapsto F_f^\delta$ $(f \in I_{c,\delta}(G))$, $f \mapsto F_f^{\delta,\sigma}$ $(f \in I_{c,\delta}(G))$, I_{pq}, $\Phi_\delta^{\sigma,\mu}$, $\psi_\delta^{\sigma,\mu}$, $[\delta:\sigma]$

Chapter 7

7.1.1 $C^*(G)$, $\hat{C}^*(G)$, $\mathrm{spt}\,(U)$, $\hat{G}$, $\hat{G}_r$
7.1.2 $\mathscr{Q}_P$, $\hat{G}_P$, $\check{G}_P$, $C_*^\infty(G)$, F_q, $\hat{U}_q$
7.2.1 μ
7.2.2 J_-, J_+, $\xi_{\lambda_-,\lambda_+}$, $\Lambda_{-,+}$, U^{λ_-,λ_+}, T^{λ_-,λ_+}

Chapter 8

8.1.1 j^μ $(j \in \mathsf{J})$, ξ_ρ, ξ_α, α^θ, α^σ; $\diamondsuit(j)$ $(= \diamondsuit_\mathsf{J}(j))$, $\diamondsuit_I(j)$, $\diamondsuit_R(j)$, $\diamondsuit_{R,C}(j)$; $\Delta(j)$ $(= \Delta_\mathsf{J}(j))$, $r(R:j)$, $\epsilon_R(j)$, $\epsilon_R(w)$, $\epsilon_w(j_K)$
8.1.2 $\sigma(x)$ $(x \in G)$
8.1.3 $\frown$, $\mathbf{c}_G$, ρ_P $(P = \mathsf{M}J_\mathfrak{p}N^+)$, $\nabla_\mathfrak{l}$, $\pi_{R,C}$, Δ_I, $\Delta_{R,C}$
8.1.4 G_γ
8.1.5 $\|x\|_\theta$ $(= \|x\|)$
8.1.6 $\mathfrak{g}(r)$ $(r > 0)$
8.2.1 $f(p_1; D_1 : p_2; D_2)$, $v \mapsto f_v$, τ_T, $f(x:p) = f(x\cdot p)$, $\mathfrak{I}(D)$, V_v, $\tilde{T}$, G_p, D_T
8.2.2 $f(x:X) = f(x\cdot X)$ $(x \in G,\ X \in \mathfrak{g})$, $U_X(D)$, Γ_X, $\pi = \prod_{\alpha>0}\alpha$, $\mathfrak{I}_H(D) = \mathfrak{I}(D)_H$, $\mathfrak{P}_I(\mathfrak{g})$, $\mathfrak{J}(\mathfrak{g})$, $\mathfrak{P}_I(\mathfrak{j})$, $\bar{\mathfrak{I}}$, $\delta_\mathfrak{h}(X)$, $'\mathfrak{h}$, $\mathfrak{I}_\mathfrak{h}(D)$
8.2.3 $f(x; X)$, $f(X; x)$, $\mathfrak{G}$, $U_x(D)$, Γ_x, $f(x:y) = f(xyx^{-1})$ $(x, y \in G)$, $\mathfrak{I}(D)_j$, η_α $(\alpha \in \Phi)$, $\mathfrak{R}$; δ_γ, $\mathfrak{g}_\gamma$, G_γ, $'G_\gamma$, $\Omega_\gamma = \gamma^{-1}\Omega \cap {}'G_\gamma$, $\mathfrak{I}_\gamma(D)$
8.2.4 $\Upsilon(X)$, $\phi \mapsto f_\phi$, $\nabla(D)$, $Z \mapsto p_Z$ $(Z \in \mathfrak{Z})$, $T \mapsto \sigma_T$
8.3.1 $T \leftrightarrow F_T$, δ_Γ, $'\mathfrak{g}_\Gamma$, Ω_Γ, F_σ, R_p, ∇_p
8.3.2 $F_{T,\mathfrak{j}}$, $\nabla_\mathfrak{g}$, $\mathfrak{F}_T = \nabla_\mathfrak{g}F_T$, $\mathfrak{j}'(R)$
8.3.3 $T \leftrightarrow F_T$
8.3.4 $\Omega_\gamma = \gamma^{-1}\Omega \cap {}'G_\gamma$, Υ_γ, c_γ, $\diamondsuit_R$, $\mathsf{J}'(R)$, $F_{T,\mathsf{J}}$, $\Pi = \prod_{\alpha>0} H_\alpha$, ∇_G
8.3.5 $\mathscr{C}(E)$, $\hat{T}$, $\hat{f}$, $p^\sharp$, Ξ, $Y(t)$
8.3.6 f_g
8.3.7 $\vdash\!\circ\!\dashv$, $\mathscr{C}(\mathscr{O})$, $\mathscr{C}(G)$, $\mathscr{C}(\mathsf{G})$, ${}_{D_1}|\cdot|_{r,D_2}$, $\mathfrak{A}$, $\mathfrak{A}_R^+$, $\mathfrak{A}_L^+$
8.3.8 $F_{T,\mathsf{J}}$, $\mathsf{J}^\pm$, T_δ $(\delta \in \hat{K})$, f_g
8.3.9 ϕ_z, $\Delta = \omega + 2\omega_-$
8.4.1 G_Γ, Φ_R^+, π_R, ϵ_R, ϕ_f $(f \in C_c^\infty(\mathfrak{g}))$, Φ_{SI}^+, π_{SI}, $\mathfrak{j}'(SI)$, $\mathbf{M}_\mathfrak{g}$, d_l
8.4.2 ψ_f, ψ_f^+, ψ_f^-, ϕ_f^+, ϕ_f^-, $\mathfrak{l}$, $\mathfrak{h}^-$, $\mathfrak{h}^+$, $\mathfrak{l}^-$, $\mathfrak{l}^+$
8.4.3 $|\cdot|_m$, $\|\cdot\|_m$, f_K, f_P, $f_{K,P}$, $p_\mathfrak{l}$, $\phi_{f_{K,P}}^\mathfrak{l}$
8.4.4 $\Phi_{SI}^+(H_0)$, $\mathfrak{h}_\Gamma^\pm$, $\mathsf{H}_\Gamma^\pm$, $\mathscr{O}_\Gamma$, ψ_f^+, ψ_f^-, $v_f^\pm$, ϕ_f^+, ϕ_f^-, π_+, π_-
8.4.5 $\mathbf{M}_\mathfrak{g}$, $\mathbf{m}_\mathfrak{g}$, $(2\pi)^r$, $T_{H_0}(f)$, F_{H_0}, $T_\mathscr{C}$, η, $\mathfrak{m}(\mathfrak{j}_\mathfrak{k})$, $\phi_f^\mathfrak{l}$, π_{SI}, π_C, r_{SI}, r_C
8.5.1 Φ_f, G_γ, G_γ, $\dot{G}_\gamma$, J_γ, Φ_γ, π_γ, $\mathfrak{j}_\gamma'$, $\diamondsuit_\gamma$, ϵ_γ, $\epsilon_{\gamma,R}$, Δ_γ, v_f, D_γ, $\mathsf{J}'(SI)$, $\Pi = \prod_{\alpha>0} H_\alpha$, $\mathbf{M}_G$, $\mathbf{m}_G$, Γ
8.5.2 d_P, $\vdash\!\circ\!\dashv_L$, $H(x) = H_1(x) + H_2(x)$, $\lhd(H)$, ρ_1, ρ_2
8.5.3 $\mathfrak{L}$, $\mathfrak{Q}^+$, $D_\mathfrak{l}$, $\mu_P = \mu_{\mathfrak{g}/\mathfrak{l}}$, ρ_P, $D^\flat$, ${}^\flat D$, f_K, f_P, $f_{K,P}$

8.5.4 Φ_f^L, $f_{K,P}^a$ $(a \in \mathsf{Z}(J_\mathfrak{p}))$
8.5.5 $\nu(f)$, $\nu_L(f)$
8.5.6 D_J, $\mathscr{E} = G(T)$, $\psi_\mathscr{E}$, $\Delta_{R,C}$, $I_c^\infty(G)$
8.5.7 $I(G)$, $I^+(G)$
8.5.8 $\mathsf{J}'(SI)$, $\mathscr{C}(\mathsf{J}'(SI))$

Chapter 9

9.1.1 m_λ, $\mathscr{R}(\lambda)$, $\mathscr{I}(\lambda)$, $\mu > \nu$, $\mu \gg \nu$, L, $\mathscr{C}_c$, $\mathfrak{a}_{\mathfrak{p}_c}(\epsilon, \eta)$, $\mathfrak{S}(\mu, \kappa)$, Ψ, Ψ^L, $\mathscr{E}(\Psi)$, $\mathscr{E}_0(\Psi)$, E_M, $\rho = 2^{-1} \sum_{\lambda>0} m(\lambda)\lambda$, $\Psi(a : \nu : x)$
9.1.2 $\mathfrak{q}$, $\mathfrak{Q}$, $A'_\mathfrak{p}$, $D^x = xD$ $(x \in G)$, γ_-, μ_-, W_-, $I_-(\mathfrak{a}_c)$, $\mathfrak{Z}_-$, $f_\alpha^\pm$ $(\alpha \in P_+)$, $\mathfrak{F}$, $\mathfrak{F}_0$, $\mu_-^*(.) = (\mu_-(.))^\flat$, $X_\alpha = Y_\alpha + Z_\alpha$ $(Y_\alpha \in \mathfrak{k}_c$, $Z_\alpha \in \mathfrak{p}_c)$, $\mathfrak{S}(\delta_1, \delta_2, \kappa)$, $\mathfrak{D}(A'_\mathfrak{p})$, $\mathfrak{A}$, Γ_h, $\wp_{A_\mathfrak{p},h}(D)$, $\wp_{A_\mathfrak{p}}(D)$, Δ, $\deg(\Delta)$, $\deg_t(\Delta)$, Δ_h, ω, $\omega_\mathfrak{l}$, $\omega_\mathfrak{m}$, $\wp_{A_\mathfrak{p}}(\omega)$, $\mathfrak{M}$
9.1.3 $\mathfrak{S}(\mu, \kappa)$, $\mathfrak{e}^\lambda$, $\Omega(D)$, $\Omega(D : \lambda)$, $\omega(D)$, $\omega(D : \lambda)$
9.1.4 $\langle \nu, H \rangle = \langle H, \nu \rangle$, $\gamma(T)$ $(T \in \mathrm{Hom}_\mathbf{C}(E_\mathsf{M}, E_\mathsf{M}))$, Γ_λ, Γ, $\gamma(\Gamma) = \Gamma^\gamma$, $\Psi(\nu : h)$, $\Phi(\nu : h)$, $\mathfrak{T}$, L', $\{\gamma_1, \ldots, \gamma_r\}$, $\tau_{\lambda,i}$, $\sigma_{\lambda,i}$, $\mathfrak{a}_{\mathfrak{p}_c}(\eta)$, $\mathfrak{a}_{\mathfrak{p}_c}(\epsilon, \eta)$, p_λ $(\lambda \in \mathsf{L}')$, L_ω, P_ω, $\Upsilon_\lambda = P_\omega \Gamma_\lambda$, $\psi(\nu : H)$, $\mathfrak{t}$, $\mathfrak{t}(\mathsf{M})$, ψ_t, $A_\mathfrak{p}(\eta)$, $\mathbf{t}$, $\mathbf{t}(\nu : D : t)$, $\mathbf{t}(\nu : D : t : \lambda)$
9.1.5 Ψ $(a : \nu : x)$, $\Gamma(\mathbf{c})$, $\Gamma'(\mathbf{c})$, $\mathbf{c}_w(\nu)$, W_{H_0}, $I_{H_0}(\mathfrak{a}_c)$, $I_{H_0}^+(\mathfrak{a}_c)$, $\mathfrak{E}(H_0)$, $\mathfrak{E}_\mathfrak{p}(H_0)$, $\mathfrak{E}_\mathfrak{p}(H_0 : E)$, E_M^i, $\rho_- = 2^{-1} \sum_{\alpha \in P_-} \alpha$, $\mathfrak{E}_\mathfrak{p}(\Lambda + \Lambda_i + \rho_- : E_\mathsf{M}^i)$, $H(w)$ $(w \in \complement W_\sigma)$, $G^+ = K \exp(\mathscr{C})K$, $\mathfrak{z}^+(\mu, \kappa)$, $\mathfrak{S}^+(\mu, \kappa)$, $\Phi_0(a : \nu)$, $\Phi_0(a : \nu)$
9.1.6 $T(w, \nu, \delta)$, $\mathsf{C}(w)$, $\Gamma_w(\nu)$, f_ν, $\mathbf{c}(\nu)$, $\mathbf{I}(\nu)$, $\mathbf{c}_\lambda(\nu)$, $\mathbf{c}_w(\nu) = \prod_{\lambda \in \Delta_0(w)} \mathbf{c}_\lambda(\nu)$
9.1.7 $\phi_\nu(x)$, Γ_λ, $\mathbf{c}$, $\hat{\Psi}$, $\hat{\Gamma}_\lambda$, γ_λ, $\hat{\gamma}_\lambda$, Δ, $\hat{M}(\nu)$, $\hat{m}(\nu)$, $\hat{d}$, $\hat{D}$, d, D
9.2.1 $I_c^\infty(G)$, $I^p(G)$ $(1 \leqslant p < \infty)$, $I^1(G)$, $\phi(\nu : x)$ $(= \phi_\nu(x))$, $\mathfrak{C}_\rho$, $\mathfrak{T}_\rho$, Spec $(I^1(G))$, $\hat{f}(\phi)$, $\hat{f}(\nu)$, $\mathbf{P}$, $\mathfrak{P}$, $\hat{f}(\nu, \dot{k})$, a_ν, L_ν, E_ν, $L_{G/K}$
9.2.2 $\mathscr{I}(G)$, $\mathscr{I}(A_\mathfrak{p})$, $I_c^\infty(G)$, F_f $(f \in \mathscr{I}(G))$, $\mathbf{c}_{N^+}$, ϕ_a, $\pi = \pi_\Sigma$, $\mathscr{I}_0(G)$, T_ν, $\boldsymbol{\beta}$, $\mathscr{C}(\mathfrak{F})$, $\mathscr{I}(\mathfrak{F})$
9.2.3 $f_\lambda(h)$
9.2.4 $I^1(G)$, $\mathfrak{C}_\rho$, $\mathfrak{T}_\rho$, $\mathfrak{T}_\rho^*$, $\mathscr{C}^\vee$, $\mathfrak{H}(\mathfrak{T}_\rho^*)$, $\mathsf{I}(G)$
9.3.1 $\mapsto\!\dashv_r$, ι, $\mathscr{S}^+$, Π_γ, $\mathbf{M}_\gamma$
9.3.2 $\mathfrak{p}^*$, G^*
9.3.3 $F_\mathscr{O}(T)$, $\mathscr{O}_T$, $\mathfrak{Z}$, $\mathfrak{Z}_\mathfrak{l}$, $E_\mathfrak{l}$, $\mathbf{E}$, $\mathbf{E}^+$, $\mathbf{E}^-$, $\mathbf{E}^0$, Γ, Ψ, Ψ^+, Ψ^-, Ψ^0, $\boldsymbol{\mu}$, Ψ_Z, $\mapsto\!\dashv_L$, Θ_H, Ψ_P, Ψ_P

Chapter 10

10.1.1 L_T, L'_T, ξ_τ $(\tau \in \mathsf{L}_T)$, γ_λ, Θ_τ, θ_τ, $\mathfrak{Z}$, $\mathfrak{Z}'$, $G(\epsilon)$, Ψ_τ
10.1.2 $\Theta_{\tau,\delta}$, $\mathscr{C}(G)$
10.2.1 T_U
10.2.2 $\mathscr{E}_U$, $\mathscr{C}_U(G)$, F_U, $F_{U,T}$, $F_{\bar{U}}$, $F_{\bar{U},T}$
10.2.3 $\hat{G}_d(\tau)$, $W(\tau)$, $\Theta_{\tau,T}$, $\mathbf{T}(\tau)$
10.2.4 $\hat{G}_d$ $\hat{U}_\tau$, $\epsilon(\tau)$, $d_{\hat{U}_\tau}$, m_δ, $c(\tau)$

Appendices

A1 δ_G, δ_H, ρ, μ_ρ, s_ρ, ρ_H, μ_H, s_H, $\mu^x(x \in G)$
A2.1 $C^\infty(M)$, $C^\infty(p)(p \in M)$, $C_c^\infty(M)$, $C_\omega^\infty(M)$, $C_c^p(M)(0 \leqslant p \leqslant \infty)$, $\mathfrak{D}_F = \mathfrak{D}_F(M)$, $\mathfrak{D}(M)$, $D_1 \circ D_2$, $\{D_1, D_2\}$, $T(f)$, $\langle f, T \rangle$, $\int_M f(p)\, dT(p)$, $|\cdot|_D$, $|\cdot|_{D\,\omega}$, $f \cdot T$, $D^\dagger$, $D \cdot T$
A2.2 $E \hat{\otimes} F$, $E \bar{\otimes} F$, $C^\infty(M; E)$, $|\cdot|_{D,\omega}^\alpha$
A2.3 $\mathscr{S}_T$, $\pi^* f$, $\pi_* T$
A2.4 $f * T$, $T * f$, $T_1 * T_2$, $f^x = xf$, $T^x = xT$, $D^x = xD$, $\bar{X}$, T_D, f', f'', T', T'', $G_p(p \in M)$
A3.1 $\mathbf{C}^l$, $\mathbf{R}^l$, $\mathbf{N}^l$, $\langle \mathbf{z}, \bar{\mathbf{z}} \rangle$, $\|\mathbf{z}\|$, $\mathscr{R}(\mathbf{z})$, $\mathbf{z}^\mathbf{n}$, $|\mathbf{n}|$, $\mathbf{n}!$, $D^\mathbf{n}$, $\mathbf{C}^l(\eta)$, $\mathbf{C}^l(\epsilon, \eta)$, $\|p\|$, $\mathscr{E}(\Psi)$, $\mathscr{E}_0(\Psi)$
A3.2 None
A3.3 $H(z, \lambda)$
A3.4 None
A3.5 $\{p_\mathbf{n} : \mathbf{n} \in \mathbf{N}^l\}$, $P(\mathbf{C}^l; E)$, $(\partial/\partial z_i + \gamma)^{-1}$, $(\partial/\partial z_i + T)^{-1}$, $v_j(\mathbf{n} : z)$, $\Delta(\mathbf{z})$

Guide to the Literature

Chapter 1

1.1.1 Adams [1], Helgason [2], Hochschild [1], Jacobson [3], Loos [1], Serre [3], [4]

1.1.2 Aribaud [1], Bourbaki [5], Serre [4], Steinberg [5]

1.1.3 Araki [1], Goto and Kobayashi [1], Harish-Chandra [15], [27], Helgason [2], Kunze and Stein [3], Murakami [1], Satake [1], Schattschneider [1]

1.1.4 Borel and Harish-Chandra [1], Harish-Chandra [8], [17], [29], Kunze and Stein [3], Langlands [5], Schiffmann [2]

1.1.5 Bourbaki [4], Helgason [2], Jacobson [3]

1.2.1 Bourbaki [5], Carter [1], Coxeter and Moser [1], Dornhoff [1], Tits [1], [2]

1.2.2 Aribaud [1], Borel [4], Griffiths and Schmid [1], Helgason [2], Serre [4], Wang [1]

1.2.3 Aomoto [1], Borel and Tits [1], Bruhat [1], Chevalley [3], Furstenberg [1], Harish-Chandra [12], Helgason [12], Iwahori and Matsumoto [1], Kunze and Stein [3], C. C. Moore [2], Satake [1]

1.2.4 Borel and Tits [1], Bruhat [1], Harish-Chandra [15], [31], Langlands [2], C. C. Moore [2], Oniščik [1], [2], Satake [1], Takeuchi [1]

1.3.1 Harish-Chandra [13], Kostant [1], Sugiura [1], Rothschild [1]

1.3.2 Harish-Chandra [15], [25]

1.3.3 Harish-Chandra [15]

1.3.4 Harish-Chandra [15], [16], [25], Serre [4]

1.3.5 Borel and Harish-Chandra [1], Bourbaki [4], Harish-Chandra [26], Jacobson [3], Kostant [2], Rallis [1], Richardson [1], Vinberg [1]

1.4.1 Harish-Chandra [13], [25], Steinberg [4]

1.4.2 Harish-Chandra [15], [16], [26], [27], Hirai [7], [8]

1.4.3 Borel [4], Borel and Harish-Chandra [1], Borel and Mostow [1], Harish-Chandra [27], [29], Steinberg [4]

Chapter 2

2.1.1 Aribaud [1], Bourbaki [4], Harish-Chandra [6], Jacobson [3]

2.1.2 Bourbaki [4], Dixmier [1], Hadziev [1], Harish-Chandra [6], Nagata [1]

2.1.3 Bourbaki [5], Chevalley [5], Dress [1], Flatto [1], [2], Flatto and Wiener [1], Harish-Chandra [18], Helgason [2], Séminaire Sophus Lie [1], Shephard and Todd [1], Solomon [1], Steinberg [3], [5]

2.1.4 Harish-Chandra [24], Séminaire Sophus Lie [1]

2.1.5 Borel and Chevalley [1], Bourbaki [5], Chevalley [2], Coleman [1], Coxeter [1], Harish-Chandra [5], [15], [18], [24], Helgason [2], [3], [5], Kostant [2], [4], Rallis [1], Steinberg [5], Varadarajan [1]

2.2.1 Harish-Chandra [3], [5]

2.2.2 Dixmier [1], Harish-Chandra [3], [5], [6], Rallis [1]

2.3.1 Bourbaki [4], Harish-Chandra [1], [2], [5], Jacobson [3], Séminaire Sophus Lie [1]

2.3.2 Harish-Chandra [1]

2.3.3 Gelfand [3], Harish-Chandra [3], [13], [27], Kostant [4], Séminaire Sophus Lie [1]

2.4.1 Bouwer [1], Harish-Chandra [3], Jacobson [3], Kimura [1], Lemire [1], Séminaire Sophus Lie [1]
2.4.2 Aribaud [1], Jacobson [3], Kostant [3], Séminaire Sophus Lie [1], Serre [4]
2.4.3 Harish-Chandra [3], [6], Séminaire Sophus Lie [1]
2.5.1 Aribaud [1], Bourbaki [4], Hattori [1], Hochschild and Serre [1], Koszul [1]
2.5.2 Aribaud [1], Bott [1], Cartier [3], Kostant [3]

Chapter 3
3.1.1 Harish-Chandra [2], Hochschild [1], Séminaire Sophus Lie [1], Serre [4]
3.1.2 Aribaud [1], Bott [1], Griffiths and Schmid [1], Hirzebruch [3], [4], Kostant [3], G. Segal [1], Wallach [1], [2]
3.2.1 Harish-Chandra [14], [27]
3.2.2 Harish-Chandra [30], Séminaire Sophus Lie [1]
3.3.1 Harish-Chandra [18], Helgason [14], Sugiura [2], [3]
3.3.2 Harish-Chandra [18], [30]

Chapter 4
4.1.1 Bourbaki [3], Bruhat [1], R. T. Moore [2]
4.1.2 Bourbaki [3], Bruhat [1], R. T. Moore [2]
4.1.3 Bruhat [1]
4.2.1 Dixmier [10], Fell [3], [4], [8], Godement [7], Jacobson [2], Naimark [2]
4.2.2 Berezin [1], Fell [4], Godement [7], de Leeuw and Glicksberg [1], Shiga [1]
4.3.1 Dixmier [15], Hewitt and Ross [1], Kaplansky et al. [1], Langlands [2], Loebel [1], Loomis [1], Mackey [1], [5], [9], [11], Naimark [1], [2]
4.3.2 Bichteler [1], Dixmier [15], Segal and von Neumann [1], van der Waerden [1]
4.4.1 Blattner [1], Bruhat [1], Fell [3], [4], Gårding [1], Harish-Chandra [30], Segal [6], Stein [1]
4.4.2 Harish-Chandra [30], Taylor [1]
4.4.3 Harish-Chandra [29], [30]
4.4.4 Bers, John, and Schechter [1], Goodman [2], [4], [7], John [1], Nelson [1], Nelson and Stinespring [1], Segal [5], [11], Sternheimer [1]
4.4.5 Browder [1], Cartier and Dixmier [1], Gaffney [1], Gårding [2], Godement [7], Harish-Chandra [5], Nelson [1]
4.4.6 Goodman [1], [2], Kotake and Narasimhan [1], R.T. Moore [1], Nelson [1], Tits and Waelbroeck [1]
4.4.7 Goodman (unpublished; cf. Zeitlin [1]), Nelson and Stinespring [1]
4.5.1 Dieudonné [3], Dixmier [15], Fell [4], [8], Godement [7], Harish-Chandra [7], Kaplansky [2], Mackey [9]
4.5.2 Fell [8], Godement [7], Harish-Chandra [5], [6], [7], [21], Rader [1]
4.5.3 Fell [2], [4], [6], [8]
4.5.4 Fell [8], Harish-Chandra [11]
4.5.5 Dixmier [11], [17], Fell [4], Godement [7], Harish-Chandra [5]
4.5.6 Fell [8]
4.5.7 Dixmier [15], Godement [7], Harish-Chandra [5], [7], Nelson and Stinespring [1], Stinespring [1]
4.5.8 Fell [8], Godement [6], [13], [14], Harish-Chandra [6], [7], [8], [13], [27]
4.5.9 Bargmann [1], Dixmier [15], Godement [1], [2], Harish-Chandra [10], Kunze [3], Kunze and Stein [1], Mackey [9], Rieffel [2]

Chapter 5
5.1.1 Auslander, Green, and Hahn [1], Blattner [1], [2], Bruhat [1], Dauns

and Hoffmann [1], Fell [7], [8], Kleppner [1], Mackey [2], [3], [4], [5], [9], Mautner [1], [2], [3], [4], Poulsen [1], [2], Rieffel [1], Rigelhof [1]

5.1.2 Blattner [3], Fell [8]

5.1.3 Mackey [3], Fell [8]

5.2.1 Bruhat [1]

5.2.2 Bruhat [1]

5.2.3 Bruhat [1]

5.2.4 Bruhat [1], Fell [8]

5.3.1 Bruhat [1]

5.3.2 Bruhat [1], Fell [8]

5.3.3 Bruhat [1], Mackey [3], [4], Mautner [4], C.C. Moore [1], Weil [1]

5.3.4 Mackey [3]

5.4.1 Fell [8], Mackey [9], [11]

5.4.2 Auslander, Green, and Hahn [1], Fell [8], Gindikin [2], Kohari [1], Mackey [9], [11], Naimark [1], Nelson and Stinespring [1], Pukanszky [2]

5.5.1 Gross [1], Harish-Chandra [5], [6], [21], Kunze and Stein [3], Parthasarathy, K.R., Ranga Rao, R. and Varadarajan, V.S. [1], Rader [1], Sakai [1], [2], Takahashi [1]

5.5.2 Arthur [1], Bruhat [1], Harish-Chandra [8], [32], [36], Knapp and Stein [1], [4], Kostant [5], Wallach [3]

5.5.3 Harish-Chandra [8], Hirai [6], Lipsman [8]

5.5.4 Harish-Chandra [34], Lipsman [6]

Chapter 6

6.1.1 Dieudonné [3], Fell [4], [8], Godement [7], Harish-Chandra [7], [21], Rader [1], Sakai [1], [2], Vilenkin [1], [2]

6.1.2 Godement [7], Harish-Chandra [7], Rader [1]

6.2.1 Fell [8], Godement [7]

6.2.2 H. Glover (unpublished; cf. Rader [1]), Harish-Chandra [18], [19], [21], Helgason [2], Takahashi [1], [2]

Chapter 7

7.1.1 Dixmier [15], [19], Fell [1], Heyer [1], Schochetman [1], Takenouchi [1], S.P. Wang [1]

7.1.2 Fell [1], Lipsman [6]

7.2.1 Dixmier [15], Harish-Chandra [4], [11], [34], [36], Hirai [4], [8], Okamoto [1], Pukanszky [1], Romm [1], I.E. Segal [4], [10], Takahashi [1], [2]

7.2.2 Dixmier [3], Duflo [2], Gelfand and Naimark [1], Gutkin [1], Harish-Chandra [8]

Chapter 8

8.1.1 Harish-Chandra [27], [28], [29], Hirai [7]

8.1.2 Harish-Chandra [15], [28], [30], Helgason [2]

8.1.3 Harish-Chandra [15], [18], [29], Helgason [2]

8.1.4 Harish-Chandra [17]

8.1.5 Harish-Chandra [15], [26]

8.1.6 Harish-Chandra [17], [26], [27]

8.2.1 Berezin [1], Harish-Chandra [13], [14], [18], [23], [24], [27], Helgason [8], [14], Methée [1], Tengstrand [1]

8.2.2 Harish-Chandra [14], [23], [24]

8.2.3 Harish-Chandra [13], [27]

8.2.4 Harish-Chandra [27]

8.3.1 Harish-Chandra [13], [23], [26]

8.3.2 Harish-Chandra [26]

8.3.3 Harish-Chandra [13], [27]

8.3.4 Harish-Chandra [27]

8.3.5 Harish-Chandra [14], [15], [16], [25], [27], [29], de Rham [1], Schwartz [1], [2], Treves [2]

8.3.6 Harish-Chandra [28]

8.3.7 Harish-Chandra [18], [30], Rader (unpublished; cf. Harish-Chandra [34])

8.3.8 Harish-Chandra [30], [34]
8.3.9 Harish-Chandra [30]
8.4.1 Harish-Chandra [15], [16], [25]
8.4.2 Harish-Chandra [16], [25]
8.4.3 Harish-Chandra [15]
8.4.4 Harish-Chandra [15], [16], [25]
8.4.5 Harish-Chandra [15], [16], [25], [28]
8.5.1 Harish-Chandra [17], [26], [27], [28], [29], [30]
8.5.2 Harish-Chandra [30]
8.5.3 Harish-Chandra [27], [29], [30]
8.5.4 Harish-Chandra [27]
8.5.5 Harish-Chandra [17], [27]
8.5.6 Harish-Chandra [29]
8.5.7 Harish-Chandra [30], [34]
8.5.8 Harish-Chandra [30]

Chapter 9

9.1.1 Harish-Chandra (unpublished; cf. Harish-Chandra [21])
9.1.2 Harish-Chandra (unpublished; cf. Harish-Chandra [18])
9.1.3 Harish-Chandra (unpublished; cf. Harish-Chandra [18])
9.1.4 Harish-Chandra (unpublished; cf. Harish-Chandra [18]), Helgason [10], Miller and Simms [1], [2]
9.1.5 Harish-Chandra (unpublished; cf. Harish-Chandra [18])
9.1.6 Bhanu Murthy [1], [2], Gindikin and Karpelevič [1], [3], Harish-Chandra [18], [34], [36], Schiffmann [2]
9.1.7 Gangolli [3], Harish-Chandra [18], [19], Helgason [5], [14]
9.2.1 Ehrenpreis and Mautner [1], Godement [15], Harish-Chandra [18], [19], [30], Helgason [8], [9], [14], Helgason and Johnson [1], Takahashi [1]
9.2.2 Harish-Chandra [18], [19], [30], Helgason [5]
9.2.3 Ehrenpreis and Mautner [1], [2], [3], Gangolli [3], Helgason [5], [14], Y. Shimizu [1]
9.2.4 Harish-Chandra [18], Helgason [14], Helgason and Johnson [1], Trombi and Varadarajan [1]
9.3.1 Harish-Chandra [30], Langlands [1]
9.3.2 Harish-Chandra [30]
9.3.3 Harish-Chandra [30]

Chapter 10

10.1.1 Harish-Chandra [28]
10.1.2 Harish-Chandra [30], [34]
10.2.1 Harish-Chandra [30]
10.2.2 Harish-Chandra [30]
10.2.3 Harish-Chandra [30]
10.2.4 Harish-Chandra [30]

Appendices

A1 Bourbaki [3], Bruhat [1], Helgason [2], Mackey [3]
A2.1 Harish-Chandra [13], [23], Helgason [2], Peetre [1], Schwartz [1], Treves [1], [2]
A2.2 Grothendieck [1], Treves [2]
A2.3 Bruhat [1], Schwartz [3], [4]
A2.4 Bruhat [1], Godement [7], Harish-Chandra [13], [23], [24], Helgason [2]
A3.1 Harish-Chandra (unpublished; cf. Harish-Chandra [21])
A3.2 Harish-Chandra [18]
A3.3 Harish-Chandra (unpublished; cf. Harish-Chandra [21])
A3.4 Harish-Chandra (unpublished; cf. Harish-Chandra [18])
A3.5 Harish-Chandra (unpublished; cf. Harish-Chandra [21])

Bibliography

Adams, J. F.: [1] Lectures on Lie groups, Benjamin Inc., New York, 1969.

Aomoto, K.: [1] On some double coset decompositions of complex semi-simple Lie groups, J. Math. Soc. Japan, vol. 18 (1966), pp. 1–44.

— [2] L'analyse harmonique sur les espaces riemanniens à courbure riemannienne negative I, J. Fac. Sci. Univ. of Tokyo, vol. 13 (1966), pp. 85–107.

— [3] Sur les transformations d'horisphere et les équations integrales qui s'y rattachent, J. Fac. Sci. Univ. of Tokyo, vol. 14 (1967), pp. 1–23.

Araki, S.: [1] On root systems and an infinitesimal classification of irreducible symmetric spaces, J. of Math., Osaka City Univ., vol. 13 (1962), pp. 1–34.

Aribaud, F.: [1] Une nouvelle démonstration d'un théorème de R. Bott et B. Kostant, Bull. Math. Soc. France, vol. 95 (1967), pp. 205–242.

Arthur, J. G.: [1] Harmonic analysis of tempered distributions on semi-simple Lie groups of real rank one, Ph.D. Thesis, Yale University, 1970.

Atiyah, M. and Bott, R.: [1] A Lefschetz fixed point formula for elliptic complexes I, Ann. of Math., vol. 86 (1967), pp. 374–407.

— [2] A Lefschetz fixed point formula for elliptic complexes II, Ann. of Math., vol. 88 (1968), pp. 451–491.

Atiyah, M. and Segal, G.: [1] The index of elliptic operators II, Ann. of Math., vol. 87 (1968), pp. 531–545.

Atiyah, M. and Singer, I.: [1] The index of elliptic operators I, Ann. of Math., vol. 87 (1968), pp. 484–530.

— [2] The index of elliptic operators III, Ann. of Math., vol. 87 (1968), pp. 546–604.

Auslander, L., Green, L. and Hahn, F.: [1] Flows on homogeneous spaces, Princeton University Press, Princeton, New Jersey, 1963.

Auslander, L. and Kostant, B.: [1] Quantization and representations of solvable Lie groups, Bull. Amer. Math. Soc., vol. 73 (1967), pp. 692–695.

Auslander, L. and Moore, C. C.: [1] Unitary representations of solvable Lie groups, Memoirs Amer. Math. Soc., n°62, Providence, Rhode Island, 1966.

Baggett, L.: [1] Hilbert-Schmidt representations of groups, Proc. Amer. Math. Soc., vol. 21 (1969), pp. 502–506.

Bargmann, V.: [1] Irreducible unitary representations of the Lorentz group, Ann. of Math., vol. 48 (1947), pp. 568–640.

— [2] On unitary ray representations of continuous groups, Ann. of Math., vol. 59 (1954), pp. 1–46.

Berezin, F. A.: [1] Laplace operators on semi-simple Lie groups, Amer. Math. Soc. Translations (2), vol. 21 (1962), pp. 239–339.

Bernat, P.: [1] Sur les représentations unitaires des groupes de Lie résolubles, Ann. Sci. École. Norm. Sup., vol. 82 (1965), pp. 37–99.

Bernat, P. and Dixmier, J.: [1] Sur le dual d'un groupe de Lie, C. R. Acad. Sc. Paris, vol. 250 (1960), pp. 1778–1779.

Bers, L., John, F. and Schechter, M.: [1] Partial differential equations, Lectures in Applied Mathematics, vol. III, Interscience, New York, 1964.

Bhanu, Murthy, T. S.: [1] Plancherel's measure for the factor space **SL**(n, **R**)/**SO**(n, **R**), Doklady Akad. Nauk. SSSR, vol. 133 (1960), pp. 503–506.
— [2] The asymptotic behavior of zonal spherical functions on the Siegel upper half-plane, Doklady Akad. Nauk. SSSR, vol. 135 (1960), pp. 1027–1029.

Bichteler, K.: [1] On the existence of noncontinuous irreducible representations of a locally compact group, Inventiones Math., vol. 6 (1968), pp. 159–162.

Blattner, R. J.: [1] On induced representations, Amer. J. Math., vol. 83 (1961), pp. 79–98.
— [2] On induced representations II: Infinitesimal induction, Amer. J. Math., vol. 83 (1961), pp. 499–512.
— [3] Positive definite measures, Proc. Amer. Math. Soc., vol. 14 (1963), pp. 423–428.
— [4] Group extension representations and the structure space, Pacific J. Math., vol. 15 (1965), pp. 1101–1113.

Borel, A.: [1] Les fonctions automorphes de plusieurs variables complexes, Bull. Soc. Math. France, vol. 80 (1952), pp. 167–182.
— [2] Kählerian coset spaces of semi-simple Lie groups, Proc. Nat. Acad. Sci. U.S.A., vol. 40 (1954), pp. 1147–1151.
— [3] Topology of Lie groups and characteristic classes, Bull. Amer. Math. Soc., vol. 61 (1955), pp. 397–432.
— [4] Groupes linéaires algébriques, Ann. of Math., vol. 64 (1956), pp. 20–82.
— [5] On the curvature tensor of the Hermitian symmetric manifolds, Ann. of Math., vol. 71 (1960), pp. 508–521.
— [6] Density properties for certain subgroups of semi-simple groups without compact components, Ann. of Math., vol. 72 (1960), pp. 179–188.
— [7] Ensembles fondamentaux pour les groupes arithmétiques, Colloq. des Groupes Algébriques, Bruxelles, 1962, pp. 23–40.
— [8] Arithmetic properties of linear algebraic groups, Proc. Int. Congr. of Math., pp. 10–22, Stockholm, 1962.
— [9] Compact Clifford-Klein forms of symmetric spaces, Topology, vol. 2 (1963), pp. 111–122.
— [10] Introduction aux groupes arithmétiques, Publications de l'Institut Mathématique de l'Université de Strasbourg XV, Hermann, Paris, 1969.

Borel, A. and Chevalley, C.: [1] The Betti numbers of the exceptional groups, Memoirs Amer. Math. Soc., n°14, pp. 1–9, Providence, Rhode Island, 1955.

Borel, A. and Harish-Chandra: [1] Arithmetic subgroups of algebraic groups, Ann. of Math., vol. 75 (1962), pp. 485–535.

Borel, A. and Mostow, G. D.: [1] On semi-simple automorphisms of Lie algebras, Ann. of Math., vol. 61 (1955), pp. 389–405.

Borel, A. and de Siebenthal, J.: [1] Les sous-groupes fermés de rang maximum des groupes de Lie clos, Comment. Math. Helv., vol. 23 (1949), pp. 200–221.

Borel, A. and Tits, J.: [1] Groupes réductifs, Inst. Hautes Études Sci. Publ. Math., vol. 27 (1965), pp. 55–152.

Bott, R.: [1] Homogeneous vector bundles, Ann. of Math., vol. 66 (1957), pp. 203–248.

Bourbaki, N.: [1] Éléments de mathématique, vol. VI, Intégration, Chapters I–IV, Hermann, Paris, 1952.
— [2] Éléments de mathématique, Vol. VI, Intégration, Chapter VI, Hermann, Paris, 1959.
— [3] Éléments de mathématique, Vol. VI, Intégration, Chapters VII–VIII, Hermann, Paris, 1963.
— [4] Éléments de mathématique, Groupes et algèbres de Lie, Chapter I, Hermann, Paris, 1960.

— [5] Éléments de mathématique, Groupes et algèbres de Lie, Chapters IV–VI, Hermann, Paris, 1968.

Bouwer, I. Z.: [1] Standard representations of simple Lie algebras, Canad. J. Math., vol. 20 (1968), pp. 344–361.

Brezin, J.: [1] Unitary representation theory for solvable Lie groups, Memoirs Amer. Math. Soc., n°79, Providence, Rhode Island, 1968.
— [2] Harmonic analysis on nilmanifolds, Trans. Amer. Math. Soc., vol. 150 (1970), pp. 611–618.

Browder, F. I.: [1] Analyticity and partial differential equations I, Amer. J. Math., vol. 84 (1962), pp. 666–710.

Bruhat, F.: [1] Sur les représentations induites des groupes de Lie, Bull. Soc. Math. France, vol. 84 (1956), pp. 97–205.

Butzer, P. L. and Berens, H.: [1] Semi-groups of operators and approximation, Springer-Verlag, New York, 1967.

Carmona, J.: [1] Représentations unitaires induites holomorphes des groupes de Lie; application aux groupes de Lie semi-simples connexes reels, Ph.D. Thesis, Marseille, 1968.

Cartan, H.: [1] Fonctions automorphes, Séminaire, 1957–1958, Paris.

Carter, R.: [1] Simple groups and simple Lie algebras, J. London Math. Soc., vol. 40 (1965), pp. 193–240.

Cartier, P.: [1] Développements de fonctions arbitraires suivant les fonctions propres d'un opérateur différential, Séminaire Bourbaki, n° 102 (1954), Paris.
— [2] On H. Weyl's character formula, Bull. Amer. Math. Soc., vol. 67 (1961), pp. 228–230.
— [3] Remarks on "Lie algebra cohomology and the generalized Borel-Weil theorem," by B. Kostant, Ann. of Math., vol. 74 (1961), pp. 388–390.

Cartier P. and Dixmier, J.: [1] Vecteurs analytiques dans les représentations des groupes de Lie, Amer. J. Math., vol. 80 (1958), pp. 131–145.

Chevalley, C.: [1] Theory of Lie groups, vol. 1, Princeton University Press, Princeton, New Jersey, 1946.
— [2] The Betti numbers of the exceptional simple Lie groups, Proc. Int. Congr. of Math., vol. 2, pp. 21–24, Harvard, 1950.
— [3] Sur certains groupes simples, Tôhoku Math. J., vol. 7 (1955), pp. 14–66.
— [4] Theorie des groupes de Lie, vol. 3, Hermann, Paris, 1955.
— [5] Invariants of finite groups generated by reflections, Amer. J. Math., vol. 77 (1955), pp. 778–782.
— [6] Fundamental concepts of algebra, Academic Press, New York, 1956.

Cohn, L.: [1] The dimension of spaces of automorphic forms on a certain two dimensional complex domain, Ph.D. Thesis, University of Chicago, 1969.

Coifman, R. R. and Weiss, G.: [1] Representations of compact groups and spherical harmonics, L'Enseignement Mathématique, vol. 14 (1969), pp. 121–173.

Coleman, A. J.: [1] The Betti numbers of the simple Lie groups, Canad. J. Math., vol. 10 (1958), pp. 349–356.

Coxeter, H. S. M.: [1] The product of the generators of a finite group generated by reflections, Duke Math. J., vol. 18 (1951), pp. 765–782.

Coxeter, H. S. M. and Moser, W.: [1] Generators and relations for discrete groups, Berlin, 1965.

Curtis, C. and Reiner, I.: [1] Representation theory of finite groups and associative algebras, Interscience, New York, 1962.

Dauns, J. and Hofmann, K.: [1] Representations of rings by sections, Memoirs Amer. Math. Soc., n°83, Providence, Rhode Island, 1968.

Dieudonné, J.: [1] Sur le théorème de Lebesgue-Nikodym (III), Ann. Univ. Grenoble (NS), vol. 23 (1948), pp. 25–53.
— [2] Foundations of modern analysis, Academic Press, New York, 1960.
— [3] Note sur les fonctions sphériques, J. Math. Pures Appl., vol. 41 (1962), pp. 233–240.
— [4] Éléments d'analyse, Tome II, Gauthier-Villars, Paris, 1968.

Dixmier, J.: [1] Sur un théorème d'Harish-Chandra, Acta Sci. Math. Szeged., vol. 14 (1952), pp. 145–156.
— [2] Les algèbres d'operateurs dans l'espace Hilbertien (algèbres de von Neumann), Gauthier-Villars, Paris, 1957.
— [3] Sur les représentations unitaires des groupes de Lie nilpotents II, Bull. Soc. Math. France, vol. 85 (1957), pp. 325–388.
— [4] Sur les représentations unitaires des groupes de Lie algébriques, Ann. Inst. Fourier (Grenoble), vol. 7 (1957), pp. 315–328.
— [5] Sur les représentations unitaires des groupes de Lie nilpotents III, Canad. J. Math., vol. 10 (1958), pp. 321–348.
— [6] Sur les représentations unitaires des groupes de Lie nilpotents I, Amer. J. Math., vol. 81 (1959), pp. 160–170.
— [7] Sur les représentations unitaires des groupes de Lie nilpotents IV, Canad. J. Math., vol. 11 (1959), pp. 321–344.
— [8] Sur les représentations unitaires des groupes de Lie nilpotents V, Bull. Soc. Math. France, vol. 87 (1959), pp. 65–79.
— [9] Sur les représentations unitaires des groupes de Lie nilpotents VI, Canad. J. Math., vol. 12 (1960), pp. 324–352.
— [10] Opérateurs de rang fini dans les représentations unitaires, Inst. Hautes Études Sci. Publ. Math., vol. 6 (1960), pp. 13–25.
— [11] Représentations intégrables du groupe de De Sitter, Bull. Soc. Math. France, vol. 89 (1961), pp. 9–41.
— [12] Points isolés dans le dual d'un groupe localement compact, Bull. Sc. Math., vol. 85 (1961), pp. 91–96.
— [13] Sur les représentations unitaires des groupes de Lie résolubles, Math. J. Okayama Univ., vol. 11 (1962), pp. 1–18.
— [14] Représentations irréductibles des algèbres de Lie nilpotentes, Anais. Acad. Brasil. Ciencias, vol. 35 (1963), pp. 491–519.
— [15] Les C^*-algèbres et leurs représentations, Gauthier-Villars, Paris, 1964.
— [16] Représentations irréductibles des algèbres de Lie résolubles, J. Math. Pures et Appl., vol. 45 (1966), pp. 1–66.
— [17] Sur le noyau infinitesimal d'une représentation unitaire d'un groupe résoluble, C. R. Acad. Sc. Paris, vol. 262 (1966), Serie A, pp. 483–486.
— [18] Représentations induites holomorphes des groupes résolubles algébriques, Bull. Soc. Math. France, vol. 94 (1966), pp. 181–206.
— [19] Espace dual d'une algèbre ou d'un groupe localement compact, Proc. Int. Congr. of Math., pp. 357–366, Moscow, 1966.
— [20] Sur le dual d'un groupe de Lie nilpotent, Bull. Sc. Math., vol. 90 (1966), pp. 113–118.
— [21] Sur le centre de l'algèbre enveloppante d'une algèbre de Lie, C. R. Acad. Sc. Paris, vol. 265 (1967), Serie A, pp. 408–410.
— [22] Représentations induites des algèbres de Lie, L'Enseignement Mathématique, vol. 16 (1970), pp. 169–175.

Dopplicher, S., Kastler, D. and Robinson, D.: [1] Covariance algebras in field theory and statistical mechanics, Comm. Math. Phys., vol. 3 (1966), pp. 1–28.

Dornhoff, L.: [1] Linear and reflection groups, Lecture Notes, Yale University, 1967.

Dress, A.: [1] On finite groups generated by pseudoreflections, J. of Algebra, vol. 11 (1969), pp. 1–5.

Duflo, M.: [1] Caractères des groupes et des algèbres de Lie résolubles, Ann. Scientifiques L'École Normale Supérieure, vol. 3 (1970), pp. 23–74.

— [2] Fundamental series representations of a semi-simple Lie group, Func. Analysis and Applic., vol. 4 (1970), pp. 122–126.

Dunford, N. and Schwartz, J. T.: [1] Linear operators, Part I, Interscience, New York, 1958.

— [2] Linear operators, Part II, Interscience, New York, 1963.

Dunkl, C. F.: [1] Functions that operate in the Fourier algebra of a compact group, Proc. Amer. Math. Soc., vol. 21 (1969), pp. 540–544.

Effros, E.: [1] Transformation groups and C^*-algebras, Ann. of Math., vol. 81 (1965), pp. 38–55.

Effros, E. and Hahn, F.: [1] Locally compact transformation groups and C^*-algebras, Memoirs Amer. Math. Soc., n°75, Providence, Rhode Island, 1967.

Ehrenpreis, L. and Mautner, F. I.: [1] Some properties of the Fourier transform on semi-simple Lie groups I, Ann. of Math., vol. 61 (1955), pp. 406–439.

— [2] Some properties of the Fourier transform on semi-simple Lie groups II, Trans. Amer. Math. Soc., vol. 84 (1957), pp. 1–55.

— [3] Some properties of the Fourier transform on semi-simple Lie groups III, Trans. Amer. Math. Soc., vol. 90 (1959), pp. 431–484.

Eymard, P.: [1] L'algèbre de Fourier d'un groupe localement compact, Bull. Soc. Math. France, vol. 92 (1964), pp. 181–236.

Fell, J. M. G.: [1] The dual spaces of C^*-algebras, Trans. Amer. Math. Soc., vol. 94 (1960), pp. 365–403.

— [2] Weak containment and induced representations of groups, Canad. J. Math., vol. 14 (1962), pp. 237–268.

— [3] The dual spaces of Banach algebras, Trans. Amer. Math. Soc., vol. 114 (1965), pp. 227–250.

— [4] Non-unitary dual spaces of groups, Acta Math., vol. 114 (1965), pp. 267–310.

— [5] Conjugating representations and related results on semi-simple Lie groups, Trans. Amer. Math. Soc., vol. 127 (1967), pp. 405–426.

— [6] An extension of Mackey's method to algebraic bundles over finite groups, Amer. J. Math., vol. 91 (1969), pp. 203–238.

— [7] An extension of Mackey's method to Banach *-algebraic bundles, Memoirs Amer. Math. Soc., n°90, Providence, Rhode Island, 1969.

— [8] Group representation theory, To Appear.

Flatto, L.: [1] Basic sets of invariants for finite reflection groups, Bull. Amer. Math. Soc., vol. 74 (1968), pp. 730–734.

— [2] Invariants of finite reflection groups and mean value problems II, Amer. J. Math., vol. 92 (1970), pp. 552–561.

Flatto, L. and Wiener, M. M.: [1] Invariants of finite reflection groups and mean value problems, Amer. J. Math., vol. 91 (1969), pp. 591–598.

Flensted-Jensen, M.: [1] On the Fourier transform on a symmetric space of rank one, Institut Mittag-Leffler, 1970.

Freudenthal, H.: [1] Zur Berechnung der Charaktere der halbeinfachen Lieschen Gruppen I, II and III. I: Indag. Math., vol. 16 (1954), pp. 369–376. II: ibid., pp. 487–491. III: ibid., vol. 18 (1956), pp. 511–514.

Frobenius, G.: [1] Über Relationen zwischen den Charakteren einer Gruppe und denen ihrer Untergruppen, Sitz. Preus. Akad. Wiss., 1898, pp. 501–515.

Furstenberg, H.: [1] A Poisson formula for semi-simple Lie groups, Ann. of Math., vol. 77 (1963), pp. 335–386.

— [2] Translation-invariant cones of functions on semi-simple Lie groups, Bull. Amer. Math. Soc., vol. 71 (1965), pp. 271– 326.
— [3] Poisson boundaries and envelopes of discrete groups, Bull. Amer. Math. Soc., vol. 73 (1967), pp. 350–356.

Gaffney, M.: [1] The conservation property of the heat equation on Riemannian manifolds, Comm. Pure and Applied Math., vol. 12 (1959), pp. 1–11.

Gangolli, R.: [1] Isotropic infinitely divisible measures on symmetric spaces, Acta Math., vol. 111 (1964), pp. 213–246.
— [2] Asymptotic behaviour of spectra of compact quotients of certain symmetric spaces, Acta Math., vol. 121 (1968), pp. 151–192.
— [3] On the Plancherel formula and the Paley-Wiener theorem for spherical functions on semi-simple Lie groups, Ann. of Math., vol. 93 (1971), pp. 150–165.

Gårding, L.: [1] Note on continuous representations of Lie groups, Proc. Nat. Acad. Sci. U.S.A., vol. 33 (1947), pp. 331–332.
— [2] Vecteurs analytiques dans les représentations des groupes de Lie, Bull. Soc. Math. France, vol. 88 (1960), pp. 73–93.
— [3] Applications of the theory of direct integrals of Hilbert spaces to some integral and differential operators, Institute for Fluid Dynamics and Applied Mathematics Lecture Series, n° 11, University of Maryland.
— [4] Transformation de Fourier des distributions homogènes, Bull. Soc. Math. France, vol. 89 (1961), pp. 381–428.

Garland, H.: [1] The spectrum of non-compact G/Γ and the cohomology of arithmetic groups, Bull. Amer. Math. Soc., vol. 75 (1969), pp. 807–811.

Garland, H. and Goto, M.: [1] Lattices and the adjoint group of a Lie group, Trans. Amer. Math. Soc., vol. 124 (1966), pp. 450–460.

Garland, H. and Raghunathan, M. S.: [1] Fundamental domains for lattices in (**R**)-rank 1 semi-simple Lie groups, Ann. of Math., vol. 92 (1970), pp. 279–326.

Gelbart, S.: [1] Fourier analysis on matrix space, Memoirs Amer. Math. Soc., n° 108, Providence, Rhode Island, 1971.

Gelfand, I.: [1] On one-parameter groups of operators in a normed space, Doklady Akad. Nauk. SSSR (N. S.), vol. 25 (1939), pp. 713–718.
— [2] Spherical functions on symmetric spaces, Doklady Akad. Nauk. SSSR, vol. 70 (1950), pp. 5–8.
— [3] The center of an infinitesimal group algebra, Mat. Sb., vol. 26 (1950), pp. 103–112.
— [4] Integral geometry and its relation to group representations, Russ. Math. Surveys, vol. 15 (1960), pp. 143–151.
— [5] Automorphic functions and the theory of representations, Proc. Int. Congr. of Math., pp. 74–85, Stockholm, 1962.

Gelfand, I. M. and Graev, M. I.: [1] The geometry of homogeneous spaces, group representations in homogeneous spaces and questions in integral geometry related to them, Amer. Math. Soc. Translations (2), vol. 37 (1964), pp. 351–429.

Gelfand, I. M., Graev, M. I. and Pyateckii-Shapiro, I. I.: [1] Representation theory and automorphic functions, W. B. Saunders Co., Philadelphia, 1969.

Gelfand, I. M., Graev, M. I. and Shapiro, S. J.: [1] Differential forms and integral geometry, Func. Analysis and Applic., vol. 3 (1969), pp. 24–40.

Gelfand I. M. and Kirillov A. A.: [1] Sur les corps Liés aux algèbres enveloppantes des algèbres de Lie, Inst. Hautes Études Sci. Publ. Math., vol. 31 (1966), pp. 5–20.

Gelfand, I. M. and Naimark, M. A.: [1] Unitäre Darstellungen der klassischen Gruppen, Akademie Verlag, Berlin, 1957.

Gelfand, I. M. and Pyateckii-Shapiro, I. I.: [1] Theory of representations and theory of automorphic functions, Amer. Math. Soc. Translations (2), vol. 26 (1963), pp. 173–200.

Gindikin, S.: [1] Analysis in homogeneous domains, Russ. Math. Surveys, vol. 19 (1964), pp. 1–89.

— [2] Unitary representations of groups of automorphisms of Riemannian symmetric spaces, Func. Analysis and Applic., vol. 1 (1967), pp. 28–32.

Gindikin, S. G. and Karpelevič, F. I.: [1] Plancherel measure of Riemannian symmetric spaces of non-positive curvature, Doklady Akad. Nauk. SSSR, vol. 145 (1962), pp. 252–255.

— [2] On a problem in integral geometry, Chebotarev Memorial Volume, Kazan University, 1964.

— [3] On an integral connected with symmetric Riemannian spaces of negative curvature, Amer. Math. Soc. Translations (2), vol. 85 (1969), pp. 249–258.

Glimm, J.: [1] Type I C^*-algebras, Ann. of Math., vol. 73 (1961), pp. 572–612.

— [2] Locally compact transformation groups, Trans. Amer. Math. Soc., vol. 101 (1961), pp. 124–138.

— [3] Families of induced representations, Pacific J. Math., vol. 12 (1962), pp. 885–911.

Godement, R.: [1] Sur les relations d'orthogonalité de V. Bargmann I, Resultats préliminaires, C. R. Acad. Sc. Paris, vol. 225 (1947), pp. 521–523.

— [2] Sur les relations d'orthogonalité de V. Bargmann II, Demonstration générale, C. R. Acad. Sc. Paris, vol. 225 (1947), pp. 657–659.

— [3] Les fonctions de type positif et la théorie des groupes, Trans. Amer. Math. Soc., vol. 63 (1948), pp. 1–84.

— [4] Sur la transformation de Fourier dans les groupes discrets, C. R. Acad. Sc. Paris, vol. 228 (1949), pp. 627–628.

— [5] Sur la théorie des représentations unitaires, Ann. of Math., vol. 53 (1951), pp. 68–124.

— [6] Mémoire sur la théorie des caractères dans les groupes localement compacts unimodulaires, J. Math. Pures Appl., vol. 30 (1951), pp. 1–110.

— [7] A theory of spherical functions I, Trans. Amer. Math. Soc., vol. 73 (1952), pp. 496–556.

— [8] Les travaux de E. Hecke I, Séminaire Bourbaki, n° 51 (1951), Paris.

— [9] Les travaux de E. Hecke II, Séminaire Bourbaki, n° 59 (1952), Paris.

— [10] Les travaux de E. Hecke III, Séminaire Bourbaki, n° 74 (1953), Paris.

— [11] Les travaux de E. Hecke IV, Séminaire Bourbaki, n° 80 (1953), Paris.

— [12] Cohomologie des groupes discontinus, Séminaire Bourbaki, n° 90 (1954), Paris.

— [13] Théorie des caractères I: Algèbres unitaires, Ann. of Math., vol. 59 (1954), pp. 47–62.

— [14] Théorie des caractères II: Définition et propriétés générales des caractères, Ann. of Math., vol. 59 (1954), pp. 63–85.

— [15] Introduction aux travaux de A. Selberg, Séminaire Bourbaki, n° 144 (1957), Paris.

— [16] Les fonctions ζ des algèbres simples I, Séminaire Bourbaki, n° 171 (1958), Paris.

— [17] Les fonctions ζ des algèbres simples II, Séminaire Bourbaki, n° 176 (1959), Paris.

— [18] Articles in H. Cartan [1].

— [19] Groupes linéaires algébriques sur un corps parfait, Séminaire Bourbaki, n° 206 (1960), Paris.

— [20] La formule des traces de Selberg, Séminaire Bourbaki, n° 244 (1962), Paris.

— [21] Domaines fondamentaux des groupes arithmétiques, Séminaire Bourbaki, n° 257 (1963), Paris.

— [22] Analyse spectrale des fonctions modulaires, Séminaire Bourbaki, n° 278 (1964), Paris.
— [23] The decomposition of $L^2(G/\Gamma)$ for $\Gamma = SL(2, Z)$, Proc. Sympos. Pure Math., vol. 9, Amer. Math. Soc., Providence, Rhode Island, 1966, pp. 211–224.
— [24] The spectral decomposition of cusp forms, Proc. Sympos. Pure Math., vol. 9, Amer. Math. Soc., Providence, Rhode Island, 1966, pp. 225–234.
— [25] Introduction a la théorie de Langlands, Séminaire Bourbaki, n° 321 (1967), Paris.
— [26] Formes automorphes et produits Euleriens, Séminaire Bourbaki, n° 349 (1968), Paris.

Goodman, R.: [1] Analytic domination by fractional powers of a positive operator, J. of Functional Analysis, vol. 3 (1969), pp. 246–264.
— [2] Analytic and entire vectors for representations of Lie groups, Trans. Amer. Math. Soc., vol. 143 (1969), pp. 55–76.
— [3] Differential operators of infinite order on a Lie group I, J. Math. and Mechanics, vol. 19 (1970), pp. 879–894.
— [4] Some regularity theorems for operators in an enveloping algebra, To Appear.
— [5] One parameter groups generated by operators in an enveloping algebra, J. of Functional Analysis, vol. 6 (1970), pp. 218–236.
— [6] Complex Fourier analysis on nilpotent Lie groups, To Appear.
— [7] Differential operators of infinite order on a Lie group II, To Appear.

Goto, M. and Kobayashi, E.: [1] On the subgroups of the centers of simply connected simple Lie groups—classification of simple Lie groups in the large, Osaka J. Math., vol. 6 (1969), pp. 251–281.

Griffiths, P. and Schmid, W.: [1] Locally homogeneous complex manifolds, Acta Math., vol. 123 (1969), pp. 253–302.

Gross, K.: [1] The Plancherel transform on the nilpotent part of $\mathbf{G}_2$ and some applications to the representation theory of $\mathbf{G}_2$, Tulane University, 1968.

Gross, K. and Kunze, R. A.: [1] Fourier decompositions of certain representations, To Appear.

Grothendieck, A.: [1] Produits tensoriels topologiques et espaces nucléaires, Memoirs Amer. Math. Soc., n° 16, Providence, Rhode Island, 1955.

Gutkin, E. A.: [1] Representations of the principal series of a complex semi-simple Lie group, Func. Analysis and Applic., vol. 4 (1970), pp. 117–121.

Hadziev, Dž.: [1] Some questions in the theory of vector invariants, Math. of the U.S.S.R.—Sbornik, Amer. Math. Soc. Translation, vol. 1 (1967), pp. 383–396.

Harish-Chandra: [1] On representations of Lie algebras, Ann. of Math., vol. 50 (1949), pp. 900–915.
— [2] Lie algebras and the Tannaka duality theorem, Ann. of Math., vol. 51 (1950), pp. 299–330.
— [3] On some applications of the universal enveloping algebra of a semi-simple Lie algebra, Trans. Amer. Math. Soc., vol. 70 (1951), pp. 28–96.
— [4] Plancherel formula for the 2×2 real unimodular group, Proc. Nat. Acad. Sci. U.S.A., vol. 38 (1952), pp. 337–342.
— [5] Representations of semi-simple Lie groups I, Trans. Amer. Math. Soc., vol. 75 (1953), pp. 185–243.
— [6] Representations of semi-simple Lie groups II, Trans. Amer. Math. Soc., vol. 76 (1954), pp. 26–65.
— [7] Representations of semi-simple Lie groups III, Trans. Amer. Math. Soc., vol. 76 (1954), pp. 234–253.
— [8] The Plancherel formula for complex semi-simple Lie groups, Trans. Amer. Math. Soc., vol. 76 (1954), pp. 485–528.

— [9] Representations of semi-simple Lie groups IV, Amer. J. Math., vol. 77 (1955), pp. 743–777.
— [10] Representations of semi-simple Lie groups V, Amer. J. Math., vol. 78 (1956), pp. 1–41.
— [11] Representations of semi-simple Lie groups VI, Amer. J. Math., vol. 78 (1956), pp. 564–628.
— [12] On a lemma of F. Bruhat, J. Math. Pures Appl., vol. 35 (1956), pp. 203–210.
— [13] The characters of semi-simple Lie groups, Trans. Amer. Math. Soc., vol. 83 (1956), pp. 98–163.
— [14] Differential operators on a semi-simple Lie algebra, Amer. J. Math., vol. 79 (1957), pp. 87–120.
— [15] Fourier transforms on a semi-simple Lie algebra I, Amer. J. Math., vol. 79 (1957), pp. 193–257.
— [16] Fourier transforms on a semi-simple Lie algebra II, Amer. J. Math., vol. 79 (1957), pp. 653–686.
— [17] A formula for semi-simple Lie groups, Amer. J. Math., vol. 79 (1957), pp. 733–760.
— [18] Spherical functions on a semi-simple Lie group I, Amer. J. Math., vol. 80 (1958), pp. 241–310.
— [19] Spherical functions on a semi-simple Lie group II, Amer. J. Math., vol. 80 (1958), pp. 553–613.
— [20] Automorphic forms on a semi-simple Lie group, Proc. Nat. Acad. Sci. U.S.A., vol. 45 (1959), pp. 570–573.
— [21] Some results on differential equations and their applications, Proc. Nat. Acad. Sci. U.S.A., vol. 45 (1959), pp. 1763–1764.
— [22] Invariant eigendistributions on semi-simple Lie groups, Bull. Amer. Math. Soc., vol. 69 (1963), pp. 117–123.
— [23] Invariant distributions on Lie algebras, Amer. J. Math., vol. 86 (1964), pp. 271–309.
— [24] Invariant differential operators and distributions on a semi-simple Lie algebra, Amer. J. Math., vol. 86 (1964), pp. 534–564.
— [25] Some results on an invariant integral on a semi-simple Lie algebra, Ann. of Math., vol. 80 (1964), pp. 551–593.
— [26] Invariant eigendistributions on a semi-simple Lie algebra, Inst. Hautes Études Sci. Publ. Math., vol. 27 (1965), pp. 5–54.
— [27] Invariant eigendistributions on a semi-simple Lie group, Trans. Amer. Math. Soc., vol. 119 (1965), pp. 457–508.
— [28] Discrete series for semi-simple Lie groups I, Acta Math., vol. 113 (1965), pp. 241–318.
— [29] Two theorems on semi-simple Lie groups, Ann. of Math., vol. 83 (1966), pp. 74–128.
— [30] Discrete series for semi-simple Lie groups II, Acta Math., vol. 116 (1966), pp. 1–111.
— [31] Automorphic forms on semi-simple Lie groups, Springer-Verlag, Lecture Notes in Mathematics, vol. 62, 1968.
— [32] Eisenstein series over finite fields, Functional Analysis and Related Fields, pp. 76–88, Springer-Verlag, Berlin, 1970.
— [33] Some applications of the Schwartz space of a semi-simple Lie group, Lectures in Modern Analysis and Applications, Georgetown, 1969. [Lecture Notes in Mathematics, Springer-Verlag, vol. 140, 1970.]
— [34] Harmonic analysis on semi-simple Lie groups, Bull. Amer. Math. Soc., vol. 76 (1970), pp. 529–551.
— [35] Harmonic analysis on reductive $\mathfrak{p}$-adic groups, Springer-Verlag, Lecture Notes in Mathematics, vol. 162, 1970.
— [36] On the theory of the Eisenstein integral, To Appear.

Hattori, A.: [1] On 1-cohomology groups of infinite dimensional representations of semi-simple Lie algebras, J. Math. Soc. Japan, vol. 16 (1964), pp. 226–229.

Helgason, S.: [1] Differential operators on homogeneous spaces, Acta Math., vol. 102 (1959), pp. 239–299.

— [2] Differential geometry and symmetric spaces, Academic Press, New York, 1962.

— [3] Invariants and fundamental functions, Acta Math., vol. 109 (1963), pp. 241–258.

— [4] Duality and Radon transform for symmetric spaces, Amer. J. Math., vol. 85 (1963), pp. 667–692.

— [5] Fundamental solutions of invariant differential operators on symmetric spaces, Amer. J. Math., vol. 86 (1964), pp. 565–601.

— [6] A duality in integral geometry; some generalizations of the Radon transform, Bull. Amer. Math. Soc., vol. 70 (1964), pp. 435–446.

— [7] The Radon transform on Euclidean spaces, compact two-point homogeneous spaces and Grassmann manifolds, Acta Math., vol. 113 (1965), pp. 153–180.

— [8] Radon-Fourier transforms on symmetric spaces and related group representations, Bull. Amer. Math. Soc., vol. 71 (1965), pp. 757–763.

— [9] A duality in integral geometry on symmetric spaces, Proc. U.S.-Japan Seminar in Differential Geometry, Kyoto, 1965.

— [10] An analogue of the Paley-Wiener theorem for the Fourier transform on certain symmetric spaces, Math. Ann., vol. 165 (1966), pp. 297–308.

— [11] Totally geodesic spheres in compact symmetric spaces, Math. Ann., vol. 165 (1966), pp. 309–317.

— [12] Lie groups and symmetric spaces, Battelle Rencontres, pp. 1–71, Benjamin, New York, 1968.

— [13] Applications of the Radon transform to representations of semi-simple Lie groups, Proc. Nat. Acad. Sci. U.S.A., vol. 63 (1969), pp. 643–647.

— [14] A duality for symmetric spaces with applications to group representations, Advances in Math., vol. 5 (1970), pp. 1–154.

Helgason, S. and Johnson, K.: [1] The bounded spherical functions on symmetric spaces, Advances in Math., vol. 3 (1969), pp. 586–593.

Helgason, S. and Korányi, A.: [1] A Fatou-type theorem for harmonic functions on symmetric spaces, Bull. Amer. Math. Soc., vol. 74 (1968), pp. 258–263.

Herz, C. S.: [1] Bessel functions of matrix argument, Ann. of Math., vol. 61 (1955), pp. 474–523.

— [2] Sur le phénomène de Kunze-Stein, C. R. Acad. Sc. Paris, vol. 271 (1970), Serie A, pp. 491–493.

Hewitt, E.: [1] A certain algebra of functions on a compact group, Queen's Papers on Pure and Applied Math., Queen's University, Kingston, Ontario, 1967.

Hewitt, E. and Ross, K. A.: [1] Abstract harmonic analysis, vol. I, Springer-Verlag, Berlin, 1963.

— [2] Abstract harmonic analysis, vol. II, Springer-Verlag, Berlin, 1970.

Heyer, H.: [1] Dualität lokalkompakter Gruppen, Springer-Verlag, Lecture Notes in Mathematics, vol. 150, 1970.

Hirai, T.: [1] On infinitesimal operators of irreducible representations of the Lorentz group of n-th order, Proc. Japan Acad., vol. 38 (1962), pp. 83–87.

— [2] On irreducible representations of the Lorentz group of n-th order, Proc. Japan Acad., vol. 38 (1962), pp. 258–262.

— [3] The characters of irreducible representations of the Lorentz group of n-th order, Proc. Japan Acad., vol. 41 (1965), pp. 526–531.

— [4] The Plancherel formula for the Lorentz group of n-th order, Proc. Japan Acad., vol. 42 (1965), pp. 323–326.

— [5] Classification and the characters of irreducible representations of **SU**(p, 1), Proc. Japan Acad., vol. 42 (1966), pp. 907–912.
— [6] The characters of some induced representations of semi-simple Lie groups, J. Math. Kyoto Univ., vol. 8 (1968), pp. 313–363.
— [7] Invariant eigendistributions of Laplace operators on real simple Lie groups I, To Appear.
— [8] The Plancherel formula for **SU**(p, q), J. Math. Soc. Japan, vol. 22 (1970), pp. 134–179.

Hirzebruch, F.: [1] Characteristic numbers of homogeneous domains, Seminars on Analytic Functions, vol. 2, pp. 92–104, Institute for Advanced Study, 1957.
— [2] Automorphe Formen und der Satz von Riemann-Roch, Symp. Intern. Top. Alg., 1956, pp. 129–144, Universidad de Mexico, 1958.
— [3] Topological methods in algebraic geometry, Third Edition, Springer-Verlag, New York, 1966.
— [4] Elliptische Differentialoperatoren auf Mannigfaltigkeiten, Weierstrass Festband, Westdeutscher Verlag, Opladen, 1966.

Hochschild, G.: [1] The structure of Lie groups, Holden-Day Inc., San Francisco, 1965.

Hochschild, G. and Serre, J-P.: [1] Cohomology of Lie algebras, Ann. of Math., vol. 57 (1953), pp. 591–603.

Hörmander, L.: [1] Linear partial differential operators, Springer-Verlag, Berlin, 1963.
— [2] Complex analysis in several variables, Van Nostrand, New York, 1966.

Hotta, R.: [1] Elliptic complexes on some homogeneous spaces, Osaka J. Math., vol. 7 (1970), pp. 117–160.
— [2] On realization of the discrete series for semi-simple Lie groups, Proc. Japan Acad., vol. 46 (1970), pp. 993–996.

Howe, R.: [1] On Frobenius reciprocity for unipotent algebraic groups over **Q**, To Appear.
— [2] On representations of discrete, finitely generated, torsion-free, nilpotent groups, To Appear.

Ise, M.: [1] Generalized automorphic forms and certain holomorphic vector bundles, Amer. J. Math., vol. 86 (1964), pp. 70–108.

Iwahori, N.: [1] On real irreducible representations of Lie algebras, Nagoya Math. J., vol. 14 (1959), pp. 59–83.
— [2] On the structure of a Hecke ring of a Chevalley group over a finite field, J. Fac. Sci. Univ. Tokyo, vol. 10 (1964), pp. 215–236.

Iwahori, N. and Matsumoto, H.: [1] On some Bruhat decompositions and the structure of the Hecke rings of $\mathfrak{p}$-adic Chevalley groups, Inst. Hautes Études Sci. Publ. Math., vol. 25 (1965), pp. 5–48.

Jacobson, N.: [1] Lectures in abstract algebra, vol. I, Van Nostrand, New York, 1951.
— [2] Lectures in abstract algebra, vol. II, Van Nostrand, New York, 1953.
— [3] Lie algebras, Interscience, New York, 1962.
— [4] Structure of rings, Amer. Math. Soc. Colloquium Publication, vol. 37, Providence, Rhode Island, 1964.

Jacquet, H.: [1] Fonctions de Whittaker associées aux groupes de Chevalley, Bull. Soc. Math. France, vol. 95 (1967), pp. 243–309.

Jeanquartier, P.: [1] Distributions et opérateurs différentiels homogènes et invariants, Comment. Math. Helv., vol. 39 (1964), pp. 205–252.

John, F.: [1] Plane waves and spherical means, Interscience, New York, 1955.

Kaneyuki, S. and Nagano, T.: [1] On the first Betti numbers of compact quotient spaces of complex semi-simple Lie groups by discrete groups, Sci. Papers Coll. Gen. Ed. Univ. Tokyo, vol. 12 (1962), pp. 1–11.
— [2] On certain quadratic forms related to symmetric Riemannian spaces, Osaka Math. J., vol. 14 (1962), pp. 1–20.

Kaplansky, I.: [1] Primary ideals in group algebras, Proc. Nat. Acad. Sci. U.S.A., vol. 35 (1949), pp. 133–136.
— [2] Groups with representations of bounded degree, Canad. J. Math., vol. 1 (1949), pp. 105–112.

Kaplansky, I. et al.: [1] Operator theory and group representations, National Academy of Sciences - National Research Council, Publication 387, Washington, D. C., 1955.

Karpelevič, F. I.: [1] Orispherical radial parts of Laplace operators on symmetric spaces, Soviet Mathematics, vol. 3 (1962), pp. 528–531.
— [2] The geometry of geodesics and the eigenfunctions of the Laplace-Beltrami operator on symmetric spaces, Trans. Moscow Math. Soc., vol. 14 (1965), pp. 48–185.

Kazdan, D.: [1] Connection of the dual space of a group with the structure of its closed subgroups, Func. Analysis and Applic., vol. 1 (1967), pp. 63–65.

Kazdan, D. A. and Margolies, G. A.: [1] A proof of Selberg's hypothesis, Mat. Sb. (N. S.), vol. 75 (1968), pp. 163–168.

Kelley, J. L. and Namioka, I.: [1] Linear topological spaces, Van Nostrand, Princeton, New Jersey, 1963.

Kimura, H.: [1] On some infinite dimensional representations of semi-simple Lie algebras, Nagoya Math. J., vol. 25 (1965), pp. 211–220.

Kirillov, A. A.: [1] Unitary representations of nilpotent Lie groups, Russ. Math. Surveys, vol. 17 (1962), pp. 53–104.
— [2] Plancherel measure for nilpotent Lie groups, Func. Analysis and Applic., vol. 1 (1967), pp. 330–331.
— [3] Method of orbits in the theory of unitary representations of Lie groups, Func. Analysis and Applic., vol. 2 (1968), pp. 90–93.
— [4] The characters of unitary representations of Lie groups, Func. Analysis and Applic., vol. 2 (1968), pp. 133–146.

Kleppner, A.: [1] Intertwining forms for summable induced representations, Trans. Amer. Math. Soc., vol. 112 (1964), pp. 164–183.

Klimyk, A.: [1] Decomposition of a direct product of irreducible representations of a semi-simple Lie algebra into a direct sum of irreducible representations, Amer. Math. Soc. Translations (2), vol. 76 (1968), pp. 63–73.

Knapp, A. W. and Okamoto, K.: [1] Limits of holomorphic discrete series, To Appear.

Knapp, A. W. and Stein, E. M.: [1] Singular integrals and the principal series, Proc. Nat. Acad. Sci. U.S.A., vol. 63 (1969), pp. 281–284.
— [2] Existence of complementary series, Problems in Analysis: Symposium in Honor of Salomon Bochner, Princeton University Press, Princeton, New Jersey, 1970.
— [3] Singular integrals and the principal series II, Proc. Nat. Acad. Sci. U.S.A., vol. 66 (1970), pp. 13–17.
— [4] Intertwining operators for semi-simple groups, Ann. of Math., vol. 93 (1971), pp. 489–578.

Kohari, A.: [1] Harmonic analysis on the group of linear transformations of the straight line, Proc. Jap. Acad., vol. 37 (1962), pp. 250–254.

Korányi, A. and Wolf, J. A.: [1] Realization of Hermitian symmetric spaces as generalized half-planes, Ann. of Math., vol. 81 (1965), pp. 265–288.

Kostant, B.: [1] On the conjugacy of real Cartan subalgebras I, Proc. Nat. Acad. Sci. U.S.A., vol. 41 (1955), pp. 967–970.
— [2] The principal three-dimensional subgroup and the Betti numbers of a complex simple Lie group, Amer. J. Math., vol. 81 (1959), pp. 973–1032.
— [3] Lie algebra cohomology and the generalized Borel-Weil theorem, Ann. of Math., vol. 74 (1961), pp. 329–387.
— [4] Lie group representations on polynomial rings, Amer. J. Math., vol. 85 (1963), pp. 327–404.
— [5] On the existence and irreducibility of certain series of representations, Bull. Amer. Math. Soc., vol. 75 (1969), pp. 627–642.
— [6] Quantization and unitary representations, Lectures in Modern Analysis and Applications, Georgetown, 1969. [Lecture Notes in Mathematics, Springer-Verlag, vol. 170, 1970.]

Kostant, B. and Rallis, S.: [1] On orbits associated with symmetric spaces, Bull. Amer. Math. Soc., vol. 75 (1969), pp. 879–883.
— [2] On representations associated with symmetric spaces, Bull. Amer. Math. Soc., vol. 75 (1969), pp. 884–888.

Koszul, J-L.: [1] Homologie et cohomologie des algèbres de Lie, Bull. Soc. Math. France, vol. 78 (1950), pp. 65–127.

Kotake, T. and Narasimhan, M. S.: [1] Regularity theorems for fractional powers of a linear operator, Bull. Soc. Math. France, vol. 90 (1962), pp. 449–471.

Kubota, T.: [1] Introduction to Selberg's theory, Lecture Notes, Nagoya Univ., 1966.
— [2] An elementary theory of Eisenstein series, Tokyo Univ. Seminar Lecture Notes, vol. 18, 1968.
— [3] On automorphic functions and the reciprocity law in a number field, Kinokuniya Book Store Co., Tokyo, Japan, 1969.

Kuga, M.: [1] Topological analysis and its applications in weakly symmetric Riemannian spaces, Sûgaku, vol. 9 (1957/58), pp. 166–185.
— [2] On a uniformity of distribution of O-cycles and the eigenvalues of Hecke's operators I, Sci. Papers Coll. Gen. Ed. Univ. Tokyo, vol. 10 (1960), pp.1–16.
— [3] On a uniformity of distribution of O-cycles and the eigenvalues of Hecke's operators II, Sci. Papers Coll. Gen. Ed. Univ. Tokyo, vol. 10 (1960), pp. 171–186.
— [4] Fiber varieties over a symmetric space whose fibers are abelian varieties, University of Chicago Lecture Notes, 1963-1964.

Kuga, M. and Shimura, G.: [1] On vector differential forms attached to automorphic forms, J. Math. Soc. Japan, vol. 12 (1960), pp. 258–270.

Kunze, R. A.: [1] Positive definite operator-valued kernels and unitary representations, Functional Analysis (Proc. Conf., Irvine, Calif., 1966), pp. 235–247, Thompson Book Co., Washington, D. C., 1967.
— [2] Analytic continuation of intertwining operators, To Appear.
— [3] A note on square integrable representations, J. Functional Analysis, vol. 6 (1970), pp. 454–459.

Kunze, R. A. and Stein, E. M.: [1] Uniformly bounded representations and harmonic analysis on the 2×2 real unimodular group, Amer. J. Math., vol. 82 (1960), pp. 1–62.
— [2] Uniformly bounded representations II: Analytic continuation of the principal series of representations of the $n \times n$ complex unimodular group, Amer. J. Math., vol. 83 (1961), pp. 723–786.
— [3] Uniformly bounded representations III: Intertwining operators for the principal series on semi-simple groups, Amer. J. Math., vol. 89 (1967), pp. 385–442.

— [4] Uniformly bounded representations IV: Analytic continuation of the principal series for complex classical groups of types $\mathbf{B}_n$, $\mathbf{C}_n$, $\mathbf{D}_n$, To Appear.

Langlands, R. P.: [1] The dimension of spaces of automorphic forms, Amer. J. Math., vol. 85 (1963), pp. 99–125.
— [2] On the functional equations satisfied by Eisenstein series, unpublished.
— [3] Eisenstein series, Proc. Sympos. Pure Math., vol. 9, Amer. Math. Soc., Providence, Rhode Island, 1966, pp. 235–252.
— [4] Dimension of spaces of automorphic forms, Proc. Sympos. Pure Math., vol. 9, Amer. Math. Soc., Providence, Rhode Island, 1966, pp. 253–257.
— [5] Euler products, Lecture Notes, Yale University, 1967.
— [6] Representations of abelian algebraic groups, Lecture Notes, Yale University, 1968.
— [7] Problems in the theory of automorphic forms, Lectures in Modern Analysis and Applications, Georgetown, 1969. [Lecture Notes in Mathematics, Springer-Verlag, vol. 170, 1970.]
— [8] On the functional equation of the Artin L-functions, To Appear.

de Leeuw, K. and Glicksberg, I.: [1] The decomposition of certain group representations, J. d'Analyse Math., vol. 15 (1965), pp. 135–192.

Lemire, F. W.: [1] Irreducible representations of a simple Lie algebra admitting a one-dimensional weight space, Proc. Amer. Math. Soc., vol. 19 (1968), pp. 1161–1164.

Lipsman, R.: [1] Uniformly bounded representations of $\mathbf{SL}(2, \mathbf{C})$, Amer. J. Math., vol. 91 (1969), pp. 47–66.
— [2] An indicator diagram for locally compact unimodular groups, Duke Math. J., vol. 36 (1969), pp. 765–780.
— [3] Harmonic analysis on $\mathbf{SL}(n,\mathbf{C})$, J. of Functional Analysis, vol. 3 (1969), pp. 126–155.
— [4] Uniformly bounded representations of the Lorentz group, Amer. J. Math., vol. 91 (1969), pp. 938–962.
— [5] An indicator diagram for locally compact unimodular groups II, Duke Math. J., vol. 37 (1970), pp. 459–466.
— [6] The dual topology for the principal and discrete series on semi-simple groups, Trans. Amer. Math. Soc., vol. 152 (1970), pp. 399–417.
— [7] An explicit realization of Kostant's complementary series with applications to uniformly bounded representations, To Appear.
— [8] On the characters and equivalence of continuous series representations, To Appear.

Loebel, E. et. al.: [1] Group theory and its applications, Academic Press, New York, 1968.

Loomis, L. H.: [1] Abstract harmonic analysis, Van Nostrand, New York, 1953.
— [2] Positive definite functions and induced representations of groups, Duke Math. J., vol. 27 (1960), pp. 569–580.

Loos, O.: [1] Symmetric spaces, vols. I–II, Benjamin, New York, 1969.

Maas, H.: [1] Über eine neue Art von nichtanalytischen automorphen Funktionen, Math. Ann., vol. 121 (1949), pp. 141–183.
— [2] Spherical functions and quadratic forms, J. Indian Math. Soc., vol. 20 (1956), pp. 117–162.
— [3] Zur Theorie der Kugelfunktionen einer Matrixvariablen, Math. Ann., vol. 135 (1958), pp. 391–416.

MacDonald, I.: [1] Spherical functions on a $\mathfrak{p}$-adic Chevalley group, Bull. Amer. Math. Soc., vol. 74 (1968), pp. 520–525.

Mackey, G. W.: [1] Functions on locally compact groups, Bull. Amer. Math. Soc., vol. 56 (1950), pp. 385–412.
— [2] On induced representations of groups, Amer. J. Math., vol. 73 (1951), pp. 576–592.

— [3] Induced representations of locally compact groups I, Ann. of Math., vol. 55 (1952), pp. 101–139.
— [4] Induced representations of locally compact groups II, Ann. of Math., vol. 58 (1953), pp. 193–221.
— [5] Mimeographed notes on group representations, Univ. of Chicago, 1956.
— [6] Borel structure in groups and their duals, Trans. Amer. Math. Soc., vol. 85 (1957), pp. 134–165.
— [7] Unitary representations of group extensions I, Acta Math., vol. 99 (1958), pp. 265–311.
— [8] Point realizations of transformation groups, Illinois J. Math., vol. 6 (1962), pp. 327–335.
— [9] Infinite dimensional group representations, Bull. Amer. Math. Soc., vol. 69 (1963), pp. 628–686.
— [10] Ergodic transformation groups with a pure point spectrum, Illinois J. Math., vol. 8 (1964), pp. 593–600.
— [11] Mimeographed notes on group representations and non-commutative harmonic analysis, University of California (Berkeley), 1965.
— [12] Ergodic theory and virtual groups, Math. Ann., vol. 166 (1966), pp. 187–207.

Magnus, W. and Oberhettinger, F.: [1] Formeln und Sätze für die Speziellen Funktionen der Mathematischen Physik, Springer-Verlag, Berlin, 1948.

Matsumoto, H.: [1] Fonctions sphériques sur un groupe semi-simple $\mathfrak{p}$-adic, C. R. Acad. Sc. Paris, vol. 269 (1969), Serie A, pp. 829–832.

Matsushima, Y.: [1] On the first Betti number of compact quotient spaces of higher dimensional symmetric spaces, Ann. of Math., vol. 75 (1962), pp. 312–330.
— [2] On Betti numbers of compact, locally symmetric Riemannian manifolds, Osaka Math. J., vol. 14 (1962), pp. 1–20.
— [3] A formula for the Betti numbers of compact locally symmetric Riemannian manifolds, J. Diff. Geom., vol. 1 (1967), pp. 99–109.

Matsushima, Y. and Murakami, S.: [1] On vector bundle valued harmonic forms and automorphic forms on symmetric Riemannian manifolds, Ann. of Math., vol. 78 (1963), pp. 365–416.
— [2] On certain cohomology groups attached to Hermitian symmetric spaces, Osaka J. Math., vol. 2 (1965), pp. 1–35.
— [3] On certain cohomology groups attached to Hermitian symmetric spaces II, Osaka J. Math., vol. 5 (1968), pp. 223–241.

Matsushima, Y. and Shimura, G.: [1] On the cohomology groups attached to certain vector-valued differential forms on the product of upper half-planes, Ann. of Math., vol. 78 (1963), pp. 417–449.

Mautner, F. I.: [1] Unitary representations of locally compact groups I, Ann. of Math., vol. 51 (1950), pp. 1–25.
— [2] Unitary representations of locally compact groups II, Ann. of Math., vol. 52 (1950), pp. 528–556.
— [3] On the decomposition of unitary representations of Lie groups, Proc. Amer. Math. Soc., vol. 2 (1951), pp. 490–496.
— [4] A generalization of the Frobenius reciprocity theorem, Proc. Nat. Acad. Sci. U.S.A., vol. 37 (1951), pp. 431–435.
— [5] Fourier analysis and symmetric spaces, Proc. Nat. Acad. Sci. U.S.A., vol. 37 (1951), pp. 529–533.
— [6] Induced representations, Amer. J. Math., vol. 74 (1952), pp. 737–758.
— [7] Geodesic flows and unitary representations, Proc. Nat. Acad. Sci. U.S.A., vol. 40 (1954), pp. 33–36.
— [8] Geodesic flows on symmetric Riemann spaces, Ann. of Math., vol. 65 (1957), pp. 416–431.

— [9] Spherical functions over $\mathfrak{p}$-adic fields I, Amer. J. Math., vol. 80 (1958), pp. 441–457.
— [10] Spherical functions over $\mathfrak{p}$-adic fields II, Amer. J. Math., vol. 86 (1964), pp. 171–200.
— [11] The trace of Hecke operators, Monat. für Math., vol. 72 (1968), pp. 137–143.
— [12] Fonctions propres des opérateurs de Hecke, C. R. Acad. Sc. Paris, vol. 269 (1969), Serie A, pp. 940–943.
— [13] Fonctions propres des opérateurs de Hecke, C. R. Acad. Sc. Paris, vol. 270 (1970), Serie A, pp. 89–92.

McKean, P. and Singer, I.: [1] Curvature and the eigenvalues of the Laplacian, J. Diff. Geom., vol. 1 (1967), pp. 43–70.

Methée, P-D.: [1] Sur les distributions invariantes dans le groupe des rotations de Lorentz, Comment. Math. Helv., vol. 28 (1954), pp. 225–269.

Miller, J. and Simms, D. J.: [1] Radial limits of the rational functions $'\Gamma_\mu$ of Harish-Chandra, Proc. Royal Irish Acad. (Sec. A), vol. 68 (1969), pp. 41–47.
— [2] A difference equation satisfied by the functions $'\Gamma_\mu$ of Harish-Chandra, Amer. J. Math., vol. 92 (1970), pp. 362–368.

Minakshisundaram S. and Pleijel, A.: [1] Eigenfunctions of the Laplace operator on Riemann manifolds, Canad. J. Math., vol. 1 (1949), pp. 242–256.

Moore, C. C.: [1] On the Frobenius reciprocity theorem for locally compact groups, Pacific J. Math., vol. 12 (1962), pp. 359–365.
— [2] Compactifications of symmetric spaces, Amer. J. Math., vol. 86 (1964), pp. 201–218.
— [3] Compactifications of symmetric spaces, Amer. J. Math., vol. 86 (1964), pp. 358–378.

Moore, R. T.: [1] Exponentiation of operator Lie algebras on Banach spaces, Bull. Amer. Math. Soc., vol. 71 (1965), pp. 903–908.
— [2] Measurable, continuous and smooth vectors for semigroups and group representations, Memoirs Amer. Math. Soc., n°78, Providence, Rhode Island, 1968.

Morimoto, M: [1] Sur les transformations horosphériques généralisées dans les espaces homogènes, J. Fac. Sci. Univ. of Tokyo, vol. 13 (1966), pp. 65–83.

Mostow, G. D.: [1] Some new decomposition theorems for semi-simple groups, Memoirs Amer. Math. Soc., n° 14, pp. 31-54, Providence, Rhode Island, 1955.
— [2] Self-adjoint group, Ann. of Math., vol. 62 (1955), pp. 44–55.
— [3] Fully reducible subgroups of algebraic groups, Amer. J. Math., vol. 78 (1956), pp. 200–221.
— [4] Equivariant embeddings in Euclidean space, Ann. of Math., vol. 65 (1957), pp. 432–446.
— [5] On maximal subgroups of real Lie groups, Ann. of Math., vol. 74 (1961), pp. 503–517.
— [6] Covariant fiberings of Klein spaces II, Amer. J. Math., vol. 84 (1962), pp. 466–474.
— [7] Homogeneous spaces with finite invariant measure, Ann. of Math., vol. 75 (1962), pp. 17–37.
— [8] Representative functions on groups, Advanced Science Seminar, pp. 1–37, Bowdoin College, 1968.
— [9] Quasi-conformal mappings in n-space and the rigidity of hyperbolic space forms, Inst. Hautes Études Sci. Publ. Math., vol. 34 (1968), pp. 53–104.

Mostow, G. D. and Tamagawa, T.: [1] On the compactness of arithmetically defined homogeneous spaces, Ann. of Math., vol. 76 (1962), pp. 446-463.

Murakami, S.: [1] Sur la classification des algèbres de Lie réelles et simples, Osaka J. Math., vol. 2 (1965), pp. 291–307.

Nachbin, L.: [1] The Haar integral, Van Nostrand, New York, 1965.

Nagata, M.: [1] On the 14th problem of Hilbert, Amer. J. Math., vol. 81 (1959), pp. 766–772.

Naimark, M. A.: [1] Normed rings, Noordhoff, Groningen—The Netherlands, 1964.

— [2] Linear representations of the Lorentz group, Macmillan, New York, 1964.

Narasimhan, M. S. and Okamoto, K.: [1] An analogue of the Borel-Weil-Bott theorem for Hermitian symmetric pairs of non-compact type, Ann. of Math., vol. 91 (1970), pp. 486–511.

Nelson, E.: [1] Analytic vectors, Ann. of Math., vol. 70 (1959), pp. 572–615.

Nelson, E. and Stinespring, W. F.: [1] Representation of elliptic operators in an enveloping algebra, Amer. J. Math., vol. 81 (1959), pp. 547–560.

Neto, J. B.: [1] Spaces of vector valued real analytic functions, Trans. Amer. Math. Soc., vol. 112 (1964), pp. 381–391.

Okamoto, K.: [1] On the Plancherel formulas for some types of simple Lie groups, Osaka J. Math., vol. 2 (1965), pp. 247–282.

— [2] On induced representations, Osaka J. Math., vol. 4 (1967), pp. 85–94.

— [3] On square integrable $\bar{\partial}$-cohomology spaces attached to homogeneous symplectic manifolds, To Appear.

Okamoto, K. and Ozeki, H.: [1] On square integrable $\bar{\partial}$-cohomology spaces attached to Hermitian symmetric spaces, Osaka J. Math., vol. 4 (1967), pp. 95–110.

Oniščik, A. L.: [1] Lie groups transitive on compact manifolds, Amer. Math. Soc. Translations (2), vol. 73 (1968), pp. 59–72.

— [2] Lie groups transitive on compact manifolds II, Math. U.S.S.R.-Sbornik, vol. 3 (1967), pp. 373–388.

Osima, M.: [1] Note on the Kronecker product of representations of a group, Proc. Imperial Acad. Tokyo, vol. 17 (1941), pp. 411–413.

Palais, R.: [1] Imbedding of compact, differentiable transformation groups in orthogonal representations, J. Math. Mech., vol. 6 (1957), pp. 673–678.

— [2] Seminar on the Atiyah-Singer Index Theorem, Princeton University Press, Princeton, New Jersey, 1965.

Paley, R. and Wiener, N : [1] Fourier transforms in the complex domain, Amer. Math. Soc., New York, 1934.

Parthasarathy, K. R., Ranga Rao, R. and Varadarajan, V. S.: [1] Representations of complex semi-simple Lie groups and Lie algebras, Ann. of Math., vol. 85 (1967), pp. 383–429.

Peetre, J : [1] Rectification à l'article "Une charactérisation abstraite des opérateurs differentials," Math. Scand., vol. 8 (1960), pp. 116–120.

Poulsen, N.: [1] Regularity aspects of the theory of infinite dimensional representations of Lie groups, Ph.D. Thesis, Massachusetts Institute of Technology, 1970.

— [2] On C^∞-vectors and intertwining bilinear forms for representations of Lie groups, To Appear.

Pukanszky, L.: [1] The Plancherel formula for the universal covering group of $\mathbf{SL}(2, \mathbf{R})$, Math. Ann., vol. 156 (1964), pp. 96–143.

— [2] Leçons sur les représentations des groupes, Dunod, Paris, 1967.

— [3] On the theory of exponential groups, Trans. Amer. Math. Soc., vol. 126 (1967), pp. 487–507.

— [4] On the characters and the Plancherel formula of nilpotent groups, J. of Functional Analysis, vol. 1 (1967), pp. 255–280.

— [5] On the unitary representations of exponential groups, J. of Functional Analysis, vol. 2 (1968), pp. 73–113.
— [6] Characters of algebraic solvable groups, J. of Functional Analysis, vol. 3 (1969), pp. 435–494.

Pyateckii-Shapiro, I. I.: [1] The geometry of classical domains and the theory of automorphic functions, Moscow, 1961.

Rader, C.: [1] Spherical functions on a semi-simple Lie group, Ph.D. Thesis, University of Washington, 1971.

Rallis, S.: [1] Lie group representations associated to symmetric spaces, Ph.D. Thesis, Massachusetts Institute of Technology, 1968.

de Rham, G.: [1] Solution élémentaire d'opérateurs différentiels du second ordre, Ann. de l'Institut Fourier, vol. 8 (1958), pp. 337–366.

Rhie, Y : [1] A density theorem on spectra of discrete subgroups of semi-simple Lie groups, Illinois J. Math., vol. 10 (1966), pp. 147–158.

Richardson Jr., R. W.: [1] Conjugacy classes in Lie algebras and algebraic groups, Ann. of Math., vol. 86 (1967), pp. 1–15.

Rickart, C. E.: [1] General theory of Banach algebras, Van Nostrand, New York, 1960.

Rieffel, M. A.: [1] Induced Banach representations of Banach algebras and locally compact groups, J. of Functional Analysis, vol. 1 (1967), pp. 443–491.
— [2] Square integrable representations of Hilbert algebras, J. of Functional Analysis, vol. 3 (1969), pp. 265–300.

Riesz, F. and Sz-Nagy, B.: [1] Functional analysis, Ungar, New York, 1955.

Rigelhof, R.: [1] Induced representations of locally compact groups, Acta Math., vol. 125 (1970), pp. 155–187.

Roelcke, W.: [1] Über die Wellengleichung bei Grenzkreisgruppen erster Art, Sitz. Ber. Heidelberger Ak. d. Wiss., Math.-nat. Kl. 1955, 4. Abh.
— [2] Das Eigenwertproblem der automorphen Formen in der hyperbolischen Ebene I, Math. Ann., vol. 167 (1966), pp. 292–337.
— [3] Das Eigenwertproblem der automorphen Formen in der hyperbolischen Ebene II, Math. Ann., vol. 168 (1967), pp. 261–324.

Romm, B. D.: [1] Analogue to the Plancherel formula for the real unimodular group of the nth order, Amer. Math. Soc. Translations (2), vol. 58 (1966), pp. 155–215.
— [2] Fully reducible representations of a semi-simple Lie algebra, Soviet Math., vol. 175 (1967), pp. 861–864.

Rothschild, L.: [1] Conjugacy classes of real Cartan subalgebras revisited, A.M.S. Notices, vol. 18 (1971), Abstract 682-22-9.

Sakai, S.: [1] On the representations of semi-simple Lie groups, Proc. Japan Acad., vol. 30 (1954), pp. 14–18.
— [2] On infinite-dimensional representations of semi-simple Lie algebras and some functionals on the universal enveloping algebras I, Proc. Japan Acad., vol. 30 (1954), pp. 305–312.

Sally, P. and Shalika, J. [1] Characters of the discrete series of representations of **SL**(2) over a local field, Proc. Nat. Acad. Sci. U.S.A., vol. 61 (1968), pp. 1231–1237.
— [2] The Plancherel formula for **SL**(2) over a local field, Proc. Nat. Acad. Sci. U.S.A., vol. 63 (1969), pp. 661–667.

Satake, I.: [1] On representations and compactifications of symmetric Riemannian spaces, Ann. of Math., vol. 71 (1960), pp. 77–110.
— [2] On compactifications of the quotient space for arithmetically defined discontinuous groups, Ann. of Math., vol. 72 (1960), pp. 555–580.

— [3] Theory of spherical functions on reductive algebraic groups over 𝔭-adic fields, Inst. Hautes Études Sci. Publ. Math., vol. 18 (1963), pp. 5–69.

Schattschneider, D.: [1] On restricted roots of semi-simple algebraic groups, J. Math. Soc. Japan, vol. 21 (1969), pp. 94–115.

Schiffmann, G.: [1] Distributions centrales de type positif sur un groupe de Lie nilpotent, Bull. Soc. Math. France, vol. 96 (1968), pp. 347–355.

— [2] Intégrales d'entrelacement et fonctions de Whittaker, Bull. Soc. Math. France, vol. 99 (1971), pp. 3–72.

Schmid, W.: [1] Homogeneous complex manifolds and representations of semi-simple Lie groups, Ph.D. Thesis, University of California—Berkeley, 1967.

— [2] Die Randwerte holomorpher Funktionen auf hermitesch symmetrischen Räumen, Inventiones Math., vol. 9 (1969), pp. 61–80.

— [3] On a conjecture of Langlands, Ann. of Math., vol. 93 (1971), pp. 1–42.

— [4] Some properties of square integrable representations of semi-simple Lie groups, To Appear.

Schochetman, I.: [1] Topology and the duals of certain locally compact groups, Trans. Amer. Math. Soc., vol. 150 (1970), pp. 477–489.

Schwartz, L : [1] Théorie des distributions, vol. I, Hermann, Paris, 1957.

— [2] Théorie des distributions, vol. II, Hermann, Paris, 1957.

— [3] Distributions à valeurs vectorielles I, Ann. Inst. Fourier Grenoble, vol. 7 (1957), pp. 1–141.

— [4] Distributions à valeurs vectorielles II, Ann. Inst. Fourier Grenoble, vol. 8 (1959), pp. 1–207.

Segal, G.: [1] The representation ring of a compact Lie group, Inst. Hautes Études Sci. Publ. Math., vol. 34 (1968), pp. 113–128.

Segal, I. E.: [1] The group algebra of a locally compact group, Trans. Amer. Math. Soc., vol. 61 (1947), pp. 69–105.

— [2] Irreducible representations of operator algebras, Bull. Amer. Math. Soc., vol. 53 (1947), pp. 73–88.

— [3] The two-sided regular representation of a unimodular locally compact group, Ann. of Math., vol. 51 (1950), pp. 293–298.

— [4] An extension of Plancherel's formula to separable unimodular groups, Ann. of Math., vol. 52 (1950), pp. 272–292.

— [5] A class of operator algebras which are determined by groups, Duke Math. J., vol. 18 (1951), pp. 221–265.

— [6] Hypermaximality of certain operators on Lie groups, Proc. Amer. Math. Soc., vol. 3 (1952), pp. 13–15.

— [7] A non-commutative extension of abstract integration, Ann. of Math., vol. 57 (1953), pp. 401–457.

— [8] A theorem on the measurability of group-invariant operators, Duke Math. J., vol. 26 (1959), pp. 549–552.

— [9] Infinite dimensional irreducible representations of compact semi-simple groups, Bull. Amer. Math. Soc., vol. 70 (1964), pp. 155–160.

— [10] Algebraic integration theory, Bull. Amer. Math. Soc., vol. 71 (1965), pp. 419–489.

— [11] An extension of a theorem of L. O'Raifeartaigh, J. of Functional Analysis, vol. 1 (1967), pp. 1–21.

Segal, I. E. and von Neumann, J.: [1] A theorem on unitary representations of semi-simple Lie groups, Ann. of Math., vol. 52 (1950), pp. 509–517.

Selberg, A.: [1] Harmonic analysis and discontinuous groups in weakly symmetric Riemannian spaces with applications to Dirichlet series, J. Indian Math. Soc., vol. 20 (1956), pp. 47–87.

— [2] Automorphic functions and integral operators, Seminars on Analytic Functions, Institute for Advanced Study, Princeton, New Jersey, vol. 2 (1957), pp. 152–161.

— [3] A new type of zeta function connected with quadratic forms, Report of the Institute in the Theory of Numbers, pp. 207–210, Boulder, Colorado, 1959.
— [4] On discontinuous groups in higher dimensional symmetric spaces, Contributions to Function Theory, pp. 147–164, Bombay, 1960.
— [5] Discontinuous groups and harmonic analysis, Proc. Int. Congr. of Math., pp. 177–189, Stockholm, 1962.
— [6] On the estimation of Fourier coefficients of modular forms, Proc. Sympos. Pure Math., vol. 8, Amer. Math. Soc., Providence, Rhode Island, 1965, pp. 1–15.
— [7] Recent developments in the theory of discontinuous groups of motions of symmetric spaces, Proceedings of the 15th Scandinavian Congress, Oslo, 1968. [Lecture Notes in Mathematics, vol. 118, 1970, Springer-Verlag.]

Séminaire Henri Cartan: [1] Fonctions automorphes, Paris, 1957–58.

Séminaire Sophus Lie: [1] Théorie des algèbres de Lie, Topologie des groupes de Lie, Paris, 1955.

Semyanistyi, V. I.: [1] Homogeneous functions and some problems in integral geometry in spaces of constant curvature, Soviet Math., vol. 2 (1961), pp. 59–62.

Serre, J-P.: [1] Faisceaux algébriques cohérents, Ann. of Math., vol. 61 (1955), pp. 197–278.
— [2] Une théorème de dualité, Comment. Math. Helv., vol. 29 (1955), pp. 9–26.
— [3] Lie algebras and Lie groups, Benjamin, New York, 1965.
— [4] Algèbres de Lie semi-simples complexes, Benjamin, New York, 1966.

Shephard, G. and Todd, J.: [1] Finite unitary reflection groups, Canad. J. Math., vol. 6 (1954), pp. 274–304.

Sherman, T.: [1] A weight theory for unitary representations, Canad. J. Math., vol. 18 (1966), pp. 159–168.

Shiga, K.: [1] Representations of a compact group on a Banach space, J. Math. Soc. Japan, vol. 7 (1955), pp. 224–248.

Shimizu, H.: [1] On discontinuous groups operating on the product of upper half planes, Ann. of Math., vol. 77 (1963), pp. 33–71.
— [2] On traces of Hecke operators, J. Fac. of Sci. Univ. of Tokyo, Sec. 1, vol. 10 (1963), pp. 1–19.
— [3] On zeta functions of quaternion algebras, Ann. of Math., vol. 81 (1965), pp. 166–193.

Shimizu, Y.: [1] An analogue of the Paley-Wiener theorem for certain function spaces on the generalized Lorentz group, J. Fac. of Sci. Univ. of Tokyo, vol. 16 (1969), pp. 13–51.

Shimura, G.: [1] Sur les intégrales attachées aux formes automorphes, J. Math. Soc. Japan, vol. 11 (1959), pp. 291–311.

Shoda, K.: [1] Über die monomialen Darstellungen einer endlichen Gruppe, Proc. Physico-Mathematical Soc. Japan, vol. 15 (1933), pp. 249–257.

de Siebenthal, J.: [1] Sur certains modules dans une algèbre de Lie semi-simple, Comment. Math. Helv., vol. 44 (1969), pp. 1–44.

Sobolev, S.: [1] Applications of functional analysis in mathematical physics, Translations of Math. Mono., vol. 7, Amer. Math. Soc., Providence, Rhode Island, 1963.

Solomon, L.: [1] Invariants of finite reflection groups, Nagoya Math. J., vol. 22 (1963), pp. 57–64.

Solomon, L. and Verma, D. N.: [1] Sur le corps des quotients de l'algèbre enveloppante d'une algèbre de Lie, C. R. Acad. Sc. Paris, vol. 264 (1967), Serie A, pp. 985–986.

Springer, T. A.: [1] Weyl's character formula for algebraic groups, Inventiones Math., vol. 5 (1968), pp. 85–105.

Stein, E. M.: [1] A survey of representations of non-compact groups, Lect. Sem. on High Energy Physics and Elementary Particles (Trieste, 1965), International Atomic Energy Agency, Vienna, 1965, pp. 563–584.

— [2] Analysis in matrix spaces and some new representations of $\mathbf{SL}(n, \mathbf{C})$, Ann. of Math., vol. 86 (1967), pp. 461–490.

— [3] Analytic continuation of group representations, Advances in Math., vol. 4 (1970), pp. 172–207.

Stein, E. M. and Weiss, G.: [1] Generalization of the Cauchy-Riemann equations and representations of the rotation group, Amer. J. Math., vol. 90 (1968), pp. 163–196.

Steinberg, R.: [1] A general Clebsch-Gordan theorem, Bull. Amer. Math. Soc., vol. 67 (1961), pp. 406–407.

— [2] Representations of algebraic groups, Nagoya Math. J., vol. 22 (1963), pp. 33–56.

— [3] Differential equations invariant under finite reflection groups, Trans. Amer. Math. Soc., vol. 112 (1964), pp. 392–400.

— [4] Regular elements of semi-simple algebraic groups, Inst. Hautes Études Sci. Publ. Math., vol. 25(1965), pp. 49–80.

— [5] Lectures on Chevalley groups, Lecture Notes, Yale University, 1967.

Stern, A. I.: [1] Completely irreducible class I representations of real semisimple Lie groups, Soviet Math., vol. 10 (1969), pp. 1254–1257.

Sternheimer, D.: [1] Propriétés spectrales dans les représentations de groupes de Lie, J. Math. Pures et Appl., vol. 47 (1968), pp. 289–319.

Stinespring, W. F.: [1] A semi-simple matrix group is of type I, Proc. Amer. Math. Soc., vol. 9 (1958), pp. 965–967.

— [2] Integrability of Fourier transforms for unimodular Lie groups, Duke Math. J., vol. 26 (1959), pp. 123–131.

Stone, M.: [1] Linear transformations in Hilbert space, Amer. Math. Soc. Colloquium Publication, vol. 15, New York, 1932.

Sugiura, M.: [1] Conjugate classes of Cartan subalgebras in real semi-simple Lie algebras, J. Math. Soc. Japan, vol. 11 (1959), pp. 374–434.

— [2] Spherical functions and representation theory of compact Lie groups, Sci. Papers Coll. Gen. Ed. Univ. Tokyo, vol. 10 (1960), pp. 187–193.

— [3] Representations of compact groups realized by spherical functions on symmetric spaces, Proc. Japan Acad., vol. 38 (1962), pp. 111–113.

Takahashi, R.: [1] Sur les fonctions sphériques et la formule de Plancherel dans le groupe hyperbolique, Jap. J. Math., vol. 31 (1961), pp. 55–90.

— [2] Sur les représentations unitaires des groupes de Lorentz généralisés, Bull. Soc. Math. France, vol. 91 (1963), pp. 289–433.

Takenouchi, O.: [1] Sur une classe de fonctions continues de type positif sur un groupe localement compact, Math. J. Okayama U., vol. 4 (1955), pp. 143–173.

— [2] Sur les facteurs représentations d'un groupe de Lie résoluble de type E, Math. J. Okayama U., vol. 7 (1957), pp. 151–161.

Takesaki, M.: [1] Covariant representations of C^*-algebras and their locally compact automorphism groups, Acta Math., vol. 119 (1967), pp. 273–303.

Takeuchi, M.: [1] Cell decompositions and Morse equalities on certain symmetric spaces, J. Fac. of Sci. Univ. of Tokyo, vol. 12 (1965), pp. 81–191.

— [2] On orbits in a compact Hermitian symmetric space, Amer. J. Math., vol. 90 (1968), pp. 657–680.

Tamagawa, T.: [1] On Selberg's trace formula, J. Fac. of Sci. Univ. of Tokyo, vol. 8 (1960), pp. 363–386.

— [2] On the ζ-functions of a division algebra, Ann. of Math., vol. 77 (1963), pp. 387–405.
— [3] Harmonic analysis on adele groups, Advanced Science Seminar, pp. 1–39, Bowdoin College, 1968.

Tanaka, S.: [1] Selberg's trace formula and spectrum, Osaka J. Math., vol. 3 (1966), pp. 205–216.

Tatsuuma, N.: [1] Regularity of orbit spaces on semi-simple Lie groups, Proc. Japan Acad., vol. 42 (1966), pp. 84–86.

Taylor, M.: [1] Fourier series on compact Lie groups, Proc. Amer. Math. Soc., vol. 19 (1968), pp. 1103–1105.

Tengstrand, A.: [1] Distributions invariant under an orthogonal group of arbitrary signature, Math. Scand., vol. 8 (1960), pp. 201–218.

Tirao, J. and Wolf, J.: [1] Homogeneous holomorphic vector bundles, J. Math. and Mechanics, vol. 20 (1970), pp. 15–31.

Titchmarsh, E. C.: [1] Eigenfunction expansions, Oxford University Press, London, 1946.

Tits, J.: [1] Théorème de Bruhat et sous-groupes paraboliques, C. R. Acad. Sci. Paris, vol. 254 (1962), pp. 2910-2912.
— [2] Algebraic and abstract simple groups, Ann. of Math., vol. 80 (1964), pp. 313–329.
— [3] Tabellen zu den einfachen Lie Gruppen und ihren Darstellungen, Springer-Verlag, Lecture Notes in Mathematics, vol. 40, 1967.

Tits, J. and Waelbroeck, L.: [1] The integration of a Lie algebra representation, Pacific J. of Math., vol. 26 (1968), pp. 595–600.

Treves, F.: [1] Linear partial differential equations with constant coefficients, Gordon and Breach, New York, 1966.
— [2] Topological vector spaces, distributions and kernels, Academic Press, New York, 1967.

Trombi, P. C. and Varadarajan, V. S.: [1] Spherical transforms on semi-simple Lie groups, Ann. of Math., vol. 94 (1971), pp. 246–303.
— [2] Asymptotic behaviour of eigenfunctions on a semi-simple Lie group: The discrete spectrum, To Appear.

Tsuchikawa, M.: [1] On the representations of **SL**(3, **C**) I, Proc. Japan Acad., vol. 43 (1967), pp. 852–855.
— [2] On the representations of **SL**(3, **C**) II, Proc. Japan Acad., vol. 44 (1968), pp. 127–129.
— [3] On the representations of **SL**(3, **C**) III, Proc. Japan Acad., vol. 44 (1968), pp. 130–132.

van Dijk, G.: [1] On symmetry of group algebras of motion groups, Math. Ann., vol. 179 (1969), pp. 219–226.

van der Waerden, B. L.: [1] Stetigketissätze der halbeinfachen Lieschen Gruppen, Math. Zeit., vol. 36 (1933), pp. 780–786.

Varadarajan, V. S.: [1] On the ring of invariant polynomials on a semi-simple Lie algebra, Amer. J. Math., vol. 90 (1968), pp. 308–317.

Vilenkin, N.: [1] On the theory of associated spherical functions on Lie groups, Mat. Sb., vol. 42 (1957), pp. 485–496.
— [2] Special functions and the theory of group representations, Translations of Math. Mono., vol. 22, Amer. Math. Soc., Providence, Rhode Island, 1968.

Vinberg, E. B.: [1] The Borel-Morosow theorem for real Lie groups, Soviet Math., vol. 2 (1961), pp. 1416–1419.

Wallach, N.: [1] Induced representations of Lie algebras and a theorem of Borel-Weil, Trans. Amer. Math. Soc., vol. 136 (1969), pp. 181–187.

— [2] Induced representations of Lie algebras II, Proc. Amer. Math. Soc., vol. 21 (1969), pp. 161–166.
— [3] Cyclic vectors and irreducibility for principal series representations, To Appear.

Wang, H. C.: [1] Closed manifolds with homogeneous complex structure, Amer. J. Math., vol. 76 (1954), pp. 1–32.
— [2] On a maximality property of discrete subgroups with fundamental domain of finite measure, Amer. J. Math., vol. 89 (1967), pp. 124–132.

Wang, S. P.: [1] The dual space of semi-simple Lie groups, Amer. J. Math., vol. 91 (1969), pp. 921–937.

Weil, A.: [1] L'intégration dans les groupes topologiques et ses applications, Hermann, Paris, 1940.
— [2] On discrete subgroups of Lie groups I, Ann. of Math., vol. 72 (1960), pp. 369–384.
— [3] On discrete subgroups of Lie groups II, Ann. of Math., vol. 75 (1962), pp. 578–602.
— [4] Remarks on the cohomology of groups, Ann. of Math., vol. 80 (1964), pp. 149–157.

Weyl, H.: [1] The classical groups, Princeton University Press, Princeton, New Jersey, 1939.

Whitney, H.: [1] Elementary structure of real algebraic varieties, Ann. of Math., vol. 66 (1957), pp. 545–556.

Whittaker, E. T. and Watson, G. N.: [1] Modern analysis, Cambridge Press, London, 1927.

Wolf, J. A.: [1] The geometry and structure of isotropy irreducible homogeneous spaces, Acta Math., vol. 120 (1968), pp. 59–148.
— [2] The action of a real semi-simple group on a complex flag manifold I: Orbit structure and holomorphic arc components, Bull. Amer. Math. Soc., vol. 75 (1969), pp. 1121–1237.
— [3] The action of a real semi-simple group on a complex flag manifold II: Unitary representations on partially holomorphic cohomology spaces, To Appear.

Wolf, J. A. and Gray, A.: [1] Homogeneous spaces defined by Lie group automorphisms, J. Diff. Geom., vol. 2 (1968), pp. 77–159.

Wolf, J. A. and Korányi, A.: [1] Generalized Cayley transformations of bounded symmetric domains, Amer. J. Math., vol. 87 (1965), pp. 899–939.

Zeitlin, J.: [1] Correspondence between Lie algebra invariant subspaces and Lie group invariant subspaces of representations of Lie groups, Ph. D. Thesis, University of California-Los Angeles, 1969.

Želobenko, D. P.: [1] Infinitely differentiable vectors in the theory of representations, Vestnik. Moskov. Univ. Ser I. Math. Meh. 1965, pp. 3–10.
— [2] Harmonic analysis of functions on semi-simple Lie groups I, Amer. Math. Soc. Translations (2), vol. 54 (1966), pp. 177–230.
— [3] Analogue of the Cartan-Weyl theory for infinite dimensional representations of a complex semi-simple Lie group, Soviet Math., vol. 8 (1967), pp. 798–802.
— [4] Symmetry in a class of elementary representations of a semi-simple complex Lie group, Func. Analysis and Applic., vol. 1 (1967), pp. 15–38.
— [5] The analysis of irreducibility in the class of elementary representations of a complex semi-simple Lie group, Math. U.S.S.R.-Izvestija, vol. 2 (1968), pp. 105–128.
— [6] Operational calculus on a complex semi-simple Lie group, Math. U.S.S.R.-Izvestija, vol. 3 (1969), pp. 881–915.

— [7] On the harmonic analysis of functions on semi-simple Lie groups II, Math. U.S.S.R.-Izvestija, vol. 3 (1969), pp. 1183–1217.

— [8] On the irreducible representations of a complex semi-simple Lie group, Func. Analysis and Applic., vol. 4 (1970), pp. 163–165.

Želobenko, D. P. and Naimark, M. A.: [1] Description of the completely irreducible representations of a complex semi-simple Lie group, Math. U.S.S.R.-Izvestija, vol. 4 (1970), pp. 59–83.

Subject Index to Volumes I and II

Page numbers in *italics* refer to Volume I

Die Grundlehren der mathematischen Wissenschaften in Einzeldarstellungen mit besonderer Berücksichtigung der Anwendungsgebiete

Eine Auswahl

67. Byrd/Friedman: Handbook of Elliptic Integrals for Engineers and Scientists
68. Aumann: Reelle Funktionen
74. Boerner: Darstellungen von Gruppen
76. Tricomi: Vorlesungen über Orthogonalreihen
77. Behnke/Sommer: Theorie der analytischen Funktionen einer komplexen Veränderlichen
78. Lorenzen: Einführung in die operative Logik und Mathematik
86. Richter: Wahrscheinlichkeitstheorie
87. van der Waerden: Mathematische Statistik
94. Funk: Variationsrechnung und ihre Anwendung in Physik und Technik
97. Greub: Linear Algebra
99. Cassels: An Introduction to the Geometry of Numbers
104. Chung: Markov Chains with Stationary Transition Probabilities
107. Köthe: Topologische lineare Räume
114. MacLane: Homology
116. Hörmander: Linear Partial Differential Operators
117. O'Meara: Introduction to Quadratic Forms
120. Collatz: Funktionalanalysis und numerische Mathematik
121./122. Dynkin: Markov Processes
123. Yosida: Functional Analysis
124. Morgenstern: Einführung in die Wahrscheinlichkeitsrechnung und mathematische Statistik
125. Itô/McKean jr.: Diffusion Processes and their sample Paths
126. Lehto/Virtanen: Quasikonforme Abbildungen
127. Hermes: Enumerability, Decidability, Computability
128. Braun/Koecher: Jordan-Algebren
129. Nikodým: The Mathematical Apparatus for Quantum Theories
130. Morrey jr.: Multiple Integrals in the Calculus of Variations
131. Hirzebruch: Topological Methods in Algebraic Geometry
132. Kato: Perturbation Theory for Linear Operators
133. Haupt/Künneth: Geometrische Ordnungen
134. Huppert: Endliche Gruppen I
135. Handbook for Automatic Computation. Vol. 1/Part a: Rutishauser: Description of ALGOL 60
136. Greub: Multilinear Algebra
137. Handbook for Automatic Computation. Vol. 1/Part b: Grau/Hill/Langmaack: Translation of ALGOL 60
138. Hahn: Stability of Motion
139. Mathematische Hilfsmittel des Ingenieurs. 1. Teil
140. Mathematische Hilfsmittel des Ingenieurs. 2. Teil
141. Mathematische Hilfsmittel des Ingenieurs. 3. Teil
142. Mathematische Hilfsmittel des Ingenieurs. 4. Teil
143. Schur/Grunsky: Vorlesungen über Invariantentheorie
144. Weil: Basic Number Theory
145. Butzer/Berens: Semi-Groups of Operators and Approximation

146. Treves: Locally Convex Spaces and Linear Partial Differential Equations
147. Lamotke: Semisimpliziale algebraische Topologie.
148. Chandrasekharan: Introduction to Analytic Number Theory
149. Sario/Oikawa: Capacity Functions
150. Iosifescu/Theodorescu: Random Processes and Learning
151. Mandl: Analytical Treatment of One-dimensional Markov Processes
152. Hewitt/Ross: Abstract Harmonic Analysis. Vol. 2: Structure and Analysis for Compact Groups. Analysis on Locally Compact Abelian Groups
153. Federer: Geometric Measure Theory
154. Singer: Bases in Banach Spaces I
155. Müller: Foundations of the Mathematical Theory of Electromagnetic Waves
156. van der Waerden: Mathematical Statistics
157. Prohorov/Rozanov: Probability Theory
159. Köthe: Topological Vector Spaces I
160. Agrest/Maksimov: Theory of Incomplete Cylindrical Functions and Their Applications
161. Bhatia/Szegö: Stability Theory of Dynamical Systems
162. Nevanlinna: Analytic Functions
163. Stoer/Witzgall: Convexity and Optimization in Finite Dimensions I
164. Sario/Nakai: Classification Theory of Riemann Surfaces
165. Mitrinović/Vasić: Analytic Inequalities
166. Grothendieck/Dieudonné: Eléments de Géométrie Algébrique I
167. Chandrasekharan: Arithmetical Functions
168. Palamodov: Linear Differential Operators with Constant Coefficients
170. Lions: Optimal Control Systems Governed by Partial Differential Equations
171. Singer: Best Approximation in Normed Linear Spaces by Elements of Linear Subspaces
172. Bühlmann: Mathematical Methods in Risk Theory
173. F. Maeda/S. Maeda: Theory of Symmetric Lattices
174. Stiefel/Scheifele: Linear and Regular Celestial Mechanics. Perturbed Two-body Motion—Numerical Methods—Canonical Theory
175. Larsen: An Introduction of the Theory of Multipliers
176. Grauert/Remmert: Analytische Stellenalgebren
177. Flügge: Practical Quantum Mechanics I
178. Flügge: Practical Quantum Mechanics II
179. Giraud: Cohomologie non abélienne
180. Landkoff: Foundations of Modern Potential Theory
181. Lions/Magenes: Non-Homogeneous Boundary Value Problems and Applications I
182. Lions/Magenes: Non-Homogeneous Boundary Value Problems and Applications II
183. Lions/Magenes: Non-Homogeneous Boundary Value Problems and Applications III. In preparation
184. Rosenblatt: Markov Processes. Structure and Asymptotic Behavior
185. Rubinowicz: Sommerfeldsche Polynommethode
186. Wilkinson/Reinsch: Handbook for Automatic Computation II, Linear Algebra
187. Siegel/Moser: Lectures on Celestial Mechanics
188. Warner: Harmonic Analysis on Semi-Simple Lie Groups I
189. Warner: Harmonic Analysis on Semi-Simple Lie Groups II
190. Faith: Algebra: Rings, Modules, and Categories I. In preparation
191. Faith: Algebra: Rings, Modules, and Categories II. In preparation
192. Maltsev: Algebraic Systems. In preparation
193. Pólya/Szegö: Problems and Theorems in Analysis. Vol. 1. In preparation
194. Igusa: Theta Functions